OXFORD CLASSIC TEXTS IN THE PHYSICAL SCIENCES

THE MATHEMATICAL THEORY OF BLACK HOLES

S. CHANDRASEKHAR

CLARENDON PRESS · OXFORD

This book has been printed digitally and produced in a standard specification in order to ensure its continuing availability

OXFORD
UNIVERSITY PRESS

Great Clarendon Street, Oxford OX2 6DP

Oxford University Press is a department of the University of Oxford.
It furthers the University's objective of excellence in research, scholarship,
and education by publishing worldwide in

Oxford New York

Auckland Cape Town Dar es Salaam Hong Kong Karachi
Kuala Lumpur Madrid Melbourne Mexico City Nairobi
New Delhi Shanghai Taipei Toronto

With offices in

Argentina Austria Brazil Chile Czech Republic France Greece
Guatemala Hungary Italy Japan South Korea Poland Portugal
Singapore Switzerland Thailand Turkey Ukraine Vietnam

Published in the United States
by Oxford University Press Inc., New York

Reprinted 2009

ISBN 978-0-19-850370-5

TO THE READER

First, my fear; then, my curtsy; last, my speech. My fear is, your displeasure; my curtsy, my duty; and my speech, to beg your pardons. If you look for a good speech now, you undo me; for what I have to say is of mine own making; and what indeed I should say will, I doubt, prove mine own marring. But to the purpose, and so to the venture.

W. Shakespeare
(*Henry IV, Part II*)

Contents. The table of contents has been made sufficiently detailed that the topics covered and their organization can be gathered from it. Also, the opening section of each chapter provides the prospect and the closing section, often, a retrospect.

Arrangement. The section numbers run serially through the entire book; and so do those of the tables and the figures. References to equation numbers are to those in the same chapter unless qualified by the prefix of the number of the chapter to which it belongs.

Notation. The choice of a notation which will be strictly consistent throughout the book proved to be impossible: all the alphabets that are available are simply not adequate. Besides, one is constrained not to mar the appearance of an equation or a formula and obscure its meaning by a random assortment of symbols. A compromise had to be made. While certain symbols, easily recognizable from the context, maintain their meanings throughout, many are transient. It is hoped that going to the beginning of the chapter (if not the section) will resolve most of the ambiguities.

Bibliographical Notes. Since the entire subject matter (including the mathematical developments) has been written (or, worked out) *ab initio,* independently of the origins, the author has not made any serious search of the literature. The bibliographical notes at the end of each chapter provide no more than the sources of his information.

The book is an expression of the author's perspective with the limitations which that implies.

ACKNOWLEDGMENTS

In my study of the mathematical theory of black holes, I have greatly benefited from my association with several colleagues. I am particularly indebted

to John L. Friedman, for his critical judgement, perceptive comments, and constant encouragement, besides his help with Chapters 10 (§ 102) and 11 (§ 114);

to Steven Detweiler, for his active collaboration in several investigations (incorporated in the book) and for his ever-readiness to be helpful both in matters requiring clarification and in matters requiring computations—Tables IV, VI, IX, X, and XI, as well as the tables included in the Appendix, are due to him;

to Garret Toomey, for his assistance with the sections dealing with the geodesics in the Schwarzschild, the Reissner-Nordström, and the Kerr space-times and for providing the beautiful illustrations included in Chapters 3, 5, and 7; and

to Robert Wald, Robert Geroch, and James Hartle, for acting as referees on diverse matters.

Of my indebtedness to Basilis Xanthopoulos, I can make no adequate acknowledgement. He undertook, gladly, the formidable task of reading the entire book in several stages of the manuscript and checking the mathematical developments. But his help went much beyond that: besides his original contributions included in Chapters 4 (§ 25), 5 (§ 42), 6 (§ 57), 7 (§ 60), and 11(§ 109), his critical appraisal was always invaluable.

I am also grateful to Ms. Mavis Takeuchi-Lozano for her assistance in the preparation of the manuscript and in its passage through the press.

The writing of this book (during the period March 1980–January 1982) was supported in part by the National Science Foundation under grant PHY-78-24275 with the University of Chicago. Also, during the three-month period (March–June 1980) I had the support of a Regents Fellowship of the Smithsonian Institution (Washington, D.C.) at the Center for Astrophysics, Harvard College Observatory.

And finally, I am grateful to the Clarendon Press for bringing to this book (as to two earlier books of mine) that excellence of craftsmanship and typography which is characteristic of all their work.

S.C.

I have taken this occasion of the third printing of this book to scrutinize the entire developments formula by formula (in effect rederiving them all *ab initio*) for possible misprints. Apart from many errors in Chapter 1, all derived from characters which should have been bold face appearing as light face (and conversely) the actual number of misprints in formulae (including even obvious ones) do not exceed some forty; but none of the errors that were found were substantive; and none propagated. It is of course idle to expect that a book of this size and complexity can be totally free of errors; but I hope that there are none that are of substance.

S. Chandrasekhar

September 1985

CONTENTS

Roy Patrick Kerr (1934–   )

Karl Schwarzschild (1873–1916)

PROLOGUE

The black holes of nature are the most perfect macroscopic objects there are in the universe: the only elements in their construction are our concepts of space and time. And since the general theory of relativity provides only a single unique family of solutions for their descriptions, they are the simplest objects as well.

The unique two-parameter family of solutions which describes the space-time around black holes is the Kerr family discovered by Roy Patrick Kerr in July, 1963. The two parameters are the mass of the black hole and the angular momentum of the black hole. The static solution, with zero angular momentum, was discovered by Karl Schwarzschild in December, 1915. A study of the black holes of nature is then a study of these solutions. It is to this study that this book is devoted.

1

MATHEMATICAL PRELIMINARIES

1. Introduction

In this chapter, we shall provide an account of the analytical methods that lie at the base of much of the developments that are to be described in this book.These consist of Cartan's calculus of differential forms and the tetrad and the Newman–Penrose formalisms. While none of these matters are novel in themselves, they cannot all be found in a coherent treatment in one place; and the account is included to make the book as self-contained as is possible. The presentation of the elements of differential geometry in §§2–6 is, however, not intended to replace the standard accounts of the subject available elsewhere: the presentation is confined to the barest essentials leading to Cartan's equations of structure.

2. The elements of differential geometry

Differential geometry deals with *manifolds*. Manifolds are essentially spaces that are locally Euclidean in a sense which we shall first make precise.

We recall that an Euclidean space of n-dimensions, $\mathbb{R}_n$, is the set of all n-tuples, $(x^1, \ldots, x^n)$ $(-\infty < x^i < +\infty)$, with open and closed sets (or, neighbourhoods) defined in the usual way. A manifold, M, is locally identical to Euclidean space in the sense that M is covered (i.e., a union of) neighbourhoods, $\mathcal{U}_\alpha$, and that associated with each $\mathcal{U}_\alpha$ there is a one-one map, ϕ_α, which images each point $p \in \mathcal{U}_\alpha$ to a point in an open neighbourhood of $\mathbb{R}_n$ (onto which $\mathcal{U}_\alpha$ is imaged by ϕ_α) with the coordinates $(x^1, \ldots, x^n)$. Further, if two neighbourhoods, $\mathcal{U}_\alpha$ and $\mathcal{V}_\alpha$ of M, intersect and have points in common (i.e., $\mathcal{U}_\alpha \cap \mathcal{V}_\alpha \neq 0$), and if ϕ_α and ψ_α are the associated maps onto neighbourhoods in $\mathbb{R}_n$, then the map $\phi_\alpha \circ \psi_\alpha^{-1}$ images a point $\psi_\alpha(p)$ $(p \in \mathcal{U}_\alpha \cap \mathcal{V}_\alpha)$ with the coordinates $(x^1, \ldots, x^n)$, say, to the point $\phi_\alpha(p)$ with the coordinates $(\bar{x}^1, \ldots, \bar{x}^n)$, then as a part of the definition of M, it is required that x^i $(i = 1, \ldots, n)$ are *smooth functions* of $(\bar{x}^1, \ldots, \bar{x}^n)$. (Smooth functions are those which have continuous partial derivatives of all orders.)

The *Cartesian product* $M \times N$ of two manifolds M and N is, in the first instance, the ordered pair of points, (p, q), where $p \in M$ and $q \in N$; further, if $\mathcal{U}_\alpha$ and $\mathcal{V}_\alpha$ are neighbourhoods in M and in N, ϕ_α and ψ_β are the associated maps, and $\phi_\alpha(p) = (x^1, \ldots, x^n)$ and $\psi_\beta(q) = (y^1, \ldots, y^m)$ (m not necessarily equal to n), then the map,

$$(\phi_\alpha \times \psi_\beta)\,(p, q) = (x^1, \ldots, x^n, y^1, \ldots, y^m),$$

suffices to complete the definition of $M \times N$ as a manifold of $(m+n)$-dimensions.

We now consider a function f on M defined by a map f: $M \to \mathbb{R}^1$. We shall suppose that the combined map $f \circ \phi_\alpha^{-1}$ which images a point $(x^1, \ldots, x^n)$ in $\mathbb{R}^n$ on to the reals, $\mathbb{R}^1$, is a smooth function of the coordinates $(x^1, \ldots, x^n)$. We define a smooth curve λ on M by the map

$$\lambda: \quad \text{an interval } I(a < t < b) \text{ in } \mathbb{R}^1 \to \lambda(t) = p \in M$$

such that

$$(\phi_\alpha \circ \lambda)(t) = [x^1(t), \ldots, x^n(t)]; \tag{1}$$

and we require that $x^i(t)$ $(i = 1, \ldots, n)$ are smooth functions of t. Finally, we may note that a function f, defined on the manifold, enables us to define the function, $f \circ \lambda$ on the curve λ. With the aid of the map $\phi_\alpha \circ \lambda$ we are led to consider the function $f(\lambda(t)) = f(x^1(t), \ldots, x^n(t))$ where $(x^1(t), \ldots, x^n(t))$ are the coordinates of $p = \lambda(t)$ by the map ϕ_α.

(a) Tangent vectors

With the definition of $f(x^1(t), \ldots, x^n(t))$ on a curve λ on M, which we have just given, consider

$$\left(\frac{\partial f}{\partial t}\right)_{\lambda(t)}\Bigg|_{t=t_0} = \underset{\varepsilon \to 0}{\text{limit}}\, \frac{1}{\varepsilon}\Big\{f(\lambda(t_0+\varepsilon)) - f(\lambda(t_0))\Big\}$$

$$= \sum_{j=1}^{n} \frac{\mathrm{d}x^j(t)}{\mathrm{d}t}\Bigg|_{t_0} \left(\frac{\partial f}{\partial x^j}\right)_{\lambda(t_0)} = \left(\frac{\mathrm{d}x^j}{\mathrm{d}t}\frac{\partial f}{\partial x^j}\right)_{t_0}, \tag{2}$$

where, in the last step, summation over repeated indices is assumed. (This summation convention will be adopted throughout the book.)

It is now clear that by considering various curves λ passing through a given point p, we can define a linear vector-space (at p) consisting of linear combinations of the coordinate derivatives $\partial/\partial x^j$ of the forms

$$\boldsymbol{X} = X^j \frac{\partial}{\partial x^j}, \tag{3}$$

where the X^j's are any set of n numbers. These *tangent vectors* arise by considering the curves λ defined by

$$x^j(t) = x^j(p) + X^j t \qquad (j = 1, \ldots, n) \tag{4}$$

for t in some small interval $-\varepsilon < t < +\varepsilon$.

The tangent vectors at p form a linear vector-space over $\mathbb{R}^1$ spanned by the coordinate derivatives, since the requirement for a linear vector-space, namely,

$$(\alpha \boldsymbol{X} + \beta \boldsymbol{Y})f = \alpha(\boldsymbol{X}f) + \beta(\boldsymbol{Y}f), \tag{5}$$

is satisfied for all vectors X and Y, numbers α and β, and functions f. Moreover, the vectors $(\partial/\partial x^j)_p$ are linearly independent; for, otherwise, there should exist numbers $X^j (j = 1, \ldots, n)$, not all zero, such that $X = X^j \partial/\partial x^j$ applied to any smooth function is identically zero; but the application of X to the coordinate functions $x^k (k = 1, \ldots, n)$ would lead to $X^k = 0$ for all k; and this is a contradiction.

Finally, the definition of X by

$$Xf = X^j \frac{\partial f}{\partial x^j} = X^j f_{,j}, \tag{6}$$

for every smooth function f clearly satisfies the Leibnitz rule when operating on products of functions, thus

$$X(fg)|_{\lambda(t)} = (fXg + gXf)|_{\lambda(t)}. \tag{7}$$

Tangent vectors may, in fact, be considered as directional derivatives. (In equation (6) we have introduced the notation of denoting derivatives with respect to x^j by the index j, following a comma, both as subscripts.)

The space of tangent vectors (or *contravariant vectors*, as they are also called) to an n-dimensional manifold M at p, denoted by $T_p(M)$ or simply T_p, is an n-dimensional vector-space. This space, which may be visualized as the set of all 'directions' at p, is called the *tangent space* at p.

Instead of a basis determined by local coordinates, we may choose any other n linearly independent vectors $e_a (a = 1, \ldots, n)$ (say). There must, then, exist linear relations of the form

$$e_a = \Phi_a{}^k \frac{\partial}{\partial x^k}, \tag{8}$$

where the determinant of the matrix formed by $\Phi_a{}^k$ must be non-zero. The inverse relation is then given by

$$\frac{\partial}{\partial x^j} = \Phi^b{}_j e_b \tag{9}$$

where $[\Phi^b{}_j]$ is the inverse of the matrix $[\Phi_a{}^k]$:

$$\Phi_a{}^k \Phi^a{}_j = \delta^k{}_j \quad \text{and} \quad \Phi_a{}^k \Phi^b{}_k = \delta_a{}^b. \tag{10}$$

Given any basis (e_j), we can express any tangent vector at p in the form

$$X = X^j e_j. \tag{11}$$

The X^j's are the components of X relative to the basis (e_j).

(b) *One-forms (or, cotangent or covariant vectors)*

A *one-form*, ω, at p is a linear mapping of the tangent space T_p on to the reals:

$$\omega: \quad T_p \to \mathbb{R}^1. \tag{12}$$

In other words, given any tangent vector X at p, the one-form ω associates uniquely with it a number $\omega(X)$ which is also written as

$$\omega(X) = \langle \omega, X \rangle. \tag{13}$$

The required linearity of the map is expressed by the relation

$$\langle \omega, \alpha X + \beta Y \rangle = \alpha \langle \omega, X \rangle + \beta \langle \omega, Y \rangle, \tag{14}$$

where X and Y are any two tangent vectors and α and β are any two real numbers. We, further, define multiplication of forms by real numbers and sums of forms by the rules that, for any $X \in T_p$ and any real number α,

$$(\alpha\omega)(X) = \alpha\langle \omega, X \rangle \qquad \text{and} \qquad (\omega + \pi)(X) = \langle \omega, X \rangle + \langle \pi, X \rangle, \tag{15}$$

where ω and π are two one-forms. By these rules, one-forms span a vector space which we denote by $T^*{}_p$; it is called the *cotangent space* at p and the *dual* of the tangent space. For this reason one-forms are also called *cotangent vectors* (or, *covariant vectors*).

We shall now verify that a basis for $T^*{}_p$, associated with a basis (e_j) for T_p, is provided by the one-forms $(e^i)(i = 1, \ldots, n)$ which map any tangent vector $X = X^j e_j$ to its components; thus,

$$e^i(X) = \langle e^i, X^j e_j \rangle = X^i \qquad (i = 1, \ldots, n). \tag{16}$$

From this last equation it follows that

$$e^i(e_j) = \langle e^i, e_j \rangle = \delta^i{}_j. \tag{17}$$

The expression of any arbitrary one-form ω as a linear combination of the e^i's is obtained by observing that

$$\langle \omega, X \rangle = \langle \omega, X^i e_i \rangle = X^i \langle \omega, e_i \rangle. \tag{18}$$

Now letting

$$\omega_i = \langle \omega, e_i \rangle = \omega(e_i), \tag{19}$$

be the numbers to which ω maps the basis vectors (e_i) of the tangent space T_p at p, we may write

$$\begin{aligned} \langle \omega, X \rangle = \omega_i X^i &= \omega_i \langle e^i, X^j e_j \rangle \\ &= \langle \omega_i e^i, X \rangle. \end{aligned} \tag{20}$$

Since this last equation is valid for any $X \in T_p$, it follows that

$$\omega = \omega_i e^i; \tag{21}$$

and this is the required expression of ω as a linear combination of the e^i's. That the vectors e^i are linearly independent is manifest from its definition. The bases (e_i) and (e^i) are said to provide *dual bases* for the tangent and the cotangent spaces at p.

If in place of the dual bases $(\boldsymbol{e}_i)$ and $(\boldsymbol{e}^j)$ we should choose different bases,

$$\boldsymbol{e}_{i'} = \Phi_{i'}{}^{j}\boldsymbol{e}_j \qquad \text{and} \qquad \boldsymbol{e}^{j'} = \Phi^{j'}{}_{j}\boldsymbol{e}^j, \tag{22}$$

obtained by non-singular linear transformations represented by $\Phi_{i'}{}^{j}$ and $\Phi^{j'}{}_{j}$, then the condition, that the new bases $(\boldsymbol{e}_{i'})$ and $(\boldsymbol{e}^{j'})$ continue to be dual, requires

$$\begin{aligned}\delta^{j'}{}_{i'} = \langle \boldsymbol{e}^{j'}, \boldsymbol{e}_{i'} \rangle &= \Phi^{j'}{}_{j}\Phi_{i'}{}^{k}\langle \boldsymbol{e}^j, \boldsymbol{e}_k \rangle \\ &= \Phi^{j'}{}_{j}\Phi_{i'}{}^{k}\delta^{j}{}_{k} = \Phi^{j'}{}_{j}\Phi^{j}{}_{i'};\end{aligned} \tag{23}$$

in other words, the matrices $[\Phi_{i'}{}^{i}]$ and $[\Phi^{j'}{}_{j}]$ are the inverses of one another.

Finally, we may note that if we change from one set of local coordinates (x^i) to another set $(x^{i'})$, then the corresponding expressions for $\Phi_{i'}{}^{i}$ and $\Phi^{j'}{}_{j}$ are

$$\Phi_{i'}{}^{i} = \left(\frac{\partial x^i}{\partial x^{i'}}\right)_p \qquad \text{and} \qquad \Phi^{j'}{}_{j} = \left(\frac{\partial x^{j'}}{\partial x^j}\right)_p. \tag{24}$$

Associated with any function f on the manifold, one defines a one-form $\mathrm{d}f$ by requiring that

$$\mathrm{d}f(\boldsymbol{X}) = \langle \mathrm{d}f, \boldsymbol{X} \rangle = \boldsymbol{X}f, \tag{25}$$

for any vector $\boldsymbol{X} \in T_p$. In a local coordinate basis,

$$\boldsymbol{X} = X^i \frac{\partial}{\partial x^i}; \tag{26}$$

and by definition,

$$\langle \mathrm{d}f, \boldsymbol{X} \rangle = X^i \frac{\partial f}{\partial x^i} = X^i f_{,i}; \tag{27}$$

in particular,

$$\langle \mathrm{d}x^j, \frac{\partial}{\partial x^i} \rangle = \delta^j{}_i. \tag{28}$$

Hence, the one-forms $(\mathrm{d}x^j)$ provide a local coordinate basis for the cotangent vectors which is dual to the local coordinate basis provided by the tangent vectors $(\partial_j = \partial/\partial x^j)$ for the tangent space. The bases (∂_j) and $(\mathrm{d}x^j)$ are sometimes referred to as the *canonical bases* for the tangent and the cotangent spaces.

We note that if

$$\mathrm{d}f = \alpha_j \mathrm{d}x^j, \tag{29}$$

then it follows from

$$\begin{aligned}X^i f_{,i} = \langle \mathrm{d}f, \boldsymbol{X} \rangle &= \langle \alpha_j \mathrm{d}x^j, X^i \partial_i \rangle \\ &= \alpha_j X^i \langle \mathrm{d}x^j, \partial_i \rangle = \alpha_i X^i,\end{aligned} \tag{30}$$

that

$$\alpha_i = f_{,i} \qquad \text{and} \qquad \mathrm{d}f = f_{,i}\,\mathrm{d}x^i; \tag{31}$$

and this last equation is consistent with the conventional meaning one attaches to d*f*.

(*c*) *Tensors and tensor products*

Let

$$\Pi^s{}_r = \underbrace{T^*{}_p \times T^*{}_p \times \ldots \times T^*{}_p}_{r \text{ factors}} \times \underbrace{T_p \times T_p \times \ldots \times T_p}_{s \text{ factors}} \tag{32}$$

represent the Cartesian product or r cotangent spaces and s tangent spaces at some point p of a manifold, i.e., the space of ordered sets of r one-forms and s tangent vectors: $(\boldsymbol{\omega}^1, \ldots, \boldsymbol{\omega}^r, \boldsymbol{X}_1, \ldots, \boldsymbol{X}_s)$. And consider a *multilinear* mapping, $\boldsymbol{T}$, of the manifold $\Pi^s{}_r$ to the reals:

$$\boldsymbol{T}: \qquad \Pi^s{}_r \rightarrow \mathbb{R}^1. \tag{33}$$

Precisely, what the mapping provides is an association (in some unique manner) of any given ordered set of r one-forms and s tangent vectors to a real number:

$$\boldsymbol{T}(\boldsymbol{\omega}^1, \ldots, \boldsymbol{\omega}^r, \boldsymbol{X}_1, \ldots, \boldsymbol{X}_s) = \text{a real number.} \tag{34}$$

The condition that the map is multilinear requires that

$$\begin{aligned} &\boldsymbol{T}(\boldsymbol{\omega}^1, \ldots, \boldsymbol{\omega}^r, \alpha \boldsymbol{X} + \beta \boldsymbol{Y}, \boldsymbol{X}_2, \ldots, \boldsymbol{X}_s) \\ &\quad = \alpha \boldsymbol{T}(\boldsymbol{\omega}^1, \ldots, \boldsymbol{\omega}^r, \boldsymbol{X}, \boldsymbol{X}_2, \ldots, \boldsymbol{X}_s) + \beta \boldsymbol{T}(\boldsymbol{\omega}^1, \ldots, \boldsymbol{\omega}^r, \boldsymbol{Y}, \boldsymbol{X}_2, \ldots, \boldsymbol{X}_s), \end{aligned} \tag{35}$$

for all α, $\beta \in \mathbb{R}^1$ and $\boldsymbol{X}$, $\boldsymbol{Y} \in T_p$; and for similar replacements of all the other forms and vectors. A multilinear mapping so defined is said to be a *tensor of type* (r, s). Linear combinations of tensors of a given type (r, s) are defined by the rule

$$\begin{aligned} &(\alpha \boldsymbol{T} + \beta \boldsymbol{S})(\boldsymbol{\omega}^1, \ldots, \boldsymbol{\omega}^r, \boldsymbol{X}_1, \ldots, \boldsymbol{X}_s) \\ &\quad = \alpha \boldsymbol{T}(\boldsymbol{\omega}^1, \ldots, \boldsymbol{\omega}^r, \boldsymbol{X}_1, \ldots, \boldsymbol{X}_s) + \beta \boldsymbol{S}(\boldsymbol{\omega}^1, \ldots, \boldsymbol{\omega}^r, \boldsymbol{X}_1, \ldots, \boldsymbol{X}_s), \end{aligned} \tag{36}$$

for all $\alpha, \beta \in \mathbb{R}^1$, $\boldsymbol{\omega}^i \in T^*{}_p$, and $\boldsymbol{X}_j \in T_p$ $(i = 1, \ldots, r;\ j = 1, \ldots, s)$. By these rules, tensors of a given type (r, s) span a linear vector-space of dimension n^{r+s}. And the space of such tensors is called the space of *tensor products* and denoted by

$$T^r{}_s(p) = \underbrace{T_p \otimes \ldots \otimes T_p}_{r \text{ factors}} \otimes \underbrace{T^*{}_p \otimes \ldots \otimes T^*{}_p}_{s \text{ factors}} \tag{37}$$

We shall presently verify that a basis for tensor products of type (r, s) is provided by the n^{r+s} special mappings

$$\begin{aligned} &\boldsymbol{e}_{i_1 \ldots i_r}{}^{j_1 \ldots j_s}(\boldsymbol{\omega}^1, \ldots, \boldsymbol{\omega}^r, \boldsymbol{X}_1, \ldots, \boldsymbol{X}_s) \\ &\quad = \boldsymbol{e}_{i_1 \ldots i_r}{}^{j_1 \ldots j_s}(\omega^1{}_{k_1} \boldsymbol{e}^{k_1}, \ldots, \omega^r{}_{k_r} \boldsymbol{e}^{k_r}, X_1{}^{l_1} \boldsymbol{e}_{l_1}, \ldots, X_s{}^{l_s} \boldsymbol{e}_{l_s}) \\ &\quad = \omega^1{}_{i_1} \ldots \omega^r{}_{i_r} X_1{}^{j_1} \ldots X_s{}^{j_s}. \end{aligned} \tag{38}$$

These mappings are clearly linear in every argument and are tensors of type (r, s). An equivalent way of defining these mappings is

$$e_{i_1 \ldots i_r}{}^{j_1 \ldots j_s}(e^{k_1}, \ldots, e^{k_r}, e_{l_1}, \ldots, e_{l_s}) = \delta^{k_1}{}_{i_1}\delta^{k_2}{}_{i_2} \ldots \delta^{k_r}{}_{i_r}\delta^{j_1}{}_{l_1} \ldots \delta^{j_s}{}_{l_s}. \tag{39}$$

That any tensor of type (r, s) can be expressed as a linear combination of the mappings (39) follows from noting

$$\begin{aligned} T(\omega^1, \ldots, \omega^r, X_1, \ldots, X_s) &= T(\omega^1{}_{i_1}e^{i_1}, \ldots, \omega^r{}_{i_r}e^{i_r}, X_1{}^{j_1}e_{j_1}, \ldots, X_s{}^{j_s}e_{j_s}) \\ &= \omega^1{}_{i_1} \ldots \omega^r{}_{i_r}X_1{}^{j_1} \ldots X_s{}^{j_s}T(e^{i_1}, \ldots, e^{i_r}, e_{j_1}, \ldots, e_{j_s}), \end{aligned} \tag{40}$$

and letting

$$T(e^{i_1}, \ldots, e^{i_r}, e_{j_1}, \ldots, e_{j_s}) = T^{i_1 \ldots i_r}{}_{j_1 \ldots j_s}; \tag{41}$$

for, we can then write (cf. equation (38))

$$T = T^{i_1 \ldots i_r}{}_{j_1 \ldots j_s}e_{i_1 \ldots i_r}{}^{j_1 \ldots j_s}; \tag{42}$$

and this is the required expression for T as a linear combination of the mappings (39). It is manifest that the mappings (39) are linearly independent; they, therefore, provide a basis for tensors of type (r, s). The number of these basis elements, $e_{i_1 \ldots i_r}{}^{j_1 \ldots j_s}$, is n^{r+s} (which is the dimension of the $T^r{}_s$).

The coefficients, $T^{i_1 \ldots i_r}{}_{j_1 \ldots j_s}$, in the expansion (42), are said to be the *components* of T relative to the chosen basis.

One generally writes

$$e_{i_1 \ldots i_r}{}^{j_1 \ldots j_s} = e_{i_1} \otimes \ldots \otimes e_{i_r} \otimes e^{j_1} \otimes \ldots \otimes e^{j_s}, \tag{43}$$

as representing the tensor product of the dual bases (e_i) and (e^j) of T_p and $T^*{}_p$. In this notation, the tensor product

$$Y_1 \otimes \ldots \otimes Y_r \otimes \Omega^1 \otimes \ldots \otimes \Omega^s \tag{44}$$

of r tangent vectors and s one-forms is that element of $T^r{}_s$ which maps $(\omega^1 \ldots, \omega^r, X_1, \ldots, X_s)$ to the number

$$\langle \omega^1, Y_1 \rangle \ldots \langle \omega^r, Y_r \rangle \langle \Omega^1, X_1 \rangle \ldots \langle \Omega^s, X_s \rangle. \tag{45}$$

In particular,

$$\begin{aligned} (e_{i_1} \otimes \ldots \otimes e_{i_r} \otimes e^{j_1} \otimes \ldots \otimes e^{j_s})&(\omega^1, \ldots, \omega^r, X_1, \ldots, X_s) \\ &= \langle \omega^1, e_{i_1} \rangle \ldots \langle \omega^r, e_{i_r} \rangle \langle e^{j_1}, X_1 \rangle \ldots \langle e^{j_s}, X_s \rangle \\ &= \omega^1{}_{i_1} \ldots \omega^r{}_{i_r}X_1{}^{j_1} \ldots X_s{}^{j_s} = e_{i_1 \ldots i_r}{}^{j_1 \ldots j_s}(\omega^1, \ldots, \omega^r, X_1, \ldots, X_s); \end{aligned} \tag{46}$$

and this justifies the notation.

If instead of the dual bases (e_i) and (e^j), we choose different dual bases $(e_{i'})$ and $(e^{j'})$, then it follows from equations (22) that the components of T, relative to the new basis,

$$e_{i'_1} \otimes \ldots \otimes e_{i'_r} \otimes e^{j'_1} \otimes \ldots \otimes e^{j'_s},$$

are given by

$$T^{i'_1 \dots i'_r}{}_{j'_1 \dots j'_s} = \Phi^{i'_1}{}_{i_1} \dots \Phi^{i'_r}{}_{i_r} \Phi_{j'_1}{}^{j_1} \dots \Phi_{j'_s}{}^{j_s} T^{i_1 \dots i_r}{}_{j_1 \dots j_s}. \tag{47}$$

The *contraction* of a tensor of type (r, s) with the components $T^{i_1 \dots i_r}{}_{j_1 \dots j_s}$, with respect to a chosen contravariant index i_p and a chosen covariant index j_q, is defined as the following tensor of type $(r-1, s-1)$:

$$T^{i_1 \dots i_{p-1} k i_{p+1} \dots i_r}{}_{j_1 \dots j_{q-1} k j_{q+1} \dots j_s} e_{i_1} \otimes \dots \otimes e_{i_{p-1}} \otimes e_{i_{p+1}} \otimes \dots \otimes e_{i_r} \otimes e^{j_1} \otimes \dots \otimes e^{j_{q-1}} \otimes e^{j_{q+1}} \otimes \dots \otimes e^{j_s}, \tag{48}$$

where, as the notation indicates, summation over all values of k (of the i_p-contravariant index and of the j_q-covariant index) is to be effected. It can be readily verified, with the aid of equations (23) and (47), that the process of contraction is independent of the chosen basis.

A tensor of type $(0, 2)$ is said to be *symmetric* or *antisymmetric* if

$$T(X, Y) = T(Y, X) \quad \text{or} \quad T(X, Y) = -T(Y, X) \quad \text{for all } X \text{ and } Y \text{ in } T_p. \tag{49}$$

In terms of components, in an arbitrary basis, symmetry or antisymmetry implies

$$T_{ij} = T_{ji} \quad \text{or} \quad T_{ij} = -T_{ji}. \tag{50}$$

More generally, a tensor of type (r, s) is said to be symmetric or antisymmetric in its covariant indices i and j if

$$T(\omega^1, \dots, \omega^i, \dots, \omega^j, \dots, \omega^r, X_1, \dots X_s) = \pm T(\omega^1, \dots, \omega^j, \dots, \omega^i, \dots, \omega^r, X_1, \dots, X_s), \tag{51}$$

for all ω's and X's. Symmetry or antisymmetry with respect to chosen contravariant indices are similarly defined.

3. The calculus of forms

A particularly important class of tensors of type $(0, s)$ is the class of *totally antisymmetric tensors*, i.e., covariant tensors which are antisymmetric in every pair of their arguments, i.e.,

$$T(X_1, \dots, X_i, \dots, X_j, \dots, X_s) = -T(X_1, \dots, X_j, \dots, X_i, \dots, X_s), \tag{52}$$

for all pairs of indices i and j and for all X's. Tensors of this kind can be constructed out of a general tensor T of type $(0, s)$ by applying to it the alternating operator A whose effect on it is to give the linear combination defined by

$$AT(X_1, \dots, X_s) = \frac{1}{s!} \sum_{j_1, \dots, j_s} \operatorname{sgn}(j_1, \dots, j_s)\, T(X_{j_1}, \dots, X_{j_s}), \tag{53}$$

where the summation is extended over all $s!$ permutations of the s integers $(1, \ldots, s)$ and $\operatorname{sgn}(j_1, \ldots, j_s) = \pm 1$, according as $(j_1, \ldots, j_s)$ is an even or an odd permutation of $(1, \ldots, s)$; and equation (53) is to be valid for every $(X_1, \ldots, X_s)$.

It is clear that if T is already totally antisymmetric, the effect of A on it is, simply, to reproduce T. Also, if $s > n$ (the dimension of the vector space) the effect of A on $T(X_1, \ldots, X_s)$ is to reduce it to zero; in other words, there can be no totally antisymmetric tensor of type $(0, s)$ for $s > n$.

Totally antisymmetric tensors of type $(0, s)$ are called *s-forms*. Since they must vanish when any two of their arguments coincide, it follows that the s-forms span a vector space of dimension $n!/s!\,(n-s)!$. This space is denoted by $\Lambda^s T^*_p$.

If $T_{j_1 \ldots j_s}$ are the components of a tensor of type $(0, s)$ relative to the basis,

$$e^{j_1} \otimes \ldots \otimes e^{j_s},$$

and if the tensor should be totally antisymmetric, then its $n!/s!(n-s)!$ distinct components can be distinguished by arranging its indices in a strictly descending sequence in the manner:

$$T_{j_1 \ldots j_s} \qquad \text{where} \qquad j_1 > j_2 > \ldots > j_s. \tag{54}$$

A basis for $\Lambda^s T^*_p$ can be obtained by applying the alternating operator A to the basis elements of the tensor product:

$$A(e^{j_1} \otimes \ldots \otimes e^{j_s}).$$

The resulting basis elements are written as the *exterior* or the *wedge product* of the e^j's in the manner:

$$e^{j_1} \wedge e^{j_2} \wedge \ldots \wedge e^{j_s} \qquad (j_1 > j_2 > \ldots > j_s). \tag{55}$$

A general s-form can then be written as

$$\boldsymbol{\Omega} = \Omega_{j_1 \ldots j_s} e^{j_1} \wedge e^{j_2} \wedge \ldots \wedge e^{j_s} \qquad (j_1 > j_2 > \ldots > j_s), \tag{56}$$

where the summation is now extended only over strictly descending sequences.

Since interchanging a pair of indices is equivalent to interchanging the corresponding elements in the wedge product, it follows that interchanging the elements in a wedge product must be accompanied by a change of sign; thus

$$e^j \wedge e^k = -e^k \wedge e^j. \tag{57}$$

In a local coordinate basis, the expression for an s-form is

$$\boldsymbol{\Omega} = \Omega_{j_1 \ldots j_s} \mathrm{d}x^{j_1} \wedge \ldots \wedge \mathrm{d}x^{j_s}. \tag{58}$$

Given any p-form $\boldsymbol{\Omega}^1$ and a q-form $\boldsymbol{\Omega}^2$, we can form their wedge (or, exterior) product by the rule

$$\boldsymbol{\Omega}^1 \wedge \boldsymbol{\Omega}^2 = A(\boldsymbol{\Omega}^1 \otimes \boldsymbol{\Omega}^2) \tag{59}$$

to obtain a $(p+q)$-form. (It must accordingly vanish identically if $p+q>n$.)

Wedge products of forms clearly obey the associative and the distributive laws, but they are not in general commutative. For, by definition,

$$\boldsymbol{\Omega}^1 \wedge \boldsymbol{\Omega}^2 = (\Omega^1{}_{j_1 \ldots j_p} \boldsymbol{e}^{j_1} \wedge \ldots \wedge \boldsymbol{e}^{j_p}) \wedge (\Omega^2{}_{k_1 \ldots k_q} \boldsymbol{e}^{k_1} \wedge \ldots \wedge \boldsymbol{e}^{k_q}), \quad (60)$$

where $(j_1, \ldots, j_p)$ and $(k_1, \ldots, k_q)$ are strictly descending sequences. Accordingly,

$$\begin{aligned} \boldsymbol{\Omega}^1 \wedge \boldsymbol{\Omega}^2 &= (-1)^{pq} (\Omega^2{}_{k_1 \ldots k_q} \boldsymbol{e}^{k_1} \wedge \ldots \wedge \boldsymbol{e}^{k_q}) \wedge (\Omega^1{}_{j_1 \ldots j_p} \boldsymbol{e}^{j_1} \wedge \ldots \wedge \boldsymbol{e}^{j_p}) \\ &= (-1)^{pq} \boldsymbol{\Omega}^2 \wedge \boldsymbol{\Omega}^1, \end{aligned} \quad (61)$$

since each of the q basis elements $\boldsymbol{e}^{k_1}, \ldots, \boldsymbol{e}^{k_q}$ must suffer p interchanges before $\boldsymbol{\Omega}^1 \wedge \boldsymbol{\Omega}^2$ can be brought to the form required of $\boldsymbol{\Omega}^2 \wedge \boldsymbol{\Omega}^1$.

So far, we have considered tensors and forms defined at a point on the manifold. We shall now enlarge the basic definitions in a way which enables us to envisage *fields* defined on M. Thus, a smooth *tensor-field* $T^r{}_s(M)$ (or, simply $T^r{}_s$) of type (r, s) on M is an assignment of an element of $T^r{}_s(p)$ at each point $p \in M$ in such a way that the components of $T^r{}_s$ relative to any local coordinate basis are smooth functions of the coordinates. This enlargement of the basic definitions is necessary if we are to formulate notions of differentiation.

In the future we shall be concerned only with smooth tensor fields; and this is to be understood even if the qualifying words 'smooth' and 'field' are omitted.

(*a*) *Exterior differentiation*

Exterior differentiation is effected by an operator d applied to forms. It converts p-forms to $(p+1)$-forms consistently with the following rules:

(a) The operator d, applied to functions (or zero-forms) f, yields a one-form $\mathrm{d}f$ defined by

$$\mathrm{d}f(\boldsymbol{X}) = \langle \mathrm{d}f, \boldsymbol{X} \rangle = \boldsymbol{X}f$$

for every $\boldsymbol{X} \in T^1{}_0$. In particular, in a local coordinate basis

$$\mathrm{d}f = \frac{\partial f}{\partial x^j} \mathrm{d}x^j.$$

(b) If $\boldsymbol{A}_1$ and $\boldsymbol{A}_2$ are two p-forms,

$$\mathrm{d}(\alpha \boldsymbol{A}_1 + \beta \boldsymbol{A}_2) = \alpha \mathrm{d}\boldsymbol{A}_1 + \beta \mathrm{d}\boldsymbol{A}_2 \qquad (\alpha, \beta \in \mathbb{R}^1).$$

(c) If $\boldsymbol{A}$ is a p-form and $\boldsymbol{B}$ is a q-form,

$$\mathrm{d}(\boldsymbol{A} \wedge \boldsymbol{B}) = \mathrm{d}\boldsymbol{A} \wedge \boldsymbol{B} + (-1)^p \boldsymbol{A} \wedge \mathrm{d}\boldsymbol{B}.$$

(d) Poincaré's lemma, which requires that

$$\mathrm{d}(\mathrm{d}\boldsymbol{A}) = 0,$$

for every p-form $\boldsymbol{A}$.

To clarify that the operator d subject to the foregoing rules is well-defined, consider, first, the exterior derivative

$$\mathrm{d}A = \mathrm{d}(A_{j_1 \ldots j_p} \mathrm{d}x^{j_1} \wedge \ldots \wedge \mathrm{d}x^{j_p})$$

of a p-form A. By rules (a), (b), and (d)

$$\begin{aligned} \mathrm{d}A &= \mathrm{d}A_{j_1 \ldots j_p} \wedge \mathrm{d}x^{j_1} \wedge \ldots \wedge \mathrm{d}x^{j_p} \\ &= \frac{\partial A_{j_1 \ldots j_p}}{\partial x^k} \mathrm{d}x^k \wedge \mathrm{d}x^{j_1} \wedge \ldots \wedge \mathrm{d}x^{j_p}. \end{aligned} \tag{62}$$

It is important to verify at this point that $\mathrm{d}A$, as given by equation (62), is independent of the choice of the local coordinate system. For, if instead of the local coordinates (x^j) we had chosen a different set of local coordinates $(x^{j'})$, then by equations (24) and (47),

$$A_{j'_1 \ldots j'_p} = A_{j_1 \ldots j_p} \frac{\partial x^{j_1}}{\partial x^{j'_1}} \cdots \frac{\partial x^{j_p}}{\partial x^{j'_p}}, \tag{63}$$

and we should conclude that

$$\begin{aligned} &\mathrm{d}(A_{j'_1 \ldots j'_p} \mathrm{d}x^{j'_1} \wedge \ldots \wedge \mathrm{d}x^{j'_p}) \\ &= \mathrm{d}\left(A_{j_1 \ldots j_p} \frac{\partial x^{j_1}}{\partial x^{j'_1}} \cdots \frac{\partial x^{j_p}}{\partial x^{j'_p}} \mathrm{d}x^{j'_1} \wedge \ldots \wedge \mathrm{d}x^{j'_p} \right) \\ &= \frac{\partial x^{j_1}}{\partial x^{j'_1}} \cdots \frac{\partial x^{j_p}}{\partial x^{j'_p}} \mathrm{d}A_{j_1 \ldots j_p} \wedge \mathrm{d}x^{j'_1} \wedge \ldots \wedge \mathrm{d}x^{j'_p} \\ &\quad + \frac{\partial^2 x^{j_1}}{\partial x^{k'} \partial x^{j'_1}} \frac{\partial x^{j_2}}{\partial x^{j'_2}} \cdots \frac{\partial x^{j_p}}{\partial x^{j'_p}} A_{j_1 \ldots j_p} \mathrm{d}x^{k'} \wedge \mathrm{d}x^{j'_1} \wedge \ldots \wedge \mathrm{d}x^{j'_p} + \ldots \\ &\quad \ldots + \frac{\partial x^{j_1}}{\partial x^{j'_1}} \cdots \frac{\partial x^{j_{p-1}}}{\partial x^{j'_{p-1}}} \frac{\partial^2 x^{j_p}}{\partial x^{k'} \partial x^{j'_p}} A_{j_1 \ldots j_p} \mathrm{d}x^{k'} \wedge \mathrm{d}x^{j'_1} \ldots \wedge \mathrm{d}x^{j'_p}. \end{aligned} \tag{64}$$

All the terms on the right-hand side of equation (64) involving the second derivatives of $x^{j_i} (i = 1, \ldots, p)$ vanish on account of their symmetry in k' and j'_i and the antisymmetry of the basis elements, in these same indices, in the wedge product; and the sole surviving term is the first one which is clearly the same as

$$\mathrm{d}A_{j_1 \ldots j_p} \wedge \mathrm{d}x^{j_1} \wedge \ldots \wedge \mathrm{d}x^{j_p} = \mathrm{d}(A_{j_1 \ldots j_p} \mathrm{d}x^{j_1} \wedge \ldots \wedge \mathrm{d}x^{j_p}).$$

Accordingly,

$$\mathrm{d}(A_{j'_1 \ldots j'_p} \mathrm{d}x^{j'_1} \wedge \ldots \wedge \mathrm{d}x^{j'_p}) = \mathrm{d}(A_{j_1 \ldots j_p} \mathrm{d}x^{j_1} \wedge \ldots \wedge \mathrm{d}x^{j_p}); \tag{65}$$

and this is what we set out to verify.

We next verify that the rule (c) is consistent with the expression (62) for $\mathrm{d}A$.

For, by the rules (a), (b), and (d)

$$d(\boldsymbol{A} \wedge \boldsymbol{B}) = d(A_{j_1 \dots j_p} dx^{j_1} \wedge \dots \wedge dx^{j_p} \wedge B_{k_1 \dots k_q} dx^{k_1} \wedge \dots \wedge dx^{k_q})$$

$$= \frac{\partial A_{j_1 \dots j_p}}{\partial x^i} dx^i \wedge dx^{j_1} \wedge \dots \wedge dx^{j_p} \wedge B_{k_1 \dots k_q} dx^{k_1} \wedge \dots \wedge dx^{k_q}$$

$$+ A_{j_1 \dots j_p} \frac{\partial B_{k_1 \dots k_q}}{\partial x^i} dx^i \wedge dx^{j_1} \wedge \dots \wedge dx^{j_p} \wedge dx^{k_1} \wedge \dots \wedge dx^{k_q}$$

$$= d\boldsymbol{A} \wedge \boldsymbol{B} + (-1)^p A_{j_1 \dots j_p} dx^{j_1} \wedge \dots \wedge dx^{j_p} \wedge \left(\frac{\partial B_{k_1 \dots k_q}}{\partial x^i} dx^i \wedge dx^{k_1} \wedge \dots \wedge dx^{k_q} \right)$$

$$= d\boldsymbol{A} \wedge \boldsymbol{B} + (-1)^p \boldsymbol{A} \wedge d\boldsymbol{B}. \tag{66}$$

And finally, to establish the consistency of rule (d), we need only observe that

$$d(d\boldsymbol{A}) = d\left(\frac{\partial A_{j_1 \dots j_p}}{\partial x^k} dx^k \wedge dx^{j_1} \wedge \dots \wedge dx^{j_p} \right)$$

$$= \frac{\partial^2 A_{j_1 \dots j_p}}{\partial x^i \partial x^k} dx^i \wedge dx^k \wedge dx^{j_1} \wedge \dots \wedge dx^{j_p} \equiv 0. \tag{67}$$

This completes the demonstration that the operation d is, indeed, well-defined.

(*b*) *Lie bracket and Lie differentiation*

Given any two vector fields, $\boldsymbol{X}$ and $\boldsymbol{Y}$, their *Lie bracket*, $[\boldsymbol{X}, \boldsymbol{Y}]$, is defined by its action on any function f; and it is given by

$$[\boldsymbol{X}, \boldsymbol{Y}] f = (\boldsymbol{XY} - \boldsymbol{YX}) f = \boldsymbol{X}(\boldsymbol{Y} f) - \boldsymbol{Y}(\boldsymbol{X} f). \tag{68}$$

The Lie bracket of any two tangent vectors is, again, a tangent vector, since

$$[\boldsymbol{X}, \boldsymbol{Y}](\alpha f + \beta g) = \alpha [\boldsymbol{X}, \boldsymbol{Y}] f + \beta [\boldsymbol{X}, \boldsymbol{Y}] g \tag{69}$$

and

$$[\boldsymbol{X}, \boldsymbol{Y}](fg) = g[\boldsymbol{X}, \boldsymbol{Y}] f + f[\boldsymbol{X}, \boldsymbol{Y}] g, \tag{70}$$

where f and g are any two functions and α and β are any two real numbers. The first of these relations is manifest while the second follows quite readily:

$$\begin{aligned}
[\boldsymbol{X}, \boldsymbol{Y}](fg) &= \boldsymbol{X}(\boldsymbol{Y} fg) - \boldsymbol{Y}(\boldsymbol{X} fg) \\
&= \boldsymbol{X}(g\boldsymbol{Y} f + f\boldsymbol{Y} g) - \boldsymbol{Y}(g\boldsymbol{X} f + f\boldsymbol{X} g) \\
&= g\boldsymbol{XY} f + (\boldsymbol{X} g)(\boldsymbol{Y} f) + (\boldsymbol{X} f)(\boldsymbol{Y} g) + f\boldsymbol{XY} g \\
&\quad - \{g\boldsymbol{YX} f + (\boldsymbol{X} f)(\boldsymbol{Y} g) + (\boldsymbol{Y} f)(\boldsymbol{X} g) + f\boldsymbol{YX} g\} \\
&= g[\boldsymbol{X}, \boldsymbol{Y}] f + f[\boldsymbol{X}, \boldsymbol{Y}] g.
\end{aligned} \tag{71}$$

The relation (69) establishes the Lie bracket as a linear operator while the relation (70) establishes it as a differentiation.

As may be readily verified, the Lie bracket satisfies the *Jacobi identity*,

$$[[X, Y], Z]+[[Y, Z], X]+[[Z, X], Y]=0. \tag{72}$$

We have seen that the Lie bracket of X and Y is a tangent vector. Its components, relative to a local coordinate basis, can be obtained by its action on x^j. Thus,

$$\begin{aligned}[X, Y]^j &= (XY-YX)x^j = XY^j - YX^j \\ &= X^k Y^j{}_{,k} - Y^k X^j{}_{,k},\end{aligned} \tag{73}$$

where (as indicated earlier) a comma preceding an index denotes partial differentiation with respect to the local coordinate with that same index.

In a local coordinate basis, the Lie bracket $[\partial_k, \partial_j]$ clearly vanishes.

Considered as a differentiation, $[X, Y]$ is called the *Lie derivative of* Y *in the direction* X and is written as

$$\mathscr{L}_X Y = [X, Y] = -[Y, X] = -\mathscr{L}_Y X. \tag{74}$$

More generally, we define the Lie derivative, $\mathscr{L}_X T$, of a tensor field T of a given type, as a tensor of the same type which satisfies the following rules:

(a) Its action on a scalar field f is given by

$$\mathscr{L}_X f = Xf = \mathrm{d}f(X). \tag{75}$$

(b) Its action on a tangent vector Y, as we have already defined, is given by

$$\mathscr{L}_X Y = [X, Y]. \tag{76}$$

And (c) it operates linearly on tensor fields and satisfies the Liebnitz rule when acting on tensor products:

$$\mathscr{L}_X(S \otimes T) = \mathscr{L}_X S \otimes T + S \otimes \mathscr{L}_X T, \tag{77}$$

where S and T are arbitrary tensor fields.

The last of the foregoing rules enables us to derive the effect of $\mathscr{L}_X$ on a tensor of arbitrary type. Thus, its effect on a one-form ω can be determined by considering, for any arbitrary vector-field Y, the contracted version of the relation

$$\mathscr{L}_X(\omega \otimes Y) = (\mathscr{L}_X \omega) \otimes Y + \omega \otimes (\mathscr{L}_X Y), \tag{78}$$

namely,

$$\mathscr{L}_X \langle \omega, Y \rangle = \langle \mathscr{L}_X \omega, Y \rangle + \langle \omega, \mathscr{L}_X Y \rangle. \tag{79}$$

Writing out this last equation explicitly, we have

$$X^k(\omega_j Y^j)_{,k} = (\mathscr{L}_X \omega)_j Y^j + \omega_j (\mathscr{L}_X Y)^j \tag{80}$$

or, making use of equation (73), we obtain

$$\begin{aligned}(\mathscr{L}_X \omega)_j Y^j &= X^k(\omega_{j,k} Y^j + \omega_j Y^j{}_{,k}) - \omega_j(X^k Y^j{}_{,k} - Y^k X^j{}_{,k}) \\ &= (X^k \omega_{j,k} + \omega_k X^k{}_{,j}) Y^j.\end{aligned} \tag{81}$$

Since this last equation is valid for an arbitrary Y, we conclude that

$$(\mathscr{L}_X\omega)_j = \omega_{j,k}X^k + \omega_k X^k{}_{,j}. \tag{82}$$

We may write equation (79) alternatively in the form

$$\mathscr{L}_X[\omega(Y)] = (\mathscr{L}_X\omega)(Y) + \omega(\mathscr{L}_X Y). \tag{83}$$

By rule (c), equation (83) admits of generalization to a tensor of type (r, s). We have

$$\begin{aligned}\mathscr{L}_X[T(\omega^1, \ldots, \omega^r, Y_1, \ldots, Y_s)] &= (\mathscr{L}_X T)(\omega^1, \ldots, \omega^r, Y_1, \ldots, Y_s)\\ &+ T(\mathscr{L}_X\omega^1, \omega^2, \ldots, \omega^r, Y_1, \ldots, Y_s) + \ldots\\ &+ T(\omega^1, \ldots, \omega^r, Y_1, \ldots, \mathscr{L}_X Y_s),\end{aligned} \tag{84}$$

where all the terms in this equation, except the first one on the right-hand side, can be evaluated in terms of the known results (73), (75), and (82). The components of $\mathscr{L}_X T$ can, therefore, be deduced from equation (84).

For later use, we shall derive here a simple identity relating the exterior derivative of a one-form to Lie derivatives. By making use of equation (82), we have

$$\begin{aligned}&\langle \mathscr{L}_X\omega, Y\rangle - Y\langle\omega, X\rangle\\ &= (\omega_{j,k}X^k + \omega_k X^k{}_{,j})Y^j - Y^j(\omega_{k,j}X^k + \omega_k X^k{}_{,j})\\ &= (\omega_{j,k} - \omega_{k,j})X^k Y^j = 2\,\mathrm{d}\omega(X, Y).\end{aligned} \tag{85}$$

Now, substituting for $\langle \mathscr{L}_X\omega, Y\rangle$ from equation (79), we obtain the required result:

$$\mathrm{d}\omega(X, Y) = \tfrac{1}{2}\{X\langle\omega, Y\rangle - Y\langle\omega, X\rangle - \langle\omega, [X, Y]\rangle\} \tag{86}$$

where we have written the Lie bracket $[X, Y]$ in place $\mathscr{L}_X Y$.

4. Covariant differentiation

We shall now define a type of differentiation which, unlike exterior and Lie differentiation, requires that the manifold be endowed with an additional structure. This additional structure is an *affine connection*, ∇, which assigns to each vector field X on M a differential operator, ∇_X, which maps an arbitrary vector-field, Y, into a vector field $\nabla_X Y$. Consistent with these requirements, we impose the conditions,

(a) $\nabla_X Y$ is linear in the argument X, i.e.,

$$\nabla_{fX+gY}Z = f\nabla_X Z + g\nabla_Y Z \quad (X, Y, Z \in T^1{}_0), \tag{87}$$

where f and g are any two arbitrary functions defined on M;

(b) $\nabla_X Y$ is linear in the argument Y, i.e.,

$$\nabla_X(Y+Z) = \nabla_X Y + \nabla_X Z \qquad (X, Y, Z \in T^1{}_0), \tag{88}$$

(c)

$$\nabla_X f = Xf, \tag{89}$$

where f is any function on M; and, finally,

(d)
$$\nabla_X(fY) = (\nabla_X f)Y + f\nabla_X Y. \tag{90}$$

It should be noted that, according to equation (89), in a local coordinate basis (∂_k), ∇_{∂_k}, when acting on functions, coincides with partial differentiation with respect to x^k.

With the action of ∇_X on vector fields $Y(\in T^1{}_0)$ specified by the rules (a)–(d), we now define the covariant derivative, ∇Y, of Y as a tensor field of type (1, 1) which maps the contravariant vector-field X to $\nabla_X Y$, i.e.,

$$\nabla Y(X) = \langle \nabla Y, X \rangle = \nabla_X Y, \tag{91}$$

for every $X \in \mathrm{T}^1{}_0$. In this notation, we can rewrite equation (90) in the form

$$\nabla(fY) = \mathrm{d}f \otimes Y + f\nabla Y. \tag{92}$$

To clarify what the assignment of a connection precisely means, it will be useful to rewrite $\nabla_X Y$ relative to some chosen dual bases (e_i) and (e^j). Thus, making use of the rules (a)–(d), we have

$$\nabla_X Y = \nabla_X(Y^j e_j) = (XY^j)e_j + Y^j \nabla_X e_j. \tag{93}$$

Since $\nabla_X e_j$, for a particular e_j, is a tensor field of type (1,0) we must have a representation, in the chosen basis, of the form

$$\nabla_X e_j = \omega^l{}_j(X)e_l, \tag{94}$$

where $\omega^l{}_j$ (depending on l and j) are one-forms. Accordingly, we may write

$$\nabla_X Y = (XY^j)e_j + Y^j \omega^l{}_j(X)e_l. \tag{95}$$

Alternatively, we may also rewrite equation (93) in the form

$$\begin{aligned} \nabla_X Y &= (XY^j)e_j + Y^j \nabla_{X^k e_k} e_j \\ &= (XY^j)e_j + Y^j X^k \nabla_{e_k} e_j, \end{aligned} \tag{96}$$

or, in conformity with the definition (94),

$$\nabla_X Y = (XY^j)e_j + Y^j X^k \omega^l{}_j(e_k)e_l. \tag{97}$$

Letting

$$\omega^l{}_j(e_k) = \omega^l{}_{jk} \tag{98}$$

be the coefficient of e^k in the expansion of $\omega^l{}_j$ in the basis (e^k), we conclude that a connection ∇ is specified by the n^2 one-forms $\omega^l{}_j$ or, equivalently, by the n^3 scalar fields $\omega^l{}_{jk}$.

Returning to equation (95) and rewriting it in the form

$$\nabla_X Y = [XY^j + \omega^j{}_l(X)Y^l]e_j, \tag{99}$$

we infer that

$$(\nabla_X Y)^j = XY^j + \omega^j{}_l(X)Y^l. \tag{100}$$

In a local coordinate basis $(\partial_k, \mathrm{d}x^l)$, equation (100) gives

$$(\nabla_{\partial_k} Y)^j = \partial_k Y^j + Y^l \omega^j{}_{lk} = Y^j{}_{,k} + Y^l \omega^j{}_{lk}. \tag{101}$$

In a local coordinate basis, it is customary to write

$$\Gamma^j{}_{lk} \text{ in a place of } \omega^j{}_{lk}; \tag{102}$$

and using semicolons to indicate covariant derivatives (in contrast to commas which indicate ordinary partial derivatives), we obtain the standard formula

$$Y^j{}_{;k} = Y^j{}_{,k} + Y^l \Gamma^j{}_{lk}. \tag{103}$$

The definition of covariant derivatives of vector fields can be extended to tensor fields, in general, by requiring that the operation of ∇ satisfies the Leibnitz rule when acting on tensor products. Thus, we require that

$$\nabla(\boldsymbol{S} \otimes \boldsymbol{T}) = \nabla \boldsymbol{S} \otimes \boldsymbol{T} + \boldsymbol{S} \otimes \nabla \boldsymbol{T}, \tag{104}$$

where $\boldsymbol{S}$ and $\boldsymbol{T}$ are two arbitrary tensor-fields. An immediate consequence of this requirement is (cf. equation (84))

$$\begin{aligned} \nabla_X\{\boldsymbol{T}(\boldsymbol{\omega}^1, \ldots, \boldsymbol{\omega}^r, \boldsymbol{Y}_1, \ldots, \boldsymbol{Y}_s)\} &= (\nabla_X \boldsymbol{T})(\boldsymbol{\omega}^1, \ldots, \boldsymbol{\omega}^r, \boldsymbol{Y}_1, \ldots, \boldsymbol{Y}_s) \\ &+ \boldsymbol{T}(\nabla_X \boldsymbol{\omega}^1, \boldsymbol{\omega}^2, \ldots, \boldsymbol{\omega}^r, \boldsymbol{Y}_1, \ldots, \boldsymbol{Y}_s) + \ldots \\ &+ \boldsymbol{T}(\boldsymbol{\omega}^1, \ldots, \boldsymbol{\omega}^r, \boldsymbol{Y}_1, \ldots, \boldsymbol{Y}_{s-1}, \nabla_X \boldsymbol{Y}_s). \end{aligned} \tag{105}$$

Thus, if $\boldsymbol{\Omega}$ is a one-form, then, for every vector field $\boldsymbol{Y}$, the foregoing equation gives

$$\nabla_X(\boldsymbol{\Omega}(\boldsymbol{Y})) = (\nabla_X \boldsymbol{\Omega})(\boldsymbol{Y}) + \boldsymbol{\Omega}(\nabla_X \boldsymbol{Y}), \tag{106}$$

or, in terms of a local basis $(\boldsymbol{e}_i)$ and $(\boldsymbol{e}^j)$, we have

$$\nabla_X(\Omega_j Y^j) = (\nabla_X \boldsymbol{\Omega})_j Y^j + \Omega_j (\nabla_X Y)^j. \tag{107}$$

Now making use of rule (c) and equation (100), we find

$$\begin{aligned} (\nabla_X \boldsymbol{\Omega})_j Y^j &= (X\Omega_j) Y^j + \Omega_j (XY^j) - \Omega_j [XY^j + Y^l \omega^j{}_l(X)] \\ &= (X\Omega_j) Y^j - \Omega_l \omega^l{}_j(X) Y^j. \end{aligned} \tag{108}$$

We conclude that

$$(\nabla_X \boldsymbol{\Omega})_j = X\Omega_j - \Omega_l \omega^l{}_j(X), \tag{109}$$

or, alternatively,

$$\nabla_X \boldsymbol{\Omega} = [X\Omega_j - \Omega_l \omega^l{}_j(X)] \boldsymbol{e}^j. \tag{110}$$

Specializing this last equation to the case when $\boldsymbol{\Omega} = \boldsymbol{e}^j$, we obtain the formula

$$\nabla_X \boldsymbol{e}^j = -\omega^j{}_l(X) \boldsymbol{e}^l, \tag{111}$$

which is to be contrasted with the earlier formula (94). Equation (111) shows that a knowledge of the n^2 one-forms $\boldsymbol{\omega}^l{}_j$ suffices to determine the covariant derivatives of one-forms, as well, once we accept the Leibnitz rule for tensor products.

Also, we may note that in a local coordinate basis, equation (109) gives

$$\Omega_{j;k} = \Omega_{j,k} - \Omega_l \Gamma^l{}_{jk}. \tag{112}$$

An important result follows from equations (109) and (112) when applied to the one-form $\mathrm{d}f$. Since the components of $\mathrm{d}f$ in a local coordinate basis are $f_{,j}$, we obtain from equation (112), in this case,

$$f_{,j;k} = f_{,j,k} - f_{,l}\Gamma^l{}_{jk}; \tag{113}$$

and by permuting the indices j and k in this equation, we obtain

$$f_{,k;j} = f_{,k,j} - f_{,l}\Gamma^l{}_{kj}. \tag{114}$$

Since partial differentiations applied to functions permute, we find, on taking the difference of equations (113) and (114),

$$f_{,j;k} - f_{,k;j} = -f_{,l}(\Gamma^l{}_{jk} - \Gamma^l{}_{kj}). \tag{115}$$

The right-hand side of equation (115) is non-vanishing only for non-symmetric connections. On this account, it is customary to write

$$T^l{}_{jk} = -(\Gamma^l{}_{jk} - \Gamma^l{}_{kj}). \tag{116}$$

From the occurrence of this quantity in equation (115), it is clear that $T^l{}_{jk}$ are the components of a tensor of type (1, 2). It is called the *torsion tensor*. We define the torsion tensor more generally in §5; meantime, we may note that in terms of it, we can write equation (115) in the form

$$f_{,j;k} - f_{,k;j} = T^l{}_{jk} f_{,l}. \tag{117}$$

Returning to equation (105), we now observe that, with the aid of equations (99) and (109), we can readily write down the covariant derivative of an arbitrary tensor-field. Thus,

$$S^{ij}{}_{k;l} = S^{ij}{}_{k,l} + S^{mj}{}_k \Gamma^i{}_{ml} + S^{im}{}_k \Gamma^j{}_{ml} - S^{ij}{}_m \Gamma^m{}_{kl}. \tag{118}$$

(a) *Parallel displacements and geodesics*

Let $\boldsymbol{Y}$ represent a contravariant vector-field. Consider its variation along a curve λ on M. The change $\delta \boldsymbol{Y}$ in $\boldsymbol{Y}$ caused by a displacement along λ resulting from an increment δt in t (which parametrizes λ) is, in a local coordinate system, given by

$$(\delta \boldsymbol{Y})^j = Y^j{}_{,k} \frac{\mathrm{d}x^k(\lambda(t))}{\mathrm{d}t} \delta t. \tag{119}$$

In Euclidean geometry and in a Cartesian system of coordinates, one would say that $\boldsymbol{Y}$ is '*parallely propagated*' along λ if $\delta \boldsymbol{Y} = 0$. In a general differentiable manifold with a connection, one defines, analogously, that a vector $\boldsymbol{Y}$ is parallely propagated along λ, if

$$(DY)^j = (\nabla_{\partial_k} Y)^j \frac{\mathrm{d}x^k(\lambda(t))}{\mathrm{d}t}\delta t = Y^j{}_{;k}\frac{\mathrm{d}x^k(\lambda(t))}{\mathrm{d}t}\delta t = 0, \qquad (120)$$

or, alternatively, if

$$(Y^j{}_{,k} + Y^l\Gamma^j{}_{lk})\frac{\mathrm{d}x^k(\lambda(t))}{\mathrm{d}t}\delta t = 0. \qquad (121)$$

In other words, for parallel propagation of Y along λ, we require that (cf. equation (119))

$$(\delta Y)^j = -Y^l\Gamma^j{}_{lk}\frac{\mathrm{d}x^k(\lambda(t))}{\mathrm{d}t}\delta t. \qquad (122)$$

In particular, for the tangent vector to the curve λ, $\mathrm{d}x^j(\lambda(t))/\mathrm{d}t$ parallely propagated along λ,

$$\delta\left(\frac{\mathrm{d}x^j(\lambda(t))}{\mathrm{d}t}\right) = -\Gamma^j{}_{lk}\frac{\mathrm{d}x^l(\lambda(t))}{\mathrm{d}t}\frac{\mathrm{d}x^k(\lambda(t))}{\mathrm{d}t}\delta t. \qquad (123)$$

A curve λ on M is said to be a *geodesic* if the tangent vector to λ, parallely propagated, remains a multiple of itself. This condition, for λ to be a geodesic, is, clearly,

$$\frac{\mathrm{d}x^j(\lambda(t))}{\mathrm{d}t} - \Gamma^j{}_{lk}\frac{\mathrm{d}x^l(\lambda(t))}{\mathrm{d}t}\frac{\mathrm{d}x^k(\lambda(t))}{\mathrm{d}t}\delta t$$
$$= [1-\phi(t)\delta t]\left(\frac{\mathrm{d}x^j(\lambda(t))}{\mathrm{d}t} + \frac{\mathrm{d}^2x^j(\lambda(t))}{\mathrm{d}t^2}\delta t\right), \qquad (124)$$

where $\phi(t)$ is some function of t. In the limit $\delta t \to 0$, the equation for a geodesic becomes

$$\frac{\mathrm{d}^2x^j}{\mathrm{d}t^2} + \Gamma^j{}_{lk}\frac{\mathrm{d}x^l}{\mathrm{d}t}\frac{\mathrm{d}x^k}{\mathrm{d}t} = \phi(t)\frac{\mathrm{d}x^j}{\mathrm{d}t}. \qquad (125)$$

It can be readily verified that if we reparametrize the curve λ by the variable

$$s = \int^t \mathrm{d}t'' \exp\left\{\int^{t''}\mathrm{d}t'\phi(t')\right\}, \qquad (126)$$

equation (125) becomes

$$\frac{\mathrm{d}^2x^j}{\mathrm{d}s^2} + \Gamma^j{}_{lk}\frac{\mathrm{d}x^l}{\mathrm{d}s}\frac{\mathrm{d}x^k}{\mathrm{d}s} = 0; \qquad (127)$$

and when the equation for a geodesic is reduced to this form, we say that it is *affinely parametrized*. It should be noticed that the only freedom we have in the choice of s is its origin and its scale.

5. Curvature forms and Cartan's equations of structure

For a manifold endowed with a connection, we define the two mappings

$$T(X, Y) = \nabla_X Y - \nabla_Y X - [X, Y] \tag{128}$$

and

$$R(X, Y) = \nabla_X \nabla_Y - \nabla_Y \nabla_X - \nabla_{[X, Y]} \tag{129}$$

where X and Y are two contravariant vector-fields. These mappings are called *torsion* and *curvature*, respectively. As defined, both are antisymmetric in their arguments.

Considering torsion first, we readily verify that T is linear in the arguments X and Y; thus

$$T(X+Y, Z) = T(X, Z) + T(Y, Z) \qquad (X, Y, Z \in T^1{}_0); \tag{130}$$

also

$$T(fX, Y) = f\,T(X, Y), \tag{131}$$

where f is any function. (In proving the second of these relations, we must make use of the identity

$$[fX, Y] = f[X, Y] - (Yf)X.)$$

The relations (130) and (131) clearly imply that the mapping

$$T\colon \qquad T^1{}_0 \times T^1{}_0 \to T^1{}_0 \tag{132}$$

is multilinear. Accordingly, T is a tensor field of type (1, 2).

Let (e_j) and (e^i) provide dual bases for T_p and $T^*{}_p$. Then, as we have shown (equation (100)),

$$(\nabla_X Y)^j = XY^j + Y^l \omega^j{}_l(X) = Xe^j(Y) + e^l(Y)\omega^j{}_l(X). \tag{133}$$

Therefore,

$$\nabla_X Y - \nabla_Y X = [Xe^j(Y) + \omega^j{}_l(X)e^l(Y) - Ye^j(X) - \omega^j{}_l(Y)e^l(X)]e_j. \tag{134}$$

Hence,

$$\begin{aligned} T^j(X, Y) &= \langle e^j, \nabla_X Y - \nabla_Y X - [X, Y]\rangle \\ &= X\langle e^j, Y\rangle - Y\langle e^j, X\rangle - \langle e^j, [X, Y]\rangle \\ &\quad + \omega^j{}_l(X)e^l(Y) - \omega^j{}_l(Y)e^l(X), \end{aligned} \tag{135}$$

or, making use of the general identity (86), we have

$$\tfrac{1}{2}T^j(X, Y) = (\mathrm{d}e^j + \omega^j{}_l \wedge e^l)(X, Y). \tag{136}$$

Since this equation is valid for arbitrary X and Y, we conclude that

$$\tfrac{1}{2}T^j = \mathrm{d}e^j + \omega^j{}_l \wedge e^l = \Omega^j \quad \text{(say)}. \tag{137}$$

This is the first of *Cartan's equations of structure*. In the important special case, when the torsion is zero, equation (137) reduces to

$$de^j + \omega^j{}_l \wedge e^l = 0. \tag{137'}$$

In a local coordinate basis, $de^j = 0$ (since $e^j = dx^j$) and equation (137) reduces to

$$T^j = 2\Gamma^j{}_{lk} dx^k \wedge dx^l = (\Gamma^j{}_{lk} - \Gamma^j{}_{kl}) dx^k \wedge dx^l, \tag{138}$$

in agreement with our earlier definition of this quantity in equation (116).

Turning next to the curvature, by definition, we have

$$\boldsymbol{R}(X, Y)Z = \nabla_X \nabla_Y Z - \nabla_Y \nabla_X Z - \nabla_{[X,Y]} Z. \tag{139}$$

The expression on the right-hand side of equation (139) is manifestly linear in X, Y, and Z. And, moreover, it can also be verified that

$$\left.\begin{aligned} &\boldsymbol{R}(fX, Y)Z = \boldsymbol{R}(X, fY)Z = f\boldsymbol{R}(X, Y)Z \\ \text{and}\quad &\boldsymbol{R}(X, Y)fZ = f\boldsymbol{R}(X, Y)Z, \end{aligned}\right\} \tag{140}$$

where f is an arbitrary function. Hence, the mapping

$$\boldsymbol{R}: \quad T^1{}_0 \times T^1{}_0 \times T^1{}_0 \to T^1{}_0, \tag{141}$$

is a multilinear function of the arguments. Consequently, $\boldsymbol{R}$ is a tensor field of type (1, 3): it is called the *Riemann tensor*.

Now, making use of known relations, we obtain

$$\begin{aligned} \nabla_X \nabla_Y Z &= \nabla_X \{Ye^j(Z) + \omega^j{}_k(Y)e^k(Z)\} e_j \\ &= [Ye^l(Z) + \omega^l{}_k(Y)e^k(Z)] \nabla_X e_l + [XYe^j(Z) + X\{\omega^j{}_l(Y)e^l(Z)\}] e_j \\ &= [Ye^l(Z) + \omega^l{}_k(Y)e^k(Z)] \omega^j{}_l(X) e_j \\ &\quad + [XYe^j(Z) + e^l(Z)X\omega^j{}_l(Y) + \omega^j{}_l(Y)Xe^l(Z)] e_j \\ &= [XYe^j(Z) + \omega^j{}_l(Y)Xe^l(Z) + e^l(Z)X\omega^j{}_l(Y) \\ &\quad + \omega^j{}_l(X)Ye^l(Z) + \omega^j{}_l(X)\omega^l{}_k(Y)e^k(Z)] e_j. \end{aligned} \tag{142}$$

Consequently,

$$\begin{aligned} \nabla_X \nabla_Y Z - \nabla_Y \nabla_X Z = \{&[X\omega^j{}_l(Y) - Y\omega^j{}_l(X) + \omega^j{}_k(X)\omega^k{}_l(Y) \\ &- \omega^j{}_k(Y)\omega^k{}_l(X)] e^l(Z) + [X, Y] e^j(Z)\} e_j. \end{aligned} \tag{143}$$

We also have

$$\nabla_{[X,Y]} Z = \{[X, Y] e^j(Z) + \omega^j{}_l([X, Y]) e^l(Z)\} e_j. \tag{144}$$

Now combining equations (143) and (144), we obtain

$$\begin{aligned} \boldsymbol{R}(X, Y)Z = \{&X \langle \omega^j{}_l, Y \rangle - Y \langle \omega^j{}_l, X \rangle - \langle \omega^j{}_l, [X, Y] \rangle \\ &+ \omega^j{}_k(X)\omega^k{}_l(Y) - \omega^j{}_k(Y)\omega^k{}_l(X)\} e^l(Z) e_j, \end{aligned} \tag{145}$$

or, making use of the identity (86), we have

$$\tfrac{1}{2}\boldsymbol{R}(X,Y)\boldsymbol{Z} = (\mathrm{d}\omega^j{}_l + \omega^j{}_k \wedge \omega^k{}_l)(X,Y)e^l(\boldsymbol{Z})\boldsymbol{e}_j. \tag{146}$$

Accordingly, if ω is any one-form,

$$\begin{aligned}\boldsymbol{R}(\omega,\boldsymbol{Z},\boldsymbol{X},\boldsymbol{Y}) &= R^j{}_{lkm}[\boldsymbol{e}_j \otimes e^l \otimes (e^k \wedge e^m)](\omega,\boldsymbol{Z},\boldsymbol{X},\boldsymbol{Y}) \\ &= (R^j{}_{lkm}e^k \wedge e^m)(X,Y)e^l(\boldsymbol{Z})\boldsymbol{e}_j(\omega).\end{aligned} \tag{147}$$

From a comparison of equations (146) and (147), we obtain the relation

$$\tfrac{1}{2}R^j{}_{lkm}e^k \wedge e^m = \mathrm{d}\omega^j{}_l + \omega^j{}_k \wedge \omega^k{}_l. \tag{148}$$

Defining the two-form,

$$\boldsymbol{\Omega}^j{}_l = \mathrm{d}\omega^j{}_l + \omega^j{}_k \wedge \omega^k{}_l, \tag{149}$$

we have *Cartan's second equation of structure*:

$$\tfrac{1}{2}R^j{}_{lkm}e^k \wedge e^m = \boldsymbol{\Omega}^j{}_l. \tag{150}$$

In a local coordinate basis

$$\omega^j{}_l = \Gamma^j{}_{lm}\mathrm{d}x^m; \tag{151}$$

therefore

$$\mathrm{d}\omega^j{}_l = \Gamma^j{}_{lm,n}\mathrm{d}x^n \wedge \mathrm{d}x^m = \tfrac{1}{2}(\Gamma^j{}_{lm,n} - \Gamma^j{}_{ln,m})\mathrm{d}x^n \wedge \mathrm{d}x^m. \tag{152}$$

Also,

$$\begin{aligned}\omega^j{}_k \wedge \omega^k{}_l &= \Gamma^j{}_{kn}\Gamma^k{}_{lm}\mathrm{d}x^n \wedge \mathrm{d}x^m \\ &= \tfrac{1}{2}(\Gamma^j{}_{kn}\Gamma^k{}_{lm} - \Gamma^j{}_{km}\Gamma^k{}_{ln})\mathrm{d}x^n \wedge \mathrm{d}x^m.\end{aligned} \tag{153}$$

Accordingly, Cartan's second equation of structure is equivalent to the definition

$$R^j{}_{lnm} = \Gamma^j{}_{lm,n} - \Gamma^j{}_{ln,m} + \Gamma^j{}_{kn}\Gamma^k{}_{lm} - \Gamma^j{}_{km}\Gamma^k{}_{ln}; \tag{154}$$

and this is the Riemann tensor as conventionally defined.

It is of interest to relate the foregoing treatment of the Riemann tensor, following Cartan, to the more customary treatment of it, by evaluating the right-hand side of equation (139), *ab initio*, in a local coordinate basis. Thus, by making use of equation (103), we have

$$\begin{aligned}\nabla_X\nabla_Y\boldsymbol{Z} &= \nabla_X(Y^kZ^j{}_{;k}\boldsymbol{e}_j) = Y^k\nabla_X(Z^j{}_{;k}\boldsymbol{e}_j) + Z^j{}_{;k}\boldsymbol{e}_j\nabla_XY^k \\ &= Y^kX^lZ^j{}_{;k;l}\boldsymbol{e}_j + Y^k{}_{;l}X^lZ^j{}_{;k}\boldsymbol{e}_j.\end{aligned} \tag{155}$$

Accordingly,

$$\begin{aligned}\nabla_X\nabla_Y\boldsymbol{Z} - \nabla_Y\nabla_X\boldsymbol{Z} &= Y^kX^l(Z^j{}_{;k;l} - Z^j{}_{;l;k})\boldsymbol{e}_j + [Y^k{}_{;l}X^l - X^k{}_{;l}Y^l]Z^j{}_{;k}\boldsymbol{e}_j \\ &= Y^kX^l(Z^j{}_{;k;l} - Z^j{}_{;l;k})\boldsymbol{e}_j + [X,Y]^kZ^j{}_{;k}\boldsymbol{e}_j + X^lY^n(\Gamma^k{}_{nl} - \Gamma^k{}_{ln})Z^j{}_{;k}\boldsymbol{e}_j;\end{aligned} \tag{156}$$

or, since

$$\nabla_{[X,Y]}\boldsymbol{Z} = [X, Y]^k Z^j{}_{;k}\boldsymbol{e}_j, \tag{157}$$

$$\begin{aligned} \boldsymbol{R}(X, Y)\boldsymbol{Z} &= (Z^j{}_{;k;l} - Z^j{}_{;l;k} + T^n{}_{lk} Z^j{}_{;n}) X^l Y^k \boldsymbol{e}_j \\ &= R^j{}_{ilk} Z^i X^l Y^k \boldsymbol{e}_j. \end{aligned} \tag{158}$$

We thus obtain the relation

$$Z^j{}_{;k;l} - Z^j{}_{;l;k} = -R^j{}_{ikl} Z^i + T^n{}_{kl} Z^j{}_{;n}. \tag{159}$$

This is the Ricci identity; it is the customary starting point for the introduction of the Riemann tensor when the torsion is zero.

It is of interest to contrast equation (159) with the equation,

$$f_{,k;l} - f_{,l;k} = T^n{}_{kl} f_{,n}, \tag{160}$$

which we derived earlier in §4 (equation (117)).

A further result of some importance which follows from equation (129) is (cf. equation (114))

$$\begin{aligned} \boldsymbol{R}(X, Y)f &= \nabla_X(Y^j f_{,j}) - \nabla_Y X^j f_{,j}) - [X, Y]^j f_{,j} \\ &= (X^l Y^j{}_{;l} - Y^l X^j{}_{;l}) f_{,j} + Y^j X^l (f_{,j;l} - f_{,l;j}) - [X, Y]^j f_{,j} \\ &= X^l Y^n (\Gamma^j{}_{nl} - \Gamma^j{}_{ln}) f_{,j} + X^l Y^j (-\Gamma^n{}_{jl} + \Gamma^n{}_{lj}) f_{,n} = 0. \end{aligned} \tag{161}$$

So far, we have considered the effect of $\boldsymbol{R}(X, Y)$ on contravariant vector-fields and scalar fields only. We shall now consider its effect on arbitrary tensor-fields.

By virtue of the Leibnitz rule satisfied by covariant differentiation of tensor products, we readily verify that

$$\boldsymbol{R}(X, Y)(\boldsymbol{P} \otimes \boldsymbol{Q}) = \boldsymbol{R}(X, Y)\boldsymbol{P} \otimes \boldsymbol{Q} + \boldsymbol{P} \otimes \boldsymbol{R}(X, Y)\boldsymbol{Q}, \tag{162}$$

where $\boldsymbol{P}$ and $\boldsymbol{Q}$ are arbitrary tensor-fields. With the aid of equation (162) we can, for example, find the effect of $\boldsymbol{R}(X, Y)$ on a one-form $\boldsymbol{\Omega}$. Thus, if $\boldsymbol{Z}$ is any contravariant vector-field, it follows from equation (162) that

$$\boldsymbol{R}(X, Y)(\Omega_j Z^j) = [\boldsymbol{R}(X, Y)\boldsymbol{\Omega}]_j Z^j + \Omega_j [\boldsymbol{R}(X, Y)\boldsymbol{Z}]^j. \tag{163}$$

Since $\boldsymbol{R}(X, Y)$ acting on a scalar field vanishes (by equation (161)), we conclude:

$$\begin{aligned} [\boldsymbol{R}(X, Y)\boldsymbol{\Omega}]_j Z^j &= -\Omega_j R^j{}_{ilk} Z^i X^l Y^k \\ &= -\Omega_i R^i{}_{jlk} Z^j X^l Y^k. \end{aligned} \tag{164}$$

Hence,

$$[\boldsymbol{R}(X, Y)\boldsymbol{\Omega}]_j = -R^i{}_{jlk} \Omega_i X^l Y^k. \tag{165}$$

On the other hand, by evaluating the effect of $\boldsymbol{R}(X, Y)$ on $\boldsymbol{\Omega}$ by the same procedure that was followed in deriving equation (158), we now obtain

$$\boldsymbol{R}(X, Y)\boldsymbol{\Omega} = X^l Y^k [\Omega_{j;k;l} - \Omega_{j;l;k} + T^n{}_{lk} \Omega_{j;n}] \boldsymbol{e}^j. \tag{166}$$

Now combining equations (165) and (166), we have

$$\Omega_{j;k;l} - \Omega_{j;l;k} = R^i{}_{jkl}\Omega_i + T^n{}_{kl}\Omega_{j;n}. \tag{167}$$

Again, considering the effect of $\boldsymbol{R}(\boldsymbol{X}, \boldsymbol{Y})$ on a tensor of type (2, 0), we have (on making use of equations (161) and (162))

$$\begin{aligned}\boldsymbol{R}(\boldsymbol{X}, \boldsymbol{Y})[S^{ij}\boldsymbol{e}_i \otimes \boldsymbol{e}_j] &= \boldsymbol{R}(\boldsymbol{X}, \boldsymbol{Y})S^{ij}\boldsymbol{e}_i \otimes \boldsymbol{e}_j + \boldsymbol{e}_i \otimes \boldsymbol{R}(\boldsymbol{X}, \boldsymbol{Y})S^{ij}\boldsymbol{e}_j \\ &= (R^i{}_{lmn}S^{lj} + R^j{}_{lmn}S^{il})X^m Y^n \boldsymbol{e}_i \otimes \boldsymbol{e}_j;\end{aligned} \tag{168}$$

and we conclude that

$$S^{ij}{}_{;k;l} - S^{ij}{}_{;l;k} = -R^i{}_{mkl}S^{mj} - R^j{}_{mkl}S^{im} + T^n{}_{kl}S^{ij}{}_{;n}. \tag{169}$$

In a similar fashion, we find

$$S_{ij;k;l} - S_{ij;l;k} = R^m{}_{jkl}S_{im} + R^m{}_{ikl}S_{mj} + T^n{}_{kl}S_{ij;n}. \tag{170}$$

It is now manifest how we can write down corresponding formulae for tensors of arbitrary type.

Finally, we note that by contracting the Riemann tensor (154) with respect to the second (or, the third) covariant index, we obtain the *Ricci tensor* (or, its negative); thus

$$R^j{}_{ljm} = -R^j{}_{lmj} = R_{lm}. \tag{171}$$

In a local coordinate basis, the expression for the components of the Ricci tensor is (cf. equation (154))

$$R_{lm} = \Gamma^j{}_{lm,j} - \Gamma^j{}_{lj,m} + \Gamma^j{}_{kj}\Gamma^k{}_{lm} - \Gamma^j{}_{km}\Gamma^k{}_{lj}. \tag{172}$$

(*a*) *The cyclic and the Bianchi identities in case the torsion is zero*

In case the torsion is zero (cf. equation (128)),

$$\nabla_X \boldsymbol{Q} - \nabla_Q \boldsymbol{X} = [\boldsymbol{X}, \boldsymbol{Q}] \qquad (\boldsymbol{X}, \boldsymbol{Q} \in T^1{}_0), \tag{173}$$

the curvature tensor satisfies two important identities which we shall now establish.

First, we verify that by virtue of the relation (173),

$$\begin{aligned}&\nabla_X[\boldsymbol{Y}, \boldsymbol{Z}] + \nabla_Y[\boldsymbol{Z}, \boldsymbol{X}] + \nabla_Z[\boldsymbol{X}, \boldsymbol{Y}] \\ &\quad = (\nabla_X\nabla_Y - \nabla_Y\nabla_X)\boldsymbol{Z} + (\nabla_Z\nabla_X - \nabla_X\nabla_Z)\boldsymbol{Y} + (\nabla_Y\nabla_Z - \nabla_Z\nabla_Y)\boldsymbol{X}.\end{aligned} \tag{174}$$

Therefore,

$$\begin{aligned}&\boldsymbol{R}(\boldsymbol{X}, \boldsymbol{Y})\boldsymbol{Z} + \boldsymbol{R}(\boldsymbol{Z}, \boldsymbol{X})\boldsymbol{Y} + \boldsymbol{R}(\boldsymbol{Y}, \boldsymbol{Z})\boldsymbol{X} \\ &\quad = \nabla_X[\boldsymbol{Y}, \boldsymbol{Z}] + \nabla_Y[\boldsymbol{Z}, \boldsymbol{X}] + \nabla_Z[\boldsymbol{X}, \boldsymbol{Y}] - \nabla_{[X,Y]}\boldsymbol{Z} - \nabla_{[Y,Z]}\boldsymbol{X} - \nabla_{[Z,X]}\boldsymbol{Y}.\end{aligned} \tag{175}$$

On the other hand, by writing $\boldsymbol{Q} = [\boldsymbol{Y}, \boldsymbol{Z}]$ in equation (173), we obtain

$$\nabla_X[\boldsymbol{Y}, \boldsymbol{Z}] - \nabla_{[Y,Z]}\boldsymbol{X} = [\boldsymbol{X}, [\boldsymbol{Y}, \boldsymbol{Z}]]. \tag{176}$$

Accordingly, equation (175) reduces to

$$\begin{aligned}&\boldsymbol{R}(\boldsymbol{X},\boldsymbol{Y})\boldsymbol{Z}+\boldsymbol{R}(\boldsymbol{Z},\boldsymbol{X})\boldsymbol{Y}+\boldsymbol{R}(\boldsymbol{Y},\boldsymbol{Z})\boldsymbol{X}\\&\quad=[\boldsymbol{X},[\boldsymbol{Y},\boldsymbol{Z}]]+[\boldsymbol{Y},[\boldsymbol{Z},\boldsymbol{X}]]+[\boldsymbol{Z},[\boldsymbol{X},\boldsymbol{Y}]]=0,\end{aligned}\tag{177}$$

by the Jacobi identity. In a local coordinate basis, equation (177) provides the *cyclic identity*

$$R^j{}_{lkm}+R^j{}_{kml}+R^j{}_{mlk}=0.\tag{178}$$

Next, consider the exterior derivative of Cartan's two-form $\Omega^j{}_l$ (cf. equation (149)). We have

$$\begin{aligned}\mathrm{d}\boldsymbol{\Omega}^j{}_l&=\mathrm{d}\boldsymbol{\omega}^j{}_k\wedge\boldsymbol{\omega}^k{}_l-\boldsymbol{\omega}^j{}_k\wedge\mathrm{d}\boldsymbol{\omega}^k{}_l\\&=(\boldsymbol{\Omega}^j{}_k-\boldsymbol{\omega}^j{}_m\wedge\boldsymbol{\omega}^m{}_k)\wedge\boldsymbol{\omega}^k{}_l-\boldsymbol{\omega}^j{}_k\wedge(\boldsymbol{\Omega}^k{}_l-\boldsymbol{\omega}^k{}_m\wedge\boldsymbol{\omega}^m{}_l).\end{aligned}\tag{179}$$

The triple wedge-products of the one-forms which occur in the second line of equation (179) are seen to cancel; and we are left with

$$\mathrm{d}\boldsymbol{\Omega}^j{}_l-\boldsymbol{\Omega}^j{}_k\wedge\boldsymbol{\omega}^k{}_l+\boldsymbol{\omega}^j{}_k\wedge\boldsymbol{\Omega}^k{}_l=0;\tag{180}$$

and this equation expresses the *Bianchi identities*. We can obtain them in their standard forms by rewriting equation (180) in a local coordinate basis when

$$\boldsymbol{\Omega}^j{}_l=\tfrac{1}{2}R^j{}_{lpq}\mathrm{d}x^p\wedge\mathrm{d}x^q\qquad\text{and}\qquad\boldsymbol{\omega}^k{}_l=\Gamma^k{}_{lr}\mathrm{d}x^r.\tag{181}$$

With these substitutions, equation (180) gives

$$(R^j{}_{lpq,r}-R^j{}_{kpq}\Gamma^k{}_{lr}+\Gamma^j{}_{kr}R^k{}_{lpq})\mathrm{d}x^p\wedge\mathrm{d}x^q\wedge\mathrm{d}x^r=0.\tag{182}$$

Since the connection $\Gamma^j{}_{kr}$ is symmetric in k and r, when the torsion is zero, the additional terms we include in the following equation do not affect its validity:

$$\begin{aligned}&(R^j{}_{lpq,r}-R^j{}_{kpq}\Gamma^k{}_{lr}+\Gamma^j{}_{kr}R^k{}_{lpq}\\&\quad-R^j{}_{lkq}\Gamma^k{}_{pr}-R^j{}_{lpk}\Gamma^k{}_{qr})\mathrm{d}x^p\wedge\mathrm{d}x^q\wedge\mathrm{d}x^r=0.\end{aligned}\tag{183}$$

But the quantity in parentheses in equation (183) is precisely the covariant derivative of $R^j{}_{lpq}$ with respect to x^r. We conclude that

$$R^j{}_{lpq;r}+R^j{}_{lqr;p}+R^j{}_{lrp;q}=0;\tag{184}$$

and this is the Bianchi identity in its standard form.

6. The metric and the connection derived from it. Riemannian geometry and the Einstein field-equation

A *metric tensor* $\boldsymbol{g}$ is a non-singular symmetric tensor-field of type (0, 2). Thus,

$$\left.\begin{aligned}&\text{(a)}\quad \boldsymbol{g}:T^1{}_0\times T^1{}_0\to\mathbb{R}^1;\\&\text{(b)}\quad \boldsymbol{g}(\boldsymbol{X},\boldsymbol{Y})=\boldsymbol{g}(\boldsymbol{Y},\boldsymbol{X})\text{ for every }\boldsymbol{X},\boldsymbol{Y}\in T^0{}_1;\\&\text{(c)}\quad \boldsymbol{g}(\boldsymbol{X},\boldsymbol{Y})=0\text{ for every }\boldsymbol{Y}\in T^0{}_1\text{ implies that }\boldsymbol{X}=0.\end{aligned}\right\}\tag{185}$$

The condition (b) ensures the symmetry of $\boldsymbol{g}$ while the condition (c) its non-singular nature.

In a local basis, we may write

$$\boldsymbol{g} = g_{ij}\boldsymbol{e}^i \otimes \boldsymbol{e}^j \qquad \text{and} \qquad g_{ij} = g_{ji}; \tag{186}$$

and, similarly, in a local coordinate basis,

$$\boldsymbol{g} = g_{ij}\mathrm{d}x^i \otimes \mathrm{d}x^j \qquad \text{and} \qquad g_{ij} = g_{ji}. \tag{187}$$

In terms of its components, the requirement that the metric tensor be non-singular is equivalent to the requirement that the determinant g of the matrix $[g_{ij}]$ is non-zero at every point of the manifold. The matrix $[g_{ij}]$ has then a unique inverse. We denote the elements of the inverse matrix by g^{ij} so that

$$g^{ij}g_{jk} = \delta^i{}_k. \tag{188}$$

This last equation guarantees that we may, in fact, regard g^{ij} as the components of a tensor field, $\boldsymbol{g}^{-1}$, of type $(2, 0)$ whose representation in the basis $\boldsymbol{e}_i \otimes \boldsymbol{e}_j$ (dual to the basis used in equation (186)) is given by

$$\boldsymbol{g}^{-1} = g^{ij}\boldsymbol{e}_i \otimes \boldsymbol{e}_j. \tag{189}$$

One uses the metric tensor to define a path length L, along a curve λ on M, from $\lambda(a)$ to $\lambda(b)$ (for example) by the formula

$$L = \int_a^b \left| g_{ij} \frac{\mathrm{d}x^i(\lambda(t))}{\mathrm{d}t} \frac{\mathrm{d}x^j(\lambda(t))}{\mathrm{d}t} \right|^{1/2} \mathrm{d}t. \tag{190}$$

In conformity with this definition, it is customary to write

$$\mathrm{d}s^2 = g_{ij}\mathrm{d}x^i\mathrm{d}x^j \tag{191}$$

and consider $\mathrm{d}s^2$ as giving the square of the interval $\mathrm{d}s$ between neighbouring points of the manifold.

Given a tensor field $\boldsymbol{T}$ of type (r, s), we may contract $\boldsymbol{g} \otimes \boldsymbol{T}$ and $\boldsymbol{g}^{-1} \otimes \boldsymbol{T}$ with respect to one of the indices of the metric tensor (or its inverse) to obtain tensors of types $(r-1, s+1)$ and $(r+1, s-1)$, respectively. The components of the contracted tensor are written as

$$\left.\begin{aligned} &g_{ij}T^{ab\ldots i\ldots p}{}_{cd\ldots q} = T^{ab\ldots}{}_{j}{}^{\ldots p}{}_{cd\ldots q} \\ \text{and}\quad & \\ &g^{ij}T^{ab\ldots p}{}_{cd\ldots i\ldots q} = T^{ab\ldots p}{}_{cd\ldots}{}^{j}{}_{\ldots q}. \end{aligned}\right\} \tag{192}$$

The process can clearly be repeated. We regard tensors derived by such *raising* and *lowering* of indices as representing the same geometric quantity since by raising an index and subsequently lowering it, we recover the original tensor.

An important notion concerning the metric tensor is its *signature*. It is defined as the difference in the number of coefficients that are positive and the number of coefficients that are negative when g_{ij} (at some point) is brought to its diagonal form. It can be shown that the signature so defined is the same at all points of a (connected) manifold. An Euclidean metric is one for which the signature is, numerically, equal to the dimension of the manifold. And the metric is said to be Lorentzian or Minkowskian if the signature is $\pm(n-2)$-, the plus or minus sign being a matter of convention.

(a) *The connection derived from a metric*

So far, we have considered the introduction of the metric as independent of whatever connection we may or may not have endowed the manifold. We shall now show that, associated with a metric, we can endow the manifold with a unique torsion-free connection by the requirement that

$$\nabla \boldsymbol{g} = 0. \tag{193}$$

With such a connection

$$\nabla(\boldsymbol{g} \otimes \boldsymbol{T}) = \boldsymbol{g} \otimes \nabla \boldsymbol{T}, \tag{194}$$

where $\boldsymbol{T}$ is any tensor field. The principal advantage of such a connection is that the operation of raising or lowering of indices commutes with the operation of covariant differentiation.

To deduce the torsion-free connection which follows from the requirement (193), we evaluate $\nabla_X \boldsymbol{g}$ with $\boldsymbol{g}$ expressed in the form (186). Thus, we require

$$\nabla_X g_{ij} \boldsymbol{e}^i \otimes \boldsymbol{e}^j = X g_{ij} \boldsymbol{e}^i \otimes \boldsymbol{e}^j + g_{ij}[(\nabla_X \boldsymbol{e}^i) \otimes \boldsymbol{e}^j + \boldsymbol{e}^i \otimes (\nabla_X \boldsymbol{e}^j)] = 0, \tag{195}$$

or, making use of equation (111), we have

$$[X g_{ij} - g_{lj}\omega^l{}_i(X) - g_{il}\omega^l{}_j(X)]\boldsymbol{e}^i \otimes \boldsymbol{e}^j = 0. \tag{196}$$

We conclude that

$$(\mathrm{d}g_{ij} - g_{lj}\omega^l{}_i - g_{il}\omega^l{}_j)(X) = 0. \tag{197}$$

Letting $X = \partial_k$ in a local coordinate basis, we obtain from equation (197) the requirement

$$g_{ij,k} = g_{lj}\Gamma^l{}_{ik} + g_{il}\Gamma^l{}_{jk}, \tag{198}$$

where by our assumption of zero torsion, the Γ-symbols are symmetric in their 'covariant' indices. From equation (198) we derive, in the usual fashion, that

$$g_{il}\Gamma^l{}_{jk} = \frac{1}{2}\left(\frac{\partial g_{ij}}{\partial x^k} + \frac{\partial g_{ik}}{\partial x^j} - \frac{\partial g_{jk}}{\partial x^i}\right), \tag{199}$$

or, equivalently,

$$\Gamma^i{}_{jk} = \frac{1}{2}g^{il}\left(\frac{\partial g_{lj}}{\partial x^k} + \frac{\partial g_{lk}}{\partial x^j} - \frac{\partial g_{jk}}{\partial x^l}\right). \tag{200}$$

The connection is thus uniquely specified; and this connection underlies *Riemannian geometry*.

The Γ-symbols (200), appropriate to a metric $\boldsymbol{g}$, are called the *Christoffel symbols*; and the connection itself is called the *Christoffel connection.*

Two elementary consequences of the Christoffel connection are the following.

The first is that the *scalar product*, $(\boldsymbol{X}\cdot\boldsymbol{Y})$, of two contravariant vector-fields, X and Y, defined by

$$\boldsymbol{g}(X, Y) = (X\cdot Y) = g_{ij}X^iY^j, \tag{201}$$

remains unchanged as X and Y are parallely propagated along a curve λ on M. For (by equations (120) and (121))

$$D(g_{ij}X^iY^j) = (Dg_{ij})X^iY^j + g_{ij}(Y^jDX^i + X^iDY^j) = 0, \tag{202}$$

since Dg_{ij} vanishes by virtue of the condition $\nabla\boldsymbol{g} = 0$ (from which the connection was derived) and DX^i and DY^j vanish by the assumption of parallel propagation along λ.

A second consequence is that the geodesic equation (127), derived in §4 from the requirement that the tangent vector to the curve λ, as it is parallely propagated along λ, remains a multiple of itself, now emerges as the Euler-Lagrange equation of an extremal problem. Thus, consider the integral (cf. equation (190))

$$I = \int_a^b L\,\mathrm{d}s, \quad \text{where} \quad L = g_{ij}\frac{\mathrm{d}x^i(\lambda(s))}{\mathrm{d}s}\frac{\mathrm{d}x^j(\lambda(s))}{\mathrm{d}s}, \tag{203}$$

and the curve λ is parametrized by the arc length, s, along λ. The Euler-Lagrange equation, for the extremal problem associated with the integral I, is

$$\frac{\mathrm{d}}{\mathrm{d}s}\left(\frac{\partial L}{\partial \dot{x}^j}\right) - \frac{\partial L}{\partial x^j} = 0 \quad \text{where} \quad \dot{x}^j = \frac{\mathrm{d}x^j(\lambda(s))}{\mathrm{d}s}. \tag{204}$$

Since

$$\frac{\partial L}{\partial \dot{x}^j} = 2g_{ij}\dot{x}^i \qquad \text{and} \qquad \frac{\partial L}{\partial x^j} = g_{ik,j}\dot{x}^i\dot{x}^k, \tag{205}$$

the Euler-Lagrange equation reduces to

$$g_{ij}\ddot{x}^i + (g_{ij,k} - \tfrac{1}{2}g_{ik,j})\dot{x}^i\dot{x}^k = 0, \tag{206}$$

or, alternatively,

$$g_{ij}\ddot{x}^i + \tfrac{1}{2}(g_{ij,k} + g_{kj,i} - g_{ik,j})\dot{x}^i\dot{x}^k = 0. \tag{207}$$

Contracting this last equation with g^{jl}, we obtain

$$\ddot{x}^l + \Gamma^l{}_{ik}\dot{x}^i\dot{x}^k = 0. \tag{208}$$

An alternative form of the foregoing equation, which we shall find useful, is to express it in terms of the '*velocity*'

$$u^j = \dot{x}^j = \frac{\mathrm{d}x^j}{\mathrm{d}s}. \tag{209}$$

Then

$$\ddot{x}^j = u^j{}_{,k} u^k, \tag{210}$$

and the equation for the geodesic takes the form

$$(u^j{}_{,k} + \Gamma^j{}_{ik} u^i) u^k = u^j{}_{;k} u^k = 0. \tag{211}$$

If the geodesic is not affinely parametrized, it will take the form (cf. equation (125))

$$u^j{}_{;k} u^k = \phi u^j, \tag{212}$$

where ϕ is some scalar function.

As may be directly verified (by contracting with u_j), equation (211) allows the integral $(\boldsymbol{u} \cdot \boldsymbol{u}) =$ constant, consistently with the fact that $\boldsymbol{u}$ is parallely propagated along the geodesic.

(b) Some consequences of the Christoffel connection for the Riemann and the Ricci tensors

When the connection is that compatible with a metric, the Riemann and the Ricci tensors have additional symmetries. Thus, equation (170) with g_{ij} substituted for S_{ij} gives

$$g_{im} R^m{}_{jkl} + g_{mj} R^m{}_{ikl} = 0; \tag{213}$$

or, with the index m lowered, we have

$$R_{ijkl} + R_{jikl} = 0. \tag{214}$$

Hence, the completely covariant Riemann-tensor is antisymmetric in the first pair of indices, as well. We deduce a further symmetry by the following sequence of transformations. Starting from the cyclic identity (cf. equation (178)),

$$R_{jkmn} + R_{jmnk} + R_{jnkm} = 0, \tag{215}$$

making use of the known antisymmetry in the first and the second pair of indices and of the cyclic identity as well, we find successively,

$$\begin{aligned} R_{jkmn} &= -(R_{jmnk} + R_{jnkm}) = R_{mjnk} + R_{njkm} \\ &= -(R_{mnkj} + R_{mkjn}) - (R_{nkmj} + R_{nmjk}) \\ &= 2\, R_{mnjk} + (R_{kmjn} + R_{knmj}) \\ &= 2\, R_{mnjk} - R_{kjnm} = 2\, R_{mnjk} - R_{jkmn}. \end{aligned} \tag{216}$$

Hence

$$R_{jkmn} = R_{mnjk}. \tag{217}$$

The Riemann tensor is therefore unchanged by the simultaneous interchange of the first pair with the second pair of its covariant indices. From these various symmetries, one concludes that the number of independent components of the Riemann tensor is $n^2(n^2 - 1)/12$ (i.e., 20 for the Riemann tensor of a four-dimensional manifold).

From the symmetries of the Riemann tensor, the symmetry of the Ricci tensor follows; for

$$R_{ij} = g^{kl} R_{ikjl} = g^{lk} R_{ljki} = R_{ji}. \tag{218}$$

We may note in this connection the explicit expression for the Ricci tensor in a local coordinate basis. But first we observe that

$$\begin{aligned}\Gamma^j{}_{jk} &= \tfrac{1}{2} g^{jl} (g_{lj,k} + g_{lk,j} - g_{jk,l}) = \tfrac{1}{2} g^{jl} g_{lj,k} \\ &= (\lg \sqrt{|g|})_{,k}\end{aligned} \tag{219}$$

where $|g|$ denotes the absolute value of the determinant of $[g_{ij}]$. With the foregoing expression for the contracted Christoffel symbol, equation (172) gives

$$R_{lm} = \Gamma^j{}_{lm,j} - \frac{\partial^2 \lg \sqrt{|g|}}{\partial x^l \partial x^m} + \frac{\partial \lg \sqrt{|g|}}{\partial x^k} \Gamma^k{}_{lm} - \Gamma^j{}_{km} \Gamma^k{}_{lj}. \tag{220}$$

(c) The Einstein tensor

The *Einstein tensor* G_{ij} is related to the Ricci tensor by

$$G_{ij} = R_{ij} - \tfrac{1}{2} g_{ij} R, \tag{221}$$

where

$$R = R^j{}_j = g^{ij} R_{ij} \tag{222}$$

is the contracted Ricci tensor (or, as it is usually called, the *scalar curvature*).

The most important property of the Einstein tensor is that its covariant divergence vanishes:

$$G^i{}_{j;i} = 0. \tag{223}$$

This identity follows from the Bianchi identity (184). Thus, contracting the relation,

$$R^i{}_{jkl;m} - g^{ip} R_{jplm;k} + R^i{}_{jmk;l} = 0, \tag{224}$$

with respect to the indices i and k, we obtain

$$R_{jl;m} - g^{kp} R_{jplm;k} - R_{jm;l} = 0. \tag{225}$$

Now raising the index j and contracting it with m, we obtain

$$R^j{}_{l;j} + R^k{}_{l;k} - R_{,l} = 0, \tag{226}$$

or

$$(R^j{}_l - \tfrac{1}{2} \delta^j{}_l R)_{;j} = 0, \tag{227}$$

which is the required identity.

(d) The Weyl tensor

We have seen that the Reimann tensor, R_{ijkl}, is antisymmetric in both pairs of indices (ij) and (kl) and it is, moreover, unchanged by a simultaneous interchange of the two pairs of indices (ij) and (kl). By virtue of these symmetries, the only non-trivial contraction we can make, by raising one of its indices and contracting it with any one of the three remaining covariant indices, is that leading to the Ricci tensor. Accordingly, it will be convenient to separate the Riemann tensor into a 'trace-free' part and a 'Ricci' part. This separation is accomplished by the Weyl tensor,

$$C_{ijkl} = R_{ijkl} - \frac{1}{(n-2)}(g_{ik}R_{jl} + g_{jl}R_{ik} - g_{jk}R_{il} - g_{il}R_{jk})$$

$$+\frac{1}{(n-1)(n-2)}(g_{ik}g_{jl} - g_{il}g_{jk})R. \tag{228}$$

This tensor has manifestly all the symmetries of the Riemann tensor; but

$$g^{jl}C_{ijkl} = 0 \qquad \text{in contrast to} \qquad g^{jl}R_{ijkl} = R_{ik}. \tag{229}$$

A further distinction is that while the Riemann tensor can be defined in a manifold endowed only with a connection, the Weyl tensor can be defined only when a metric is also defined.

The importance of the Weyl tensor for the deeper problems of differential geometry (and of general relativity) is the conformal invariance of $C^i{}_{jkl}$, i.e., its invariance to the transformation $\boldsymbol{g} \to \Omega^2 \boldsymbol{g}$ where Ω is some scalar function; but we shall not be concerned with those problems in this book.

(e) *Space-time as a four-dimensional manifold; matters of notation and Einstein's field-equation*

In our account of differential geometry, we have, up to the present, made no restrictions on the dimensionality of the manifold (or, of its signature). But space–time, in the general theory of relativity, is considered as a four-dimensional differentiable manifold with a Lorentzian signature; and our subsequent discussion in this chapter (as in the rest of the book) will be restricted to this case. In this section, we shall recapitulate the basic equations of the theory in the notation and with the conventions (regarding signs) we shall adopt.

First, with regard to the signature, our convention will be that when the metric is brought to its diagonal form (at any point of space–time), the difference in the number of coefficients that are positive and the number of coefficients that are negative is -2; in particular, in flat space, the Minkowskian metric will be taken to have the form

$$ds^2 = c^2 dt^2 - dx^2 - dy^2 - dz^2, \tag{230}$$

where c denotes the velocity of light (which we shall generally set equal to 1 by a choice of units).

With its Lorentzian signature, the metric of space–time is not positive-definite in the sense that $\boldsymbol{g}(\boldsymbol{X},\boldsymbol{X})$ can be positive, zero, or negative. And we call vectors as *time-like*, *null*, or *space-like* according as $\boldsymbol{g}(\boldsymbol{X},\boldsymbol{X}) > 0, = 0$, or < 0. Material particles (with finite rest-mass) can describe only *time-like trajectories* (i.e., curves along which the tangent vector is always time-like), while massless particles (such as photons, gravitons, and neutrinos) describe *null trajectories* (i.e., curves along which the tangent vector is always null).

Freely falling particles describe geodesics—time-like if they have finite rest-mass and null if they are massless. Null geodesics cannot clearly be parametrized by the arc length even though they can be affinely parametrized by a change of variables (as shown in §4, equation (126)). The equations for the geodesics (time-like or null), when expressed in terms of the *four-velocity*, $\boldsymbol{u}$, can always be reduced to the forms given in equations (211) and (212). Along time-like geodesics $(\boldsymbol{u}\cdot\boldsymbol{u}) = 1$, while along null geodesics $(\boldsymbol{u}\cdot\boldsymbol{u}) = 0$.

We note that by our definition of the Riemann tensor, the Ricci identity has the form

$$R^{i}{}_{jkl}Z_i = Z_{j;k;l} - Z_{j;l;k}. \tag{231}$$

Further, the Ricci tensor is obtained by the contraction

$$g^{jl}R_{ijkl} = R_{ik}. \tag{232}$$

And the Weyl tensor is now related to the Riemann and the Ricci tensors by

$$\begin{aligned} C_{ijkl} = R_{ijkl} &- \tfrac{1}{2}(g_{ik}R_{jl} + g_{jl}R_{ik} - g_{jk}R_{il} - g_{il}R_{jk}) \\ &+ \tfrac{1}{6}(g_{ik}g_{jl} - g_{jk}g_{il})R. \end{aligned} \tag{233}$$

The Riemann tensor has twenty distinct components while the Weyl and the Ricci tensors have ten components each. And the Bianchi identity,

$$R_{ij[kl;m]} = \tfrac{1}{3}(R_{ijkl;m} + R_{ijlm;k} + R_{ijmk;l}) = 0, \tag*{(234)*}$$

includes 24 distinct equations corresponding to the six distinct index-pairs $(i,j,i \neq j)$, each of which can be associated with the four choices for $k \neq l \neq m$. However, only 20 of these 24 equations are linearly independent since, as may be directly verified, the following four linear combinations identically vanish by virtue of the various symmetries of the Riemann tensor (including that of the cyclical relation):

$$\left.\begin{aligned} &R_{12[34;1]} - R_{13[41;2]} + R_{14[12;3]} = 0, \\ &R_{23[41;2]} - R_{24[12;3]} + R_{21[23;4]} = 0, \\ &R_{34[12;3]} - R_{31[23;4]} + R_{32[34;1]} = 0, \\ \text{and}\quad &R_{41[23;4]} - R_{42[34;1]} + R_{43[41;2]} = 0. \end{aligned}\right\} \tag{235}$$

* We are here adopting the notation that enclosing a group of indices in square brackets signifies that the quantity in question has been 'antisymmetrized' by applying the alternating operator A to them in the manner of equation (53).

Einstein's equation governing the geometry of space-time is

$$G_{ij} = R_{ij} - \tfrac{1}{2} g_{ij} R = \frac{8\pi G}{c^4} T_{ij}, \tag{236}$$

where T_{ij} denotes the *energy-momentum tensor* of the matter and of the fields (other than gravitational) which may be present and G is the constant of gravitation (which, like c, we shall generally set equal to 1 by a choice of units). An alternative form of equation (236) is

$$R_{ij} = \frac{8\pi G}{c^4} (T_{ij} - \tfrac{1}{2} g_{ij} T) \qquad (T = g^{ij} T_{ij}). \tag{236'}$$

In a *vacuum* (i.e., in regions of space-time in which $T_{ij} = 0$), Einstein's equation reduces to

$$G_{ij} = 0, \qquad \text{or, equivalently,} \qquad R_{ij} = 0; \tag{237}$$

and the Riemann and the Weyl tensors coincide.

One generally considers Einstein's equation as an equation for determining the metric (more particularly, the ten *metric coefficients* g_{ij}). In view of the identity (equation (223)),

$$G^{i}{}_{j;i} = 0, \tag{238}$$

it is clear that any solution of Einstein's equation must involve four arbitrary functions. The freedom we thus have in the choice of these functions is described as *gauge freedom* and traced to the 'general covariance' of the theory. And this gauge freedom allows us to impose four *coordinate conditions* on the metric.

And finally, it should be noted that, in the vacuum, the number of Bianchi identities is reduced from 20 to 16 since four of the identities included in equation (238) are identically satisfied by virtue of the field equation (237).

7. The tetrad formalism

The standard way of treating problems in the general theory of relativity used to be to consider the Einstein field-equation in a local coordinate basis adapted to the problem on hand. But in recent years, it has appeared advantageous, in some contexts, to proceed somewhat differently by choosing a suitable *tetrad basis* of four linearly independent vector-fields, projecting the relevant quantities on to the chosen basis, and considering the equations satisfied by them. This is the *tetrad formalism.*

In the applications of the tetrad formalism, the choice of the tetrad basis depends on the underlying symmetries of the space–time we wish to grasp and is, to some extent, a part of the problem. Besides, it is not always clear what the 'relevant' equations are and what the relations among them may be. On these accounts, we shall present the basic ideas of the theory without any prior commitments and derive the various equations which will later appear as the relevant ones of the formalism for the applications we have in view.

(a) The tetrad representation

We set up at each point of space–time a basis of four contravariant vectors,

$$e_{(a)}{}^{i} \qquad (a = 1, 2, 3, 4), \tag{239}$$

where enclosure in parentheses distinguishes the *tetrad indices* from the tensor indices (which are not so enclosed). (Also, we shall reserve the earlier letters of the Latin alphabet (a, b, etc.,) for the tetrad indices and the later letters (i, j, etc.) for the tensor indices.) Associated with the contravariant vectors (239) we have the covariant vectors,

$$e_{(a)i} = g_{ik}\, e_{(a)}{}^{k}, \tag{240}$$

where g_{ik} denotes the metric tensor. In addition, we also define the inverse, $e^{(b)}{}_{i}$, of the matrix $[e_{(a)}{}^{i}]$ (with the tetrad index labelling the rows and the tensor index labelling the columns) so that

$$e_{(a)}{}^{i} e^{(b)}{}_{i} = \delta^{(b)}{}_{(a)} \qquad \text{and} \qquad e_{(a)}{}^{i} e^{(a)}{}_{j} = \delta^{i}{}_{j}, \tag{241}$$

where the summation convention with respect to the indices of the two sorts, independently, is assumed (here and elsewhere). Further, as a part of the definitions, we shall also assume that

$$e_{(a)}{}^{i} e_{(b)i} = \eta_{(a)(b)}, \tag{242}$$

where

$$\eta_{(a)(b)} \qquad \text{is a constant symmetric matrix.} \tag{243}$$

One most often supposes that the basis vectors, $e_{(a)}{}^{i}$, are *orthonormal* in which case $\eta_{(a)(b)}$ represents a diagonal matrix with the diagonal elements, $+1$, -1, -1, -1. We shall not make this assumption though it should be stated that a formalism, more general than the one we shall describe in detail, can be developed in which the $\eta_{(a)(b)}$'s are allowed to be functions on the manifold. In some contexts this further generalization may commend itself. We consider it briefly in §(e) below; but, for the present, we shall proceed on the assumption that the $\eta_{(a)(b)}$'s are constants.

Returning to equation (242), let $\eta^{(a)(b)}$ be the inverse of the matrix $[\eta_{(a)(b)}]$; then

$$\eta^{(a)(b)} \eta_{(b)(c)} = \delta^{(a)}{}_{(c)}. \tag{244}$$

As a consequence of the various definitions,

$$\eta_{(a)(b)} e^{(a)}{}_{i} = e_{(b)i}, \qquad \eta^{(a)(b)} e_{(a)i} = e^{(b)}{}_{i}; \tag{245}$$

and most importantly,

$$e_{(a)i}\, e^{(a)}{}_j = g_{ij}. \tag{246}$$

Given any vector or tensor field, we project it onto the tetrad frame to obtain its *tetrad components*. Thus,

$$\left.\begin{aligned} A_{(a)} &= e_{(a)j}\, A^j &&= e_{(a)}{}^j\, A_j, \\ A^{(a)} &= \eta^{(a)(b)}\, A_{(b)} &&= e^{(a)}{}_j\, A^j = e^{(a)j}\, A_j, \\ A^i &= e_{(a)}{}^i\, A^{(a)} &&= e^{(a)i}\, A_{(a)}; \end{aligned}\right\} \tag{247}$$

and more generally,

$$\left.\begin{aligned} T_{(a)(b)} &= e_{(a)}{}^i e_{(b)}{}^j T_{ij} = e_{(a)}{}^i T_{i(b)}, \\ T_{ij} &= e^{(a)}{}_i e^{(b)}{}_j T_{(a)(b)} = e^{(a)}{}_i T_{(a)j}. \end{aligned}\right\} \tag{248}$$

It is clear from equations (241), (242), (244), and (245) that (a) we can pass freely from the tensor indices to the tetrad indices and vice versa; (b) raise and lower the tetrad indices with $\eta^{(a)(b)}$ and $\eta_{(a)(b)}$ even as we can raise and lower the tensor indices with the metric tensor; (c) there is no ambiguity in having quantities in which the indices of both sorts occur; and (d) the result of contracting a tensor is the same whether it is carried out with respect to its tensor or tetrad indices.

(b) Directional derivatives and the Ricci rotation-coefficients

The contravariant vectors $\boldsymbol{e}_{(a)}$, considered as tangent vectors, define the directional derivatives (cf. equation (8))

$$\boldsymbol{e}_{(a)} = e_{(a)}{}^i \frac{\partial}{\partial x^i}; \tag{249}$$

and we shall write

$$\phi_{,(a)} = e_{(a)}{}^i \frac{\partial \phi}{\partial x^i} = e_{(a)}{}^i \phi_{,i}, \tag{250}$$

where ϕ is any scalar field. More generally, we define

$$\begin{aligned} A_{(a),(b)} &= e_{(b)}{}^i \frac{\partial}{\partial x^i} A_{(a)} = e_{(b)}{}^i \frac{\partial}{\partial x^i} e_{(a)}{}^j A_j \\ &= e_{(b)}{}^i \nabla_{\partial_i}[e_{(a)}{}^j A_j] = e_{(b)}{}^i [e_{(a)}{}^j A_{j;i} + A_k e_{(a)}{}^k{}_{;i}]. \end{aligned} \tag{251}$$

We thus obtain

$$A_{(a),(b)} = e_{(a)}{}^j A_{j;i} e_{(b)}{}^i + e_{(a)k;i} e_{(b)}{}^i e_{(c)}{}^k A^{(c)}, \tag{252}$$

making use of the various rules enunciated at the end of the preceding section and of the fact that the raising and the lowering of tensor indices permutes with the operation of covariant differentiation.

With the definition

$$\gamma_{(c)(a)(b)} = e_{(c)}{}^{k} e_{(a)k;i} e_{(b)}{}^{i}, \tag{253}$$

we can rewrite equation (252) in the form

$$A_{(a),(b)} = e_{(a)}{}^{j} A_{j;i} e_{(b)}{}^{i} + \gamma_{(c)(a)(b)} A^{(c)}. \tag{254}$$

The quantities, $\gamma_{(c)(a)(b)}$, which we have defined in equation (253) are called the *Ricci rotation-coefficients*. An equivalent definition of these coefficients is

$$e_{(a)k;i} = e^{(c)}{}_{k} \gamma_{(c)(a)(b)} e^{(b)}{}_{i}. \tag{255}$$

The Ricci rotation-coefficients are antisymmetric in the first pair of indices:

$$\gamma_{(c)(a)(b)} = -\gamma_{(a)(c)(b)} \tag{256}$$

—a fact which follows from expanding the identity

$$0 = \eta_{(a)(b),i} = [e_{(a)k} e_{(b)}{}^{k}]_{;i}. \tag{257}$$

(Notice that we could not have deduced this antisymmetry if the $\eta_{(a)(b)}$'s had not been constants.) By virtue of this antisymmetry, equation (255) can also be written in the form

$$e_{(a)}{}^{k}{}_{;i} = -\gamma_{(a)}{}^{k}{}_{i}. \tag{258}$$

Returning to equation (254), we write it in the alternative form

$$e_{(a)}{}^{i} A_{i;j} e_{(b)}{}^{j} = A_{(a),(b)} - \eta^{(n)(m)} \gamma_{(n)(a)(b)} A_{(m)}. \tag{259}$$

The quantity on the right-hand side of this equation is called the *intrinsic derivative* of $A_{(a)}$ in the direction $\boldsymbol{e}_{(b)}$ and written $A_{(a)|(b)}$:

$$A_{(a)|(b)} = e_{(a)}{}^{i} A_{i;j} e_{(b)}{}^{j} \qquad (\text{or, } A_{i;j} = e^{(a)}{}_{i} A_{(a)|(b)} e^{(b)}{}_{j}). \tag{260}$$

We thus have the formula,

$$A_{(a)|(b)} = A_{(a),(b)} - \eta^{(n)(m)} \gamma_{(n)(a)(b)} A_{(m)}, \tag{261}$$

relating the directional and the intrinsic derivatives.

It is clear from the definitions (260) that we can pass freely from intrinsic derivatives to covariant derivatives and vice versa.

The notion of the intrinsic derivative of vector fields is readily extended to tensor fields in an obvious fashion. Thus, the intrinsic derivative of the Riemann tensor is given by

$$R_{(a)(b)(c)(d)|(f)} = R_{ijkl;m} e_{(a)}{}^{i} e_{(b)}{}^{j} e_{(c)}{}^{k} e_{(d)}{}^{l} e_{(f)}{}^{m}. \tag{262}$$

Now expanding

$$R_{(a)(b)(c)(d),(f)} = [R_{ijkl} e_{(a)}{}^{i} e_{(b)}{}^{j} e_{(c)}{}^{k} e_{(d)}{}^{l}]_{;m} e_{(f)}{}^{m}, \tag{263}$$

and replacing the covariant derivatives of the different basis-vectors by the respective rotation-coefficients (in accordance with the relation (258)) we find (analogous to equation (261))

$$R_{(a)(b)(c)(d)\|(f)} = R_{(a)(b)(c)(d),(f)} - \eta^{(n)(m)}[\gamma_{(n)(a)(f)}R_{(m)(b)(c)(d)} + \gamma_{(n)(b)(f)}R_{(a)(m)(c)(d)} + \gamma_{(n)(c)(f)}R_{(a)(b)(m)(d)} + \gamma_{(n)(d)(f)}R_{(a)(b)(c)(m)}]. \tag{264}$$

Finally, it is important to observe that the evaluation of the rotation coefficients does not require the evaluation of covariant derivatives (and, therefore, of the Christoffel symbols). For, defining

$$\lambda_{(a)(b)(c)} = e_{(b)i,j}[e_{(a)}{}^i e_{(c)}{}^j - e_{(a)}{}^j e_{(c)}{}^i], \tag{265}$$

and rewriting in the form

$$\lambda_{(a)(b)(c)} = [e_{(b)i,j} - e_{(b)j,i}]\, e_{(a)}{}^i e_{(c)}{}^j, \tag{266}$$

we observe that we can replace the ordinary derivatives of $e_{(b)i}$ and $e_{(b)j}$ by the corresponding covariant derivatives (in accordance with equation (115) for symmetric connections) and write

$$\begin{aligned}\lambda_{(a)(b)(c)} &= [e_{(b)i;j} - e_{(b)j;i}]\, e_{(a)}{}^i e_{(c)}{}^j \\ &= \gamma_{(a)(b)(c)} - \gamma_{(c)(b)(a)}.\end{aligned} \tag{267}$$

By virtue of this relation, we have

$$\gamma_{(a)(b)(c)} = \tfrac{1}{2}[\lambda_{(a)(b)(c)} + \lambda_{(c)(a)(b)} - \lambda_{(b)(c)(a)}] \tag{268}$$

and as is manifest from equation (266) the evaluation of $\lambda_{(a)(b)(c)}$ requires only the evaluation of ordinary derivatives. Notice also that the λ-symbols are antisymmetric in the first and the third indices:

$$\lambda_{(a)(b)(c)} = -\lambda_{(c)(b)(a)}. \tag{269}$$

(c) *The commutation relations and the structure constants*

The Lie bracket, $[e_{(a)}, e_{(b)}]$, plays an important role in the theory; and being a tangent vector itself, can be expanded in terms of the same basis, $e_{(a)}$. Thus, we must have an expansion of the form

$$[e_{(a)}, e_{(b)}] = C^{(c)}{}_{(a)(b)}\, e_{(c)}. \tag{270}$$

The coefficients, $C^{(c)}{}_{(a)(b)}$, in this expansion are called the *structure constants*; they are antisymmetric in the indices (a) and (b) and there are 24 of them.

The structure constants can be expressed in terms of the rotation coefficients as follows. Consider the effect of the Lie bracket on a scalar field f. We have

$$\begin{aligned}[e_{(a)}, e_{(b)}]f &= e_{(a)}{}^i[e_{(b)}{}^j f_{,j}]_{,i} - e_{(b)}{}^i[e_{(a)}{}^j f_{,j}]_{,i} \\ &= [e_{(a)}{}^i e_{(b)}{}^j{}_{;i} - e_{(b)}{}^i e_{(a)}{}^j{}_{;i}]f_{,j} \\ &= [-\gamma_{(b)}{}^j{}_{(a)} + \gamma_{(a)}{}^j{}_{(b)}]f_{,j} \\ &= [-\gamma_{(b)}{}^{(c)}{}_{(a)} + \gamma_{(a)}{}^{(c)}{}_{(b)}]e_{(c)}{}^j f_{,j}.\end{aligned} \tag{271}$$

Comparison with equation (270) now yields the relation

$$C^{(c)}{}_{(a)(b)} = \gamma^{(c)}{}_{(b)(a)} - \gamma^{(c)}{}_{(a)(b)}. \tag{272}$$

Equation (270), written out explicitly with the structure constants expressed in terms of the rotation coefficients, provides the *commutation relations*; there are twenty-four of them.

(*d*) *The Ricci and the Bianchi identities*

Projecting the Ricci identity (equation (231)),

$$e_{(a)i;k;l} - e_{(a)i;l;k} = R_{mikl}\, e_{(a)}{}^{m}, \tag{273}$$

on to the tetrad frame, we have

$$\begin{aligned} R_{(a)(b)(c)(d)} &= R_{mikl}\, e_{(a)}{}^{m} e_{(b)}{}^{i} e_{(c)}{}^{k} e_{(d)}{}^{l} \\ &= \{ -[\gamma_{(a)(f)(g)} e^{(f)}{}_{i} e^{(g)}{}_{k}]_{;l} \\ &\quad + [\gamma_{(a)(f)(g)} e^{(f)}{}_{i} e^{(g)}{}_{l}]_{;k} \} e_{(b)}{}^{i} e_{(c)}^{(k)} e_{(d)}{}^{l}. \end{aligned} \tag{274}$$

Now expanding the quantities in the square brackets, on the right-hand side, and replacing, once again, the covariant derivatives of the basis vectors by the respective rotation-coefficients, we obtain

$$\begin{aligned} R_{(a)(b)(c)(d)} = &-\gamma_{(a)(b)(c),(d)} + \gamma_{(a)(b)(d),(c)} \\ &+ \gamma_{(b)(a)(f)}[\gamma_{(c)}{}^{(f)}{}_{(d)} - \gamma_{(d)}{}^{(f)}{}_{(c)}] \\ &+ \gamma_{(f)(a)(c)} \gamma_{(b)}{}^{(f)}{}_{(d)} - \gamma_{(f)(a)(d)} \gamma_{(b)}{}^{(f)}{}_{(c)}. \end{aligned} \tag{275}$$

Because of the antisymmetry of the rotation coefficients in the first pair of their indices and the manifest antisymmetry of the operation by which the tetrad components of the Riemann tensor are constructed from them, it is clear that there are 36 equations of the kind (275).

And finally, the Bianchi identity (234), expressed in terms of intrinsic derivatives and tetrad components, takes the form

$$\begin{aligned} R_{(a)(b)[(c)(d)|(f)]} = \tfrac{1}{6} \sum_{[(c)(d)(f)]} \{ & R_{(a)(b)(c)(d),(f)} \\ & - \eta^{(n)(m)}[\gamma_{(n)(a)(f)} R_{(m)(b)(c)(d)} + \gamma_{(n)(b)(f)} R_{(a)(m)(c)(d)} \\ & + \gamma_{(n)(c)(f)} R_{(a)(b)(m)(d)} + \gamma_{(n)(d)(f)} R_{(a)(b)(c)(m)}] \}; \end{aligned} \tag{276}$$

and, as we have noted earlier in § 6(*e*), there are 20 linearly-independent equations of this kind.

The basic equations of the tetrad formalism are the 24 commutation relations (270), the 36 Ricci identities (275), and the 20 linearly-independent Bianchi identities. It is not clear how many of these equations are independent, how they are to be ordered or used and, indeed, what they are *for*. These are all questions which we shall confront in due course.

(e) *A generalized version of the tetrad formalism*

In the foregoing account of the tetrad formalism, we have explicitly assumed that $\eta_{(a)(b)}$, as defined in equation (242), represents a constant symmetric matrix. If this should not be the case and the $\eta_{(a)(b)}$'s are functions in space–time, then (as we have already noted) we cannot infer the antisymmetry of the rotation coefficients in their first two indices; instead, we shall have (cf. equation (257))

$$\gamma_{(a)(b)(c)} - \gamma_{(b)(a)(c)} = \eta_{(a)(b),\,(c)}. \tag{277}$$

And even more important, the process of raising and lowering of tetrad indices by $\eta^{(a)(b)}$ and $\eta_{(a)(b)}$ does not permute with the operation of directional or of intrinsic differentiation. A careful scrutiny of the analysis of the preceding sections shows that while the correct expression for the structure constant, in place of equation (272), is

$$C^{(c)}{}_{(a)(b)} = \gamma_{(a)}{}^{(c)}{}_{(b)} - \gamma_{(b)}{}^{(c)}{}_{(a)}, \tag{278}$$

the Ricci and the Bianchi identities as written in equations (275) and (276) continue to be valid.

8. The Newman–Penrose formalism

The Newman–Penrose formalism is a tetrad formalism with a special choice of the basis vectors. The choice that is made is a tetrad of null vectors $\boldsymbol{l}$, $\boldsymbol{n}$, $\boldsymbol{m}$, and $\bar{\boldsymbol{m}}$ of which $\boldsymbol{l}$ and $\boldsymbol{n}$ are real and $\boldsymbol{m}$ and $\bar{\boldsymbol{m}}$ are complex conjugates of one another. The novelty of the formalism, when it was first proposed by Newman and Penrose in 1962, was precisely in their choice of a null basis:* it was a departure from the choice of an orthonormal basis which was customary till then. The underlying motivation for the choice of a null basis was Penrose's strong belief that the essential element of a space–time is its light-cone structure which makes possible the introduction of a spinor basis. And it will appear that the light-cone structure of the space–times of the black-hole solutions of general relativity is exactly of the kind that makes the Newman–Penrose formalism most effective for grasping the inherent symmetries of these space–times and revealing their analytical richness. But it may be stated, here already, that the special adaptability of the Newman–Penrose formalism to the black-hole solutions of general relativity derives from their "type-D" character and the Goldberg–Sachs theorem—matters which will be considered later in this chapter (§9) and subsequently.

* Penrose was originally led to consider the introduction of a null basis from his interest in incorporating in general relativity spinor analysis in an essential way. We briefly consider this alternative approach to the Newman–Penrose formalism in Chapter 10 (§102).

(a) The null basis and the spin coefficients

As we have already stated, underlying the Newman-Penrose formalism is the choice of a null basis consisting of a pair of real null-vectors, $\boldsymbol{l}$ and $\boldsymbol{n}$, and a pair of complex-conjugate null-vectors $\boldsymbol{m}$ and $\bar{\boldsymbol{m}}$. They are required to satisfy the *orthogonality conditions*,

$$\boldsymbol{l}\cdot\boldsymbol{m} = \boldsymbol{l}\cdot\bar{\boldsymbol{m}} = \boldsymbol{n}\cdot\boldsymbol{m} = \boldsymbol{n}\cdot\bar{\boldsymbol{m}} = 0, \tag{279}$$

besides the requirements,

$$\boldsymbol{l}\cdot\boldsymbol{l} = \boldsymbol{n}\cdot\boldsymbol{n} = \boldsymbol{m}\cdot\boldsymbol{m} = \bar{\boldsymbol{m}}\cdot\bar{\boldsymbol{m}} = 0, \tag{280}$$

that the vectors be null. It is customary to impose on the basis vectors the further *normalization conditions*,

$$\boldsymbol{l}\cdot\boldsymbol{n} = 1 \qquad \text{and} \qquad \boldsymbol{m}\cdot\bar{\boldsymbol{m}} = -1. \tag{281}$$

It is strictly not necessary to impose these further conditions (cf. §7(e)). Indeed, in some of Penrose's later work, he prefers not to impose the conditions (281) as more in harmony with his views on the null-cone structure of space–time. But for the purposes of this book, this additional freedom in the choice of the basis vectors does not have any advantage that overcomes the disadvantage of having rotation coefficients which are not antisymmetric in their first two indices, and the still more serious disadvantage of not being able to permute the operation of the raising and the lowering of the tetrad indices with the operation of directional and intrinsic differentiations. On these accounts, in this book, we shall retain the normalization conditions (281). Then, the fundamental matrix represented by $\eta_{(a)(b)}$ is a constant symmetric matrix of the form

$$[\eta_{(a)(b)}] = [\eta^{(a)(b)}] = \begin{vmatrix} 0 & 1 & 0 & 0 \\ 1 & 0 & 0 & 0 \\ 0 & 0 & 0 & -1 \\ 0 & 0 & -1 & 0 \end{vmatrix}, \tag{282*}$$

with the correspondence (cf. §7(a)),

$$\boldsymbol{e}_1 = \boldsymbol{l}, \qquad \boldsymbol{e}_2 = \boldsymbol{n}, \qquad \boldsymbol{e}_3 = \boldsymbol{m}, \qquad \text{and} \qquad \boldsymbol{e}_4 = \bar{\boldsymbol{m}}. \tag{283}$$

The corresponding covariant basis is given by

$$\boldsymbol{e}^1 = \boldsymbol{e}_2 = \boldsymbol{n};\quad \boldsymbol{e}^2 = \boldsymbol{e}_1 = \boldsymbol{l};\quad \boldsymbol{e}^3 = -\boldsymbol{e}_4 = -\bar{\boldsymbol{m}},\quad \text{and}\quad \boldsymbol{e}^4 = -\boldsymbol{e}_3 = -\boldsymbol{m}. \tag{284}$$

The basis vectors, considered as directional derivatives, are designated by

* In the sequel, we shall dispense with the distinguishing parentheses for the tetrad indices so long as there is no ambiguity as to what is intended.

special symbols:

$$e_1 = e^2 = D; \quad e_2 = e^1 = \triangle; \quad e_3 = -e^4 = \delta; \quad \text{and} \quad e_4 = -e^3 = \delta^*. \tag{285}$$

The various Ricci rotation-coefficients, now called the *spin coefficients*, are, similarly, designated by special symbols:

$$\left.\begin{aligned} &\kappa = \gamma_{311}; \quad \rho = \gamma_{314}; \quad \varepsilon = \tfrac{1}{2}(\gamma_{211}+\gamma_{341});\\ &\sigma = \gamma_{313}; \quad \mu = \gamma_{243}; \quad \gamma = \tfrac{1}{2}(\gamma_{212}+\gamma_{342});\\ &\lambda = \gamma_{244}; \quad \tau = \gamma_{312}; \quad \alpha = \tfrac{1}{2}(\gamma_{214}+\gamma_{344});\\ &\nu = \gamma_{242}; \quad \pi = \gamma_{241}; \quad \beta = \tfrac{1}{2}(\gamma_{213}+\gamma_{343}). \end{aligned}\right\} \tag{286}$$

It is clear that the complex conjugate of any quantity can be obtained by replacing the index 3, wherever it occurs, by the index 4, and conversely. This is a general rule.

(b) *The representation of the Weyl, the Ricci, and the Riemann tensors*

The Weyl tensor is the trace-free part of the Riemann tensors; and its tetrad components are given by (cf. equation (233))

$$\begin{aligned} R_{abcd} = C_{abcd} &+ \tfrac{1}{2}(\eta_{ac}R_{bd} - \eta_{bc}R_{ad} - \eta_{ad}R_{bc} + \eta_{bd}R_{ac})\\ &- \tfrac{1}{6}(\eta_{ac}\eta_{bd} - \eta_{ad}\eta_{bc})R, \end{aligned} \tag{287}$$

where R_{bd} denotes the tetrad components of the Ricci tensor and R the scalar curvature:

$$R_{ac} = \eta^{bd}R_{abcd} \quad \text{and} \quad R = \eta^{ab}R_{ab} = 2(R_{12} - R_{34}). \tag{288}$$

The fact that C_{abcd} is trace-free requires that

$$\eta^{ad}C_{abcd} = C_{1bc2} + C_{2bc1} - C_{3bc4} - C_{4bc3} = 0. \tag{289}$$

Besides, we must also require that

$$C_{1234} + C_{1342} + C_{1423} = 0. \tag{290}$$

The condition (289), written out explicitly for $b = c$, gives

$$C_{1314} = C_{2324} = C_{1332} = C_{1442} = 0, \tag{291}$$

while the condition for $b \neq c$ together with the cyclic requirement (290) give

$$\left.\begin{aligned} &C_{1231} = C_{1334};\ C_{1241} = C_{1443};\ C_{1232} = C_{2343};\ C_{1242} = C_{2434},\\ &C_{1212} = C_{3434};\ \text{and}\ C_{1342} = \tfrac{1}{2}(C_{1212} - C_{1234}) = \tfrac{1}{2}(C_{3434} - C_{1234}). \end{aligned}\right\} \tag{292}$$

Making use of the foregoing results, we find that the various components of the Riemann tensor are related to those of the Weyl and the Ricci tensors by

$$\left.\begin{aligned}
&R_{1212} = C_{1212}+R_{12}-\tfrac{1}{6}R;\ R_{1324} = C_{1324}+\tfrac{1}{12}R;\ R_{1234} = C_{1234};\\
&R_{3434} = C_{3434}-R_{34}-\tfrac{1}{6}R;\ R_{1313} = C_{1313};\ R_{2323} = C_{2323};\\
&R_{1314} = \tfrac{1}{2}R_{11};\ R_{2324} = \tfrac{1}{2}R_{22};\ R_{3132} = -\tfrac{1}{2}R_{33};\\
&R_{1213} = C_{1213}+\tfrac{1}{2}R_{13};\qquad R_{1334} = C_{1334}+\tfrac{1}{2}R_{13};\\
&R_{1223} = C_{1223}-\tfrac{1}{2}R_{23};\qquad R_{2334} = C_{2334}+\tfrac{1}{2}R_{23};
\end{aligned}\right\} \tag{293}$$

and the additional complex-conjugate relations obtained by replacing the index 3 by the index 4, and vice versa.

In the Newman–Penrose formalism, the ten independent components of the Weyl tensor are represented by the five complex scalars,

$$\left.\begin{aligned}
\Psi_0 &= -C_{1313} = -C_{pqrs}l^p m^q l^r m^s.\\
\Psi_1 &= -C_{1213} = -C_{pqrs}l^p n^q l^r m^s,\\
\Psi_2 &= -C_{1342} = -C_{pqrs}l^p m^q \bar{m}^r n^s,\\
\Psi_3 &= -C_{1242} = -C_{pqrs}l^p n^q \bar{m}^r n^s,\\
\Psi_4 &= -C_{2424} = -C_{pqrs}n^p \bar{m}^q n^r \bar{m}^s.
\end{aligned}\right\} \tag{294}$$

While it is clear on general grounds that the Weyl tensor is completely specified by the five complex scalars $\Psi_0, \ldots, \Psi_4$, it will be convenient to have a general formula which expresses the different components of the Weyl tensor explicitly in terms of the five scalars. Such a formula can be obtained in the following manner.

First, we shall introduce a symbol which indicates the construction out of products of four quantities a linear combination of them which will have all the principal symmetries of the Riemann and the Weyl tensors. Thus, we shall let

$$\begin{aligned}
\{w_p x_q y_r z_s\} = {} & w_p x_q y_r z_s - w_p x_q z_r y_s - x_p w_q y_r z_s + x_p w_q z_r y_s\\
& + y_p z_q w_r x_s - y_p z_q x_r w_s - z_p y_q w_r x_s + z_p y_q x_r w_s.
\end{aligned} \tag{295}$$

The linear combination so constructed is manifestly antisymmetric in (p, q) and in (r, s) and unchanged for the simultaneous interchange of the index pairs (p, q) and (r, s). In this notation, it is clear that the tensor components, C_{pqrs}, of the Weyl tensor must have the representation

$$\begin{aligned}
C_{pqrs} = {} & Q_{1212}\{l_p n_q l_r n_s\} + Q_{3434}\{m_p \bar{m}_q m_r \bar{m}_s\} + Q_{1234}\{l_p n_q m_r \bar{m}_s\}\\
& + Q_{1314}\{l_p m_q l_r \bar{m}_s\} + Q_{2324}\{n_p m_q n_r \bar{m}_s\}\\
& + [Q_{1313}\{l_p m_q l_r m_s\} + Q_{2323}\{n_p m_q n_r m_s\} + Q_{1213}\{l_p n_q l_r m_s\}\\
& + Q_{1223}\{l_p n_q n_r m_s\} + Q_{1323}\{l_p m_q n_r m_s\} + Q_{1324}\{l_p m_q n_r \bar{m}_s\}\\
& + Q_{1334}\{l_p m_q m_r \bar{m}_s\} + Q_{2334}\{n_p m_q m_r \bar{m}_s\} + \text{complex conjugates}],
\end{aligned} \tag{296}$$

where $Q_{....}$ are coefficients, unspecified for the present, and the 'complex conjugates' of the eight quantities in the square brackets on the right-hand side are to be obtained by writing $\bar{\boldsymbol{m}}$ wherever $\boldsymbol{m}$ occurs and index 4 wherever the index 3 occurs and vice versa.

The coefficients $Q_{....}$ in the representation (296) can be obtained by contracting both sides of the equation by appropriate products of the components of $\boldsymbol{l}, \boldsymbol{n}, \boldsymbol{m}$, and $\bar{\boldsymbol{m}}$. Thus, by contracting with $l^p m^q l^r m^s$, the left-hand side, by definition, yields $-\Psi_0$, while the only surviving term on the right-hand side has the coefficient Q_{2424}; and, accordingly, $Q_{2424} = -\Psi_0$. Similarly, by contracting the equation with $l^p m^q l^r \bar{m}^s$, the only surviving term on the right-hand side has the coefficient Q_{2423}; and this coefficient must vanish since the left-hand side yields C_{1314} which we have shown to be zero (cf. equation (291)). Proceeding in this fashion, systematically making use of the various definitions and requirements included in equations (291), (292), and (294), we find

$$\begin{aligned}
&Q_{1324} = \Psi_2;\; Q_{1212} = Q_{3434} = -(\Psi_2 + \Psi_2^*);\; Q_{1234} = (\Psi_2 - \Psi_2^*);\\
&Q_{1313} = -\Psi_4;\; Q_{2424} = -\Psi_0;\; Q_{1213} = -Q_{1334} = \Psi_3;\\
&Q_{1224} = Q_{2443} = -\Psi_1;\; \text{and}\quad Q_{1314} = Q_{2324} = Q_{1323} = Q_{1424} = 0.
\end{aligned} \tag{297}$$

The required formula is, therefore,

$$\begin{aligned}
C_{pqrs} = &-(\Psi_2 + \Psi_2^*)[\{l_p n_q l_r n_s\} + \{m_p \bar{m}_q m_r \bar{m}_s\}] + (\Psi_2 - \Psi_2^*)\{l_p n_q m_r \bar{m}_s\}\\
&+ [\![-\Psi_0 \{n_p \bar{m}_q n_r \bar{m}_s\} - \Psi_4 \{l_p m_q l_r m_s\} + \Psi_2 \{l_p m_q n_r \bar{m}_s\}\\
&-\Psi_1 [\{l_p n_q n_r \bar{m}_s\} + \{n_p \bar{m}_q \bar{m}_r m_s\}]\\
&+\Psi_3 [\{l_p n_q l_r m_s\} - \{l_p m_q m_r \bar{m}_s\}] + \text{complex conjugates}]\!].
\end{aligned} \tag{298}$$

In particular,

$$C_{1334} = \Psi_1;\; C_{2443} = \Psi_3;\; C_{1212} = C_{3434} = -(\Psi_2 + \Psi_2^*);$$

and
$$C_{1234} = (\Psi_2 - \Psi_2^*). \tag{299}$$

Including the four components which vanish and allowing for complex conjugation, we have specified in equations (294) and (299) all the distinct components of the Weyl tensor.

And finally, the ten components of the Ricci tensor are defined in terms of the following four real and three complex scalars:

$$\left.\begin{aligned}
&\Phi_{00} = -\tfrac{1}{2}R_{11};\; \Phi_{22} = -\tfrac{1}{2}R_{22};\; \Phi_{02} = -\tfrac{1}{2}R_{33};\; \Phi_{20} = -\tfrac{1}{2}R_{44};\\
&\Phi_{11} = -\tfrac{1}{4}(R_{12} + R_{34});\; \Phi_{01} = -\tfrac{1}{2}R_{13};\; \Phi_{10} = -\tfrac{1}{2}R_{14};\\
&\quad \Lambda = \tfrac{1}{24}R = \tfrac{1}{12}(R_{12} - R_{34});\; \Phi_{12} = -\tfrac{1}{2}R_{23};\; \Phi_{21} = -\tfrac{1}{2}R_{24}.
\end{aligned}\right\} \tag{300}$$

(c) The commutation relations and the structure constants

We shall now proceed to write down the explicit forms of the various equations of the theory.

We consider first the commutation relation (cf. equations (270) and (272))

$$[\boldsymbol{e}_{(a)}, \boldsymbol{e}_{(b)}] = (\gamma_{cba} - \gamma_{cab})\boldsymbol{e}^c = C^c{}_{ab}\boldsymbol{e}_c. \tag{301}$$

As an example, consider

$$\begin{aligned}[\triangle, D] = [\boldsymbol{n}, \boldsymbol{l}] = [\boldsymbol{e}_2, \boldsymbol{e}_1] &= (\gamma_{c12} - \gamma_{c21})\boldsymbol{e}^c \\ &= -\gamma_{121}\boldsymbol{e}^1 + \gamma_{212}\boldsymbol{e}^2 + (\gamma_{312} - \gamma_{321})\boldsymbol{e}^3 + (\gamma_{412} - \gamma_{421})\boldsymbol{e}^4 \\ &= -\gamma_{121}\triangle + \gamma_{212}D - (\gamma_{312} - \gamma_{321})\delta^* - (\gamma_{412} - \gamma_{421})\delta,\end{aligned} \tag{302}$$

or, giving the spin coefficients their designated symbols, we obtain

$$\triangle D - D\triangle = (\gamma + \gamma^*)D + (\varepsilon + \varepsilon^*)\triangle - (\tau^* + \pi)\delta - (\tau + \pi^*)\delta^*. \tag{303}$$

In similar fashion, we obtain:

$$\delta D - D\delta = (\alpha^* + \beta - \pi^*)D + \kappa\triangle - (\rho^* + \varepsilon - \varepsilon^*)\delta - \sigma\delta^*, \tag{304}$$

$$\delta\triangle - \triangle\delta = -\nu^* D + (\tau - \alpha^* - \beta)\triangle + (\mu - \gamma + \gamma^*)\delta + \lambda^*\delta^*, \tag{305}$$

$$\delta^*\delta - \delta\delta^* = (\mu^* - \mu)D + (\rho^* - \rho)\triangle + (\alpha - \beta^*)\delta + (\beta - \alpha^*)\delta^*. \tag{306}$$

By expressing the foregoing relations in the manner of equation (301), we find that the structure constants are related to the spin coefficients as in the accompanying tabulation:

$$\begin{aligned}
&C^1{}_{21} = +(\gamma + \gamma^*);\ C^1{}_{31} = +(\alpha^* + \beta - \pi^*);\ C^1{}_{32} = -\nu^*;\quad C^1{}_{43} = \mu^* - \mu,\\
&C^2{}_{21} = +(\varepsilon + \varepsilon^*);\ C^2{}_{31} = +\kappa;\quad C^2{}_{32} = \tau - \alpha^* - \beta;\ C^2{}_{43} = \rho^* - \rho,\\
&C^3{}_{21} = -(\tau^* + \pi);\ C^3{}_{31} = -(\rho^* + \varepsilon - \varepsilon^*);\ C^3{}_{32} = \mu - \gamma + \gamma^*;\ C^3{}_{43} = \alpha - \beta^*,\\
&C^4{}_{21} = -(\tau + \pi^*);\ C^4{}_{31} = -\sigma;\quad C^4{}_{32} = +\lambda^*;\quad C^4{}_{43} = \beta - \alpha^*.
\end{aligned} \tag{307}$$

(d) The Ricci identities and the eliminant relations

We shall now write down the explicit forms which the Ricci identities take in the Newman–Penrose formalism. Considering the (1313)-component of equation (275), for example, we have (making use of the relations (293))

$$\begin{aligned}-\Psi_0 = C_{1313} = R_{1313} = {}&\gamma_{133,1} - \gamma_{131,3}\\ &+ \gamma_{133}(\gamma_{121} + \gamma_{431} - \gamma_{413} + \gamma_{431} + \gamma_{134})\\ &- \gamma_{131}(\gamma_{433} + \gamma_{123} - \gamma_{213} + \gamma_{231} + \gamma_{132});\end{aligned} \tag{308}$$

or, substituting for the directional derivatives and the spin coefficients their designated symbols, we obtain

$$D\sigma - \delta\kappa = \sigma(3\varepsilon - \varepsilon^* + \rho + \rho^*) + \kappa(\pi^* - \tau - 3\beta - \alpha^*) + \Psi_0. \tag{309}$$

As we have noted in the context of the standard tetrad formalism in §7(*d*), we can write down a total of 36 equations by considering the various different components of equation (275). But in the context of the Newman–Penrose formalism, it will suffice to write down only half the number of equations (by omitting to write down the complex conjugate of an equation). We list below a set of 18 equations, as originally derived by Newman and Penrose in 1962; and we have indicated in each case the component of the Riemann tensor which gives rise to the equation.

$$D\rho - \delta^*\kappa = (\rho^2 + \sigma\sigma^*) + \rho(\varepsilon + \varepsilon^*) - \kappa^*\tau - \kappa(3\alpha + \beta^* - \pi) + \Phi_{00}, \qquad [R_{1314}] \qquad \text{(a)}$$

$$D\sigma - \delta\kappa = \sigma(\rho + \rho^* + 3\varepsilon - \varepsilon^*) - \kappa(\tau - \pi^* + \alpha^* + 3\beta) + \Psi_0, \qquad [R_{1313}] \qquad \text{(b)}$$

$$D\tau - \Delta\kappa = \rho(\tau + \pi^*) + \sigma(\tau^* + \pi) + \tau(\varepsilon - \varepsilon^*) - \kappa(3\gamma + \gamma^*) + \Psi_1 + \Phi_{01}, \qquad [R_{1312}] \qquad \text{(c)}$$

$$D\alpha - \delta^*\varepsilon = \alpha(\rho + \varepsilon^* - 2\varepsilon) + \beta\sigma^* - \beta^*\varepsilon - \kappa\lambda - \kappa^*\gamma + \pi(\varepsilon + \rho) + \Phi_{10}, \qquad [\tfrac{1}{2}(R_{3414} - R_{1214})] \qquad \text{(d)}$$

$$D\beta - \delta\varepsilon = \sigma(\alpha + \pi) + \beta(\rho^* - \varepsilon^*) - \kappa(\mu + \gamma) - \varepsilon(\alpha^* - \pi^*) + \Psi_1, \qquad [\tfrac{1}{2}(R_{1213} - R_{3413})] \qquad \text{(e)}$$

$$D\gamma - \Delta\varepsilon = \alpha(\tau + \pi^*) + \beta(\tau^* + \pi) - \gamma(\varepsilon + \varepsilon^*) - \varepsilon(\gamma + \gamma^*) + \tau\pi - \nu\kappa + \Psi_2 + \Phi_{11} - \Lambda, \qquad [\tfrac{1}{2}(R_{1212} - R_{3412})] \qquad \text{(f)}$$

$$D\lambda - \delta^*\pi = (\rho\lambda + \sigma^*\mu) + \pi(\pi + \alpha - \beta^*) - \nu\kappa^* - \lambda(3\varepsilon - \varepsilon^*) + \Phi_{20}, \qquad [R_{2441}] \qquad \text{(g)}$$

$$D\mu - \delta\pi = (\rho^*\mu + \sigma\lambda) + \pi(\pi^* - \alpha^* + \beta) - \mu(\varepsilon + \varepsilon^*) - \nu\kappa + \Psi_2 + 2\Lambda, \qquad [R_{2431}] \qquad \text{(h)}$$

$$D\nu - \Delta\pi = \mu(\pi + \tau^*) + \lambda(\pi^* + \tau) + \pi(\gamma - \gamma^*) - \nu(3\varepsilon + \varepsilon^*) + \Psi_3 + \Phi_{21}, \qquad [R_{2421}] \qquad \text{(i)}$$

$$\Delta\lambda - \delta^*\nu = -\lambda(\mu + \mu^* + 3\gamma - \gamma^*) + \nu(3\alpha + \beta^* + \pi - \tau^*) - \Psi_4, \qquad [R_{2442}] \qquad \text{(j)}$$

$$\delta\rho - \delta^*\sigma = \rho(\alpha^* + \beta) - \sigma(3\alpha - \beta^*) + \tau(\rho - \rho^*) + \kappa(\mu - \mu^*) - \Psi_1 + \Phi_{01}, \qquad [R_{3143}] \qquad \text{(k)}$$

$$\delta\alpha - \delta^*\beta = (\mu\rho - \lambda\sigma) + \alpha\alpha^* + \beta\beta^* - 2\alpha\beta + \gamma(\rho - \rho^*) + \varepsilon(\mu - \mu^*) - \Psi_2 + \Phi_{11} + \Lambda, \qquad [\tfrac{1}{2}(R_{1234} - R_{3434})] \qquad \text{(l)}$$

$$\delta\lambda - \delta^*\mu = \nu(\rho - \rho^*) + \pi(\mu - \mu^*) + \mu(\alpha + \beta^*) + \lambda(\alpha^* - 3\beta) - \Psi_3 + \Phi_{21}, \qquad [R_{2443}] \qquad \text{(m)}$$

$$\delta\nu - \Delta\mu = (\mu^2 + \lambda\lambda^*) + \mu(\gamma + \gamma^*) - \nu^*\pi + \nu(\tau - 3\beta - \alpha^*) + \Phi_{22}, \qquad [R_{2423}] \qquad \text{(n)}$$

$$\delta\gamma - \Delta\beta = \gamma(\tau - \alpha^* - \beta) + \mu\tau - \sigma\nu - \varepsilon\nu^* - \beta(\gamma - \gamma^* - \mu) + \alpha\lambda^* + \Phi_{12}, \qquad [\tfrac{1}{2}(R_{1232} - R_{3432})] \qquad \text{(o)}$$

$$\delta\tau - \Delta\sigma = (\mu\sigma + \lambda^*\rho) + \tau(\tau + \beta - \alpha^*) - \sigma(3\gamma - \gamma^*) - \kappa\nu^* + \Phi_{02}, \qquad [R_{1332}] \qquad \text{(p)}$$

$$\Delta\rho - \delta^*\tau = -(\rho\mu^* + \sigma\lambda) + \tau(\beta^* - \alpha - \tau^*) + \rho(\gamma + \gamma^*) + \nu\kappa - \Psi_2 - 2\Lambda; \qquad [R_{1324}] \qquad \text{(q)}$$

$$\Delta\alpha - \delta^*\gamma = \nu(\rho + \varepsilon) - \lambda(\tau + \beta) + \alpha(\gamma^* - \mu^*) + \gamma(\beta^* - \tau^*) - \Psi_3. \qquad [\tfrac{1}{2}(R_{1242} - R_{3442})] \qquad \text{(r)}$$

(310)

While each of the foregoing equations contains one or more of the Weyl and the Ricci scalars, it is clear that one should be able to obtain from them a total of sixteen real equations which are independent of them and involve only the spin coefficients. The reason is that, while the right-hand side of equation (275) (from which all of the foregoing equations were derived) is antisymmetric in the first pair and in the second pair of indices, it has not the invariance of the Riemann tensor (on the left-hand side) for the simultaneous interchange of the first and the second index of pairs. It is precisely this latter invariance (together with the cyclic identity) which reduces the number of independent components of the Riemann tensor from thirty-six to twenty. For this same reason, we must be able to eliminate the Riemann tensor, altogether, from sixteen of the thirty-six equations included in equation (275) (and (310)). This elimination can be carried out directly from the equations listed. Thus, the imaginary part of equation (310,a) will not contain Φ_{00} since it is real. Similarly, from equations (310,c), (310,e), and the complex conjugate of equation (310,d) we can readily eliminate Ψ_1 and Φ_{01} ($= \Phi_{10}^*$), and obtain a single complex equation involving only the spin coefficients. By such systematic eliminations, we obtain the following set of four real and six complex *eliminant relations*:

$$D(\rho - \rho^*) + \delta\kappa^* - \delta^*\kappa = (\rho - \rho^*)(\rho + \rho^* + \varepsilon + \varepsilon^*) + \kappa(\tau^* + \pi - 3\alpha - \beta^*) - \kappa^*(\tau + \pi^* - 3\alpha^* - \beta), \qquad \text{(a)}$$

$$D(\mu - \mu^*) + \delta(\alpha + \beta^* - \pi) - \delta^*(\alpha^* + \beta - \pi^*) = (\gamma + \gamma^*)(\rho - \rho^*) + \alpha(\pi^* - 2\beta) - \alpha^*(\pi - 2\beta^*) + \kappa^*\nu^* - \kappa\nu + \beta\pi - \beta^*\pi^* + (\rho + \rho^*)(\mu - \mu^*) \qquad \text{(b)}$$

$$D(\mu - \mu^* - \gamma + \gamma^*) + \Delta(\varepsilon - \varepsilon^*) - \delta\pi + \delta^*\pi^* = (\varepsilon + \varepsilon^*)(\mu^* - \mu)$$

$$+\tau^*(\alpha^*+\pi^*-\beta)-\tau(\alpha+\pi-\beta^*)+\lambda\sigma-\lambda^*\sigma^*+\rho^*\mu-\rho\mu^*+2(\epsilon\gamma-\epsilon^*\gamma^*), \tag{c}$$

$$\begin{aligned}\triangle(\mu^*-\mu)+\delta\nu-\delta^*\nu^* = (\mu-\mu^*)(\mu+\mu^*+\gamma+\gamma^*)+\nu(\tau-3\beta-\alpha^*+\pi^*)\\ -\nu^*(\tau^*+\pi-3\beta^*-\alpha),\end{aligned} \tag{d}$$

$$\begin{aligned}D(\tau-\alpha^*-\beta)-\triangle\kappa+\delta(\epsilon+\epsilon^*) = \rho(\tau+\pi^*)+\kappa^*\lambda^*+\sigma(\tau^*-\alpha-\beta^*)\\ +\epsilon(\tau-\pi^*)-\rho^*(\beta+\alpha^*+\pi^*)+\epsilon^*(2\alpha^*+2\beta-\tau-\pi^*)+\kappa(\mu-2\gamma),\end{aligned} \tag{e}$$

$$\begin{aligned}\delta(\rho-\epsilon+\epsilon^*)-\delta^*\sigma+D(\beta-\alpha^*) = \rho(\alpha^*+\beta+\tau)-\rho^*(\tau-\beta+\alpha^*+\pi^*)\\ +(\epsilon^*-\epsilon)(2\alpha^*-\pi^*)+\sigma(\pi-2\alpha)+\kappa(\gamma^*-\gamma-\mu^*)+\kappa^*\lambda^*,\end{aligned} \tag{f}$$

$$\begin{aligned}D\lambda+\triangle\sigma^*-\delta^*(\tau^*+\pi) = \sigma^*(3\gamma^*-\gamma+\mu-\mu^*)+(\pi+\tau^*)(\pi-\tau^*+\alpha)\\ +\lambda(\rho-\rho^*-3\epsilon+\epsilon^*)-\beta\pi-\tau^*\beta^*,\end{aligned} \tag{g}$$

$$\begin{aligned}D\nu+\triangle(\alpha+\beta^*-\pi)-\delta^*(\gamma+\gamma^*) = \nu(\rho-2\epsilon)+\lambda(\pi^*-\alpha^*-\beta)+\mu(\pi+\tau^*)\\ -\mu^*(\alpha+\beta^*+\tau^*)+\gamma(\pi-\tau^*)+\gamma^*(2\alpha+2\beta^*-\pi-\tau^*)+\sigma^*\nu^*,\end{aligned} \tag{h}$$

$$\begin{aligned}\triangle(\beta^*-\alpha)+\delta\lambda+\delta^*(\gamma-\gamma^*-\mu) = \nu(\epsilon^*-\epsilon-\rho^*)+\lambda(\tau-2\beta)+\alpha(\mu+\mu^*)\\ -\mu^*(\pi+\tau^*+\beta^*)+\mu(\pi+\beta^*)+(\gamma-\gamma^*)(\tau^*-2\beta^*)+\sigma^*\nu^*,\end{aligned} \tag{i}$$

$$\begin{aligned}D\mu+\triangle\rho-\delta\pi-\delta^*\tau = \rho^*\mu-\rho\mu^*+\pi(\pi^*-\alpha^*+\beta)+\tau(\beta^*-\alpha-\tau^*)\\ +\rho(\gamma+\gamma^*)-\mu(\epsilon+\epsilon^*).\end{aligned} \tag{j}$$

(311)

(e) *The Bianchi identities*

As we have stated earlier in § 7(*d*), there are, altogether, twenty linearly-independent Bianchi-identities. A complete set is provided by the eight complex identities

$$\left.\begin{aligned} R_{13[13|4]}=0;\quad R_{13[21|4]}=0;\quad R_{13[13|2]}=0;\quad R_{13[43|2]}=0;\\ R_{42[13|4]}=0;\quad R_{42[21|4]}=0;\quad R_{42[13|2]}=0;\quad R_{42[43|2]}=0;\end{aligned}\right\} \tag{312}$$

and the four real identities which follow from

$$\eta^{bc}(R_{ab}-\tfrac{1}{2}\eta_{ab}R)_{|c}=0. \tag{313}$$

Explicitly written out, equation (313) provides the two real equations,

$$\left.\begin{aligned} R_{11|2}+R_{34|1}-R_{13|4}-R_{14|3}=0,\\ \text{and}\qquad\qquad\\ R_{22|1}+R_{34|2}-R_{23|4}-R_{24|3}=0,\end{aligned}\right\} \tag{314}$$

and the one complex equation

$$R_{33|4}+R_{12|3}-R_{31|2}-R_{32|1}=0. \tag{315}$$

We shall now write down the explicit forms of the various identities in terms of the spin coefficients and the Weyl and the Ricci scalars. As an example, consider the first of the identities listed in equation (312). We have

$$R_{1313|4}+R_{1334|1}+R_{1341|3}=0. \tag{316}$$

By making use of the relations given in equations (293), we can write instead,

$$C_{1313|4}+(C_{1334}+\tfrac{1}{2}R_{13})_{|1}-\tfrac{1}{2}R_{11|3}=0. \tag{317}$$

Considering the terms in the Weyl tensor, we have

$$\begin{aligned} C_{1313|4} &= C_{1313,4}-\eta^{nm}(\gamma_{n14}C_{m313}+\gamma_{n34}C_{1m13} \\ &\qquad +\gamma_{n14}C_{13m3}+\gamma_{n34}C_{131m}) \\ &= C_{1313,4}-2(\gamma_{214}+\gamma_{344})C_{1313}+2\gamma_{314}(C_{1213}+C_{4313}) \\ &= -\delta^*\Psi_0+4\alpha\Psi_0-4\rho\Psi_1, \end{aligned} \tag{318}$$

and

$$\begin{aligned} C_{1334|1} &= C_{1334,1}-\eta^{nm}[\gamma_{n11}C_{m334}+\gamma_{n31}(C_{1m34}+C_{13m4}) \\ &\qquad +\gamma_{n41}C_{133m}] \\ &= C_{1334,1}-[(\gamma_{211}+\gamma_{341})C_{1334}+\gamma_{131}(C_{1234}-C_{3434}) \\ &\qquad +\gamma_{231}C_{1314}+\gamma_{141}C_{1332}+\gamma_{131}C_{1324}+\gamma_{241}C_{1331}] \\ &= D\Psi_1-2\varepsilon\Psi_1+3\kappa\Psi_2-\pi\Psi_0. \end{aligned} \tag{319}$$

Similarly, we find

$$\begin{aligned} \tfrac{1}{2}(R_{13|1}-R_{11|3}) = {} & -D\Phi_{01}+\delta\Phi_{00}+2(\varepsilon+\rho^*)\Phi_{01}+2\sigma\Phi_{10} \\ & -2\kappa\Phi_{11}-\kappa^*\Phi_{02}+(\pi^*-2\alpha^*-2\beta)\Phi_{00}. \end{aligned} \tag{320}$$

Combining equations (318), (319), and (320), we obtain the required explicit form of the identity (316). The remaining identities can be derived in similar fashion. We give below the explicit forms of the eight complex identities (312):

$$\begin{aligned} &-\delta^*\Psi_0+D\Psi_1+(4\alpha-\pi)\Psi_0-2(2\rho+\varepsilon)\Psi_1+3\kappa\Psi_2 \\ &\quad +[\text{Ricci}]=0; \qquad R_{13[13|4]}=0, \qquad \text{(a)} \end{aligned}$$

$$\begin{aligned} &+\delta^*\Psi_1-D\Psi_2-\lambda\Psi_0+2(\pi-\alpha)\Psi_1+3\rho\Psi_2-2\kappa\Psi_3 \\ &\quad +[\text{Ricci}]=0; \qquad R_{13[21|4]}=0, \qquad \text{(b)} \end{aligned}$$

$$\begin{aligned} &-\delta^*\Psi_2+D\Psi_3+2\lambda\Psi_1-3\pi\Psi_2+2(\varepsilon-\rho)\Psi_3+\kappa\Psi_4 \\ &\quad +[\text{Ricci}]=0; \qquad R_{42[13|4]}=0, \qquad \text{(c)} \end{aligned}$$

$$\begin{aligned} &+\delta^*\Psi_3-D\Psi_4-3\lambda\Psi_2+2(2\pi+\alpha)\Psi_3-(4\varepsilon-\rho)\Psi_4 \\ &\quad +[\text{Ricci}]=0; \qquad R_{42[21|4]}=0, \qquad \text{(d)} \end{aligned}$$

$$\begin{aligned} &-\triangle\Psi_0+\delta\Psi_1+(4\gamma-\mu)\Psi_0-2(2\tau+\beta)\Psi_1+3\sigma\Psi_2 \\ &\quad +[\text{Ricci}]=0; \qquad R_{13[13|2]}=0, \qquad \text{(e)} \end{aligned}$$

$$-\Delta\Psi_1+\delta\Psi_2+\nu\Psi_0+2(\gamma-\mu)\Psi_1-3\tau\Psi_2+2\sigma\Psi_3 +[\text{Ricci}]=0; \qquad R_{13[43|2]}=0, \tag{f}$$

$$-\Delta\Psi_2+\delta\Psi_3+2\nu\Psi_1-3\mu\Psi_2+2(\beta-\tau)\Psi_3+\sigma\Psi_4 +[\text{Ricci}]=0; \qquad R_{42[13|2]}=0, \tag{g}$$

$$-\Delta\Psi_3+\delta\Psi_4+3\nu\Psi_2-2(\gamma+2\mu)\Psi_3-(\tau-4\beta)\Psi_4 +[\text{Ricci}]=0; \qquad R_{42[43|2]}=0, \tag{h}$$

(321)

where the terms in the Ricci tensors (enclosed in square brackets), in the respective equations, are:

$$-D\Phi_{01}+\delta\Phi_{00}+2(\varepsilon+\rho^*)\Phi_{01}+2\sigma\Phi_{10}-2\kappa\Phi_{11}-\kappa^*\Phi_{02} +(\pi^*-2\alpha^*-2\beta)\Phi_{00}, \tag{a}$$

$$+\delta^*\Phi_{01}-\Delta\Phi_{00}-2(\alpha+\tau^*)\Phi_{01}+2\rho\Phi_{11}+\sigma^*\Phi_{02} -(\mu^*-2\gamma-2\gamma^*)\Phi_{00}-2\tau\Phi_{10}-2D\Lambda, \tag{b}$$

$$-D\Phi_{21}+\delta\Phi_{20}+2(\rho^*-\varepsilon)\Phi_{21}-2\mu\Phi_{10}+2\pi\Phi_{11}-\kappa^*\Phi_{22} -(2\alpha^*-2\beta-\pi^*)\Phi_{20}-2\delta^*\Lambda, \tag{c}$$

$$-\Delta\Phi_{20}+\delta^*\Phi_{21}+2(\alpha-\tau^*)\Phi_{21}+2\nu\Phi_{10}+\sigma^*\Phi_{22}-2\lambda\Phi_{11} -(\mu^*+2\gamma-2\gamma^*)\Phi_{20}, \tag{d}$$

$$-D\Phi_{02}+\delta\Phi_{01}+2(\pi^*-\beta)\Phi_{01}-2\kappa\Phi_{12}-\lambda^*\Phi_{00}+2\sigma\Phi_{11} +(\rho^*+2\varepsilon-2\varepsilon^*)\Phi_{02}, \tag{e}$$

$$\Delta\Phi_{01}-\delta^*\Phi_{02}+2(\mu^*-\gamma)\Phi_{01}-2\rho\Phi_{12}-\nu^*\Phi_{00}+2\tau\Phi_{11} +(\tau^*-2\beta^*+2\alpha)\Phi_{02}+2\delta\Lambda, \tag{f}$$

$$-D\Phi_{22}+\delta\Phi_{21}+2(\pi^*+\beta)\Phi_{21}-2\mu\Phi_{11}-\lambda^*\Phi_{20}+2\pi\Phi_{12} +(\rho^*-2\varepsilon-2\varepsilon^*)\Phi_{22}-2\Delta\Lambda, \tag{g}$$

$$\Delta\Phi_{21}-\delta^*\Phi_{22}+2(\mu^*+\gamma)\Phi_{21}-2\nu\Phi_{11}-\nu^*\Phi_{20}+2\lambda\Phi_{12} +(\tau^*-2\alpha-2\beta^*)\Phi_{22}. \tag{h}$$

(321′)

The explicit forms of the contracted identities (314) and (315) are:

$$\begin{aligned}\delta^*\Phi_{01}+\delta\Phi_{10}-D(\Phi_{11}+3\Lambda)-\Delta\Phi_{00} &=\kappa^*\Phi_{12}+\kappa\Phi_{21}+(2\alpha+2\tau^*-\pi)\Phi_{01}+(2\alpha^*+2\tau-\pi^*)\Phi_{10}\\ &\quad-2(\rho+\rho^*)\Phi_{11}-\sigma^*\Phi_{02}-\sigma\Phi_{20}+[\mu+\mu^*-2(\gamma+\gamma^*)]\Phi_{00},\end{aligned} \tag{i}$$

$$\begin{aligned}\delta^*\Phi_{12}+\delta\Phi_{21}-\Delta(\Phi_{11}+3\Lambda)-D\Phi_{22} &=-\nu\Phi_{01}-\nu^*\Phi_{10}+(\tau^*-2\beta^*-2\pi)\Phi_{12}+(\tau-2\beta-2\pi^*)\Phi_{21}\\ &\quad+2(\mu+\mu^*)\Phi_{11}-(\rho+\rho^*-2\varepsilon-2\varepsilon^*)\Phi_{22}+\lambda\Phi_{02}+\lambda^*\Phi_{20},\end{aligned} \tag{j}$$

$$\delta(\Phi_{11}-3\Lambda)-D\Phi_{12}-\triangle\Phi_{01}+\delta^*\Phi_{02}$$
$$=\kappa\Phi_{22}-\nu^*\Phi_{00}+(\tau^*-\pi+2\alpha-2\beta^*)\Phi_{02}-\sigma\Phi_{21}+\lambda^*\Phi_{10}$$
$$+2(\tau-\pi^*)\Phi_{11}-(2\rho+\rho^*-2\varepsilon^*)\Phi_{12}+(2\mu^*+\mu-2\gamma)\Phi_{01}. \qquad \text{(k)}$$

(322)

In the vacuum, the Ricci scalars vanish and the relevant Bianchi identities are given by the eight complex equations (321, a–h) with the Ricci terms set equal to zero; also, in this case, we need not concern ourselves with the contracted identities (322). In general, however, the Ricci terms in equations (321, a–h) must be included; and in this formalism they are replaced by the components of the energy–momentum tensor in accordance with Einstein's equation,

$$R_{ij}=-\frac{8\pi G}{c^4}(T_{ij}-\tfrac{1}{2}g_{ij}T). \qquad (323)$$

The basic equations of the Newman–Penrose formalism are comprised in the commutation relations (303)–(306), the Ricci identities (310, a–r), the eliminant relations (311, a–j), and the Bianchi identities (equations (321) and (322)). As we have remarked earlier, it is not clear what these equations are for and in what sense they replace Einstein's equation or are equivalent to it.

(*f*) *Maxwell's equations*

In the Newman–Penrose formalism, the antisymmetric Maxwell-tensor, F_{ij}, is replaced by the three complex scalars

$$\left.\begin{aligned}\phi_0&=F_{13}=F_{ij}l^im^j,\\ \phi_1&=\tfrac{1}{2}(F_{12}+F_{43})=\tfrac{1}{2}F_{ij}(l^in^j+\bar{m}^im^j),\\ \text{and}\qquad \phi_2&=F_{42}=F_{ij}\bar{m}^in^j;\end{aligned}\right\} \qquad (324)$$

and Maxwell's equations,

$$F_{[ij;k]}=0 \qquad \text{and} \qquad g^{ik}F_{ij;k}=0, \qquad (325)$$

expressed in terms of tetrad components and intrinsic derivatives, namely,

$$F_{[ab|c]}=0 \qquad \text{and} \qquad \eta^{nm}F_{an|m}=0, \qquad (326)$$

are replaced by the equations

$$\left.\begin{aligned}\phi_{1|1}-\phi_{0|4}&=0, \qquad \phi_{2|1}-\phi_{1|4}=0,\\ \phi_{1|3}-\phi_{0|2}&=0, \qquad \phi_{2|3}-\phi_{1|2}=0.\end{aligned}\right\} \qquad (327)$$

The explicit forms of these equations are readily found. Thus,

$$\begin{aligned}\phi_{1|1}&=\tfrac{1}{2}[F_{12,1}-\eta^{nm}(\gamma_{n11}F_{m2}+\gamma_{n21}F_{1m})\\ &\qquad+F_{43,1}-\eta^{nm}(\gamma_{n41}F_{m3}+\gamma_{n31}F_{4m})]\\ &=\phi_{1,1}-(\gamma_{131}F_{42}+\gamma_{241}F_{13})\\ &=D\phi_1+\kappa\phi_2-\pi\phi_0;\end{aligned} \qquad (328)$$

and, similarly,

$$\phi_{0|4} = \delta^* \phi_0 - 2\alpha\phi_0 + 2\rho\phi_1. \tag{329}$$

The first of Maxwell's equations (327), therefore, takes the form

$$D\phi_1 - \delta^* \phi_0 = (\pi - 2\alpha)\phi_0 + 2\rho\phi_1 - \kappa\phi_2. \tag{330}$$

And the remaining equations are

$$D\phi_2 - \delta^* \phi_1 = -\lambda\phi_0 + 2\pi\phi_1 + (\rho - 2\varepsilon)\phi_2, \tag{331}$$

$$\delta\phi_1 - \triangle\phi_0 = (\mu - 2\gamma)\phi_0 + 2\tau\phi_1 - \sigma\phi_2, \tag{332}$$

$$\delta\phi_2 - \triangle\phi_1 = -\nu\phi_0 + 2\mu\phi_1 + (\tau - 2\beta)\phi_2. \tag{333}$$

Turning next to the energy-momentum tensor of the Maxwell field, we have

$$T_{ab} = \eta^{cd} F_{ac} F_{bd} - \tfrac{1}{4}\eta_{ab} F_{ef} F^{ef}; \tag{334}$$

and in terms of the Maxwell scalars ϕ_0, ϕ_1, and ϕ_2, we readily find that

$$\begin{aligned} -\tfrac{1}{2}T_{11} &= \phi_0\phi_0^*; & -\tfrac{1}{2}T_{13} &= \phi_0\phi_1^*; \\ -\tfrac{1}{4}(T_{12}+T_{34}) &= \phi_1\phi_1^*; & -\tfrac{1}{2}T_{23} &= \phi_1\phi_2^*; \\ -\tfrac{1}{2}T_{22} &= \phi_2\phi_2^*; & -\tfrac{1}{2}T_{33} &= \phi_0\phi_2^*; \end{aligned} \tag{335}$$

and the trace of T_{ab} is, of course, zero. In accordance with equations (300) and (323), we can, therefore

$$\textit{set} \quad \Lambda = 0 \quad \textit{and replace} \quad \Phi_{mn} \quad \textit{by} \quad \phi_m\phi_n^*, \tag{336}$$

apart from the constant of proportionality, $-8\pi G/c^4$.

In the Bianchi identities listed in equations (321, a–h), the Ricci terms (321′, a–h) simplify considerably when use is made of Maxwell's equations (330)–(334). Thus, considering the terms in the derivatives of the Ricci scalars in equation (321′, a), we have

$$\begin{aligned} -D\Phi_{01} + \delta\Phi_{00} \to & -D(\phi_0\phi_1^*) + \delta(\phi_0\phi_0^*) \\ &= -\phi_0(D\phi_1^* - \delta\phi_0^*) - \phi_1^* D\phi_0 + \phi_0^* \delta\phi_0 \\ &= -\phi_0\{(\pi^* - 2\alpha^*)\phi_0^* + 2\rho^*\phi_1^* - \kappa^*\phi_2^*\} \\ &\quad -\phi_1^* D\phi_0 + \phi_0^*\delta\phi_0; \end{aligned} \tag{337}$$

and inserting this last expression in (321′, a), we find that we are left with

$$-\phi_1^* D\phi_0 + \phi_0^* \delta\phi_0 + 2(\varepsilon\phi_0\phi_1^* + \sigma\phi_1\phi_0^* - \kappa\phi_1\phi_1^* - \beta\phi_0\phi_0^*). \tag{338}$$

The other terms listed in (321′) allow of similar simplifications; and the Ricci terms to be included in equations (321, a–h) for an Einstein–Maxwell field take the following forms (apart from a constant of proportionality):

$$
\begin{aligned}
&-\phi_1{}^* D\phi_0+\phi_0{}^*\delta\phi_0+2(\varepsilon\phi_0\phi_1{}^*+\sigma\phi_1\phi_0{}^*-\kappa\phi_1\phi_1{}^*-\beta\phi_0\phi_0{}^*), &&\text{(a)}\\
&+\phi_1{}^*\delta^*\phi_0-\phi_0{}^*\triangle\phi_0+2(-\alpha\phi_0\phi_1{}^*+\rho\phi_1\phi_1{}^*+\gamma\phi_0\phi_0{}^*-\tau\phi_1\phi_0{}^*), &&\text{(b)}\\
&-\phi_1{}^* D\phi_2+\phi_0{}^*\delta\phi_2+2(-\varepsilon\phi_2\phi_1{}^*-\mu\phi_1\phi_0{}^*+\beta\phi_2\phi_0{}^*+\pi\phi_1\phi_1{}^*), &&\text{(c)}\\
&-\phi_0{}^*\triangle\phi_2+\phi_1{}^*\delta^*\phi_2+2(\alpha\phi_2\phi_1{}^*+\nu\phi_1\phi_0{}^*-\gamma\phi_2\phi_0{}^*-\lambda\phi_1\phi_1{}^*), &&\text{(d)}\\
&-\phi_2{}^* D\phi_0+\phi_1{}^*\delta\phi_0+2(-\kappa\phi_1\phi_2{}^*-\beta\phi_0\phi_1{}^*+\sigma\phi_1\phi_1{}^*+\varepsilon\phi_0\phi_2{}^*), &&\text{(e)}\\
&+\phi_1{}^*\triangle\phi_0-\phi_2{}^*\delta^*\phi_0+2(-\rho\phi_1\phi_2{}^*-\gamma\phi_0\phi_1{}^*+\tau\phi_1\phi_1{}^*+\alpha\phi_0\phi_2{}^*), &&\text{(f)}\\
&-\phi_2{}^* D\phi_2+\phi_1{}^*\delta\phi_2+2(-\varepsilon\phi_2\phi_2{}^*-\mu\phi_1\phi_1{}^*+\beta\phi_2\phi_1{}^*+\pi\phi_1\phi_2{}^*), &&\text{(g)}\\
&+\phi_1{}^*\triangle\phi_2-\phi_2{}^*\delta^*\phi_2+2(-\alpha\phi_2\phi_2{}^*-\nu\phi_1\phi_1{}^*+\gamma\phi_2\phi_1{}^*+\lambda\phi_1\phi_2{}^*). &&\text{(h)}
\end{aligned}
\tag{339}
$$

And since the vanishing of the covariant divergence of the energy–momentum tensor is automatically guaranteed by Maxwell's equations, we need not concern ourselves with equations (322).

(g) *Tetrad transformations*

Having chosen a tetrad frame—either an orthonormal frame, as in the conventional tetrad formalism, or a null frame, as in the Newman–Penrose formalism—we can subject the frame to a Lorentz transformation at some point and extend it continuously through all of space–time. Corresponding to the six parameters of the group of Lorentz transformations, we have six degrees of freedom to rotate a chosen tetrad frame. In considering the effect of such Lorentz transformations on the various Newman–Penrose quantities, we shall find it convenient to regard a general Lorentz transformation of the basis vectors $\boldsymbol{l}, \boldsymbol{n}, \boldsymbol{m}$, and $\bar{\boldsymbol{m}}$ as made up of the following three classes of rotations:

(a) *rotations of class I* which leave the vector $\boldsymbol{l}$ unchanged;
(b) *rotations of class II* which leave the vector $\boldsymbol{n}$ unchanged; and
(c) *rotations of class III* which leave the directions of $\boldsymbol{l}$ and $\boldsymbol{n}$ unchanged and rotate $\boldsymbol{m}$ (and $\bar{\boldsymbol{m}}$) by an angle θ in the $(\boldsymbol{m}, \bar{\boldsymbol{m}})$-plane.

Associated with these three classes of rotations, we have the following explicit transformations (which, as may be readily verified, preserve the underlying orthogonality and normalization conditions):

I: $\boldsymbol{l} \to \boldsymbol{l}$, $\boldsymbol{m} \to \boldsymbol{m} + a\boldsymbol{l}$, $\bar{\boldsymbol{m}} \to \bar{\boldsymbol{m}} + a^*\boldsymbol{l}$, and, $\boldsymbol{n} \to \boldsymbol{n} + a^*\boldsymbol{m} + a\bar{\boldsymbol{m}} + aa^*\boldsymbol{l}$;

II: $\boldsymbol{n} \to \boldsymbol{n}$, $\boldsymbol{m} \to \boldsymbol{m} + b\boldsymbol{n}$, $\bar{\boldsymbol{m}} \to \bar{\boldsymbol{m}} + b^*\boldsymbol{n}$, and, $\boldsymbol{l} \to \boldsymbol{l} + b^*\boldsymbol{m} + b\bar{\boldsymbol{m}} + bb^*\boldsymbol{n}$;

III: $\boldsymbol{l} \to A^{-1}\boldsymbol{l}, \boldsymbol{n} \to A\boldsymbol{n}, \boldsymbol{m} \to e^{i\theta}\boldsymbol{m}$, and $\bar{\boldsymbol{m}} \to e^{-i\theta}\bar{\boldsymbol{m}}$;

where a and b are two complex functions and A and θ are two real functions on the manifold.

The effect of a rotation of class I on the various Newman–Penrose quantities is readily found. Thus, considering the Weyl scalars Ψ_0 and Ψ_1, we have

$$\begin{aligned} -\Psi_0 = C_{1313} &\rightarrow C_{pqrs}\, l^p (m^q + a l^q) l^r (m^s + a l^s) \\ &= C_{pqrs}\, l^p m^q l^r m^s = -\Psi_0, \end{aligned} \tag{340}$$

and

$$\begin{aligned} -\Psi_1 = C_{1213} &\rightarrow C_{pqrs} l^p (n^q + a^* m^q + a\bar{m}^q + a a^* l^q) l^r (m^s + a l^s) \\ &= C_{pqrs}\, l^p (n^q + a^* m^q) l^r m^s \\ &= C_{1213} + a^* C_{1313} = -(\Psi_1 + a^* \Psi_0), \end{aligned} \tag{341}$$

where in the reduction for Ψ_1, apart from the obvious symmetries of the Weyl tensor, the fact that $C_{1413} = 0$ (cf., equation (291)) has been used. The effects of the transformation on the remaining Weyl scalars are similarly found; and the result is

$$\left.\begin{aligned} &\Psi_0 \rightarrow \Psi_0, \Psi_1 \rightarrow \Psi_1 + a^*\Psi_0, \Psi_2 \rightarrow \Psi_2 + 2a^*\Psi_1 + (a^*)^2\Psi_0, \\ &\Psi_3 \rightarrow \Psi_3 + 3a^*\Psi_2 + 3(a^*)^2\Psi_1 + (a^*)^3\Psi_0, \\ &\Psi_4 \rightarrow \Psi_4 + 4a^*\Psi_3 + 6(a^*)^2\Psi_2 + 4(a^*)^3\Psi_1 + (a^*)^4\Psi_4. \end{aligned}\right\} \tag{342}$$

In a similar fashion, we find that the spin coefficients transform as follows:

$$\begin{aligned} &\kappa \rightarrow \kappa;\ \sigma \rightarrow \sigma + a\kappa;\ \rho \rightarrow \rho + a^*\kappa;\ \varepsilon \rightarrow \varepsilon + a^*\kappa; \\ &\tau \rightarrow \tau + a\rho + a^*\sigma + aa^*\kappa;\ \pi \rightarrow \pi + 2a^*\varepsilon + (a^*)^2\kappa + Da^*; \\ &\alpha \rightarrow \alpha + a^*(\rho + \varepsilon) + (a^*)^2\kappa;\ \beta \rightarrow \beta + a\varepsilon + a^*\sigma + aa^*\kappa; \\ &\gamma \rightarrow \gamma + a\alpha + a^*(\beta + \tau) + aa^*(\rho + \varepsilon) + (a^*)^2\sigma + a(a^*)^2\kappa; \\ &\lambda \rightarrow \lambda + a^*(2\alpha + \pi) + (a^*)^2(\rho + 2\varepsilon) + (a^*)^3\kappa + \delta^* a^* + a^* Da^*; \\ &\mu \rightarrow \mu + a\pi + 2a^*\beta + 2aa^*\varepsilon + (a^*)^2\sigma + a(a^*)^2\kappa + \delta a^* + aDa^*; \\ &\nu \rightarrow \nu + a\lambda + a^*(\mu + 2\gamma) + (a^*)^2(\tau + 2\beta) + (a^*)^3\sigma + aa^*(\pi + 2\alpha) \\ &\qquad + a(a^*)^2(\rho + 2\varepsilon) + a(a^*)^3\kappa + (\triangle + a^*\delta + a\delta^* + aa^*D)a^*. \end{aligned} \tag{343}$$

And the scalars representing the Maxwell field become

$$\phi_0 \rightarrow \phi_0, \phi_1 \rightarrow \phi_1 + a^*\phi_0, \phi_2 \rightarrow \phi_2 + 2a^*\phi_1 + (a^*)^2\phi_0. \tag{344}$$

The corresponding effects of a rotation of class II on the various Newman–Penrose quantities can be readily written down from the foregoing formulae, since the effect of interchanging ***l*** and ***n*** results in the transformation

$$\left.\begin{aligned} &\Psi_0 \rightleftarrows \Psi_4^*, \Psi_1 \rightleftarrows \Psi_3^*, \Psi_2 \rightleftarrows \Psi_2^*, \phi_0 \rightleftarrows -\phi_2^*, \phi_1 \rightleftarrows -\phi_1^*, \\ &\kappa \rightleftarrows -\nu^*, \rho \rightleftarrows -\mu^*, \sigma \rightleftarrows -\lambda^*, \alpha \rightleftarrows -\beta^*, \varepsilon \rightleftarrows -\gamma^*, \quad \text{and} \quad \pi \rightleftarrows -\tau^*. \end{aligned}\right\} \tag{345}$$

In particular, the effect of a rotation of class II on the Weyl scalars is

$$\left.\begin{aligned}&\Psi_0 \to \Psi_0 + 4b\Psi_1 + 6b^2\Psi_2 + 4b^3\Psi_3 + b^4\Psi_4,\\&\Psi_1 \to \Psi_1 + 3b\Psi_2 + 3b^2\Psi_3 + b^3\Psi_4,\\&\Psi_2 \to \Psi_2 + 2b\Psi_3 + b^2\Psi_4,\ \Psi_3 \to \Psi_3 + b\Psi_4,\ \Psi_4 \to \Psi_4.\end{aligned}\right\} \tag{346}$$

And finally, we write down the effects of a rotation of class III on the Newman–Penrose quantities.

$$\begin{aligned}&\Psi_0 \to A^{-2}e^{2i\theta}\Psi_0,\ \Psi_1 \to A^{-1}\,e^{i\theta}\Psi_1,\ \Psi_2 \to \Psi_2,\\&\Psi_3 \to Ae^{-i\theta}\Psi_3, \qquad \text{and} \qquad \Psi_4 \to A^2e^{-2i\theta}\Psi_4;\\&\phi_0 \to A^{-1}e^{i\theta}\phi_0,\ \phi_1 \to \phi_1,\ \phi_2 \to Ae^{-i\theta}\phi_2;\\&\kappa \to A^{-2}e^{i\theta}\kappa,\ \sigma \to A^{-1}e^{2i\theta}\sigma,\ \rho \to A^{-1}\rho,\ \tau \to e^{i\theta}\tau,\\&\pi \to e^{-i\theta}\pi,\ \lambda \to Ae^{-2i\theta}\lambda,\ \mu \to A\mu,\ \nu \to A^2e^{-i\theta}\nu,\\&\gamma \to A\gamma - \tfrac{1}{2}\Delta A + \tfrac{1}{2}iA\,\Delta\theta,\\&\varepsilon \to A^{-1}\varepsilon - \tfrac{1}{2}A^{-2}DA + \tfrac{1}{2}iA^{-1}D\theta,\\&\alpha \to e^{-i\theta}\alpha + \tfrac{1}{2}ie^{-i\theta}\delta^*\theta - \tfrac{1}{2}A^{-1}e^{-i\theta}\delta^*A,\\&\beta \to e^{i\theta}\beta + \tfrac{1}{2}ie^{i\theta}\delta\theta - \tfrac{1}{2}A^{-1}e^{i\theta}\delta A.\end{aligned} \tag{347}$$

This completes our account of the Newman–Penrose formalism.

9. The optical scalars, the Petrov classification, and the Goldberg-Sachs theorem

The physical significance of the spin coefficients that were formally introduced in §8 in the context of the Newman–Penrose formalism becomes apparent when we consider the propagation of the basis vectors along $\boldsymbol{l}$ or $\boldsymbol{n}$. Thus, the first-order change in a basis vector $\boldsymbol{e}_{(a)}$, when it suffers an infinitesimal displacement ξ, is, by definition,

$$\begin{aligned}\delta e_{(a)i} &= e_{(a)i;j}\xi^j = e^{(b)}{}_i\gamma_{(b)(a)(c)}e^{(c)}{}_j\xi^j\\&= -\gamma_{(a)(b)(c)}e^{(b)}{}_i\xi^{(c)}.\end{aligned} \tag{348}$$

Therefore, the change, $\delta\boldsymbol{e}_{(a)}(c)$ in $\boldsymbol{e}_{(a)}$, per unit displacement along the direction c, is given by

$$\delta\boldsymbol{e}_{(a)}(c) = -\gamma_{(a)(b)(c)}\boldsymbol{e}^{(b)}. \tag{349}$$

In particular, for the change in $\boldsymbol{l}$, per unit displacement along $\boldsymbol{l}$, we have (cf. equation (284))

$$\begin{aligned}\delta\boldsymbol{l}(1) &= -\gamma_{1(b)1}\boldsymbol{e}^{(b)} = -\gamma_{121}\boldsymbol{e}^{(2)} - \gamma_{131}\boldsymbol{e}^{(3)} - \gamma_{141}\boldsymbol{e}^{(4)}\\&= -\gamma_{121}\boldsymbol{l} + \gamma_{131}\bar{\boldsymbol{m}} + \gamma_{141}\boldsymbol{m},\end{aligned} \tag{350}$$

or, giving to the spin coefficients their designated symbols, we have

$$\delta \boldsymbol{l}(1) = (\varepsilon+\varepsilon^*)\boldsymbol{l} - \kappa\bar{\boldsymbol{m}} - \kappa^*\boldsymbol{m}. \tag{351}$$

Similarly, we find

$$\delta \boldsymbol{n}(1) = -\gamma_{2(b)1}\boldsymbol{e}^{(b)} = -(\varepsilon+\varepsilon^*)\boldsymbol{n} + \pi\boldsymbol{m} + \pi^*\bar{\boldsymbol{m}} \tag{352}$$

and

$$\delta \boldsymbol{m}(1) = -\gamma_{3(b)1}\boldsymbol{e}^{(b)} = +(\varepsilon-\varepsilon^*)\boldsymbol{m} + \pi^*\boldsymbol{l} - \kappa\boldsymbol{n}. \tag{353}$$

Several important conclusions follow from equations (351)–(353). Thus, rewriting equation (351) in the alternative form (cf. equation (348)),

$$l_{i;j}l^j = (\varepsilon+\varepsilon^*)l_i - \kappa\bar{m}_i - \kappa^* m_i, \tag{354}$$

we conclude from equations (211) and (212) that *the* $\boldsymbol{l}$*-vectors form a congruence of null geodesics if, and only if,* $\kappa = 0$*; and, further, that they are affinely parametrized if, and only if, in addition* $\text{Re}\,\varepsilon = 0$. If $\kappa = 0$ the latter requirement for affine parametrization (namely, $\varepsilon = 0$) can always be met by a rotation of class III described in § 8(g) (cf., equations (347)) which will not affect the direction of $\boldsymbol{l}$ nor of an initially vanishing κ.

If the vectors $\boldsymbol{l}$ should initially be defined as a congruence of null geodesics affinely parameterized, so that $\kappa = \varepsilon = 0$, then by a rotation of class I (which will not affect $\boldsymbol{l}$ nor the initial vanishing of κ and ε), we can arrange that $\pi = 0$. After such a rotation, the newly oriented vectors $\boldsymbol{n}$, $\boldsymbol{m}$, and $\bar{\boldsymbol{m}}$ will, according to equations (352) and (353), suffer no change for displacements along $\boldsymbol{l}$. In other words, under these circumstances, *all the basis vectors* $\boldsymbol{l}$, $\boldsymbol{n}$, $\boldsymbol{m}$, *and* $\bar{\boldsymbol{m}}$ *will remain unchanged as they are parallely propagated along* $\boldsymbol{l}$.

(a) *The optical scalars*

Some further properties of a congruence of null geodesics can be derived by writing out explicitly the summations on the right-hand side of the equation,

$$l_{i;j} = e^{(a)}{}_i\,\gamma_{(a)1(b)}\,e^{(b)}{}_j. \tag{355}$$

We find

$$\begin{aligned} l_{i;j} = {} & (\varepsilon+\varepsilon^*)l_i n_j + (\gamma+\gamma^*)l_i l_j - (\alpha^*+\beta)l_i\bar{m}_j - (\alpha+\beta^*)l_i m_j \\ & -\kappa\bar{m}_i n_j - \kappa^* m_i n_j + \sigma\bar{m}_i\bar{m}_j + \sigma^* m_i m_j \\ & -\tau\bar{m}_i l_j - \tau^* m_i l_j + \rho\bar{m}_i m_j + \rho^* m_i\bar{m}_j. \end{aligned} \tag{356}$$

First, we observe that if we contract this equation with l^j we recover equation (354). Since the $\boldsymbol{l}$'s form a congruence of null geodesics affinely parametrized, $\kappa = \varepsilon = 0$, and equation (356) becomes

$$\begin{aligned} l_{i;j} = {} & (\gamma+\gamma^*)l_i l_j - (\alpha^*+\beta)l_i\bar{m}_j - (\alpha+\beta^*)l_i m_j - \tau\bar{m}_i l_j \\ & +\sigma\bar{m}_i\bar{m}_j + \sigma^* m_i m_j + \rho\bar{m}_i m_j + \rho^* m_i\bar{m}_j - \tau^* m_i l_j. \end{aligned} \tag{357}$$

From this last equation, we find

$$l_{[i;j]} = -(\alpha^* + \beta - \tau) l_{[i}\bar{m}_{j]} - (\alpha + \beta^* - \tau^*) l_{[i}m_{j]} + (\rho - \rho^*)\bar{m}_{[i}m_{j]} \tag{358}$$

and

$$l_{[i;j}l_{k]} = (\rho - \rho^*)\bar{m}_{[i}m_j l_{k]}. \tag{359}$$

From the standard theorems of the subject (see the references in the Bibliographical Notes at the end of the chapter) it follows that *the congruence of the null geodesics will be hyper-surface orthogonal* (i.e. $\boldsymbol{l}$ will be proportional to the gradient of a scalar field) *if, and only if, the spin coefficient ρ is real; and $\boldsymbol{l}$ will be equal to the gradient of a scalar field if, and only if, in addition $\alpha^* + \beta = \tau$.*

From equations (357), (358), and (359) we find

$$\tfrac{1}{2} l^i{}_{;i} = -\tfrac{1}{2}(\rho + \rho^*) = \theta \text{ (say)}, \tag{360}$$

$$\tfrac{1}{2} l_{[i;j]} l^{i;j} = -\tfrac{1}{4}(\rho - \rho^*)^2 = \omega^2 \text{ (say)}, \tag{361}$$

and

$$\tfrac{1}{2} l_{(i;j)} l^{i;j} = \theta^2 + |\sigma|^2. \tag{362}$$

The quantities θ, ω, and σ, as defined by the foregoing equations, were first introduced by R. Sachs and are called the *optical scalars.* Alternative definitions of θ and ω are

$$\theta = -\text{Re}\rho \qquad \text{and} \qquad \omega = \text{Im}\rho. \tag{363}$$

The geometrical meanings of ρ and σ are as follows. But first, we recall that $\boldsymbol{l}$ is the tangent vector to a null ray, say, N, and $\boldsymbol{m}$ is a complex vector orthogonal to $\boldsymbol{l}$. At a point P on N, the real part of $\boldsymbol{m}$ spans with $\boldsymbol{l}$ a 2-plane. Now consider a small circle with its centre at a point P on N and lying in the 2-plane orthogonal to $\boldsymbol{l}$. If we follow the rays of the congruence $\boldsymbol{l}$, which intersect the circle, into the future null-direction, the circle may become contracted or expanded, rotated or sheared (into an ellipse). The expansion (or contraction), the rotation, and the shear are measured, respectively, by $-\text{Re}\rho$, $\text{Im}\rho$, and σ (see Fig. 1); and $|\sigma|$ measures the magnitude of the shear and $\tfrac{1}{2}\arg\sigma$ the angle the minor axis of the ellipse makes with the assigned 2-plane.

The fact that $\text{Im}\rho$ is a measure of the rotation is consistent with the requirement that this quantity vanish for the congruence to be hypersurface orthogonal.

The essential roles which the spin coefficients ρ and σ play in describing the behaviour of bundles of light-rays as they traverse a gravitational field, are, further, manifested by the change experienced by $\boldsymbol{l}$, in the orthogonal direction $\boldsymbol{m}$, as it propagates. This change, in accordance with equation (349), is given by

$$\delta \boldsymbol{l}(3) = -\gamma_{1(b)3}\, \boldsymbol{e}^{(b)} = (\alpha^* + \beta)\boldsymbol{l} - \rho^* \boldsymbol{m} - \sigma \bar{\boldsymbol{m}}. \tag{364}$$

The equations governing the variation of ρ and σ, along the geodesic, are given by equations (310, a) and (310, b) in which we can now set $\kappa = \varepsilon = 0$.

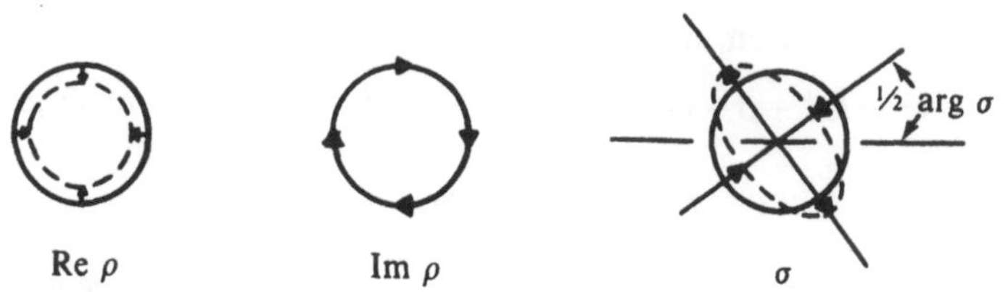

FIG. 1. The geometrical interpretation of the optical scalars in terms of the effect of propagation of a small circle perpendicular to the beam.

Thus

$$D\rho = (\rho^2 + |\sigma|^2) + \Phi_{00}. \tag{365}$$

and

$$D\sigma = \sigma(\rho + \rho^*) + \Psi_0, \tag{366}$$

where it may be recalled that

$$\Phi_{00} = -\tfrac{1}{2}R_{11} = -\tfrac{1}{2}R_{ij}l^i l^j \quad \text{and} \quad \Psi_0 = -C_{pqrs}l^p m^q l^r m^s. \tag{367}$$

Some useful alternative forms of the foregoing equations governing the optical scalars may be noted here. Thus, equation (366) may be rewritten in the form

$$D\sigma = -2\theta\sigma + \Psi_0. \tag{368}$$

An analogous equation for ω follows from taking the imaginary part of equation (365) and remembering that Φ_{00} is real. Thus,

$$D\omega = \tfrac{1}{2}(\rho + \rho^*)(\rho - \rho^*) = -2\theta\omega. \tag{369}$$

And the real part of equation (365) gives

$$D\theta = \omega^2 - \theta^2 - |\sigma|^2 - \Phi_{00}. \tag{370}$$

Equations (368), (369), and (370) are the standard forms in which they are commonly used.

(*b*) *The Petrov classification*

We have seen how, with respect to a chosen null tetrad-frame, the Weyl tensor is completely specified by the five complex scalars, $\Psi_0, \Psi_1, \ldots, \Psi_4$. But the values which these scalars take are dependent on the choice of frame and on the six-parameter group of Lorentz transformations to which it is subject. The question now arises as to which of these scalars and how many of them can be made to vanish by a suitable orientation of the frame. The answer to this question gives rise to an algebraic classification of the Weyl tensor into types called the *Petrov classification* and the *Petrov types*.

Let $\Psi_4 \neq 0$. (If Ψ_4 should happen to be zero in the chosen frame, then by a rotation of class I, we can make it non-zero so long as space is not flat conformally and not all of the Weyl scalars vanish.) Consider now a rotation of class II with a parameter b. The Weyl scalars Ψ_0, Ψ_1, etc., become (cf. equation (346))

$$\left.\begin{aligned}&\Psi_0^{(1)} = \Psi_0 + 4b\Psi_1 + 6b^2\Psi_2 + 4b^3\Psi_3 + b^4\Psi_4,\\&\Psi_1^{(1)} = \Psi_1 + 3b\Psi_2 + 3b^2\Psi_3 + b^3\Psi_4,\\&\Psi_2^{(1)} = \Psi_2 + 2b\Psi_3 + b^2\Psi_4,\ \Psi_3^{(1)} = \Psi_3 + b\Psi_4,\ \Psi_4^{(1)} = \Psi_4,\end{aligned}\right\} \tag{371}$$

where the superscript "(1)" distinguishes the new values of the scalars from the old. It is clear that by a rotation of class II, $\Psi_0^{(1)}$ can be made to vanish if b is chosen as a root of the equation

$$\Psi_4 b^4 + 4\Psi_3 b^3 + 6\Psi_2 b^2 + 4\Psi_1 b + \Psi_0 = 0. \tag{372}$$

This equation has always four roots and the corresponding new directions of $\boldsymbol{l}$, namely, $\boldsymbol{l} + b^*\boldsymbol{m} + b\bar{\boldsymbol{m}} + bb^*\boldsymbol{n}$, are called the *principal null-directions* of the Weyl tensor. If one or more of the roots coincide, the tensor is said to be algebraically special; otherwise, it is said to be *algebraically general*. And the various ways in which the roots can coincide, or be distinct, leads to the Petrov classification.

(α) *Petrov type* I—In this case, all four roots of equation (372) are distinct. Let them be b_1, b_2, b_3, and b_4. Then by a rotation of class II with parameter $b = b_1$ (say) we can make $\Psi_0 = 0$. Then by a rotation of class I (which does not affect Ψ_0) we can make Ψ_4 vanish.* With Ψ_0 and Ψ_4 made to vanish in this manner, Ψ_1, Ψ_2, and Ψ_3 will be left non-vanishing and of these scalars $\Psi_1\Psi_3$ and Ψ_2 are invariant to the remaining rotations of class III (which cannot, however, affect the vanishing of Ψ_0 and Ψ_4).

(β) *Petrov type* II—Let equation (372) allow two coincident roots $b_1 = b_2$ ($\neq b_3 \neq b_4$ and $b_3 \neq b_4$). In this case, besides equation (372), its derivative (with respect to b), namely

$$\Psi_4 b^3 + 3\Psi_3 b^2 + 3\Psi_2 b + \Psi_1 = 0, \tag{373}$$

will also be satisfied when $b = b_2$. Therefore, by a rotation of class II with parameter $b = b_1$ ($= b_2$), Ψ_0 and Ψ_1 will vanish. Then by a rotation of class I (which will not affect the initial simultaneous vanishing of Ψ_0 and Ψ_1) (see equations (342)) we can make Ψ_4 vanish. Only Ψ_2 and Ψ_3 will be left non-vanishing; and of them only Ψ_2 will be invariant to rotations of class III.

(γ) *Petrov type* D—Let equation (372) allow two distinct double roots b_1

* Note that the roots of the equation (cf. equation (342))

$$\Psi_0(a^*)^4 + 4\Psi_1(a^*)^3 + 6\Psi_2(a^*)^2 + 4\Psi_3 a^* + \Psi_4 = 0 \tag{372$'$}$$

are the reciprocals of the roots of equation (372).

and b_2. In this case, we shall show how, by a rotation of class II followed by a rotation of class I, we can make Ψ_0, Ψ_1, Ψ_3, and Ψ_4 vanish simultaneously and leave Ψ_2 as the only remaining non-vanishing scalar.

By assumption, Ψ_0, after a rotation of class II with parameter b, will have the value

$$\Psi_0^{(1)} = \Psi_4 (b-b_1)^2 (b-b_2)^2. \tag{374}$$

The values of the remaining scalars can be obtained by successively differentiating the expression for $\Psi_0^{(1)}$ and renormalizing, at each stage, to have the same coefficient Ψ_4 for the highest power of b. We, thus, find

$$\left.\begin{aligned}\Psi_1^{(1)} &= \tfrac{1}{2}\Psi_4 (b-b_1)(b-b_2)(2b-b_1-b_2)\\ \Psi_2^{(1)} &= \tfrac{1}{3}\Psi_4 [(b-b_1)(b-b_2)+\tfrac{1}{2}(2b-b_1-b_2)^2]\\ \Psi_3^{(1)} &= \tfrac{1}{2}\Psi_4 (2b-b_1-b_2), \qquad \text{and} \qquad \Psi_4^{(1)} = \Psi_4.\end{aligned}\right\} \tag{375}$$

With the choice, $b = b_1$,

$$\left.\begin{aligned}&\Psi_0^{(1)} = \Psi_1^{(1)} = 0, \qquad \Psi_2^{(1)} = \tfrac{1}{6}\Psi_4 (b_1-b_2)^2,\\ &\Psi_3^{(1)} = \tfrac{1}{2}\Psi_4 (b_1-b_2), \qquad \text{and} \qquad \Psi_4^{(1)} = \Psi_4.\end{aligned}\right\} \tag{376}$$

Now subject the frame to a rotation of class I with a parameter a^*. Then, the new values of the Weyl scalars are (cf. equation (342))

$$\left.\begin{aligned}\Psi_0^{(2)} &= \Psi_1^{(2)} = 0\\ \Psi_2^{(2)} &= \Psi_2^{(1)} = \tfrac{1}{6}\Psi_4 (b_1-b_2)^2,\end{aligned}\right\} \tag{377}$$

$$\begin{aligned}\Psi_3^{(2)} &= \Psi_3^{(1)} + 3a^*\Psi_2^{(1)} = \tfrac{1}{2}\Psi_4 (b_1-b_2) + \tfrac{1}{2}\Psi_4 a^* (b_1-b_2)^2\\ &= \tfrac{1}{2}\Psi_4 (b_1-b_2)[1+a^*(b_1-b_2)],\end{aligned} \tag{378}$$

and

$$\begin{aligned}\Psi_4^{(2)} &= \Psi_4 + 4a^*.\tfrac{1}{2}\Psi_4 (b_1-b_2) + 6(a^*)^2.\tfrac{1}{6}\Psi_4 (b_1-b_2)^2\\ &= \Psi_4 [1+a^*(b_1-b_2)]^2.\end{aligned} \tag{379}$$

Accordingly, with the choice

$$a^* = (b_2-b_1)^{-1}, \tag{380}$$

we can make $\Psi_3^{(2)}$ and $\Psi_4^{(2)}$ also vanish. Thus, by a rotation of class II with parameter b_1 followed by a rotation of class I with parameter $a^* = (b_2-b_1)^{-1}$, Ψ_0, Ψ_1, Ψ_3, and Ψ_4 have been reduced to zero and Ψ_2 is the only non-vanishing scalar. And Ψ_2 is invariant to rotations of class III.

(δ) *Petrov type* III—If three roots of equation (372) coincide and $b_1 = b_2 = b_3 \neq b_4$, then by a rotation of class II with parameter $b = b_1 (= b_2 = b_3)$ we can make Ψ_0, Ψ_1, and Ψ_2 vanish simultaneously; and by a subsequent rotation of class I we can make Ψ_4 vanish (without affecting the vanishing of Ψ_0, Ψ_1, and Ψ_2). And Ψ_3 will be the only non-vanishing scalar whose magnitude can be altered by rotations of class III.

(ε) *Petrov type* N—If all four roots coincide and there is only one distinct root b (say), then, by a rotation of class II with parameter b, we can make Ψ_0, Ψ_1, Ψ_2, and Ψ_3 vanish simultaneously and Ψ_4 will be the only non-vanishing scalar.

We shall now derive some simple conditions for $\boldsymbol{l}$ to be a principal null-direction of the Weyl tensor and for the occurrence of the various Petrov types.

Using the general expression (298) for C_{pqrs}, we find that

$$C_{pqrs}l^q l^r = (\Psi_2+\Psi_2^*)l_p l_s + \Psi_0 \bar{m}_p \bar{m}_s + \Psi_0^* m_p m_s - \Psi_1(l_p \bar{m}_s + l_s \bar{m}_p) - \Psi_1^*(l_p m_s + l_s m_p). \tag{381}$$

Accordingly,

$$C_{pqr[s}l_{t]}l^q l^r = (\Psi_0^* m_p - \Psi_1^* l_p)m_{[s}l_{t]} + (\Psi_0 \bar{m}_p - \Psi_1 l_p)\bar{m}_{[s}l_{t]} \tag{382}$$

and

$$l_{[u}C_{p]qr[s}l_{t]}l^q l^r = \Psi_0 l_{[u}m_{p]}m_{[s}l_{t]} + \Psi_0^* l_{[u}\bar{m}_{p]}\bar{m}_{[s}l_{t]}. \tag{383}$$

Consequently, the condition, that $\boldsymbol{l}$ is a non-degenerate principal null-direction belonging to a Weyl tensor of type I, is

$$l_{[u}C_{p]qr[s}l_{t]}l^q l^r = 0, \text{ if and only if } \quad \Psi_0 = 0; \tag{384}$$

and the condition, that $\boldsymbol{l}$ is a doubly-degenerate principal null-direction belonging to a Weyl tensor of type II, is

$$C_{pqr[s}l_{t]}l^q l^r = 0, \quad \text{if and only if } \quad \Psi_0 = \Psi_1 = 0. \tag{385}$$

And when Ψ_0 and Ψ_1 vanish,

$$C_{pqrs}l^q l^r = (\Psi_2+\Psi_2^*)l_p l_s. \tag{386}$$

If the Weyl tensor is of type D and $\boldsymbol{n}$ is the second doubly-degenerate null-direction, then, besides the condition (385), we must also have

$$C_{pqr[s}n_{t]}n^q n^r = 0; \tag{387}$$

and when $\Psi_4 = \Psi_3 = 0$,

$$C_{pqrs}n^q n^r = (\Psi_2+\Psi_2^*)n_p n_s. \tag{388}$$

If the Weyl tensor is of type III and $\Psi_0 = \Psi_1 = \Psi_2 = 0$, then it follows from equation (298) that

$$C_{pqrs}l^s = [\Psi_3(l_p m_q - l_q m_p) + \Psi_3^*(l_p \bar{m}_q - l_q \bar{m}_p)]l_r; \tag{389}$$

and we conclude that

$$C_{pqr[s}l_{t]}l^r = 0, \text{ if and only if } \quad \Psi_0 = \Psi_1 = \Psi_2 = 0. \tag{390}$$

And, finally, if the tensor is of type N and $\Psi_0 = \Psi_1 = \Psi_2 = \Psi_3 = 0$, then

$$C_{pqrs} = -\Psi_4\{l_p m_q l_r m_s\} - \Psi_4^*\{l_p \bar{m}_q l_r \bar{m}_s\}, \tag{391}$$

and

$$C_{pqrs}l^s = 0, \text{ if and only if } \quad \Psi_0 = \Psi_1 = \Psi_2 = \Psi_3 = 0. \tag{392}$$

(c) *The Goldberg–Sachs theorem*

We restrict ourselves to the vacuum when the Ricci tensor vanishes and the Riemann and the Weyl tensors coincide. With this restriction the theorem states:

If the Riemann tensor is of type II *and a null basis is so chosen that l is the repeated null direction and $\Psi_0 = \Psi_1 = 0$, then $\kappa = \sigma = 0$; and, conversely, if $\kappa = \sigma = 0$, then $\Psi_0 = \Psi_1 = 0$ and the Riemann tensor is of type* II.

The importance of the theorem consists in its relating, reciprocally, the type-II character of a field with the existence of a congruence of shear-free null geodesics.

To prove that, when $\Psi_0 = \Psi_1 = 0$, the congruence formed by the vector l is geodesic ($\kappa = 0$) and shear-free ($\sigma = 0$), is straightforward. Thus, when $\Psi_0 = \Psi_1 = 0$, the Bianchi identities (equations (321, a, b, c, e, f, and g)) are

$$\left.\begin{aligned} 3\kappa\Psi_2 = 0, \text{ (a)};\ D\Psi_2 &= -2\kappa\Psi_3 + 3\rho\Psi_2, \text{ (b)}; \\ D\Psi_3 - \delta^*\Psi_2 &= -\kappa\Psi_4 - 2(\varepsilon - \rho)\Psi_3 + 3\pi\Psi_2, \text{ (c)}; \\ 3\sigma\Psi_2 = 0, \text{ (e)};\ -\delta\Psi_2 &= -3\tau\Psi_2 + 2\sigma\Psi_3, \text{ (f)}; \\ \Delta\Psi_2 - \delta\Psi_3 &= -3\mu\Psi_2 - 2(\tau - \beta)\Psi_3 + \sigma\Psi_4, \text{ (g)}. \end{aligned}\right\} \tag{393}$$

If space is not conformally flat (i.e., not all of Ψ_0, Ψ_1, Ψ_2, Ψ_3, and Ψ_4 are zero), then it follows from equations (393,a,b,c) that $\kappa = 0$ and from equations (393,e,f,g) that $\sigma = 0$.

To prove the converse, namely, that if $\kappa = \sigma = 0$ then $\Psi_0 = \Psi_1 = 0$, is somewhat less direct. First, we shall suppose that by a rotation of class III (which will not affect the vanishing of κ and σ), ε has been made to vanish. Then the Ricci identities (310,a,b,c,e, and k) for $\kappa = \sigma = 0$ give

$$\left.\begin{aligned} &D\rho = \rho^2, \text{ (a)};\ \Psi_0 = 0, \text{ (b)};\ D\tau = (\tau^* + \pi)\rho + \Psi_1 \text{ (c)}; \\ &D\beta = \rho^*\beta + \Psi_1, \text{ (e)};\ \delta\rho = (\alpha^* + \beta)\rho + (\rho - \rho^*)\tau - \Psi_1, \text{ (k)}; \end{aligned}\right\} \tag{394}$$

and the Bianchi identities (equations (321,a and e), with Ψ_0 set equal to zero in accordance with equation (394,b)) are

$$D\Psi_1 = 4\rho\Psi_1, \text{ (a)}; \quad \text{and} \quad \delta\Psi_1 = 2(2\tau + \beta)\Psi_1, \text{ (e)}. \tag{395}$$

And finally, the commutation relation (304) gives

$$D\delta - \delta D = (\pi^* - \alpha^* - \beta)D + \rho^*\delta. \tag{396}$$

From equations (343) it is clear that by a rotation of class I (which will not affect Ψ_0 nor the vanishing of κ, σ, or ε), we can arrange that $\tau = 0$ *provided*

$\rho \neq 0$. (If on the other hand $\rho = 0$, it would follow from equation (394, k) that $\Psi_1 = 0$, which is the result that we wish to prove.) Assuming then, that τ has been made to vanish (and $\rho \neq 0$), we have by equations (395)

$$D \lg \Psi_1 = 4\rho \quad \text{and} \quad \delta \lg \Psi_1 = 2\beta. \tag{397}$$

Accordingly,

$$(D\delta - \delta D) \lg \Psi_1 = 2 D\beta - 4\delta\rho. \tag{398}$$

Now, using the relations (394,e and k) and remembering that τ has been made to vanish, we have

$$(D\delta - \delta D) \lg \Psi_1 = 2\rho^*\beta - 4(\alpha^* + \beta)\rho + 6\Psi_1. \tag{399}$$

On the other hand, by applying the commutation relation (396) to $\lg \Psi_1$, and making use of equations (395,a and e), we obtain

$$\begin{aligned}(D\delta - \delta D) \lg \Psi_1 &= (\pi^* - \alpha^* - \beta) D \lg \Psi_1 + \rho^* \delta \lg \Psi_1 \\ &= 4(\pi^* - \alpha^* - \beta)\rho + 2\beta\rho^*. \end{aligned} \tag{400}$$

Equating now the right-hand sides of equations (399) and (400), we obtain

$$\Psi_1 = \tfrac{2}{3}\pi^*\rho, \tag{401}$$

whereas, equation (394,c) with $\tau = 0$ requires

$$\Psi_1 = -\pi^*\rho. \tag{402}$$

Since we have assumed that $\rho \neq 0$ it follows from equations (401) and (402) that $\Psi_1 = 0$; and the proof of the theorem is completed.

A corollary to the Goldberg–Sachs theorem is that if the field is algebraically special and of Petrov type D, then the congruences formed by the two principal null-directions, $\boldsymbol{l}$ and $\boldsymbol{n}$, must both be geodesic and shear-free, i.e., $\kappa = \sigma = \nu = \lambda = 0$ when $\Psi_0 = \Psi_1 = \Psi_3 = \Psi_4 = 0$; and conversely.

It is a remarkable fact that the black-hole solutions of general relativity are all of Petrov type D and, therefore, enable their analysis in a null tetrad-frame in which the spin coefficients κ, σ, λ, and ν and all the Weyl scalars, except Ψ_2, vanish. It is this circumstance which accounts for the special effectiveness of the Newman–Penrose formalism for their consideration.

BIBLIOGRAPHICAL NOTES

§§ 1–6. There are many good books on differential geometry; but most of them are either more advanced or more extensive than is strictly necessary for our purposes. The account in these sections is adapted to our particular needs. In style and outlook, it is similar to the accounts in:

1. D. Lovelock and H. Rund, *Tensors, Differential Forms, and Variational Principles*, Appendix, 331–52, John Wiley & Sons, Inc., New York, 1975.

2. S. W. Hawking and G. F. R. Ellis, *The Large-Scale Structure of Space–Time*, chapter II, 19–55, Cambridge, England, 1973.

§ 7. Brief accounts of the tetrad formalism will be found in:

3. L. D. Landau and E. M. Lifschitz, *Classical Fields*, § 98, 291–94, Pergamon Press, Oxford, 1975.
4. S. Chandrasekhar in *General Relativity—An Einstein Centenary Survey*, chapter 7, 371–91, edited by S. W. Hawking and W. Israel, Cambridge, England, 1979.

§ 8. The Newman-Penrose formalism is central to the book; and as we shall see in Chapters 8, 9, and 10, it is peculiarly well adapted to the study of the black-hole solutions of general relativity. The basic paper on the subject is:

5. E. T. Newman and R. Penrose, *J. Math. Phys.*, **3**, 566–79 (1962).

An account which more fully motivates the underlying concepts is:

6. R. Penrose, *An Analysis of the Structure of Space–Time*, Cambridge University Adams Prize Essay, Cambridge, England, 1966.

An alternative version of the formalism which exploits the underlying symmetry with respect to the basis vectors $\boldsymbol{l}$ and $\boldsymbol{n}$ and with respect to $\boldsymbol{m}$ and $\bar{\boldsymbol{m}}$ is given in:

7. R. Geroch, A. Held, and R. Penrose, *J. Math. Phys.*, **14**, 874–81 (1973).

For certain practical considerations (explained in the Bibliographical Notes for Chapter VIII) we have preferred to adhere to the original less symmetric version. For other accounts of the Newman-Penrose formalism see:

8. F. A. E. Pirani in *Lectures on General Relativity*, Brandeis Summer Institute in Theoretical Physics, I, 249–373, Prentice Hall, Inc., Englewood, New Jersey, 1964.
9. R. A. Breuer, *Gravitational Perturbation Theory and Synchotron Radiation*, Lecture Notes in Physics, 44, Springer Verlag, Berlin, 1975.

See also reference 4.

§ 9(*a*). The optical scalars were introduced by Sachs:

10. R. K. Sachs, *Proc. Roy. Soc.* (*London*) A, **264**, 309–37 (1961).
11. ———, ibid. **270**, 103–26 (1962).

See also:

12. R. K. Sachs, *Relativity, Groups, and Topology*, 523–62, Les Houches Summer School in Theoretical Physics, Gordon and Breach Science Publishers, New York, 1964.

For a condensed and an illuminating account see:

13. R. Penrose in *Perspectives in Geometry and Relativity*, 259–74, edited by B. Hoffmann, Indiana University Press, Bloomington, Indiana, 1966.

The significance of the optical scalars, as spin coefficients, emerged, of course, only after the Newman–Penrose formalism had been developed. (See also reference 6.)

§ 9(*b*). The original papers in which the Petrov classification (sometimes also called the Petrov-Pirani classification) is described are:

14. A. Z. Petrov, *Sci. Nat. Kazan State University*, **114**, 55–69 (1954).
15. F. A. E. Pirani, *Phys. Rev.*, **105**, 1089–99 (1957).

For more extensive accounts see:

16. A. Z. Petrov, *Einstein Spaces*, translated by R. E. Kelleher and J. Woodrow, Pergamon Press, Oxford, 1969.
17. F. A. E. Pirani in *Gravitation: An Introduction to Current Research*, 119–226, edited by L. Witten, John Wiley & Sons, Inc., New York, 1962.

See also reference 8.

The account in the text is an elaboration of the one given in reference 5. (See also reference 4.)

§9(*c*). The Goldberg-Sachs theorem is proved in:

18. J. N. Goldberg and R. K. Sachs, *Acta Phys. Polonica, Supp.* 13, **22**, 13–23 (1962).

But the proof given follows reference 5.

2

A SPACE-TIME OF SUFFICIENT GENERALITY

10. Introduction

The black-hole solutions of general relativity are static and spherically symmetric (as the Schwarzschild and the Reissner–Nordström solutions are) or stationary and axisymmetric (as the Kerr and the Kerr–Newman solutions are). The perturbed state of these solutions, which are of equal concern to us, will be neither stationary nor axisymmetric. As a basis for the treatment of the various problems that are suggested in these latter contexts, it will be convenient to have on hand the Einstein and the Maxwell equations in a space-time of sufficient generality which will encompass the different situations we shall encounter in the course of our study. In this chapter, we shall consider a space-time which appears to have the requisite generality; and we shall provide expressions for the basic tensors which will enable us to write down, without any ado, the relevant equations appropriate to the different contexts.

11. Stationary axisymmetric space-times and the dragging of inertial frames

We consider first the metric appropriate to stationary axisymmetric space-times as a preliminary to our consideration of more general space-times which are neither stationary nor axisymmetric.

For the description of a stationary axisymmetric space-time, it is convenient to take as two of the coordinates the time $t(=x^0)$ and the azimuthal angle $\varphi(=x^1)$ about the axis of symmetry. The stationary and the axisymmetric character of the space-time require that the metric coefficients be independent of t and φ and that

$$g_{ij} = g_{ij}(x^2, x^3), \tag{1}$$

where x^2 and x^3 are the two remaining spatial coordinates.

Besides stationarity and axisymmetry, we shall also require that the space-time is invariant to the simultaneous inversion of the time t and the angle φ, i.e., to the transformation, $t \to -t$ and $\varphi \to -\varphi$. The physical meaning of this additional requirement is that the source of the gravitational field, whatever that may be, has motions that are purely rotational about the axis of symmetry; for, in that case, the energy-momentum tensor of the source will have the stated invariance. In other words, the space-time we are considering is that associated with a 'rotating body'. In any event, the assumed invariance requires that

$$g_{02} = g_{03} = g_{12} = g_{13} = 0, \tag{2}$$

since the terms in the metric with these coefficients change sign under the inversion, $t \to -t$ and $\varphi \to -\varphi$. Therefore, under the assumptions made, the metric must have the form

$$\begin{aligned} ds^2 = {} & g_{00}(dx^0)^2 + 2g_{01}\,dx^0 dx^1 + g_{11}(dx^1)^2 \\ & + [g_{22}(dx^2)^2 + 2g_{23}\,dx^2 dx^3 + g_{33}(dx^3)^2]. \end{aligned} \tag{3}$$

where all the metric coefficients are functions of x^2 and x^3 only.

A further reduction in the form of the metric can be achieved by making use of the following theorem.

THEOREM. *The metric*

$$ds^2 = g_{11}(dx^1)^2 + 2g_{12}\,dx^1 dx^2 + g_{22}(dx^2)^2 \tag{4}$$

of a two-dimensional space, (x^1, x^2), *with a positive- or a negative-definite signature* $(+, +$ *or* $-, -)$ *can always be brought to the diagonal form,*

$$ds^2 = \pm e^{2\mu}[(dx^1)^2 + (dx^2)^2] \tag{5}$$

by a coordinate transformation, where $e^{2\mu}$ *is some function of* x^1 *and* x^2.

Proof. To prove the theorem, it will clearly suffice to show that there exist transformations,

$$x^{1'} = \phi(x^1, x^2) \qquad \text{and} \qquad x^{2'} = \psi(x^1, x^2), \tag{6}$$

which will reduce the contravariant form of the metric,

$$ds^2 = g^{11}(dx_1)^2 + 2g^{12}\,dx_1 dx_2 + g^{22}(dx_2)^2, \tag{7}$$

to a diagonal form with equal coefficients for $(dx_1)^2$ and $(dx_2)^2$.

For a transformation to achieve the stated ends, it is necessary and sufficient that

$$g^{1'2'} = g^{11}\phi_{,1}\psi_{,1} + g^{12}(\phi_{,1}\psi_{,2} + \phi_{,2}\psi_{,1}) + g^{22}\phi_{,2}\psi_{,2} = 0 \tag{8}$$

and

$$\begin{aligned} g^{1'1'} - g^{2'2'} = {} & g^{11}(\phi_{,1}{}^2 - \psi_{,1}{}^2) + 2g^{12}(\phi_{,1}\phi_{,2} - \psi_{,1}\psi_{,2}) \\ & + g^{22}(\phi_{,2}{}^2 - \psi_{,2}{}^2) = 0 \end{aligned} \tag{9}$$

We readily verify that equation (8) is identically satisfied by the substitutions

$$\phi_{,1} = \kappa(g^{21}\psi_{,1} + g^{22}\psi_{,2}) \quad \text{and} \quad \phi_{,2} = -\kappa(g^{11}\psi_{,1} + g^{12}\psi_{,2}), \tag{10}$$

where κ is any arbitrary function. But the same substitutions in equation (9) yield the condition

$$[\kappa^2\{g^{11}g^{22}-(g^{12})^2\}-1]\,[g^{11}\psi_{,1}{}^2+2g^{12}\psi_{,1}\psi_{,2}+g^{22}\psi_{,2}{}^2]=0 \qquad (11)$$

For a positive- or a negative-definite metric, the term in square brackets in equation (11) cannot vanish; and the equation can be satisfied only by the choice

$$\kappa^2=\frac{1}{g^{11}g^{22}-(g^{12})^2}=g_{11}g_{22}-(g_{12})^2=g, \qquad (12)$$

where g denotes the determinant of the covariant form of the metric. We conclude that equations (8) and (9) are satisfied by the substitutions

$$\phi_{,1}=+g^{\frac{1}{2}}(g^{21}\psi_{,1}+g^{22}\psi_{,2})=+g^{\frac{1}{2}}g^{2k}\psi_{,k}$$

and

$$\phi_{,2}=-g^{\frac{1}{2}}(g^{11}\psi_{,1}+g^{12}\psi_{,2})=-g^{\frac{1}{2}}g^{1k}\psi_{,k}. \qquad (13)$$

The integrability condition for these equations requires

$$(g^{\frac{1}{2}}g^{ik}\psi_{,k})_{,i}=0. \qquad (14)$$

In other words, ψ can be *any* solution of Laplace's equation in the 2-space considered. The existence of transformations which will reduce the metric to a diagonal form with equal coefficients for $(dx_1)^2$ and $(dx_2)^2$ (and, therefore, also for $(dx^1)^2$ and $(dx^2)^2$ in the covariant form of the metric) is established.

It is worth noticing that by rewriting equations (13) in the covariant form,

$$\phi_{,i}=g^{\frac{1}{2}}\varepsilon_{ij}g^{jk}\psi_{,k}, \qquad (15)$$

(where ε_{ij} is the alternating symbol in two indices), the gradients of ϕ and of ψ appear as 'duals' of one another.

It follows from the theorem we have proved that the part of the metric in square brackets in equation (3) can be reduced to the form

$$\pm e^{2\mu}[(dx^2)^2+(dx^3)^2], \qquad (16)$$

where μ is some function of x^2 and x^3.

With the various simplifications available, we shall write the metric appropriate to a stationary axisymmetric space-time in the form

$$ds^2=e^{2\nu}(dt)^2-e^{2\psi}(d\varphi-\omega\,dt)^2-e^{2\mu_2}(dx^2)^2-e^{2\mu_3}(dx^3)^2, \qquad (17)$$

where ν, ψ, ω, μ_2, and μ_3 are functions of x^2 and x^3. It will be noticed that we have not specialized the metric as fully as we might have, within the requirements of our basic assumptions of stationarity, axisymmetry, and invariance under simultaneous inversion of t and φ, since we have left the functions μ_2 and μ_3 to be different. We have done so in order that we may, at a later stage, avail ourselves of this gauge-freedom to restrict μ_2 and μ_3 by a coordinate condition which may appear advantageous.

(*a*) *The dragging of the inertial frame*

Associated with the metric (17), we can define a tetrad frame by the basis vectors:

$$\begin{aligned} e_{(0)i} &= (e^{\nu}, \quad 0, \quad 0, \quad 0),\\ e_{(1)i} &= (\omega e^{\psi}, \quad -e^{\psi}, \quad 0, \quad 0),\\ e_{(2)i} &= (0, \quad 0, \quad -e^{\mu_2}, \quad 0),\\ e_{(3)i} &= (0, \quad 0, \quad 0, \quad -e^{\mu_3}). \end{aligned} \tag{18}$$

The corresponding contravariant basis-vectors are:

$$\begin{aligned} e_{(0)}{}^{i} &= (e^{-\nu}, \quad \omega e^{-\nu}, \quad 0, \quad 0),\\ e_{(1)}{}^{i} &= (0, \quad e^{-\psi}, \quad 0, \quad 0),\\ e_{(2)}{}^{i} &= (0, \quad 0, \quad e^{-\mu_2}, \quad 0),\\ e_{(3)}{}^{i} &= (0, \quad 0, \quad 0, \quad e^{-\mu_3}). \end{aligned} \tag{19}$$

For a tetrad frame so defined

$$e_{(a)}{}^{i} e_{(b)i} = \eta_{(a)(b)} = \begin{vmatrix} 1 & 0 & 0 & 0\\ 0 & -1 & 0 & 0\\ 0 & 0 & -1 & 0\\ 0 & 0 & 0 & -1 \end{vmatrix}. \tag{20}$$

Thus, in the chosen frame, the metric is Minkowskian: it, therefore, represents locally an *inertial frame.*

Now consider a point in space-time which is assigned the four-velocity,

$$u^0 = \frac{dt}{ds} = \frac{e^{-\nu}}{\sqrt{(1-V^2)}}, \quad u^1 = \frac{d\varphi}{ds} = \Omega u^0, \quad u^\alpha = \frac{dx^\alpha}{ds} = u^0 v^\alpha \quad (\alpha = 2, 3), \tag{21}$$

where

$$\Omega = \frac{d\varphi}{dt}, \quad v^\alpha = \frac{dx^\alpha}{dt} \quad (\alpha = 2, 3), \tag{22}$$

and

$$V^2 = e^{2\psi - 2\nu}(\Omega - \omega)^2 + e^{2\mu_2 - 2\nu}(v^2)^2 + e^{2\mu_3 - 2\nu}(v^3)^2. \tag{23}$$

The same point will be assigned, in the local inertial frame, the four-velocity,

$$u^{(a)} = e^{(a)}{}_{i} u^{i} = \eta^{(a)(b)} e_{(b)i} u^{i}, \tag{24}$$

or, as one readily finds,

$$u^{(0)} = \frac{1}{\sqrt{(1-V^2)}}, \quad u^{(1)} = \frac{e^{\psi-\nu}(\Omega-\omega)}{\sqrt{(1-V^2)}}, \quad \text{and} \quad u^{(\alpha)} = \frac{e^{\mu_\alpha - \nu} v^\alpha}{\sqrt{(1-V^2)}} \quad (\alpha = 2, 3). \tag{25}$$

Therefore, a point, considered as describing a circular orbit (with the proper circumference πe^{ψ}) with an angular velocity Ω in the chosen coordinate frame, will be assigned an angular velocity,

$$e^{\psi-\nu}(\Omega-\omega), \tag{26}$$

in the local inertial frame. Accordingly, a point which is considered as at rest in the local inertial frame (i.e., $u^{(1)} = u^{(2)} = u^{(3)} = 0$), will be assigned an angular velocity ω in the coordinate frame. On this account, the non-vanishing of ω is said to describe *a 'dragging' of the inertial frame*. In Chapter 6, we shall show that in an asymptotically flat space-time with the metric (17),

$$\omega \to 2Jr^{-3} \tag{27}$$

where J is a constant which will be interpreted as the angular momentum of the source.

12. A space-time of requisite generality

We now wish to generalize the line element (17) to encompass situations which are non-stationary and non-axisymmetric. Our principal interest in these generalizations is to be able to treat, in full generality, perturbations of space-times which are either static and spherically symmetric or stationary and axisymmetric.

In the first instance, we shall restrict ourselves to space-times which retain their axisymmetry at all times, i.e., we shall allow the metric to be dependent on t but restrict it to be independent of φ. Let the contravariant form of the metric be

$$g = g^{ij}\partial_i\partial_j. \tag{28}$$

Then by assumption, all the components, g^{ij}, of the metric are functions only of x^0, x^2, and x^3. We shall now show how, by a local coordinate-transformation involving only x^0, x^2, and x^3, we can bring the 3×3-matrix, $[g^{ij}]$, $(i, j = 0, 2, 3)$, to a diagonal form.

THEOREM (COTTON-DARBOUX). *The metric*

$$g = g^{ij}\partial_i\partial_j \qquad (i, j = 0, 1, 2), \tag{29}$$

in a three-dimensional space, (x^0, x^1, x^2), *can always be brought to a diagonal form by a local coordinate-tranformation.*

Proof. First, we observe that by the choice of a geodesic system of coordinates, the metric can be brought to the form

$$g = e^0\partial_0 + g^{\alpha\beta}\partial_\alpha\partial_\beta \quad (e^0 = \pm 1;\ \alpha,\ \beta = 1, 2). \tag{30}$$

[A geodesic system of coordinates is constructed by considering a surface

$f(x^0, x^1, x^2)$ such that $g^{ij} f_{,i} f_{,j} \neq 0$, letting the geodesics, normal to $f = 0$, be the coordinate lines x^0, and choosing the coordinates x^1 and x^2 on the surfaces geodesically parallel to f.]

Consider the coordinate transformation,

$$x^{i'} = \phi^{i'}(x^0, x^1, x^2) \qquad (i' = 0, 1, 2), \quad (31)$$

where $\phi^{i'}$ are functions of x^0, x^1, and x^2 which we wish to determine by the conditions that the metric (30), in the new coordinate system, is diagonal. The conditions clearly are

$$g^{i'j'} = e^0 \frac{\partial \phi^{i'}}{\partial x^0} \frac{\partial \phi^{j'}}{\partial x^0} + g^{\alpha\beta} \frac{\partial \phi^{i'}}{\partial x^\alpha} \frac{\partial \phi^{j'}}{\partial x^\beta} = 0 \quad [(i', j') = (0, 1), (1, 2)\ (2, 0)], \quad (32)$$

or, written out explicitly,

$$\left.\begin{aligned} e^0 \frac{\partial \phi^1}{\partial x^0} \frac{\partial \phi^2}{\partial x^0} &= -g^{\alpha\beta} \frac{\partial \phi^1}{\partial x^\alpha} \frac{\partial \phi^2}{\partial x^\beta} = K^0, \text{ (say)}, \\ e^0 \frac{\partial \phi^2}{\partial x^0} \frac{\partial \phi^0}{\partial x^0} &= -g^{\alpha\beta} \frac{\partial \phi^2}{\partial x^\alpha} \frac{\partial \phi^0}{\partial x^\beta} = K^1, \text{ (say)}, \\ e^0 \frac{\partial \phi^0}{\partial x^0} \frac{\partial \phi^1}{\partial x^0} &= -g^{\alpha\beta} \frac{\partial \phi^0}{\partial x^\alpha} \frac{\partial \phi^1}{\partial x^\beta} = K^2, \text{ (say)}. \end{aligned}\right\} \quad (33)$$

Hence,

$$\frac{\partial \phi^0}{\partial x^0} = \left(\frac{K^1 K^2}{e^0 K^0}\right)^{1/2}, \frac{\partial \phi^1}{\partial x^0} = \left(\frac{K^2 K^0}{e^0 K^1}\right)^{1/2}, \quad \text{and} \quad \frac{\partial \phi^2}{\partial x^0} = \left(\frac{K^0 K^1}{e^0 K^2}\right)^{1/2}. \quad (34)$$

Now, suppose that ϕ^0, ϕ^1, and ϕ^2, as functions of the two variables x^1 and x^2, are specified on a surface $x^0 = 0$ (say); and, further, that K^0, K^1, and K^2 are nowhere zero on the surface and satisfy the usual conditions of smoothness. Then, by the Cauchy–Kowalewski theorem, there exist unique functions, ϕ^0, ϕ^1, and ϕ^2 which satisfy the system of equations (33) and which reduce to the values specified on $x^0 = 0$. The existence of local coordinate-transformations, which will bring the metric in a 3-space to the diagonal form, is thus established.

Returning to the consideration of a metric appropriate for a non-stationary axisymmetric space-time, we can, in accordance with the theorem we have established, set

$$g^{02} = g^{03} = g^{23} = 0. \quad (35)$$

And we shall write the diagonal coefficients in the forms

$$g^{00} = e^{-2\nu}, \quad g^{22} = -e^{-2\mu_2}, \quad \text{and} \quad g^{33} = -e^{-2\mu_3}, \quad (36)$$

where ν, μ_2, and μ_3 are functions of x^0, x^2, and x^3. Let the remaining coefficients of g^{ij} be

and
$$\left.\begin{aligned} g^{01} = \omega e^{-2\nu}, \quad g^{12} = -q_2 e^{-2\mu_2}, \quad g^{13} = -q_3 e^{-2\mu_3} \\ g^{11} = \omega^2 e^{-2\nu} - e^{-2\psi} - q_2^2 e^{-2\mu_2} - q_3^2 e^{-2\mu_3}, \end{aligned}\right\} \tag{37}$$

where ω, q_2, q_3, and ψ are further functions of x^0, x^2, and x^3.

With the contravariant form of the metric, chosen in the manner we have described, the covariant form of the metric is

$$\begin{aligned} ds^2 = e^{2\nu}(dt)^2 - e^{2\psi}(d\varphi - q_2 dx^2 - q_3 dx^3 - \omega\, dt)^2 \\ - e^{2\mu_2}(dx^2)^2 - e^{2\mu_3}(dx^3)^2. \end{aligned} \tag{38}$$

Comparison with the metric (17) shows that the metric (38) is, in some sense, a natural generalization of it to allow for non-stationarity. Also, it should be noted that we do not now have the freedom of gauge to impose on μ_2 and μ_3 any additional coordinate condition—a freedom we did have in the context of the stationary metric (17).

The chosen form of the metric (38) involves seven functions, namely, ν, ψ, μ_2, μ_3, ω, q_2, and q_3. On the other hand, since Einstein's equation is covariant and provides only six independent equations for the metric coefficients, it should be the case that these seven functions occur in the field equations only in six independent combinations. As to how this might arise can be inferred from considering the coordinate transformation

$$x^1 = x^{1'} + f(x^{0'}, x^{2'}, x^{3'}) \qquad \text{and} \qquad x^i = x^{i'} \quad (i = 0, 2, 3). \tag{39}$$

Under this transformation, only the metric coefficients $g_{i1}\,(i = 0, 2, 3)$ will change; thus,

$$g_{i'1'} = g_{i1} + g_{11} f_{,i} \qquad \text{or} \qquad (g_{i'1'} - g_{i1})/g_{11} = f_{,i} \quad (i = 0, 2, 3). \tag{40}$$

From the integrability conditions of these relations, it follows that

$$\omega_{,2} - q_{2,0}, \quad \omega_{,3} - q_{3,0}, \qquad \text{and} \qquad q_{2,3} - q_{3,2}, \tag{41}$$

are invariant to the transformation (39). And it is a fact, which we shall verify, that the functions ω, q_2, and q_3 occur in the field equations only in the combinations (41). But among the combinations (41), we have the identity,

$$(\omega_{,2} - q_{2,0})_{,3} - (\omega_{,3} - q_{3,0})_{,2} + (q_{2,3} - q_{3,2})_{,0} = 0. \tag{42}$$

In this manner, the seven functions, in terms of which we have expressed the metric coefficients, occur in the field equations (as we have indicated) only as six independent quantities.

The form of the metric (38), which we have derived for non-stationary axisymmetric space-times, can be applied to a class of non-axisymmetric space-times whose metric coefficients are separable in the variable x^1 ($= \varphi$), i.e., when

$$g^{ij}(x^0, x^1, x^2, x^3) = g^{ij}(x^0, x^2, x^3)h(x^1), \tag{43}$$

where $h(x^1)$ is some function of x^1. For, in this case, it is clear that by a coordinate transformation, involving only x^0, x^2, and x^3 and leaving x_1 unchanged, we can make g^{02}, g^{03}, and g^{23} vanish: the common factor, $h(x^1)$, which will occur in equation (33) will not, of course, make any difference. To allow for this possible generalization, we shall formally let the seven functions, ν, ψ, μ_2, μ_3, q_2, q_3, and ω be functions of x^1 as well. It will appear in later chapters that the form of the metric has the requisite flexibility and generality to encompass all the situations to which we shall apply the equations and the expressions appropriate for the metric (38) with the metric coefficients allowed to be functions of all four variables x^0, x^1, x^2, and x^3. Also, it should be remarked that, even with this generalization, the seven functions in terms of which the metric coefficients are expressed occur in the field equations only in six independent combinations (see §§13 and 14).

The remaining sections of this chapter are devoted to writing down the basic expressions and equations which should be used in conjunction with the chosen form of the metric.

13. Equations of structure and the components of the Riemann tensor

For the purposes of writing down the explicit form of Einstein's equation for the most general form of the metric chosen in §12, we shall obtain the components of the Riemann tensor via Cartan's equations of structure. In deriving the relevant equations, we can preserve maximum symmetry by writing

$$dt = -i\mathrm{d}x^4, \quad \partial_t = i\partial_4, \quad \nu = \mu_4, \quad \text{and} \quad \omega = iq_4. \tag{44}$$

The metric (38) then takes the form

$$ds^2 = -\sum_A e^{2\mu_A}(\mathrm{d}x^A)^2 - e^{2\psi}(\mathrm{d}x^1 - \sum_A q_A \mathrm{d}x^A)^2, \tag{45}$$

where the capital Latin indices ($A, B, \ldots$) and summations over them are restricted to the values 2, 3, and 4. In the form (45), the metric is entirely symmetric in the indices 2, 3, and 4. But it should be noted that in this complex version, its signature is $(-, -, -, -)$. Also, it should be recalled that ψ, μ_A, and q_A are functions of all four variables x^1 and $x^A (A = 2, 3, 4)$.

To avoid ambiguity, summation over a repeated capital Latin index (restricted to its range) will not be assumed: summation will be explicitly indicated (as in equation (45)) whenever it is needed,

In applying Cartan's method, we shall take

$$\omega^A = e^{\mu_A}\mathrm{d}x^A \quad \text{and} \quad \omega^1 = e^{\psi}(\mathrm{d}x^1 - \sum_A q_A \mathrm{d}x^A) \tag{46}$$

as our basis of one-forms (denoted by e^i in Ch. 1, §5). These forms correspond to the choice of an orthonormal tetrad-frame with signature $(-, -, -, -)$. The inverse relations expressing the one-forms dx^A and dx^1 in terms of ω^A and ω^1 are

$$dx^A = e^{-\mu_A}\omega^A \quad \text{and} \quad dx^1 = e^{-\psi}\omega^1 + \sum_A e^{-\mu_A} q_A \omega^A. \tag{47}$$

The first step in deriving Cartan's equations of structure (Ch. 1, equations (137′) and (149)) is to express the exterior derivatives of the ω's in terms of the basis of the two-forms, namely, $\omega^i \wedge \omega^j$ $(i \neq j,\ i, j = 1, 2, 3, 4)$. We find

$$\begin{aligned} d\omega^A &= \sum_B e^{\mu_A} \mu_{A,B} dx^B \wedge dx^A + e^{\mu_A}\mu_{A,1} dx^1 \wedge dx^A \\ &= \sum_B e^{-\mu_B} \mu_{A,B} \omega^B \wedge \omega^A + \mu_{A,1}\Big[e^{-\psi}\omega^1 + \sum_B e^{-\mu_B} q_B \omega^B\Big] \wedge \omega^A \\ &= \sum_B e^{-\mu_B}(\mu_{A,B} + q_B \mu_{A,1})\omega^B \wedge \omega^A + e^{-\psi}\mu_{A,1}\omega^1 \wedge \omega^A. \end{aligned} \tag{48}$$

In writing compactly equations such as the foregoing, we shall find it convenient to define a *colon derivative* of a function $f(x^1, x^2, x^3, x^4)$ with respect to $x^A (A = 2, 3, 4)$ by

$$f_{:A} = f_{,A} + q_A f_{,1}. \tag{49}$$

It is a differentiation, since it satisfies the Leibnitz rule:

$$(fg)_{:A} = f g_{:A} + g f_{:A}. \tag{50}$$

In this notation, equation (48) takes the form

$$d\omega^A = -\sum_B e^{-\mu_B}\mu_{A:B}\omega^A \wedge \omega^B - e^{-\psi}\mu_{A,1}\omega^A \wedge \omega^1. \tag{51}$$

Similarly, we find

$$d\omega^1 = \sum_A e^{-\mu_A}(\psi_{:A} + q_{A,1})\omega^A \wedge \omega^1 - \sum_{A,B} e^{\psi - \mu_A - \mu_B} q_{B:A}\omega^A \wedge \omega^B. \tag{52}$$

Now, when the torsion is zero, Cartan's first equation of structure (Ch. 1, equation (137)) gives

$$d\omega^1 = -\sum_A \omega^1{}_A \wedge \omega^A \tag{53}$$

and

$$d\omega^A = -\sum_B \omega^A{}_B \wedge \omega^B - \omega^A{}_1 \wedge \omega^1; \tag{54}$$

and these equations determine the connection one-forms $\omega^1{}_A$ and $\omega^A{}_B$ from our knowledge of $d\omega^1$ and $d\omega^A$. Also, since the one-forms ω^1 and ω^A provide the basis for an orthonormal tetrad-frame, it follows from Ch. 1, equation (197) that

$$\omega^i{}_j = -\omega^j{}_i \qquad (i, j = 1, 2, 3, 4). \tag{55}$$

From a comparison of equations (51) and (52) with equations (53) and (54), we conclude that

$$\omega^1{}_A = -\omega^A{}_1 = e^{-\mu_A}(\psi_{:A} + q_{A,1})\omega^1 - e^{-\psi}\mu_{A,1}\omega^A + \tfrac{1}{2}\sum_B e^{\psi-\mu_A-\mu_B}Q_{AB}\omega^B \tag{56}$$

and

$$\omega^A{}_B = -\omega^B{}_A = -\tfrac{1}{2}e^{\psi-\mu_A-\mu_B}Q_{AB}\omega^1 + e^{-\mu_B}\mu_{A:B}\omega^A - e^{-\mu_A}\mu_{B:A}\omega^B, \tag{57}$$

where

$$Q_{AB} = q_{A:B} - q_{B:A}. \tag{58}$$

Now, letting

$$\Psi_A = \psi_{:A} + q_{A,1}, \tag{59}$$

we have the following explicit expressions for the different connection forms.

$$\begin{aligned}
\omega^1{}_2 &= e^{-\mu_2}\Psi_2\omega^1 - e^{-\psi}\mu_{2,1}\omega^2 + \tfrac{1}{2}e^{\psi-\mu_2-\mu_3}Q_{23}\omega^3 + \tfrac{1}{2}e^{\psi-\mu_2-\mu_4}Q_{24}\omega^4,\\
\omega^1{}_3 &= e^{-\mu_3}\Psi_3\omega^1 - e^{-\psi}\mu_{3,1}\omega^3 + \tfrac{1}{2}e^{\psi-\mu_3-\mu_4}Q_{34}\omega^4 + \tfrac{1}{2}e^{\psi-\mu_3-\mu_2}Q_{32}\omega^2,\\
\omega^1{}_4 &= e^{-\mu_4}\Psi_4\omega^1 - e^{-\psi}\mu_{4,1}\omega^4 + \tfrac{1}{2}e^{\psi-\mu_4-\mu_2}Q_{42}\omega^2 + \tfrac{1}{2}e^{\psi-\mu_4-\mu_3}Q_{43}\omega^3,\\
\omega^2{}_3 &= -\tfrac{1}{2}e^{\psi-\mu_2-\mu_3}Q_{23}\omega^1 + e^{-\mu_3}\mu_{2:3}\omega^2 - e^{-\mu_2}\mu_{3:2}\omega^3,\\
\omega^3{}_4 &= -\tfrac{1}{2}e^{\psi-\mu_3-\mu_4}Q_{34}\omega^1 + e^{-\mu_4}\mu_{3:4}\omega^3 - e^{-\mu_3}\mu_{4:3}\omega^4,\\
\omega^4{}_2 &= -\tfrac{1}{2}e^{\psi-\mu_4-\mu_2}Q_{42}\omega^1 + e^{-\mu_2}\mu_{4:2}\omega^4 - e^{-\mu_4}\mu_{2:4}\omega^2.
\end{aligned} \tag{60}$$

To obtain the components of the Riemann tensor from Cartan's second equation of structure, namely,

$$\tfrac{1}{2}R^i{}_{jkl}\omega^k \wedge \omega^l = \Omega^i{}_j = d\omega^i{}_j + \omega^i{}_k \wedge \omega^k{}_j, \tag{61}$$

we have to evaluate the exterior derivatives of the connection forms listed in equations (60). (Actually, it will appear that, by taking into account the underlying symmetry of the metric in the indices 2, 3, and 4, we can write down all the components of the Riemann tensor from a knowledge only of, for example, $\Omega^1{}_2$ and $\Omega^2{}_3$.)

The required exterior derivatives of $\omega^i{}_j$ can be written down with the aid of the following readily verifiable lemmas.

LEMMAS. *If F is any function of* x^1, x^2, x^3, *and* x^4, *then*

$$d(F\omega^1) = \sum_A e^{-\psi-\mu_A}\mathscr{D}_A(Fe^\psi)\omega^A\wedge\omega^1 + \tfrac{1}{2}\sum_{A,B} Fe^{\psi-\mu_A-\mu_B}Q_{AB}\omega^A\wedge\omega^B, \tag{62}$$

and

$$d(F\omega^A) = \sum_B e^{-\mu_A-\mu_B}(e^{\mu_A}F)_{:B}\omega^B\wedge\omega^A + e^{-\psi-\mu_A}(e^{\mu_A}F)_{,1}\,\omega^1\wedge\omega^A, \tag{63}$$

where $\mathscr{D}_A$ *is an operator, which acting on any function* $f(x^1, \ldots, x^4)$, *yields*

$$\mathscr{D}_A f = f_{:A} + q_{A,1}f = f_{,A} + (q_Af)_{,1}. \tag{64}$$

(Note that $\mathscr{D}_A$ is *not* a differentiation; and also that (cf. equation (59))

$$e^{-\psi}\mathscr{D}_A(e^\psi) = \psi_{:A} + q_{A,1} = \Psi_A. \tag{65}$$

Now making use of the foregoing lemmas in conjunction with the connection forms (60), we find that

$$\begin{aligned}
d\omega^1{}_2 = {} & \sum_A e^{-\psi-\mu_A}\mathscr{D}_A(e^{\psi-\mu_2}\Psi_2)\omega^A\wedge\omega^1 - e^{-\psi-\mu_2}(e^{\mu_2-\psi}\mu_{2,1})_{,1}\,\omega^1\wedge\omega^2\\
& -\tfrac{1}{2}e^{-\psi-\mu_3}(e^{\psi-\mu_2}Q_{23})_{,1}\,\omega^3\wedge\omega^1 - \tfrac{1}{2}e^{-\psi-\mu_4}(e^{\psi-\mu_2}Q_{24})_{,1}\,\omega^4\wedge\omega^1\\
& +\omega^2\wedge\omega^3\,\{e^{\psi-2\mu_2-\mu_3}\Psi_2Q_{23} + \tfrac{1}{2}e^{-\mu_2-\mu_3}(e^{\psi-\mu_2}Q_{23})_{:2}\\
& +e^{-\mu_2-\mu_3}(e^{-\psi+\mu_2}\mu_{2,1})_{:3}\}\\
& +\omega^2\wedge\omega^4\,\{e^{\psi-2\mu_2-\mu_4}\Psi_2Q_{24} + \tfrac{1}{2}e^{-\mu_2-\mu_4}(e^{\psi-\mu_2}Q_{24})_{:2}\\
& +e^{-\mu_2-\mu_4}(e^{-\psi+\mu_2}\mu_{2,1})_{:4}\}\\
& +\omega^3\wedge\omega^4\,\{e^{\psi-\mu_2-\mu_3-\mu_4}\Psi_2Q_{34} + \tfrac{1}{2}e^{-\mu_3-\mu_4}(e^{\psi-\mu_2}Q_{24})_{:3}\\
& -\tfrac{1}{2}e^{-\mu_4-\mu_3}(e^{\psi-\mu_2}Q_{23})_{:4}\},
\end{aligned} \tag{66}$$

$$\begin{aligned}
\omega^1{}_3\wedge\omega^2{}_3 = {} & [e^{-2\mu_3}\Psi_3\mu_{2:3} - \tfrac{1}{4}e^{2\psi-2\mu_2-2\mu_3}Q_{23}^2]\omega^1\wedge\omega^2\\
& -[e^{-\mu_2-\mu_3}\Psi_3\mu_{3:2} + \tfrac{1}{2}e^{-\mu_2-\mu_3}\mu_{3,1}Q_{23}]\omega^1\wedge\omega^3\\
& +[\tfrac{1}{2}e^{\psi-2\mu_2-\mu_3}Q_{23}\mu_{3:2} + e^{-\psi-\mu_3}\mu_{3,1}\mu_{2:3}]\omega^2\wedge\omega^3\\
& +\tfrac{1}{4}e^{2\psi-2\mu_3-\mu_2-\mu_4}Q_{34}Q_{23}\omega^1\wedge\omega^4 - \tfrac{1}{2}e^{\psi-2\mu_3-\mu_4}Q_{34}\mu_{2:3}\omega^2\wedge\omega^4\\
& +\tfrac{1}{2}e^{\psi-\mu_2-\mu_3-\mu_4}Q_{34}\mu_{3:2}\,\omega^3\wedge\omega^4,
\end{aligned} \tag{67}$$

$$\begin{aligned}
\omega^1{}_4\wedge\omega^2{}_4 = {} & [e^{-2\mu_4}\Psi_4\mu_{2:4} - \tfrac{1}{4}e^{2\psi-2\mu_2-2\mu_4}Q_{24}^2]\omega^1\wedge\omega^2\\
& -[e^{-\mu_2-\mu_4}\Psi_4\mu_{4:2} + \tfrac{1}{2}e^{-\mu_2-\mu_4}\mu_{4,1}Q_{24}]\omega^1\wedge\omega^4
\end{aligned}$$

$$+[\tfrac{1}{2}e^{\psi-2\mu_2-\mu_4}Q_{24}\mu_{4:2}+e^{-\psi-\mu_4}\mu_{4,1}\mu_{2:4}]\omega^2\wedge\omega^4$$
$$-\tfrac{1}{2}e^{\psi-2\mu_4-\mu_3}Q_{43}\mu_{2:4}\omega^2\wedge\omega^3+\tfrac{1}{2}e^{\psi-\mu_2-\mu_3-\mu_4}Q_{43}\mu_{4:2}\omega^4\wedge\omega^3$$
$$+\tfrac{1}{4}e^{2\psi-2\mu_4-\mu_2-\mu_3}Q_{43}Q_{24}\omega^1\wedge\omega^3; \tag{68}$$

and

$$d\omega^2{}_3=-\tfrac{1}{2}\sum_A e^{-\psi-\mu_A}\mathscr{D}_A(e^{2\psi-\mu_2-\mu_3}Q_{23})\omega^A\wedge\omega^1$$
$$-e^{-\psi-\mu_2}(e^{\mu_2-\mu_3}\mu_{2:3})_{,1}\,\omega^2\wedge\omega^1+e^{-\psi-\mu_3}(e^{\mu_3-\mu_2}\mu_{3:2})_{,1}\,\omega^3\wedge\omega^1$$
$$+\omega^2\wedge\omega^3\{-\tfrac{1}{2}e^{2\psi-2\mu_2-2\mu_3}Q_{23}^2-e^{-\mu_2-\mu_3}(e^{\mu_2-\mu_3}\mu_{2:3})_{:3}$$
$$-e^{-\mu_2-\mu_3}(e^{\mu_3-\mu_2}\mu_{3:2})_{:2}\}$$
$$+\omega^2\wedge\omega^4\{-\tfrac{1}{2}e^{2\psi-2\mu_2-\mu_3-\mu_4}Q_{23}Q_{24}-e^{-\mu_2-\mu_4}(e^{\mu_2-\mu_3}\mu_{2:3})_{:4}\}$$
$$+\omega^3\wedge\omega^4\{-\tfrac{1}{2}e^{2\psi-2\mu_3-\mu_2-\mu_4}Q_{23}Q_{34}+e^{-\mu_3-\mu_4}(e^{\mu_3-\mu_2}\mu_{3:2})_{:4}\}, \tag{69}$$

$$\omega^1{}_2\wedge\omega^1{}_3=[\tfrac{1}{2}e^{\psi-\mu_2-\mu_3-\mu_4}\Psi_2Q_{34}-\tfrac{1}{2}e^{\psi-\mu_2-\mu_3-\mu_4}\Psi_3Q_{24}]\omega^1\wedge\omega^4$$
$$+[\tfrac{1}{2}e^{\psi-2\mu_2-\mu_3}\Psi_2Q_{32}+e^{-\psi-\mu_3}\Psi_3\mu_{2,1}]\omega^1\wedge\omega^2$$
$$+[-\tfrac{1}{2}e^{\psi-\mu_2-2\mu_3}\Psi_3Q_{23}-e^{-\psi-\mu_2}\Psi_2\mu_{3,1}]\omega^1\wedge\omega^3$$
$$+[\tfrac{1}{4}e^{2\psi-2\mu_2-2\mu_3}Q_{23}^2+e^{-2\psi}\mu_{2,1}\mu_{3,1}]\omega^2\wedge\omega^3$$
$$+[\tfrac{1}{4}e^{2\psi-2\mu_2-\mu_3-\mu_4}Q_{24}Q_{23}-\tfrac{1}{2}e^{-\mu_3-\mu_4}Q_{34}\mu_{2,1}]\omega^2\wedge\omega^4$$
$$+[\tfrac{1}{4}e^{2\psi-2\mu_3-\mu_2-\mu_4}Q_{23}Q_{34}+\tfrac{1}{2}e^{-\mu_2-\mu_4}Q_{24}\mu_{3,1}]\omega^3\wedge\omega^4, \tag{70}$$

$$\omega^3{}_4\wedge\omega^4{}_2=[-\tfrac{1}{2}e^{\psi-\mu_2-\mu_3-\mu_4}Q_{34}\mu_{4:2}-\tfrac{1}{2}e^{\psi-\mu_2-\mu_3-\mu_4}Q_{42}\mu_{4:3}]\omega^1\wedge\omega^4$$
$$+\tfrac{1}{2}e^{\psi-\mu_3-2\mu_4}Q_{34}\mu_{2:4}\omega^1\wedge\omega^2+\tfrac{1}{2}e^{\psi-2\mu_4-\mu_2}Q_{42}\mu_{3:4}\omega^1\wedge\omega^3$$
$$+e^{-2\mu_4}\mu_{3:4}\mu_{2:4}\omega^2\wedge\omega^3-e^{-\mu_3-\mu_4}\mu_{4:3}\mu_{2:4}\omega^2\wedge\omega^4$$
$$+e^{-\mu_2-\mu_4}\mu_{3:4}\mu_{4:2}\omega^3\wedge\omega^4. \tag{71}$$

Now (cf. equations (55) and (61))

$$\tfrac{1}{2}R^1{}_{2kl}\omega^k\wedge\omega^l=\Omega^1{}_2=d\omega^1{}_2-\omega^1{}_3\wedge\omega^2{}_3-\omega^1{}_4\wedge\omega^2{}_4. \tag{72}$$

Therefore, combining equations (66), (67), and (68) in accordance with the foregoing expression for $\Omega^1{}_2$, we can read of the components of $R^1{}_{2kl}$ of the Riemann tensor by simply collecting the coefficients of $\omega^k\wedge\omega^l$. Thus, $R^1{}_{212}$, which is the coefficient of $\omega^1\wedge\omega^2$ in $\Omega^1{}_2$, is given by

$$R^1{}_{212}=-e^{-\psi-\mu_2}\mathscr{D}_2(e^{\psi-\mu_2}\Psi_2)-e^{-2\mu_3}\Psi_3\mu_{2:3}-e^{-\psi-\mu_2}(e^{\mu_2-\psi}\mu_{2,1})_{,1}$$
$$+\tfrac{1}{4}e^{2\psi-2\mu_2-2\mu_3}Q_{23}^2+\tfrac{1}{4}e^{2\psi-2\mu_2-2\mu_4}Q_{24}^2-e^{-2\mu_4}\Psi_4\mu_{2:4}; \tag{73}$$

and, lowering the index, we obtain equation (a) in the list of equations (75) given below. Equations (g), (j), and (s) in the list are, similarly, the coefficients of $\omega^1 \wedge \omega^3$, $\omega^2 \wedge \omega^3$, and $\omega^3 \wedge \omega^4$ in $\Omega^1{}_2$. In the same way, equations (d) and (p) are obtained by combining equations (69), (70), and (71) in accordance with the definition

$$\tfrac{1}{2}R^2{}_{3kl}\omega^k \wedge \omega^l = \Omega^2{}_3 = \mathrm{d}\omega^2{}_3 - \omega^1{}_2 \wedge \omega^1{}_3 - \omega^3{}_4 \wedge \omega^4{}_2. \tag{74}$$

The remaining components of the Riemann tensor are obtained by suitable permutations of the indices in equations (a), (d), (j), (p), and (s) by making use of the underlying symmetry in the indices (2), (3), and (4).

$$-R_{1212} = -e^{-\psi-\mu_2}\mathscr{D}_2(e^{\psi-\mu_2}\Psi_2) - e^{-2\mu_3}\Psi_3\mu_{2:3} - e^{-2\mu_4}\Psi_4\mu_{2:4}$$
$$-e^{-\psi-\mu_2}(e^{\mu_2-\psi}\mu_{2,1})_{,1} + \tfrac{1}{4}e^{2\psi-2\mu_2}[e^{-2\mu_3}Q_{23}^2 + e^{-2\mu_4}Q_{24}^2]; \tag{a}$$

$$-R_{1313} = -e^{-\psi-\mu_3}\mathscr{D}_3(e^{\psi-\mu_3}\Psi_3) - e^{-2\mu_2}\Psi_2\mu_{3:2} - e^{-2\mu_4}\Psi_4\mu_{3:4}$$
$$-e^{-\psi-\mu_3}(e^{\mu_3-\psi}\mu_{3,1})_{,1} + \tfrac{1}{4}e^{2\psi-2\mu_3}[e^{-2\mu_2}Q_{23}^2 + e^{-2\mu_4}Q_{34}^2]; \tag{b}$$

$$-R_{1414} = -e^{-\psi-\mu_4}\mathscr{D}_4(e^{\psi-\mu_4}\Psi_4) - e^{-2\mu_2}\Psi_2\mu_{4:2} - e^{-2\mu_3}\Psi_3\mu_{4:3}$$
$$-e^{-\psi-\mu_4}(e^{\mu_4-\psi}\mu_{4,1})_{,1} + \tfrac{1}{4}e^{2\psi-2\mu_4}[e^{-2\mu_2}Q_{24}^2 + e^{-2\mu_3}Q_{34}^2]; \tag{c}$$

$$-R_{2323} = -e^{-\mu_2-\mu_3}[(e^{\mu_2-\mu_3}\mu_{2:3})_{:3} + (e^{\mu_3-\mu_2}\mu_{3:2})_{:2}] - e^{-2\mu_4}\mu_{3:4}\mu_{2:4}$$
$$-\tfrac{3}{4}e^{2\psi-2\mu_2-2\mu_3}Q_{23}^2 - e^{-2\psi}\mu_{2,1}\mu_{3,1}; \tag{d}$$

$$-R_{2424} = -e^{-\mu_2-\mu_4}[(e^{\mu_2-\mu_4}\mu_{2:4})_{:4} + (e^{\mu_4-\mu_2}\mu_{4:2})_{:2}] - e^{-2\mu_3}\mu_{4:3}\mu_{2:3}$$
$$-\tfrac{3}{4}e^{2\psi-2\mu_2-2\mu_4}Q_{24}^2 - e^{-2\psi}\mu_{2,1}\mu_{4,1}; \tag{e}$$

$$-R_{3434} = -e^{-\mu_3-\mu_4}[(e^{\mu_3-\mu_4}\mu_{3:4})_{:4} + (e^{\mu_4-\mu_3}\mu_{4:3})_{:3}] - e^{-2\mu_2}\mu_{4:2}\mu_{3:2}$$
$$-\tfrac{3}{4}e^{2\psi-2\mu_3-2\mu_4}Q_{34}^2 - e^{-2\psi}\mu_{3,1}\mu_{4,1}; \tag{f}$$

$$-R_{1213} = -e^{-\psi-\mu_3}\mathscr{D}_3(e^{\psi-\mu_2}\Psi_2) + e^{-\mu_2-\mu_3}\Psi_3\mu_{3:2}$$
$$+\tfrac{1}{4}e^{2\psi-2\mu_4-\mu_2-\mu_3}Q_{43}Q_{42}$$
$$+\tfrac{1}{2}e^{-\mu_2-\mu_3}[Q_{23,1} + Q_{23}(\mu_3-\mu_2+\psi)_{,1}]; \tag{g}$$

$$-R_{1214} = -e^{-\psi-\mu_4}\mathscr{D}_4(e^{\psi-\mu_2}\Psi_2) + e^{-\mu_2-\mu_4}\Psi_4\mu_{4:2}$$
$$+\tfrac{1}{4}e^{2\psi-2\mu_3-\mu_2-\mu_4}Q_{34}Q_{32}$$
$$+\tfrac{1}{2}e^{-\mu_2-\mu_4}[Q_{24,1} + Q_{24}(\mu_4-\mu_2+\psi)_{,1}]; \tag{h}$$

$$-R_{1314} = -e^{-\psi-\mu_4}\mathscr{D}_4(e^{\psi-\mu_3}\Psi_3) + e^{-\mu_3-\mu_4}\Psi_4\mu_{4:3}$$
$$+\tfrac{1}{4}e^{2\psi-2\mu_2-\mu_3-\mu_4}Q_{24}Q_{23}$$
$$+\tfrac{1}{2}e^{-\mu_3-\mu_4}[Q_{34,1} + Q_{34}(\mu_4-\mu_3+\psi)_{,1}]; \tag{i}$$

$$-R_{1223} = e^{\psi-2\mu_2-\mu_3}Q_{23}(\Psi_2 - \tfrac{1}{2}\mu_{3:2}) + \tfrac{1}{2}e^{-\mu_2-\mu_3}(e^{\psi-\mu_2}Q_{23})_{:2}$$
$$+\tfrac{1}{2}e^{\psi-2\mu_4-\mu_3}Q_{43}\mu_{2:4}$$

$$+e^{-\mu_2-\mu_3}(e^{-\psi+\mu_2}\mu_{2,1})_{:3}-e^{-\psi-\mu_3}\mu_{3,1}\mu_{2:3}; \qquad \text{(j)}$$

$$-R_{1224}=e^{\psi-2\mu_2-\mu_4}Q_{24}(\Psi_2-\tfrac{1}{2}\mu_{4:2})+\tfrac{1}{2}e^{-\mu_2-\mu_4}(e^{\psi-\mu_2}Q_{24})_{:2}$$
$$+\tfrac{1}{2}e^{\psi-2\mu_3-\mu_4}Q_{34}\mu_{2:3}$$
$$+e^{-\mu_2-\mu_4}(e^{-\psi+\mu_2}\mu_{2,1})_{:4}-e^{-\psi-\mu_4}\mu_{4,1}\mu_{2:4}; \qquad \text{(k)}$$

$$-R_{1334}=e^{\psi-2\mu_3-\mu_4}Q_{34}(\Psi_3-\tfrac{1}{2}\mu_{4:3})+\tfrac{1}{2}e^{-\mu_3-\mu_4}(e^{\psi-\mu_3}Q_{34})_{:3}$$
$$+\tfrac{1}{2}e^{\psi-2\mu_2-\mu_4}Q_{24}\mu_{3:2}$$
$$+e^{-\mu_3-\mu_4}(e^{-\psi+\mu_3}\mu_{3,1})_{:4}-e^{-\psi-\mu_4}\mu_{4,1}\mu_{3:4}; \qquad \text{(l)}$$

$$-R_{1332}=e^{\psi-2\mu_3-\mu_2}Q_{32}(\Psi_3-\tfrac{1}{2}\mu_{2:3})+\tfrac{1}{2}e^{-\mu_3-\mu_2}(e^{\psi-\mu_3}Q_{32})_{:3}$$
$$+\tfrac{1}{2}e^{\psi-2\mu_4-\mu_2}Q_{42}\mu_{3:4}$$
$$+e^{-\mu_3-\mu_2}(e^{-\psi+\mu_3}\mu_{3,1})_{:2}-e^{-\psi-\mu_2}\mu_{2,1}\mu_{3:2}; \qquad \text{(m)}$$

$$-R_{1442}=e^{\psi-2\mu_4-\mu_2}Q_{42}(\Psi_4-\tfrac{1}{2}\mu_{2:4})+\tfrac{1}{2}e^{-\mu_4-\mu_2}(e^{\psi-\mu_4}Q_{42})_{:4}$$
$$+\tfrac{1}{2}e^{\psi-2\mu_3-\mu_2}Q_{32}\mu_{4:3}$$
$$+e^{-\mu_4-\mu_2}(e^{-\psi+\mu_4}\mu_{4,1})_{:2}-e^{-\psi-\mu_2}\mu_{2,1}\mu_{4:2}; \qquad \text{(n)}$$

$$-R_{1443}=e^{\psi-2\mu_4-\mu_3}Q_{43}(\Psi_4-\tfrac{1}{2}\mu_{3:4})+\tfrac{1}{2}e^{-\mu_4-\mu_3}(e^{\psi-\mu_4}Q_{43})_{:4}$$
$$+\tfrac{1}{2}e^{\psi-2\mu_2-\mu_3}Q_{23}\mu_{4:2}$$
$$+e^{-\mu_4-\mu_3}(e^{-\psi+\mu_4}\mu_{4,1})_{:3}-e^{-\psi-\mu_3}\mu_{3,1}\mu_{4:3}; \qquad \text{(o)}$$

$$-R_{2334}=e^{-\mu_2-\mu_4}[\mu_{3:2:4}+\mu_{3:2}(\mu_3-\mu_2)_{:4}-\mu_{3:4}\mu_{4:2}]$$
$$-\tfrac{3}{4}e^{2\psi-\mu_2-2\mu_3-\mu_4}Q_{23}Q_{34}-\tfrac{1}{2}e^{-\mu_2-\mu_4}Q_{24}\mu_{3,1}; \qquad \text{(p)}$$

$$-R_{3224}=e^{-\mu_3-\mu_4}[\mu_{2:3:4}+\mu_{2:3}(\mu_2-\mu_3)_{:4}-\mu_{2:4}\mu_{4:3}]$$
$$-\tfrac{3}{4}e^{2\psi-\mu_3-2\mu_2-\mu_4}Q_{32}Q_{24}-\tfrac{1}{2}e^{-\mu_3-\mu_4}Q_{34}\mu_{2,1}; \qquad \text{(q)}$$

$$-R_{3442}=e^{-\mu_3-\mu_2}[\mu_{4:3:2}+\mu_{4:3}(\mu_4-\mu_3)_{:2}-\mu_{4:2}\mu_{2:3}]$$
$$-\tfrac{3}{4}e^{2\psi-\mu_3-2\mu_4-\mu_2}Q_{34}Q_{42}-\tfrac{1}{2}e^{-\mu_3-\mu_2}Q_{32}\mu_{4,1}; \qquad \text{(r)}$$

$$-R_{1234}=\tfrac{1}{2}e^{-\mu_3-\mu_4}[(e^{\psi-\mu_2}Q_{24})_{:3}-(e^{\psi-\mu_2}Q_{23})_{:4}]$$
$$+\tfrac{1}{2}e^{\psi-\mu_2-\mu_3-\mu_4}Q_{34}(2\Psi_2-\mu_{3:2}-\mu_{4:2}); \qquad \text{(s)}$$

$$-R_{1423}=\tfrac{1}{2}e^{-\mu_2-\mu_3}[(e^{\psi-\mu_4}Q_{43})_{:2}-(e^{\psi-\mu_4}Q_{42})_{:3}]$$
$$+\tfrac{1}{2}e^{\psi-\mu_2-\mu_3-\mu_4}Q_{23}(2\Psi_4-\mu_{2:4}-\mu_{3:4}); \qquad \text{(t)}$$

$$-R_{1342}=\tfrac{1}{2}e^{-\mu_4-\mu_2}[(e^{\psi-\mu_3}Q_{32})_{:4}-(e^{\psi-\mu_3}Q_{34})_{:2}]$$
$$+\tfrac{1}{2}e^{\psi-\mu_2-\mu_3-\mu_4}Q_{42}(2\Psi_3-\mu_{4:3}-\mu_{2:3}). \qquad \text{(u)}$$

(75)

It should be noted that we can obtain formally different expressions for the same component of the Riemann tensor (modulo its symmetries) by differently ordering the permutations of the indices 2, 3, and 4. For example, by combining equations (68), (70), and (71), as required by the definition of ${\Omega^2}_3$, we obtain

$$
\begin{aligned}
-R_{2313} = {} & \tfrac{1}{2}e^{-\psi-\mu_3}\mathscr{D}_3(e^{2\psi-\mu_2-\mu_3}Q_{23}) - e^{-\psi-\mu_3}(e^{\mu_3-\mu_2}\mu_{3:2})_{,1} \\
& +\tfrac{1}{2}e^{\psi-\mu_2-2\mu_3}\Psi_3 Q_{23} + e^{-\psi-\mu_2}\Psi_2\mu_{3,1} - \tfrac{1}{2}e^{\psi-2\mu_4-\mu_2}Q_{42}\mu_{3:4},
\end{aligned} \tag{76}
$$

as the coefficient of $\omega^1 \wedge \omega^3$. It is not manifest that this expression for R_{2313} is, apart from sign, the same as the expression for R_{1332} given in equation (75, m). Nevertheless, by expanding the two expressions, we can verify that, in fact, they agree as required if proper use is made of our definitions of the colon derivative and of the operator $\mathscr{D}_A$. The establishment of other similar equalities is not always easy: they often depend on identities which are not obvious, as the following:

$$
Q_{[23:4]} - Q_{[23}q_{4],1} = 0. \tag{77}
$$

Turning next to the components of the Ricci and the Einstein tensors, we can obtain them by combining appropriately the components of the Riemann tensor. Thus,

$$
\left.
\begin{aligned}
-R_{11} &= R_{1212}+R_{1313}+R_{1414}, \text{ etc.,} \\
+R_{12} &= +G_{12} = R_{1332}+R_{1442}, \text{ etc.,} \\
+\tfrac{1}{2}R &= R_{1212}+R_{1313}+R_{1414}+R_{2323}+R_{2424}+R_{3434}, \\
\text{and} \qquad G_{11} &= R_{2323}+R_{2424}+R_{3434}, \text{ etc.}
\end{aligned}
\right\} \tag{78}
$$

We shall not write out these expressions explicitly: we shall have occasions to consider various special cases in later chapters.

Finally, we may note that we can revert from the (complex) coordinate x^4 to the (real) space-time coordinate $x^0 (= t)$ by the correspondence

$$
-i^4 \to {}^0,; \quad {}_{,4} \to -i_{,0,} \quad \mu_4 \to \nu; \qquad \text{and} \qquad q_4 \to -i\omega. \tag{79}
$$

14. The tetrad frame and the rotation coefficients

In some contexts, as in obtaining the explicit forms of Maxwell's equations in §15, it is convenient to apply the tetrad formalism to the space-time with the metric chosen in §12. For such applications, a knowledge of the Ricci rotation-coefficients in a suitable tetrad-frame is, of course, essential. The tetrad frame we shall select is the real version, with a signature $(+, -, -, -)$, of the complex frame with the signature $(-, -, -, -)$, used in §13. Explicitly, the covariant vectors, which provide the basis for the orthonormal frame selected, are (cf. equations (18) and (19))

$$\begin{aligned} e_{(0)i} &= (e^{\nu}, \quad 0, \quad 0, \quad 0), \\ e_{(1)i} &= (\omega e^{\psi}, \quad -e^{\psi}, \quad q_2 e^{\psi}, \quad q_3 e^{\psi}), \\ e_{(2)i} &= (0, \quad 0, \quad -e^{\mu_2}, \quad 0), \\ e_{(3)i} &= (0, \quad 0, \quad 0, \quad -e^{\mu_3}). \end{aligned} \tag{80}$$

The corresponding contravariant vectors are

$$\begin{aligned} e_{(0)}{}^{i} &= (e^{-\nu}, \quad \omega e^{-\nu}, \quad 0, \quad 0), \\ e_{(1)}{}^{i} &= (0, \quad e^{-\psi}, \quad 0, \quad 0), \\ e_{(2)}{}^{i} &= (0, \quad q_2 e^{-\mu_2}, \quad e^{-\mu_2}, \quad 0), \\ e_{(3)}{}^{i} &= (0, \quad q_3 e^{-\mu_3}, \quad 0, \quad e^{-\mu_3}). \end{aligned} \tag{81}$$

As explained in Chapter 1 §8(*b*), the most convenient method of evaluating the rotation coefficients is via the symbols (Ch. 1, equation (266))

$$\lambda_{(a)(b)(c)} = e_{(a)}{}^{i}[e_{(b)i,j} - e_{(b)j,i}]e_{(c)}{}^{j}. \tag{82}$$

The evaluation of these symbols is direct and straightforward. The results can be expressed in compact forms in the notation of 'colon derivatives' introduced in §13. However, since we do not wish to introduce the imaginary coordinate, $x^4 = ix^0$, we shall now restrict the capital Latin letters (A, B, etc.) to the indices 0, 2, and 3. Also we shall write

$$q_0 = \omega. \tag{83}$$

DEFINITIONS. (1) *The colon derivative of a function* $f(x^0, x^1, x^2, x^3)$ *with respect to* $x^A (A = 0, 2, 3)$ *is given by*

$$f_{:A} = f_{,A} + q_A f_{,1}. \tag{84}$$

(2) *The operator* $\mathscr{D}_A$ *acting on* $f(x^0, x^1, x^2, x^3)$ *gives*

$$\mathscr{D}_A f = f_{:A} + q_{A,1} f = f_{,A} + (q_A f)_{,1}, \tag{85}$$

(3)
$$Q_{AB} = q_{A:B} - q_{B:A}, \tag{86}$$

and

(4)
$$\Psi_A = \psi_{:A} + q_{A,1} = e^{-\psi}\,\mathscr{D}_A(e^{\psi}). \tag{87}$$

It is of interest to notice in this connection that the colon derivatives are simply related to the directional derivatives along the tangent vectors $e_{(0)}$. Thus,

$$e_{(0)} f = e^{-\nu} f_{:0}, \quad e_{(2)} f = e^{-\mu_2} f_{:2}, \quad e_{(3)} f = e^{-\mu_3} f_{:3}$$

and

$$e_{(1)} f = e^{-\psi} f_{,1}. \tag{88}$$

With the foregoing definitions we find that the λ-symbols have the values listed below:

$$\begin{aligned}
&\lambda_{100} = -\nu_{,1}e^{-\psi}, && \lambda_{110} = -\Psi_0 e^{-\nu}, && \lambda_{223} = -\mu_{2:3}e^{-\mu_3},\\
&\lambda_{200} = -\nu_{:2}e^{-\mu_2}, && \lambda_{112} = -\Psi_2 e^{-\mu_2}, && \lambda_{233} = +\mu_{3:2}e^{-\mu_2},\\
&\lambda_{300} = -\nu_{:3}e^{-\mu_3}, && \lambda_{113} = -\Psi_3 e^{-\mu_3}, && \lambda_{210} = Q_{20}e^{\psi-\mu_2-\nu},\\
&\lambda_{122} = \mu_{2,1}e^{-\psi}, && \lambda_{220} = -\mu_{2:0}e^{-\nu}, && \lambda_{310} = Q_{30}e^{\psi-\mu_3-\nu},\\
&\lambda_{133} = \mu_{3,1}e^{-\psi}, && \lambda_{330} = -\mu_{3:0}e^{-\nu}, && \lambda_{213} = Q_{23}e^{\psi-\mu_2-\mu_3}.
\end{aligned} \tag{89}$$

The rotation coefficients $\gamma_{(a)(b)(c)}$ are now determined in terms of $\lambda_{(a)(b)(c)}$ by the formula (cf. Ch. 1, equation (268))

$$\gamma_{(a)(b)(c)} = \tfrac{1}{2}[\lambda_{(a)(b)(c)} + \lambda_{(c)(a)(b)} - \lambda_{(b)(c)(a)}]. \tag{90}$$

We find:

$$\begin{aligned}
&\gamma_{100} = -\nu_{,1}e^{-\psi}, && \gamma_{101} = -\Psi_0 e^{-\nu},\\
&\gamma_{200} = -\nu_{:2}e^{-\mu_2}, && \gamma_{121} = -\Psi_2 e^{-\mu_2},\\
&\gamma_{300} = -\nu_{:3}e^{-\mu_3}, && \gamma_{131} = -\Psi_3 e^{-\mu_3},\\
&\gamma_{122} = \mu_{2,1}e^{-\psi}, && \gamma_{202} = -\mu_{2:0}e^{-\nu},\\
&\gamma_{133} = \mu_{3,1}e^{-\psi}, && \gamma_{303} = -\mu_{3:0}e^{-\nu},\\
&\gamma_{233} = \mu_{3:2}e^{-\mu_2}, && \gamma_{232} = -\mu_{2:3}e^{-\mu_3},\\
&\gamma_{102} = \gamma_{201} = -\gamma_{120} = \tfrac{1}{2}Q_{20}e^{\psi-\mu_2-\nu},\\
&\gamma_{103} = \gamma_{301} = -\gamma_{130} = \tfrac{1}{2}Q_{30}e^{\psi-\mu_3-\nu},\\
&\gamma_{132} = \gamma_{231} = -\gamma_{123} = \tfrac{1}{2}Q_{23}e^{\psi-\mu_2-\mu_3}.
\end{aligned} \tag{91}$$

15. Maxwell's equations

We conclude this chapter by deriving Maxwell's equations in the space-time with the metric we have chosen. The derivation is most conveniently carried out in the tetrad frame of §14.

Maxwell's equations,

$$\eta^{(n)(m)}F_{(a)(n)|(m)} = 0 \quad \text{and} \quad F_{[(a)(b)|(c)]} = 0, \tag{92}$$

in the orthonormal frame of §14, become

$$\begin{aligned}
e^{-\nu}F_{a0:0} - e^{-\psi}F_{a1,1} - e^{-\mu_2}F_{a2:2} - e^{-\mu_3}F_{a3:3}&\\
= -F_{01}(\gamma_{0a1}-\gamma_{1a0}) - F_{02}(\gamma_{0a2}-\gamma_{2a0}) - F_{03}(\gamma_{0a3}-\gamma_{3a0})&\\
+F_{12}(\gamma_{1a2}-\gamma_{2a1}) + F_{23}(\gamma_{2a3}-\gamma_{3a2}) + F_{31}(\gamma_{3a1}-\gamma_{1a3})&\\
+F_{a1}(-\gamma_{100}+\gamma_{122}+\gamma_{133}) + F_{a2}(-\gamma_{200}+\gamma_{211}+\gamma_{233})&\\
+F_{a3}(-\gamma_{300}+\gamma_{311}+\gamma_{322}) + F_{a0}(\gamma_{101}+\gamma_{202}+\gamma_{303})&
\end{aligned} \tag{93}$$

and

$$F_{[a,b,c]}+\sum_{[a,b,c]}(-\gamma_{0ac}F_{0b}+\gamma_{1ac}F_{1b}+\gamma_{2ac}F_{2b}+\gamma_{3ac}F_{3b}$$
$$-\gamma_{0bc}F_{a0}+\gamma_{1bc}F_{a1}+\gamma_{2bc}F_{a2}+\gamma_{3bc}F_{a3})=0. \tag{94}$$

The particular form of the left-hand side of equation (93) derives from the fact that the directional derivatives, along the contravariant basis-vectors (81), are simply related to the colon derivatives (as we have noted in equations (88)).

Now, substituting in equations (93) and (94) the values for the rotation coefficients given in equations (91), we find, after some lengthy but straightforward reductions, the following set of eight equations:

$$\mathscr{D}_3(e^{\psi+\mu_2}F_{12})-\mathscr{D}_2(e^{\psi+\mu_3}F_{13})+(e^{\mu_2+\mu_3}F_{23})_{,1}=0, \qquad \text{(a)}$$
$$\mathscr{D}_2(e^{\psi+\nu}F_{01})+\mathscr{D}_0(e^{\psi+\mu_2}F_{12})-(e^{\nu+\mu_2}F_{02})_{,1}=0, \qquad \text{(b)}$$
$$\mathscr{D}_3(e^{\psi+\nu}F_{01})+\mathscr{D}_0(e^{\psi+\mu_3}F_{13})-(e^{\nu+\mu_3}F_{03})_{,1}=0, \qquad \text{(c)}$$
$$(e^{\mu_2+\mu_3}F_{01})_{:0}+(e^{\nu+\mu_3}F_{12})_{:2}+(e^{\nu+\mu_2}F_{13})_{:3}$$
$$=e^{\psi+\mu_3}F_{02}Q_{02}+e^{\psi+\mu_2}F_{03}Q_{03}-e^{\psi+\nu}F_{23}Q_{23}, \qquad \text{(d)}$$
$$\mathscr{D}_2(e^{\psi+\mu_3}F_{02})+\mathscr{D}_3(e^{\psi+\mu_2}F_{03})+(e^{\mu_2+\mu_3}F_{01})_{,1}=0, \qquad \text{(e)}$$
$$-\mathscr{D}_2(e^{\psi+\nu}F_{23})+\mathscr{D}_0(e^{\psi+\mu_2}F_{03})-(e^{\mu_2+\nu}F_{13})_{,1}=0, \qquad \text{(f)}$$
$$+\mathscr{D}_3(e^{\psi+\nu}F_{23})+\mathscr{D}_0(e^{\psi+\mu_3}F_{02})-(e^{\mu_3+\nu}F_{12})_{,1}=0, \qquad \text{(g)}$$
$$(e^{\nu+\mu_2}F_{02})_{:3}-(e^{\nu+\mu_3}F_{03})_{:2}+(e^{\mu_2+\mu_3}F_{23})_{:0}$$
$$=e^{\psi+\nu}F_{01}Q_{23}+e^{\psi+\mu_2}F_{12}Q_{03}-e^{\psi+\mu_3}F_{13}Q_{02}. \qquad \text{(h)} \tag{95}$$

The foregoing eight equations are not all linearly independent; and that they are not depends on the following commutation rules (which are readily verified):

$$\left.\begin{aligned}&\mathscr{D}_A(f_{,1})=f_{:A,1}\\ \text{and}\qquad &(\mathscr{D}_A\mathscr{D}_B-\mathscr{D}_B\mathscr{D}_A)f=-(Q_{AB}f)_{,1}.\end{aligned}\right\} \tag{96}$$

Thus, by applying the operators $\mathscr{D}_0$, $-\mathscr{D}_3$, and $\mathscr{D}_2$ to equations (a), (b), and (c), respectively, adding and simplifying with the aid of the commutation rules, we obtain

$$[-Q_{03}e^{\psi+\mu_2}F_{12}+Q_{02}e^{\psi+\mu_3}F_{13}-Q_{23}e^{\psi+\nu}F_{01}$$
$$+(e^{\mu_2+\mu_3}F_{23})_{:0}+(e^{\nu+\mu_2}F_{02})_{:3}-(e^{\nu+\mu_3}F_{03})_{:2}]_{,1}=0. \tag{97}$$

But this equation is identically satisfied by virtue of equation (h). Similarly, by applying the operators $\mathscr{D}_0$, $-\mathscr{D}_3$, and $-\mathscr{D}_2$ to equations (e), (f), and (g), respectively, adding and simplifying with the aid of the commutation rules (96), we find that the resulting equation is identically satisfied by virtue of equation (d).

We have now the means to write down the explicit forms of the Einstein and the Einstein-Maxwell equations for the non-stationary non-axisymmetric space-time with the metric (38).

BIBLIOGRAPHICAL NOTES

§11. Bardeen appears to have been the first to recognize the special appropriateness of the form of the metric chosen in this section to describe stationary axisymmetric space-times:

1. J. M. BARDEEN, *Astrophys. J.*, 161, 103–9 (1970).

See also:

2. J. M. BARDEEN, W. H. PRESS, and S. A. TEUKOLSKY, ibid., 178, 347–69 (1972).

§12. In spite of its elementary character, the theorem (which we have designated as the Cotton–Darboux theorem) that in a 3-space one can always set up locally, i.e., in finite open neighbourhoods, a triply orthogonal system of curvilinear coordinates, seems to have escaped mention in any of the standard sources on differential geometry. Indeed, the reference to

3. E. COTTON, *Ann. Fac. Sci. Toulouse, Ser.* 2, 1, 385–438 (1899) (see particularly p. 410)

was traced by Professor A. Trautman to an 'example' in:

4. A. Z. PETROV, *Einstein Spaces*, translated by R. E. Kelleher and J. Woodrow, Pergamon Press, Oxford, 1969. See Example 5 on p. 44.

We have adjoined Darboux's name with that of Cotton, in designating the theorem, since Cotton's proof is modelled on one given by Darboux for Euclidean spaces:

5. G. DARBOUX, *Lecons sur les Systèmes Orthogonaux et les Coordinnèes Curvilignes*, Gauthier–Villars, Paris, 1898. See pp. 1–2.

I am grateful to Professor Trautman for his interest and advice in these and related questions.

§13. The tetrad components of the Riemann tensor for a metric of the same form but under conditions of axisymmetry are given in:

6. S. CHANDRASEKHAR and J. L. FRIEDMAN, *Astrophys. J.*, 175, 379–405 (1972).

§§14–15. The analysis in these sections generalizes the treatment in:

7. S. CHANDRASEKHAR and B. C. XANTHOPOULOS, *Proc. Roy. Soc.* (*London*) A, 367, 1–14 (1979).

3

THE SCHWARZSCHILD SPACE-TIME

16. Introduction

A full understanding of the Schwarzschild space-time, as consisting of an event horizon and an essential singularity at the centre, was achieved only comparatively recently. In this chapter, we shall bypass the common historical route and provide, instead, a derivation of the Schwarzschild metric (in essence, due to Synge) which addresses itself directly to the essential features of the space-time. Also, since the geodesics in the Schwarzschild space-time illuminate some basic aspects of space-times with event horizons, we shall include an account of them – perhaps more complete than is strictly necessary. We conclude the chapter with a description of the Schwarzschild space-time in a Newman–Penrose formalism and a proof of its type-D character.

17. The Schwarzschild metric

The Schwarzschild metric is a spherically symmetric solution of Einstein's equation for the vacuum. Following Synge, we shall define a spherically symmetric space-time as a manifold which is the Cartesian product, $S_2 \times U_2$, of a unit two-dimensional sphere S_2 and a two-dimensional manifold U_2 with an indefinite metric. On S_2, we take the usual polar coordinates, (θ, φ), with the metric

$$d\Omega^2 = (d\theta)^2 + (d\varphi)^2 \sin^2\theta. \tag{1}$$

On U_2, since it is characterized by an indefinite metric, null lines

$$u = \text{constant} \qquad \text{and} \qquad v = \text{constant} \tag{2}$$

must exist. We shall take them as a basis for a coordinate system on U_2. With this choice of coordinates on S_2 and U_2, we can write the most general metric for a spherically symmetric space-time in the form

$$ds^2 = 4f\,du\,dv - e^{2\mu_3}[(d\theta)^2 + (d\varphi)^2 \sin^2\theta], \tag{3}$$

where f and μ_3 are functions of u and v.

The metric (3) can be brought to a form in which it becomes a special case of the general metric considered in Chapter 2 by the substitutions

$$u = \tfrac{1}{2}(x^0 + x^2) \qquad \text{and} \qquad v = \tfrac{1}{2}(x^0 - x^2); \tag{4}$$

for, the metric, then, takes the form

$$ds^2 = f[(dx^0)^2 - (dx^2)^2] - e^{2\mu_3}[(d\theta)^2 + (d\varphi)^2 \sin^2\theta]. \tag{5}$$

In order that we may transcribe the formulae of Chapter 2 directly, it is convenient to write

$$f = e^{2\mu_2},\ e^{2\psi} = e^{2\mu_3}\sin^2\theta,\ dx^1 = d\varphi,\ dx^3 = d\theta, \quad \text{and} \quad dx^4 = i\,dx^0, \tag{6}$$

when the metric becomes

$$-ds^2 = e^{2\mu_2}(dx^4)^2 + e^{2\psi}(dx^1)^2 + e^{2\mu_2}(dx^2)^2 + e^{2\mu_3}(dx^3)^2. \tag{7}$$

The metric is now of the form considered in Chapter 2, §13 with

$$q_2 = q_3 = q_4 = 0,\ \mu_4 = \mu_2, \quad \text{and} \quad e^{\psi} = e^{\mu_3}\sin\theta. \tag{8}$$

Accordingly, for the present choice of the metric,

$$\Psi_4 = \mu_{3,4},\ \Psi_2 = \mu_{3,2}, \quad \text{and} \quad \Psi_3 = \cot\theta, \tag{9}$$

and μ_2 and μ_3 are functions of x^2 and x^4 only.

We can now run down the list of the components of the Riemann tensor given in Chapter 2 (equations (75)) for the special case we are presently considering. We find that the only non-vanishing components of the Riemann tensor are:

$$\begin{aligned}
-R_{1212} = -R_{2323} &= -e^{-\mu_3-\mu_2}(e^{\mu_3-\mu_2}\mu_{3,2})_{,2} - e^{-2\mu_2}\mu_{3,4}\mu_{2,4} \\
&= -e^{-2\mu_2}[\mu_{3,2,2} + \mu_{3,2}(\mu_3-\mu_2)_{,2} + \mu_{3,4}\mu_{2,4}], && \text{(a)} \\
-R_{1414} = -R_{3434} &= -e^{-\mu_3-\mu_2}(e^{\mu_3-\mu_2}\mu_{3,4})_{,4} - e^{-2\mu_2}\mu_{3,2}\mu_{2,2} \\
&= -e^{-2\mu_2}[\mu_{3,4,4} + \mu_{3,4}(\mu_3-\mu_2)_{,4} + \mu_{3,2}\mu_{2,2}], && \text{(b)} \\
-R_{1214} = +R_{2334} &= -e^{-\mu_3-\mu_2}(e^{\mu_3-\mu_2}\mu_{3,2})_{,4} + e^{-2\mu_2}\mu_{3,4}\mu_{2,2} \\
&= -e^{-2\mu_2}[\mu_{3,2,4} + \mu_{3,2}(\mu_3-\mu_2)_{,4} - \mu_{3,4}\mu_{2,2}], && \text{(c)} \\
-R_{1313} &= e^{-2\mu_3} - e^{-2\mu_2}[(\mu_{3,2})^2 + (\mu_{3,4})^2], && \text{(d)} \\
-R_{2424} &= -e^{-2\mu_2}(\mu_{2,4,4} + \mu_{2,2,2}). && \text{(e)}
\end{aligned}$$

(10)

The field equations now require that the components of the Ricci tensor vanish. These conditions give (cf. Ch. 2, equations (78))

$$-R_{22} = R_{1212} + R_{2323} + R_{2424} = 2R_{1212} + R_{2424} = 0, \tag{11}$$

$$-R_{44} = R_{1414} + R_{2424} + R_{3434} = 2R_{1414} + R_{2424} = 0, \tag{12}$$

$$-R_{11} = R_{1212} + R_{1313} + R_{1414} = 0, \tag{13}$$

$$-R_{33} = R_{1313} + R_{2323} + R_{3434} = 0; \tag{14}$$

and the only non-diagonal component of the Ricci tensor, which does not vanish identically, is

$$R_{24} = R_{2114} + R_{2334} = 0. \tag{15}$$

Equations (10, b) and (11)–(15) are readily seen to require

$$\left.\begin{aligned} R_{1212} = R_{1414} = -\tfrac{1}{2}R_{2424} = -\tfrac{1}{2}R_{1313} \\ \text{and} \qquad R_{2334} = R_{1214} = 0. \end{aligned}\right\} \tag{16}$$

Reverting to the space-time variable x^0, we have

$$\left.\begin{aligned} R_{1212} = -R_{1010} = \tfrac{1}{2}R_{2020} = -\tfrac{1}{2}R_{1313} \\ \text{and} \qquad R_{1210} = 0, \end{aligned}\right\} \tag{17}$$

where

$$\begin{aligned}
R_{1212} &= e^{-2\mu_2}[\mu_{3,2,2} + \mu_{3,2}(\mu_3 - \mu_2)_{,2} - \mu_{3,0}\mu_{2,0}], && \text{(a)} \\
R_{1010} &= e^{-2\mu_2}[\mu_{3,0,0} + \mu_{3,0}(\mu_3 - \mu_2)_{,0} - \mu_{3,2}\mu_{2,2}], && \text{(b)} \\
R_{1210} &= e^{-2\mu_2}[\mu_{3,2,0} + \mu_{3,2}(\mu_3 - \mu_2)_{,0} - \mu_{3,0}\mu_{2,2}], && \text{(c)} \\
R_{1313} &= -e^{-2\mu_3} + e^{-2\mu_2}[(\mu_{3,2})^2 - (\mu_{3,0})^2], && \text{(d)} \\
R_{2020} &= e^{-2\mu_2}(\mu_{2,0,0} - \mu_{2,2,2}). && \text{(e)}
\end{aligned} \tag{18}$$

We consider equations (17) in the combinations

$$\begin{gathered} R_{1212} + R_{1010} = 0, \quad R_{1210} = 0, \\ R_{1212} + R_{1313} - R_{1010} = 0, \qquad \text{and} \qquad R_{0202} + R_{1313} = 0. \end{gathered} \tag{19}$$

Inserting from equations (18), for the components of the Riemann tensor, in the foregoing equations, we find that they can be reduced to the forms

$$[(e^{\mu_3})_{,2}e^{-\mu_2}]_{,2} + [(e^{\mu_3})_{,0}e^{-\mu_2})]_{,0} + (e^{\mu_3})_{,0}(e^{-\mu_2})_{,0} + (e^{\mu_3})_{,2}(e^{-\mu_2})_{,2} = 0, \tag{20}$$

$$(e^{\mu_3})_{,0,2} - (e^{\mu_3})_{,2}\mu_{2,0} - (e^{\mu_3})_{,0}\mu_{2,2} = 0, \tag{21}$$

$$e^{2\mu_2} + \{e^{\mu_3}(e^{\mu_3})_{,0,0} - e^{\mu_3}(e^{\mu_3})_{,2,2} + [(e^{\mu_3})_{,0}]^2 - [(e^{\mu_3})_{,2}]^2\} = 0, \tag{22}$$

and

$$e^{-2\mu_2}(\mu_{2,0,0} - \mu_{2,2,2}) - e^{-2\mu_3} + e^{-2\mu_2-2\mu_3}\{[(e^{\mu_3})_{,2}]^2 - [(e^{\mu_3})_{,0}]^2\} = 0. \tag{23}$$

Reverting to the variable

$$f = e^{2\mu_2} \quad \text{and letting} \quad Z = e^{\mu_3}, \tag{24}$$

we can rewrite equations (20)–(23) in the forms

$$f(Z_{,2,2} + Z_{,0,0}) - (Z_{,2}f_{,2} + Z_{,0}f_{,0}) = 0, \tag{25}$$

$$2fZ_{,0,2}-(Z_{,2}f_{,0}+Z_{,0}f_{,2})=0, \tag{26}$$

$$f+Z(Z_{,0,0}-Z_{,2,2})+(Z_{,0})^2-(Z_{,2})^2=0, \tag{27}$$

$$\frac{1}{2f}[(\lg f)_{,0,0}-(\lg f)_{,2,2}]-\frac{1}{Z^2}+\frac{1}{Z^2 f}[(Z_{,2})^2-(Z_{,0})^2]=0. \tag{28}$$

Equations (25) and (26) can be combined to give

$$f(Z_{,0,0}\pm 2Z_{,0,2}+Z_{,2,2})-(Z_{,2}\pm Z_{,0})(f_{,2}\pm f_{,0})=0; \tag{29}$$

while equation (28) can be simplified, with the aid of equation (27), to give

$$\tfrac{1}{2}[(\lg f)_{,0,0}-(\lg f)_{,2,2}]+\frac{1}{Z}(Z_{,0,0}-Z_{,2,2})=0. \tag{30}$$

Finally, reverting to the original variables u and v, we have the equations

$$fZ_{,u,u}-Z_{,u}f_{,u}=0, \tag{31}$$

$$fZ_{,v,v}-Z_{,v}f_{,v}=0, \tag{32}$$

$$f+ZZ_{,u,v}+Z_{,u}Z_{,v}=0, \tag{33}$$

and

$$\tfrac{1}{2}(\lg f)_{,u,v}+\frac{1}{Z}Z_{,u,v}=0. \tag{34}$$

(*a*) *The solution of the equations*

Rewriting equation (31) in the form

$$\frac{Z_{,u,u}}{Z_{,u}}=\frac{f_{,u}}{f}, \tag{35}$$

we conclude that

$$f=B(v)Z_{,u}, \tag{36}$$

where $B(v)$ is an arbitrary function of v. Similarly, from equation (32), we conclude that

$$f=A(u)Z_{,v}, \tag{37}$$

where $A(u)$ is an arbitrary function of u. Now writing equation (33) in the two alternative forms,

$$f+(ZZ_{,v})_{,u}=0 \qquad \text{and} \qquad f+(ZZ_{,u})_{,v}=0, \tag{38}$$

and substituting for f from equations (36) or (37), we obtain

$$[B(v)Z+ZZ_{,v}]_{,u}=0 \qquad \text{and} \qquad [A(u)Z+ZZ_{,u}]_{,v}=0. \tag{39}$$

Therefore,

$$Z_{,u}=-A(u)+\frac{F(u)}{Z} \qquad \text{and} \qquad Z_{,v}=-B(v)+\frac{G(v)}{Z}, \tag{40}$$

where $F(u)$ and $G(v)$ are further arbitrary functions of the arguments specified. By making use of equations (36), (37), and (40), we obtain

$$Z_{,u}Z_{,v} = -A(u)Z_{,v} + F(u)\frac{Z_{,v}}{Z} = -f + F(u)\frac{Z_{,v}}{Z}$$

$$= -B(v)Z_{,u} + G(v)\frac{Z_{,u}}{Z} = -f + G(v)\frac{Z_{,u}}{Z}. \tag{41}$$

Accordingly,

$$F(u)\frac{Z_{,v}}{Z} = G(v)\frac{Z_{,u}}{Z}. \tag{42}$$

Together with equations (36) and (37), equation (42) requires

$$\frac{Z_{,u}}{Z_{,v}} = \frac{F(u)}{G(v)} = \frac{A(u)}{B(v)}. \tag{43}$$

Therefore,

$$\frac{F(u)}{A(u)} = \frac{G(v)}{B(v)} = \text{a constant} = 2M \quad \text{(say)}. \tag{44}$$

Thus, we obtain the solution

$$Z_{,u} = -A(u)\left(1 - \frac{2M}{Z}\right) \quad \text{and} \quad Z_{,v} = -B(v)\left(1 - \frac{2M}{Z}\right), \tag{45}$$

and

$$f = -A(u)B(v)\left(1 - \frac{2M}{Z}\right). \tag{46}$$

It can now be verified that the solution of equations (31)–(33) expressed by equations (45) and (46) satisfies the remaining equation (34).

The meaning of the solution for Z given by equation (45) becomes clearer in the form

$$\mathrm{d}Z = -\left(1 - \frac{2M}{Z}\right)[A(u)\,\mathrm{d}u + B(v)\,\mathrm{d}v]. \tag{47}$$

According to this equation, Z is not a function of u and of v independently: it depends jointly on them in the sense that we can draw curves of constant Z in the (u, v)-plane. The latter statement is essentially the content of Birkhoff's theorem to which we shall return in §(c) below and also in §18.

Finally, we may note that the solution for the metric itself is given by

$$\mathrm{d}s^2 = -4\left(1 - \frac{2M}{Z}\right)A(u)B(v)\,\mathrm{d}u\,\mathrm{d}v - Z^2\,\mathrm{d}\Omega^2. \tag{48}$$

(b) The Kruskal frame

The occurrence of the two arbitrary functions, $A(u)$ and $B(v)$, in the solution (48) for the metric, is entirely consistent with the freedom we have in defining, as coordinates in U_2, u and v or some function of u and some function of v. The availability of this gauge freedom is in accord with the fact that the two arbitrary functions, $A(u)$ and $B(v)$, occur in the combinations $A(u)\,\mathrm{d}u$ and $B(v)\,\mathrm{d}v$ both in the metric (48) and in the solution (47) for Z. However, some care must be exercised in the choice of the functions of u and v which we wish to designate as coordinates in U_2 if we are not to introduce spurious coordinate-singularities in the metric (particularly at $Z = 2M$, as is manifest from the form of the metric). The occurrence of singularities might only signify that the coordinate functions chosen are valid in certain well-defined neighbourhoods of the manifold and not valid globally.

A choice of the coordinate functions that avoids spurious singularities is obtained b› letting

$$-A(u)\,\mathrm{d}u \to 2M\,\mathrm{d}\lg u \qquad \text{and} \qquad -B(v)\,\mathrm{d}v \to 2M\,\mathrm{d}\lg v. \tag{49}$$

Then

$$\mathrm{d}Z = 2M\left(1 - \frac{2M}{Z}\right)\mathrm{d}[\lg(uv)], \tag{50}$$

and

$$\mathrm{d}s^2 = -\frac{16M^2}{uv}\left(1 - \frac{2M}{Z}\right)\mathrm{d}u\,\mathrm{d}v - Z^2\,\mathrm{d}\Omega^2. \tag{51}$$

On integration, equation (50) gives

$$|uv| = C|Z - 2M|e^{Z/2M}, \tag{52}$$

where C is a constant of integration. With the choice

$$C = (2M)^{-1}, \tag{53}$$

and replacing Z by the customary radial-coordinate r, the metric (51) takes the *Kruskal form*

$$\mathrm{d}s^2 = -\frac{32M^3}{r}e^{-r/2M}\,\mathrm{d}u\,\mathrm{d}v - r^2\,\mathrm{d}\Omega^2, \tag{54}$$

where r is related to uv by

$$-uv = (1 - r/2M)e^{r/2M}. \tag{55}$$

We shall now examine the domain in the (u, v)-plane in which the metric is well-defined.

First, we observe that the (r, uv)-relation (55) is double-valued if we allow negative values for r. The two branches of the relation, for $r > 0$ and for $r < 0$, must, therefore, refer to entirely distinct physical situations. Since reversing the sign of r has the same effect as reversing the sign of M, it is clear that the branch

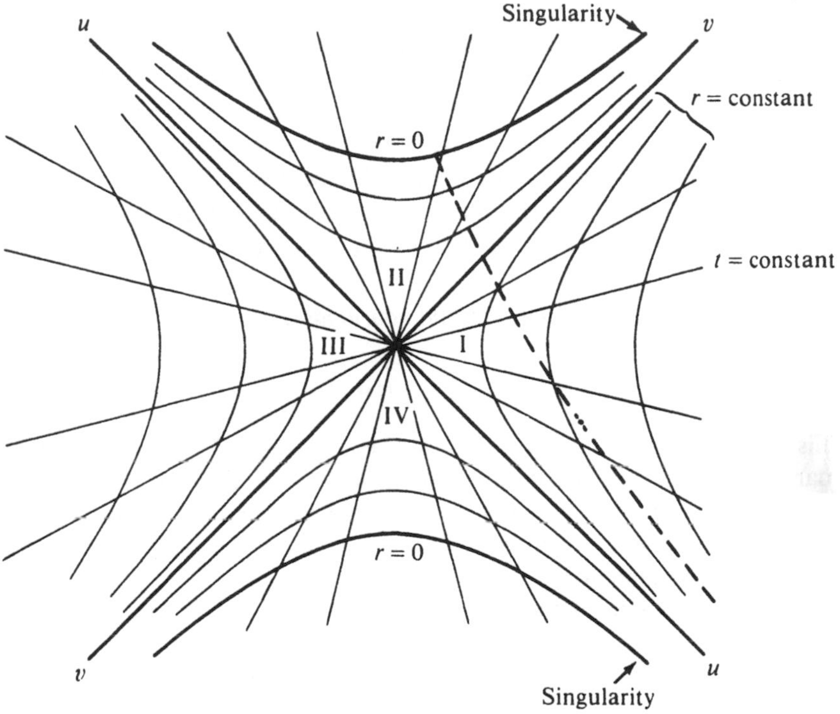

FIG. 2. The Schwarzschild geometry in the u- and v- coordinates (equivalent to the Kruskal coordinates). The light cones at any point are defined by the null geodesics, u = constant and v = constant, passing through that point. The curves of constant t and constant r are the radial lines from the origin and the hyperbolae, respectively. The dashed line is a possible time-like trajectory crossing the horizon.

of the relation (55) for $r < 0$ is that appropriate to a negative value of M; it is, therefore, not of much interest for, as we shall presently see, M has the physical meaning of mass. We shall, accordingly, restrict ourselves to $r > 0$ and $M > 0$.

In the (u, v)-plane, the curves of constant r are rectangular hyperbolae with the coordinate axes as asymptotes. Also, the 'iso-radial' curves for $r > 2M$ are confined to the quadrants I and III (in Fig. 2), while the curves for $r < 2M$ are confined to the quadrants II and IV; and the axes themselves define the loci for $r = 2M$. The light cones at any point are defined by the null geodesics, u = constant and v = constant, passing through that point. It is clear from the disposition of the light cones in Fig. 2 that no time-like (or, null) trajectory, in the direction of increasing r in the quadrant II, can ever emerge into the quadrant I. These geometrically obvious facts imply (as we shall see more explicitly in §§19 and 20 below) that the surface $r = 2M$ is an *event horizon*.

A further essential feature of the space-time which Fig. 2 exhibits, is that a future-directed time-like trajectory (in the direction of decreasing r), once it crosses the horizon (at $r = 2M$) into the quadrant II, cannot avoid intersecting the locus $r = 0$ where the element of proper volume vanishes. In other words, anything which crosses the horizon cannot escape being crushed to 'nothing' at $r = 0$. For this reason, the centre emerges as a singular line of the Schwarzschild space-time—a fact which is further exemplified by all the non-vanishing components of the Riemann tensor (in a local inertial-frame) diverging at $r = 0$ and nowhere else (see equations (75) below).

(c) *The transition to the Schwarzschild coordinates*

Let

$$v/u = e^{t/2M}. \tag{56}$$

This substitution is strictly valid in the quadrants I and III. In these quadrants, the curves of constant t are straight lines through the origin. In particular, $t = 0$ along $u = +v$, and $t \to +\infty$ along the v-axis (approached in the counter-clockwise direction) and $t \to -\infty$ along the u-axis (approached in the clockwise direction).

Restricting our considerations to the quadrants I and III, we have

$$dt = 2M\left(\frac{du}{u} - \frac{dv}{v}\right). \tag{57}$$

We also have (replacing Z by r in equation (50))

$$\frac{r\,dr}{r-2M} = 2M\left(\frac{du}{u} + \frac{dv}{v}\right). \tag{58}$$

From these equations, it follows that

$$(dt)^2 - \frac{r^2(dr)^2}{(r-2M)^2} = -16M^2\frac{du\,dv}{uv}; \tag{59}$$

and the expression (51) for the metric becomes

$$ds^2 = \left(1 - \frac{2M}{r}\right)(dt)^2 - \frac{(dr)^2}{1-2M/r} - r^2[(d\theta)^2 + (d\varphi)^2\sin^2\theta]. \tag{60}$$

This is the Schwarzschild metric in its most familiar form; it is also the form in which Schwarzschild originally wrote it. The present derivation, however, emphasizes that it is strictly applicable only in the quadrants I and III. But the form (60) has several advantages. It makes it manifest, for example, that the radial coordinate r is a 'luminosity distance' in the sense that the surface area of a sphere of 'radius' r is $4\pi r^2$ and that t is a measure of the proper time for an observer at rest at infinity. Further, the asymptotic form of the metric for $r \to \infty$ shows that it represents the space-time external to a spherical distribution of inertial mass M. And finally, since the metric coefficients are explicitly independent of time and there is, in addition, no dragging of the inertial frame (cf. Ch. 2, §11(a)) the space-time is *static* as experienced by an observer external to the horizon. And this is Birkhoff's theorem as commonly stated.

Returning to the (u, v)-plane, we now observe that any time-like or null trajectory, crossing into quadrant II from quadrant I, will, according to any observer stationed in quadrant I, take an infinite time to do so. (We shall provide illustrations of this phenomenon in §§19 and 20.) Also, since no time-like or null trajectory can ever emerge from quadrant II into quadrant I, events in the space interior to $r = 2M$ are inaccessible to all observers in quadrant I: the events taking place interior to $r = 2M$ are incommunicable to the space outside. It is in this sense that the surface $r = 2M$ is an event horizon; and it is also the sense in which the Schwarzschild metric represents the space-time of a black hole.

18. An alternative derivation of the Schwarzschild metric

In view of the important role which the Schwarzschild metric in its standard form (60) plays in subsequent developments, it will be useful to have an *ab initio* derivation of it starting with the coordinates r and t as they eventually emerged in §17(c). Besides, it will provide the occasion to obtain directly the various components of the Riemann tensor, in these coordinates, which we shall need.

The arguments, which led us to postulate a metric of the form (1) for a spherically symmetric space-time, will permit us to write

$$ds^2 = f_{00}(dt)^2 + 2f_{02}\,dr\,dt + f_{22}(dr)^2 + g\,d\Omega^2 \tag{61}$$

where f_{00}, f_{02}, f_{22}, and g are functions of the variables r and t only. By a coordinate transformation involving these variables, we can arrange that f_{02} vanishes and g is replaced by the square of the luminosity distance r. Such a transformation is clearly possible in some open neighbourhood; but it is not evident that it may be possible globally. Indeed, as our discussion in §17 has shown, such a transformation is not, in fact, possible for the entire space-time. With this understanding, the metric can be reduced to the form

$$ds^2 = e^{2\nu}(dt)^2 - e^{2\mu_2}(dr)^2 - r^2[(d\theta)^2 + (d\varphi)^2 \sin^2\theta], \tag{62}$$

where ν and μ_2 are functions of $t (= x^0 = -ix^4)$ and $r (= x^2)$. The metric (62) is again of the general form considered in Chapter 2. In the present instance

$$\left.\begin{aligned} &\nu = \mu_4, \quad \mu_3 = \lg r, \quad \psi = \lg r + \lg \sin\theta, \quad \mu_{3,2} = r^{-1}, \\ &\Psi_4 = 0, \; \Psi_1 = 0, \quad \Psi_2 = r^{-1}, \quad \text{and} \quad \Psi_3 = \cot\theta. \end{aligned}\right\} \tag{63}$$

Running down the list of the components of the Riemann tensor given in Chapter 2 (equations (75)), we find that the only non-vanishing components of the Riemann tensor are

$$-R_{1212} = -R_{2323} = +\frac{e^{-2\mu_2}}{r}\mu_{2,r},$$

$$-R_{1414} = -R_{3434} = -\frac{e^{-2\mu_2}}{r}\nu_{,r},$$

$$-R_{2334} = +R_{1214} = -\frac{e^{-\nu-\mu_2}}{r}\mu_{2,4}, \tag{64}$$

$$-R_{1313} = r^{-2}(1-e^{-2\mu_2}),$$

and

$$-R_{2424} = -e^{-\mu_2-\nu}[(e^{\mu_2-\nu}\mu_{2,4})_{,4} + (e^{\nu-\mu_2}\nu_{,r})_{,r}],$$

where it should be remembered that these are the tetrad components in a frame with the basis vectors

$$e_{(a)i} = \begin{vmatrix} e^{\nu} & 0 & 0 & 0 \\ 0 & -r\sin\theta & 0 & 0 \\ 0 & 0 & -e^{\mu_2} & 0 \\ 0 & 0 & 0 & -r \end{vmatrix} \tag{65}$$

The field equations now follow from setting the various components of the Ricci tensor equal to zero. Thus,

$$-R_{42} = R_{1412} + R_{3432} = -2\frac{e^{-\nu-\mu_2}}{r}\mu_{2,4} = 0. \tag{66}$$

Hence,

$$\mu_2 \equiv \mu_2(r). \tag{67}$$

Considering next the equation,

$$-R_{44} + R_{22} = R_{1414} + R_{3434} - (R_{1212} + R_{2323}) = 0, \tag{68}$$

we find

$$\frac{2}{r}e^{-2\mu_2}(\nu + \mu_2)_{,r} = 0. \tag{69}$$

Therefore,

$$\nu = -\mu_2(r) + f(t), \tag{70}$$

where $f(t)$ is an arbitrary function of t. There is no loss of generality in setting $f(t) = 0$, since it can be 'absorbed' in the definition of t, i.e., by replacing $e^{f(t)}dt$ by dt. With this redefinition of the time coordinate

$$\nu = -\mu_2, \tag{71}$$

and both ν and μ_2 are functions of r only. The fact that the metric coefficients, originally allowed to be functions of t as well, are now seen to be independent of t is a further demonstration of Birkhoff's theorem: *the space-time external to a spherical mass is necessarily static.*

The solution can now be completed by considering the equation

$$-R_{33} = R_{1313} + R_{2323} + R_{4343} = 0. \tag{72}$$

Substituting for the components of the Riemann tensor from equations (64), we find

$$-2\mu_{2,r}\frac{e^{-2\mu_2}}{r} = \frac{1}{r^2}(1 - e^{-2\mu_2}). \tag{73}$$

The solution of this equation is

$$e^{-2\mu_2} = 1 - 2M/r (= e^{2\nu}), \tag{74}$$

where M is a constant. It can now be verified that the vanishing of the remaining components of the Riemann tensor is ensured by the solution (74). We have thus recovered the solution (60) for the metric.

Finally, we note that, with the solution for μ_2 and ν given by equation (74), the non-vanishing components (64) of the Riemann tensor reduce to

$$\left.\begin{aligned} -R_{(1)(3)(1)(3)} = R_{(2)(0)(2)(0)} = 2Mr^{-3} \\ \text{and} \quad R_{(1)(2)(1)(2)} = R_{(2)(3)(2)(3)} = -R_{(1)(0)(1)(0)} = -R_{(3)(0)(3)(0)} = Mr^{-3}, \end{aligned}\right\} \tag{75}$$

where we have enclosed the indices in parentheses to emphasize that these are tetrad components in a local inertial-frame specified by the basis vectors (65). We observe that all these components of the Riemann tensor diverge at $r = 0$—a manifestation of the singular nature of the Schwarzschild space-time at $r = 0$.

To obtain the covariant tensor-components, in the coordinate frame in which the Schwarzschild metric is expressed, we must subject the tetrad components (with respect to each of the indices) to the transformation (65); and applying this transformation to the components listed in equation (75), we find

$$\left.\begin{aligned} R_{0101} &= -Mr^{-1}e^{2\nu}\sin^2\theta, & R_{0303} &= -Mr^{-1}e^{2\nu}, \\ R_{1212} &= +Mr^{-1}e^{-2\nu}\sin^2\theta, & R_{2323} &= +Mr^{-1}e^{-2\nu}, \\ R_{1313} &= -2Mr\sin^2\theta, & R_{0202} &= 2Mr^{-3}, \end{aligned}\right\} \tag{76}$$

where it may be recalled that

$$e^{2\nu} = 1 - 2M/r, \tag{77}$$

and, further, that the indices 0, 1, 2, and 3 refer to the coordinates t, φ, r, and θ, respectively.

19. The geodesics in the Schwarzschild space-time: the time-like geodesics

We have shown in Chapter 1 (§6(*a*), equation (203)) that the equations governing the geodesics in a space-time with the line element,

$$ds^2 = g_{ij}\,dx^i\,dx^j, \tag{78}$$

can be derived from the Lagrangian

$$2\mathscr{L} = g_{ij}\frac{dx^i}{d\tau}\frac{dx^j}{d\tau}, \tag{79}$$

where τ is some affine parameter along the geodesic. For time-like geodesics, τ may be identified with the proper time, s, of the particle describing the geodesic.

For the Schwarzschild space-time, the Lagrangian is

$$\mathscr{L} = \frac{1}{2}\left[\left(1-\frac{2M}{r}\right)\dot{t}^2 - \frac{\dot{r}^2}{1-2M/r} - r^2\dot{\theta}^2 - (r^2\sin^2\theta)\dot{\varphi}^2\right], \tag{80}$$

where the dot denotes differentiation with respect to τ. The corresponding canonical momenta are

$$\left.\begin{aligned} p_t &= \frac{\partial\mathscr{L}}{\partial\dot{t}} = \left(1-\frac{2M}{r}\right)\dot{t}, \quad p_\varphi = -\frac{\partial\mathscr{L}}{\partial\dot{\varphi}} = (r^2\sin^2\theta)\dot{\varphi}, \\ p_r &= -\frac{\partial\mathscr{L}}{\partial\dot{r}} = \left(1-\frac{2M}{r}\right)^{-1}\dot{r}, \quad \text{and} \quad p_\theta = -\frac{\partial\mathscr{L}}{\partial\dot{\theta}} = r^2\dot{\theta}. \end{aligned}\right\} \tag{81}$$

The resulting Hamiltonian is

$$\mathscr{H} = p_t\dot{t} - (p_r\dot{r} + p_\theta\dot{\theta} + p_\varphi\dot{\varphi}) - \mathscr{L} = \mathscr{L}. \tag{82}$$

The equality of the Hamiltonian and the Lagrangian signifies that there is no 'potential energy' in the problem: the energy is derived solely from the 'kinetic energy' as is, indeed, manifest from the expression (79) for the Lagrangian. The constancy of the Hamiltonian and of the Lagrangian follows from this fact:

$$\mathscr{H} = \mathscr{L} = \text{constant}. \tag{83}$$

By rescaling the affine parameter τ, we can arrange that $2\mathscr{L}$ has the value $+1$ for time-like geodesics. For null geodesics, $\mathscr{L}$ has the value zero. (We shall not be concerned with space-like geodesics.)

Further integrals of the motion follow from the equations

$$\frac{dp_t}{d\tau} = \frac{\partial\mathscr{L}}{\partial t} = 0 \quad \text{and} \quad \frac{dp_\varphi}{d\tau} = -\frac{\partial\mathscr{L}}{\partial\varphi} = 0. \tag{84}$$

Thus,

$$p_t = \left(1 - \frac{2M}{r}\right)\frac{\mathrm{d}t}{\mathrm{d}\tau} = \text{constant} = E \quad \text{(say)} \tag{85}$$

and

$$p_\varphi = r^2 \sin^2\theta \frac{\mathrm{d}\varphi}{\mathrm{d}\tau} = \text{constant}. \tag{86}$$

Moreover, from the equation of motion,

$$\frac{\mathrm{d}p_\theta}{\mathrm{d}\tau} = \frac{\mathrm{d}}{\mathrm{d}\tau}(r^2\dot{\theta}) = -\frac{\partial \mathscr{L}}{\partial \theta} = (r^2 \sin\theta\cos\theta)\left(\frac{\mathrm{d}\varphi}{\mathrm{d}\tau}\right)^2, \tag{87}$$

it follows that if we choose to assign the value $\pi/2$ to θ when $\dot{\theta}$ is zero, then $\ddot{\theta}$ will also be zero; and θ will remain constant at the assigned value. We conclude that the geodesic is described in an invariant plane which we may distinguish by $\theta = \pi/2$. Equation (86) then gives

$$p_\varphi = r^2 \frac{\mathrm{d}\varphi}{\mathrm{d}\tau} = \text{constant} = L \quad \text{(say)}, \tag{88}$$

where L denotes the angular momentum about an axis normal to the invariant plane.

With $\dot{t}$ and $\dot{\varphi}$ given by equations (85) and (88), the constancy of the Lagrangian gives

$$\frac{E^2}{1-2M/r} - \frac{\dot{r}^2}{1-2M/r} - \frac{L^2}{r^2} = 2\mathscr{L} = +1 \quad \text{or} \quad 0, \tag{89}$$

depending on whether we are considering time-like or null geodesics.

In this section we shall restrict ourselves to time-like geodesics. (Null geodesics are considered in the following section.) For time-like geodesics, equations (88) and (89) can be rewritten in the forms

$$\left(\frac{\mathrm{d}r}{\mathrm{d}\tau}\right)^2 + \left(1 - \frac{2M}{r}\right)\left(1 + \frac{L^2}{r^2}\right) = E^2 \tag{90}$$

and

$$\frac{\mathrm{d}\varphi}{\mathrm{d}\tau} = \frac{L}{r^2}. \tag{91}$$

By considering r as a function of φ (instead of τ), we obtain the equation

$$\left(\frac{\mathrm{d}r}{\mathrm{d}\varphi}\right)^2 = (E^2 - 1)\frac{r^4}{L^2} + \frac{2M}{L^2}r^3 - r^2 + 2Mr. \tag{92}$$

Letting

$$u = r^{-1}, \tag{93}$$

as in the analysis of the Keplerian orbit in the Newtonian theory, we obtain the basic equation of the problem:

$$\left(\frac{\mathrm{d}u}{\mathrm{d}\varphi}\right)^2 = 2Mu^3 - u^2 + \frac{2M}{L^2}u - \frac{1-E^2}{L^2}. \tag{94}$$

This equation determines the geometry of the geodesics in the invariant plane. Once equation (94) has been solved for $u(\varphi)$, the solution can be completed by direct quadratures of the equations,

$$\frac{\mathrm{d}\tau}{\mathrm{d}\varphi} = \frac{1}{Lu^2} \quad \text{and} \quad \frac{\mathrm{d}t}{\mathrm{d}\varphi} = \frac{E}{Lu^2(1-2Mu)}. \tag{95}$$

(*a*) *The radial geodesics*

The radial geodesics of zero angular momentum, while simple, illustrate some essential features of the space-time to which references were made in §17. The equations governing these geodesics are (cf. equations (85) and (90))

$$\left(\frac{\mathrm{d}r}{\mathrm{d}\tau}\right)^2 = \frac{2M}{r} - (1-E^2) \quad \text{and} \quad \frac{\mathrm{d}t}{\mathrm{d}\tau} = \frac{E}{1-2M/r}. \tag{96}$$

We shall consider the trajectories of particles which start from rest at some finite distance r_i and fall towards the centre. The starting distance is related to the constant E by

$$r_i = \frac{2M}{1-E^2} \quad (r = r_i \quad \text{when} \quad \dot{r} = 0). \tag{97}$$

The equations of motion are most conveniently integrated in terms of a variable η where

$$r = \frac{M}{1-E^2}(1+\cos\eta) = \frac{2M}{1-E^2}\cos^2\tfrac{1}{2}\eta = r_i\cos^2\tfrac{1}{2}\eta. \tag{98}$$

Clearly, $\eta = 0$ when $r = r_i$; and the values of η when r crosses the horizon at $r = 2M$ and arrives at the singularity at $r = 0$ are

$$\eta = \eta_H = 2\sin^{-1}E \quad \text{when} \quad r = 2M$$

and

$$\eta = \pi \quad \text{when} \quad r = 0. \tag{99}$$

In terms of η, the equations to be integrated are

$$\left(\frac{\mathrm{d}r}{\mathrm{d}\tau}\right)^2 = (1-E^2)\tan^2\tfrac{1}{2}\eta \quad \text{and} \quad \frac{\mathrm{d}t}{\mathrm{d}\tau} = \frac{E\cos^2\frac{1}{2}\eta}{\cos^2\frac{1}{2}\eta - \cos^2\frac{1}{2}\eta_H}, \tag{100}$$

together with

$$\frac{\mathrm{d}r}{\mathrm{d}\eta} = -r_i\sin\tfrac{1}{2}\eta\cos\tfrac{1}{2}\eta. \tag{101}$$

Since we are considering infalling particles,

$$\frac{dr}{d\tau} = -(1-E^2)^{\frac{1}{2}}\tan\tfrac{1}{2}\eta = -\left(\frac{2M}{r_i}\right)^{1/2}\tan\tfrac{1}{2}\eta. \tag{102}$$

From equations (101) and (102), we now obtain

$$\frac{d\tau}{d\eta} = \left(\frac{r_i^3}{2M}\right)^{1/2}\cos^2\tfrac{1}{2}\eta = \left(\frac{r_i^3}{8M}\right)^{1/2}(1+\cos\eta). \tag{103}$$

Therefore,

$$\tau = \left(\frac{r_i^3}{8M}\right)^{1/2}(\eta+\sin\eta), \tag{104}$$

where we have assumed that $\tau = 0$ at the starting point $\eta = 0$. From equation (104) it follows that the particle crosses the horizon and arrives at the singularity at the finite proper times,

$$\tau_{\mathrm{H}} = \left(\frac{r_i^3}{8M}\right)^{1/2}(\eta_{\mathrm{H}}+\sin\eta_{\mathrm{H}}) \qquad \text{and} \qquad \tau_0 = \left(\frac{r_i^3}{8M}\right)^{1/2}\pi. \tag{105}$$

The situation, as we shall see presently, is very different when we consider the equation of the trajectory in coordinate time, t. The equation to be integrated to obtain t is (cf. equations (100) and (103))

$$\frac{dt}{d\eta} = E\left(\frac{r_i^3}{2M}\right)^{1/2}\frac{\cos^4\frac{1}{2}\eta}{\cos^2\frac{1}{2}\eta-\cos^2\frac{1}{2}\eta_{\mathrm{H}}}. \tag{106}$$

On integration, this equation gives

$$t = E\left(\frac{r_i^3}{2M}\right)^{1/2}\left[\tfrac{1}{2}(\eta+\sin\eta)+(1-E^2)\eta\right] + 2M\lg\left[\frac{\tan\frac{1}{2}\eta_{\mathrm{H}}+\tan\frac{1}{2}\eta}{\tan\frac{1}{2}\eta_{\mathrm{H}}-\tan\frac{1}{2}\eta}\right]. \tag{107}$$

According to this equation,

$$t \to \infty \qquad \text{as} \qquad \eta \to \eta_{\mathrm{H}} - 0, \tag{108}$$

in sharp contrast with the behaviour of the proper time τ. This is an example of what we concluded on general grounds in §17(c), namely, that with respect to an observer stationed at 'infinity', a particle describing a time-like trajectory will take an infinite time to reach the horizon even though by its own proper time it will cross the horizon in a finite time. And after crossing the horizon, the particle will arrive at the singularity, again, at finite proper time. These facts are illustrated in Fig. 3 in terms of the solutions (104) and (108).

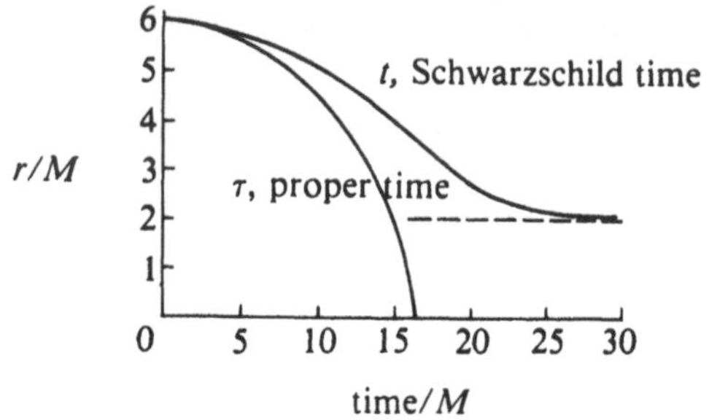

FIG. 3. The variation of the coordinate time (t) and the proper time (τ) along a time-like radial-geodesic described by a test particle, starting at rest at $r = 6M$ and falling towards the singularity.

(b) *The bound orbits* ($E^2 < 1$)

The consideration of equation (94), governing the geometry of the orbits described in the invariant plane, is conveniently separated into two parts, pertaining, respectively, to $E^2 < 1$ and $E^2 \geqslant 1$. These two classes of orbits are characterized by energies (exclusive of the rest energy) which are negative or positive (and zero). As we should expect, this distinction is one which will determine whether the orbits are bound or unbound (i.e., whether along the orbits r remains bounded or not). In this section (*b*), we shall restrict ourselves to bound orbits ($E^2 < 1$). These orbits are governed by the equation

$$\left(\frac{du}{d\varphi}\right)^2 = f(u), \tag{109}$$

where

$$f(u) = 2Mu^3 - u^2 + \frac{2M}{L^2}u - \frac{1-E^2}{L^2} \quad (E^2 < 1). \tag{110}$$

It is clear that the geometry of the geodesics will be determined by the disposition of the roots of the equation $f(u) = 0$. Since $f(u)$ is a cubic in u, there are two possibilities: *either*, all the roots are real *or* one of them is real and the two remaining are a complex-conjugate pair. Letting u_1, u_2, and u_3 denote the roots of the cubic equation $f(u) = 0$, we have

$$u_1 u_2 u_3 = (1 - E^2)/2ML^2 \tag{111}$$

and

$$u_1 + u_2 + u_3 = 1/2M. \tag{112}$$

Since we have assumed that $(1 - E^2) > 0$, the equation $f(u) = 0$ must always allow a positive real root. From the further facts, $f < 0$ for $u = 0$ and $f(u) \to$

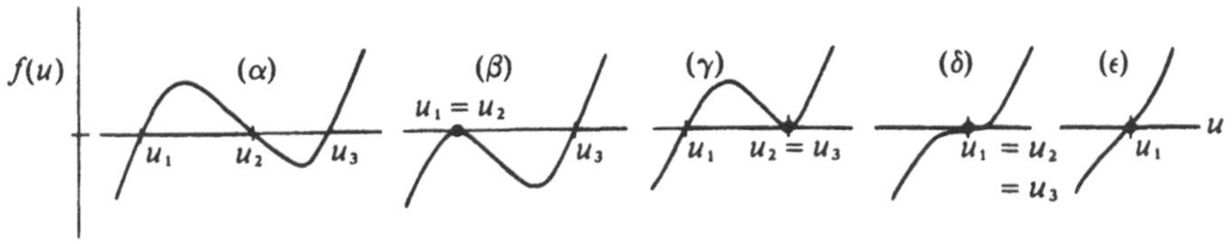

FIG. 4. The disposition of the roots of a cubic $f(u) = 0$ for $E^2 < 1$. The various cases (α), (β), etc., are distinguished in the text.

$\pm\infty$ for $u \to \pm\infty$, it follows that we must consider the five cases distinguished in Fig. 4. These different cases lead to the following possibilities.*

Case (α): For every pair of values E and L, which allows for u three real roots $0 < u_1 < u_2 < u_3$, there exist two distinct orbits confined, respectively, to the intervals $u_1 \leqslant u \leqslant u_2$ and $u \geqslant u_3$, i.e., an orbit which oscillates between two extreme values of $r(= u_1^{-1}$ and $u_2^{-1})$ and an orbit, which, starting at a certain aphelion distance $(= u_3^{-1})$, plunges into the singularity at $r = 0$ (i.e., as $u \to \infty$). We shall call these two classes of orbits as of the *first* and *second kinds*, respectively. The orbits of the first kind are the relativistic analogues of the Keplerian orbits and to which they tend in the Newtonian limit. The orbits of the second kind have no Newtonian analogues. However, we shall find that the orbits of both kinds (and the unbound orbits, as well, as we shall see in §(c) below) are most conveniently parametrized by an eccentricity, $e \geqslant 0$ and a *latus rectum*, l, even as the Newtonian orbits are.

Case (β): In this case the orbit of the first kind is a stable circular orbit (of zero eccentricity) while the orbit of the second kind, even though labelled as of zero eccentricity, still plunges into the singularity.

Case (γ): In this case, the orbit of the first kind starts at a certain aphelion distance, u_1^{-1}, and approaches the circle of radius u_3^{-1}, asymptotically, by spiralling around it an infinite number of times. The orbit of the second kind is, in some sense, a continuation of the orbit of the first kind in that it spirals away from the same circle (towards the centre) to plunge eventually into the singularity.

Case (δ): In this case all three roots coincide; and the principal difference with case (ε) below is that the equations also allow an unstable circular orbit of radius $u_1^{-1}(= u_2^{-1} = u_3^{-1})$.

* By combining the relations (111) and (112) with the further relation

$$u_1 u_2 + u_2 u_3 + u_3 u_1 = 1/L^2,$$

it can be shown that the equation $f(u) = 0$ for $E^2 < 1$ does not allow two negative real roots beside a positive root. All the allowed possibilities are distinguished in Fig. 4.

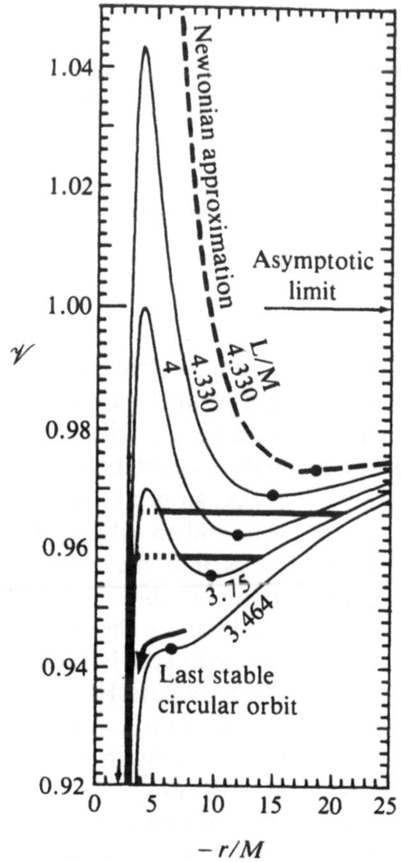

FIG. 5. The effective potentials appropriate for time-like trajectories (cf. equation (113)). The minima in the potentials correspond to the stable circular orbits while the maxima correspond to unstable circular orbits. At the point of inflexion the last stable circular orbit occurs.

Case (ε): In this case, we have only one class of orbits: they all plunge into the singularity after starting from certain finite aphelion distances. We shall find that these orbits are most conveniently parametrized by an imaginary eccentricity; otherwise, they are similar to the radial geodesics.

The different cases we have distinguished can also be inferred by interpreting

$$\mathscr{V}^2 = \left(1-\frac{2M}{r}\right)\left(1+\frac{L^2}{r^2}\right), \tag{113}$$

which occurs on the left-hand side of equation (90), as a 'potential energy' in the sense that, together with $\dot{r}^2$, interpreted as 'kinetic energy', it is a constant of the motions. (See Fig. 5 in which we have displayed the potential-energy curves for some typical values of the parameters E and L.)

(i) *Orbits of the first kind*

Orbits of the first kind occur in the cases (α), (β), (γ), and (δ); and these cases require that $f(u) = 0$ allows three real roots, all of which are positive; and we shall write them as

$$u_1 = \frac{1}{l}(1-e), \quad u_2 = \frac{1}{l}(1+e), \qquad \text{and} \qquad u_3 = \frac{1}{2M} - \frac{2}{l}, \tag{114}$$

where the latus rectum, l, is some positive constant and the eccentricity e is less than 1 for $u_1 > 0$, as required by the condition $E^2 < 1$:

$$0 \leqslant e < 1 \qquad \text{for} \qquad E^2 < 1. \tag{115}$$

(We have assigned to u_3 the value consistent with the requirement (112)).

It is important to note that to be in conformity with the ordering, $u_1 \leqslant u_2 \leqslant u_3$, we must require

$$\frac{1}{2M} - \frac{2}{l} \geqslant \frac{1}{l}(1+e) \qquad \text{or} \qquad l \geqslant 2M(3+e). \tag{116}$$

Defining

$$\mu = M/l, \tag{117}$$

we have the important inequality

$$\mu \leqslant \frac{1}{2(3+e)} \qquad \text{or} \qquad 1 - 6\mu - 2\mu e \geqslant 0. \tag{118}$$

The condition, that

$$f(u) = 2M\left(u - \frac{1-e}{l}\right)\left(u - \frac{1+e}{l}\right)\left(u - \frac{1}{2M} + \frac{2}{l}\right) \tag{119}$$

agrees with its definition (110), yields the relations,

$$\frac{M}{L^2} = \frac{1}{l^2}[l - M(3+e^2)] \quad \text{and} \quad \frac{1-E^2}{L^2} \equiv \frac{1}{l^3}(l-4M)(1-e^2), \tag{120}$$

or, in terms of μ,

$$\frac{1}{L^2} = \frac{1}{lM}[1 - \mu(3+e^2)] \quad \text{and} \quad \frac{1-E^2}{L^2} = \frac{1}{l^2}(1-4\mu)(1-e^2). \tag{121}$$

It follows from these relations that

$$\mu < (3+e^2)^{-1} \qquad \text{and} \qquad \mu < \tfrac{1}{4}. \tag{122}$$

It can be verified that these inequalities are guaranteed by (118).

An alternative expression for E^2 may be noted here:

$$\frac{E^2}{L^2} = \frac{1}{lM}[(2\mu-1)^2 - 4\mu^2 e^2]. \tag{123}$$

Returning to equation (109), we now make the substitution

$$u = \frac{1}{l}(1 + e\cos\chi), \tag{124}$$

where χ is a new variable which we may call the 'relativistic anomaly', following Darwin. According to equation (124)

$$\left.\begin{array}{lll} \text{at aphelion, where} & u = (1-e)/l, & \chi = \pi \\ \text{and} & & \\ \text{at perihelion, where} & u = (1+e)/l, & \chi = 0. \end{array}\right\} \tag{125}$$

We readily verify that the substitution (124) reduces equation (109) to the simple form

$$\begin{aligned} \left(\frac{d\chi}{d\varphi}\right)^2 &= [1 - 2\mu(3 + e\cos\chi)] \\ &= [(1 - 6\mu + 2\mu e) - 4\mu e\cos^2\tfrac{1}{2}\chi], \end{aligned} \tag{126}$$

or, alternatively,

$$\pm\frac{d\chi}{d\varphi} = (1 - 6\mu + 2\mu e)^{1/2}(1 - k^2\cos^2\tfrac{1}{2}\chi)^{1/2}, \tag{127}$$

where

$$k^2 = \frac{4\mu e}{1 - 6\mu + 2\mu e}. \tag{128}$$

The inequality (118) guarantees that

$$k^2 \leqslant 1 \qquad \text{and} \qquad 1 - 6\mu + 2\mu e > 0. \tag{129}$$

It is now apparent that the solution for φ can be expressed in terms of the Jacobian elliptic integral

$$F(\psi, k) = \int_0^\psi \frac{d\gamma}{\sqrt{(1 - k^2\sin^2\gamma)}}, \tag{130}$$

where

$$\psi = \tfrac{1}{2}(\pi - \chi). \tag{131}$$

Thus, we may write

$$\varphi = \frac{2}{(1 - 6\mu + 2\mu e)^{1/2}} F(\tfrac{1}{2}\pi - \tfrac{1}{2}\chi, k), \tag{132}$$

where the origin of φ has been chosen at aphelion passage where $\chi = \pi$. The perihelion passage occurs at $\chi = 0$ where $\psi = \pi/2$. (Orbits derived on the basis of equation (132) are illustrated in Fig. 7*a*, (a), (b), (c), pp. 116 and 117.)

The solution can now be completed by direct integration of the equations (95) for t and τ. Thus,

$$\tau = \frac{1}{L}\int \frac{d\varphi}{u^2} = \frac{1}{L}\int \frac{d\varphi}{d\chi}\frac{d\chi}{u^2}, \tag{133}$$

and

$$t = \frac{E}{L}\int \frac{d\varphi}{d\chi}\frac{d\chi}{u^2(1-2Mu)}. \tag{134}$$

By making use of equations (121), (123), (124), and (126), we obtain

$$\tau = \frac{l^{3/2}}{M^{1/2}}[1-\mu(3+e^2)]^{1/2}\int_\chi^\pi d\chi(1+e\cos\chi)^{-2}[1-2\mu(3+e\cos\chi)]^{-1/2} \tag{135}$$

and

$$t = \frac{l^{3/2}}{M^{1/2}}[(2\mu-1)^2-4\mu^2e^2]^{1/2}\int_\chi^\pi d\chi(1+e\cos\chi)^{-2}$$
$$\times[1-2\mu(3+e\cos\chi)]^{-1/2}[1-2\mu(1+e\cos\chi)]^{-1}. \tag{136}$$

Expressing t and τ in units of the Newtonian period,

$$T_{\text{Newton}} = \left[\frac{4\pi^2 l^3}{(1-e^2)^3 GM}\right]^{1/2}, \tag{137}$$

of a Kepler orbit with the same eccentricity and latus rectum, we find that the factors in front of the integrals on the right-hand sides of equations (135) and (136) are, respectively,

$$\frac{1}{2\pi}T_{\text{Newton}}(1-e^2)^{3/2}[1-\mu(3+e^2)]^{1/2}$$

and

$$\frac{1}{2\pi}T_{\text{Newton}}(1-e^2)^{3/2}[(2\mu-1)^2-4\mu^2e^2]^{1/2}. \tag{138}$$

(With T_{Newton} given by equations (137), the integrals on the right-hand side of equations (135) and (136), with the foregoing factors, give t and τ in seconds: the units in which $c = 1$ and $G = 1$, which we have adopted hitherto, is abandoned in this instance for comparison with the results of the Newtonian theory.)

We now consider two special cases: the case $e = 0$ when the two roots u_1 and u_2 coincide (the case (β) of Fig. 4) and the case $2\mu(3+e) = 1$ when the two roots u_2 and u_3 coincide (the case (γ) of Fig. 4).

(α) *The case* $e = 0$: In this case the orbit is a circle with the radius

$$r_c = l \qquad \text{and} \qquad \mu = M/r_c. \tag{139}$$

Equations (121) and (123), relating the angular momentum L and the energy E of the orbit to the parameters e and l of the orbit, now give

$$\frac{1}{L^2} = \frac{1 - 3M/r_c}{r_c M} \qquad \text{and} \qquad \frac{E^2}{L^2} = \frac{(2M/r_c - 1)^2}{r_c M}. \tag{140}$$

Rewriting the first of these equations in the form

$$r_c^2 - \frac{L^2}{M} r_c + 3L^2 = 0, \tag{141}$$

we conclude (as is evident from Fig. 5) that an orbit of zero eccentricity is compatible with the one or the other of the two roots,

$$r_c = \frac{L^2}{2M}[1 \pm (1 - 12M^2/L^2)^{1/2}]; \tag{142}$$

and, further, that no circular orbit is possible for

$$L/M < 2\sqrt{3}. \tag{143}$$

For the minimum allowed value of L/M,

$$r_c = 6M \qquad \text{and} \qquad E^2 = \tfrac{8}{9} \qquad \text{for} \qquad L/M = 2\sqrt{3}. \tag{144}$$

It is clear that the larger of the two roots of equation (142) (for $L/M > 2\sqrt{3}$) locates the minimum of the potential-energy curve $\mathscr{V}(r)$ defined by equation (113), while the smaller root locates the maximum of the potential-energy curve. Therefore, the circular orbit of the larger radius will be stable in contrast to the circular orbit of the smaller radius which will be unstable. The allowed ranges for the radii of these two classes of orbits are

$$6M < r_c(\text{stable}) < \infty \qquad \text{and} \qquad 3M \leqslant r_c\ (\text{unstable}) \leqslant 6M. \tag{144'}$$

Of the two orbits of zero eccentricity allowed for $L/M > 2\sqrt{3}$, it is the stable one with the larger radius (smaller u) that is included among the orbits of the first kind we are presently considering.

The periods for one complete revolution of these circular orbits, measured in proper time and in coordinate time, are (cf. equations (135)–(138))

$$\tau_{\text{period}} = T_{\text{Newton}}\left(\frac{1-3\mu}{1-6\mu}\right)^{1/2} \qquad \text{and} \qquad t_{\text{period}} = T_{\text{Newton}}(1-6\mu)^{-1/2}. \tag{145}$$

(Notice that $t_{\text{period}} \to \infty$ when $\mu \to \frac{1}{6}$ and $r_c \to 6M$.)

(β) *The case* $2\mu(3+e)=1$: In this case, the perihelion, r_p, and the aphelion r_{ap}, distances are given by

$$r_p=\frac{l}{1+e}=2M\frac{3+e}{1+e} \qquad \text{and} \qquad r_{ap}=2M\frac{3+e}{1-e}. \tag{146}$$

It follows that the perihelion distances for these orbits are restricted to the range

$$4M\leqslant r_p<6M. \tag{147}$$

Also, for these orbits

$$\frac{L^2}{M^2}=4\frac{(3+e)^2}{(3-e)(1+e)} \qquad \text{and} \qquad 1-E^2=\frac{1-e^2}{9-e^2}. \tag{148}$$

The modulus, k, of the elliptic integral, in terms of which the solution (132) is expressed, becomes 1 in this case and it is convenient to go back to equation (126); it gives

$$\left(\frac{d\chi}{d\varphi}\right)^2=4\mu e\sin^2\tfrac{1}{2}\chi \qquad \text{or} \qquad \frac{d\chi}{d\varphi}=-2(\mu e)^{1/2}\sin\tfrac{1}{2}\chi, \tag{149}$$

where the negative sign has been chosen so that φ may increase (from zero) when χ decreases from its aphelion value π (to its perihelion value 0). The required solution of equation (149) is

$$\varphi=-\frac{1}{\sqrt{(\mu e)}}\lg(\tan\tfrac{1}{4}\chi) \quad (\varphi=0 \quad \text{when} \quad \chi=\pi). \tag{150}$$

Equation (150) shows that $\varphi\to\infty$ when $\chi\to 0$ and the perihelion is approached. In other words, the orbit approaches the circle at r_p, asymptotically, spiralling around it (in the counter-clockwise direction) an infinite number of times. In §(ii, β) below, we shall show that this orbit 'continues' into the interior of the circle as an orbit of the second kind to plunge eventually into the singularity. (See Fig. 7*a*, (d), p. 117.)

(γ) *The post-Newtonian approximation*: The first-order correction, of relativistic origin, to the Keplerian orbits of the Newtonian theory can be readily deduced from equation (126) by noting that under conditions of normal occurrence, $\mu=M/l$ is a very small quantity: it is essentially the ratio of the Schwarzschild radius (~ 2 km) to the major axis of a planetary or a binary-star orbit ($\sim 10^6$–10^8 km). Expanding, then, equation (126), to the first order in μ, we obtain

$$-d\varphi=d\chi(1+3\mu+\mu e\cos\chi); \tag{151}$$

or, in integrated form,

$$-\varphi=(1+3\mu)\chi+\mu e\sin\chi+\text{constant}. \tag{152}$$

From equation (152) we infer that the change in φ, after one complete revolution during which χ changes by 2π, is $2(1+3\mu)\pi$. Therefore, the *advance* in the perihelion, $\Delta\varphi$, per revolution is

$$\Delta\varphi = 6\pi\frac{M}{l} = 6\pi\frac{M}{a(1-e^2)}, \tag{153}$$

where a denotes the semi-major axis of the Keplerian ellipse; and this is the standard result (first derived by Einstein).

(ii) *Orbits of the second kind*

As we have already explained, orbits of the second kind have their aphelions at u_3^{-1} and eventually plunge into the singularity at $r = 0$. It is important to observe that, since $u_1+u_2+u_3 = 1/2M$ and $u_1+u_2 > 0$, $u_3 < 1/2M$; and that, therefore, all these orbits start outside the horizon.

To obtain the solution for these orbits, we now make the substitution

$$u = \left(\frac{1}{2M}-\frac{2}{l}\right)+\left(\frac{1}{2M}-\frac{3+e}{l}\right)\tan^2\tfrac{1}{2}\xi, \tag{154}$$

in place of (124). By this substitution,

$$\left.\begin{aligned} u = u_3 = \frac{1}{2M}-\frac{2}{l} \quad &\text{when} \quad \xi = 0 \\ \text{and} \qquad u \to \infty \quad &\text{as} \quad \xi \to \pi. \end{aligned}\right\} \tag{155}$$

We further find that equation (109) now reduces to

$$\left(\frac{d\xi}{d\varphi}\right)^2 = (1-6\mu+2\mu e)(1-k^2\sin^2\tfrac{1}{2}\xi), \tag{156}$$

with the same definition of k^2 as for the orbits of the first kind. The solution for φ can, accordingly, be expressed in terms of the same elliptic integral (130). Thus, we may now write (cf. equation (132))

$$\varphi = \frac{2}{(1-6\mu+2\mu e)^{1/2}}F(\tfrac{1}{2}\xi, k). \tag{157}$$

At aphelion, $\xi = 0$ and $\varphi = 0$; and at the singularity $\xi \to \pi$ and φ takes the finite value

$$\varphi_0 = \frac{2}{(1-6\mu+2\mu e)^{1/2}}K(k), \tag{158}$$

where $K(k)$ denotes the complete elliptic integral,

$$K(k) = \int_0^{\pi/2}\frac{d\gamma}{\sqrt{(1-k^2\sin^2\gamma)}}. \tag{159}$$

(Examples of orbits derived on the basis of equation (157) are illustrated in Fig. 7*a*, (a), (b), (c), pp. 116 and 117.)

The solution for the proper time τ and the coordinate time t can be obtained by integrating the equations

$$\frac{d\tau}{d\xi} = \frac{1}{Lu^2}\frac{d\varphi}{d\xi} \quad \text{and} \quad \frac{dt}{d\xi} = \frac{E}{Lu^2(1-2Mu)}\frac{d\varphi}{d\xi}. \tag{160}$$

The occurrence of the factor $(1-2Mu)^{-1}$ on the right-hand side of the equation for $dt/d\xi$ shows that the integral for t diverges as $u \to 1/2M$. The parts of these orbits for $r < 2M$ are, therefore, inaccessible to an observer stationed outside the horizon; and this is a manifestation of the same phenomenon which occurred in the case of the radial geodesics: it is a further example of what was deduced on general grounds in §§17(*b*) and (*c*). It is also important to observe that all these orbits which cross the horizon necessarily end at the centre—a manifestation of the singular nature of the space-time at this point.

We now consider the two special cases, $e = 0$ and $2\mu(3+e) = 1$.

(α) *The case* $e = 0$: In this case (since k^2 is also zero) equation (156) integrates to give

$$\xi = (1-6\mu)^{1/2}(\varphi - \varphi_0), \tag{161}$$

where φ_0 is a constant of integration. The corresponding solution for u is

$$u = \frac{1}{l} + \left(\frac{1}{2M} - \frac{3}{l}\right)\sec^2\left[\tfrac{1}{2}(1-6\mu)^{1/2}(\varphi-\varphi_0)\right]. \tag{162}$$

This orbit of 'zero eccentricity' is not a circle! Starting at an aphelion distance u_3^{-1} (in the range $3M \leqslant u_3^{-1} \leqslant 6M$), when $\varphi = \varphi_0$, it arrives at the singularity at $r = 0$, when

$$\varphi - \varphi_0 = \pi/(1-6\mu)^{1/2}, \tag{163}$$

after circling one or more times depending on how close μ is to 1/6. The circle at u_3^{-1} is the envelope of these solutions, so that the circular orbit, predicted for this radius, is a singular solution of the equations of motion. (See Fig. 7*a*, (e), p. 118.)

The case $e = 0$ and $\mu = 1/6$ must be treated separately. In this case, all three roots of $f(u) = 0$ coincide (case (δ) of Fig. 4) and $u_1 = u_2 = 1/6M$. As we have shown in §(i, α), a circular orbit of radius $6M$ is allowed by the equations of motion as a special solution; it is, in fact, the lower bound for the radii of stable circular orbits (see equations (144′) and (146)). The general solution can be obtained from the equation appropriate to this case, namely,

$$\left(\frac{du}{d\varphi}\right)^2 = 2M\left(u - \frac{1}{6M}\right)^3; \tag{164}$$

and the solution of this equation is

$$u = \frac{1}{6M} + \frac{2}{M(\varphi - \varphi_0)^2}. \tag{165}$$

This orbit approaches the circle at $6M$, asymptotically, by spiralling around it an infinite number of times (see Fig. 7*a*, (f), p. 118).

(β) *The case* $2\mu(3+e) = 1$: We cannot obtain the solution for this case by simply letting $\mu = 1/2(3+e)$ in the analysis leading to equation (156): the coefficient of $\tan^2 \frac{1}{2}\xi$, in the initial substitution (152), vanishes. We must, therefore, consider this case *ab initio*.

When $2\mu(3+e) = 1$, the roots of $f(u) = 0$ are

$$u_1 = \frac{1-e}{l} \quad \text{and} \quad u_2 = u_3 = \frac{1}{4M} - \frac{1-e}{2l} = \frac{1+e}{l}; \tag{166}$$

and the substitution that is suggested is

$$u = \frac{1}{l}(1 + e + 2e \tan^2 \tfrac{1}{2}\xi). \tag{167}$$

By this substitution,

$$u = u_2 = u_3 = (1+e)/l \quad \text{when} \quad \xi = 0, \quad \text{and} \quad u \to \infty \quad \text{when} \quad \xi = \pi. \tag{168}$$

We further find that equation (109) reduces to

$$\left(\frac{d\xi}{d\varphi}\right)^2 = 4\mu e \sin^2 \tfrac{1}{2}\xi. \tag{169}$$

We observe that this is exactly the same equation we obtained when considering the orbits of the first kind in the same context, $u_2 = u_3$ (see equation (149)). And, as before, we shall write

$$\varphi = -\frac{1}{\sqrt{(\mu e)}} \lg (\tan \tfrac{1}{4}\xi). \tag{170}$$

Along this orbit, $\varphi = 0$ when $\xi = \pi$ and $r \to 0$; and $\varphi \to \infty$ as $\xi \to 0$ and we approach the aphelion at $r = l/(1+e)$. In other words, the orbit approaches the circle at $l/(1+e)$, asymptotically, by spiralling around it (in the counter-clockwise direction) an infinite number of times. This behaviour is the same as of the orbits of the first kind. But there is one important difference: the circle at $l/(1+e)$ is the perihelion for the orbits of the first kind, while it is the aphelion for the orbits of the second kind. If we wish to consider the orbits of the two kinds as continuations of one another (when $u_2 = u_3$), we must suppose that the circle at $l/(1+e)$ is approached by the orbit of the first kind by spiralling around it an infinite number of times and then spiralling inwards before its ultimate fall into the singularity. (See Fig. 7*a*, (d), p. 117.)

(iii) *The orbits with imaginary eccentricities*

Finally, we have to consider the case when $f(u) = 0$ allows only one real root (which is necessarily positive for the bound orbits we are presently considering) and a pair of complex-conjugate roots (the case (ε) of Fig. 4). It is evident that the corresponding orbits, while starting at some finite aphelion distances, will fall into the singularity though they may circle the origin one or more times before doing so. We shall characterize these orbits by an imaginary eccentricity ie ($e > 0$) and write the roots of $f(u) = 0$ as

$$u_1 = \frac{1}{2M} - \frac{2}{l}, \quad u_2 = \frac{1}{l}(1 + ie), \qquad \text{and} \qquad u_3 = \frac{1}{l}(1 - ie). \tag{171}$$

It is manifest that with the present definitions, we can obtain the equations which replace equations (121) and (123) by simply writing $-e^2$ wherever $+e^2$ occurs and conversely. We thus obtain

$$\left.\begin{aligned} \frac{1}{L^2} = \frac{1}{lM}[1 - \mu(3 - e^2)], \quad \frac{1 - E^2}{L^2} &= \frac{1}{l^2}(1 - 4\mu)(1 + e^2), \\ \text{and} \qquad \frac{E^2}{L^2} = \frac{1}{lM}[(2\mu - 1)^2 + 4\mu^2 e^2]. & \end{aligned}\right\} \tag{172}$$

The last of these equations shows, incidentally, that l must be taken to be positive. Moreover, the fact that we are presently considering bound orbits with $(1 - E^2) > 0$, requires that $\mu < \frac{1}{4}$; and this inequality guarantees

$$1 - 3\mu + \mu e^2 > 0. \tag{173}$$

Also, we observe that we can set no upper limit to e^2.

With the assumption (171) regarding the roots of $f(u) = 0$, the equation we have to consider is

$$\left(\frac{du}{d\varphi}\right)^2 = 2M\left(u - \frac{1}{2M} + \frac{2}{l}\right)\left[\left(u - \frac{1}{l}\right)^2 + \frac{e^2}{l^2}\right]. \tag{174}$$

And the substitution we now make is

$$u = \frac{1}{l}(1 + e\tan\tfrac{1}{2}\xi). \tag{175}$$

Since the range of u is

$$\frac{1}{2M} - \frac{2}{l} \leqslant u < \infty, \tag{176}$$

the corresponding range of ξ is

$$\xi_0 \leqslant \xi < \pi, \tag{177}$$

where

$$\tan\tfrac{1}{2}\xi_0 = -\frac{6\mu-1}{2\mu e}, \tag{178}$$

or, equivalently,

$$\sin\tfrac{1}{2}\xi_0 = -\frac{6\mu-1}{\Delta} \qquad \text{and} \qquad \cos\tfrac{1}{2}\xi_0 = \frac{2\mu e}{\Delta}, \tag{179}$$

where

$$\Delta = [(6\mu-1)^2 + 4\mu^2 e^2]^{1/2}. \tag{180}$$

We find that with the substitution (175), equation (174) reduces to

$$\frac{d\xi}{d\varphi} = \pm\sqrt{2}[(6\mu-1) + 2\mu e\sin\xi + (6\mu-1)\cos\xi]^{1/2}. \tag{181}$$

By a standard formula in the theory of elliptic integrals, the solution for φ can be expressed in terms of the Jacobian elliptic integral. Thus,

$$\pm\varphi = \frac{1}{\sqrt{\Delta}}\int^{\psi}\frac{d\gamma}{\sqrt{(1-k^2\sin^2\gamma)}}, \tag{182}$$

where

$$k^2 = \frac{1}{2\Delta}(\Delta + 6\mu - 1), \tag{183}$$

and

$$\begin{aligned}\sin^2\psi &= \frac{1}{\Delta+6\mu-1}\{\Delta - 2\mu e\sin\xi - (6\mu-1)\cos\xi)\}\\ &= \frac{1}{\Delta+6\mu-1}\{\Delta+6\mu-1-2[2\mu e\sin\tfrac{1}{2}\xi + (6\mu-1)\cos\tfrac{1}{2}\xi]\cos\tfrac{1}{2}\xi\}.\end{aligned} \tag{184}$$

From equations (179) and (184), it follows that

$$\sin^2\psi = 1 \qquad \text{both when} \qquad \xi = \xi_0 \quad \text{(at aphelion)}$$
$$\text{and} \qquad \xi = \pi \quad \text{(at the singularity)}. \tag{185}$$

Moreover,

$$\sin^2\psi = 0 \qquad \text{when} \qquad \xi = \tan^{-1}\frac{2\mu e}{6\mu-1}. \tag{186}$$

Therefore, ψ assumes the value zero within the range (177) of ξ. We conclude that the range of ψ associated with the range of ξ is

$$-\pi/2 \leqslant \psi \leqslant +\pi/2 \qquad (\xi_0 \leqslant \xi \leqslant \pi). \tag{187}$$

Accordingly, we may write the solution for φ as

$$\varphi = \frac{1}{\sqrt{\Delta}}\{K(k) - F(\psi, k)\}, \tag{188}$$

where $K(k)$ denotes the complete elliptic integral and $F(\psi, k)$ (as usual) the incomplete Jacobian integral. In writing the solution for φ in the form (188), we have assumed that $\varphi = 0$ at the singularity where $\xi = \pi$ and $\psi = \pi/2$. The value of φ at aphelion, where $\xi = \xi_0$ and $\psi = -\pi/2$, is

$$\varphi_{ap} = 2K(k)/\sqrt{\Delta}. \tag{189}$$

(See Fig. 7*a*, (g), (h), p. 119, where examples of orbits derived on the basis of equation (188) are illustrated.)

This completes our discussion of the bound orbits. We now turn to the unbound orbits with $E^2 \geqslant 1$.

(c) *The unbound orbits* ($E^2 > 1$)

When $E^2 > 1$, the constant term in $f(u)$ is positive. The equation $f(u) = 0$ must, therefore, allow a negative root; and the cases we must distinguish (fewer than for bound orbits) are those shown in Fig. 6. [Again, it can be shown that the case of all three real roots being negative cannot arise (see footnote on p. 101).]

So long as there are three real roots and two of them are positive (distinct or coincident), we must continue to distinguish between orbits of two kinds: orbits of the first kind restricted to the interval, $0 < u \leqslant u_2$ (which are the analogues of the hyperbolic orbits of the Newtonian theory) and the orbits of the second kind with $u \geqslant u_3$ (which are, in essence, no different from the bound orbits of the second kind). When $u_2 = u_3$, the two kinds of orbits coalesce as they approach, asymptotically, a common circle from opposite sides by spiralling round it an infinite number of times. And finally, when the equation $f(u) = 0$ allows a pair of complex-conjugate roots (besides a negative real root), the resulting orbits can be considered as belonging to imaginary eccentricities; and they differ from the bound orbits considered in §(*b*, iii) only in that they are not bound!

(i) *Orbits of the first and second kinds*

When all three roots are real, we shall continue to express them in terms of an eccentricity e as in the case of the bound orbits, with the only difference that now $e \geqslant 1$. Thus, we shall write (cf. equation (114))

$$u_1 = -\frac{1}{l}(e-1), \quad u_2 = \frac{1}{l}(e+1), \quad \text{and} \quad u_3 = \frac{1}{2M} - \frac{2}{l} \quad (e \geqslant 1). \tag{190}$$

The inequality (118), namely,

$$1 - 6\mu - 2\mu e \geqslant 0, \tag{191}$$

continues to hold, since it is a consequence only of the assumed ordering of the roots: $u_1 < u_2 \leqslant u_3$. The relations (121) also continue to hold, again with the

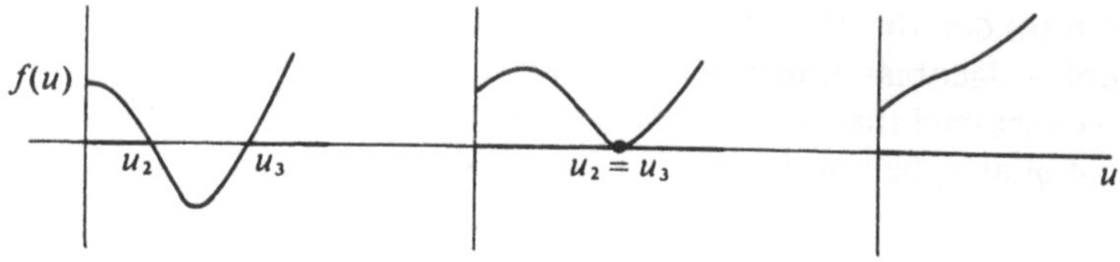

FIG. 6. The disposition of the roots of the cubic equation $f(u) = 0$ for $E^2 > 1$.

difference that now $e \geqslant 1$. We, therefore, write

$$\frac{1}{L^2} = \frac{1}{lM}[1 - \mu(3 + e^2)] \qquad \text{and} \qquad \frac{E^2 - 1}{L^2} = \frac{1}{l^2}(1 - 4\mu)(e^2 - 1). \tag{192}$$

Since $L^2 > 0$ and $E^2 - 1 \geqslant 0$ (by assumption),

$$1 - \mu(3 + e^2) > 0 \tag{193}$$

and

$$\mu \leqslant \tfrac{1}{4}. \tag{194}$$

The requirement, $\mu \leqslant \frac{1}{4}$, is guaranteed by both the inequalities (191) and (193). With regard to these latter two inequalities, one readily verifies that (193) ensures (191) for $e < 3$ while (191) ensures (193) for $e \geqslant 3$. In this connection, we may note that when $2\mu(3 + e) = 1$ (i.e., when the two roots u_2 and u_3 coincide) the relations (192) become (cf. equation (148))

$$\frac{L^2}{M^2} = 4\frac{(3 + e)^2}{(3 - e)(e + 1)} \qquad \text{and} \qquad E^2 - 1 = \frac{e^2 - 1}{9 - e^2}. \tag{195}$$

Accordingly, for these special orbits, the allowed range of e is

$$1 \leqslant e < 3; \tag{196}$$

and the corresponding perihelion-distances must lie in the range (cf. the range (147) for the bound orbits)

$$3M < r_p \leqslant 4M. \tag{197}$$

Quite generally, we may also note that since the orbits of the first kind (and the orbits belonging to imaginary eccentricities, as well) 'arrive from infinity', we can associate with them an impact parameter, D, and a velocity at infinity, V; these are related to L and E by

$$D^2 = \frac{L^2}{V^2} = \frac{L^2 E^2}{E^2 - 1} \qquad \left(E^2 = \frac{1}{1 - V^2}\right). \tag{198}$$

Turning to the analytical representation of these orbits, we shall make the same substitution as before, namely,

$$u = \frac{1}{l}(1 + e\cos\chi). \tag{199}$$

However, since $e \geqslant 1$,

$$u = 0 \quad \text{when} \quad \chi = \cos^{-1}(-e^{-1}) = \chi_\infty \quad \text{(say)}. \tag{200}$$

But the perihelion passage still occurs when $\chi = 0$. The allowed range of χ is, therefore,

$$0 \leqslant \chi < \chi_\infty = \cos^{-1}(-e^{-1}). \tag{201}$$

Apart from this restriction on the range of χ, the analysis in § (*b*, i) is applicable as it stands; and the solution for χ can be expressed in terms of the elliptic integral (130) with the same modulus k. Keeping in mind the restriction in the range of χ, we may now write the solution for φ in the form (cf. equation (132))

$$\varphi = \frac{2}{(1 - 6\mu + 2\mu e)^{1/2}}\left[K(k) - F(\tfrac{1}{2}\pi - \tfrac{1}{2}\chi, k)\right]. \tag{202}$$

In writing the solution for φ in the form (202), we have taken the origin of φ at perihelion passage where $\chi = 0$. The trajectory goes off to infinity, asymptotically, along the direction

$$\varphi = \varphi_\infty = \frac{2}{(1 - 6\mu + 2\mu e)^{1/2}}\left[K(k) - F(\psi_\infty, k)\right] \quad (\psi_\infty = \tfrac{1}{2}\cos^{-1}e^{-1}). \tag{203}$$

The limiting case of these orbits, when the two roots u_2 and u_3 coincide, is described by the same equation (150) (except, again, for the restriction in the range of χ). And as was stated in the context of the bound orbits, the unbound orbits approach the circle at r_p, asymptotically, by spiralling around it an infinite number of times.

The discussion of the bound orbits of the second kind in §(*b*, ii) applies to the unbound orbits without any change; only we are now concerned with values of $e \geqslant 1$.

(ii) The orbits with imaginary eccentricities

As we have shown in §(*b*, iii), the bound orbits which fall into the singularity are best characterized by imaginary eccentricities. The unbound orbits, which similarly fall into the singularity, can likewise be characterized by imaginary eccentricities; and the analysis of § (*b*, iii) can be adapted to the present context with only minor changes to allow for the orbits arriving from infinity rather than from finite aphelion-distances.

The relations (172) for unbound orbits are

$$\frac{1}{L^2} = \frac{1}{lM}[1 - \mu(3 - e^2)] \quad \text{and} \quad \frac{E^2 - 1}{L^2} = \frac{1}{l^2}(4\mu - 1)(1 + e^2). \tag{204}$$

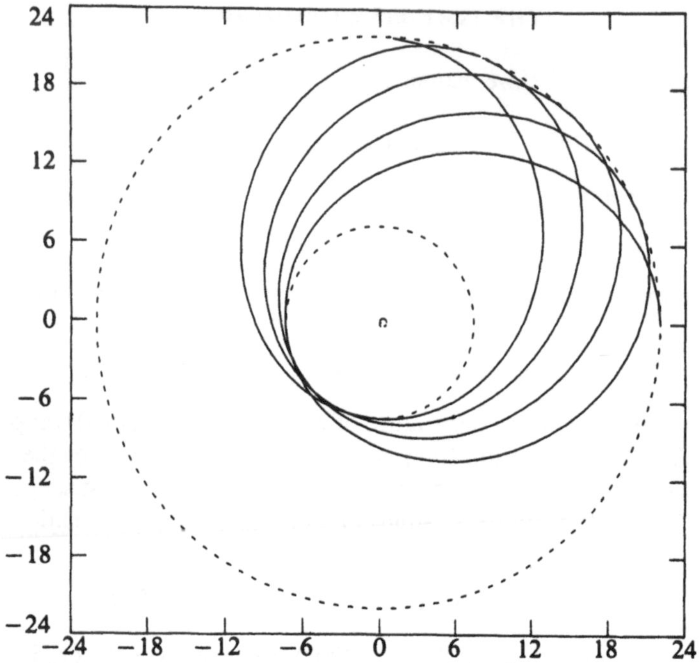

(a) $e = ½$, $\ell = 11$, $M = \frac{3}{14}$

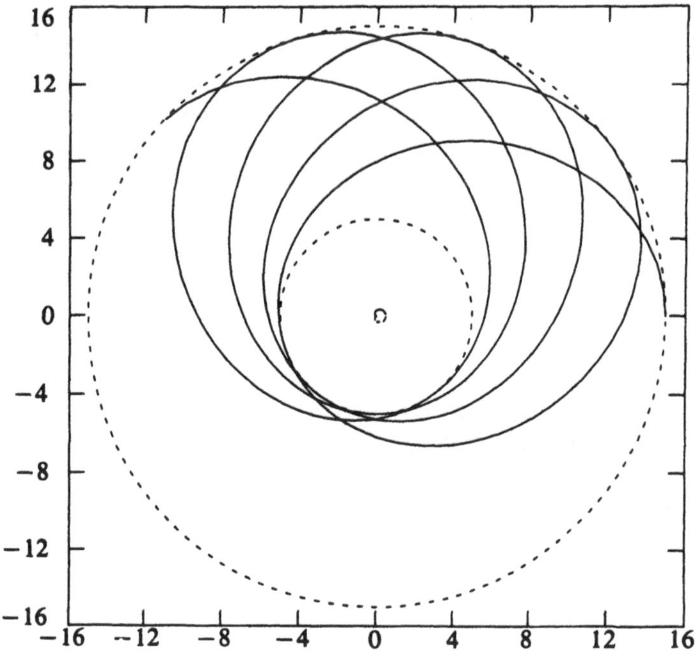

(b) $e = ½$, $\ell = 7.5$, $M = \frac{3}{14}$

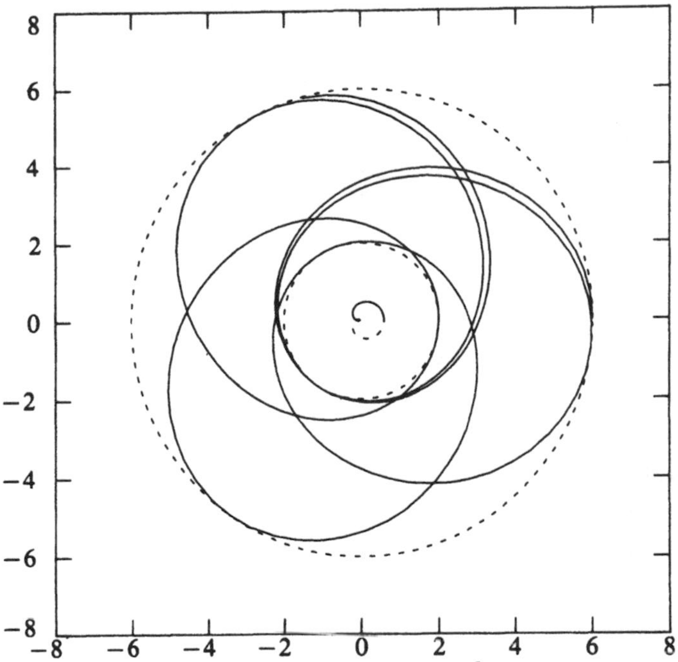

(c) $e = ½$, $\ell = 3$, $M = \dfrac{3}{14}$

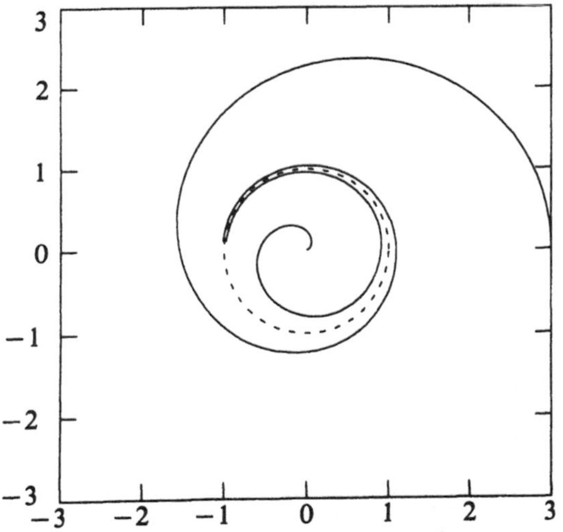

(d) $e = ½$, $\ell = \dfrac{3}{2}$, $M = \dfrac{3}{14}$

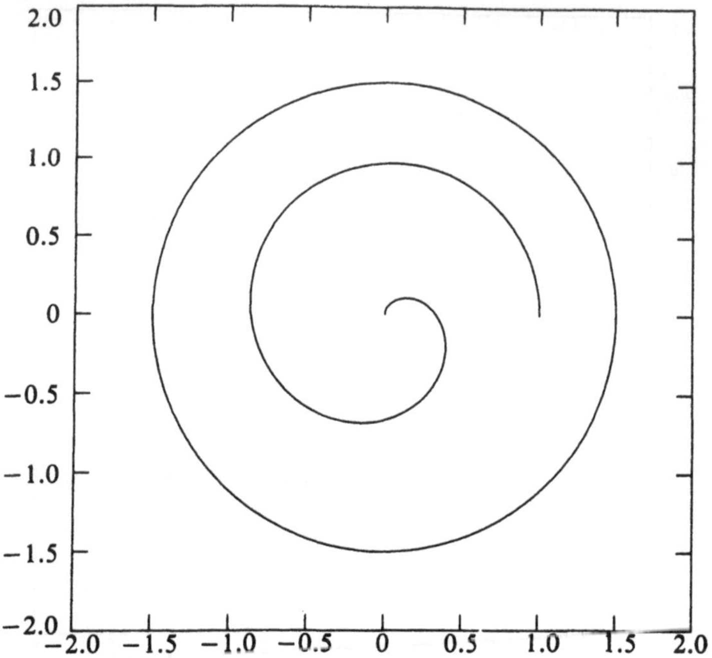

(e) $e = 0,\ \ell = \frac{3}{2},\ M = \frac{3}{14}$

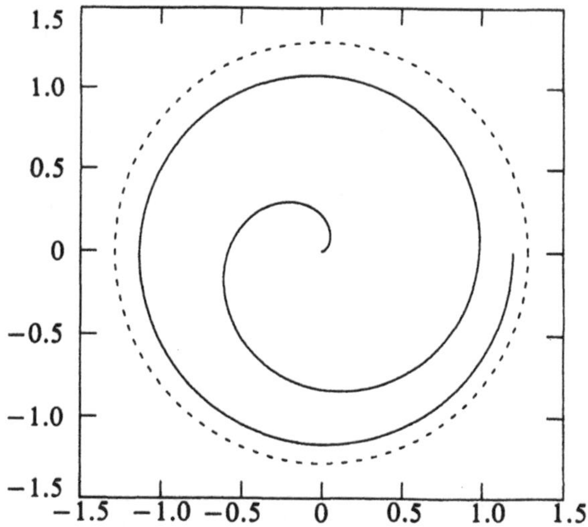

(f) $e = 0,\ \mu = \frac{1}{6},\ M = \frac{3}{14}$

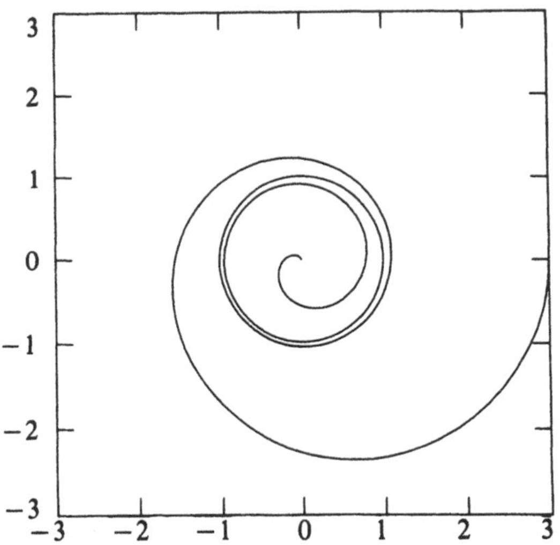

(g) $e = 0.01\,i,\ \ell = 1,\ M = \dfrac{3}{14}$

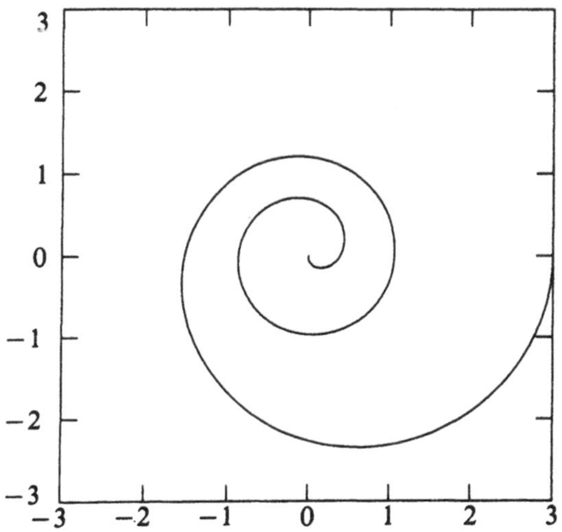

(h) $e = 0.1\,i,\ \ell = 1,\ M = \dfrac{3}{14}$

FIG. 7*a*. Various classes of time-like geodesics described by a test particle with $E^2 < 1$: (a), (b), (c): geodesics of the first and the second kind with eccentricity $e = 1/2$ and latera recta, $l = 11$, 7.5 and 3 respectively; (d): an example of a trajectory in which the orbits of the first and the second kind coalesce ($e = 1/2$, $l = 3/2$) for which $2\mu(3+e) = 1$); (e): an example of a circular orbit ($e = 0$, $l = 3/2$) and the associated orbit of the second kind; (f): the last unstable circular orbit when the orbit of the second kind spirals out of the orbit of the first kind ($e = 0$, $\mu = 1/6$); (g), (h): bound orbits with $l = 1$ and imaginary eccentricities $e = 0.01\,i$ and $0.1\,i$. (In all these figures $M = 3/14$ in the scale along the coordinate axes.)

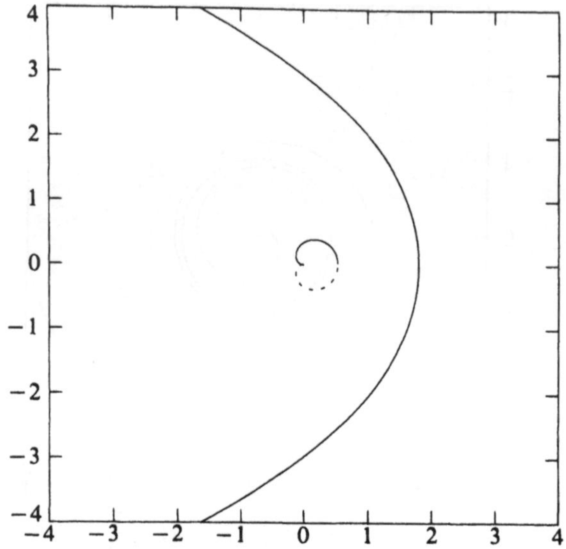

(a) $e=\frac{3}{2}$, $\ell=4.5$, $M=\frac{3}{14}$

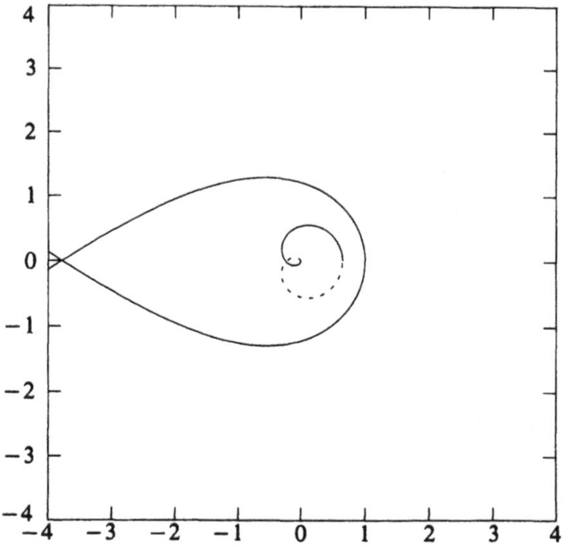

(b) $e=\frac{3}{2}$, $\ell=\frac{5}{2}$, $M=\frac{3}{14}$

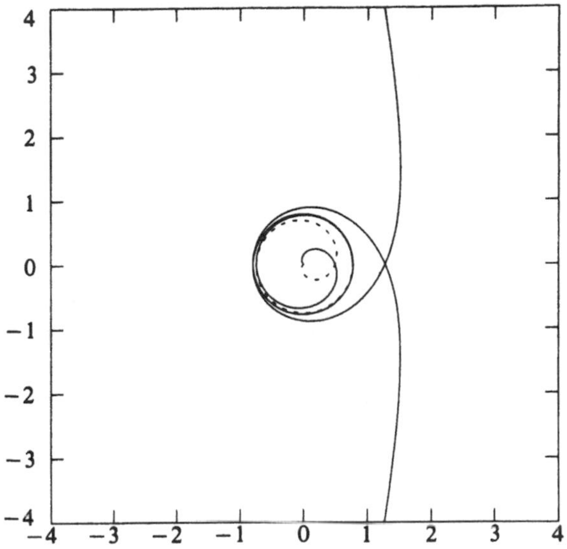

(c) $e=\frac{3}{2}$, $\ell=1.94$, $M=\frac{3}{14}$

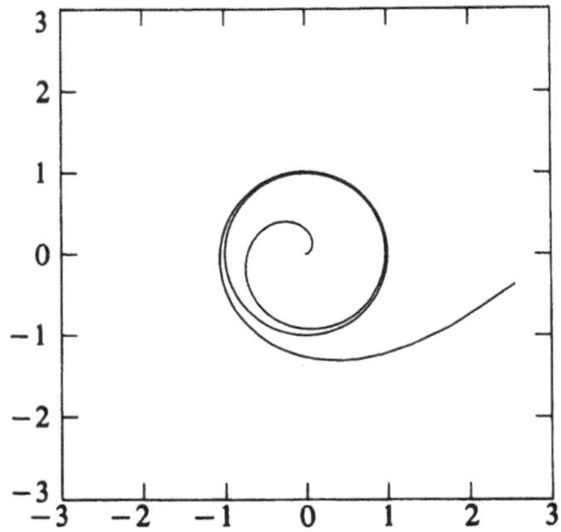

(d) $e=0.001i$, $\ell=1$, $M=0.3$

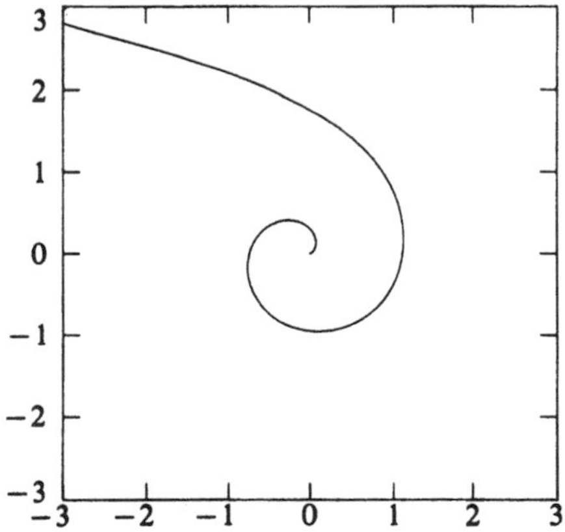

(e) $e=0.1i$, $\ell=1$, $M=0.3$

FIG. 7*b*. Various classes of time-like geodesics described by a test particle with $E^2 > 1$: (a), (b), (c): orbits of the first and the second kind with eccentricity $e = 3/2$ and latera recta, 4.5, 2.5, and 1.94 respectively ($M = 3/14$ in the scale along the coordinate axes); (d), (e): unbound orbits with $l = 1$ and with imaginary eccentricities $e = 0.001\,i$ and $0.1\,i$ ($M = 0.3$ in the scale along the coordinate axes).

Accordingly, we must now require

$$\mu \geqslant \tfrac{1}{4} \qquad \text{and} \qquad 1-3\mu+\mu e^2 > 0. \tag{205}$$

For $\frac{1}{3} \geqslant \mu \geqslant \frac{1}{4}$, the second inequality imposes no restrictions on e^2; but for $\mu > \frac{1}{3}$, it is necessary to impose the restriction

$$e^2 > 3 - 1/\mu \quad (\mu > \tfrac{1}{3}). \tag{206}$$

For obtaining the solution for these orbits, we make the same substitution (175) as before. However, in view of present requirement, $\mu \geqslant \frac{1}{4}$, the range of ξ must be terminated at ξ_∞ (*before* ξ_0 as defined in equation (179)) where

$$\tan \tfrac{1}{2}\xi_\infty = -e^{-1},$$

or,

$$\sin \tfrac{1}{2}\xi_\infty = -\frac{1}{\sqrt{(1+e^2)}} \qquad \text{and} \qquad \cos \tfrac{1}{2}\xi_\infty = \frac{e}{\sqrt{(1+e^2)}}. \tag{207}$$

For, when $\xi = \xi_\infty$, u becomes zero and when $\xi < \xi_\infty$, u becomes negative. Therefore, as $\xi \to \xi_\infty + 0$, $u \to 0$ and $r \to \infty$. But the upper limit on ξ, namely π (when $u \to \infty$ and $r \to 0$), is unaffected. The allowed range of ξ is, therefore,

$$\xi_\infty < \xi \leqslant \pi. \tag{208}$$

Apart from these changes, the solution for φ can again be written as (cf. equations (183), (184), and (188))

$$\varphi = \frac{1}{\sqrt{\Delta}}\{K(k) - F(\psi, k)\}. \tag{209}$$

The origin of φ (as in the earlier solution) is at the singularity (where $\xi = \pi$ and $\psi = +\pi/2$). But the lower limit on ψ, say ψ_∞, as the orbit goes off to infinity, can be obtained from equation (184) by inserting for $\sin \frac{1}{2}\xi$ and $\cos \frac{1}{2}\xi$ the values appropriate for ξ_∞ given in equations (207). We find

$$\sin^2 \psi_\infty = \frac{1}{\Delta + 6\mu - 1}\left(\Delta + 6\mu - 1 - 2e^2\frac{4\mu - 1}{e^2+1}\right). \tag{210}$$

(Note that for $\mu > \frac{1}{4}$, the right-hand side of equation (210) is less than 1; and that it is equal to 1 for $\mu = \frac{1}{4}$, and $\psi_\infty = -\pi/2$ as in the case for the bound orbits.)

In Fig. 7*b*, a number of examples of the different classes of unbound orbits are illustrated.

This completes our discussion of the time-like geodesics in the Schwarzschild space-time.

20. The geodesics in the Schwarzschild space-time: the null geodesics

As we have already explained in §19, the Lagrangian must be equated to zero for the null geodesics. Therefore, equation (89) must now be replaced by

$$\frac{E^2}{1-2M/r}-\frac{\dot{r}^2}{1-2M/r}-\frac{L^2}{r^2}=0, \tag{211}$$

or

$$\left(\frac{\mathrm{d}r}{\mathrm{d}\tau}\right)^2+\frac{L^2}{r^2}\left(1-\frac{2M}{r}\right)=E^2. \tag{212}$$

This equation must be considered together with

$$\left(1-\frac{2M}{r}\right)\frac{\mathrm{d}t}{\mathrm{d}\tau}=E \qquad \text{and} \qquad \frac{\mathrm{d}\varphi}{\mathrm{d}\tau}=\frac{L}{r^2}. \tag{213}$$

Again, by considering r as a function of φ and replacing r by $u = 1/r$ as the independent variable, we obtain the equation (cf. equation (94))

$$\left(\frac{\mathrm{d}u}{\mathrm{d}\varphi}\right)^2=2Mu^3-u^2+\frac{1}{D^2}=f(u) \quad \text{(say)}, \tag{214}$$

where

$$D=L/E \tag{215}$$

denotes the impact parameter. We observe that in contrast to the earlier definition, $f(u)$, as now defined, has no term linear in u; and, in addition, the constant term is always positive.

(a) *The radial geodesics*

We begin our consideration of the null geodesics with the radial geodesics. The relevant equations are (cf. equations (212) and (213))

$$\frac{\mathrm{d}r}{\mathrm{d}\tau}=\pm E \qquad \text{and} \qquad \left(1-\frac{2M}{r}\right)\frac{\mathrm{d}t}{\mathrm{d}\tau}=E. \tag{216}$$

Accordingly,

$$\frac{\mathrm{d}r}{\mathrm{d}t}=\pm\left(1-\frac{2M}{r}\right), \tag{217}$$

or, in the integrated form,

$$t=\pm r_*+\text{constant}_\pm, \tag{218}$$

where

$$r_*=r+2M\lg\left(\frac{r}{2M}-1\right). \tag{219}$$

The variable r_* to which we have now been led plays an extremely important role in subsequent developments. It is customarily defined by the equation (cf. equation (74))

$$\frac{\mathrm{d}}{\mathrm{d}r_*} = \frac{\Delta}{r^2}\frac{\mathrm{d}}{\mathrm{d}r}, \tag{220}$$

where

$$\Delta = r^2 - 2\,Mr = r^2 e^{2\nu} \tag{221}$$

is the *horizon function*. With r_* as defined in equation (219)

$$r_* \to -\infty \quad \text{as} \quad r \to 2M + 0 \qquad \text{and} \qquad r_* \to r \quad \text{as} \quad r \to +\infty. \tag{222}$$

The importance of the variable r_* arises from this fact: its range from $-\infty$ to $+\infty$ exhausts the entire part of the space-time that is accessible to observers outside the horizon.

Equation (218) must be contrasted with the equation

$$r = \pm E\tau + \text{constant}_{\pm} \tag{223}$$

which relates r to the proper time τ. Equations (218) and (223) show that while the radial geodesic crosses the horizon in its own proper time without ever noticing it, it takes an infinite coordinate-time even to arrive at the horizon: its 'crossing the bar' is not within the realm of experience of a mere observer outside the horizon. All of this is made manifest in Fig. 8 where the null cones included between the geodesics (217) are illustrated.

The tangent vectors associated with the radial geodesics (217) are

$$\frac{\mathrm{d}t}{\mathrm{d}\tau} = \frac{r^2}{\Delta}E, \quad \frac{\mathrm{d}r}{\mathrm{d}\tau} = \pm E, \quad \frac{\mathrm{d}\theta}{\mathrm{d}\tau} = 0, \qquad \text{and} \qquad \frac{\mathrm{d}\varphi}{\mathrm{d}\tau} = 0. \tag{224}$$

We shall find in §21, that these null vectors provide the basis for constructing a null tetrad-frame towards a description of the Schwarzschild space-time in a Newman–Penrose formalism.

(*b*) *The critical orbits*

Returning to the general equation (214), we first consider the different cases that must be distinguished. They clearly relate to the disposition of the roots of the cubic equation

$$f(u) = 2Mu^3 - u^2 + \frac{1}{D^2} = 0. \tag{225}$$

The sum and the product of the roots u_1, u_2, and u_3 of this equation are given by

$$u_1 + u_2 + u_3 = \frac{1}{2M} \qquad \text{and} \qquad u_1 u_2 u_3 = -\frac{1}{2MD^2}. \tag{226}$$

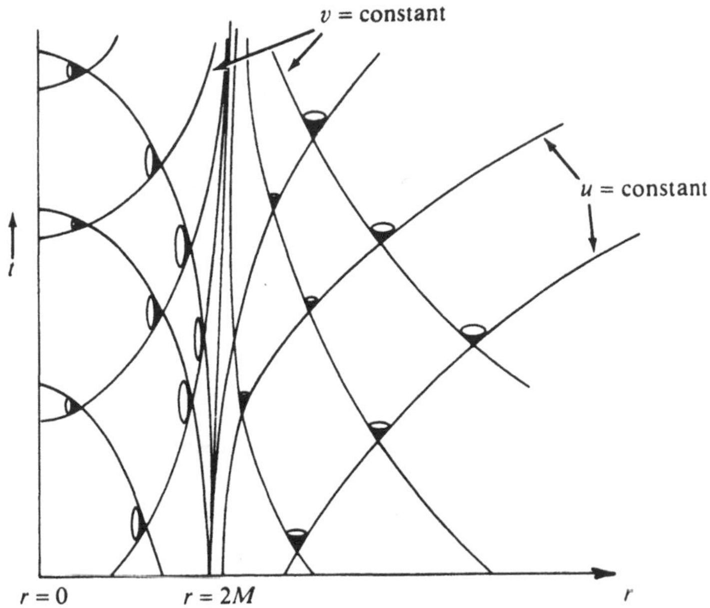

FIG. 8. Illustrating the ingoing and outgoing radial null-geodesics in the Schwarzschild coordinates.

Clearly the equation $f(u) = 0$ must allow a negative real root; and the two remaining roots may either be real (distinct or coincident) or be a complex-conjugate pair. The different cases that must be distinguished are, therefore, essentially the same as those for the unbound time-like geodesics. In the present instance, the case when two of the positive real roots* coincide plays a specially decisive role in the discrimination of the null geodesics. On this account, we shall consider this critical case first.

The conditions for the occurrence of two coincident roots can be obtained as follows. The derivative of equation (225), namely,

$$f'(u) = 6Mu^2 - 2u = 0, \tag{227}$$

allows $u = (3M)^{-1}$ as a root; and $u = (3M)^{-1}$ will be a root of equation (225) (indeed, a double root) if

$$D^2 = 27M^2 \qquad \text{or} \qquad D = (3\sqrt{3})M. \tag{228}$$

* The impossibility of three real negative roots follows from the equation $u_1u_2 + u_2u_3 + u_3u_1 = 0$.

From the condition on the product of the roots, we infer that the roots of $f(u) = 0$ are

$$u_1 = -1/6M \quad \text{and} \quad u_2 = u_3 = 1/3M \quad \text{and} \quad D = (3\sqrt{3})M. \tag{229}$$

Moreover when D has the value (228), $du/d\varphi$ vanishes when $u = (3M)^{-1}$. Therefore, *a circular orbit of radius 3M is an allowed null-geodesic*. This circular orbit cannot, however, be a stable one. Its place in the family of the null geodesics can be understood by considering the full equation, appropriate to this case, namely,

$$\left(\frac{du}{d\varphi}\right)^2 = 2M\left(u + \frac{1}{6M}\right)\left(u - \frac{1}{3M}\right)^2. \tag{230}$$

We verify that this equation is satisfied by the substitution

$$u = -\frac{1}{6M} + \frac{1}{2M}\tanh^2\tfrac{1}{2}(\varphi - \varphi_0), \tag{231}$$

where φ_0 is a constant of integration. If φ_0 is so chosen that

$$\tanh^2\tfrac{1}{2}\varphi_0 = \tfrac{1}{3}, \tag{232}$$

then

$$u = 0 \quad \text{and} \quad r \to \infty \quad \text{when} \quad \varphi = 0. \tag{233}$$

But we also notice that

$$u = \frac{1}{3M} \quad \text{when} \quad \varphi \to \infty. \tag{234}$$

Therefore, a null geodesic arriving from infinity with an impact parameter $D = (3\sqrt{3})M$ approaches the circle of radius $3M$, asymptotically, by spiralling around it.

Associated with the orbit (231), we must have an 'orbit of the second kind' which, originating at the singularity, approaches, from the opposite side, the same circle at $r = 3M$, asymptotically, by spiralling around it. Such an orbit can be obtained by the substitution

$$u = \frac{1}{3M} + \frac{1}{2M}\tan^2\tfrac{1}{2}\xi \tag{235}$$

in the same equation (230). We find that the equation then reduces to

$$\left(\frac{d\xi}{d\varphi}\right)^2 = \sin^2\tfrac{1}{2}\xi; \tag{236}$$

and we may take

$$\varphi = 2\,\lg\,(\tan\tfrac{1}{4}\xi) \quad \text{or} \quad \tan\tfrac{1}{4}\xi = e^{\varphi/2} \tag{237}$$

as the appropriate solution. Inserting this solution in equation (235), we obtain

$$u = \frac{1}{3M} + \frac{2e^{\varphi}}{M(e^{\varphi}-1)^2}. \tag{238}$$

Along this orbit

$$\left.\begin{aligned} &u \to \infty \quad \text{and} \quad r \to 0 \quad \text{when} \quad \varphi \to 0 \\ \text{and} \quad &u \to (3M)^{-1} \quad \text{as} \quad \varphi \to \infty; \end{aligned}\right\} \tag{239}$$

it has accordingly the attributes that were stated. As we have explained in a similar context in §19(*b*, ii), the solution (239) with the sign of φ reversed may be considered as a 'continuation' of the solution (231).

In Fig. 9 the various classes of null geodesics that can occur are illustrated.

(i) *The cone of avoidance*

At any point we can define a '*cone of avoidance*' whose generators are the null rays, described by the solution (231) of equation (230), passing through that point, since, as is clear on general grounds and as we shall establish analytically in §(*c*) below, light rays, included in the cone, must necessarily cross the horizon and get trapped.

If ψ denotes the half-angle of the cone (directed inward at large distances), then

$$\cot\psi = +\frac{1}{r}\frac{d\tilde{r}}{d\varphi}, \tag{240}$$

where

$$d\tilde{r} = (1-2M/r)^{-1/2}\,dr \tag{241}$$

is an element of proper length along the generators of the cone. Therefore,

$$\cot\psi = +\frac{1}{r(1-2M/r)^{1/2}}\frac{dr}{d\varphi} = -\frac{1}{u(1-2Mu)^{1/2}}\frac{du}{d\varphi}, \tag{242}$$

where $u = 1/r$. In equation (242) we may substitute for $du/d\varphi$ from equation (230). In this manner we obtain

$$\cot\psi = -\frac{1}{(r/2M-1)^{1/2}}\left(1-\frac{r}{3M}\right)\left(1+\frac{r}{6M}\right)^{1/2} \tag{243}$$

or

$$\tan\psi = \frac{(r/2M-1)^{1/2}}{(r/3M-1)(r/6M+1)^{1/2}}. \tag{244}$$

From this last equation it follows that

$$\left.\begin{aligned} &\psi \sim \frac{3\sqrt{3}}{r}M \quad \text{as} \quad r\to\infty; \quad \psi = \tfrac{1}{2}\pi \quad \text{for} \quad r = 3M, \\ &\text{and} \quad \psi = \pi \quad \text{for} \quad r = 2M. \end{aligned}\right\} \tag{245}$$

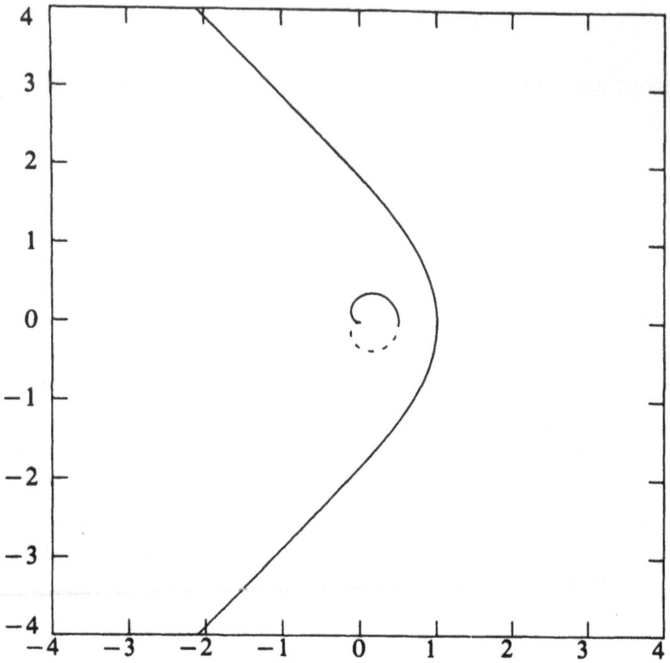

(a) $P = 1,\ M = \dfrac{3}{14}$

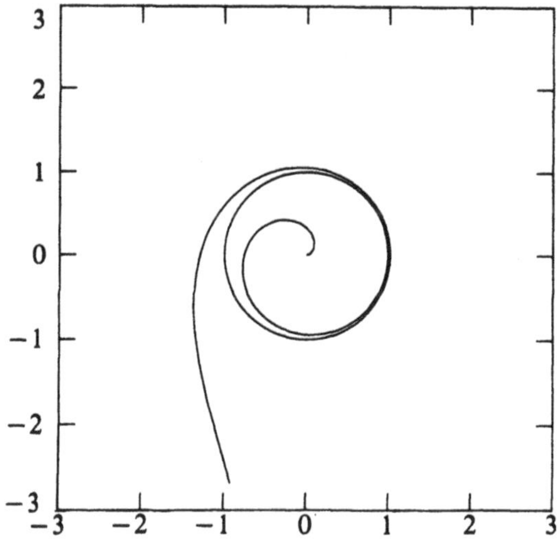

(c) $e = 0.001i,\ \ell = 1,\ M = \dfrac{1}{3}$

FIG. 9. Various classes of null geodesics in the Schwarzschild metric: (*a*): a null geodesic with $P = 1$ (cf. equation (251)) illustrating orbits of the first and the second kind; (*b*): the critical null-geodesic, with $D = D_c = 3\sqrt{3}.M$, for which the orbits of the two kinds spiral towards the unstable

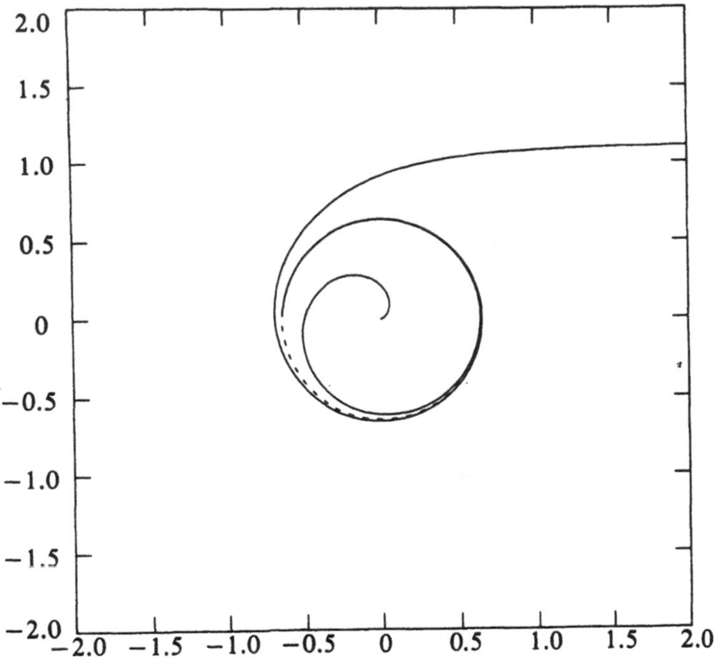

(b) $D = D_c = 3\sqrt{3}.M,\ M = \frac{3}{14}$

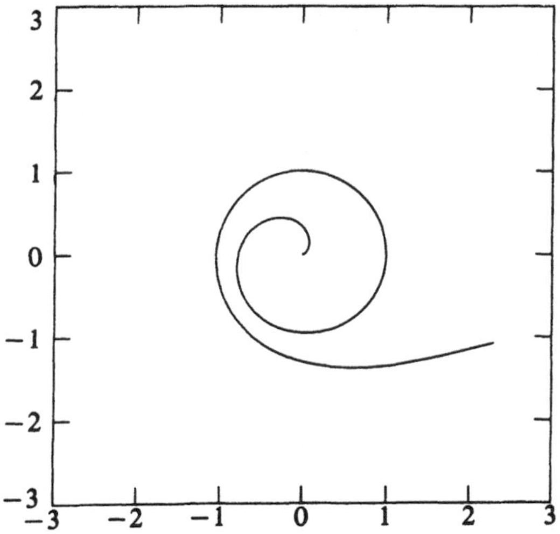

(d) $e = 0.01i,\ \ell = 1,\ M = \frac{1}{3}$

circular orbit at $3M$ ($M = 3/14$ in the scale along the coordinate axes for (*a*) and (*b*)); (*c*), (*d*): orbits with imaginary eccentricities ($e = 0.001\,i$ and $0.01\,i$ with $l = 1$) ($M = 1/3$ in the scale along the coordinate axes).

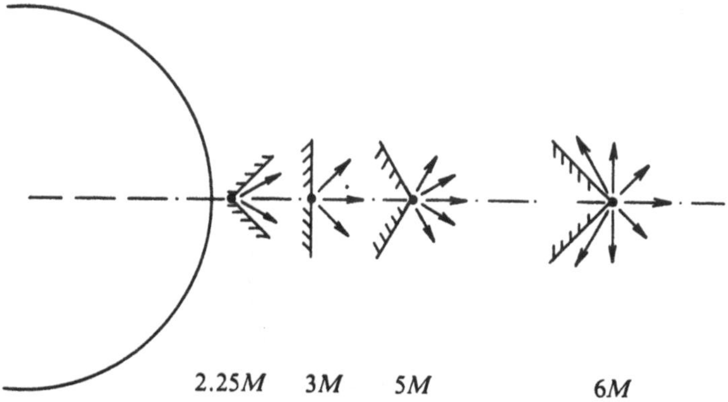

FIG. 10. The cone of avoidance at different distances from the centre.

Therefore, the cone of avoidance is narrow and, at large distances, its angle is that subtended by a disc of radius $(3\sqrt{3})M$. The cone opens out fully at $r = 3M$; and for $r < 3M$, it is directed outward and becomes narrower as $r \to 2M$; and at $r = 2M$, everything is blotted out (see Fig. 10).

(c) *The geodesics of the first kind*

We now consider the case when all the roots of the cubic equation $f(u) = 0$ are real and the two positive roots are distinct. Let the roots be

$$u_1 = \frac{P-2M-Q}{4MP} (<0), \quad u_2 = \frac{1}{P}, \quad \text{and} \quad u_3 = \frac{P-2M+Q}{4MP}, \tag{246}$$

where P denotes the perihelion distance and Q is a constant to be specified presently. The sum of the roots has been arranged to be equal to $1/2M$ as required (cf. equation (226)). Also, it should be noted that the ordering of the roots, $u_1 < u_2 < u_3$, requires that

$$Q + P - 6M > 0. \tag{247}$$

Next, evaluating

$$f(u) = 2M(u-u_1)(u-u_2)(u-u_3) \tag{248}$$

with u_1, u_2, and u_3 as specified in equations (246), and comparing the result with the expression (225), we obtain the relations

$$Q^2 = (P-2M)(P+6M) \tag{249}$$

and

$$\frac{1}{D^2} = \frac{1}{8MP^3}[Q^2 - (P-2M)^2]. \tag{250}$$

With the aid of the first of these relations, the second simplifies to give

$$D^2 = \frac{P^3}{P-2M}. \tag{251}$$

Combined with the relation (249), the inequality (247) gives

$$(P-2M)(P+6M) > (P-6M)^2, \tag{252}$$

or, after simplification,

$$P > 3M. \tag{253}$$

From equation (251), we now obtain the further inequality

$$D > (3\sqrt{3})M = D_c \quad \text{(say)}. \tag{254}$$

Therefore, the orbits we are now considering have impact parameters in excess of the critical value leading to the special solutions considered in §(*b*) above. Moreover, they are entirely outside the circle $r = 3M$.

We now make the substitution

$$u - \frac{1}{P} = -\frac{Q-P+6M}{8MP}(1+\cos\chi), \tag{255}$$

or, equivalently,

$$u + \frac{Q-P+2M}{4MP} = \frac{Q-P+6M}{8MP}(1-\cos\chi). \tag{256}$$

By this substitution,

$$\left.\begin{aligned} u = \frac{1}{P} \quad &\text{when } \chi = \pi, \text{ and} \\ u = 0 \text{ and } r \to \infty \quad &\text{when } \sin^2\tfrac{1}{2}\chi = \frac{Q-P+2M}{Q-P+6M} = \sin^2\tfrac{1}{2}\chi_\infty \quad \text{(say)}. \end{aligned}\right\} \tag{257}$$

We further find that the substitutions (255) and (256) reduce equation (214) to

$$\left(\frac{d\chi}{d\varphi}\right)^2 = \frac{Q}{P}(1 - k^2\sin^2\tfrac{1}{2}\chi), \tag{258}$$

where

$$k^2 = \frac{1}{2Q}(Q-P+6M). \tag{259}$$

The solution for φ can, therefore, be expressed in terms of the Jacobian elliptic integral in the form

$$\varphi = 2\left(\frac{P}{Q}\right)^{1/2}[K(k) - F(\tfrac{1}{2}\chi, k)], \tag{260}$$

where the origin of φ has been chosen at perihelion passage when $\chi = \pi$ (see equations (257)). The asymptotic value of φ, as $r \to \infty$, is given by

$$\varphi_\infty = 2\left(\frac{P}{Q}\right)^{1/2} [K(k) - F(\tfrac{1}{2}\chi_\infty, k)], \tag{261}$$

where χ_∞ is specified in (257). (See Fig. 9(a), p. 128.)

(i) *The asymptotic behaviours of φ_∞ for $P \to 3M$ and for $P/3M \gg 1$*

It is of interest to obtain the asymptotic behaviours of φ_∞ for $P \to 3M$ and for $P/3M \to \infty$.

First, we observe that

$$\left.\begin{aligned} &\text{when}\quad P = 3M, \quad Q = 3M, \quad D = (3\sqrt{3})M, \quad k^2 = 1, \\ &\sin^2 \tfrac{1}{2}\chi_\infty = \tfrac{1}{3}, \qquad \text{and} \qquad F(\tfrac{1}{2}\chi_\infty, 1) = \tfrac{1}{2}\lg\frac{\sqrt{3}+1}{\sqrt{3}-1}. \end{aligned}\right\} \tag{262}$$

Next, we find from equations (249), (251), and (259) that

$$\begin{aligned} &\text{if}\quad P = M(3+\delta), \qquad Q = M(3+\tfrac{5}{3}\delta), \quad k'^2 = 1-k^2 = \tfrac{4}{9}\delta, \\ &\text{and} \\ &\qquad D = D_c + \delta D \qquad \text{where} \qquad \delta D = (\tfrac{1}{2}\sqrt{3})\, M\,\delta^2. \end{aligned} \tag{263}$$

Also, we have the known asymptotic relation,

$$K(k) \to \lg(4/k') = \lg 6 - \tfrac{1}{2}\lg\delta \quad (k' \to 0). \tag{264}$$

Inserting these various relations in equation (261), we find that

$$\varphi_\infty = \tfrac{1}{2}\lg\left[\frac{6^4\sqrt{3}}{2}\frac{(\sqrt{3}-1)^2}{(\sqrt{3}+1)^2}\right] - \tfrac{1}{2}\lg\frac{\delta D}{M}, \tag{265}$$

or

$$\frac{\delta D}{M} \to \frac{6^4\sqrt{3}}{2}\frac{(\sqrt{3}-1)^2}{(\sqrt{3}+1)^2}e^{-2\varphi_\infty}. \tag{266}$$

Letting

$$\varphi_\infty = \tfrac{1}{2}(\pi + \Theta), \tag{267}$$

we obtain

$$\frac{\delta D}{M} = 648\sqrt{3}\frac{(\sqrt{3}-1)^2}{(\sqrt{3}+1)^2}e^{-\pi}e^{-\Theta} = 3.4823\,e^{-\Theta}; \tag{268}$$

and this is the required relation.

It should be noted that geodesics, that are deflected by an angle Θ given by equation (268), include not only the one that has suffered this deflection, but also the ones which have been deflected by $\Theta + 2n\pi$ ($n = 1, 2, \ldots$): they derive from geodesics with the impact parameters

$$D_n = D_c + 3.4823\,M\,e^{-(\Theta+2n\pi)}. \tag{269}$$

By similar (and simpler) arguments, it can be shown that for $P \gg 3M$, the deflection Θ (as defined in equation (267)) is given by

$$\Theta \to \frac{4M}{D} \quad \text{and} \quad D \to P\left(1+\frac{M}{P}\right). \tag{270}$$

The first of these relations is the celebrated one which provided the basis for one of the early (and spectacular) confirmations of the general theory of relativity.

(*d*) *The geodesics of the second kind*

To obtain the null geodesics, whose ranges are $u_3 \leqslant u < \infty$, we make the substitution

$$u = \frac{1}{P} + \frac{Q+P-6M}{4MP}\sec^2 \tfrac{1}{2}\chi, \tag{271}$$

in place of (255). By this substitution,

$$\left.\begin{array}{l} u \text{ is at aphelion when} \quad u = u_3 = \dfrac{1}{4MP}(Q+P-2M) \quad \text{and} \quad \chi = 0 \\ \qquad \text{and} \quad u \to \infty \quad \text{and} \quad r \to 0 \quad \text{when} \quad \chi = \pi. \end{array}\right\} \tag{272}$$

The substitution (271) reduces equation (214) to the same form (258) and with the same value of k^2; and we may now write (cf. equation (260))

$$\varphi = 2\left(\frac{P}{Q}\right)^{1/2} F(\tfrac{1}{2}\chi, k), \tag{273}$$

where the origin of φ is now at aphelion passage.

(*e*) *The orbits with imaginary eccentricities and impact parameters less than* $(3\sqrt{3})\,M$

Finally, we consider the orbits when the equation $f(u) = 0$ has a pair of complex-conjugate roots, besides a negative real root. As in §19(*b*, iii), we shall now write the roots in terms of an imaginary eccentricity ie as in equations (171). On evaluating

$$f(u) = 2M\left(u - \frac{1}{2M} + \frac{2}{l}\right)\left[\left(u - \frac{1}{l}\right)^2 + \frac{e^2}{l^2}\right], \tag{274}$$

and comparing it with $f(u)$ given in equation (214), we obtain the relations

$$l - M(3-e^2) = 0 \quad \text{and} \quad \frac{1}{2MD^2} = \left(\frac{2}{l} - \frac{1}{2M}\right)\frac{1+e^2}{l^2} \tag{275}$$

or, alternatively, in terms of $\mu = M/l$, we have

$$e^2 = \frac{1}{\mu}(3\mu - 1) \quad \text{and} \quad \frac{D^2}{M^2} = \frac{1}{\mu(4\mu-1)^2}. \tag{276}$$

These relations require

$$\mu > \tfrac{1}{3} \quad \text{and} \quad D < (3\sqrt{3})M. \tag{277}$$

Therefore, these orbits have impact parameters less than the value which distinguishes the orbits considered in §(*c*) above.

The analysis of §19(*b*, iii) applies equally to the present with the only proviso that e^2 and μ are no longer to be considered as independent parameters but related: $e^2 = 3 - \mu^{-1}$. Since these orbits are unbounded, the solution for φ is given by the same equation (209) as in §19(*c*, ii). In particular,

$$\varphi_\infty = \frac{1}{\sqrt{\Delta}}\{K(k) - F(\psi_\infty, k)\}, \tag{278}$$

where ψ_∞ (on substituting for e^2 its present value, $3 - \mu^{-1}$) is now given by

$$\sin^2 \psi_\infty = \frac{\Delta + 1}{\Delta + 6\mu - 1}, \tag{279}$$

where it should be remembered that $\mu > \frac{1}{3}$; and further that

$$\Delta = (48\mu^2 - 16\mu + 1)^{1/2}. \tag{280}$$

Examples of orbits derived on the basis of equations (230), (238), (260), (273), and (278) are illustrated in Fig. 9, (c) and (d) p. 129.

This completes our considerations relating to the geodesics in the Schwarzschild space-time.

21. The description of the Schwarzschild space-time in a Newman–Penrose formalism

We shall conclude our consideration of the Schwarzschild space-time by a description of it in a Newman–Penrose formalism. For the construction of a null-tetrad frame needed for such a description, we start with the null vectors (224) representing radial null-geodesics. Precisely, for the real null-vectors $\boldsymbol{l}$ and $\boldsymbol{n}$ of the Newman–Penrose formalism we shall take

$$\left.\begin{aligned} l^i &= (l^t, l^r, l^\theta, l^\varphi) = \frac{1}{\Delta}(r^2, +\Delta, 0, 0) \\ \text{and} \qquad & \\ n^i &= (n^t, n^r, n^\theta, n^\varphi) = \frac{1}{2r^2}(r^2, -\Delta, 0, 0), \end{aligned}\right\} \tag{281}$$

where the vector, associated with the negative sign for $\mathrm{d}r/\mathrm{d}\tau$ in (224), has been multiplied by $\Delta/2r^2$ to satisfy the required orthogonality condition

$$\boldsymbol{l}\cdot\boldsymbol{n} = 1. \tag{282}$$

On this account, the vector $\boldsymbol{n}$, unlike the vector $\boldsymbol{l}$, is not affinely parametrized.

We now complete the tetrad basis by adjoining to $\boldsymbol{l}$ and $\boldsymbol{n}$ the complex null-vector,

$$m^i = (m^t, m^r, m^\theta, m^\varphi) = \frac{1}{r\sqrt{2}}(0, 0, 1, i \operatorname{cosec} \theta), \tag{283}$$

which is orthogonal to $\boldsymbol{l}$ and $\boldsymbol{n}$ and satisfies the further normalization condition

$$\boldsymbol{m} \cdot \bar{\boldsymbol{m}} = -1. \tag{284}$$

The contravariant vectors l^i, n^i, m^i, and $\bar{m}^i$ provide the required basis. The corresponding covariant vectors are

$$\left.\begin{aligned} l_i &= (1, -\frac{r^2}{\Delta}, 0, 0), \\ n_i &= \frac{1}{2r^2}(\Delta, +r^2, 0, 0), \\ m_i &= \frac{1}{r\sqrt{2}}(0, 0, -r^2, -ir^2 \sin\theta). \end{aligned}\right\} \tag{285}$$

The spin coefficients (as defined in Chapter 1, equations (286)) with respect to the chosen basis are most conveniently evaluated via the λ-symbols (Chapter 1, equation (266)) even as the rotation coefficients (in a different basis) were evaluated in Chapter 2, §14. We now find that the non-vanishing λ symbols are

$$\lambda_{122} = -\frac{M}{r^2}, \quad \lambda_{243} = -\frac{r-2M}{2r^2}, \quad \lambda_{341} = -\frac{1}{r}, \quad \text{and} \quad \lambda_{334} = \frac{\cot\theta}{r\sqrt{2}}; \tag{286}$$

and the spin coefficients, derived with their aid, are

$$\kappa = \sigma = \lambda = \nu = \varepsilon = \pi = \tau = 0, \tag{287}$$

and

$$\rho = -\frac{1}{r}, \quad \beta = -\alpha = \frac{1}{2\sqrt{2}}\frac{\cot\theta}{r}, \quad \mu = -\frac{r-2M}{2r^2}, \quad \text{and} \quad \gamma = \frac{M}{2r^2}. \tag{288}$$

The fact that the spin coefficients κ, σ, λ, and ν vanish shows that the congruences of the null geodesics, $\boldsymbol{l}$ and $\boldsymbol{n}$, are shear-free as is, indeed, obvious from the purely radial character of these null geodesics in a spherically symmetric background. From the shear-free character of these congruences, we can conclude on the basis of the Goldberg–Sachs theorem that *the Schwarzschild space-time is of Petrov type-*D.

The Goldberg–Sachs theorem also allows us to conclude that in the chosen basis, the Weyl scalars, Ψ_0, Ψ_1, Ψ_3, and Ψ_4 vanish; and that the only non-vanishing scalar is Ψ_2. These conclusions can be directly verified by contracting the non-vanishing components of the Riemann tensor listed in equations (76) with the vectors $\boldsymbol{l}$, $\boldsymbol{n}$, $\boldsymbol{m}$, and $\bar{\boldsymbol{m}}$ in accordance with the definitions of the Weyl scalars in Chapter 1, equations (294) (remembering that the tensor indices 0, 1, 2, and 3 correspond to the coordinate indices t, φ, r, and θ, respectively). The vanishing of the scalars Ψ_0, Ψ_1, Ψ_3, and Ψ_4 is readily verified. Considering the scalar Ψ_2, we have

$$\begin{aligned}\Psi_2 &= R_{ijkl}l^i m^j n^k \bar{m}^l \\ &= \frac{r^2}{2\Delta}(R_{0j0l} - e^{2\nu}R_{0j2l} - e^{2\nu}R_{2j0l} - e^{4\nu}R_{2j2l})m^j\bar{m}^l \\ &= \frac{e^{-2\nu}}{4r^2}(R_{0101}\operatorname{cosec}^2\theta + R_{0303} - e^{4\nu}R_{2323} - e^{4\nu}R_{1212}\operatorname{cosec}^2\theta), \qquad (289)\end{aligned}$$

or, inserting for the components of the Riemann tensor their values given in equations (76), we find

$$\Psi_2 = -Mr^{-3}. \qquad (290)$$

With the specification of the null basis, the spin coefficients, and the Weyl scalars, we have completed the description of the Schwarzschild space-time in a Newman–Penrose formalism which exemplifies its special algebraic character.

BIBLIOGRAPHICAL NOTES

K. Schwarzschild's (1873–1916) original derivation of the solution, known by his name, was published in:

1. K. SCHWARZSCHILD, *Berliner Sitzungsbesichte* (*Phys. Math. Klasse*), 189–96, 3 Feb. 1916 (Mitt. Jan. 13).

See also:

2. K. SCHWARZSCHILD, *Berliner Sitzungsbesichte* (*Phys. Math. Klasse*), 424–34, 23 Mar. 1916 (Mitt. Feb. 24).

There can hardly be a book on general relativity which does not include some discussion of Schwarzschild's solution. Nevertheless, the heroic circumstances under which the solution was discovered seem hardly to be known. The first published account (so far as the present author is aware) is in:

3. S. CHANDRASEKHAR, *Notes and Records of the Royal Society of London*, 30, 249–60 (1976).

And the story is retold in:

4. W. SULLIVAN, *Black Holes*, 61–2, Anchor Press, Doubleday, Garden City, New York, 1979.

Since these accounts may easily have been overlooked by most students of relativity, the account given in reference 3 is briefly abstracted below:

> Schwarzschild's paper, in which he derived his solution, was communicated to the Berlin Academy by Einstein on 13 January 1916 just about two months after Einstein himself had published the basic equations of his theory in a short communication. In acknowledging the manuscript of Schwarzschild's paper on 9 January 1916, Einstein wrote: 'Ihre Arbeit habe ich mit grössten Interesse durchgesehen. Ich hätte nicht

erwartet, dass man so einfach die strenge Lössung der Aufgabe formulieren konnte. Die rechenrische Behandlung des Gegenstandes gefält mir ausgezeichnet.' [I have read your paper with the greatest interest. I had not expected that one could formulate the exact solution of the problem so simply. The analytical treatment of the problem appears to me splendid.]

The circumstances under which Schwarzschild derived his now famous solution are the following:

During the spring and summer of 1915, Schwarzschild was serving in the German army at the eastern front. While at the eastern front with a small technical staff, Schwarzschild contracted pemphigus—a fatal disease; and he died on 11 May 1916. It was during this period of illness that Schwarzschild wrote his two papers on relativity [besides a fundamental one on the Bohr-Sommerfeld theory].

Among the many books which discuss Schwarzschild's solutions, the reader can find excellent accounts in:

5. L. D. Landau and E. M. Lifschitz, *Classical Fields*, §§100–1, 299–321, Pergamon Press, Oxford, 1975.
6. C. W. Misner, K. S. Thorne, and J. A. Wheeler, *Gravitation*, chapter 25, 655–78, and chapters 31–32, 819–41, W. H. Freeman and Co., San Francisco, 1970.

Figs. 3 and 5 in the text are taken from reference 6.

§§17–18. It is customary (as, for example, in references 5 and 6) to derive Schwarzschild's solution in the coordinate system (t, r, θ, φ) (of §18) and then transform it to the 'Eddington–Finklestein' or the 'Kruskal' frame to show that there is no real singularity at the Schwarzschild-radius, $R_S = 2GM/c^2$, and to clarify the nature of the surface at R_S as an event horizon. We have inverted this common procedure and obtained the solution *ab initio* in the Kruskal frame and then transformed it to the Schwarzschild frame as a matter of convenience. The manner of presentation is due to Synge:

7 J. L. Synge, *Annali di Matematica pura ed Applicata*, 98, 239–55 (1974).

As Synge remarks: "If the problem of spherical symmetry had been attacked originally in this way, it would never have occurred to anyone to think of a singularity here [i.e., at $r = R_S$]." However, the basic idea that the nature of the Schwarzschild space-time is best clarified in a system of null coordinates is due to:

8. A. S. Eddington, *Nature*, 113, 192 (1924).
9. D. Finklestein, *Phys. Rev.*, 110, 965–7 (1958).
10. M. D. Kruskal, ibid., 110, 1743–5 (1960).

For fuller references to the literature see reference 6.

§§19–20. The geodesics in the Schwarzschild space-time have been discussed widely in the literature ever since Einstein's first evaluation, in a post-Newtonian approximation, of the deflection of light and the precession of the Kepler orbit in the field of a central spherically symmetric source of gravitation. The interest in the geodesics as a means of understanding the nature of the space-time itself is, however, relatively recent; and the accounts in references 5 and 6 are adequate.

On the purely analytical side, an exceptionally complete treatment of the geodesics is contained in an early investigation by:

11. Y. Hagihara, *Jap. J. Astron. Geophys*, 8, 67–175 (1931).

But Hagihara's treatment makes the subject much more complicated than is necessary. The account in the text is essentially a completion of the program set out by:

12. C. G. Darwin, *Proc. Roy. Soc.* (*London*) A, 249, 180–94 (1958).
13. ———, ibid., 263, 39–50 (1961).

In particular, by classifying the orbits into those of 'the first kind' and those of 'the second kind,' and allowing the eccentricity to be imaginary, we are able to treat all cases in a coherent and a unified manner.
The geodesics illustrated in Figs. 7, 8, and 9 have all been evaluated and drawn by Mr. Garret Toomey to whom the author is most grateful.
For a complete listing of references bearing on geodesics in the Schwarzschild space-time see:

14. N. A. Sharp, *General Relativity and Gravitation*, 10, 659–70 (1979).

For a beautiful pictorial visualization of the Schwarzschild black-hole with a luminous accretion disc see:

15. J. P. Luminet, *Astron. Astrophys.*, 75, 228–35 (1979).

§21. The first applications of the Newman–Penrose formalism for a study of the Schwarzschild space-time are due to:

16. R. H. Price, *Phys. Rev. D*, 5, 2419–38 and 2439–54 (1972).
17. J. M. Bardeen and W. H. Press, *J. Math. Phys.*, 14, 7–19 (1972).

The actual applications considered in these papers, however, are in the context of the next chapter.

4

THE PERTURBATIONS OF THE SCHWARZSCHILD BLACK-HOLE

22. Introduction

This chapter is devoted to the study of the perturbations of the Schwarzschild black-hole. On the physical side, the principal question to which the study is addressed is the manner in which gravitational waves, incident on the black-hole, are scattered and absorbed. The answer to this question has clearly some astrophysical interest. On the theoretical side, the answer has a more transcendent interest: it provides insight, in its simplest and purest context, into the deeper aspects of space and time as conceived in general relativity; and it reveals the analytical richness of the theory.

There are, at present, two avenues of approach to the study of the perturbations of space-times. One can either study, directly, the perturbations in the metric coefficients via the Einstein or the Einstein–Maxwell equations linearized about the unperturbed space-times; or, one can study the perturbations in the Weyl and in the Maxwell scalars via the equations of the Newman–Penrose formalism. While the latter avenue appears specially suitable to the study of the perturbations of the space-times around black holes (on account of their special algebraic character), it will emerge that the theories developed along both avenues complement each other very effectively in disclosing inner relationships which will have remained shrouded otherwise.

23. The Ricci and the Einstein tensors for non-stationary axisymmetric space-times

In studying the perturbations of any spherically symmetric system, one can, without any loss of generality, restrict oneself to axisymmetric modes of perturbations. For, non-axisymmetric modes of perturbations with an $e^{im\varphi}$-dependence on the azimuthal angle φ (where m is an integer positive or negative) can be deduced from modes of axisymmetric perturbations with $m = 0$ by suitable rotations since there are no preferred axes in a spherically symmetric background. Thus, an axisymmetric mode, evaluated at a point (θ, φ) on the sphere with respect to a chosen polar axis, when expressed in another frame with its polar axis pointing in a direction (θ', φ') will be assigned a polar angle Θ given by

$$\cos\Theta = \cos\theta\cos\theta' + \sin\theta\sin\theta'\cos(\varphi' - \varphi). \qquad (1)$$

When the chosen axisymmetric mode is subjected to this transformation, it will be decomposed into non-axisymmetric components, each of which may be considered as representing different non-axisymmetric modes; and we may do this by virtue of the linearity of the underlying perturbation theory.

In practice, the above mentioned decomposition will occur in the following way. The equations governing the perturbations of a spherically symmetric system will be separable in all four of the variables t, r, θ, and φ. Ignoring the dependence on t and r, we may expect the dependence of an axisymmetric mode on the angle θ to be expressible in terms of the Legendre functions $P_l(\cos\theta)$. When a mode belonging to a particular $P_l(\cos\theta)$ is subjected to the transformation (1), it will become

$$P_l(\cos\Theta) = \sum_{m=-l}^{m=+l} P_l^m(\cos\theta)e^{im\varphi}P_l^m(\cos\theta')e^{-im\varphi'}, \tag{2}$$

where $P_l^m(\cos\theta)$ are the associated Legendre functions. Thus an axisymmetric mode belonging to a particular l is decomposed into a complete set of $(2l+1)$ modes with angular dependences given by $P_l^m(\cos\theta)e^{im\varphi}$. The functions describing the radial dependence will, of course, be unaffected by this decomposition. It is this same reason which ensures the independence of the radial wave-function of an electron in a central field on the magnetic quantum number m and its dependence only on the orbital angular momentum specified by l.

In considering the perturbations of the Schwarzschild black-hole, we shall, in accordance with the foregoing remarks, restrict ourselves to time-dependent axisymmetric modes. It will suffice then to consider the Schwarzschild solution as a special, spherically symmetric time-independent solution of the field equations appropriate to the line element (Ch. 2, equation (38)),

$$\begin{aligned} ds^2 = e^{2\nu}(dt)^2 - e^{2\psi}(d\varphi - \omega\, dt - q_2\, dx^2 - q_3\, dx^3)^2 \\ - e^{2\mu_2}(dx^2)^2 - e^{2\mu_3}(dx^3)^2, \end{aligned} \tag{3}$$

and obtain the relevant perturbation equations by linearizing the field equations about the Schwarzschild solution. For this purpose (and other purposes, as well) it will be convenient to have the explicit expressions for the various components of the Ricci and the Einstein tensors for the line element (3) when ν, ψ, μ_2, μ_3, ω, q_2, and q_3 are functions only of t, x^2, and x^3. These components can be readily obtained by suitable contractions of the Riemann tensor whose components (when the metric functions are dependent on φ, as well) are listed in Chapter 2 (equations (75)). The results of such contractions are given below.

$$
\begin{aligned}
-R_{00} = {} & e^{-2\nu}[(\psi+\mu_2+\mu_3)_{,0,0}+\psi_{,0}(\psi-\nu)_{,0}+\mu_{2,0}(\mu_2-\nu)_{,0} \\
& +\mu_{3,0}(\mu_3-\nu)_{,0}] \\
& -e^{-2\mu_2}[\nu_{,2,2}+\nu_{,2}(\psi+\nu-\mu_2+\mu_3)_{,2}] \\
& -e^{-2\mu_3}[\nu_{,3,3}+\nu_{,3}(\psi+\nu+\mu_2-\mu_3)_{,3}] \\
& +\tfrac{1}{2}e^{2\psi-2\nu}[e^{-2\mu_2}Q_{20}^2+e^{-2\mu_3}Q_{30}^2], \qquad \text{(a)}
\end{aligned}
$$

$$
\begin{aligned}
-R_{11} = {} & e^{-2\mu_2}[\psi_{,2,2}+\psi_{,2}(\psi+\nu+\mu_3-\mu_2)_{,2}] \\
& +e^{-2\mu_3}[\psi_{,3,3}+\psi_{,3}(\psi+\nu+\mu_2-\mu_3)_{,3}] \\
& -e^{-2\nu}[\psi_{,0,0}+\psi_{,0}(\psi-\nu+\mu_2+\mu_3)_{,0}]-\tfrac{1}{2}e^{2\psi-2\mu_2-2\mu_3}Q_{23}^2 \\
& +\tfrac{1}{2}e^{2\psi-2\nu}[e^{-2\mu_3}Q_{30}^2+e^{-2\mu_2}Q_{20}^2], \qquad \text{(b)}
\end{aligned}
$$

$$
\begin{aligned}
-R_{22} = {} & e^{-2\mu_2}[(\psi+\nu+\mu_3)_{,2,2}+\psi_{,2}(\psi-\mu_2)_{,2} \\
& +\mu_{3,2}(\mu_3-\mu_2)_{,2}+\nu_{,2}(\nu-\mu_2)_{,2}] \\
& +e^{-2\mu_3}[\mu_{2,3,3}+\mu_{2,3}(\psi+\nu+\mu_2-\mu_3)_{,3}] \\
& -e^{-2\nu}[\mu_{2,0,0}+\mu_{2,0}(\psi-\nu+\mu_2+\mu_3)_{,0}] \\
& +\tfrac{1}{2}e^{2\psi-2\mu_2}[e^{-2\mu_3}Q_{23}^2-e^{-2\nu}Q_{20}^2], \qquad \text{(c)}
\end{aligned}
$$

$$
-R_{01} = \tfrac{1}{2}e^{-2\psi-\mu_2-\mu_3}[(e^{3\psi-\nu-\mu_2+\mu_3}Q_{20})_{,2}+(e^{3\psi-\nu-\mu_3+\mu_2}Q_{30})_{,3}], \qquad \text{(d)}
$$

$$
-R_{12} = \tfrac{1}{2}e^{-2\psi-\nu-\mu_3}[(e^{3\psi+\nu-\mu_2-\mu_3}Q_{32})_{,3}-(e^{3\psi-\nu+\mu_3-\mu_2}Q_{02})_{,0}], \qquad \text{(e)}
$$

$$
\begin{aligned}
-R_{02} = {} & e^{-\mu_2-\nu}[(\psi+\mu_3)_{,2,0}+\psi_{,2}(\psi-\mu_2)_{,0}+\mu_{3,2}(\mu_3-\mu_2)_{,0} \\
& -(\psi+\mu_3)_{,0}\nu_{,2})]-\tfrac{1}{2}e^{2\psi-\nu-2\mu_3-\mu_2}Q_{23}Q_{30}, \qquad \text{(f)}
\end{aligned}
$$

$$
\begin{aligned}
-R_{23} = {} & e^{-\mu_2-\mu_3}[(\psi+\nu)_{,2,3}-(\psi+\nu)_{,2}\mu_{2,3}-(\psi+\nu)_{,3}\mu_{3,2}+\psi_{,2}\psi_{,3} \\
& +\nu_{,2}\nu_{,3}]-\tfrac{1}{2}e^{2\psi-2\nu-\mu_2-\mu_3}Q_{20}Q_{30}, \qquad \text{(g)}
\end{aligned}
$$

$$
\begin{aligned}
G_{00} = {} & -e^{-2\mu_2}[(\psi+\mu_3)_{,2,2}+\psi_{,2}(\psi-\mu_2+\mu_3)_{,2}+\mu_{3,2}(\mu_3-\mu_2)_{,2}] \\
& -e^{-2\mu_3}[(\psi+\mu_2)_{,3,3}+\psi_{,3}(\psi-\mu_3+\mu_2)_{,3}+\mu_{2,3}(\mu_2-\mu_3)_{,3}] \\
& +e^{-2\nu}[\psi_{,0}(\mu_2+\mu_3)_{,0}+\mu_{3,0}\mu_{2,0}]-\tfrac{1}{4}e^{2\psi-2\nu}[e^{-2\mu_2}Q_{20}^2+e^{-2\mu_3}Q_{30}^2] \\
& -\tfrac{1}{4}e^{2\psi-2\mu_2-2\mu_3}Q_{23}^2, \qquad \text{(h)}
\end{aligned}
$$

$$
\begin{aligned}
G_{11} = {} & e^{-2\mu_2}[(\nu+\mu_3)_{,2,2}+\nu_{,2}(\nu-\mu_2+\mu_3)_{,2}+\mu_{3,2}(\mu_3-\mu_2)_{,2}] \\
& +e^{-2\mu_3}[(\nu+\mu_2)_{,3,3}+\nu_{,3}(\nu-\mu_3+\mu_2)_{,3}+\mu_{2,3}(\mu_2-\mu_3)_{,3}] \\
& -e^{-2\nu}[(\mu_2+\mu_3)_{,0,0}+\mu_{2,0}(\mu_2-\nu)_{,0}+\mu_{3,0}(\mu_3-\nu)_{,0}+\mu_{2,0}\mu_{3,0}] \\
& +\tfrac{3}{4}e^{2\psi}[e^{-2\mu_2-2\mu_3}Q_{23}^2-e^{-2\mu_2-2\nu}Q_{20}^2-e^{-2\mu_3-2\nu}Q_{30}^2], \qquad \text{(i)}
\end{aligned}
$$

$$
\begin{aligned}
G_{22} = {} & e^{-2\mu_3}[(\psi+\nu)_{,3,3}+(\psi+\nu)_{,3}(\nu-\mu_3)_{,3}+\psi_{,3}\psi_{,3}] \\
& +e^{-2\mu_2}[\nu_{,2}(\psi+\mu_3)_{,2}+\psi_{,2}\mu_{3,2}] \\
& -e^{-2\nu}[(\psi+\mu_3)_{,0,0}+(\psi+\mu_3)_{,0}(\mu_3-\nu)_{,0}+\psi_{,0}\psi_{,0}] \\
& -\tfrac{1}{4}e^{2\psi}[e^{-2\mu_2-2\mu_3}Q_{23}^2-e^{-2\mu_2-2\nu}Q_{20}^2+e^{-2\mu_3-2\nu}Q_{30}^2], \qquad \text{(j)}
\end{aligned}
$$

(4)

where

$$Q_{AB}=q_{A,B}-q_{B,A} \quad \text{and} \quad Q_{A0}=q_{A,0}-\omega_{,A} \quad (A, B=2, 3). \tag{5}$$

The components R_{33}, R_{13}, R_{03}, and G_{33} are not listed; they can be obtained by interchanging the indices 2 and 3 in the components R_{22}, R_{12}, R_{02}, and G_{22} which are listed.

24. The metric perturbations

The Schwarzschild line-element considered as a special case of the line element (3) has the metric coefficients

$$e^{2\nu}=e^{-2\mu_2}=1-2M/r=\Delta/r^2, \quad e^{\mu_3}=r, \quad e^{\psi}=r\sin\theta, \tag{6}$$

and

$$\omega=q_2=q_3=0 \quad (\Delta=r^2-2Mr;\ x^2=r;\ x^3=\theta). \tag{7}$$

Consequently, a general perturbation of the Schwarzschild black-hole will result in

$$\omega, q_2, \quad \text{and} \quad q_3 \tag{8}$$

becoming small quantities of the first order and the functions ν, μ_2, μ_3, and ψ experiencing small increments,

$$\delta\nu, \delta\mu_2, \delta\mu_3, \quad \text{and} \quad \delta\psi. \tag{9}$$

It is clear that the perturbations leading to non-vanishing values of ω, q_2, and q_3 and perturbations leading to increments in ν, μ_2, μ_3, and ψ are of very different kinds: the former induce a dragging of the inertial frame and impart a rotation to the black hole while the latter impart no such rotation. For this reason we shall call them *axial* and *polar* perturbations, respectively. The terminology is justified when we consider the effect of a reversal in the sign of φ on the metric. It has no effect when the perturbations are polar while, when the perturbations are axial, the signs of ω, q_2, and q_3 must also be reversed if the metric is to remain unchanged. As one may expect from these different behaviours of the perturbations of the two kinds, they must decouple in the sense that they can be considered independently of each other. We shall find that this is indeed the case.

(*a*) *Axial perturbations*

As we have stated, axial perturbations are characterized by the non-vanishing of ω, q_2, and q_3. The equations governing these quantities are given by

$$R_{12} = R_{13} = 0. \tag{10}$$

From the expression (4, e) for R_{12}, for example, it is apparent that in equations (10), we may insert for ν, μ_2, μ_3, and ψ their unperturbed values (6). The resulting equations are

$$(e^{3\psi+\nu-\mu_2-\mu_3}Q_{23})_{,3} = -e^{3\psi-\nu+\mu_3-\mu_2}Q_{02,0} \quad (\delta R_{12} = 0), \tag{11}$$

$$(e^{3\psi+\nu-\mu_2-\mu_3}Q_{23})_{,2} = +e^{3\psi-\nu+\mu_2-\mu_3}Q_{03,0} \quad (\delta R_{13} = 0). \tag{12}$$

Letting

$$Q(t, r, \theta) = \Delta Q_{23}\sin^3\theta = \Delta(q_{2,3} - q_{3,2})\sin^3\theta, \tag{13}$$

and substituting for ν, μ_2, μ_3, and ψ their unperturbed values (6), we obtain the pair of equations

$$\frac{1}{r^4\sin^3\theta}\frac{\partial Q}{\partial\theta} = -(\omega_{,2} - q_{2,0})_{,0}, \tag{14}$$

$$\frac{\Delta}{r^4\sin^3\theta}\frac{\partial Q}{\partial r} = +(\omega_{,3} - q_{3,0})_{,0}. \tag{15}$$

In our further considerations, we shall assume that the perturbations have the time-dependence

$$e^{i\sigma t}, \tag{16}$$

where σ is a constant (generally real). This assumption corresponds to a Fourier analysis of the perturbations and considering the Fourier component with the frequency $-\sigma$. Retaining the same symbols for the amplitudes of the perturbations with the foregoing time-dependent factor, we can rewrite equations (14) and (15) in the forms

$$\left.\begin{aligned} \frac{1}{r^4\sin^3\theta}\frac{\partial Q}{\partial\theta} &= -i\sigma\omega_{,2} - \sigma^2 q_2, \\ \text{and} \qquad \frac{\Delta}{r^4\sin^3\theta}\frac{\partial Q}{\partial r} &= +i\sigma\omega_{,3} + \sigma^2 q_3. \end{aligned}\right\} \tag{17}$$

Eliminating ω from these equations, we obtain

$$r^4\frac{\partial}{\partial r}\left(\frac{\Delta}{r^4}\frac{\partial Q}{\partial r}\right) + \sin^3\theta\frac{\partial}{\partial\theta}\left(\frac{1}{\sin^3\theta}\frac{\partial Q}{\partial\theta}\right) + \sigma^2\frac{r^4}{\Delta}Q = 0. \tag{18}$$

The variables r and θ in equation (18) can be separated by the substitution

$$Q(r,\theta) = Q(r) C_{l+2}^{-3/2}(\theta), \tag{19}$$

where C_n^ν denotes the Gegenbauer function governed by the equation

$$\left[\frac{d}{d\theta}\sin^{2\nu}\theta\frac{d}{d\theta} + n(n+2\nu)\sin^{2\nu}\theta\right] C_n^\nu(\theta) = 0. \tag{20}$$

It may be noted here that the Gegenbauer function $C_{l+2}^{-3/2}(\theta)$ is related to the Legendre function $P_l(\theta)$ by the formula

$$C_{l+2}^{-3/2}(\theta) = \sin^3\theta \frac{d}{d\theta}\frac{1}{\sin\theta}\frac{dP_l(\theta)}{d\theta}, \tag{21}*$$

or

$$C_{l+2}^{-3/2}(\theta) = (P_{l,\theta,\theta} - P_{l,\theta}\cot\theta)\sin^2\theta. \tag{22}$$

With the substitution (19) in equation (18), we obtain the radial equation

$$\Delta\frac{d}{dr}\left(\frac{\Delta}{r^4}\frac{dQ}{dr}\right) - \mu^2\frac{\Delta}{r^4}Q + \sigma^2 Q = 0, \tag{23}$$

where

$$\mu^2 = 2n = (l-1)(l+2) \tag{24}$$

specifies the associated angular dependence. Changing to the variable (cf. Ch. 3, equation (219))

$$r_* = r + 2M\lg(r/2M - 1) \quad \left(\frac{d}{dr_*} = \frac{\Delta}{r^2}\frac{d}{dr}\right), \tag{25}$$

and further letting

$$Q(r) = rZ^{(-)}, \tag{26}$$

we find that $Z^{(-)}$ satisfies the one-dimensional Schrodinger wave-equation

$$\left(\frac{d^2}{dr_*^2} + \sigma^2\right)Z^{(-)} = V^{(-)}Z^{(-)}, \tag{27}$$

where the potential $V^{(-)}$ is given by

$$V^{(-)} = \frac{\Delta}{r^5}[(\mu^2+2)r - 6M]. \tag{28}$$

Equation (27) governing the axial perturbation was first derived (though by an entirely different procedure) by Regge and Wheeler; and it is often referred to as the *Regge–Wheeler equation.*

* With the usual normalization of the Gegenbauer and the Legendre functions, there is an additional factor $3[(l-1)l(l+1)(l+2)]^{-1}$ on the right-hand side of this equation.

TABLE I

The potential barrier, $V^{(-)}$, for axial perturbations for different values of l and n

r/M	$l=2$; $n=2$	$l=3$; $n=5$	$l=4$; $n=9$	r/M	$l=2$; $n=2$	$l=3$; $n=5$	$l=4$; $n=9$
2.0	0.	0.	0.	3.8	0.14503	0.34185	0.60428
2.1	0.03394	0.09872	0.18511	4.0	0.14063	0.32813	0.57813
2.2	0.06147	0.17417	0.32443	4.5	0.12803	0.29264	0.51212
2.3	0.08362	0.23156	0.42881	5.0	0.11520	0.25920	0.45120
2.4	0.10127	0.27488	0.50637	6.0	0.09259	0.20370	0.35185
2.5	0.11520	0.30720	0.56320	7.0	0.07497	0.16243	0.27905
2.6	0.12605	0.33087	0.60397	8.0	0.06152	0.13184	0.22559
2.7	0.13435	0.34773	0.63224	9.0	0.05121	0.10882	0.18564
2.8	0.14057	0.35923	0.65077	10.0	0.04320	0.09120	0.15520
2.9	0.14506	0.36647	0.66169	12.0	0.03183	0.06655	0.11285
3.0	0.14815	0.37037	0.66667	14.0	0.02436	0.05060	0.08559
3.2	0.15106	0.37079	0.66376	16.0	0.01923	0.03973	0.06708
3.4	0.15086	0.36458	0.64954	18.0	0.01555	0.03201	0.05396
3.6	0.14861	0.35437	0.62871	20.0	0.01283	0.02633	0.04433

In our subsequent work, we shall find it convenient to introduce the operators

$$\Lambda_{\pm} = \frac{d}{dr_*} \pm i\sigma \quad \text{and} \quad \Lambda^2 = \Lambda_+\Lambda_- = \Lambda_-\Lambda_+ = \frac{d^2}{dr_*^2} + \sigma^2. \tag{29}$$

In this notation, the equation satisfied by $Z^{(-)}$ is

$$\Lambda^2 Z^{(-)} = V^{(-)} Z^{(-)}. \tag{30}$$

Table I provides a brief tabulation of the potential $V^{(-)}$ for $l = 2$, 3, and 4; and they are exhibited in Fig. 11.

(b) *The polar perturbations*

Polar perturbations are characterized by non-vanishing increments in the metric functions ν, μ_2, μ_3, and ψ. Examining the expressions (4*f*, *g*, *b*, and *j*) for R_{02}, R_{23}, R_{11}, and G_{22}, we find that the Q_{AB}'s occur quadratically in them; they can, accordingly, be ignored in a linear perturbation-theory. Thus, as expected, the equations governing the axial and the polar perturbations do decouple.

Linearizing the expressions for R_{02}, R_{03}, R_{23}, R_{11}, and G_{22} about the Schwarzschild values, we obtain the equations

$$(\delta\psi + \delta\mu_3)_{,r} + \left(\frac{1}{r} - \nu_{,r}\right)(\delta\psi + \delta\mu_3) - \frac{2}{r}\delta\mu_2 = 0 \qquad (\delta R_{02} = 0), \tag{31}$$

$$(\delta\psi + \delta\mu_2)_{,\theta} + (\delta\psi - \delta\mu_3)\cot\theta = 0 \qquad (\delta R_{03} = 0), \tag{32}$$

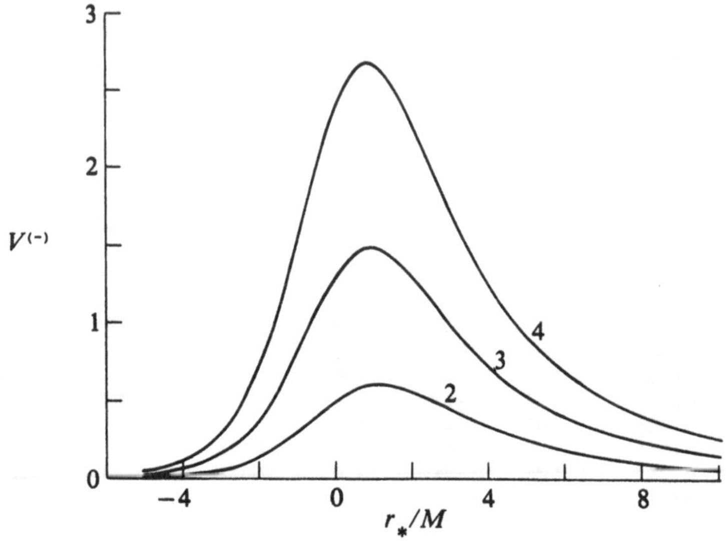

FIG. 11. The potential barriers surrounding the Schwarzschild black-holes for axial perturbations. The curves are labelled by the values of l to which they belong.

$$(\delta\psi+\delta\nu)_{,r,\theta}+(\delta\psi-\delta\mu_3)_{,r}\cot\theta+\left(\nu_{,r}-\frac{1}{r}\right)\delta\nu_{,\theta}$$
$$-\left(\nu_{,r}+\frac{1}{r}\right)\delta\mu_{2,\theta}=0 \qquad (\delta R_{23}=0), \tag{33}$$

$$e^{-2\mu_2}\left[\frac{2}{r}\delta\nu_{,r}+\left(\frac{1}{r}+\nu_{,r}\right)(\delta\psi+\delta\mu_3)_{,r}-2\delta\mu_2\left(\frac{2}{r}\nu_{,r}+\frac{1}{r^2}\right)\right]$$
$$+\frac{1}{r^2}[(\delta\psi+\delta\nu)_{,\theta,\theta}+(2\delta\psi+\delta\nu-\delta\mu_3)_{,\theta}\cot\theta+2\delta\mu_3]$$
$$-e^{-2\nu}(\delta\psi+\delta\mu_3)_{,0,0}=0 \qquad (\delta G_{22}=0), \tag{34}$$

and

$$e^{+2\nu}\left\{\delta\psi_{,r,r}+2\left(\frac{1}{r}+\nu_{,r}\right)\delta\psi_{,r}+\frac{1}{r}(\delta\psi+\delta\nu-\delta\mu_2+\delta\mu_3)_{,r}\right.$$
$$\left.-2\frac{\delta\mu_2}{r}\left(\frac{1}{r}+2\nu_{,r}\right)\right\}+\frac{1}{r^2}\{\delta\psi_{,\theta,\theta}+(2\delta\psi+\delta\nu+\delta\mu_2-\delta\mu_3)_{,\theta}\cot\theta+2\delta\mu_3\}$$
$$-e^{-2\nu}\delta\psi_{,0,0}=0 \qquad (\delta R_{11}=0). \tag{35}$$

We shall assume once again that the perturbations have a time-dependence $e^{i\sigma t}$ so that "$_{,0}$" is replaced by the factor $i\sigma$; and we shall suppress the factor $e^{i\sigma t}$ in our subsequent considerations.

The variables r and θ in equations (31)–(35) can be separated by the substitutions (due to J. Friedman)

$$\delta\nu = N(r)P_l(\cos\theta), \tag{36}$$

$$\delta\mu_2 = L(r)P_l(\cos\theta), \tag{37}$$

$$\delta\mu_3 = [T(r)P_l + V(r)P_{l,\theta,\theta}], \tag{38}$$

and

$$\delta\psi = [T(r)P_l + V(r)P_{l,\theta}\cot\theta]. \tag{39}$$

We may note that, according to these substitutions,

$$\delta\psi + \delta\mu_3 = [2T - l(l+1)V]P_l, \tag{40}$$

$$\delta\psi_{,\theta} + (\delta\psi - \delta\mu_3)\cot\theta = (T - V)P_{l,\theta}, \tag{41}$$

and

$$\begin{aligned}\delta\psi_{,\theta,\theta} + (2\delta\psi - \delta\mu_3)_{,\theta}\cot\theta + 2\delta\mu_3 &= [2 - l(l+1)]TP_l \\ &= -2nTP_l.\end{aligned} \tag{42}$$

Returning to a consideration of equations (31)–(35), we first observe that equation (32), by the substitutions (37) and (41), leads to the relation

$$T - V + L = 0 \qquad (\delta R_{03} = 0). \tag{43}$$

Accordingly, only three of the four radial functions we have defined are linearly independent. We shall choose N, L, and V as the independent functions.

Making use of equations (40) and (41), the (0, 2)- and the (2, 3)-components of the field equations—equations (31) and (33)—give

$$\left(\frac{\mathrm{d}}{\mathrm{d}r} + \frac{1}{r} - \nu_{,r}\right)[2T - l(l+1)V] - \frac{2}{r}L = 0 \tag{44}$$

and

$$(T - V + N)_{,r} - \left(\frac{1}{r} - \nu_{,r}\right)N - \left(\frac{1}{r} + \nu_{,r}\right)L = 0, \tag{45}$$

or, after the elimination of T, we have

$$N_{,r} - L_{,r} = \left(\frac{1}{r} - \nu_{,r}\right)N + \left(\frac{1}{r} + \nu_{,r}\right)L \tag{46}$$

and

$$L_{,r} + \left(\frac{2}{r} - \nu_{,r}\right)L = -\left[X_{,r} + \left(\frac{1}{r} - \nu_{,r}\right)X\right], \tag{47}$$

where we have replaced V by

$$X = nV = \tfrac{1}{2}(l-1)(l+2)V. \tag{48}$$

Similarly, we find that by making use of equations (40) and (42), equation (34) gives

$$\frac{2}{r}N_{,r}+\left(\frac{1}{r}+\nu_{,r}\right)\Big[2T-l(l+1)V\Big]_{,r}-\frac{2}{r}\left(\frac{1}{r}+2\nu_{,r}\right)L$$
$$-l(l+1)\frac{e^{-2\nu}}{r^2}N-2n\frac{e^{-2\nu}}{r^2}T+\sigma^2e^{-4\nu}[2T-l(l+1)V]=0, \qquad (49)$$

or, after the elimination of T, we have

$$\frac{2}{r}N_{,r}-l(l+1)\frac{e^{-2\nu}}{r^2}N-\frac{2}{r}\left(\frac{1}{r}+2\nu_{,r}\right)L-2\left(\frac{1}{r}+\nu_{,r}\right)(L+nV)_{,r}$$
$$-2n\frac{e^{-2\nu}}{r^2}(V-L)-2\sigma^2e^{-4\nu}(L+nV)=0. \qquad (50)$$

Finally, considering equation (35), we find that it consists of terms in P_l and $P_{l,\theta}\cot\theta$. These terms vanish separately by virtue of the preceding equations; and the terms in $P_{l,\theta}\cot\theta$ give

$$V_{,r,r}+2\left(\frac{1}{r}+\nu_{,r}\right)V_{,r}+\frac{e^{-2\nu}}{r^2}(N+L)+\sigma^2e^{-4\nu}V=0. \qquad (51)$$

This equation, as we have stated, is not independent of equations (46), (47), and (50); it is, nevertheless, useful in this form in our further reductions.

It will be observed that equations (46), (47), and (50) provide three linear first-order equations for the three radial functions L, N, and V (or X). By suitably combining them, we can express the derivatives of each of them as linear combinations of themselves. Thus, we find

$$N_{,r}=aN+bL+cX, \qquad (52)$$

$$L_{,r}=\left(a-\frac{1}{r}+\nu_{,r}\right)N+\left(b-\frac{1}{r}-\nu_{,r}\right)L+cX, \qquad (53)$$

$$X_{,r}=-\left(a-\frac{1}{r}+\nu_{,r}\right)N-\left(b+\frac{1}{r}-2\nu_{,r}\right)L-\left(c+\frac{1}{r}-\nu_{,r}\right)X, \qquad (54)$$

where, for the sake of brevity, we have written

$$\left.\begin{aligned}
a&=\frac{n+1}{r-2M},\quad \nu_{,r}=\frac{M}{r(r-2M)},\\
b&=-\frac{1}{r}-\frac{n}{r-2M}+\frac{M}{r(r-2M)}+\frac{M^2}{r(r-2M)^2}+\sigma^2\frac{r^3}{(r-2M)^2},\\
c&=-\frac{1}{r}+\frac{1}{r-2M}+\frac{M^2}{r(r-2M)^2}+\sigma^2\frac{r^3}{(r-2M)^2}.
\end{aligned}\right\} \qquad (55)$$

We shall find the following alternative forms of equations (47) and (54) useful:

$$(L+X)_{,r} = -\left(\frac{2}{r}-\nu_{,r}\right)L-\left(\frac{1}{r}-\nu_{,r}\right)X$$

$$= -\frac{1}{r(r-2M)}[(2r-5M)L+(r-3M)X] \tag{56}$$

and

$$X_{,r} = -\frac{nr+3M}{r(r-2M)}N-\left[\frac{1}{r-2M}-\frac{M}{r(r-2M)}+\frac{M^2+\sigma^2r^4}{r(r-2M)^2}\right](L+X)$$
$$+\frac{n+1}{r-2M}L. \tag{57}$$

(i) *The reduction of the equations to a one-dimensional wave equation*

It is a remarkable fact that

$$Z^{(+)} = \frac{r^2}{nr+3M}\left(\frac{3M}{nr}X-L\right), \tag{58}$$

by virtue of equations (52)–(54), satisfies a one-dimensional wave equation similar to $Z^{(-)}$. This reduction of the third-order system of equations (52)–(54) to a single second-order equation emerged empirically. Its origin must, however, lie in some deeper fact at the base of equations (52)–(54). As to what this may be is clarified in §25 below. But first, we shall verify that $Z^{(+)}$, as defined in equation (58), does in fact satisfy a one-dimensional wave equation.

Rewriting $Z^{(+)}$ in the form

$$Z^{(+)} = rV-\frac{r^2}{nr+3M}(L+X), \tag{59}$$

differentiating it with respect to r_* and substituting for $(L+X)_{,r}$ from equation (56), we find

$$Z^{(+)}_{,r_*} = \left(1-\frac{2M}{r}\right)Z_{,r}$$

$$= (r-2M)V_{,r}+\frac{3M(r-2M)}{r(nr+3M)}V+\frac{nr^2-3nMr-3M^2}{(nr+3M)^2}(L+X). \tag{60}$$

Differentiating this expression with respect to r and making use, once again, of equation (56), we obtain

$$Z^{(+)}_{,r_*,r} = (r-2M)V_{,r,r}+V_{,r}+\frac{3M(r-2M)}{r(nr+3M)}V_{,r}$$
$$+3M\frac{-nr^2+4Mnr+6M^2}{r^2(nr+3M)^2}V+\frac{3Mn}{(nr+3M)^3}[(2+n)r-M](L+X)$$
$$-\frac{nr^2-3Mnr-3M^2}{r(r-2M)(nr+3M)^2}[(2r-5M)L+(r-3M)X]. \tag{61}$$

On substituting for $V_{,r,r}$ from equation (51), we shall be left with an expression, which, besides L, X, and N, involves only the derivative of V for which we can substitute from equation (57). After making these substitutions and simplifying, we find that we are left with (after some remarkable cancellations)

$$\left(\frac{d^2}{dr_*^2}+\sigma^2\right)Z^{(+)} = V^{(+)}Z^{(+)}, \tag{62}$$

where

$$V^{(+)} = \frac{2\Delta}{r^5(nr+3M)^2}[n^2(n+1)r^3+3Mn^2r^2+9M^2nr+9M^3]. \tag{63}$$

Equation (62) was first derived (by an entirely different procedure) by Zerilli; and it is often called the *Zerilli equation*.

We shall rewrite equation (62) in the alternative form (cf. equation (30))

$$\Lambda^2 Z^{(+)} = V^{(+)}Z^{(+)}. \tag{64}$$

Table II provides a brief tabulation of the potential $V^{(+)}$ for $l = 2$, 3, and 4; and they are exhibited in Fig. 12. A comparison of the tabulations of $V^{(+)}$ and $V^{(-)}$ shows that they differ remarkably little over the entire range of r_* even though their analytical forms (28) and (63) are so different.

(ii) *The completion of the solution*

With the reduction of the third-order system of equations (52)–(54) to the single second-order equation (61) for $Z^{(+)}$, it is clear that the solution for L, X,

TABLE II

The potential barrier, $V^{(+)}$, for polar perturbations for different values of l and n

r/M	$l=2$; $n=2$	$l=3$; $n=5$	$l=4$; $n=9$	r/M	$l=2$; $n=2$	$l=3$ $n=5$	$l=4$; $n=9$
2.0	0.	0.	0.	3.8	0.13970	0.33866	0.60225
2.1	0.04183	0.10289	0.18766	4.0	0.13443	0.32458	0.57591
2.2	0.07311	0.18018	0.32809	4.5	0.12079	0.28876	0.50974
2.3	0.09647	0.23805	0.43272	5.0	0.10787	0.25542	0.44891
2.4	0.11383	0.28106	0.51004	6.0	0.08617	0.20054	0.34997
2.5	0.12660	0.31263	0.56639	7.0	0.06974	0.15993	0.27758
2.6	0.13584	0.33536	0.60656	8.0	0.05734	0.12988	0.22444
2.7	0.14235	0.35122	0.63420	9.0	0.04786	0.10728	0.18475
2.8	0.14673	0.36171	0.65211	10.0	0.04050	0.08997	0.15449
2.9	0.14946	0.36802	0.66246	12.0	0.03002	0.06575	0.11239
3.0	0.15089	0.37106	0.66691	14.0	0.02310	0.05005	0.08527
3.2	0.15094	0.37003	0.66314	16.0	0.01832	0.03934	0.06685
3.4	0.14848	0.36272	0.64827	18.0	0.01487	0.03172	0.05379
3.6	0.14452	0.35172	0.62355	20.0	0.01231	0.02611	0.04420

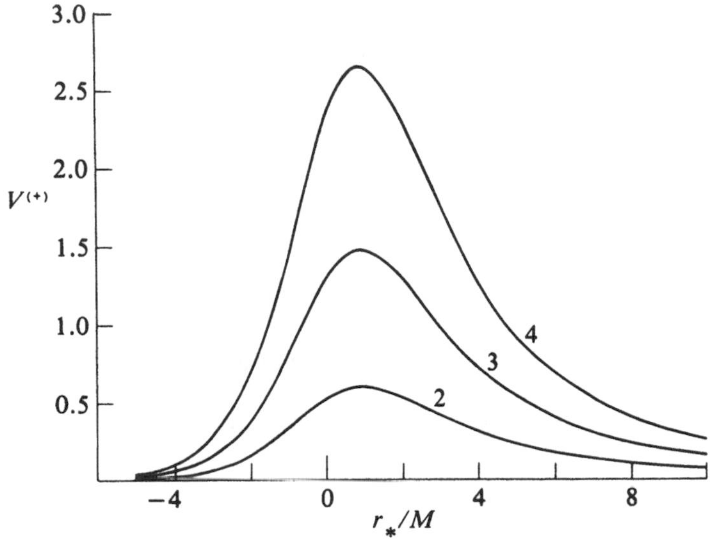

FIG. 12. The potential barriers surrounding the Schwarzschild black-holes for polar perturbations. The curves are labelled by the values of l to which they belong.

and N will require a further quadrature. Thus, rewriting equation (47) in the form

$$\frac{\mathrm{d}}{\mathrm{d}r}(r^2 e^{-\nu} L) = -nr\frac{\mathrm{d}}{\mathrm{d}r}(re^{-\nu} V) \tag{65}$$

and replacing V by

$$V = \frac{nr+3M}{3Mr} Z^{(+)} + \frac{r}{3M} L, \tag{66}$$

we obtain the equation

$$\left(1+\frac{nr}{3M}\right)\frac{\mathrm{d}}{\mathrm{d}r}(r^2 e^{-\nu} L) = -\frac{nr}{3M}\frac{\mathrm{d}}{\mathrm{d}r}[e^{-\nu}(nr+3M)Z^{(+)}]. \tag{67}$$

This equation yields the integral relation

$$r^2 e^{-\nu} L = -\int \frac{nr}{(nr+3M)}\frac{\mathrm{d}}{\mathrm{d}r}[e^{-\nu}(nr+3M)Z^{(+)}]\,\mathrm{d}r, \tag{68}$$

or, after an integration by parts

$$r^2 e^{-\nu} L = -nre^{-\nu} Z^{(+)} + 3Mn\int \frac{e^{-\nu}}{nr+3M} Z^{(+)}\,\mathrm{d}r. \tag{69}$$

Defining

$$\Phi = ne^{\nu}\int\frac{e^{-\nu}Z^{(+)}}{nr+3M}\,dr, \tag{70}$$

we have the solution

$$L = -\frac{n}{r}Z^{(+)}+\frac{3M}{r^2}\Phi. \tag{71}$$

With this solution for L, equation (66) gives

$$X = \frac{n}{r}(Z^{(+)}+\Phi). \tag{72}$$

As a consequence of these solutions for L and X,

$$L+X = \frac{1}{r^2}(nr+3M)\Phi. \tag{73}$$

To obtain the solution for N, we return to equation (60) and substitute for $V_{,r}(=X_{,r}/n)$ on the right-hand side from equation (57). In this manner, we obtain

$$Z_{,r_*} = \frac{1}{n}(r-2M)\left\{-\frac{nr+3M}{r(r-2M)}N-\left[\frac{1}{r-2M}-\frac{M}{r(r-2M)}+\frac{M^2+\sigma^2r^4}{r(r-2M)^2}\right]\right.$$
$$\left.\times(L+X)+\frac{n+1}{r-2M}L\right\}+\frac{3M(r-2M)}{nr(nr+3M)}X+\frac{nr^2-3Mnr-3M^2}{(nr+3M)^2}(L+X). \tag{74}$$

Simplifying this last equation and substituting for L and $L+X$ their solutions (71) and (72), we find

$$N = -\frac{nr}{nr+3M}Z^{(+)}_{,r_*}-\frac{n}{(nr+3M)^2}\left[\frac{6M^2}{r}+3Mn+n(n+1)r\right]Z^{(+)}$$
$$+\left(M-\frac{M^2+\sigma^2r^4}{r-2M}\right)\frac{\Phi}{r^2}. \tag{75}$$

This completes the formal solution of the basic equations. We shall return to a consideration of the wave equations for $Z^{(+)}$ and $Z^{(-)}$ in §§26 and 27.

25. A theorem relating to the particular integrals associated with the reducibility of a system of linear differential equations

In our treatment of the polar perturbations in §24, we saw that the system of three linear equations (equations (52)–(54)) governing N, L, and X could be reduced to a single second-order equation for a particular combination of them which we denoted by $Z^{(+)}$. This reducibility of the system was a verifiable but an empirically discovered fact; and in view of the complexity of the potential in the wave equation for $Z^{(+)}$, this is a remarkable coincidence shrouded in mysterious cancellations. In this section, we shall try to locate the origin of the mystery.

In principle, the reducibility of a system of linear equations to one of lower order is, of itself, nothing extraordinary: it can always be achieved if one has a prior knowledge of some particular solutions of the equations. The reduction in §24 was, however, achieved without any such prior knowledge. A general question which occurs is whether, from a knowledge of the reducibility of a system of linear equations to one of lower order, we can, conversely, devise an algorism for deriving the particular solutions which made the reduction, in the first place, possible. Such an algorism was recently devised by Xanthopoulos. We shall give an account of it, since the procedure has applications in other contexts which we shall encounter.

We consider a system of n linear equations,

$$\frac{\mathrm{d}X_j}{\mathrm{d}x} = A_{jk}X_k \quad (j = 1, \ldots, n), \tag{76}$$

where the A_{jk}'s are some known functions of x and summation over repeated indices is assumed. Letting $\boldsymbol{X}$ denote the column vector (with components X_j) and $\boldsymbol{A}$ the (n, n)-matrix (whose elements are A_{jk}) we can write equation (76) in the form

$$\frac{\mathrm{d}\boldsymbol{X}}{\mathrm{d}x} = \boldsymbol{A}\boldsymbol{X}. \tag{77}$$

We define the associated *adjoint* system of equations by

$$\frac{\mathrm{d}\tilde{\boldsymbol{X}}}{\mathrm{d}x} = -\boldsymbol{A}^{\dagger}\tilde{\boldsymbol{X}} \quad \left(\text{or } \frac{\mathrm{d}\tilde{X}_j}{\mathrm{d}x} = -A_{kj}\tilde{X}_k\right), \tag{78}$$

where $\boldsymbol{A}^{\dagger}$ denotes the transposed of the matrix $\boldsymbol{A}$.

If X_j and $\tilde{X}_j$ are solutions of equations (77) and (78), then it readily follows that

$$\frac{\mathrm{d}}{\mathrm{d}x}(X_j\tilde{X}_j) = 0 \qquad \text{or} \qquad X_j\tilde{X}_j = \text{constant}. \tag{79}$$

We define the adjoint of the operators

$$\boldsymbol{A}\frac{\mathrm{d}}{\mathrm{d}x} \qquad \text{and} \qquad \boldsymbol{A}\frac{\mathrm{d}^2}{\mathrm{d}x^2}, \tag{80}$$

in the customary manner by

$$-\boldsymbol{A}^{\dagger}\frac{\mathrm{d}}{\mathrm{d}x} - \frac{\mathrm{d}\boldsymbol{A}^{\dagger}}{\mathrm{d}x} \qquad \text{and} \qquad \boldsymbol{A}^{\dagger}\frac{\mathrm{d}^2}{\mathrm{d}x^2} + 2\frac{\mathrm{d}\boldsymbol{A}^{\dagger}}{\mathrm{d}x}\frac{\mathrm{d}}{\mathrm{d}x} + \frac{\mathrm{d}^2\boldsymbol{A}^{\dagger}}{\mathrm{d}x^2}, \tag{81}$$

respectively. Also, we shall find it convenient to write equation (76) in the form

$$EX = \left(I\frac{d}{dx} - A\right)X = \left(\delta_{jk}\frac{d}{dx} - A_{jk}\right)X_k = 0, \tag{82}$$

where I denotes the unit (diagonal) matrix.

Suppose that there exists a linear combination,

$$Z = L_j X_j = LX, \tag{83}$$

of the X_j's which, by virtue of equation (77), satisfies the second-order equation

$$OZ = \left(\frac{d^2}{dx^2} + B\frac{d}{dx} + C\right)L_j X_j = 0, \tag{84}$$

where B and C are some known functions of x. The problem now is to devise an algorism which will enable us to discover the particular solutions of equations (77) which made this reduction possible.

More generally, one may suppose that there are m combinations

$$Z_\alpha = L_{\alpha j} X_j \quad (\alpha = 1, \ldots, m;\ j = 1, \ldots n;\ 2m < n) \tag{85}$$

which satisfy the set of m equations

$$\left(\delta_{\alpha\beta}\frac{d^2}{dx^2} + B_{\alpha\beta}\frac{d}{dx} + C_{\alpha\beta}\right)Z_\beta = 0 \quad (\alpha = 1, \ldots, m), \tag{86}$$

where $B_{\alpha\beta}$ and $C_{\alpha\beta}$ are certain known functions of x. We shall, however, restrict ourselves to the simpler case when we have only one combination to consider. No essentially new ideas are required to treat the general case; but a suitable notation must be devised.

The fact that equation (84) is satisfied, by virtue of equation (82), implies a certain operator identity which can be derived as follows:

$$\frac{dZ}{dx} = \frac{d}{dx}L_j X_j = \left(\frac{dL_j}{dx} + L_k A_{kj}\right)X_j = \Gamma_j X_j \quad \text{(say)}, \tag{87}$$

where

$$\Gamma_j = \frac{dL_j}{dx} + L_k A_{kj}, \qquad \text{or} \qquad \Gamma = \frac{dL}{dx} + LA. \tag{88}$$

Differentiating equation (87), we find

$$\frac{d^2 Z}{dx^2} = \left(\frac{d\Gamma_j}{dx} + \Gamma_k A_{kj}\right)X_j. \tag{89}$$

Inserting the foregoing expressions for the derivatives of Z in equation (84), we obtain

$$OZ = \left(\frac{d\Gamma_j}{dx} + \Gamma_k A_{kj} + B\Gamma_j + CL_j\right)X_j = 0; \tag{90}$$

and we infer the identity

$$\frac{d\Gamma}{dx}+\Gamma A+B\Gamma+CL=0. \tag{91}$$

We now seek an operator $\boldsymbol{S}$ which will satisfy the identity

$$\boldsymbol{SE}=\boldsymbol{O}L, \tag{92}$$

or, explicitly,

$$S_k\left(\delta_{kj}\frac{d}{dx}-A_{kj}\right)X_j=\left(\frac{d^2}{dx^2}+B\frac{d}{dx}+C\right)L_jX_j. \tag{93}$$

To determine $\boldsymbol{S}$, we evaluate the right-hand side of equation (93) when X_j satisfies the inhomogeneous equation

$$\frac{dX_j}{dx}=A_{jk}X_k+\Lambda_j. \tag{94}$$

Then (cf. equation (87))

$$\frac{dZ}{dx}=\Gamma_jX_j+L_j\Lambda_j, \tag{95}$$

and

$$\frac{d^2Z}{dx^2}=\left(\frac{d\Gamma_j}{dx}+\Gamma_kA_{kj}\right)X_j+\Gamma_k\Lambda_k+\frac{d}{dx}(L_j\Lambda_j). \tag{96}$$

Therefore, by making use of the identity (91), we now find

$$\begin{aligned} OZ=OL_jX_j&=\left(\frac{dL_j}{dx}+L_j\frac{d}{dx}+BL_j+\Gamma_j\right)\Lambda_j\\ &=\left(\frac{dL_j}{dx}+L_j\frac{d}{dx}+BL_j+\Gamma_j\right)\left(\delta_{jk}\frac{d}{dx}-A_{jk}\right)X_k. \end{aligned} \tag{97}$$

Comparison with equation (92) now shows that

$$\boldsymbol{S}=L\frac{d}{dx}+\frac{dL}{dx}+BL+\Gamma. \tag{98}$$

The adjoint of this operator, $\boldsymbol{S}$, is

$$\boldsymbol{S}^\dagger=-L^\dagger\frac{d}{dx}+BL^\dagger+\Gamma^\dagger. \tag{99}$$

Now consider

$$\tilde{X}_j=S_j^\dagger\phi=-L_j\frac{d\phi}{dx}+(BL_j+\Gamma_j)\phi\quad(j=1,\ldots,n), \tag{100}$$

where ϕ is some function of x. We shall now suppose that the foregoing n

equations determine ϕ and $\phi_{,x}$ uniquely. This is possible if, and only if, the $(3, n)$-matrix

$$\begin{vmatrix} \tilde{X}_1 & +L_1 & -(BL_1+\Gamma_1) \\ \tilde{X}_2 & +L_2 & -(BL_2+\Gamma_2) \\ \cdot & \cdot & \cdot \\ \cdot & \cdot & \cdot \\ \cdot & \cdot & \cdot \\ \tilde{X}_n & +L_n & -(BL_n+\Gamma_n) \end{vmatrix}, \tag{101}$$

or, equivalently, the matrix,

$$\begin{vmatrix} \tilde{X}_1 & L_1 & \Gamma_1 \\ \tilde{X}_2 & L_2 & \Gamma_2 \\ \cdot & \cdot & \cdot \\ \cdot & \cdot & \cdot \\ \cdot & \cdot & \cdot \\ \tilde{X}_n & L_n & \Gamma_n \end{vmatrix}, \tag{102}$$

is of rank 2, i.e., all the third-order determinants

$$D_{ijk} = \begin{vmatrix} \tilde{X}_i & L_i & \Gamma_i \\ \tilde{X}_j & L_j & \Gamma_j \\ \tilde{X}_k & L_k & \Gamma_k \end{vmatrix}, \qquad (i \neq j \neq k \neq i), \tag{103}$$

formed out of any three rows selected from the matrix (102) vanish. We shall now show that the vanishing of these determinants is preserved for the $\tilde{X}_j$'s varying in accordance with the adjoint system of equations

$$\frac{d\tilde{X}_j}{dx} = -A_{kj}\tilde{X}_k. \tag{104}$$

To show that this is the case, we differentiate the determinant D_{ijk} and replace the derivatives of $\tilde{X}$, L, and Γ in accordance with equations (104), (88), and (91), respectively. We thus obtain

$$\frac{d}{dx}D_{ijk} = \begin{vmatrix} -A_{li}\tilde{X}_l & L_i & \Gamma_i \\ -A_{lj}\tilde{X}_l & L_j & \Gamma_j \\ -A_{lk}\tilde{X}_l & L_k & \Gamma_k \end{vmatrix} + \begin{vmatrix} \tilde{X}_i & (\Gamma_i - L_l A_{li}) & \Gamma_i \\ \tilde{X}_j & (\Gamma_j - L_l A_{lj}) & \Gamma_j \\ \tilde{X}_k & (\Gamma_k - L_l A_{lk}) & \Gamma_k \end{vmatrix}$$
$$+ \begin{vmatrix} \tilde{X}_i & L_i & -(\Gamma_l A_{li} + B\Gamma_i + CL_i) \\ \tilde{X}_j & L_j & -(\Gamma_l A_{lj} + B\Gamma_j + CL_j) \\ \tilde{X}_k & L_k & -(\Gamma_l A_{lk} + B\Gamma_k + CL_k) \end{vmatrix}$$

$$= -A_{li}\begin{vmatrix} \tilde{X}_l & L_l & \Gamma_l \\ \tilde{X}_j & L_j & \Gamma_j \\ \tilde{X}_k & L_k & \Gamma_k \end{vmatrix} - A_{lj}\begin{vmatrix} \tilde{X}_i & L_i & \Gamma_i \\ \tilde{X}_l & L_l & \Gamma_l \\ \tilde{X}_k & L_k & \Gamma_k \end{vmatrix} - A_{lk}\begin{vmatrix} \tilde{X}_i & L_i & \Gamma_i \\ \tilde{X}_j & L_j & \Gamma_j \\ \tilde{X}_l & L_l & \Gamma_l \end{vmatrix}$$
$$-BD_{ijk}. \tag{105}$$

Therefore,

$$\frac{d}{dx}D_{ijk} = -A_{li}D_{ljk} - A_{lj}D_{ilk} - A_{lk}D_{ijl} - BD_{ijk}. \tag{106}$$

We conclude that *if all the third-order determinants, D_{ijk}'s, vanish, their derivatives also vanish.*

Without loss of generality, we may assume that ϕ and $\phi_{,x}$ are determined by the first pair of equations, $j = 1$ and 2, in (100). Consider the $(n-2)$ determinants

$$D_{12\alpha} \quad (\alpha = 3, \ldots, n). \tag{107}$$

Since any determinant D_{ijk} can be expressed as a linear combination of the $D_{12\alpha}$'s (with coefficients independent of the $\tilde{X}_j$'s), we can rewrite equation (106) in the form

$$\frac{d}{dx}D_{12\alpha} = \mathcal{G}_{\alpha\beta}D_{12\beta} \quad (\alpha = 3, \ldots, n), \tag{108}$$

where $\mathcal{G}_{\alpha\beta}$ are coefficients which are independent of the $\tilde{X}_j$'s and are explicitly known. Equation (108) guarantees that the vanishing of all the $D_{12\alpha}$'s is preserved by the adjoint system (104).

Equation (108) provides a linear system of $(n-2)$ first-order equations for the $D_{12\alpha}$'s. We shall suppose that the integrating factors for these equations can be found and the solutions expressed in the forms

$$\mathcal{D}_\alpha = k_{\alpha\beta}D_{12\beta} = C_\alpha = \text{constant} \quad (\alpha = 3, \ldots, n), \tag{109}$$

where the $k_{\alpha\beta}$'s are some known functions. Since the determinants, $D_{12\beta}$, are themselves linear combinations of the $\tilde{X}_j$'s, the solutions (109), written out explicitly in terms of the $\tilde{X}_j$'s, will be of the forms

$$\mathcal{D}_\alpha = X_{(\alpha)j}\tilde{X}_j = C_\alpha \quad (\alpha = 3, \ldots, n), \tag{110}$$

where the $X_{(\alpha)j}$'s are determinate linear combinations of the $k_{\alpha\beta}$'s. Equation (79) now enables us to conclude that $X_{(\alpha)j}$ $(j = 1, \ldots, n)$, for each α, represents a particular solution of the original system of equations (77). It is the existence of these particular solutions that underlies the original reducibility of the system. We have thus solved the problem, in principle, which we set ourselves.

It is of interest to note that in the particular case of an initial third-order system, reducible to a second-order equation, equation (106) takes the particularly simple form

$$\frac{dD_{123}}{dx} = -(B + A_{11} + A_{22} + A_{33})D_{123} \tag{111}$$

—an equation which allows of immediate integration.

(*a*) *The particular solution of the system of equations (52)–(54)*

We have seen in §24 that the system of equations (52)–(54) for N, L, and X is reducible to an equation of the second order for

$$Z^{(+)} = \frac{r^2}{nr+3M}\left(-L+\frac{3M}{nr}X\right). \tag{112}$$

To apply the algorism we have developed, we let

$$x \to r, \quad \text{and} \quad X_1, X_2, X_3 \to N, L, X. \tag{113}$$

With this correspondence

$$L_1 = 0, \quad L_2 = -\frac{r^2}{nr+3M}, \quad \text{and} \quad L_3 = \frac{r^2}{nr+3M}\frac{3M}{nr}. \tag{114}$$

Also, the wave equation satisfied by $Z^{(+)}$, in terms of the variable r, is

$$\frac{d^2Z^{(+)}}{dr^2} + \frac{2M}{r(r-2M)}\frac{dZ^{(+)}}{dr} + \frac{r^2}{(r-2M)^2}(\sigma^2 - V^{(+)})Z^{(+)} = 0. \tag{115}$$

We next consider equation (111) satisfied by

$$D_{123} = \frac{r^2}{nr+3M}\begin{vmatrix} \tilde{N} & 0 & \Gamma_1 \\ \tilde{L} & -1 & \Gamma_2 \\ \tilde{X} & 3M/nr & \Gamma_3 \end{vmatrix}. \tag{116}$$

According to equations (52)–(54) and (114)

$$A_{11} + A_{22} + A_{33} = a + b - c - \frac{2}{r} = \frac{5M-2r}{r(r-2M)}, \tag{117}$$

and

$$B = \frac{2M}{r(r-2M)}. \tag{118}$$

The equation for D_{123}, therefore, becomes

$$\frac{dD_{123}}{dr} = \frac{2r-7M}{r(r-2M)}D_{123}. \tag{119}$$

The integrated form of this equation is

$$\frac{e^{3\nu}}{r^2}D_{123} = \text{constant} \quad (e^{2\nu} = 1 - 2M/r). \tag{120}$$

It remains to express D_{123} in terms of $\tilde{N}$, $\tilde{L}$, and $\tilde{X}$. By equation (116),

$$D_{123} = \frac{r^2}{nr+3M}\left\{-\left(\Gamma_3 + \frac{3M}{nr}\Gamma_2\right)\tilde{N} + \Gamma_1\left(\frac{3M}{nr}\tilde{L} + \tilde{X}\right)\right\}, \tag{121}$$

where (cf. equation (88))

$$\left.\begin{aligned}\Gamma_1 &= L_2 A_{21} + L_3 A_{31},\\ \Gamma_2 &= L_{2,r} + L_2 A_{22} + L_3 A_{32},\\ \Gamma_3 &= L_{3,r} + L_2 A_{23} + L_3 A_{33}.\end{aligned}\right\} \tag{122}$$

and

The coefficients A_{21}, etc., in the foregoing equations can be read off from equations (52)–(54). On evaluating the various expressions, we find that we are finally left with

$$D_{123} = -\frac{e^{-2\nu}}{n^2}\left(p\tilde{N} + 3M\tilde{L} + nr\tilde{X}\right), \tag{123}$$

where

$$p = M - \frac{M^2 + \sigma^2 r^4}{r - 2M}. \tag{124}$$

The integral (120) now takes the form

$$\frac{e^\nu}{r^2}(p\tilde{N} + 3M\tilde{L} + nr\tilde{X}) = \text{constant}. \tag{125}$$

Therefore

$$N = f = p\frac{e^\nu}{r^2}, \quad L = g = 3M\frac{e^\nu}{r^2}, \quad \text{and} \quad X = h = nr\frac{e^\nu}{r^2} \tag{126}$$

represent particular solutions of equations (52)–(54) which were discovered by Xanthopoulos by this method.

With the aid of the particular solution (126), equations (52)–(54) can be reduced to one of order two by the substitutions,

$$N = f\nu, \quad L = g\lambda, \quad \text{and} \quad X = h\chi, \tag{127}$$

when they become

$$\left.\begin{aligned}\nu_{,r} &= A_{12}\frac{g}{f}(\lambda - \nu) + A_{13}\frac{h}{f}(\chi - \nu),\\ \lambda_{,r} &= A_{21}\frac{f}{g}(\nu - \lambda) + A_{23}\frac{h}{g}(\chi - \lambda),\\ \chi_{,r} &= A_{31}\frac{f}{h}(\nu - \chi) + A_{32}\frac{g}{h}(\lambda - \chi).\end{aligned}\right\} \tag{128}$$

By letting

$$P = \nu - \lambda, \quad Q = \lambda - \chi, \quad \text{and} \quad R = \chi - \nu \quad (P + Q + R = 0), \tag{129}$$

the equations take the forms

$$\left.\begin{aligned}\frac{dP}{dr} &= -\left(A_{12}\frac{g}{f}+A_{21}\frac{f}{g}\right)P+A_{23}\frac{h}{g}Q+A_{13}\frac{h}{f}R,\\ \frac{dQ}{dr} &= -\left(A_{23}\frac{h}{g}+A_{32}\frac{g}{h}\right)Q+A_{31}\frac{f}{h}R+A_{21}\frac{f}{g}P,\\ \frac{dR}{dr} &= -\left(A_{31}\frac{f}{h}+A_{13}\frac{h}{f}\right)R+A_{12}\frac{g}{f}P+A_{32}\frac{g}{h}Q;\end{aligned}\right\} \tag{130}$$

and these equations are consistent with the requirement (cf. equation (129))

$$P+Q+R=0. \tag{131}$$

The reducibility of the system to one of second order is now manifest.

Since

$$Z^{(+)} = \frac{r^2}{nr+3M}\left(\frac{3M}{nr}X-L\right) = \frac{r^2}{nr+3M}g(\chi-\lambda) = -\frac{3Me^{\nu}}{nr+3M}Q, \tag{132}$$

it is clear that the solution can be completed and expressed in terms of $Z^{(+)}$ and its derivative in the manner of §24(ii).

26. The relations between $V^{(+)}$ and $V^{(-)}$ and $Z^{(+)}$ and $Z^{(-)}$

In §24, we reduced the equations governing the polar and the axial perturbations to one-dimensional Schrodinger equations for two functions $Z^{(-)}$ and $Z^{(+)}$ with potentials $V^{(-)}$ and $V^{(+)}$. It is a remarkable fact that, in spite of appearance, the two potentials $V^{(-)}$ and $V^{(+)}$, given by equations (28) and (63), are very simply related. They are in fact given by (as one may directly verify)

$$V^{(\pm)} = \pm\beta\frac{df}{dr_*}+\beta^2 f^2+\kappa f, \tag{133}$$

where

$$\beta = \text{constant} = 6M, \quad \kappa = \text{constant} = 4n(n+1) = \mu^2(\mu^2+2), \tag{134}$$

and

$$f = \frac{\Delta}{r^3(\mu^2 r+6M)} = \frac{\Delta}{2r^3(nr+3M)}. \tag{135}$$

It should be particularly noted that in equation (133) β and κ are constants and f is a function which vanishes both at the horizon ($r=2M$) and at infinity (with an inverse-square r^{-2}-behaviour).

There is no obvious reason, at this stage, to have expected (or, indeed, to expect) that the potentials $V^{(+)}$ and $V^{(-)}$ are simply related in this fashion. The origin of the relation will emerge when we come to treat the perturbation problem via the Newman–Penrose formalism in §§28 and 29. Meantime, we shall accept the relation (133) as a directly verifiable fact and show that it implies a very simple relation between the solutions $Z^{(+)}$ and $Z^{(-)}$.

Quite generally, we shall consider the two wave equations

$$\frac{d^2Z_1}{dx^2}+\sigma^2 Z_1 = V_1 Z_1 = \left(+\beta\frac{df}{dx}+\beta^2 f^2+\kappa f\right)Z_1 \tag{136}$$

and

$$\frac{d^2Z_2}{dx^2}+\sigma^2 Z_2 = V_2 Z_2 = \left(-\beta\frac{df}{dx}+\beta^2 f^2+\kappa f\right)Z_2, \tag{137}$$

where β and κ are some real constants and f is an arbitrary smooth function which, together with its derivatives of all orders, vanish for both $x \to +\infty$ and $x \to -\infty$, and whose integral over the entire range of x is finite. (For convenience of notation, we are temporarily replacing r_* by x and $Z^{(+)}$ and $Z^{(-)}$ by Z_1 and Z_2.)

There is clearly no restriction to supposing that, given a solution Z_2 of equation (137),

$$Z_1 = pZ_2 + qZ_2' \tag{138}$$

is a solution of equation (136) where Z_2' denotes the derivative of Z_2 (with respect to x) and p and q are certain suitably chosen functions. The equations that must govern p and q in order that Z_1, given by equation (138), is a solution of equation (136), can be derived as follows.

Differentiating equation (138) and making use of equation (137) satisfied by Z_2, we obtain

$$Z_1' = [p' + q(V_2-\sigma^2)]Z_2 + (p+q')Z_2'. \tag{139}$$

Differentiating this equation, once again, we similarly obtain

$$\begin{aligned} Z_1'' = {} & [p'' + (p+2q')(V_2-\sigma^2) + qV_2']Z_2 \\ & + [p' + q(V_2-\sigma^2) + p' + q'']Z_2'; \end{aligned} \tag{140}$$

and this expression must identically be the same that follows from equations (136) and (138), namely,

$$Z_1'' = (pZ_2 + qZ_2')(V_1-\sigma^2). \tag{141}$$

Equating the coefficients of Z_2 and Z_2' on the right-hand sides of equations (140) and (141), we find

$$q(V_1-\sigma^2) = 2p' + q'' + q(V_2-\sigma^2) \tag{142}$$

and

$$p(V_1-\sigma^2) = p'' + (p+2q')(V_2-\sigma^2) + qV_2'; \tag{143}$$

or, alternatively,

$$q(V_1 - V_2) = 2p' + q'' \tag{144}$$

and

$$p(V_1 - V_2) = p'' + 2q'(V_2-\sigma^2) + qV_2'. \tag{145}$$

Eliminating $(V_1 - V_2)$ between these two equations, we obtain

$$2pp' + pq'' - p''q - 2qq'(V_2 - \sigma^2) - q^2 V_2' = 0. \qquad (146)$$

This equation provides the integral

$$p^2 + (pq' - p'q) - q^2(V_2 - \sigma^2) = \text{constant} = C^2 \quad \text{(say).} \qquad (147)$$

Equations (144) and (147) are the equations which p and q must satisfy if the combination $pZ_2 + qZ_2'$ is to be a solution of equation (136). In general (i.e., for arbitrarily specified V_1 and V_2) one cannot expect to solve these equations explicitly. But for the case on hand, we can! Indeed, for V_1 and V_2 of the forms specified, one can readily verify that

$$q = 2\beta \ (= \text{constant}) \qquad \text{and} \qquad p = \kappa + 2\beta^2 f, \qquad (148)$$

do satisfy equations (144) and (147) with

$$C^2 = \kappa^2 + 4\beta^2\sigma^2. \qquad (149)$$

Accordingly, the solutions Z_1 and Z_2 of equations (136) and (137) are related in the manner

$$(\kappa + 2i\sigma\beta)Z_1 = (\kappa + 2\beta^2 f)Z_2 + 2\beta Z_2', \qquad (150)$$

where we have chosen a relative normalization of Z_1 and Z_2 such that the inverse relation in the same normalization is given by

$$(\kappa - 2i\sigma\beta)Z_2 = (\kappa + 2\beta^2 f)Z_1 - 2\beta Z_1'. \qquad (151)$$

For the particular case of $Z^{(+)}$ and $Z^{(-)}$ when β, κ, and f are given by equations (134) and (135), the relations (150) and (151) take the forms

$$[\mu^2(\mu^2+2) + 12i\sigma M]Z^{(+)} = \left[\mu^2(\mu^2+2) + 72M^2 \frac{\Delta}{r^3(\mu^2 r + 6M)}\right]Z^{(-)} + 12MZ^{(-)}_{,r_*} \qquad (152)$$

and

$$[\mu^2(\mu^2+2) - 12i\sigma M]Z^{(-)} = \left[\mu^2(\mu^2+2) + 72M^2 \frac{\Delta}{r^3(\mu^2 r + 6M)}\right]Z^{(+)} - 12MZ^{(+)}_{,r_*}. \qquad (153)$$

We shall return in §27 to the implications of the foregoing relations and to the further consequences of the potentials being of the forms (133).

An interesting consequence of the relations (152) and (153) is the following. By inserting for L, X, and N their solutions (71), (72), and (75) in the expression

$$\mathscr{X} = \frac{1}{r}(nr + 3M)\left[N - \frac{(n+1)r}{3M}L\right] + \left[(n+1) - \frac{p}{r}\right](L + X), \qquad (154)$$

we find that the terms in Φ cancel (which is the reason for considering the expression $\mathscr{X}$ in the first place) and we are left with

$$\mathscr{X} = -nZ^{(+)}_{,r_*} + \left[\frac{n^2(n+1)}{3M} + 3Mn\frac{r-2M}{r^2(nr+3M)}\right]Z^{(+)}. \tag{155}$$

Therefore,

$$\frac{12M}{n}\mathscr{X} = -12MZ^{(+)}_{,r_*} + \left[4n(n+1) + 36M^2\frac{\Delta}{r^3(nr+3M)}\right]Z^{(+)}. \tag{156}$$

From a comparison of this last equation with equation (153) (and remembering that $\mu^2 = 2n$), we conclude that the expression for $\mathscr{X}$, formed out of the radial functions describing the *polar perturbations*, will satisfy the wave equation appropriate to the *axial perturbations*!

27. The problem of reflexion and transmission

We return to the wave equations satisfied by $Z^{(+)}$ and $Z^{(-)}$ to consider the nature of the solutions of these equations and the quantities of physical interest which they determine.

We first observe that the potentials $V^{(+)}$ and $V^{(-)}$ are smooth functions, integrable over the range of r_*, $(+\infty, -\infty)$, and positive everywhere. Besides, they have an inverse-square behaviour for $r_* \to +\infty$ and vanish exponentially as we approach the horizon and $r_* \to -\infty$. Thus,

$$\left.\begin{aligned} &V^{(\pm)} \to 2(n+1)r^{-2} \qquad \text{as} \qquad r \to r_* \to +\infty \\ \text{and} \qquad & \\ &V^{(\pm)} \to (\text{constant})_{\pm}\, e^{r_*/2M} \qquad \text{as} \qquad r_* \to -\infty. \end{aligned}\right\} \tag{157}$$

Since $V^{(\pm)}$ fall off more rapidly than r_*^{-1} for $r_* \to \pm\infty$, the asymptotic behaviour of the solutions, $Z^{(\pm)}$, for $r_* \to \pm\infty$, is given by

$$e^{\pm i\sigma r_*} \quad (r_* \to \pm\infty). \tag{158}$$

For real σ, the solutions, therefore, represent ingoing and outgoing waves at $\pm\infty$. The underlying physical problem is, therefore, one of reflexion and transmission of incident waves (from $+$ or $-\infty$) by the one-dimensional potential-barriers, $V^{(\pm)}$. This is the problem of the penetration of one-dimensional potential-barriers with which we are familiar in elementary quantum theory. Precisely, in the context of the problem on hand, we must seek solutions of the wave equations which satisfy the boundary conditions,

$$\begin{aligned} Z^{(\pm)} &\to e^{+i\sigma r_*} + R^{(\pm)}(\sigma)e^{-i\sigma r_*} \quad (r_* \to +\infty), \\ &\to T^{(\pm)}(\sigma)e^{+i\sigma r_*} \quad (r_* \to -\infty). \end{aligned} \tag{159}$$

These boundary conditions correspond to an *incident wave* of unit amplitude from $+\infty$ giving rise to a *reflected wave* of amplitude $R^{(\pm)}(\sigma)$ at $+\infty$ and a *transmitted wave* of amplitude $T^{(\pm)}(\sigma)$ at $-\infty$. (Note that, in conformity with physical requirements, the boundary conditions we have imposed do not allow for waves emerging from the horizon.)

Since the potentials are real, the complex conjugate of the solution, satisfying the boundary conditions (159), will satisfy the complex-conjugate boundary conditions,

$$\begin{aligned} Z^{(\pm)*} &\to e^{-i\sigma r_*} + R^{(\pm)*}(\sigma)e^{+i\sigma r_*} \quad (r_* \to +\infty), \\ &\to \qquad T^{(\pm)*}(\sigma)e^{-i\sigma r_*} \quad (r_* \to -\infty). \end{aligned} \tag{160}$$

Since the Wronskian,

$$[Z^{(\pm)}, Z^{(\pm)*}] = Z^{(\pm)}_{,r_*}Z^{(\pm)*} - Z^{(\pm)}Z^{(\pm)*}_{,r_*}, \tag{161}$$

of the two (independent) solutions, $Z^{(\pm)}$ and $Z^{(\pm)*}$, must be a constant, we obtain by evaluating at $+\infty$ and $-\infty$,

$$-2i\sigma(|R^{(\pm)}(\sigma)|^2 - 1) = +2i\sigma|T^{(\pm)}(\sigma)|^2, \tag{162}$$

or,

$$\mathbb{R}^{(\pm)}(\sigma) + \mathbb{T}^{(\pm)}(\sigma) = 1, \tag{163}$$

where

$$\mathbb{R}^{(\pm)}(\sigma) = |R^{(\pm)}(\sigma)|^2 \qquad \text{and} \qquad \mathbb{T}^{(\pm)}(\sigma) = |T^{(\pm)}(\sigma)|^2 \tag{164}$$

are the *reflexion* and the *transmission coefficients.*

(a) *The equality of the reflexion and the transmission coefficients for the axial and the polar perturbations*

Consider, quite generally, the problem of reflexion and transmission associated with the pair of wave equations (136) and (137). From equations (150) and (151) relating the solutions of these two equations, it follows, from our assumption that f vanishes for $x \to \pm\infty$, that solutions for Z_1, derived from solutions for Z_2 having the asymptotic behaviours

$$Z_2 \to e^{+i\sigma x} \qquad \text{and} \qquad Z_2 \to e^{-i\sigma x} \quad (x \to \pm\infty), \tag{165}$$

have, respectively, the asymptotic behaviours

$$Z_1 \to e^{+i\sigma x} \qquad \text{and} \qquad Z_1 \to \frac{\kappa - 2i\sigma\beta}{\kappa + 2i\sigma\beta} e^{-i\sigma x} \quad (x \to \pm\infty). \tag{166}$$

Accordingly, for the associated problems of reflexion and transmission of incident waves by the potential barriers V_1 and V_2,

$$T_1(\sigma) = T_2(\sigma) \qquad \text{and} \qquad R_1(\sigma) = \frac{\kappa - 2i\sigma\beta}{\kappa + 2i\sigma\beta} R_2(\sigma). \tag{167}$$

TABLE III

The reflexion coefficient ($\mathbb{R}$) *and the phases of the reflected amplitudes,* $\delta^{(+)}$ *and* $\delta^{(-)}$ *in radians, for the axial and the polar waves for* $l = 2$ *and for various frequencies* (σ)

σ/M	$\mathbb{R}$	$\delta^{(+)}$	$\delta^{(-)}$
0.10	1.0000	3.172	3.272
0.20	0.9991	2.462	2.661
0.30	0.9945	1.198	1.496
0.32	0.8895	0.854	1.171
0.34	0.7929	0.471	0.808
0.36	0.6491	0.057	0.413
0.38	0.4754	5.909	0.001
0.40	0.3102	5.482	5.877
0.42	0.1841	5.080	5.494
0.44	0.1027	4.710	5.143
0.46	0.05533	4.371	4.823
0.48	0.02935	4.061	4.532
0.50	0.01548	3.773	4.263
0.52	0.00817	3.504	4.013
0.54	0.00433	3.250	3.777
0.56	0.00230	3.007	3.553
0.58	0.00123	2.775	3.340
0.60	0.00065	2.552	3.135
0.62	0.00036	2.337	2.938
0.64	0.00019	2.127	2.746
0.66	0.00010	1.923	2.560
0.68	0.00006	1.724	2.379
0.70	0.00003	1.530	2.203

The reflexion coefficients ($\mathbb{R}$) *for values of* $l > 2$

$\sigma M/l$	$l = 3$	$l = 4$	$l = 5$	$l = 6$
0.130	0.9995			
0.140	0.9985			
0.150	0.9953			
0.160	0.9863	0.9973		
0.170	0.9604	0.9888		
0.180	0.8935	0.9563	0.9787	0.9886
0.190	0.7469	0.8471	0.8934	0.9192
0.200	0.5137	0.5890	0.6109	0.6064
0.210	0.2774	0.2745	0.2312	0.1768
0.220	0.1236	0.0924	0.0556	0.0296
0.230	0.0471	0.0272	0.0117	0.0046
0.240	0.0192	0.0074	0.0023	0.0008
0.250	0.0077	0.0018	0.0005	
0.260	0.0029	0.0009		

Thus, *while the amplitudes of the transmitted waves are identically the same for the two potentials, V_1 and V_2, the reflected amplitudes differ only in their phases*:

$$R_1(\sigma) = e^{i\delta} R_2(\sigma) \qquad \text{where} \qquad e^{i\delta} = \frac{\kappa - 2i\sigma\beta}{\kappa + 2i\sigma\beta}. \tag{168}$$

The equality of the reflexion and the transmission coefficients follows from these facts.

For the particular case of $Z^{(+)}$ and $Z^{(-)}$ on hand, the phase difference in the reflected amplitudes is given by (cf. equation (134)).

$$e^{i\delta} = \frac{\mu^2(\mu^2+2) - 12i\sigma M}{\mu^2(\mu^2+2) + 12i\sigma M}. \tag{169}$$

In Table III we provide a brief listing of the reflexion and transmission coefficients (common for both the axial and the polar perturbations) and the difference in phase of the reflected waves given by equation (169).

It should be further noted that, from the relations (150) and (151) between the solutions Z_1 and Z_2, it also follows that should V_1 allow a discrete set of states (with solutions falling off exponentially as $x \to \pm\infty$), then V_2 will also allow the same set of discrete states. However, since the potentials, $V^{(+)}$ and $V^{(-)}$, are positive everywhere, they do not allow any discrete states.

28. The elements of the theory of one-dimensional potential-scattering and a necessary condition that two potentials yield the same transmission amplitude

In §27 we have seen how the pair of potentials (133) yield the same transmission amplitude for all σ. The question arises whether there are some necessary conditions that will characterize potentials with this property. The question is related to the theory of 'inverse scattering' dealing with the problem of determining the potential from a knowledge of the S-matrix (see equation (179) below) and its ramifications in the theory of solitons and the Korteweg–de Vries equation. Besides, it will emerge from our study of the associated problem in the context of the Kerr black-hole in Chapter 9 that we shall have to deal with cases of complex potentials and potentials with singularities. For these cases there is, as yet, no comprehensive theory. For this reason, we shall give a brief account of the elements of the extant theory and obtain necessary conditions for two potentials to yield the same transmission amplitude.

The theory of one-dimensional potential-scattering is concerned with the solution of Schrodinger's wave-equation

$$\left(\frac{d^2}{dx^2} + \sigma^2\right) f = V f \quad (-\infty < x < +\infty), \tag{170}$$

where $V(x)$ is a smooth function of x. We shall also suppose that all polynomials, constructed out of V and of its derivatives of all orders, are integrable over the entire range, $(-\infty, +\infty)$, of x. With these restrictions on V, the asymptotic behaviours of the solutions of equation (170), for $x \to \pm\infty$, are given by

$$e^{\pm i\sigma x}. \tag{171}$$

We now consider two particular solutions, $f_1(x, \sigma)$ and $f_2(x, \sigma)$, with the asymptotic behaviours

$$\left.\begin{aligned} f_1(x,\sigma) &\to e^{-i\sigma x} \qquad (x \to +\infty) \\ \text{and} \qquad f_2(x,\sigma) &\to e^{+i\sigma x} \qquad (x \to -\infty). \end{aligned}\right\} \tag{172}$$

Now $f_1(x, \sigma)$ and $f_1(x, -\sigma)$ are two solutions of equation (170) which are independent, since their Wronskian

$$\begin{aligned} [f_1(x,\sigma), f_1(x,-\sigma)] &= \{f_1{}'(x,\sigma)f_1(x,-\sigma) - f_1(x,\sigma)f_1{}'(x,-\sigma)\} = \text{constant} \\ &= \lim_{x\to+\infty} \{e^{i\sigma x}(-i\sigma)e^{-i\sigma x} - e^{-i\sigma x}(i\sigma)e^{+i\sigma x}\} \\ &= -2i\sigma \neq 0. \end{aligned} \tag{173}$$

Similarly,

$$[f_2(x,\sigma), f_2(x,-\sigma)] = +2i\sigma \quad (\neq 0); \tag{174}$$

and $f_2(x, \sigma)$ and $f_2(x, -\sigma)$ are also independent solutions of equation (170). It follows that there exist unique functions $R_1(\sigma)$, $R_2(\sigma)$, $T_1(\sigma)$, and $T_2(\sigma)$ such that

$$f_2(x,\sigma) = \frac{R_1(\sigma)}{T_1(\sigma)} f_1(x,\sigma) + \frac{1}{T_1(\sigma)} f_1(x,-\sigma) \tag{175}$$

and

$$f_1(x,\sigma) = \frac{R_2(\sigma)}{T_2(\sigma)} f_2(x,\sigma) + \frac{1}{T_2(\sigma)} f_2(x,-\sigma), \tag{176}$$

for $\sigma \neq 0$. From the assumed asymptotic behaviours $f_1(x, \sigma)$ and $f_2(x, \sigma)$, it follows from the foregoing representations that

$$\lim_{x\to+\infty} f_2(x,\sigma) = \frac{R_1(\sigma)}{T_1(\sigma)} e^{-i\sigma x} + \frac{1}{T_1(\sigma)} e^{+i\sigma x} \tag{177}$$

and

$$\lim_{x\to-\infty} f_1(x,\sigma) = \frac{R_2(\sigma)}{T_2(\sigma)} e^{+i\sigma x} + \frac{1}{T_2(\sigma)} e^{-i\sigma x}. \tag{178}$$

Accordingly, $T_1(\sigma)f_2(x, \sigma)$ corresponds to an incident wave of unit amplitude from $+\infty$ (cf. equation (159)) giving rise to a reflected wave of amplitude $R_1(\sigma)$ and a transmitted wave of amplitude $T_1(\sigma)$. Similarly, $T_2(\sigma)f_1(x, \sigma)$

corresponds to an incident wave of unit amplitude from $-\infty$ giving rise to reflected and transmitted waves of amplitudes $R_2(\sigma)$ and $T_2(\sigma)$, respectively.

We define

$$\boldsymbol{S}(\sigma) = \begin{vmatrix} T_1(\sigma) & R_2(\sigma) \\ R_1(\sigma) & T_2(\sigma) \end{vmatrix}, \tag{179}$$

for $\sigma \neq 0$, as the *scattering* or the ***S**-matrix*.

From the representations (175) and (176) and the relations (173) and (174), it follows that

$$\left.\begin{aligned} \frac{1}{T_1(\sigma)} &= -\frac{1}{2i\sigma}[f_1(x,\sigma), f_2(x,\sigma)] = \frac{1}{T_2(\sigma)}, \\ \frac{R_1(\sigma)}{T_1(\sigma)} &= -\frac{1}{2i\sigma}[f_2(x,\sigma), f_1(x,-\sigma)], \\ \text{and} \qquad \frac{R_2(\sigma)}{T_2(\sigma)} &= -\frac{1}{2i\sigma}[f_2(x,-\sigma), f_1(x,\sigma)]. \end{aligned}\right\} \tag{180}$$

Therefore,

$$T_1(\sigma) = T_2(\sigma) = T(\sigma) \quad \text{(say)}, \tag{181}$$

$$\frac{R_1(\sigma)}{T(\sigma)} = -\frac{R_2(-\sigma)}{T(-\sigma)}, \quad \text{and} \quad \frac{R_1(-\sigma)}{T(-\sigma)} = -\frac{R_2(\sigma)}{T(\sigma)}. \tag{182}$$

Besides,

$$T^*(\sigma) = T(-\sigma), \quad R_1^*(\sigma) = R_1(-\sigma), \quad \text{and} \quad R_2^*(-\sigma) = R_2(\sigma). \tag{183}$$

Next, inserting for $f_2(x, \sigma)$ its representation (175) in equation (176) for $f_1(x, \sigma)$ and equating the coefficients of $f_1(x, \sigma)$ and $f_1(x, -\sigma)$ in the resulting expression (which we may since $f_1(x, \sigma)$ and $f_1(x, -\sigma)$ are independent solutions) we find

$$1 = \frac{R_2(\sigma)R_1(\sigma)}{T(\sigma)T(\sigma)} + \frac{1}{T(\sigma)T(-\sigma)}, \tag{184}$$

in addition to the relation (182). Making use of the relations (182) and (183), we can rewrite equation (184) in the alternative forms

$$|R_1(\sigma)|^2 + |T(\sigma)|^2 = |R_2(\sigma)|^2 + |T(\sigma)|^2 = 1. \tag{185}$$

From equation (185) it follows that

$$|R_1(\sigma)|, \quad |R_2(\sigma)|, \quad \text{and} \quad |T(\sigma)| \quad \text{are} \quad \leqslant 1; \tag{186}$$

and further that $R_1(\sigma)$ and $R_2(\sigma)$ can differ only in phase. The relations (181), (182), and (183) establish the *symmetry* and the *unitarity* of the **S**-matrix.

(a) The Jost functions and the integral equations they satisfy

In the theory of potential scattering one defines the *Jost functions*,

$$m_1(x,\sigma) = e^{+i\sigma x} f_1(x,\sigma) \quad \text{and} \quad m_2(x,\sigma) = e^{-i\sigma x} f_2(x,\sigma), \tag{187}$$

which satisfy the boundary conditions (cf. equation (172))

$$m_1(x,\sigma) \to 1 \quad \text{as} \quad x \to +\infty \quad \text{and} \quad m_2(x,\sigma) \to 1 \quad \text{as} \quad x \to -\infty. \tag{188}$$

Also, we shall now let σ be a complex variable with $\operatorname{Im}\sigma < 0$ so that the functions, $f_1(x,\sigma)$ and $f_2(x,\sigma)$, as defined in equation (172), vanish for $x \to +\infty$ and $x \to -\infty$, respectively.

In terms of the Jost functions, we can rewrite equations (175) and (176) in the forms

$$T(\sigma) m_2(x,\sigma) = R_1(\sigma) e^{-2i\sigma x} m_1(x,\sigma) + m_1(x,-\sigma) \tag{189}$$

and

$$T(\sigma) m_1(x,\sigma) = R_2(\sigma) e^{+2i\sigma x} m_2(x,\sigma) + m_2(x,-\sigma). \tag{190}$$

From the behaviours (188) it follows in particular that

$$m_2(x,\sigma) = \frac{R_1(\sigma)}{T(\sigma)} e^{-2i\sigma x} + \frac{1}{T(\sigma)} + o(1) \quad \text{as} \quad x \to +\infty \tag{191}$$

and

$$m_1(x,\sigma) = \frac{R_2(\sigma)}{T(\sigma)} e^{+2i\sigma x} + \frac{1}{T(\sigma)} + o(1) \quad \text{as} \quad x \to -\infty. \tag{192}$$

We also note that the Jost functions satisfy the differential equations

$$\left.\begin{aligned} \frac{d^2 m_1}{dx^2} - 2i\sigma \frac{dm_1}{dx} &= V m_1, \\ \text{and} \qquad \frac{d^2 m_2}{dx^2} + 2i\sigma \frac{dm_2}{dx} &= V m_2. \end{aligned}\right\} \tag{193}$$

We shall now obtain an integral equation for $m_2(x,\sigma)$. Letting

$$f_2(x,\sigma) = e^{i\sigma x} + \psi(x,\sigma) \quad (\psi \to 0 \quad \text{as} \quad x \to -\infty), \tag{194}$$

we find that ψ satisfies the differential equation

$$\left(\frac{d^2}{dx^2} + \sigma^2\right)\psi = (e^{i\sigma x} + \psi) V. \tag{195}$$

Solving this equation with the aid of the Green function,

$$G(x-x') = \frac{1}{\sigma} \sin\sigma(x-x')\,\theta(x-x'), \tag{196}$$

where $\theta(x-x')$ is the step function,

$$\theta(x-x') = 1 \quad \text{for} \quad x > x' \qquad \text{and} \qquad = 0 \quad \text{for} \quad x < x', \tag{197}$$

we obtain (remembering that $\psi \to 0$ as $x \to -\infty$),

$$\psi(x,\sigma) = \frac{1}{2i\sigma}\int_{-\infty}^{x} [e^{i\sigma(x-x')} - e^{-i\sigma(x-x')}]\,V(x')[e^{i\sigma x'} + \psi(x',\sigma)]\,dx'. \tag{198}$$

Rewriting this equation in terms of the Jost function,

$$m_2(x,\sigma) = e^{-i\sigma x} f_2(x,\sigma) = 1 + e^{-i\sigma x}\psi(x,\sigma), \tag{199}$$

we obtain

$$m_2(x,\sigma) = 1 - \frac{1}{2i\sigma}\int_{-\infty}^{x} (e^{2i\sigma(x'-x)} - 1)\,V(x')m_2(x',\sigma)\,dx'. \tag{200}$$

Equation (200), for $\sigma \neq 0$, is a Volterra integral-equation for $m_2(x,\sigma)$; and, for $\operatorname{Im}\sigma < 0$, its solution obtained by repeated iterations always converges for any smooth V which is integrable. It is also apparent that the solution obtained by such iterations provides an expansion for $m_2(x,\sigma)$ in a power series in inverse σ. Based on these facts, it can be shown that $m_2(x,\sigma)$ is an analytic function in the lower-half complex σ-plane (i.e., for $\operatorname{Im}\sigma < 0$); and, further, that it is continuous for $\operatorname{Im}\sigma \leqslant 0$ ($\sigma \neq 0$).

From equation (200) it follows that

$$\begin{aligned} m_2(x,\sigma) &= -e^{-2i\sigma x}\left[\frac{1}{2i\sigma}\int_{-\infty}^{x} e^{+2i\sigma x'}V(x')m_2(x',\sigma)\,dx'\right] \\ &\quad + \left[1 + \frac{1}{2i\sigma}\int_{-\infty}^{x} V(x')m_2(x',\sigma)\,dx'\right] \\ &= -e^{-2i\sigma x}\left[\frac{1}{2i\sigma}\int_{-\infty}^{\infty} e^{+2i\sigma x'}V(x')m_2(x',\sigma)\,dx'\right] \\ &\quad + \left[1 + \frac{1}{2i\sigma}\int_{-\infty}^{\infty} V(x')m_2(x',\sigma)\,dx'\right] + o(1) \quad \text{for} \quad x \to \infty. \end{aligned} \tag{201}$$

Comparison of this last result with equation (191) shows that

$$\frac{R_1(\sigma)}{T(\sigma)} = -\frac{1}{2i\sigma}\int_{-\infty}^{+\infty} e^{2i\sigma x}V(x)m_2(x,\sigma)\,dx \tag{202}$$

and

$$\frac{1}{T(\sigma)} = 1 + \frac{1}{2i\sigma}\int_{-\infty}^{+\infty} V(x)m_2(x,\sigma)\,dx. \tag{203}$$

It is clear that similar equations for $R_2(\sigma)/T(\sigma)$ and $1/T(\sigma)$ can be written down in terms of the Jost function $m_1(x,\sigma)$.

From equation (203), we may draw one important conclusion. For $\operatorname{Im}\sigma < 0$, we can obtain a convergent expansion for $1/T(\sigma)$, by inserting for the Jost function, $m_2(x, \sigma)$, its solution obtained from the Volterra integral-equation (200) by repeated iterations. An expansion for $1/T(\sigma)$, obtained in this fashion, will be a power series in inverse σ with coefficients which will be bounded so long as $V(x)$ satisfies the requirements we have imposed on it, namely, the boundedness of integrals of all polynomials constructed out of V and its derivatives. In particular, it follows from equation (203) that

$$T(\sigma) = 1 - \frac{1}{2i\sigma}\int_{-\infty}^{+\infty} V(x)\,\mathrm{d}x + O(\sigma^{-2}). \tag{204}$$

(*b*) *An expansion of* $\lg T(\sigma)$ *as a power series in* σ^{-1} *and a condition for different potentials to yield the same transmission amplitude*

We have seen how, by combining equations (200) and (203), we can obtain an expansion for $1/T(\sigma)$ as a power series in σ^{-1}. But it appears that to obtain the coefficients in this expansion explicitly, it is more convenient to follow a different route due to Faddeev.

With the substitution,

$$m_2(x, \sigma) = e^{w(x, \sigma)}. \tag{205}$$

equation (193) satisfied by $m_2(x, \sigma)$ becomes

$$w'' + 2i\sigma w' + (w')^2 - V = 0, \tag{206}$$

where primes are used to denote differentiations with respect to x. From the fact that $m_2 \to 1$ as $x \to -\infty$, we conclude that

$$w \to 0 \quad \text{as} \quad x \to -\infty. \tag{207}$$

Also, from equation (191) we similarly conclude that

$$w \to -\lg T(\sigma) \quad \text{as} \quad x \to +\infty \quad \text{for} \quad \operatorname{Im}\sigma < 0. \tag{208}$$

Now letting

$$w(x, \sigma) = \int_{-\infty}^{x} v(x', \sigma)\,\mathrm{d}x' \quad (w' = v), \tag{209}$$

we obtain the equation

$$v' + 2i\sigma v + v^2 - V = 0. \tag{210}$$

The requirement (208) now implies that

$$\lg T(\sigma) = -\int_{-\infty}^{+\infty} v(x, \sigma)\,\mathrm{d}x. \tag{211}$$

We now seek a solution of equation (210) by expanding v in a power series in σ^{-1}; thus

$$v = \sum_{n=1}^{\infty} \frac{v_n(x)}{(2i\sigma)^n}, \tag{212}$$

where, according to equations (204) and (211),

$$v_1 = V. \tag{213}$$

Inserting the expansion (212) in equation (211), we obtain

$$\lg T(\sigma) = -\sum_{n=1}^{\infty} c_n \sigma^{-n}, \tag{214*}$$

where

$$(2i)^n c_n = \int_{-\infty}^{+\infty} v_n(x)\,\mathrm{d}x. \tag{215}$$

Now substituting the expansion (212) in equation (210) and equating the coefficients of the different powers of $1/\sigma$, we obtain the recurrence relation,

$$v_n = -v'_{n-1} - \sum_{l=1}^{n-1} v_l v_{n-1-l}. \tag{216}$$

Using this recurrence relation, we can solve for the v_n's successively starting with

$$v_1 = V. \tag{217}$$

We find

$$\left.\begin{aligned} v_2 &= -v'_1 = -V', \\ v_3 &= -v'_2 - v_1^2 = V'' - V^2, \\ v_4 &= -v'_3 - 2v_1 v_2 = -V''' + 2(V^2)', \\ v_5 &= -v'_4 - 2v_1 v_3 - v_2^2 = V'''' - 3(V^2)'' + (V')^2 + 2V^3, \text{ etc.} \end{aligned}\right\} \tag{218}$$

The coefficients of odd orders, c_{2n+1}, in the expansion for $\lg T(\sigma)$ are, therefore,

$$2ic_1 = \int_{-\infty}^{+\infty} V\,\mathrm{d}x; \quad -(2i)^3 c_3 = \int_{-\infty}^{+\infty} V^2\,\mathrm{d}x;$$

$$(2i)^5 c_5 = \int_{-\infty}^{+\infty} (2V^3 + V'^2)\,\mathrm{d}x; \text{ etc.}, \tag{219}$$

while all the coefficients of even order vanish.

Now if two different potentials should yield the same $T(\sigma)$, then the derived series expansions for $\lg T(\sigma)$ (convergent for $\operatorname{Im}\sigma < 0$) for the two potentials must coincide term by term. In this manner, we find that the integrals of the

* I am grateful to Dr. Roza Trautman for pointing out to me that this result follows directly for the following alternative expansion for $\lg T(\sigma)$ due to V. E. Zakharov and L. D. Faddeev, *Funct. Anal. Appl.*, 5, 280 (1971):

$$\lg T(\sigma) = \frac{i}{2\pi} \sum_{n=1}^{\infty} \frac{1}{\sigma^n} \int_{-\infty}^{+\infty} (\sigma')^{n-1} \lg |T(\sigma')|^2 \,\mathrm{d}\sigma'.$$

following quantities for the two potentials should be the same:

$$\left.\begin{array}{l}\text{(i) } V;\quad \text{(ii) } V^2;\quad \text{(iii) } 2V^3+V'^2;\quad \text{(iv) } 5V^4+10VV'^2+V''^2;\\ \text{(v) } 14V^5+70V^2V'^2+14VV''^2+V'''^2;\ \text{etc.}\end{array}\right\} \tag{220}$$

The integrals we encounter here are formally the same as the conserved quantities of the Korteweg–de Vries equation,

$$u_{,t}-6uu_{,x}+u_{,x,x,x}=0. \tag{221}$$

This coincidence is not an accident; but it will take us too far afield to state the reasons. The interested reader may wish to consult the relevant literature quoted in the Bibliographical Notes at the end of the chapter.

(c) *A direct verification of the hierarchy of integral equalities for the potentials* $V^{(\pm)}=\pm\beta f'+\beta^2f^2+\kappa f$

The various integral equalities required of the potentials $V^{(+)}$ and $V^{(-)}$, as given in equations (136) and (137), can be verified, individually, by evaluating the expressions (220) for

$$V=\beta f'+\beta^2f^2+\kappa f, \tag{222}$$

and showing that the terms odd in β are expressible as derivatives of combinations of f and its derivatives and, therefore, vanish on integration (by virtue of our assumption that f and all its derivatives vanish for $x\to\pm\infty$).

The equality of the integrals of $V^{(+)}$ and $V^{(-)}$ follows, for example, from the fact that the integral of $\beta f'$ vanishes.

Considering next V^2, for the terms odd in β, we have

$$2\beta(\beta^2f^2+\kappa f)f'=\tfrac{2}{3}\beta^3(f^3)'+\kappa\beta(f^2)'. \tag{223}$$

The integral of this expression clearly vanishes; and the remaining terms, even in β, are the same for $V^{(+)}$ and $V^{(-)}$. In the same way, the terms odd in β, in the expression for $2V^3+V'^2$, can be reduced to the form

$$2\beta^3(ff'^2)'+\kappa\beta(f'^2)'+6\beta(\beta^2f^2+\kappa f)^2f'; \tag{224}$$

and the integral of this expression also vanishes. The equality of the integrals of the two remaining quantities listed in (220) can be similarly verified, although the complexity of the reduction rapidly increases.

A direct evaluation of the integrals of the expressions (220) for the potentials $V^{(+)}$ and $V^{(-)}$ appropriate for the Schwarzschild black-hole yields the following results:

(i) $\dfrac{1}{4M}(2p-3)$;

(ii) $\dfrac{1}{480M^3}(5p^2-18p+18)$;

(iii) $\dfrac{1}{26880M^5}(16p^3-83p^2+150p-87)$;

(iv) $\dfrac{1}{128M^7}\left(\dfrac{1}{168}p^4-\dfrac{53}{1386}p^3+\dfrac{263}{2772}p^2-\dfrac{147}{1430}p+\dfrac{444}{10010}\right)$;

(v) $$\frac{1}{512M^9}\left(\frac{14}{6435}p^5-\frac{41}{2574}p^4+\frac{56}{1155}p^3-\frac{2557}{34320}p^2+\frac{1203}{19448}p-\frac{723}{38896}\right), \quad (225)$$

where $p = 2(n+1)$.

29. Perturbations treated via the Newman–Penrose formalism

The treatment of the metric perturbations in the previous sections leaves one with a sense of bafflement: relations have emerged whose origins are not clear. Thus, why should the polar and the axial perturbations define the same reflexion and transmission coefficient? Why should the solutions describing them be related as simply as they are? And why, indeed, should the potentials $V^{(+)}$ and $V^{(-)}$ be derivable from the same formula with only a change of sign in one of the terms? While the deeper physical origins of these relations are still obscure, they emerged, first, from the alternative treatment of the perturbations via the Newman–Penrose formalism. And to this study we now turn.

We shall assume that the perturbations in the various quantities have a time (t) and an azimuthal angle (φ) dependence given by

$$e^{i(\sigma t + m\varphi)}, \quad (226)$$

where σ is a constant and m is an integer positive, negative, or zero. The directional derivatives D, $\triangle$, δ, and δ^* along the basis null-vectors set up in Chapter 3 (equation (281)), when acting on functions with a t- and a φ-dependence given by (226), become

$$\left.\begin{aligned} &l = D = \mathscr{D}_0, \quad n = \triangle = -\frac{\Delta}{2r^2}\mathscr{D}_0^\dagger, \\ &m = \delta = \frac{1}{r\sqrt{2}}\mathscr{L}_0^\dagger, \quad \text{and} \quad \bar{m} = \delta^* = \frac{1}{r\sqrt{2}}\mathscr{L}_0, \end{aligned}\right\} \quad (227)$$

where

$$\left.\begin{aligned}\mathscr{D}_n &= \partial_r + \frac{ir^2\sigma}{\Delta} + 2n\frac{r-M}{\Delta}, \quad \mathscr{L}_n = \partial_\theta + n\cot\theta + m\operatorname{cosec}\theta,\\ \mathscr{D}_n^\dagger &= \partial_r - \frac{ir^2\sigma}{\Delta} + 2n\frac{r-M}{\Delta}, \quad \text{and} \quad \mathscr{L}_n^\dagger = \partial_\theta + n\cot\theta - m\operatorname{cosec}\theta.\end{aligned}\right\} \tag{228}$$

It will be noticed that while $\mathscr{D}_n$ and $\mathscr{D}_n^\dagger$ are purely radial operators, $\mathscr{L}_n$ and $\mathscr{L}_n^\dagger$ are purely angular operators.

The differential operators (228), which we have defined, satisfy a number of identities which we shall find useful; they are

$$\left.\begin{aligned}\mathscr{D}_n^\dagger &= (\mathscr{D}_n)^*; \quad \mathscr{L}_n^\dagger(\theta) = -\mathscr{L}_n(\pi-\theta),\\ \Delta\mathscr{D}_{n+1} &= \mathscr{D}_n\Delta, \quad \sin\theta\,\mathscr{L}_{n+1} = \mathscr{L}_n\sin\theta.\end{aligned}\right\} \tag{229}$$

Consistent with the type-D character of the Schwarzschild space-time (and as we have verified directly in Chapter 3, §21), the Weyl scalars Ψ_0, Ψ_1, Ψ_3, and Ψ_4 and the spin coefficients, κ, σ, λ, and ν vanish in the background. The only non-vanishing Weyl scalar is Ψ_2; and it has the value (Ch. 3, equation (290))

$$\Psi_2 = -Mr^{-3}; \tag{230}$$

and the non-vanishing spin-coefficients have the values (Ch. 3, equation (288))

$$\rho = -\frac{1}{r}, \quad -\alpha = \beta = \frac{\cot\theta}{r2\sqrt{2}}, \quad \mu = -\frac{\Delta}{2r^3}, \quad \gamma = \mu + \frac{r-M}{2r^2} = \frac{M}{2r^2}. \tag{231}$$

(a) The equations that are already linearized and their reduction

Among the various equations of the Newman–Penrose formalism listed in Chapter 1 (§8(*c*), (*d*), and (*e*)) there are six equations—the four Bianchi identities (Ch. 1, equation (321) (a), (d), (e), and (h)) and the Ricci identities (Ch. 1 equation (310) (b) and (j))—which are linear and homogeneous in the quantities which vanish identically in the background. The equations are

$$\left.\begin{aligned}(\delta^* - 4\alpha + \pi)\Psi_0 - (D - 2\varepsilon - 4\rho)\Psi_1 &= 3\kappa\Psi_2,\\ (\triangle - 4\gamma + \mu)\Psi_0 - (\delta - 4\tau - 2\beta)\Psi_1 &= 3\sigma\Psi_2,\\ (D - \rho - \rho^* - 3\varepsilon + \varepsilon^*)\sigma - (\delta - \tau + \pi^* - \alpha^* - 3\beta)\kappa &= \Psi_0;\end{aligned}\right\} \tag{232}$$

and

$$\left.\begin{aligned}(D + 4\varepsilon - \rho)\Psi_4 - (\delta^* + 4\pi + 2\alpha)\Psi_3 &= -3\lambda\Psi_2,\\ (\delta + 4\beta - \tau)\Psi_4 - (\triangle + 2\gamma + 4\mu)\Psi_3 &= -3\nu\Psi_2,\\ (\triangle + \mu + \mu^* + 3\gamma - \gamma^*)\lambda - (\delta^* + 3\alpha + \beta^* + \pi - \tau^*)\nu &= -\Psi_4.\end{aligned}\right\} \tag{233}$$

The foregoing equations are already linearized in the sense that since Ψ_0, Ψ_1, Ψ_3, Ψ_4, κ, σ, λ, and ν, as perturbations, are to be considered as quantities of the first order of smallness (with a t- and a φ-dependence given by (226)). We may accordingly, replace all the other quantities (including the basis vectors and, therefore, also the directional derivatives) which occur in them by their unperturbed values given in equations (227), (230), and (231). We thus obtain

$$\left.\begin{aligned}\frac{1}{r\sqrt{2}}(\mathscr{L}_0+2\cot\theta)\Psi_0-\left(\mathscr{D}_0+\frac{4}{r}\right)\Psi_1 &= -\frac{3M}{r^3}\kappa,\\ -\frac{\Delta}{2r^2}\left(\mathscr{D}_0^\dagger+\frac{4(r-M)}{\Delta}-\frac{3}{r}\right)\Psi_0-\frac{1}{r\sqrt{2}}(\mathscr{L}_0^\dagger-\cot\theta)\Psi_1 &= -\frac{3M}{r^3}\sigma,\\ \left(\mathscr{D}_0+\frac{2}{r}\right)\sigma-\frac{1}{r\sqrt{2}}(\mathscr{L}_0^\dagger-\cot\theta)\kappa &= \Psi_0;\end{aligned}\right\}\quad(234)$$

and

$$\left.\begin{aligned}\left(\mathscr{D}_0+\frac{1}{r}\right)\Psi_4-\frac{1}{r\sqrt{2}}(\mathscr{L}_0-\cot\theta)\Psi_3 &= +\frac{3M}{r^3}\lambda,\\ \frac{1}{r\sqrt{2}}(\mathscr{L}_0^\dagger+2\cot\theta)\Psi_4+\frac{\Delta}{2r^2}\left(\mathscr{D}_0^\dagger-\frac{2(r-M)}{\Delta}+\frac{6}{r}\right)\Psi_3 &= +\frac{3M}{r^3}\nu,\\ -\frac{\Delta}{2r^2}\left(\mathscr{D}_0^\dagger-\frac{2(r-M)}{\Delta}+\frac{4}{r}\right)\lambda-\frac{1}{r\sqrt{2}}(\mathscr{L}_0-\cot\theta)\nu &= -\Psi_4.\end{aligned}\right\}\quad(235)$$

These equations take simple and symmetrical forms if we write them in terms of the variables

$$\left.\begin{aligned}&\Phi_0=\Psi_0,\quad \Phi_1=\Psi_1 r\sqrt{2},\quad k=\frac{\kappa}{r^2\sqrt{2}},\quad s=\frac{\sigma}{r};\\ &\Phi_4=\Psi_4 r^4,\quad \Phi_3=\Psi_3\frac{r^3}{\sqrt{2}},\quad l=\tfrac{1}{2}\lambda r,\quad n=\frac{1}{\sqrt{2}}\nu r^2.\end{aligned}\right\}\quad(236)$$

In this manner, we obtain the following basic set of equations:

$$\mathscr{L}_2\Phi_0-\left(\mathscr{D}_0+\frac{3}{r}\right)\Phi_1 = -6Mk,\qquad(237)$$

$$\Delta\left(\mathscr{D}_2^\dagger-\frac{3}{r}\right)\Phi_0+\mathscr{L}_{-1}^\dagger\Phi_1 = +6Ms,\qquad(238)$$

$$\left(\mathscr{D}_0+\frac{3}{r}\right)s-\mathscr{L}_{-1}^\dagger k=\frac{\Phi_0}{r};\qquad(239)$$

and

$$\left(\mathscr{D}_0-\frac{3}{r}\right)\Phi_4-\mathscr{L}_{-1}\Phi_3 = 6Ml,\qquad(240)$$

$$\mathscr{L}_2^\dagger\Phi_4+\Delta\left(\mathscr{D}_{-1}^\dagger+\frac{3}{r}\right)\Phi_3 = 6Mn,\qquad(241)$$

$$\Delta\left(\mathscr{D}_{-1}^\dagger+\frac{3}{r}\right)l+\mathscr{L}_{-1}n=\frac{\Phi_4}{r}.\qquad(242)$$

We can eliminate Φ_1 between equations (237) and (238) by applying the operator $\mathscr{L}^{\dagger}_{-1}$ to equation (237) and the operator $(\mathscr{D}_0 + 3/r)$ to equation (238) and adding. The right-hand side of the resulting equations, apart from a factor $6M$, is precisely the quantity which occurs on the left-hand side of equation (239). We thus obtain the decoupled equation

$$\left[\mathscr{L}^{\dagger}_{-1}\mathscr{L}_2 + \left(\mathscr{D}_0 + \frac{3}{r}\right)\Delta\left(\mathscr{D}^{\dagger}_{\frac{1}{2}} - \frac{3}{r}\right)\right]\Phi_0 = \frac{6M}{r}\Phi_0. \tag{243}$$

Similarly, we obtain from equations (240)–(242) (by the elimination of Φ_3) the decoupled equation

$$\left[\mathscr{L}_{-1}\mathscr{L}^{\dagger}_{\frac{1}{2}} + \Delta\left(\mathscr{D}^{\dagger}_{-1} + \frac{3}{r}\right)\left(\mathscr{D}_0 - \frac{3}{r}\right)\right]\Phi_4 = \frac{6M}{r}\Phi_4. \tag{243'}$$

The identity,

$$\begin{aligned}\Delta\left(\mathscr{D}_1 + \frac{3}{r}\right)\left(\mathscr{D}^{\dagger}_{\frac{1}{2}} - \frac{3}{r}\right) - \frac{6M}{r} &= \Delta\mathscr{D}_1\mathscr{D}^{\dagger}_{\frac{1}{2}} + \frac{6}{r}[-ir^2\sigma + (r-M)] - \frac{6\Delta}{r^2} - \frac{6M}{r} \\ &= \Delta\mathscr{D}_1\mathscr{D}^{\dagger}_{\frac{1}{2}} - 6i\sigma r,\end{aligned} \tag{244}$$

reduces equation (243) to the form

$$[\mathscr{L}^{\dagger}_{-1}\mathscr{L}_2 + (\Delta\mathscr{D}_1\mathscr{D}^{\dagger}_{\frac{1}{2}} - 6i\sigma r)]\Phi_0 = 0. \tag{245}$$

Similarly, equation (243′) reduces to the form

$$[\mathscr{L}_{-1}\mathscr{L}^{\dagger}_{\frac{1}{2}} + (\Delta\mathscr{D}^{\dagger}_{-1}\mathscr{D}_0 + 6i\sigma r)]\Phi_4 = 0. \tag{246}$$

Equations (245) and (246) allow a separation of the variables. Thus, by the substitutions,

$$\Phi_0 = R_{+2}(r)S_{+2}(\theta) \qquad \text{and} \qquad \Phi_4 = R_{-2}(r)S_{-2}(\theta), \tag{247}$$

where $R_{\pm 2}$ and $S_{\pm 2}$ are functions, respectively, of r and θ only, we obtain the two pairs of equations

$$\mathscr{L}^{\dagger}_{-1}\mathscr{L}_2 S_{+2} = -\mu^2 S_{+2}, \tag{248}$$

$$(\Delta\mathscr{D}_1\mathscr{D}^{\dagger}_{\frac{1}{2}} - 6i\sigma r)R_{+2} = +\mu^2 R_{+2}; \tag{249}$$

and

$$\mathscr{L}_{-1}\mathscr{L}^{\dagger}_{\frac{1}{2}} S_{-2} = -\mu^2 S_{-2}, \tag{250}$$

$$(\Delta\mathscr{D}^{\dagger}_{-1}\mathscr{D}_0 + 6i\sigma r)R_{-2} = +\mu^2 R_{-2}, \tag{251}$$

where μ^2 is a separation constant.

It will be noticed that we have not distinguished the separation constants that derive from equations (245) and (246). The reason is the following: considering equation (248), we first observe that μ^2 is a characteristic-value parameter that is to be determined by the requirement that $S_{+2}(\theta)$ is regular at

$\theta = 0$ and $\theta = \pi$. On the other hand, since the operator acting on S_{-2} on the left-hand side of equation (250) is the same as the operator acting on S_{+2} in equation (248), if we replace θ by $\pi - \theta$, it follows that a proper solution, $S_{+2}(\theta; \mu^2)$, of equation (248), belonging to a characteristic value μ^2, provides a proper solution of equation (250), belonging to the same value μ^2, if we replace θ by $\pi - \theta$ in $S_{+2}(\theta; \mu^2)$. In other words, equations (248) and (250) determine the same set of characteristic values for μ^2.

The characteristic values of μ^2 can be ascertained, without loss of generality, by considering the case $m = 0$. (The reasons are the same that allow us to restrict ourselves to axisymmetric perturbations of an initially spherically symmetric system.) On expanding the equation governing S_{+2}, for $m = 0$, we find

$$\frac{d^2 S_{+2}}{d\theta^2} + \cot\theta \frac{dS_{+2}}{d\theta} - 2(\cot^2\theta + \operatorname{cosec}^2\theta) S_{+2} = -\mu^2 S_{+2}. \tag{252}$$

By the substitution,

$$S_{+2}(\theta) = C(\theta) \operatorname{cosec}^2\theta, \tag{253}$$

the equation becomes

$$\sin^3\theta \frac{d}{d\theta} \frac{1}{\sin^3\theta} \frac{dC}{d\theta} + \mu^2 C = 0. \tag{254}$$

Comparison with equations (20) and (21) shows that

$$C(\theta) = C_{l+2}^{-3/2}(\theta) \qquad \text{and} \qquad \mu^2 = 2n = (l-1)(l+2). \tag{255}$$

Therefore,

$$\begin{aligned} S_{+2}(\theta) = C_{l+2}^{-3/2}(\theta) \operatorname{cosec}^2\theta &= \sin\theta \frac{d}{d\theta} \frac{1}{\sin\theta} \frac{dP_l}{d\theta} \\ &= P_{l,\theta,\theta} - P_{l,\theta} \cot\theta \quad (m = 0), \end{aligned} \tag{256}$$

where P_l, as usual, denotes the Legendre function. When $m \neq 0$, $S_{+2}(\theta)$ becomes a 'spin-weighted' spherical harmonic; but the value of μ^2 is unaffected.

Returning to the radial equation (249), we observe that $\Delta^2 R_{+2}$ satisfies the equation

$$(\Delta \mathscr{D}_{-1} \mathscr{D}_0^\dagger - 6i\sigma r)\Delta^2 R_{+2} = \mu^2 (\Delta^2 R_{+2}); \tag{257}$$

and this is the complex conjugate of equation (251) satisfied by R_{-2}.

We shall transform equation (257) to a standard recurrent form in the theory. First, we observe that

$$\mathscr{D}_0 = \frac{r^2}{\Delta} \Lambda_+ \qquad \text{and} \qquad \mathscr{D}_0^\dagger = \frac{r^2}{\Delta} \Lambda_- \tag{258}$$

where (cf. equation (29))

$$\Lambda_{\pm} = \frac{d}{dr_*} \pm i\sigma \qquad \text{and} \qquad \frac{d}{dr_*} = \frac{\Delta}{r^2}\frac{d}{dr}. \tag{259}$$

Accordingly,

$$\Delta \mathscr{D}_{-1} \mathscr{D}_0^\dagger = \Delta^2 \mathscr{D}_0 \frac{1}{\Delta} \mathscr{D}_0^\dagger = r^2 \Delta \Lambda_+ \left(\frac{r^2}{\Delta^2} \Lambda_- \right). \tag{260}$$

Also, replacing $\Delta^2 R_{+2}$ by

$$Y_{+2} = r^{-3} \Delta^2 R_{+2}, \tag{261}$$

we can rewrite equation (257) in the form

$$\Lambda_+ \left[\frac{r^2}{\Delta^2} \Lambda_- (r^3 Y_{+2}) \right] - 6i\sigma \frac{r^2}{\Delta} Y_{+2} = \mu^2 \frac{r}{\Delta} Y_{+2}. \tag{262}$$

On further simplification, equation (262) can be brought to the form

$$\Lambda^2 Y_{+2} + P \Lambda_- Y_{+2} - Q Y_{+2} = 0, \tag{263}$$

where

$$P = \frac{d}{dr_*} \lg \frac{r^8}{\Delta^2} = \frac{4}{r^2}(r - 3M) \tag{264}$$

and

$$Q = \frac{\Delta}{r^5}(\mu^2 r + 6M). \tag{265}$$

Equation (263) is the form in which we shall mostly use the equation for Ψ_0; and it is a form which we shall encounter frequently.

A similar reduction of equation (251), governing R_{-2}, will lead to the complex conjugate of equation (263). Thus, with the substitution

$$Y_{-2} = r^{-3} R_{-2}, \tag{266}$$

we find

$$\Lambda^2 Y_{-2} + P \Lambda_+ Y_{-2} - Q Y_{-2} = 0. \tag{267}$$

With Y_{+2} and Y^*_{-2} satisfying the same equation, we readily deduce that

$$\frac{r^8}{\Delta^2}(Y^*_{-2} \Lambda_- Y_{+2} - Y_{+2} \Lambda_- Y^*_{-2}) = \text{constant}. \tag{268}$$

Since

$$Y^*_{-2} \Lambda_- Y_{+2} - Y_{+2} \Lambda_- Y^*_{-2} = Y_{+2,r_*} Y^*_{-2} - Y_{+2} Y^*_{-2,r_*}, \tag{269}$$

it follows that the Wronskian, $[Y_{+2}, Y^*_{-2}]$, of the solutions Y_{+2} and Y^*_{-2}, is given by

$$[Y_{+2}, Y^*_{-2}] = \text{constant } r^{-8} \Delta^2; \tag{270}$$

or, in view of the substitutions (261) and (266),

$$[\Delta^2 R_{+2}, R^*_{-2}] = \text{constant } r^{-2}\Delta^2. \tag{271}$$

This last relation must clearly imply a 'conservation law', even as the constancy of the Wronskian, $[Z^{(\pm)}, Z^{(\pm)*}]$, led to the conservation law incorporated in equation (163). Indeed, we shall find in §32 that the implications of equation (271) are the same as those of equation (163).

(*b*) *The completion of the solution of equations (237)–(242) and the phantom gauge*

While we shall not pursue in this chapter the complete solution of all the relevant equations of the Newman–Penrose formalism—we shall do so in Chapter 9 in the more general context of the Kerr black-hole—we shall, nevertheless, complete the solutions of equations (237)–(242) since they will be needed in some of the considerations of Chapter 5.

Two preliminary observations: *first*, we are provided with only six equations for the eight unknowns which they involve—a fact which implies that their solution must involve two arbitrary functions; and *second*, equations (237)–(239) governing Φ_0, Φ_1, k and s and equations (240)–(242) governing Φ_4, Φ_3, l, and n are decoupled. This latter decoupling of the two sets of equations has far-reaching implications—implications whose consideration we shall again postpone to Chapter 9.

Returning to equations (237)–(242), we have shown how these equations lead to independent equations for Φ_0 and Φ_4. Clearly, the solutions for the remaining quantities must involve two arbitrary functions. The origin of this arbitrariness is clear. From our discussion of the effect of tetrad rotations on the various Newman–Penrose quantities in Chapter 1 (§8(g)), it follows that Ψ_0 and Ψ_4 are affected only in the *second order* by first-order infinitesimal rotations of the tetrad basis in a background in which Ψ_0, Ψ_1, Ψ_3, and Ψ_4 vanish. But this is not the case with Ψ_1 and Ψ_3: they *are* affected in the *first order* (see particularly Ch. 1, equations (342) and (346)). In other words, *in a linear perturbation-theory, Ψ_0 and Ψ_4 are gauge-invariant quantities while Ψ_1 and Ψ_3 are not.* Consequently, we may, for instance, choose a gauge (i.e., subject the tetrad basis to an infinitesimal rotation) in which Ψ_1 and Ψ_3 vanish (without affecting Ψ_0 and Ψ_4).

If a choice of gauge, in which Ψ_1 and Ψ_3 are zero, is made, then the corresponding solutions for k, s, l, and n can be directly read off from equations (237), (238), (240), and (241);

$$\left.\begin{aligned} -6Mk &= R_{+2}\mathscr{L}_2 S_{+2}, \quad 6Ms = S_{+2}\Delta(\mathscr{D}_2^\dagger - 3/r)R_{+2}; \\ \text{and} \quad +6Mn &= R_{-2}\mathscr{L}_2^\dagger S_{-2}, \quad 6Ml = S_{-2}(\mathscr{D}_0 - 3/r)R_{-2}. \end{aligned}\right\} \tag{272}$$

One important piece of information is, however, missing in the foregoing solutions: we do not, as yet, know the relative normalization of the radial functions $\Delta^2 R_{+2}$ and R_{-2}. We shall obtain this information in §32.

Other choices of gauge, besides the one in which Ψ_1 and Ψ_3 are zero, can be made; and one in particular which brings to equations (237)–(242) a symmetry which they lack. Thus, considering equations (237)–(239), the symmetry of these equations in Φ_0, k, and s is only partially present in Φ_1, k, and s. Equation (239) is, for example, the 'right' equation which allows us to obtain a decoupled equation for Φ_0 after the elimination of Φ_1 between equations (237) and (238). But a similar elimination of Φ_0 does not lead to a decoupled equation for Φ_1 since we do not have a 'right' fourth equation. However, exercising the freedom we have to subject the tetrad frame to an infinitesimal rotation, we can rectify the situation by supplying (ad hoc?) the needed fourth equation. Thus, with the additional equation,

$$\Delta\left(\mathscr{D}_2^\dagger - \frac{3}{r}\right)k + \mathscr{L}_2 s = \frac{2}{r}\Phi_1, \tag{273}$$

we can eliminate Φ_0 between equations (237) and (238) to obtain the decoupled equation

$$\left[\Delta\left(\mathscr{D}_2^\dagger - \frac{3}{r}\right)\left(\mathscr{D}_0 + \frac{3}{r}\right) + \mathscr{L}_2\mathscr{L}_{-1}^\dagger\right]\Phi_1 = 12\frac{M}{r}\Phi_1. \tag{274}$$

On expanding this equation, we find

$$[(\Delta\mathscr{D}_2^\dagger\mathscr{D}_0 - 6i\sigma r) + \mathscr{L}_2\mathscr{L}_{-1}^\dagger]\Phi_1 = 0. \tag{275}$$

Similarly, supplementing equations (240)–(242) by the additional equation

$$\left(\mathscr{D}_0 - \frac{3}{r}\right)n - \mathscr{L}_2^\dagger l = \frac{2}{r}\Phi_3, \tag{276}$$

and eliminating Φ_4 between equations (240) and (241), we obtain the decoupled equation

$$[(\Delta\mathscr{D}_1\mathscr{D}_{-1}^\dagger + 6i\sigma r) + \mathscr{L}_2^\dagger\mathscr{L}_{-1}]\Phi_3 = 0. \tag{277}$$

Equations (275) and (277) are clearly separable: by the substitutions

$$\Phi_1 = R_{+1}(r)S_{+1}(\theta) \qquad \text{and} \qquad \Phi_3 = R_{-1}(r)S_{-1}(\theta) \tag{278}$$

—we justify the designation of these functions with the subscripts $+1$ and -1 presently—we obtain the pair of equations

$$(\Delta\mathscr{D}_2^\dagger\mathscr{D}_0 - 6i\sigma r)R_{+1} = +\mu^2 R_{+1}, \quad \mathscr{L}_2\mathscr{L}_{-1}^\dagger S_{+1} = -\mu^2 S_{+1}; \tag{279}$$

and

$$(\Delta\mathscr{D}_1\mathscr{D}_{-1}^\dagger + 6i\sigma r)R_{-1} = +\mu^2 R_{-1}, \quad \mathscr{L}_2^\dagger\mathscr{L}_{-1}S_{-1} = -\mu^2 S_{-1}. \tag{280}$$

It will be noticed that in equations (279) and (280), we have used the same separation constant μ^2 as in equations (248) and (250). The reason is the functions S_{+1} and S_{-1} are simply related to the functions S_{+2} and S_{-2}. The following relations are manifestly true if the functions are suitably normalized:

$$\mathscr{L}_2 S_{+2} = +\mu S_{+1}, \quad \mathscr{L}^\dagger_{-1} S_{+1} = -\mu S_{+2}; \tag{281}$$

and

$$\mathscr{L}^\dagger_2 S_{-2} = -\mu S_{-1}, \quad \mathscr{L}_{-1} S_{-1} = +\mu S_{-2}; \tag{282}$$

and the regularity of $S_{\pm 2}$ at $\theta = 0$ and π guarantees the regularity of $S_{\pm 1}$ as well.

We shall find (cf. Chapter 5, §46) that the functions Φ_1 and Φ_3, as defined by equations (278)–(280), describe Maxwell's field in Schwarzschild's geometry*. Thus, we have, in effect, derived Maxwell's equations (appropriate for photons with spin ± 1) by finding a gauge which rectifies the truncated symmetry of equations (237)–(242) in the quantities which occur in them. The designation of the relevant radial and angular functions by the subscripts ± 1 is, therefore, justified.

Because of the apparent veiled 'awareness' of Schwarzschild's geometry to the existence of Maxwell's field, we shall call the gauge, in which equations (273) and (276) are true, the *phantom gauge*.

Finally, we may note that the solutions for the spin coefficients in the phantom gauge are

$$\left.\begin{aligned} -6Mk &= S_{+1}[+\mu R_{+2} - (\mathscr{D}_0 + 3/r)R_{+1}], \\ +6Ms &= S_{+2}[-\mu R_{+1} + \Delta(D^\dagger_2 - 3/r)R_{+2}], \\ +6Ml &= S_{-2}[-\mu R_{-1} + (\mathscr{D}_0 - 3/r)R_{-2}], \\ +6Mn &= S_{-1}[-\mu R_{-2} + \Delta(\mathscr{D}^\dagger_{-1} + 3/r)R_{-1}]. \end{aligned}\right\} \tag{283}$$

We notice that by virtue of the relations (281) and (282), the foregoing solutions for the spin coefficients continue to be separable in r and θ. We shall find that this separability plays an important role in the treatment of the perturbation of the Reissner–Nordström black-hole in Chapter 5.

30. The transformation theory

In the preceding section, we have shown how the Newman–Penrose formalism leads to a pair of complex-conjugate equations for the radial functions, $\Delta^2 R_{+2}$ and R_{-2}. As we shall see in detail in §32, there is, associated with equations (263) and (267) and the Wronskian (271), a problem of reflexion and transmission of incident gravitational waves which has the same physical content as that associated with the equations governing $Z^{(\pm)}$ and the Wronskian (161) leading to the conservation law (163). Indeed, the require-

* We treat Maxwell's equation, in the more general context of the Kerr geometry, in Chapter 8.

ment, that the theories developed along the two avenues—the Newman–Penrose formalism and the linearized Einstein equation—be consistent with one another, is ultimately the reason why the axial and the polar perturbations determine the same reflexion and transmission coefficients. A corollary to this inference is that it should be possible to express the functions $Y_{\pm 2}$ and $Z^{(\pm)}$, explicitly, in terms of one another. This possibility—rather, the certainty—that a solution, Y_{+2}, of equation (263) can be transformed, simultaneously, to provide solutions of either of the equations governing $Z^{(+)}$ or $Z^{(-)}$, implies that there is a special feature of equation (263) which will make this possible. We shall now set out to find what this special feature is.

The problem is to express the solution of an equation of the form

$$\Lambda^2 Y + P\Lambda_- Y - QY = 0 \tag{284}$$

in terms of the solution of a one-dimensional wave-equation

$$\Lambda^2 Z = VZ, \tag{285}$$

where

$$P = \frac{d}{dr_*}\left(\lg \frac{r^8}{\Delta^2}\right), \tag{286}$$

and Q and V are certain functions which we shall, for the present, leave unspecified.

Since Y and Z both satisfy equations of the second order, there is no restriction to assuming that Y is a linear combination of Z and its derivative. But instead of making this assumption simply as we did in §26 when considering a similar problem, we shall now assume that

$$Y = f\Lambda_+\Lambda_+ Z + W\Lambda_+ Z, \tag{287}$$

or, equivalently,

$$Y = fVZ + (W + 2i\sigma f)\Lambda_+ Z, \tag{288}$$

where f and W are certain functions of r_* to be determined. It may appear odd that the expression of Y, as a linear combination of Z and its derivative, is made in this oblique fashion. That it is the most appropriate one to make in the context of equations (284) and (285) emerged in a different context which will be clarified in §31 below.

Applying the operator Λ_- to both sides of equation (287) and making use of the fact that Z has been assumed to satisfy equation (285), we find

$$\Lambda_- Y = \left[\frac{d}{dr_*}(fV) + WV\right]Z + \left[fV + \frac{d}{dr_*}(W + 2i\sigma f)\right]\Lambda_+ Z; \tag{289}$$

or, with the definitions,

$$-\beta\frac{\Delta^2}{r^8} = \frac{d}{dr_*}(fV) + WV \tag{290}$$

and

$$R = fV + \frac{d}{dr_*}(W + 2i\sigma f), \tag{291}$$

we can write

$$\Lambda_- Y = -\beta \frac{\Delta^2}{r^8} Z + R\Lambda_+ Z. \tag{292}$$

Next, applying the operator Λ_- to equation (292), we obtain (again making use of the equation satisfied by Z)

$$\Lambda_-\Lambda_- Y = -\beta \frac{\Delta^2}{r^8}(\Lambda_+ - 2i\sigma)Z - \frac{d\beta}{dr_*}\frac{\Delta^2}{r^8} Z - \beta Z \frac{d}{dr_*}\left(\frac{\Delta^2}{r^8}\right) + RVZ + \frac{dR}{dr_*}\Lambda_+ Z. \tag{293}$$

On the other hand, by equation (284),

$$\Lambda_-\Lambda_- Y = -(P + 2i\sigma)\Lambda_- Y + QY, \tag{294}$$

or, substituting for Y and $\Lambda_- Y$ from equations (287) and (292), we have (cf. equation (286))

$$\Lambda_-\Lambda_- Y = -\left(2i\sigma + \frac{d}{dr_*}\lg\frac{r^8}{\Delta^2}\right)\left(-\beta\frac{\Delta^2}{r^8} Z + R\Lambda_+ Z\right) + Q[fVZ + (W + 2i\sigma f)\Lambda_+ Z]. \tag{295}$$

Since the right-hand sides of equations (293) and (295) must identically be the same, we can equate the coefficients of Z and $\Lambda_+ Z$. In this manner, we obtain

$$RV - \frac{\Delta^2}{r^8}\frac{d\beta}{dr_*} = QfV, \tag{296}$$

and

$$\frac{dR}{dr_*} - \beta\frac{\Delta^2}{r^8} = Q(W + 2i\sigma f) - \left(2i\sigma + \frac{d}{dr_*}\lg\frac{r^8}{\Delta^2}\right)R. \tag{297}$$

We shall rewrite equations (296) and (297) in the forms

$$-\frac{\Delta^2}{r^8}\frac{d\beta}{dr_*} = (Qf - R)V, \tag{298}$$

and

$$\frac{d}{dr_*}\left(\frac{r^8}{\Delta^2}R\right) = \frac{r^8}{\Delta^2}[Q(W + 2i\sigma f) - 2i\sigma R] + \beta. \tag{299}$$

It can now be verified that equations (290), (291), (298), and (299) allow the integral

$$\frac{r^8}{\Delta^2}RfV + \beta(W + 2i\sigma f) = K = \text{constant}. \tag{300}$$

This integral is the present analogue of the integral (147) which was found in §26 in a similar context.

The integral (300) enables us to write the inverse of the relations (288) and (292) in the forms

$$\left.\begin{aligned} KZ &= \frac{r^8}{\Delta^2}RY - \frac{r^8}{\Delta^2}(W + 2i\sigma\,f)\Lambda_- Y, \\ K\Lambda_+ Z &= \beta Y + \frac{r^8}{\Delta^2}\,f\,V\Lambda_- Y. \end{aligned}\right\} \tag{301}$$

(*a*) *The conditions for the existence of transformations with* $f = 1$ *and* $\beta =$ *constant; dual transformations*

We may formally consider any *four* of the five equations (290), (291), (298)–(300) as equations governing the *five* functions β, f, R, V, and W. We now ask for the conditions when the transformation equations will be compatible with

$$\beta = \text{constant} \qquad \text{and} \qquad f = 1. \tag{302}$$

We shall find that transformations compatible with the requirements (302) are possible only if Q satisfies a certain non-linear second-order differential equation even in β. By virtue of this last fact it will emerge that the requirements (302) are precisely those which will permit transformations of the equations governing Y, simultaneously, to either of the equations governing $Z^{(+)}$ and $Z^{(-)}$ with $V^{(+)}$ and $V^{(-)}$ of the forms stated in §26 (equation (133)).

When β is a constant and $f = 1$, equation (298) requires that

$$R = Q. \tag{303}$$

Equation (291) then gives

$$V = Q - \frac{\mathrm{d}W}{\mathrm{d}r_*}. \tag{304}$$

The remaining equations—which we may take to be (299) and (300)—are

$$\frac{\mathrm{d}}{\mathrm{d}r_*}\left(\frac{r^8}{\Delta^2}R\right) = \left(\frac{r^8}{\Delta^2}Q\right)W + \beta \tag{305}$$

and

$$\left(\frac{r^8}{\Delta^2}Q\right)V + \beta W = K - 2i\sigma\beta = \text{constant} = \kappa \quad \text{(say).} \tag{306}$$

Letting

$$F = \frac{r^8}{\Delta^2}Q, \tag{307}$$

we can rewrite equations (305) and (306) in the forms

$$W = \frac{1}{F}\left(\frac{\mathrm{d}F}{\mathrm{d}r_*} - \beta\right) \tag{308}$$

and

$$FV + \beta W = F\left(Q - \frac{\mathrm{d}W}{\mathrm{d}r_*}\right) + \beta W = \kappa. \tag{309}$$

Eliminating W between these last two equations, we obtain

$$FQ - F\frac{\mathrm{d}}{\mathrm{d}r_*}\left(\frac{1}{F}\frac{\mathrm{d}F}{\mathrm{d}r_*} - \frac{\beta}{F}\right) + \frac{\beta}{F}\left(\frac{\mathrm{d}F}{\mathrm{d}r_*} - \beta\right) = \kappa, \tag{310}$$

or, after some simplifications, we have

$$\frac{1}{F}\left(\frac{\mathrm{d}F}{\mathrm{d}r_*}\right)^2 - \frac{\mathrm{d}^2 F}{\mathrm{d}r_*^2} + \frac{\Delta^2}{r^8}F^2 = \frac{\beta^2}{F} + \kappa. \tag{311}$$

Since F must be assumed as given, *equation* (311) *provides a necessary and sufficient condition for the transformation equations to be compatible with the requirements* (302): *there must exist constants* β *and* κ *such that equation* (311) *is satisfied by the given* $Q = \Delta^2 F/r^8$. Since β occurs as β^2 in equation (311), two transformations, associated with $+\beta$ and $-\beta$, are possible when the equation is satisfied. We shall call them *dual transformations*; and, as we shall presently verify, they lead directly to the equations governing the axial and the polar perturbations.

Distinguishing the transformations associated with $+\beta$ and $-\beta$ by superscripts $(\pm)$, we can write

$$W^{(\pm)} = \frac{1}{F}\left(\frac{\mathrm{d}F}{\mathrm{d}r_*} \mp \beta\right) \tag{312}$$

and

$$V^{(\pm)} = Q - \frac{\mathrm{d}W^{(\pm)}}{\mathrm{d}r_*}. \tag{313}$$

Rewriting this last equation in the form

$$V^{(\pm)} = Q - \frac{\mathrm{d}}{\mathrm{d}r_*}\left(\frac{1}{F}\frac{\mathrm{d}F}{\mathrm{d}r_*} \mp \frac{\beta}{F}\right), \tag{314}$$

and making use of equation (310), we obtain

$$V^{(\pm)} = \frac{\kappa}{F} \mp \frac{\beta}{F^2}\left(\frac{\mathrm{d}F}{\mathrm{d}r_*} \mp \beta\right). \tag{315}$$

Letting

$$f = F^{-1}, \tag{316}$$

we obtain the formula

$$V^{(\pm)} = \pm\beta\frac{\mathrm{d}f}{\mathrm{d}r_*} + \beta^2 f^2 + \kappa f. \tag{317}$$

The explicit forms of the associated transformations are

$$\left.\begin{aligned} Y &= V^{(\pm)}Z^{(\pm)} + (W^{(\pm)} + 2i\sigma)\Lambda_+ Z^{(\pm)}, \\ \Lambda_- Y &= \mp\beta\frac{\Delta^2}{r^8}Z^{(\pm)} + Q\Lambda_+ Z^{(\pm)}, \end{aligned}\right\} \tag{318}$$

and

$$\begin{aligned} K^{(\pm)}Z^{(\pm)} &= \frac{r^8}{\Delta^2}[QY - (W^{(\pm)} + 2i\sigma)\Lambda_- Y], \\ K^{(\pm)}\Lambda_+ Z^{(\pm)} &= \pm\beta Y + \frac{r^8}{\Delta^2}V^{(\pm)}\Lambda_- Y, \end{aligned} \tag{319}$$

where

$$K^{(\pm)} = \kappa \pm 2i\sigma\beta. \tag{320}$$

And finally, we may note that by making use of the relations (318) and (319), we can relate the solutions for $Z^{(+)}$ and $Z^{(-)}$. Thus, letting $Z^{(-)}$ denote a solution of the wave equation with the potential $V^{(-)}$, we may use the first of the equations (319) to express it in terms of Y and $\Lambda_- Y$ and then express *these* in terms of $Z^{(+)}$ and $\Lambda_+ Z^{(+)}$ with the aid of equations (318). By this procedure, we obtain

$$\begin{aligned} \frac{\Delta^2}{r^8}K^{(-)}Z^{(-)} &= QY - (W^{(-)} + 2i\sigma)\Lambda_- Y \\ &= Q[V^{(+)}Z^{(+)} + (W^{(+)} + 2i\sigma)\Lambda_+ Z^{(+)}] \\ &\qquad - (W^{(-)} + 2i\sigma)\left[-\beta\frac{\Delta^2}{r^8}Z^{(+)} + Q\Lambda_+ Z^{(+)}\right]; \end{aligned} \tag{321}$$

or, after some regrouping of the terms, we have

$$\begin{aligned} K^{(-)}Z^{(-)} = \left[\frac{r^8}{\Delta^2}QV^{(+)} + \beta(W^{(+)} + 2i\sigma) + \beta(W^{(-)} - W^{(+)})\right]Z^{(+)} \\ + F(W^{(+)} - W^{(-)})\Lambda_+ Z^{(+)}. \end{aligned} \tag{322}$$

The expression on the right-hand side of equation (322) can be simplified by making use of equations (306), (312), (316), and (320); and we find

$$(\kappa - 2i\sigma\beta)Z^{(-)} = (\kappa + 2\beta^2 f)Z^{(+)} - 2\beta\frac{\mathrm{d}Z^{(+)}}{\mathrm{d}r_*}. \tag{323}$$

The inverse of this relation is

$$(\kappa + 2i\sigma\beta)Z^{(+)} = (\kappa + 2\beta^2 f)Z^{(-)} + 2\beta\frac{\mathrm{d}Z^{(-)}}{\mathrm{d}r_*}. \tag{324}$$

We have thus recovered the relations (150) and (151) which we derived earlier by the transformation directly relating the solutions of equations (136) and (137).

(b) The verification of the equation governing F and the values of κ and β^2

For the case on hand, Q is defined in equation (265). Therefore,

$$F = \frac{r^8}{\Delta^2} Q = \frac{r^3}{\Delta}(\mu^2 r + 6M) = \frac{1}{f}. \tag{325}$$

We shall now verify that this expression for F satisfies equation (311) for certain determinate values of β^2 and κ. Thus, considering the left-hand side of equation (311), we find

$$-F\left[\frac{\Delta}{r^2}\frac{\mathrm{d}}{\mathrm{d}r}\left(\frac{\Delta}{r^2}\frac{\mathrm{d}}{\mathrm{d}r}\lg F\right) - \frac{\Delta^2}{r^8}F\right] = \mu^2(\mu^2+2) + 36M^2\frac{\Delta}{r^3(\mu^2 r + 6M)}$$

$$= \mu^2(\mu^2+2) + \frac{36M^2}{F}; \tag{326}$$

and this must equal $\kappa + \beta^2/F$. Therefore, F as defined in equation (325) does satisfy equation (311) with

$$\beta^2 = 36M^2 \qquad \text{and} \qquad \kappa = \mu^2(\mu^2+2). \tag{327}$$

Therefore, equation (263) governing Y allows the dual transformations considered in §(*a*) above: and the expression (317) for $V^{(\pm)}$ with $\beta = 6M$ and f as defined in equation (325) are what was quoted in §26, equations (133)–(135). Besides, with κ, β, and F as presently defined, the relations (323) and (324) are the same as those derived earlier on the assumption that $V^{(+)}$ and $V^{(-)}$ have the forms that we have now derived for them.

And finally, we may note that for F given by equation (325),

$$W^{(-)} = \frac{2}{r^2}(r-3M) \quad \text{and} \quad W^{(+)} = 2\frac{\mu^2 r^2 - 3\mu^2 M r - 6M^2}{r^2(\mu^2 r + 6M)}. \tag{328}$$

31. A direct evaluation of Ψ_0 in terms of the metric perturbations

In the preceding section, we have related the Weyl scalar,

$$\Psi_0 = \frac{r^3}{\Delta^2} Y_{+2}(r) S_{+2}(\theta), \tag{329}$$

to the metric perturbations by showing how the function Y_{+2} can be expressed in terms of $Z^{(+)}$ and $Z^{(-)}$ by the dual transformations incorporated in equations (318) and (319). We shall now complete the full circle by evaluating Ψ_0 directly in terms of the metric perturbations.

By definition

$$\Psi_0 = R_{(p)(q)(r)(s)} l^{(p)} m^{(q)} l^{(r)} m^{(s)}. \tag{330}$$

We have already shown in Chapter 3 (§21) that Ψ_0 (as well as Ψ_1, Ψ_3, and Ψ_4), for the Schwarzschild space-time vanishes as required by its type-D character. Therefore, in evaluating the perturbed value of Ψ_0, it will suffice to contract the *perturbed* Riemann-tensor with the *unperturbed* basis-vectors since the contributions arising from the contractions of the *unperturbed* Riemann-tensor with the *perturbed* basis-vectors $\boldsymbol{l}$ and $\boldsymbol{m}$ (expressed, in turn, as linear combinations of the unperturbed basis-vectors) vanish by virtue of the vanishing of unperturbed Weyl-scalars Ψ_0, Ψ_1, Ψ_3 and Ψ_4. Thus, we may write

$$\Psi_0 = -\delta R_{(p)(q)(r)(s)} l^{(p)} m^{(q)} l^{(r)} m^{(s)}, \tag{331}$$

where we have enclosed the indices in parentheses to emphasize that they signify tetrad components in the frame in which we have evaluated the Riemann tensor in Chapter 2 (equations (75)).

The tetrad components of the basis vectors can be obtained by applying to the tensor components (given in Chapter 3, equation (281)) the transformation represented by

$$e^{(a)}{}_i = \begin{vmatrix} e^{\nu} & 0 & 0 & 0 \\ 0 & r\sin\theta & 0 & 0 \\ 0 & 0 & e^{-\nu} & 0 \\ 0 & 0 & 0 & r \end{vmatrix} \tag{332}$$

where $e^{2\nu} = \Delta/r^2 = (1-2M/r)$. We find

$$\left.\begin{aligned} l^{(p)} &= (l^{(t)}, l^{(\varphi)}, l^{(r)}, l^{(\theta)}) = (e^{-\nu}, 0, e^{-\nu}, 0), \\ n^{(p)} &= (n^{(t)}, n^{(\varphi)}, n^{(r)}, n^{(\theta)}) = \tfrac{1}{2}(e^{+\nu}, 0, -e^{+\nu}, 0), \\ m^{(p)} &= (m^{(t)}, m^{(\varphi)}, m^{(r)}, m^{(\theta)}) = \frac{1}{\sqrt{2}}(0, i, 0, 1). \end{aligned}\right\} \tag{333}$$

Contraction of the Riemann tensor with these vectors gives

$$\begin{aligned} \Psi_0 = &-ie^{-2\nu}(\delta R_{0301} + \delta R_{2321} + \delta R_{2301} + \delta R_{0321}) \\ &-\tfrac{1}{2}e^{-2\nu}(\delta R_{0303} + 2\delta R_{0323} + \delta R_{2323} \\ &\qquad -\delta R_{0101} - 2\delta R_{0121} - \delta R_{2121}). \end{aligned} \tag{334}$$

It will be observed that Ψ_0 is decomposed into an axial part—the first group of terms, odd in the index 1, which reverses in sign with a reversal in the sign of φ, and a polar part—the second group of terms, even in the index 1, which is invariant to a reversal in the sign of φ.

(*a*) *The axial part of* Ψ_0

The relevant perturbed components of the Riemann tensor that are included in the axial part of Ψ_0 can be read off directly from the list given in Chapter 2. We find

$$\begin{aligned}
\delta R_{0301} &= \tfrac{1}{2}e^{\psi-2\nu-\mu_3}Q_{03,0}-\tfrac{1}{2}e^{\psi-2\mu_2-\mu_3}Q_{23}\,\nu_{,r},\\
\delta R_{2321} &= e^{\psi-2\mu_2-\mu_3}\left[\left(\frac{1}{r}+\frac{1}{2}\nu_{,r}\right)Q_{23}+\frac{1}{2}Q_{23,r}\right],\\
\delta R_{2301} &= \tfrac{1}{2}e^{\psi-\nu-\mu_2-\mu_3}\left[Q_{03,2}-Q_{02,3}+Q_{03}\left(\frac{1}{r}-\nu_{,r}\right)-Q_{02}\cot\theta\right],\\
\delta R_{0321} &= -\tfrac{1}{2}e^{\psi-\nu-\mu_2-\mu_3}\left[Q_{20,3}+Q_{20}\cot\theta-Q_{23,0}+Q_{30}\left(\frac{1}{r}-\nu_{,r}\right)\right], \qquad (335)
\end{aligned}$$

where it may be recalled that we are now restricting ourselves to axisymmetric perturbations (which, as we have explained in §23, implies no loss of generality). Inserting the expressions (335) in the terms which occur with the factor $-i$ in equation (334), we find

$$\begin{aligned}
-\operatorname{Im}\Psi_0 = \Bigg\{&\frac{Q_{23}}{r}+\frac{1}{2}Q_{23,r}+\frac{1}{2}e^{-2\nu}Q_{23,0}\\
&+\frac{1}{2}e^{-2\nu}\left[Q_{03,r}+2\left(\frac{1}{r}-\nu_{,r}\right)Q_{03}\right]+\frac{1}{2}e^{-4\nu}Q_{03,0}\Bigg\}\sin\theta. \qquad (336)
\end{aligned}$$

On the other hand, according to equations (13), (14), (15), and (19),

$$Q_{23}=\frac{Q(r)C_{l+2}^{-3/2}(\theta)}{\Delta\sin^3\theta}e^{i\sigma t} \qquad (337)$$

and

$$Q_{03,0}=\frac{\Delta}{r^4\sin^3\theta}Q_{,r}C_{l+2}^{-3/2}(\theta)e^{i\sigma t}. \qquad (338)$$

Inserting these expressions in equation (336), we find

$$\begin{aligned}
-\operatorname{Im}\Psi_0 = \Bigg\{&\frac{1}{2}\frac{\mathrm{d}}{\mathrm{d}r}\left(\frac{Q}{\Delta}\right)+\frac{Q}{r\Delta}+\frac{i\sigma r^2}{2\Delta^2}Q+\frac{1}{2\Delta}Q_{,r}\\
&+\frac{r^2}{2i\sigma\Delta}\left[\frac{\mathrm{d}}{\mathrm{d}r}\left(\frac{\Delta}{r^4}\frac{\mathrm{d}Q}{\mathrm{d}r}\right)+2\frac{\Delta}{r^4}\left(\frac{1}{r}-\nu_{,r}\right)Q_{,r}\right]\Bigg\}\frac{C_{l+2}^{-3/2}(\theta)}{\sin^2\theta}, \qquad (339)
\end{aligned}$$

where we have suppressed the common factor $e^{i\sigma t}$. Simplifying this last equation with the aid of equation (23), we obtain

$$-\operatorname{Im}\Psi_0=\left\{i\sigma\left[\frac{1}{2}\frac{\mathrm{d}}{\mathrm{d}r}\left(\frac{Q}{\Delta}\right)+\frac{Q}{r\Delta}+\frac{1}{2\Delta}\frac{\mathrm{d}Q}{\mathrm{d}r}\right]-\sigma^2\frac{r^2}{\Delta^2}Q\right.$$
$$\left.+\frac{n}{r^2\Delta}Q+\frac{r-3M}{r^2\Delta}\frac{\mathrm{d}Q}{\mathrm{d}r}\right\}\frac{C_{l+2}^{-3/2}(\theta)}{i\sigma\sin^2\theta}$$
$$=\left\{i\sigma\left[\frac{1}{\Delta}\frac{\mathrm{d}Q}{\mathrm{d}r}+\frac{Q}{r\Delta}-\frac{r-M}{\Delta^2}Q\right]-\sigma^2\frac{r^2}{\Delta^2}Q\right.$$
$$\left.+\frac{n}{r^2\Delta}Q+\frac{r-3M}{r^2\Delta}\frac{\mathrm{d}Q}{\mathrm{d}r}\right\}\frac{C_{l+2}^{-3/2}(\theta)}{i\sigma\sin^2\theta}. \tag{340}$$

Now letting

$$Q=rZ^{(-)}, \tag{341}$$

as in equation (25), we find, after some further simplifications, that equation (340) becomes

$$-\operatorname{Im}\Psi_0=\left\{\frac{r}{\Delta}\left(i\sigma+\frac{r-3M}{r^2}\right)\frac{\mathrm{d}Z^{(-)}}{\mathrm{d}r}\right.$$
$$\left.+\left[\frac{(\mu^2+2)r-6M}{2r^2\Delta}+i\sigma\frac{r(r-3M)}{\Delta^2}-\sigma^2\frac{r^3}{\Delta^2}\right]Z^{(-)}\right\}\frac{C_{l+2}^{-3/2}(\theta)}{i\sigma\sin^2\theta}. \tag{342}$$

Recalling the definitions (27) and (328), namely,

$$W^{(-)}=\frac{2}{r^2}(r-3M)\qquad\text{and}\qquad V^{(-)}=\frac{\Delta}{r^5}[(\mu^2+2)r-6M], \tag{343}$$

we can now rewrite equation (342) in the form

$$-2i\sigma\operatorname{Im}\Psi_0=\frac{r^3}{\Delta^2}\left[(2i\sigma+W^{(-)})\frac{\mathrm{d}Z^{(-)}}{\mathrm{d}r_*}\right.$$
$$\left.+(V^{(-)}+i\sigma W^{(-)}-2\sigma^2)Z^{(-)}\right]\frac{C_{l+2}^{-3/2}(\theta)}{\sin^2\theta}, \tag{344}$$

or, alternatively,

$$-2i\sigma\operatorname{Im}\Psi_0=\frac{r^3}{\Delta^2}[V^{(-)}Z^{(-)}+(W^{(-)}+2i\sigma)\Lambda_+Z^{(-)}]\frac{C_{l+2}^{-3/2}(\theta)}{\sin^2\theta}. \tag{345}$$

In view of the substitutions (256) and (261), the foregoing equation is in complete accord with equation (318) which we arrived at via the Newman–Penrose formalism and the transformation theory.

(b) The polar part of Ψ_0

The relevant perturbed components of the Riemann tensor, which are included in the polar part of Ψ_0 in equation (334), can be obtained by linearizing the corresponding components listed in Chapter 2 about the unperturbed Schwarzschild-metric. We find

$$-\delta R_{0303} = -2\delta\mu_2\frac{e^{2\nu}}{r}\nu_{,r} + e^{2\nu}\left(\frac{\delta\nu_{,r}}{r} + \nu_{,r}\delta\mu_{3,r}\right) + \frac{1}{r^2}\delta\nu_{,\theta,\theta} + e^{-2\nu}\sigma^2\delta\mu_3,$$

$$-\delta R_{0101} = -2\delta\mu_2\frac{e^{2\nu}}{r}\nu_{,r} + e^{2\nu}\frac{\delta\nu_{,r}}{r} + e^{2\nu}\nu_{,r}\delta\psi_{,r} + \frac{\cot\theta}{r^2}\delta\nu_{,\theta} + \sigma^2 e^{-2\nu}\delta\psi,$$

$$-\delta R_{2323} = -\frac{1}{r^2}\delta\mu_{2,\theta,\theta} + 2\delta\mu_2 e^{2\nu}\frac{\nu_{,r}}{r} - e^{2\nu}\left[\delta\mu_{3,r,r} + \left(\frac{2}{r} + \nu_{,r}\right)\delta\mu_{3,r} - \frac{\delta\mu_{2,r}}{r}\right],$$

$$-\delta R_{1212} = 2\delta\mu_2\frac{e^{2\nu}}{r}\nu_{,r} - e^{2\nu}\left[\delta\psi_{,r,r} + \left(\frac{2}{r} + \nu_{,r}\right)\delta\psi_{,r} - \frac{\delta\mu_{2,r}}{r}\right] - \frac{\cot\theta}{r^2}\delta\mu_{2,\theta},$$

$$-\delta R_{0323} = -i\sigma\left[\delta\mu_{3,r} + \frac{1}{r}(\delta\mu_3 - \delta\mu_2) - \nu_{,r}\delta\mu_3\right],$$

$$-\delta R_{0121} = +i\sigma\left[-\delta\psi_{,r} + \frac{1}{r}(\delta\mu_2 - \delta\psi) + \nu_{,r}\delta\psi\right]. \tag{346}$$

Inserting these expressions in the terms which occur with the factor $-\frac{1}{2}e^{-2\nu}$ in equation (334), we find

$$\begin{aligned}\operatorname{Re}\Psi_0 = \tfrac{1}{2}e^{-2\nu}\Big\{&\left[e^{2\nu}\left(\frac{\partial^2}{\partial r^2} + \frac{2}{r}\frac{\partial}{\partial r}\right) + 2i\sigma\left(\frac{\partial}{\partial r} + \frac{1}{r} - \nu_{,r}\right)\right](\delta\psi - \delta\mu_3)\\ &-\sigma^2 e^{-2\nu}(\delta\psi - \delta\mu_3) + \frac{1}{r^2}\left(\frac{\partial^2}{\partial\theta^2} - \cot\theta\frac{\partial}{\partial\theta}\right)(\delta\nu - \delta\mu_2)\Big\}.\end{aligned} \tag{347}$$

In accordance with equations (36)–(39), we may substitute in these equations

$$\left.\begin{aligned}\delta\psi - \delta\mu_3 &= -V(P_{l,\theta,\theta} - P_{l,\theta}\cot\theta) = -V\frac{C_{l+2}^{-3/2}(\theta)}{\sin^2\theta}\\ \text{and}\qquad \delta\nu - \delta\mu_2 &= (N - L)P_l.\end{aligned}\right\} \tag{348}$$

We thus obtain

$$\begin{aligned}-\operatorname{Re}\Psi_0 = \frac{1}{2}\Big\{&\left[\frac{d^2}{dr^2} + \frac{2}{r}\frac{d}{dr} + 2i\sigma e^{-2\nu}\left(\frac{d}{dr} + \frac{1}{r} - \nu_{,r}\right) - e^{-4\nu}\sigma^2\right]V\\ &-\frac{e^{-2\nu}}{r^2}(N - L)\Big\}\frac{C_{l+2}^{-3/2}(\theta)}{\sin^2\theta};\end{aligned} \tag{349}$$

or, making use of equation (51) and rearranging the terms, we have

$$-\operatorname{Re}\Psi_0=\left\{-\left(\nu_{,r}V_{,r}+\sigma^2e^{-4\nu}V+\frac{e^{-2\nu}}{r^2}N\right)\right.$$
$$\left.+i\sigma e^{-2\nu}\left[V_{,r}+\left(\frac{1}{r}-\nu_{,r}\right)V\right]\right\}\frac{C_{l+2}^{-3/2}(\theta)}{\sin^2\theta}. \tag{350}$$

Simplifying the first group of terms (without the factor $i\sigma$) with the aid of equation (57), we find

$$-\operatorname{Re}\Psi_0=\left\{\frac{i\sigma r}{r-2M}\left[V_{,r}+\frac{r-3M}{r(r-2M)}V\right]\right.$$
$$+\frac{\sigma^2r^2}{(r-2M)^2}\left[\frac{M}{n(r-2M)}(L+X)-V\right]$$
$$-\frac{nr^2-3Mnr-3M^2}{nr^2(r-2M)^2}N-\frac{M[nr^2-M(2n-1)r-3M^2]}{nr^2(r-2M)^3}L$$
$$\left.+\frac{M(r^2-3Mr+3M^2)}{r^2(r-2M)^3}V\right\}\frac{C_{l+2}^{-3/2}(\theta)}{\sin^2\theta}. \tag{351}$$

After some further reductions in which we make use of equations (57) and (60), the foregoing equation can be brought to the form

$$-\operatorname{Re}\Psi_0=\frac{1}{2\Delta}\left\{\frac{2n^2(n+1)r^3+6Mn^2r^2+18M^2nr+18M^3}{r^2(nr+3M)^2}Z\right.$$
$$\left.+\left[2i\sigma\frac{r^2}{r-2M}+2\frac{nr^2-3Mnr-3M^2}{(r-2M)(nr+3M)}\right](Z_{,r_*}+i\sigma Z)\right\}\frac{C_{l+2}^{-3/2}(\theta)}{\sin^2\theta}, \tag{352}$$

or, recalling the definitions (63) and (328) of $V^{(+)}$ and $W^{(+)}$, we have

$$-\operatorname{Re}\Psi_0=\frac{r^3}{\Delta^2}[V^{(+)}Z^{(+)}+(W^{(+)}+2i\sigma)\Lambda_+Z^{(+)}]\frac{C_{l+2}^{-3/2}(\theta)}{\sin^2\theta}, \tag{353}$$

again, in complete accord with what we derived earlier via the Newman–Penrose formalism and the transformation theory.

We have now come the full circle.

32. The physical content of the theory

We stated at the outset (in §22) that, on the physical side, our principal concern in this study of the perturbations is to elucidate the manner in which incident gravitational-waves are reflected and absorbed by the black hole.

While it would appear on general grounds that the reflexion ($\mathbb{R}$) and the transmission ($\mathbb{T}$) coefficients, derived from the equations governing $Z^{(\pm)}$, are precisely the quantities which are sought, the matter requires clarification, since at no point in the analysis or in the arguments was there an occasion to mention the incidence of gravitational waves. And the matter is not a very simple one either: it will require us, as we shall find, to relate the theories, developed on the basis of the linearized Einstein-equations and the Newman–Penrose formalism, at a deeper level than we have done hitherto.

First, we observe that the 'constant' in the Wronskian relation (270) is simply related to the Wronskian $[Z, Z^*]$, whose constancy led to the conservation law, $\mathbb{R} + \mathbb{T} = 1$ in §27. Thus, considering the Wronskian,

$$K^2[Z_1, Z_2] = K^2(Z_2\Lambda_+ Z_1 - Z_1\Lambda_+ Z_2), \tag{354}$$

of any two independent solutions, Z_1 and Z_2, of equation (285), substituting for KZ and $K\Lambda_+ Z$ from equations (301), and making use of the integral (300), we find

$$K^2[Z_1, Z_2] = \frac{r^8}{\Delta^2}\left[\frac{r^8}{\Delta^2}RfV + \beta(W + 2i\sigma f)\right](Y_2\Lambda_- Y_1 - Y_1\Lambda_- Y_2)$$

$$= \frac{r^8}{\Delta^2}K[Y_1, Y_2]. \tag{355}$$

Accordingly,

$$\frac{r^8}{\Delta^2}[Y_1, Y_2] = K[Z_1, Z_2] = \text{constant}. \tag{356}$$

This last relation clearly implies that we should be able to determine the reflexion and the transmission coefficients from the equations governing Y_{+2} and Y_{-2} (or, equivalently, from the equations governing $\Delta^2 R_{+2}$ and R_{-2}). However, with our present knowledge, we cannot obtain any useful information from the Wronskian $[Y_{+2}, Y^*_{-2}]$ since Y_{+2} and Y_{-2} satisfy independent equations—a fact to which attention was drawn, already, in §29(*b*). In Chapter 9, we shall find, in the more general context of the Kerr black-hole, that the equations of the Newman–Penrose formalism, which we have not so far considered, do in fact determine the relative normalizations of the functions Y_{+2} and Y_{-2} so that the Wronskian of Y_{+2} and Y^*_{-2} can, indeed, be used to determine $\mathbb{R}$ and $\mathbb{T}$. But in the present context, we shall circumvent the need for that analysis, by obtaining the requisite relations directly from the equations provided by the transformation theory of §30.

We know that the solutions for $Z^{(+)}$ and $Z^{(-)}$ have the asymptotic behaviours $e^{\pm i\sigma r_*}$ both for $r_* \to +\infty$ and for $r_* \to -\infty$. We may use this knowledge to obtain the associated asymptotic behaviours of Y_{+2} and Y_{-2} with the aid of equations (318) and (319) and their complex conjugates.

Remembering that $V^{(\pm)}$, $W^{(\pm)}$, and Q all vanish for $r_* \to +\infty$, we find, from the first of each of the pairs of equations (318) and (319), that for

$$Z^{(\pm)} \to e^{+i\sigma r_*} \quad \text{and} \quad Z^{(\pm)} \to e^{-i\sigma r_*} \quad (r_* \to +\infty), \tag{357}$$

the asymptotic behaviours of Y_{+2} and Y_{-2} are

$$\left.\begin{aligned} &Y_{+2} \to -4\sigma^2 e^{+i\sigma r_*}, && Y_{-2} \to -\frac{K^{(\pm)*}}{4\sigma^2}\frac{e^{+i\sigma r_*}}{r^4}, \\ &Y_{+2} \to -\frac{K^{(\pm)}}{4\sigma^2}\frac{e^{-i\sigma r_*}}{r^4}, && Y_{-2} \to -4\sigma^2 e^{-i\sigma r_*}. \end{aligned}\right\} \tag{358}$$

Similarly, remembering that $V^{(\pm)}$ and Q vanish on the horizon and that (cf. equation (328))

$$W^{(\pm)} \to -(2M)^{-1} \quad \text{for} \quad r = 2M, \tag{359}$$

we find that, for

$$Z^{(\pm)} \to e^{+i\sigma r_*} \quad \text{and} \quad Z^{(\pm)} \to e^{-i\sigma r_*} \quad (r_* \to -\infty), \tag{360}$$

the behaviours of Y_{+2} and Y_{-2} are

$$\begin{aligned} &Y_{+2} \to 4i\sigma(i\sigma - 1/4M)e^{+i\sigma r_*}; \; Y_{-2} \to \frac{K^{(\pm)*}\Delta^2 e^{+i\sigma r_*}}{(2M)^8 4(i\sigma + 1/2M)(i\sigma + 1/4M)} \\ &Y_{+2} \to \frac{K^{(\pm)}\Delta^2 e^{-i\sigma r_*}}{(2M)^8 4(i\sigma - 1/2M)(i\sigma - 1/4M)}; \; Y_{-2} \to 4i\sigma(i\sigma + 1/4M)e^{-i\sigma r_*}. \end{aligned} \tag{361}$$

Making use of the asymptotic behaviours of Y_{+2} and Y_{-2}, we can write down the boundary conditions for the Weyl scalars,

$$\Phi_0 = e^{i\sigma t + im\varphi} R_{+2} S_{+2} = e^{i\sigma t + im\varphi}\frac{r^3}{\Delta^2} Y_{+2} S_{+2} \quad (= \Psi_0) \tag{362}$$

and

$$\Phi_4 = e^{i\sigma t + im\varphi} R_{-2} S_{-2} = e^{i\sigma t + im\varphi} r^3 Y_{-2} S_{-2} \quad (= r^4 \Psi_4), \tag{363}$$

corresponding to the boundary conditions (cf. equation (159))

$$\left.\begin{aligned} Z^{(\pm)} &\to e^{+i\sigma r_*} + R^{(\pm)}(\sigma) e^{-i\sigma r_*} && (r_* \to +\infty) \\ &\to T^{(\pm)}(\sigma) e^{+i\sigma r_*} && (r_* \to -\infty), \end{aligned}\right\} \tag{364}$$

appropriate for determining the reflexion and the transmission amplitudes $R^{(\pm)}(\sigma)$ and $T^{(\pm)}(\sigma)$; thus,

$$\left.\begin{aligned} \Phi_0 &\to e^{i\sigma t + im\varphi} S_{+2}(\theta)\left(Y_{+2}^{(\text{in})}\frac{e^{+i\sigma r_*}}{r} + Y_{+2}^{(\text{ref})}\frac{e^{-i\sigma r_*}}{r^5}\right) && (r_* \to +\infty) \\ &\to e^{i\sigma t + im\varphi} S_{+2}(\theta) Y_{+2}^{(\text{tr})}\frac{e^{+i\sigma r_*}}{\Delta^2} && (r_* \to -\infty) \end{aligned}\right\} \tag{365}$$

and

$$\left.\begin{aligned}\Phi_4 &\to e^{i\sigma t + im\varphi} S_{-2}(\theta)\left(Y^{(\text{in})}_{-2}\frac{e^{+i\sigma r_*}}{r} + Y^{(\text{ref})}_{-2} r^3 e^{-i\sigma r_*}\right) && (r_* \to +\infty)\\ &\to e^{i\sigma t + im\varphi} S_{-2}(\theta) Y^{(\text{tr})}_{-2}\Delta^2 e^{+i\sigma r_*} && (r_* \to -\infty),\end{aligned}\right\} \tag{366}$$

where

$$\begin{aligned} &Y^{(\text{in})}_{+2} = -4\sigma^2;\ Y^{(\text{ref})}_{+2} = -\frac{K^{(\pm)}}{4\sigma^2} R^{(\pm)}(\sigma);\\ &Y^{(\text{in})}_{-2} = -\frac{K^{(\pm)*}}{4\sigma^2};\ Y^{(\text{ref})}_{-2} = -4\sigma^2 R^{(\pm)}(\sigma);\\ &Y^{(\text{tr})}_{+2} = (2M)^3 4i\sigma(i\sigma - 1/4M)T(\sigma);\\ &Y^{(\text{tr})}_{-2} = \frac{K^{(\pm)*}T(\sigma)}{(2M)^5 4(i\sigma + 1/2M)(i\sigma + 1/4M)}.\end{aligned} \tag{367}$$

Now, it is clear that the coefficients of the leading terms, with an r^{-1}-behaviour in the asymptotic expansions of the Weyl scalars Ψ_0 and Ψ_4 at infinity, will determine the flux of energies in the incoming and the outgoing gravitational waves. The relationships are established in Chapter 9 (98(b)). The result of the analysis is that the flux of energies, per unit time and unit solid-angle (Ω), in the incoming (incident) and the outgoing (reflected) waves are given by

$$\frac{d^2E^{(\text{in})}}{dt\,d\Omega} = \frac{(S_{+2})^2}{2\pi}\frac{1}{32\sigma^2}|Y^{(\text{in})}_{+2}|^2 \tag{368}$$

and

$$\frac{d^2E^{(\text{ref})}}{dt\,d\Omega} = \frac{(S_{-2})^2}{2\pi}\frac{1}{2\sigma^2}|Y^{(\text{ref})}_{-2}|^2, \tag{369}$$

where the angular functions, S_{+2} and S_{-2}, are assumed to be normalized to unity. From the expressions given in equations (367), we find that, in agreement with equations (368) and (369),

$$\mathbb{R} = |R^{(\pm)}(\sigma)|^2 = \left|\frac{Y^{(\text{ref})}_{-2}}{Y^{(\text{in})}_{+2}}\right|^2 = \left|\frac{Y^{(\text{ref})}_{+2}}{Y^{(\text{in})}_{-2}}\right|^2 = \left|\frac{Y^{(\text{ref})}_{-2}Y^{(\text{ref})}_{+2}}{Y^{(\text{in})}_{+2}Y^{(\text{in})}_{-2}}\right|, \tag{370}$$

or

$$\mathbb{R} = \frac{256\sigma^8}{|K^{(\pm)}|^2}\left|\frac{Y^{(\text{ref})}_{+2}}{Y^{(\text{in})}_{+2}}\right|^2 = \frac{|K^{(\pm)}|^2}{256\sigma^8}\left|\frac{Y^{(\text{ref})}_{-2}}{Y^{(\text{in})}_{-2}}\right|^2. \tag{371}$$

The expressions for $\mathbb{R}$ given in equations (371) will enable us to determine the reflexion coefficient directly from the equations governing Y_{+2} or Y_{-2} (or, equivalently, from the equations governing R_{+2} or R_{-2}); and it will, of course, be in agreement with that determined by the equations governing $Z^{(\pm)}$.

The corresponding expressions for the transmission coefficient, $\mathbb{T}$, are

$$\mathbb{T} = \frac{\sigma^2}{(2M)^6(\sigma^2 + 1/16M^2)}\left|\frac{Y^{(\text{tr})}_{+2}}{Y^{(\text{in})}_{+2}}\right|^2 = \frac{(2M)^{10}}{\sigma^4}\left(\sigma^2 + \frac{1}{16M^2}\right)\left(\sigma^2 + \frac{1}{4M^2}\right)\left|\frac{Y^{(\text{tr})}_{-2}}{Y^{(\text{in})}_{-2}}\right|^2$$

$$= (2M)^2\left|\frac{Y^{(\text{tr})}_{+2}Y^{(\text{tr})}_{-2}}{Y^{(\text{in})}_{+2}Y^{(\text{in})}_{-2}}\right|\frac{(\sigma^2 + 1/4M^2)^{1/2}}{\sigma}. \tag{372}$$

While the physical meaning of $\mathbb{T}$, as the fraction of the incident flux of energy that is absorbed by the black hole, is manifest from the conservation law $\mathbb{R} + \mathbb{T} = 1$, a direct proof requires some sophisticated arguments which we postpone to Chapter 9.

(a) The implications of the unitarity of the scattering matrix

The unitarity of the scattering matrix (179) established in §28 requires, among other things, that the solution for Z satisfying the boundary conditions,

$$\left.\begin{aligned} \tilde{Z} &\to e^{-i\sigma r_*} + R(-\sigma)e^{+i\sigma r_*} \quad (r_* \to -\infty) \\ &\to T(-\sigma)e^{-i\sigma r_*} \quad (r_* \to +\infty) \end{aligned}\right\} \tag{373}$$

determines the same reflexion and transmission coefficients, $\mathbb{R}$ and $\mathbb{T}$, as the solution satisfying the boundary conditions (364). Making use of the same relations (358) and (361), we find that the corresponding boundary conditions for the Weyl scalars Φ_0 and Φ_4 are

$$\left.\begin{aligned} \Phi_0 &\to e^{i\sigma t + im\varphi}S_{+2}\left(\tilde{Y}^{(\text{in})}_{+2}e^{-i\sigma r_*} + \tilde{Y}^{(\text{ref})}_{+2}\frac{e^{+i\sigma r_*}}{\Delta^2}\right) \quad (r_* \to -\infty), \\ &\to e^{i\sigma t + im\varphi}S_{+2}\left(\tilde{Y}^{(\text{tr})}_{+2}\frac{e^{-i\sigma r_*}}{r^5}\right) \quad (r_* \to +\infty), \end{aligned}\right\} \tag{374}$$

and

$$\left.\begin{aligned} \Phi_4 &\to e^{i\sigma t + im\varphi}S_{-2}(\tilde{Y}^{(\text{in})}_{-2}e^{-i\sigma r_*} + \tilde{Y}^{(\text{ref})}_{-2}\Delta^2 e^{+i\sigma r_*}) \quad (r_* \to -\infty), \\ &\to e^{i\sigma t + im\varphi}S_{-2}(\tilde{Y}^{(\text{tr})}_{-2}r^3 e^{-i\sigma r_*}) \quad (r_* \to +\infty), \end{aligned}\right\} \tag{375}$$

where

$$\left.\begin{aligned} \tilde{Y}^{(\text{in})}_{+2} &= \frac{K}{(2M)^5 4(i\sigma - 1/2M)(i\sigma - 1/4M)}; \qquad \tilde{Y}^{(\text{tr})}_{+2} = -\frac{K}{4\sigma^2}T(-\sigma), \\ \tilde{Y}^{(\text{in})}_{-2} &= (2M)^3 4i\sigma(i\sigma + 1/4M); \qquad \tilde{Y}^{(\text{tr})}_{-2} = -4\sigma^2 T(-\sigma), \\ \tilde{Y}^{(\text{ref})}_{+2} &= (2M)^3 4i\sigma(i\sigma - 1/4M)R(-\sigma); \\ \tilde{Y}^{(\text{ref})}_{-2} &= \frac{K^* R(-\sigma)}{(2M)^5 4(i\sigma + 1/2M)(i\sigma + 1/4M)}. \end{aligned}\right\} \tag{376}$$

From these relations, we obtain

$$|R(-\sigma)|^2 = \left|\frac{\tilde{Y}_{-2}^{(\mathrm{ref})}}{\tilde{Y}_{+2}^{(\mathrm{in})}}\right|^2 = \left|\frac{\tilde{Y}_{+2}^{(\mathrm{ref})}}{\tilde{Y}_{+2}^{(\mathrm{in})}}\right|^2 \frac{|K|^2}{(2M)^{16}256\sigma^2(\sigma^2+1/16M^2)^2(\sigma^2+1/4M^2)}$$

$$= \left|\frac{\tilde{Y}_{-2}^{(\mathrm{ref})}}{\tilde{Y}_{-2}^{(\mathrm{in})}}\right|^2 \frac{256}{|K|^2}(2M)^{16}\sigma^2(\sigma^2+1/16M^2)^2(\sigma^2+1/4M^2) \tag{377}$$

$$|T(-\sigma)|^2 = \left|\frac{\tilde{Y}_{+2}^{(\mathrm{tr})}}{\tilde{Y}_{+2}^{(\mathrm{in})}}\right|^2 \frac{\sigma^4}{(2M)^{10}(\sigma^2+1/16M^2)(\sigma^2+1/4M^2)}$$

$$= \left|\frac{\tilde{Y}_{-2}^{(\mathrm{tr})}}{\tilde{Y}_{-2}^{(\mathrm{in})}}\right|^2 \frac{(2M)^6(\sigma^2+1/16M^2)}{\sigma^2}. \tag{378}$$

33. Some observations on the perturbation theory

The theory of the perturbations of the Schwarzschild black-hole, described in the preceding sections, has so many intertwining strands that it is difficult to unravel them and reveal a coherent pattern. But certain elements of the pattern do emerge.

The centre from which the pattern seems to radiate is the equation provided by the Newman–Penrose formalism for the radial function Y_{+2} (or Y_{-2}) belonging to the Weyl scalar Ψ_0 (or Ψ_4). As we have seen in §32, we can determine, in terms of the function Y_{+2} (or Y_{-2}), without any ambiguity, the reflexion and the transmission coefficients for incident gravitational waves. And every strand of the theory must lead to these same coefficients.

When we turn to the metric perturbations, their separation into an axial and a polar part is a directly manifest aspect—an aspect which is concealed in the Newman–Penrose formalism.

The axial perturbations are described by a single scalar function, $Z^{(-)}$, which satisfies a one-dimensional Schrodinger wave-equation with a potential, $V^{(-)}$—the Regge–Wheeler equation. The reflexion and the transmission coefficients one can derive by elementary methods from this equation must of course agree with what one derives from the equation governing Y_{+2}. Clearly, a transformation must exist which relates the function Y_{+2} and $Z^{(-)}$.

The polar perturbations are described by three radial functions, L, N, and X, which satisfy a coupled system of three first-order linear equations. Since these equations must also inform on the reflexion and transmission of incident gravitational-waves, they must be reducible to a wave equation of the second order. The reducibility must be guaranteed by the existence of a special integral of the equations: this is the Xanthopoulos integral (126). But, if a reduction is guaranteed, it can be effected in numerous ways. There is, for example, the combination $Z^{(+)}$ (given by equation (59)) which satisfies the wave equation with the potential, $V^{(+)}$—the Zerilli equation. However, there is another linear combination $\mathscr{Z}$ (given by equation (154)) which satisfies the Regge-Wheeler equation. This ambiguity is really a consequence of the necessity that, whatever wave equation one derives for the polar perturbation, it must lead to the same reflexion and transmission coefficients as does the Regge–Wheeler equation; and that, therefore, its solutions must be expressible, *explicitly*, as linear combinations of $Z^{(-)}$ and its derivative.

Can one, on the foregoing grounds, dismiss the wave equation governing $Z^{(+)}$? It would not seem that one can. For, when we seek for the transformation relating the solutions for Y_{+2} and for $Z^{(-)}$, we find that it is one of a dual pair with potentials of the form (cf. equation (133))

$$V^{(\pm)} = \pm\beta\frac{\mathrm{d}f}{\mathrm{d}r_*} + \beta^2 f^2 + \kappa f \qquad (\beta \text{ and } \kappa \text{ constants}), \tag{379}$$

where the minus sign leads to the Regge–Wheeler potential and the plus sign leads to the Zerilli potential. Besides, the potentials are manifestly of the forms which guarantee the equality of the reflexion and the transmission coefficients and provide simple reciprocal relations between the solutions for $Z^{(+)}$ and $Z^{(-)}$.

There is a further strand in the theory. The Newman–Penrose formalism is not, in its inception, concerned with the decomposition of the Weyl scalars Ψ_0 and Ψ_4 into their axial and polar parts. The decomposition becomes possible only if we express them, explicitly, in terms of the perturbed components of the Riemann tensor. When the decomposition is made, the two parts decouple—as they must. The two parts must, then, of necessity, satisfy the same equation—as in fact they do!

There is yet another facet to the perturbation theory which we shall consider in Chapter 9. But from the present vantage point, it would appear that the two central facts of the theory are the admissibility of the equation governing Ψ_0 to dual transformations and the associated decomposition of the perturbations into an axial and a polar part.

34. The stability of the Schwarzschild black-hole

The stability of the Schwarzschild black-hole to external perturbations—the only kind we need consider—has been the subject of extensive discussions. The question one addresses in this context is the following: given any initial perturbation confined to a finite interval of r_* (i.e., of 'compact support'), will it remain bounded, at all times, as it evolves?

We have seen that the perturbations are governed by single one-dimensional wave-equations of the form

$$\frac{\mathrm{d}^2 Z}{\mathrm{d}r_*^2} + \sigma^2 Z = VZ, \tag{380}$$

with smooth real potentials, independent of σ, and of short range (i.e., one with a bounded integral). It would appear then that the standard theorems of the quantum theory are applicable. These theorems guarantee that the wave functions, belonging to any observable, form a complete set and that any square-integrable function that can describe a state of the system can be expanded in terms of them; and, further, that the integral of the absolute square of any state function must remain constant with time.

For the problem on hand, the solutions $Z(r_*, \sigma)$ of equation (380) for real σ, satisfying the boundary conditions (159), provide the basic complete set of wave functions; and any initial perturbation that is smooth and confined to a finite interval of r_* can be expressed as an integral over the functions $Z(r_*, \sigma)$ in the form

$$\psi(r_*; 0) = \frac{1}{(2\pi)^{1/2}} \int_{-\infty}^{+\infty} \hat{\psi}(\sigma; 0)\, Z(r_*; \sigma)\, d\sigma; \tag{381}$$

and the evolution of the perturbation, at later times, is given by

$$\psi(r_*, t) = \frac{1}{(2\pi)^{1/2}} \int_{-\infty}^{+\infty} \hat{\psi}(\sigma; 0) e^{i\sigma t} Z(r_*; \sigma)\, d\sigma. \tag{382}$$

We are guaranteed that

$$\int_{-\infty}^{+\infty} |\psi(r_*; 0)|^2\, dr_* = \int_{-\infty}^{+\infty} |\hat{\psi}(\sigma; 0)|^2\, d\sigma = \int_{-\infty}^{+\infty} |\psi(r_*; t)|^2\, dr_*. \tag{383}$$

Also the boundedness of $\psi(r_*, t)$, for all $t > 0$, follows directly from a comparison of equations (381) and (382).

A weaker result, but one of general interest, follows from the time-dependent version of equation (380), namely,

$$\frac{\partial^2 Z}{\partial t^2} = \frac{\partial^2 Z}{\partial r_*^2} - VZ. \tag{384}$$

(We should have derived this equation, had we not, in the analysis of §24, replaced $\partial/\partial t$ by $i\sigma$, in accordance with our assumption (16) that the time-dependence of the perturbations is represented by the factor $e^{i\sigma t}$.) Multiplying equation (384) by $\partial Z^*/\partial t$, we obtain, after an integration by parts,

$$\int_{-\infty}^{+\infty} \left(\frac{\partial Z^*}{\partial t} \frac{\partial^2 Z}{\partial t^2} + \frac{\partial Z}{\partial r_*} \frac{\partial^2 Z^*}{\partial t\, \partial r_*} + VZ \frac{\partial Z^*}{\partial t} \right) dr_* = 0, \tag{385}$$

provided the various integrals converge. Adding to equation (385) its complex conjugate, we obtain the *energy integral*

$$\int_{-\infty}^{+\infty} \left(\left| \frac{\partial Z}{\partial t} \right|^2 + \left| \frac{\partial Z}{\partial r_*} \right|^2 + V|Z|^2 \right) dr_* = \text{constant}. \tag{386}$$

The existence of this energy integral bounds the integral of $|\partial Z/\partial t|^2$; *it, therefore, excludes an exponential growth of any bounded solution of equation* (383).

35. The quasi-normal modes of the Schwarzschild black-hole

In this last section of this chapter we shall consider what we may describe as the *pure tones* of the black hole. The situations we have in mind are the following.

A black hole can be perturbed in a variety of ways, other than by the incidence of gravitational waves: by an object falling into it, or by the accretion of matter surrounding it. Or, we may consider a black hole being formed by a slightly aspherical collapse of a star settling towards a final state described by the Schwarzschild solution. In all these cases, the evolution of the perturbations—if they can be considered as 'small'—can, in principle, be followed by expressing them as superpositions of the basic solutions, $Z\,(r_*;\sigma)$, considered in §34. However, we may expect on general grounds that any initial perturbation will, during its last stages, decay in a manner characteristic of the black hole and independently of the original cause. In other words, we may expect that during the very last stages, the black hole will emit gravitational waves with frequencies and rates of damping, characteristic of itself, in the manner of a bell sounding its last dying pure notes. These considerations underlie the formulation of the concept of the *quasi-normal modes* of a black hole.

Precisely, quasi-normal modes are defined as solutions of the perturbation equations, belonging to complex characteristic-frequencies and satisfying the boundary conditions appropriate for purely outgoing waves at infinity and purely ingoing waves at the horizon. The problem, then, is to seek solutions of the equations governing $Z^{(\pm)}$ which will satisfy the boundary conditions

$$\left.\begin{aligned} Z^{(\pm)} &\to A^{(\pm)}(\sigma)\, e^{-i\sigma r_*} \quad (r_* \to +\infty), \\ &\to \qquad\quad e^{+i\sigma r_*} \quad (r_* \to -\infty). \end{aligned}\right\} \tag{387}$$

This is clearly a characteristic-value problem for σ; and the solutions belonging to the different characteristic values define the quasi-normal modes.

First, we may observe that exponentially growing unstable quasi-normal modes, with $\operatorname{Im}\sigma < 0$, are excluded by the energy integral (386). For, if such modes should exist, solutions satisfying the boundary conditions (387) will vanish exponentially for $r_* \to \pm\infty$; the integrals, over r_* in equations (386), converge and we shall be led to a contradiction. We can, therefore, have only damped stable quasi-normal modes; but the solutions belonging to them will diverge for $r_* \to \pm\infty$—a fact which we must accept as corresponding to an implicit assumption of an 'infinite' perturbation in the remote past.

Next, we observe that the characteristic frequencies σ are the same for $Z^{(-)}$ and $Z^{(+)}$: for, if σ is a characteristic frequency and $Z^{(-)}(\sigma)$ is the solution belonging to it, then the solution $Z^{(+)}(\sigma)$ derived from $Z^{(-)}(\sigma)$ in accordance with the relation (152), will satisfy the boundary conditions (387) with (cf. equation (169))

$$A^{(+)}(\sigma) = A^{(-)}(\sigma)\frac{\mu^2(\mu^2+2)-12i\sigma M}{\mu^2(\mu^2+2)+12i\sigma M}. \tag{388}$$

It will suffice, then, to consider only the equation governing $Z^{(-)}$.

Letting

$$Z^{(-)} = \exp\left(i\int^{r_*}\phi\,\mathrm{d}r_*\right), \tag{389}$$

we find that the equation we have to solve is

$$i\phi_{,r_*}+\sigma^2-\phi^2-V^{(-)}=0; \tag{390}$$

and the appropriate boundary conditions are

$$\phi\to-\sigma \quad \text{as} \quad r_*\to+\infty \quad \text{and} \quad \phi\to+\sigma \quad \text{as} \quad r_*\to-\infty. \tag{391}$$

Solutions of equation (390), satisfying the boundary conditions (391), exist only when σ assumes one of a discrete set of values. It is not known, in general, whether the set of characteristic values is finite or an enumerable infinity.

A useful identity, which follws from integrating equation (390) and making use of the boundary conditions (391), is (cf. equation (225))

$$-2i\sigma+\int_{-\infty}^{+\infty}(\sigma^2-\phi^2)\mathrm{d}r_* = \int_{-\infty}^{+\infty}V^{(-)}\mathrm{d}r_* = \frac{1}{2M}(\mu^2+\tfrac{1}{2}). \tag{392}$$

In Table IV we list the complex characteristic-frequencies σ for different values of l.

TABLE IV

The complex characteristic-frequencies belonging to the quasi-normal modes of the Schwarzschild black-hole (σ is expressed in the unit $(2M)^{-1}$)

l	$2M\sigma$	l	$2M\sigma$
2	$0.74734+0.17792i$	4	$1.61835+0.18832i$
	$0.69687+0.54938i$		$1.59313+0.56877i$
3	$1.19889+0.18541i$		$1.12019+0.84658i$
	$1.16402+0.56231i$	5	$2.02458+0.18974i$
	$0.85257+0.74546i$	6	$2.42402+0.19053i$

(The entries in the different lines for l = 2, 3, and 4 correspond to the characteristic values belonging to different modes.)

Detailed calculations, pertaining to slightly aspherical collapse of dust clouds and to particles of finite mass falling into black holes along geodesics, do exhibit the phenomenon of the ringing of black holes with the frequencies and rates of damping of these quasi-normal modes.

BIBLIOGRAPHICAL NOTES

The theory of the perturbations of the Schwarzschild space-time was initiated in a classic paper by T. Regge and J. A. Wheeler:

1. T. REGGE and J. A. WHEELER, *Phys. Rev.*, 108, 1063–9 (1957).

Following Regge and Wheeler, several investigations along the same lines were published during the early seventies, e.g.:

2. C. V. VISHVESHWARA, *Phys. Rev. D*, 1, 2870–9 (1970).
3. F. J. ZERILLI, ibid., 2, 2141–60 (1970).
4. ———, *Phys. Rev. Letters*, 24, 737–8 (1970).
5. J. M. BARDEEN and W. H. PRESS, *J. Math. Phys.*, 14, 7–19 (1972).

But certain essential features of the theory remained unrecognized: for example, it was not even known that the axial and the polar perturbations (rather, the 'odd' and the 'even' parity perturbations as they were then called) are characterized by the same reflexion and transmission coefficients. Many of the other relations emerged only gradually and often in the context of the Kerr black-hole. For this reason, in this chapter we bring together the results and methods scattered through the following papers to present a self-contained, coherent, and unified treatment of the theory of the perturbations of the Schwarzschild black-hole:

6. S. CHANDRASEKHAR, *Proc. Roy. Soc.* (*London*) A, 343, 289–98 (1975).
7. —— and S. DETWEILER, ibid., 344, 441–52 (1975).
8. ———, ibid., 345, 145–67 (1975).
9. S. CHANDRASEKHAR, ibid., 358, 421–39 (1978).
10. ———, ibid., 441–65 (1978).
11. ———, ibid., 365, 453–65 (1979).
12. —— and B. C. XANTHOPOULOS, ibid., 367, 1–14 (1979).
13. ———, ibid., 369, 425–33 (1980).

See also:

14. S. CHANDRASEKHAR in *General Relativity—An Einstein Centenary Survey*, edited by S. W. Hawking and W. Israel, Cambridge, England, 1979.

§24. The treatment in this section is essentially the same as in reference 6. The basic relations (36)–(39), which enable a separation of the variables, were discovered by J. L. Friedman:

15. J. L. FRIEDMAN, *Proc. Roy. Soc.* (*London*) A, 335, 163–90 (1973).

§25. In this section we present a somewhat simplified version of a theory developed by:

16. B. C. XANTHOPOULOS, *Proc. Roy. Soc.* (*London*) A, 378, 61–71 (1981).

§§26–27. The relations discussed in these sections emerged only gradually. They explicitly occur for the first time in reference 13; but several related aspects of them were partially known earlier. See references 7 and 11 and:

17. J. HEADING, *J. Phys. A. Math. Gen.*, 10, 885–97 (1977).

§28. The theory of inverse scattering and its ramifications for the theory of the Korteweg–de Vries equation are far too extensive to allow an account of any depth in the brief section we have allocated to it. But the relations we have found between $V^{(+)}$

and $V^{(-)}$ (and similar relations we shall find in subsequent chapters in the contexts of the Reissner–Nordström and the Kerr black-holes) are far too reminiscent of the theory to be simply ignored. The account in this section (inclusive of §(*a*)) is largely based on the following paper by Deift and Trubowitz:

18. P. Deift and E. Trubowitz, *Communications on Pure and Applied Math.*, 32, 121–251 (1979).

I am grateful to Professor Deift for critical comments on this section.

The analysis in §(*b*) follows closely:

19. L. D. Faddeev, *Soviet Physics Dokl.*, 3, 747–51 (1958).

For the explicit form of the relations of orders higher than those listed in equation (220) see:

20. R. M. Miura, C. S. Gardner, and M. D. Kruskal, *J. Math. Phys.*, 9, 1204–9 (1968).

General references to the classical theory of the solitons and of the Korteweg-de Vries equation in particular are:

21. G. B. Whitham, *Linear and Non-Linear Waves*, §§16.14–16.16 and 17.2–17.4, John Wiley & Sons, Inc., New York, 1974.
22. A. C. Scott, F. Y. F. Chu, and D. W. McLaughlin, *Proc. IEEE*, 61, 1443–83 (1973).
23. R. M. Miura, *SIAM Rev.*, 18, 412–59 (1976).

§29. In this section, the treatment in reference 9 in the context of the Kerr space-time is adapted to the Schwarzschild space-time.

§30. The transformation theory was originally developed in reference 8 (and subsequent papers not listed here) in the context of the perturbations of the Kerr black-hole. It is adapted here for the Schwarzschild space-time. The basic ideas are implicit in reference 6.

§31. The polar part of Ψ_0 was evaluated by J. L. Friedman (reference 15); the axial part is evaluated here to complete the analysis.

§34. For alternative treatments of the stability of the Schwarzschild black-hole see:

24. V. Moncrief, *Ann. Phys.*, 88, 323–42 (1973).
25. R. M. Wald, *J. Math. Phys.*, 20, 1056–8 (1979).

The basic theorems used in the discussion in the text are stated in:

26. L. D. Landau and E. M. Lifschitz, *Quantum Mechanics* (*Non-Relativistic Theory*), §21, 60–3, Pergamon Press, Oxford, 1977.

§35. The treatment of the quasi-normal modes in this section follows reference 7. In Table IV, the numerical results given in reference 7 are supplemented by some further results in:

27. D. L. Gunter, *Phil. Trans. Roy. Soc.* (*London*) A, 296, 497–526 (1980).

The appearance of these quasi-normal modes during the last stages of the collapse of stellar masses is examined in:

28. C. T. Cunningham, R. H. Price, and V. Moncrief, *Astrophys. J.*, 224, 643–67 (1978).
29. ———, ibid., 230, 870–92 (1979).

5

THE REISSNER–NORDSTRÖM SOLUTION

36. Introduction

A spherically symmetric solution of the coupled equations of Einstein and of Maxwell is that of Reissner and Nordström. It represents a black hole with a mass M and a charge Q_*. Since one does not expect any macroscopic body to possess a net charge, the consideration of a charged black-hole would appear to be outside the realm of reality. Nevertheless, the study of the Reissner–Nordström solution contributes to our understanding of the nature of space and time. Thus, we shall find that the scattering of incident electromagnetic and gravitational waves by the Reissner–Nordström black hole is described by a symmetric unitary scattering-matrix of order four and allows for the partial conversion of energy of one kind (electromagnetic or gravitational) into energy of the other kind. Besides, the Reissner–Nordström solution provides a more general framework for the many surprising analytical features which we encountered in the study of the Schwarzschild solution in Chapter 4. And finally, the fact that the Reissner–Nordström solution has two horizons, an external event horizon and an internal "Cauchy horizon," provides a convenient bridge to the study of the Kerr solution in the subsequent chapters.

37. The Reissner–Nordström solution

Since the Reissner–Nordström solution, like the Schwarzschild solution, represents a spherically symmetric space-time, we can, as in Chapter 3, write the line-element in either of the two forms (Ch. 3, equations (3) and (5)),

$$ds^2 = 4f\,du\,dv - e^{2\mu_3}[(d\theta)^2 + (d\varphi)^2 \sin^2\theta], \tag{1}$$

or

$$ds^2 = e^{2\mu_2}[(dx^0)^2 - (dx^2)^2] - e^{2\mu_3}[(d\theta)^2 + (d\varphi)^2 \sin^2\theta], \tag{2}$$

where $f = e^{2\mu_2}$ and μ_3 are functions solely of u and v (or, equivalently, of x^0 and x^2). Accordingly, the expressions for the non-vanishing (tetrad) components of the Riemann tensor (appropriate to the metric (2)) given in Chapter 3, equations (10), continue to be valid. However, since we are now considering, not a vacuum field, but one in which an electromagnetic field prevails, we cannot set the Ricci tensor to be zero, as we did in Chapter 3, equations (11)–(14). Instead we must now set

$$R_{ab} = -2(\eta^{cd}F_{ac}F_{bd} - \tfrac{1}{4}\eta_{ab}F_{ef}F^{ef}), \tag{3}$$

where F_{ab} denotes the (tetrad) components of the Maxwell tensor which must, in turn, be determined by Maxwell's equations. (In equation (3), η_{ab} denotes the Minkowskian metric of the chosen ortho-normal tetrad-frame.)

(a) *The solution of Maxwell's equations*

It is clear that, under the assumed conditions of spherical symmetry, the axial components of the Maxwell tensor, namely, F_{01}, F_{12}, and F_{13}, must vanish; and the equations governing the polar components, F_{02}, F_{03}, and F_{23}, can be readily written down by transcribing the equations given in Chapter 2 (equations (95), (e), (g), (f), and (h)) appropriately for the metric (2). We find

$$\begin{aligned}(e^{2\mu_3}F_{02})_{,2}+e^{\mu_2+\mu_3}F_{03}\cot\theta&=0,\\(e^{2\mu_3}F_{02})_{,0}+e^{\mu_2+\mu_3}F_{23}\cot\theta&=0,\\(e^{\mu_2+\mu_3}F_{23})_{,2}-(e^{\mu_2+\mu_3}F_{03})_{,0}&=0,\end{aligned}\tag{4}$$

and

$$(e^{\mu_2+\mu_3}F_{23})_{,0}-(e^{\mu_2+\mu_3}F_{03})_{,2}=0.$$

Since none of the components of the Maxwell tensor can depend on θ (by the assumption of spherical symmetry), it follows from the foregoing equations that

$$F_{03}=F_{23}=0,\tag{5}$$

and

$$(e^{2\mu_3}F_{02})_{,2}=(e^{2\mu_3}F_{02})_{,0}=0.\tag{6}$$

Hence, the only non-vanishing component of the Maxwell tensor is F_{02} which must be of the form

$$F_{02}=-Q_*e^{-2\mu_3},\tag{7}$$

where Q_* is a constant.

With $F_{03}=F_{23}=0$ and F_{02} given by equation (7), we readily find that

$$-R_{44}=R_{00}=R_{11}=-R_{22}=R_{33}=Q_*^2e^{-4\mu_3},\text{ and }R_{42}=0.\tag{8}$$

(b) *The solution of Einstein's equations*

With the non-vanishing components of the Ricci tensor given in equations (8), Chapter 3, equations (11)–(15), must now be replaced by

$$2R_{1212}+R_{2424}=+Q_*^2e^{-4\mu_3},\tag{9}$$
$$2R_{1414}+R_{2424}=+Q_*^2e^{-4\mu_3},\tag{10}$$
$$R_{1212}+R_{1313}+R_{1414}=-Q_*^2e^{-4\mu_3},\tag{11}$$
$$R_{1313}+R_{2323}+R_{3434}=-Q_*^2e^{-4\mu_3},\tag{12}$$

and

$$R_{2114}+R_{2334}=R_{24}=0. \tag{13}$$

From equations (9) and (10), it follows that

$$R_{1212}+R_{1010}=0; \tag{14}$$

and from equations (13) and Chapter 3, equation (10c), we conclude that

$$R_{1210}=0. \tag{15}$$

Next, adding equations (9) and (10), we obtain

$$R_{1212}-R_{1010}-R_{2020}=+Q_*^2 e^{-4\mu_3}; \tag{16}$$

and rewriting equation (11) in the form

$$R_{1212}+R_{1313}-R_{1010}=-Q_*^2 e^{-4\mu_3}, \tag{17}$$

we can combine it with equation (16) to give

$$R_{1313}+R_{2020}=-2Q_*^2 e^{-4\mu_3}. \tag{18}$$

We observe that equations (14) and (15) are the same as the first pair of equations in Chapter 3, equation (19), while equations (17) and (18) replace the second pair. Treating these equations in the same fashion, we obtain the basic set of equations

$$fZ_{,u,u}-Z_{,u}f_{,u}=0, \tag{19}$$

$$fZ_{,v,v}-Z_{,v}f_{,v}=0, \tag{20}$$

$$f\left(1-\frac{Q_*^2}{Z^2}\right)+ZZ_{,u,v}+Z_{,u}Z_{,v}=0, \tag{21}$$

and

$$\frac{1}{2}\frac{\partial}{\partial v}\left(\frac{f_{,u}}{f}\right)+\frac{Z_{,u,v}}{Z}=-\frac{Q_*^2}{Z^4}f. \tag{22}$$

Equations (19)–(22) replace equations (31)–(34) of Chapter 3; and their solution can be accomplished in similar fashion. Thus, equations (19) and (20) enable us to conclude, as before, that

$$f=A(u)Z_{,v} \qquad \text{and} \qquad f=B(v)Z_{,u}, \tag{23}$$

where $A(u)$ and $B(v)$ are functions only of the arguments specified. Next writing equation (21) in the alternative forms,

$$\left[A(u)\left(Z+\frac{Q_*^2}{Z}\right)+ZZ_{,u}\right]_{,v}=0 \tag{24}$$

and

$$\left[B(v)\left(Z+\frac{Q_*^2}{Z}\right)+ZZ_{,v}\right]_{,u}=0, \tag{25}$$

we deduce that

$$Z_{,u} = -A(u)\left(1+\frac{Q_*^2}{Z^2}\right)+\frac{F(u)}{Z} \tag{26}$$

and

$$Z_{,v} = -B(v)\left(1+\frac{Q_*^2}{Z^2}\right)+\frac{G(v)}{Z}, \tag{27}$$

where $F(u)$ and $G(v)$ are further arbitrary functions of the arguments specified.

Equations (23), (26), and (27) can be combined to give

$$\begin{aligned} Z_{,u}Z_{,v} &= -f\left(1+\frac{Q_*^2}{Z^2}\right)+\frac{F(u)Z_{,v}}{Z} \\ &= -f\left(1+\frac{Q_*^2}{Z^2}\right)+\frac{G(v)Z_{,u}}{Z}. \end{aligned} \tag{28}$$

Therefore,

$$\frac{Z_{,u}}{Z_{,v}} = \frac{F(u)}{G(v)} = \frac{A(u)}{B(v)}; \tag{29}$$

and we conclude that

$$\frac{F(u)}{A(u)} = \frac{G(v)}{B(v)} = \text{a constant} = 2M \quad \text{(say)}. \tag{30}$$

Thus, we obtain the solutions

$$\left.\begin{aligned} Z_{,u} &= -A(u)\left(1-\frac{2M}{Z}+\frac{Q_*^2}{Z^2}\right), \\ Z_{,v} &= -B(v)\left(1-\frac{2M}{Z}+\frac{Q_*^2}{Z^2}\right), \end{aligned}\right\} \tag{31}$$

and

$$f = -A(u)B(v)\left(1-\frac{2M}{Z}+\frac{Q_*^2}{Z^2}\right). \tag{32}$$

It can now be verified that the solution of equations (19)–(21), expressed by equations (31) and (32), satisfies the remaining equation (22).

The meaning of the solution for Z given by equations (31) becomes clearer in the form

$$\mathrm{d}Z = -\left(1-\frac{2M}{Z}+\frac{Q_*^2}{Z^2}\right)[A(u)\mathrm{d}u+B(v)\mathrm{d}v]; \tag{33}$$

and the form of this solution again establishes, as in Chapter 3, 17 (b), that a spherically symmetric solution of the coupled Einstein-Maxwell equations is necessarily static (outside the event horizon), i.e., a generalization of Birkhoff's theorem.

Finally, we may note that the solution for the metric itself is given by

$$ds^2 = -4\left(1-\frac{2M}{Z}+\frac{Q_*^2}{Z^2}\right)A(u)B(v)\,du\,dv - Z^2\,d\Omega^2. \tag{34}$$

The occurrence of the two arbitrary functions $A(u)$ and $B(v)$ in the solution (34) (as in the corresponding solution of Chapter 3, equation (48) for the Schwarzschild metric) is entirely consistent with the freedom of choice we have in defining, as coordinates in U_2, u and v, or some function of u and some function of v.

38. The nature of the space-time

In accordance with customary usage, we shall replace Z by r, with its meaning of 'luminosity distance,' and write equations (33) and (34) in the forms

$$dr = -\frac{\Delta}{r^2}[A(u)du + B(v)dv] \tag{35}$$

and

$$ds^2 = -4\frac{\Delta}{r^2}A(u)B(v)\,du\,dv - r^2\,d\Omega^2, \tag{36}$$

where

$$\Delta = r^2 - 2Mr + Q_*^2. \tag{37}$$

Let

$$r_+ = M + \sqrt{(M^2 - Q_*^2)} \text{ and } r_- = M - \sqrt{(M^2 - Q_*^2)}, \tag{38}$$

be the roots of $\Delta = 0$. These roots will be real and distinct if $M^2 > Q_*^2$. *We shall assume that this inequality obtains.* Since

$$\Delta > 0 \quad \text{for} \quad r > r_+ \quad \text{and} \quad 0 < r < r_-$$
$$\text{and} \quad \Delta < 0 \quad \text{for} \quad r_- < r < r_+, \tag{39}$$

it is clear that we must distinguish the three regions:

$$A\text{:}\quad 0 < r < r_-;\; B\text{: } r_- < r < r_+;\text{ and } C\text{: } r > r_+. \tag{40}$$

We shall find that, while the surface $r = r_+$ is an event horizon in the same sense that $r = 2M$ is an event horizon in the Schwarzschild space-time, the surface $r = r_-$ is a 'horizon' in a different sense which we shall clarify presently.

We shall find it convenient to rewrite equation (35) in the form

$$dr_* = -[A(u)du + B(v)dv] \tag{41}$$

where

$$r_* = \int \frac{r^2}{\Delta}dr = r + \frac{r_+^2}{r_+ - r_-}\lg|r - r_+| - \frac{r_-^2}{r_+ - r_-}\lg|r - r_-|. \tag{42}$$

As defined,

$$-\infty < r_* < +\infty \quad \text{for} \quad r_+ < r < +\infty$$

and

$$+\infty > r_* > -\infty \quad \text{for} \quad r_- < r < r_+. \tag{43}$$

To clarify the nature of the space-time described by the metric (36), we shall make different choices for the functions $A(u)$ and $B(v)$ in the regions A and C and in the region B. In the regions A and C we shall choose

$$A(u) = -B(v) = \tfrac{1}{2}, \tag{44}$$

and let

$$dt = \tfrac{1}{2}(du + dv). \tag{45}$$

Then,

$$u = t - r_* \quad \text{and} \quad v = t + r_*; \tag{46}$$

and the metric takes the form

$$ds^2 = \frac{\Delta}{r^2} du\, dv - r^2 d\Omega^2, \tag{47}$$

or, equivalently,

$$ds^2 = \frac{\Delta}{r^2}(dt)^2 - \frac{r^2}{\Delta}(dr)^2 - r^2[(d\theta)^2 + (d\varphi)^2 \sin^2\theta]. \tag{48}$$

It is in this last form that the Reissner–Nordström metric is generally written and commonly known.

In the region B, we shall choose instead

$$A(u) = B(v) = -\tfrac{1}{2}, \tag{49}$$

and let

$$dt = \tfrac{1}{2}(du - dv). \tag{50}$$

Then

$$u = r_* + t \quad \text{and} \quad v = r_* - t, \tag{51}$$

and

$$ds^2 = \frac{|\Delta|}{r^2} du\, dv - r^2 d\Omega^2. \tag{52}$$

With the foregoing choices, the parts of the manifold in the three regions can be represented by the 'blocks' in Fig. 13 for constant values of the angular coordinates θ and φ. The edges of the blocks are identified as indicated. Also, to the regions A, B, and C there correspond further regions A', B', and C' obtained by applying the transformation, $u \to -u$ and $v \to -v$, which reverses the light-cone structure. A maximal analytic extension of the Reissner–Nordström solution may then be obtained by piecing together copies of the six blocks in an analytic fashion so that an overlapping edge is covered by either a $u(r)$- or a $v(r)$-system of coordinates (with the exception of the corners of each block).

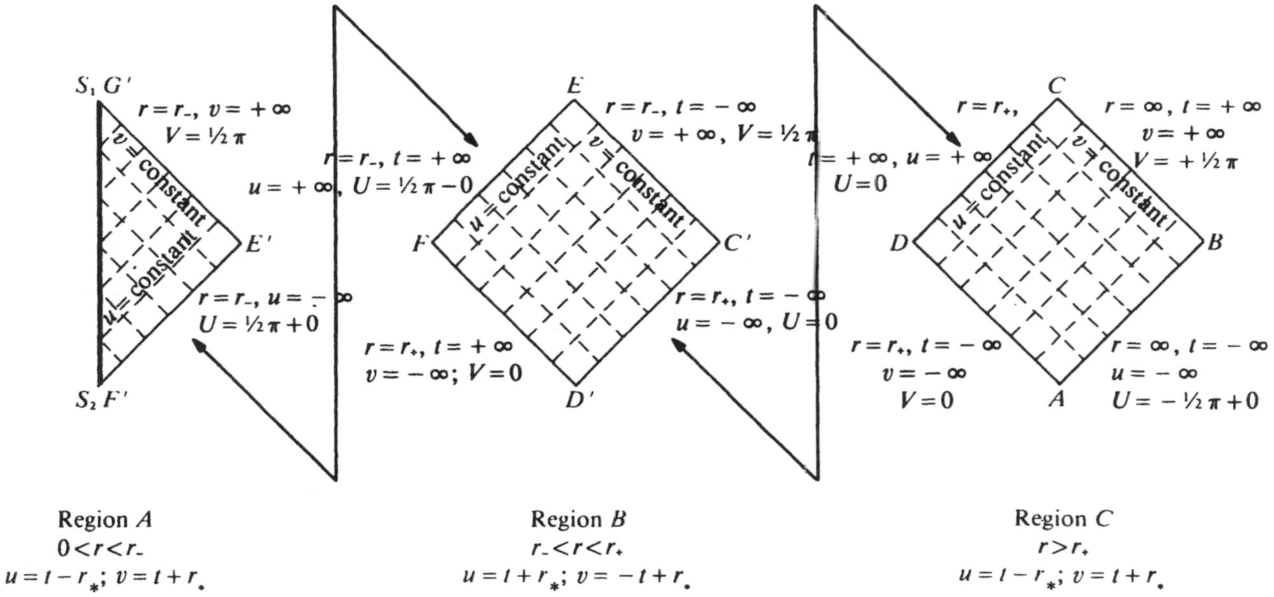

FIG. 13. The different regions in the Reissner–Nordström space-time. For explanation see text.

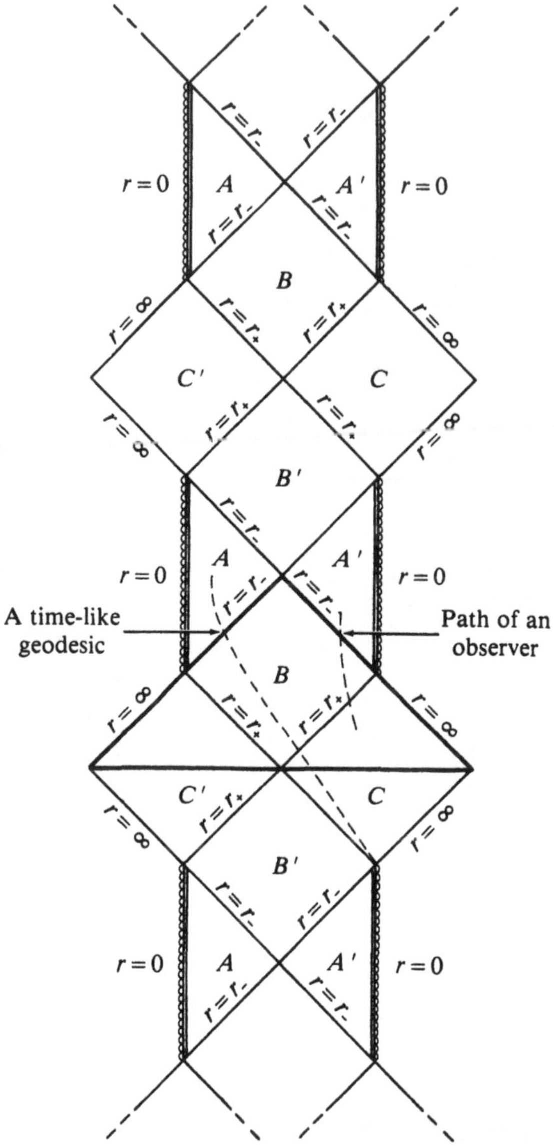

FIG. 14. The maximum analytic extension of the Reissner–Nordstrom space-time. The different regions A, B, and C of Fig. 13 are pieced together contiguously; regions A', B', and C' are obtained from A, B, and C by applying the transformation $u \to -u$ and $v \to -v$. The full extension is obtained by piecing together copies of the six blocks so that an overlapping edge is covered by a $u(r)$ or a $v(r)$-chart. A time-like geodesic from region C crossing both horizons and a possible time-like path an observer may take, to be emancipated of the past, are shown.

The resulting 'ladder' is shown in Fig. 14; it can be extended, indefinitely, in both directions. The principal reason for this extension is to make the manifold *geodesically complete*, i.e., to ensure that all geodesics, except those which terminate at the singularity, have infinite affine lengths both in the past and in the future directions.

An analytic representation of the maximally extended space-time can be obtained as follows. In the regions A and C, we set

$$\begin{aligned}\tan U &= -e^{-\alpha u} = -e^{-\alpha(t-r_*)}\\ &= -e^{-\alpha t}\{e^{+\alpha r}|r-r_+|^{1/2}|r-r_-|^{-\beta/2}\}\end{aligned}\tag{53}$$

and

$$\begin{aligned}\tan V &= +e^{+\alpha v} = +e^{+\alpha(t+r_*)}\\ &= +e^{+\alpha t}\{e^{+\alpha r}|r-r_+|^{1/2}|r-r_-|^{-\beta/2}\},\end{aligned}\tag{54}$$

where

$$\alpha = (r_+ - r_-)/2r_+^2 \quad \text{and} \quad \beta = r_-^2/r_+^2 \quad (>0 \quad \text{and} \quad <1).\tag{55}$$

In the region B, we set

$$\begin{aligned}\tan U &= +e^{+\alpha u} = +e^{+\alpha(t+r_*)}\\ &= +e^{+\alpha t}\{e^{+\alpha r}|r-r_+|^{1/2}|r-r_-|^{-\beta/2}\}\end{aligned}\tag{56}$$

and

$$\begin{aligned}\tan V &= +e^{+\alpha v} = +e^{+\alpha(-t+r_*)}\\ &= +e^{-\alpha t}\{e^{+\alpha r}|r-r_+|^{1/2}|r-r_-|^{-\beta/2}\}.\end{aligned}\tag{57}$$

By these substitutions, the metric takes the 'universal' form

$$ds^2 = -\frac{4}{\alpha^2}|r-r_+||r-r_-|\operatorname{cosec} 2U \operatorname{cosec} 2V\, dU\, dV - r^2\, d\Omega^2,\tag{58}$$

where r is implicitly defined in terms of U and V by

$$\begin{aligned}\tan U \tan V &= -e^{2\alpha r}|r-r_+||r-r_-|^{-\beta} \quad (r>r_+ \quad \text{and} \quad 0<r<r_-)\\ &= +e^{2\alpha r}|r-r_+||r-r_-|^{-\beta} \quad (r_- < r < r_+).\end{aligned}\tag{59}$$

The values of U and V, along the different edges of the blocks in Fig. 13, are indicated. The metric, as defined by equation (58), is analytic everywhere except at $r = r_-$, since $|r-r_-|$ is raised to a negative power in equation (59); but it is at least C^2 at $r = r_-$.

The delineation of the coordinate axes by null geodesics enables us to visualize, pictorially, the nature of the underlying space-time. It is clear, for example, that the space-time external to r_+ (in the region C) is very similar to the Schwarzschild space-time external to the surface $r = 2M$: any observer, who crosses the surface $r = r_+$ (represented by CD in Fig. 13), following a future-directed time-like trajectory, is forever 'lost' to an external observer (in C). Besides, any radiation transmitted by such an observer, at the instant of his crossing, will be infinitely red-shifted (because $dt/d\tau \to +\infty$; see equation (70) below) to an external observer. The surface at $r = r_+$ is, therefore, an event horizon in exactly the same sense that the surface $r = 2M$ is, in the Schwarzschild space-time. However, for one who has crossed the event horizon in the Reissner–Nordström geometry, there would appear to exist an infinite range of rich possibilities for experience, that are denied to one who crosses the event horizon in the Schwarzschild geometry. For the latter person, there is, as we have seen, no alternative to being inexorably propelled towards the singularity at $r = 0$. In contrast, the corresponding person, in region B in the Reissner–Nordström geometry, has, for example, the option to follow a future-directed time-like trajectory and cross the surface $r = r_-$ (represented by C′E) into the region A′. At the instant of such crossing, the person will witness, in a flash, a panorama of the entire history of the external world, in infinitely blue shifted (because $dt/d\tau \to -\infty$ for $r \to r_- + 0$; see equation (70) below) null rays arriving in the direction BCC′E. Once in the region A′, the person's future is no longer determined by his (or her) past history: the region A′ is outside the *domain of dependence* of the spatial slice extending across the regions C and C′ in Fig. 14. For this reason, the surface at $r = r_-$ is called a *'Cauchy' horizon.*

Alternatively, an observer in the region B has the option of following a time-like geodesic and entering the region A by crossing the surface r_- (represented by EF = E′F′). All such geodesics (except the purely radial null geodesics) skirt the singularity at $r = 0$ and cross the surface $r = r_-$ (across E′G′) to emerge, eventually, into a 'wonderfully' new asymptotically flat universe. It is clear that all these rich and varied prospects, for one who has crossed the event horizon at $r = r_+$, are essentially a consequence of the time-like character of the singularity at $r = 0$ (in contrast to the space-like character of the singularity in the Schwarzschild geometry). But a word of warning: crossing the Cauchy horizon is fraught with danger (as we shall see in §49).

39. An alternative derivation of the Reissner–Nordström metric

A derivation of the Reissner–Nordström metric, similar to the one for the Schwarzschild metric given in Chapter 3, §18, can be given. Thus, with a metric of the form Chapter 3, equation (62), we have the same expressions (Ch. 3, equation (64)) for the non-vanishing components of the Riemann tensor. And since, moreover, (cf. equations (8))

$$R_{42} = 0 \qquad \text{and} \qquad R_{22} - R_{44} = 0, \tag{60}$$

the conclusions expressed in Chapter 3, equations (67) and (72) continue to hold in the present context. However, in place of Chapter 3, equation (73), we now have (see equation (8))

$$2\mu_{2,r}\frac{e^{-2\mu_2}}{r} + \frac{1}{r^2}(1 - e^{-2\mu_2}) = R_{33} = \frac{Q_*^2}{r^4}. \tag{61}$$

The solution of this equation is

$$e^{-2\mu_2} = 1 - \frac{2M}{r} + \frac{Q_*^2}{r^2}\left(= e^{2\nu} = \frac{\Delta}{r^2}\right); \tag{62}$$

and we recover the solution in its standard form given in equation (48).

With the solution for μ_2 and ν given in equation (62), we find that the non-vanishing components of the Riemann tensor are (cf. Ch. 3, equation (75))

$$-R_{(1)(3)(1)(3)} = \frac{2Mr - Q_*^2}{r^4}; \quad R_{(2)(0)(2)(0)} = \frac{2Mr - 3Q_*^2}{r^4};$$

and (63)

$$R_{(1)(2)(1)(2)} = R_{(2)(3)(2)(3)} = -R_{(1)(0)(1)(0)} = -R_{(3)(0)(3)(0)} = \frac{Mr - Q_*^2}{r^4},$$

where we have enclosed the indices in parentheses to emphasize that these are the tetrad components in a local inertial frame. These expressions show that $r = 0$ is, indeed, a genuine space-time singularity. The corresponding tensor-components are given by (cf. Ch. 3, equations (76))

$$\left.\begin{aligned}
R_{0101} &= +R_{t\varphi t\varphi} = -\frac{Mr - Q_*^2}{r^2}e^{2\nu}\sin^2\theta,\\
R_{0303} &= +R_{t\theta t\theta} = -\frac{Mr - Q_*^2}{r^2}e^{2\nu},\\
R_{1212} &= +R_{\varphi r\varphi r} = +\frac{Mr - Q_*^2}{r^2}e^{-2\nu}\sin^2\theta,\\
R_{2323} &= +R_{r\theta r\theta} = +\frac{Mr - Q_*^2}{r^2}e^{-2\nu},\\
R_{1313} &= +R_{\varphi\theta\varphi\theta} = -(2Mr - Q_*^2)\sin^2\theta,\\
\text{and}\qquad R_{0202} &= +R_{trtr} = +\frac{2Mr - 3Q_*^2}{r^4},
\end{aligned}\right\} \tag{64}$$

where $e^{2\nu} = \Delta/r^2$.

40. The geodesics in the Reissner–Nordström space-time

Since the Reissner–Nordström metric differs from the Schwarzschild metric only in the definition of the 'horizon function' Δ, it follows that the basic equations governing geodesic motion are, with the redefinition of Δ, the same as those considered in Chapter 3, §§19 and 20. Thus, the equations governing the time-like geodesics are (Ch. 3, equations (85), (90), and (91))

$$\left(\frac{dr}{d\tau}\right)^2 + \frac{\Delta}{r^2}\left(1 + \frac{L^2}{r^2}\right) = E^2; \quad \frac{dt}{d\tau} = E\frac{r^2}{\Delta}; \quad \text{and} \quad \frac{d\varphi}{d\tau} = \frac{L}{r^2}, \tag{65}$$

where

$$\Delta = r^2 - 2Mr + Q_*^2. \tag{66}$$

By considering r as a function of φ and replacing r by $u = 1/r$ as the independent variable, we obtain the equation (cf. Ch. 3, equation (94))

$$\left(\frac{du}{d\varphi}\right)^2 = -Q_*^2 u^4 + 2Mu^3 - u^2\left(1 + \frac{Q_*^2}{L^2}\right) + \frac{2M}{L^2}u - \frac{1-E^2}{L^2} = f(u) \quad \text{(say)}. \tag{67}$$

The corresponding equations governing the null geodesics are (cf. Ch. 3, equations (212)–(215))

$$\left(\frac{dr}{d\tau}\right)^2 + L^2\frac{\Delta}{r^4} = E^2; \quad \frac{dt}{d\tau} = E\frac{r^2}{\Delta}; \qquad \text{and} \qquad \frac{d\varphi}{d\tau} = \frac{L}{r^2}; \tag{68}$$

and

$$\left(\frac{du}{d\varphi}\right)^2 = -Q_*^2 u^4 + 2Mu^3 - u^2 + \frac{1}{D^2} = f(u) \quad \text{(say)}, \tag{69}$$

where $D(= L/E)$ is, as before, the impact parameter.

Equations (67) and (69) differ from the equations considered in Chapter 3 by $f(u)$ being a biquadratic instead of a cubic. We shall find that this fact makes an essential difference only for the orbits which cross the horizon at $r = r_+$, i.e., for the orbits, which in the Schwarzschild geometry terminated at the singularity at $r = 0$; and this is in accord with the differences in the Schwarzschild and the Reissner–Nordström geometrics interior to the respective event horizons.

(a) *The null geodesics*

As in the Schwarzschild space-time, the radial null-geodesics in the Reissner–Nordström geometry provide the base for constructing a null tetrad-frame for use in a Newman–Penrose formalism. The equations governing the radial null-geodesics can be obtained by setting $L = 0$ in equations (68); thus,

$$\frac{dr}{d\tau} = \pm E, \quad \frac{dt}{d\tau} = \frac{r^2}{\Delta}E, \qquad \text{and} \qquad \frac{d\theta}{d\tau} = \frac{d\varphi}{d\tau} = 0. \tag{70}$$

Accordingly,

$$\frac{dr}{dt} = \pm\frac{\Delta}{r^2}. \tag{71}$$

The solution of this equation is

$$t = \pm r_* + \text{constant}, \tag{72}$$

where r_* is defined in equation (42). Therefore, for an in-going null-ray, the coordinate time t *increases* from $-\infty$ to $+\infty$ as r decreases from $+\infty$ to r_+, *decreases* from $+\infty$ to $-\infty$ as r further decreases from r_+ to r_-, and *increases* again from $-\infty$ to a finite limit as r decreases from r_- to zero. All the while, the proper time decreases at a constant rate E (as behoves light!). From a finite distance $r > r_+$, a co-moving observer will arrive at the singularity in a finite proper time. (We shall see, presently, that these are the only class of geodesics that terminate at the singularity: all others skirt it.) It should also be noted that as $dt/d\tau$ tends to $+\infty$ for $r \to r_+ + 0$ and to $-\infty$ as $r \to r_- + 0$, any radiation received from infinity (in the region C) will appear infinitely red-shifted at the crossing of the event horizon and infinitely blue-shifted at the crossing of the Cauchy horizon.

Turning to the consideration of equation (69) and the null geodesics in general, we first observe that the quartic equation, $f(u) = 0$, always allows two real roots: one negative (which has no physical significance) and one positive (which, one can show, occurs for $r < r_-$). We shall be concerned only with cases when the two remaining roots are *either* both real (distinct or coincident) *or* a complex-conjugate pair. Let the value of the impact parameter, D, for which $f(u) = 0$ has a double root, be denoted by D_c. Then, for all values of $D > D_c$, we shall have orbits of two kinds (as in the Schwarzschild space-time): orbits of the *first kind* which lie entirely outside the event horizon, coming from $+\infty$ and receding again to $+\infty$ after a perihelion passage; and orbits of the *second kind* which have two turning points, one outside the event horizon and one inside the Cauchy horizon. For values of the impact parameter $D < D_c$ (when $f(u) = 0$ allows only one real root, $> 1/r_-$, along the positive real u-axis) the orbit coming from $+\infty$ will cross both horizons and have a turning point for a finite value of $r < r_-$. Thus, all these orbits skirt the singularity at $r = 0$. However, from our remarks in §38, it follows that orbits which have turning points inside the Cauchy horizon cannot be assumed to continue along the symmetrically reflected time-reversed orbits: they should be considered, instead, as continuing up the 'ladder' in Fig. 14: e.g., $A \to B' \to C$ (where C is now a different asymptotically-flat universe).

The value of the impact parameter for which $f(u) = 0$ allows a double root is determined by the equations

$$f(u) = \frac{1}{D^2} - u^2(Q_*^2 u^2 - 2Mu + 1) = 0 \tag{73}$$

and

$$f'(u) = -2u(1 - 3Mu + 2Q_*^2 u^2) = 0. \tag{74}$$

Besides $u = 0$, equation (74) allows the roots

$$u = \frac{3M}{4Q_*^2}\left[1 \pm \left(1 - \frac{8Q_*^2}{9M^2}\right)^{1/2}\right]. \tag{75}$$

At the larger of these two roots, $f(u)$ has a maximum. The double root we are seeking must occur at the minimum of $f(u)$ where

$$u = \frac{3M}{4Q_*^2}\left[1-\left(1-\frac{8Q_*^2}{9M^2}\right)^{1/2}\right] = u_c \quad \text{(say)}. \tag{76}$$

The corresponding value of r is

$$r_c = 1.5M[1+(1-8Q_*^2/9M^2)^{1/2}]. \tag{77}$$

It is clear that at this radius r_c, the geodesic equations allow an unstable circular orbit.

The value of the impact parameter, D_c, associated with the double root r_c follows from equation (73). We find

$$D_c = r_c^2/\sqrt{\Delta_c}, \quad \Delta_c = r_c^2 - 2Mr_c + Q_*^2 = Mr_c - Q_*^2. \tag{78}$$

When $D = D_c$,

$$f(u) = (u-u_c)^2[-Q_*^2u^2 + 2(M-Q_*^2u_c)u + u_c(M-Q_*^2u_c)]; \tag{79}$$

and the solution for φ is given by

$$\varphi = \pm\int[-Q_*^2u^2+2(M-Q_*^2u_c)u+u_c(M-Q_*^2u_c)]^{-1/2}\frac{du}{u-u_c}. \tag{80}$$

The substitution,

$$\xi = (u-u_c)^{-1}, \tag{81}$$

reduces φ to the elementary integral,

$$\varphi = \mp\int\frac{d\xi}{(-Q_*^2+b\xi+c\xi^2)^{1/2}}, \tag{82}$$

where

$$b = 2(M-2Q_*^2u_c) \quad \text{and} \quad c = u_c(3M-4Q_*^2u_c). \tag{83}$$

We thus obtain the solution

$$\mp\varphi = \left\{\begin{array}{ll} \dfrac{1}{\sqrt{c}}\lg\{2[c(-Q_*^2+b\xi+c\xi^2)]^{1/2}+2c\xi+b\} & (c>0) \\ \dfrac{-1}{\sqrt{-c}}\sin^{-1}\left\{\dfrac{2c\xi+b}{(4Q_*^2c+b^2)^{1/2}}\right\} & (c<0) \end{array}\right\}. \tag{84}$$

The orbits of the two kinds are described by this same solution for values of the arguments in the following ranges:

$$\infty > r > r_c, \quad 0 \leqslant u \leqslant u_c, \quad \text{and} \quad -u_c^{-1} > \xi > -\infty, \tag{85}$$

for orbits of the first kind, and

$$r_c > r > r_{\min}, \quad u_c < u \leqslant u_{\max}\,(=1/r_{\min}),$$
$$\text{and} \quad +\infty > \xi > \xi_{\min}\,(=(u_{\min}-u_c)^{-1}), \tag{86}$$

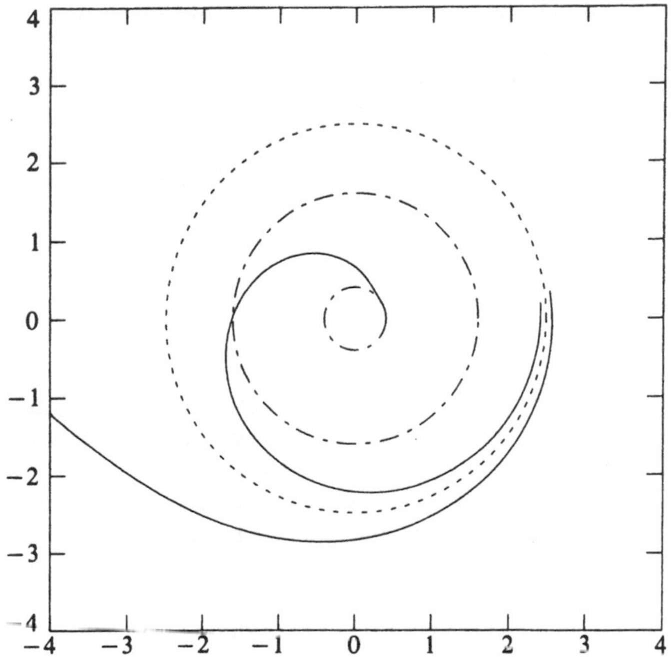

FIG. 15. The null geodesic for the critical impact-parameter D_c given by equation (78) for $Q_* = 0.8$. The inner and the outer horizons are shown by the dashed circles; the dotted circle represents the unstable circular-orbit. The trajectory actually crosses the inner horizon, but only just barely, before it terminates. The unit of length along the coordinate axes is M.

for orbits of the second kind where r_{min} is the positive root of the equation (cf. equation (79)),

$$u_c(M - Q_*^2 u_c)r^2 + 2(M - Q_*^2 u_c)r - Q_*^2 = 0. \tag{87}$$

Both these orbits approach the unstable circular orbit at $r = r_c$, asymptotically from opposite sides, by spiralling around it an infinite number of times.

An example of the solution derived from equation (84) is illustrated in Fig. 15 (cf. Fig. 9, (b) which illustrates the corresponding orbits in the Schwarzschild space-time). We can now readily visualize the nature of the null geodesics for other values of the impact parameter.

(*b*) *Time-like geodesics*

Considering first the radial geodesics, we have the governing equations

$$\left(\frac{dr}{d\tau}\right)^2 = E^2 - \frac{\Delta}{r^2} \quad \text{and} \quad \frac{dt}{d\tau} = E\frac{r^2}{\Delta}. \tag{88}$$

Since $\Delta > 0$ in the interval $0 \leqslant r < r_-$, it is clear that $(E^2 r^2 - \Delta)$ will vanish for some finite value of $0 < r < r_-$. We conclude that the trajectory will have a turning point inside the Cauchy horizon. Thus, even the radial time-like geodesics do not reach the singularity at $r = 0$: they skirt it only to emerge in other domains. It is also clear from equations (88) that the variation of the coordinate time, t, with r will exhibit singularities both for $r \to r_+$ (± 0) and for $r \to r_-$ (± 0) consistently with the horizon character of these surfaces.

The formal solutions of equations (88) can be readily written down in terms of elementary integrals. We shall not write them since they are too complicated to exhibit in any more manifest manner what can be deduced from a simple inspection of the equations.

Turning next to equation (67) and to a consideration of time-like geodesics in general, we can classify and analyze the different cases, as we did in Chapter 3, §19 in the context of the Schwarzschild geometry. It is, however, clear that the essential difference between the geodesics in the Reissner–Nordström and the Schwarzschild geometrics will be in the orbits which cross the event horizon: in the Schwarzschild geometry all such orbits must terminate in the singularity; in the Reissner–Nordström geometry they will formally terminate at some point inside the Cauchy horizon. The differences which arise on this account can be illustrated sufficiently by considering the orbits of the second kind associated with stable and unstable circular orbits.

The conditions for the occurrence of circular orbits are

$$f(u) = -Q_*^2 u^4 + 2Mu^3 - (1 + Q_*^2/L^2)u^2 + [2Mu - (1 - E^2)]/L^2 = 0, \quad (89)$$

and

$$f'(u) = -4Q_*^2 u^3 + 6Mu^2 - 2(1 + Q_*^2/L^2)u + 2M/L^2 = 0. \quad (90)$$

From these equations, it follows that the energy E and the angular momentum L of a circular orbit of radius $r_c = 1/u_c$ is given by

$$E^2 = \frac{(1 - 2Mu_c + Q_*^2 u_c^2)^2}{1 - 3Mu_c + 2Q_*^2 u_c^2} \quad (91)$$

and

$$L^2 = \frac{M - Q_*^2 u_c}{u_c(1 - 3Mu_c + 2Q_*^2 u_c^2)}. \quad (92)$$

These equations require, in particular, that

$$1 - 3Mu_c + 2Q_*^2 u_c^2 > 0. \quad (93)$$

Comparison of this inequality with equation (74) shows that the minimum radius for a time-like circular orbit is the radius of the unstable circular photon-orbit. In Fig. 16 we exhibit the dependences of E and L on the radius of the circular orbit.

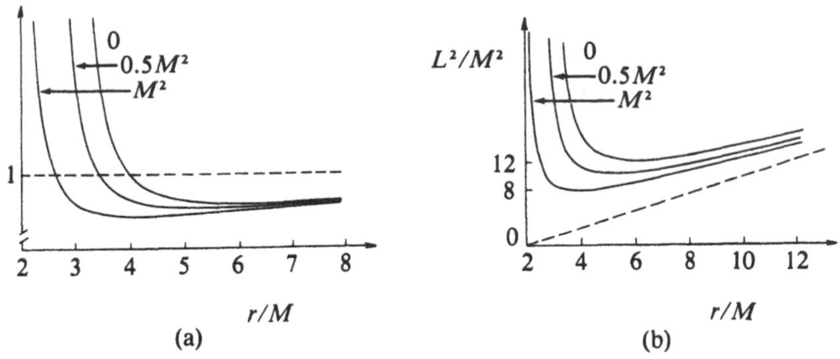

FIG. 16. The variations of E^2 (figure (*a*)) and L^2 (figure (*b*)) with the radius of the circular orbit. The curves are labelled by the values of Q_*^2 to which they belong.

When E and L have the values (91) and (92), appropriate for a circular orbit of radius $r_c = 1/u_c$, equation (67) becomes (cf. equation (79))

$$\left(\frac{\mathrm{d}u}{\mathrm{d}\varphi}\right)^2 = (u-u_c)^2[-Q_*^2u^2+2(M-Q_*^2u_c)u+(M-Q_*^2u_c-M/L^2u_c^2)u_c]. \tag{94}$$

Therefore, besides the circular orbit of radius $r_c = 1/u_c$, equation (94) provides an orbit of the second kind determined by (cf. equation (80))

$$\varphi = \pm\int[-Q_*^2u^2+2(M-Q_*^2u_c)u+(M-Q_*^2u_c-M/L^2u_c^2)u_c]^{-1/2}\frac{\mathrm{d}u}{u-u_c}. \tag{95}$$

With the substitution

$$\xi = (u-u_c)^{-1} \tag{96}$$

we obtain the same solution (84) with, however,

$$b = 2(M-2Q_*^2u_c) \qquad \text{and} \qquad c = u_c(3M-4Q_*^2u_c-M^2/L^2u_c^2). \tag{97}$$

An example of an orbit of the second kind associated with a stable circular orbit is illustrated in Fig. 17a. When the circular orbit is unstable, the orbits of both kinds can be derived from the same formula (84) (the case $c < 0$). An example of such orbits is illustrated in Fig. 17b.

The minimum radius for a stable circular orbit will occur at a point of inflexion of the function $f(u)$, i.e., we must supplement equations (89) and (90) with the further equation

$$f''(u) = -12Q_*^2u^2+12Mu-2(1+Q_*^2/L^2) = 0. \tag{98}$$

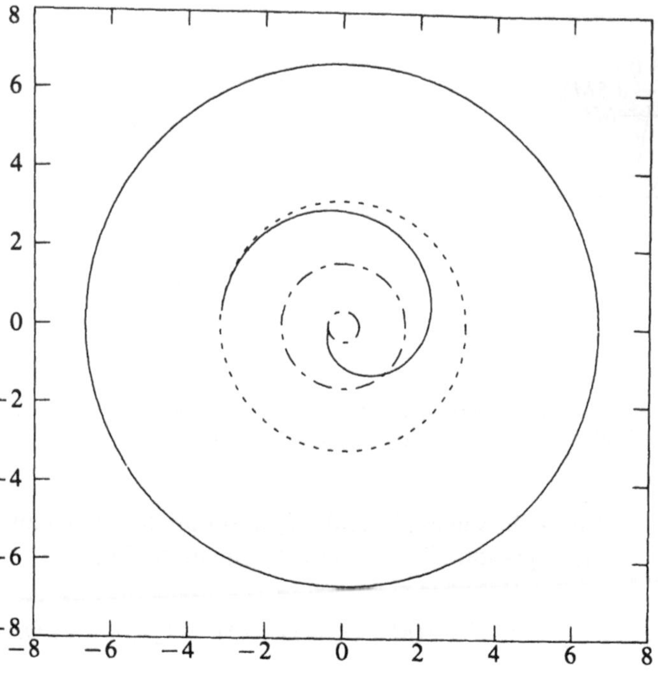

(a)

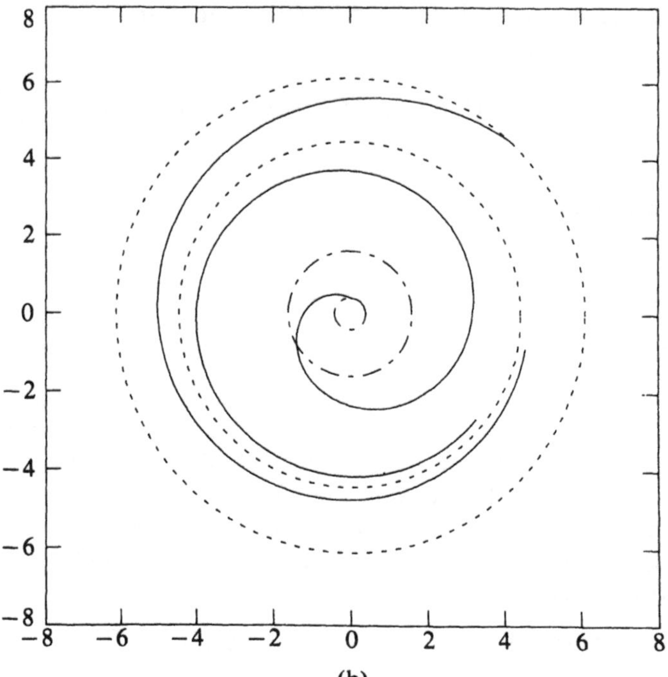

(b)

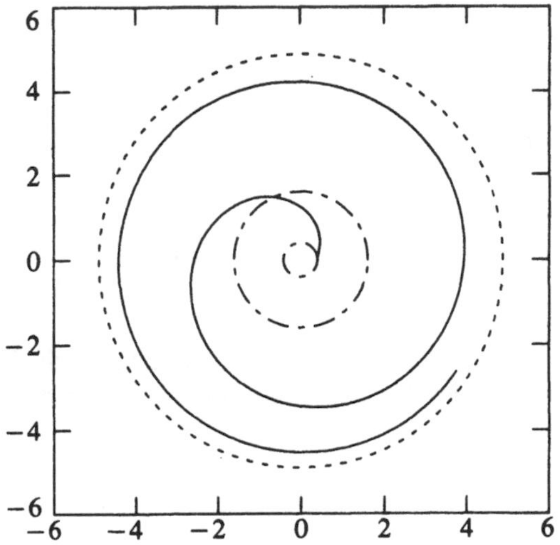

FIG. 18. The last time-like unstable circular orbit which occurs at radius 4.89 M for $Q_* = 0.8$. The horizons are indicated by the dashed circles; and the unit of length along the coordinate axes is M.

Eliminating L^2 in this equation with the aid of equation (92), we obtain the equation

$$4Q_*^4 u_c^3 - 9MQ_*^2 u_c^2 + 6M^2 u_c - M = 0, \tag{99}$$

or, alternatively,

$$r_c^3 - 6Mr_c^2 + 9Q_*^2 r_c - 4Q_*^4/M = 0. \tag{100}$$

(It may be noted here that equation (100) gives $r_c = 6M$ when $Q_*^2 = 0$, in agreement with the value for the Schwarzschild geometry.)

When all three equations, (89), (90), and (98), are satisfied, equation (67) takes the form

$$\left(\frac{\mathrm{d}u}{\mathrm{d}\varphi}\right)^2 = (u-u_c)^3(2M - 3Q_*^2 u_c - Q_*^2 u); \tag{101}$$

and the solution of this equation is (cf. Ch. 3, equation (165))

$$u = u_c + \frac{2(M - 2Q_*^2 u_c)}{(M - 2Q_*^2 u_c)^2(\varphi - \varphi_0)^2 + Q_*^2}. \tag{102}$$

An example of this orbit is illustrated in Fig. 18.

FIG. 17. (a) The orbit of the second kind associated with a stable time-like circular-orbit with a reciprocal radius $u_c = 0.15$ for $Q_* = 0.8$. (b) The case of a time-like unstable circular-orbit (with a reciprocal radius $u_c = 0.2225$) when the orbits of the two kinds coalesce. In both cases illustrated, the orbits of the second kind cross the inner horizon and, again, only just barely. The horizons are indicated by the dashed circles; and the unit of length along the coordinate axes is M.

(*c*) *The motion of charged particles*

A test particle with a net charge will not, of course, describe a geodesic in the Reissner–Nordström geometry. Since, in this geometry, the only non-vanishing component of the vector potential is A_0 ($= Q_*/r$), its motion will be determined, instead, by the Lagrangian

$$2\mathscr{L} = \left[\frac{\Delta}{r^2}\left(\frac{\mathrm{d}t}{\mathrm{d}\tau}\right)^2 - \frac{r^2}{\Delta}\left(\frac{\mathrm{d}r}{\mathrm{d}\tau}\right)^2 - r^2\left(\frac{\mathrm{d}\theta}{\mathrm{d}\tau}\right)^2 - (r^2\sin^2\theta)\left(\frac{\mathrm{d}\varphi}{\mathrm{d}\tau}\right)^2\right] + 2\frac{qQ_*}{r}\frac{\mathrm{d}t}{\mathrm{d}\tau}, \tag{103}*$$

where q denotes the charge per unit mass of the test particle. The equations of motion which follow from this Lagrangian are readily written down. They are (cf. equations (65))

$$p_t = \frac{\Delta}{r^2}\frac{\mathrm{d}t}{\mathrm{d}\tau} + \frac{qQ_*}{r} = E = \text{constant}, \quad r^2\frac{\mathrm{d}\varphi}{\mathrm{d}\tau} = L = \text{constant}, \tag{104}$$

and

$$\left(\frac{\mathrm{d}r}{\mathrm{d}\tau}\right)^2 + \frac{\Delta}{r^2}\left(1 + \frac{L^2}{r^2}\right) = \left(\frac{\Delta}{r^2}\frac{\mathrm{d}t}{\mathrm{d}\tau}\right)^2 = \left(E - \frac{qQ_*}{r}\right)^2; \tag{105}$$

and in place of equation (67) we now have

$$\left(\frac{\mathrm{d}u}{\mathrm{d}\varphi}\right)^2 = -Q_*^2u^4 + 2Mu^3 - \left[1 + Q_*^2\frac{(1-q^2)}{L^2}\right]u^2 + \frac{2}{L^2}(M - qQ_*E)u - \frac{1-E^2}{L^2}. \tag{106}$$

The only novel feature these equations present is that if a particle should have a turning point as it arrives at the event horizon, then its energy, according to equation (105), will be

$$E = qQ_*/r_+; \tag{107}$$

and this will be negative if $qQ_* < 0$. This fact gives rise to the possibility of extracting energy from the black hole, energy that is associated with its charge. We shall consider processes of this kind in Chapter 7.

41. The description of the Reissner–Nordström space-time in a Newman–Penrose formalism

For constructing a null tetrad-frame for a description of the Reissner–Nordström space-time in a Newmann–Penrose formalism, we start

* This expression for the Lagrangian follows from Lorentz's equation of motion, namely,

$$\frac{\mathrm{d}^2x^i}{\mathrm{d}\tau^2} = qF^i{}_m\frac{\mathrm{d}x^m}{\mathrm{d}\tau}.$$

with the vectors (70) specified by the radial null geodesics. Precisely, we shall take

$$\left.\begin{aligned} l^i &= (l^t, l^r, l^\theta, l^\varphi) = \frac{1}{\Delta}(r^2, +\Delta, 0, 0), \\ \text{and} \qquad & \\ n^i &= (n^t, n^r, n^\theta, n^\varphi) = \frac{1}{2r^2}(r^2, -\Delta, 0, 0), \end{aligned}\right\} \tag{108}$$

as the two real null-vectors of the basis and adjoin to them the complex null-vector,

$$m^i = (m^t, m^r, m^\theta, m^\varphi) = \frac{1}{r\sqrt{2}}(0, 0, 1, i \operatorname{cosec}\theta), \tag{109}$$

orthogonal to $\boldsymbol{l}$ and $\boldsymbol{n}$. These vectors satisfy the required normalization conditions

$$\boldsymbol{l}\cdot\boldsymbol{n} = 1 \qquad \text{and} \qquad \boldsymbol{m}\cdot\bar{\boldsymbol{m}} = -1. \tag{110}$$

The basis vectors, $(\boldsymbol{l}, \boldsymbol{n}, \boldsymbol{m}, \bar{\boldsymbol{m}})$, differ from the ones defined in the Schwarzschild space-time (Ch. 3, equations (281)–(283)) only in the definition of Δ. The evaluation of the spin coefficients proceeds exactly as before with only minor changes resulting from the present definition of Δ. We now find (cf. Ch. 3, equations (287) and (288))

$$\kappa = \sigma = \lambda = \nu = \varepsilon = \pi = \tau = 0 \tag{111}$$

and

$$\rho = -\frac{1}{r}, \quad \beta = -\alpha = \frac{1}{2\sqrt{2}}\frac{\cot\theta}{r}, \quad \mu = -\frac{\Delta}{2r^3},$$

and

$$\gamma = \mu + \frac{r-M}{2r^2} = \frac{Mr - Q_*^2}{2r^3}. \tag{112}$$

The fact, that the spin coefficients, κ, σ, λ, and ν vanish, confirms the type-D character of the space-time. The Weyl scalars, Ψ_0, Ψ_1, Ψ_3, and Ψ_4 must, therefore, vanish in the chosen basis as can, indeed, be directly verified by contracting the Riemann tensor (whose non-vanishing components are listed in equations (64)) in accordance with the definitions of the scalars. The Weyl scalar Ψ_2 does not, however, vanish. We find (cf. Ch. 3, equation (289))

$$\begin{aligned} \Psi_2 &= R_{pqrs} l^p m^q n^r \bar{m}^s \\ &= \frac{e^{-2\nu}}{4r^2}(R_{0101}\operatorname{cosec}^2\theta + R_{0303} - e^{4\nu}R_{2323} - e^{4\nu}R_{1212}\operatorname{cosec}^2\theta), \end{aligned} \tag{113}$$

or, inserting for the components of the Riemann tensor their values given in equations (64), we find

$$\Psi_2 = -(Mr - Q_*^2)r^{-4}. \tag{114}$$

The description of the Reissner–Nordström geometry in the Newman–Penrose formalism is completed by noting that the value of the only non-vanishing Maxwell-scalar is given by

$$\begin{aligned}\phi_1 &= \tfrac{1}{2} F_{pq}(l^p n^q + \bar{m}^p m^q) = \tfrac{1}{2} F_{tr}(l^t n^r - l^r n^t) \\ &= \tfrac{1}{2} Q_* r^{-2}.\end{aligned} \tag{115}$$

42. The metric perturbations of the Reissner-Nordström solution

As in the case of the Schwarzschild solution, the perturbations in the metric coefficients of the Reissner–Nordström solution can be analyzed by linearizing, about this solution, the Einstein and the Maxwell equations appropriate for the non-stationary axisymmetric metric,

$$\begin{aligned}ds^2 = e^{2\nu}(dt)^2 - e^{2\psi}(d\varphi - \omega dt - q_2 dx^2 - q_3 dx^3)^2 \\ - e^{2\mu_2}(dx^2)^2 - e^{2\mu_3}(dx^3)^2.\end{aligned} \tag{116}$$

But in contrast to the treatment of the Schwarzschild perturbations in Chapter 4, §24, we must now consider the linearization of the Maxwell equations as well. We start with Maxwell's equations.

(a) The linearized Maxwell equations

Maxwell's equations, listed in Chapter 2 (equations (95)) when specialized to the simpler case (116), take the forms

$$\left.\begin{aligned}(e^{\psi+\mu_2}F_{12})_{,3} + (e^{\psi+\mu_3}F_{31})_{,2} &= 0, \\ (e^{\psi+\nu}F_{01})_{,2} + (e^{\psi+\mu_2}F_{12})_{,0} &= 0, \\ (e^{\psi+\nu}F_{01})_{,3} + (e^{\psi+\mu_3}F_{13})_{,0} &= 0, \\ (e^{\mu_2+\mu_3}F_{01})_{,0} + (e^{\nu+\mu_3}F_{12})_{,2} + (e^{\nu+\mu_2}F_{13})_{,3} & \\ = e^{\psi+\mu_3}F_{02}Q_{02} + e^{\psi+\mu_2}F_{03}Q_{03} - e^{\psi+\nu}F_{23}Q_{23} &\end{aligned}\right\} \tag{117}$$

and

$$\left.\begin{aligned}(e^{\psi+\mu_3}F_{02})_{,2} + (e^{\psi+\mu_2}F_{03})_{,3} &= 0, \\ -(e^{\psi+\nu}F_{23})_{,2} + (e^{\psi+\mu_2}F_{03})_{,0} &= 0, \\ +(e^{\psi+\nu}F_{23})_{,3} + (e^{\psi+\mu_3}F_{02})_{,0} &= 0, \\ (e^{\nu+\mu_2}F_{02})_{,3} - (e^{\nu+\mu_3}F_{03})_{,2} + (e^{\mu_2+\mu_3}F_{23})_{,0} & \\ = e^{\psi+\nu}F_{01}Q_{23} + e^{\psi+\mu_2}F_{12}Q_{03} - e^{\psi+\mu_3}F_{13}Q_{02}, &\end{aligned}\right\} \tag{118}$$

where

$$Q_{AB} = q_{A,B} - q_{B,A} \quad \text{and} \quad Q_{A0} = q_{A,0} - \omega_{,A} \quad (A, B = 2, 3). \tag{119}$$

It will be observed that the first set of equations (117) involves only quantities (or, products of quantities) which reverse their signs when φ is replaced by $-\varphi$, while the second set of equations (118) involves only quantities which are invariant to the reversal in the sign of φ. They correspond, respectively, to what we have described as axial and polar quantities in Chapter 4, §24. Besides, in each of the two groups of equations, we can dispense with the first equation since it provides only the integrability condition for the two following equations.

Remembering that $F_{02}(=-Q_* r^{-2})$ is the only non-vanishing component of the Maxwell tensor and also that the Q_{AB}'s are quantities of the first order of smallness, we readily find that the linearized versions of equations (117) and (118) (dispensing with the first equation in each case) are

$$(re^{\nu}F_{01}\sin\theta)_{,r}+re^{-\nu}F_{12,0}\sin\theta=0, \tag{120}$$

$$re^{\nu}(F_{01}\sin\theta)_{,\theta}+r^2F_{13,0}\sin\theta=0, \tag{121}$$

$$re^{-\nu}F_{01,0}+(re^{\nu}F_{12})_{,r}+F_{13,\theta}=-Q_*(\omega_{,2}-q_{2,0})\sin\theta, \tag{122}$$

and

$$re^{-\nu}F_{03,0}=(re^{\nu}F_{23})_{,r}, \tag{123}$$

$$\delta F_{02,0}-\frac{Q_*}{r^2}(\delta\psi+\delta\mu_3)_{,0}+\frac{e^{\nu}}{r\sin\theta}(F_{23}\sin\theta)_{,\theta}=0, \tag{124}$$

$$\left[\delta F_{02}-\frac{Q_*}{r^2}(\delta\nu+\delta\mu_2)\right]_{,\theta}+(re^{\nu}F_{30})_{,r}+re^{-\nu}F_{23,0}=0. \tag{125}$$

(b) The perturbations in the Ricci tensor

In contrast to the Schwarzschild space-time, we cannot in our treatment of the Reissner–Nordström space-time set the perturbed components of the Ricci tensor to be equal to zero. Instead, we must set

$$\delta R_{(a)(b)}=-2[\eta^{(n)(m)}(\delta F_{(a)(n)}F_{(b)(m)}+F_{(a)(n)}\delta F_{(b)(m)}) \\ -\eta_{(a)(b)}Q_*\delta F_{02}/r^2]. \tag{126}$$

From this equation, we find

$$\left.\begin{aligned}
&\delta R_{00}=\delta R_{11}=-\delta R_{22}=\delta R_{33}=-2\frac{Q_*}{r^2}\delta F_{02},\\
&\delta R_{01}=-2\frac{Q_*}{r^2}F_{12},\quad \delta R_{03}=+2\frac{Q_*}{r^2}F_{23},\\
&\delta R_{12}=+2\frac{Q_*}{r^2}F_{01},\quad \delta R_{23}=+2\frac{Q_*}{r^2}F_{03},\\
&\text{and}\qquad \delta R_{02}=\delta R_{13}=0.
\end{aligned}\right\} \tag{127}$$

(c) *Axial perturbations*

The perturbation equations, which follow from the linearization of Einstein's equations about the Reissner–Nordström solution, can be obtained exactly as in the treatment of the Schwarzschild perturbations in Chapter 4, §24. In particular, the axial perturbations can be treated independently of the polar perturbations.

As in the context of the Schwarzschild solution, the axial perturbations of the Reissner–Nordström solution are characterized by the non-vanishing of ω, q_2, and q_3; and in place of Chapter 4, equations (11) and (12), we now have

$$(r^2 e^{2\nu} Q_{23} \sin^3\theta)_{,3} + r^4 Q_{02,0} \sin^3\theta = 2(r^3 e^{\nu} \sin^2\theta)\,\delta R_{12}$$
$$= 4Q_* r e^{\nu} F_{01} \sin^2\theta, \qquad (128)$$

and

$$(r^2 e^{2\nu} Q_{23} \sin^3\theta)_{,2} - r^2 e^{-2\nu} Q_{03,0} \sin^3\theta = -2(r^2 \sin^2\theta)\,\delta R_{13} = 0, \qquad (129)$$

where we have substituted for δR_{12} and δR_{13} from equations (127).

Equations (128) and (129) must be supplemented by equations (120)–(122). These latter equations can be reduced to a single equation for F_{01} by eliminating F_{12} and F_{13} from equation (122) with the aid of equations (120) and (121). We thus obtain

$$[e^{2\nu}(re^{\nu}B)_{,r}]_{,r} + \frac{e^{\nu}}{r}\left(\frac{B_{,\theta}}{\sin\theta}\right)_{,\theta} \sin\theta - re^{-\nu}B_{,0,0}$$
$$= Q_*(\omega_{,2,0} - q_{2,0,0})\sin^2\theta, \qquad (130)$$

where

$$B = F_{01}\sin\theta. \qquad (131)$$

With the substitution (cf. Ch. 4, equations (13)–(15)),

$$Q(r,\theta,t) = r^2 e^{2\nu} Q_{23} \sin^3\theta = \Delta(q_{2,3} - q_{3,2})\sin^3\theta, \qquad (132)$$

equations (128) and (129) take the forms

$$\frac{1}{r^4 \sin^3\theta}\frac{\partial Q}{\partial\theta} = -(\omega_{,2} - q_{2,0})_{,0} + \frac{4Q_*}{r^3 \sin^2\theta} e^{\nu} B \qquad (133)$$

and

$$\frac{\Delta}{r^4 \sin^3\theta}\frac{\partial Q}{\partial r} = +(\omega_{,3} - q_{3,0})_{,0}. \qquad (134)$$

Eliminating ω from these equations and assuming (as we always do) that the perturbations have a time-dependence $e^{i\sigma t}$, we obtain the equations (cf. Ch. 4, equation (18))

$$r^4\frac{\partial}{\partial r}\left(\frac{\Delta}{r^4}\frac{\partial Q}{\partial r}\right)+\sin^3\theta\frac{\partial}{\partial\theta}\left(\frac{1}{\sin^3\theta}\frac{\partial Q}{\partial\theta}\right)+\sigma^2\frac{r^4}{\Delta}Q = 4Q_* e^{\nu}r\left(\frac{B}{\sin^2\theta}\right)_{,\theta}\sin^3\theta. \quad (135)$$

Similarly, eliminating $(\omega_{,2}-q_{2,0})_{,0}$ from equation (130) with the aid of equation (133), we obtain the equation

$$[e^{2\nu}(re^{\nu}B)_{,r}]_{,r}+\frac{e^{\nu}}{r}\left(\frac{B_{,\theta}}{\sin\theta}\right)_{,\theta}\sin\theta+\left(\sigma^2re^{-\nu}-4\frac{Q_*^2}{r^3}e^{\nu}\right)B = -Q_*\frac{Q_{,\theta}}{r^4\sin\theta}. \quad (136)$$

The variables r and θ in equations (135) and (136) can be separated by the substitutions (cf. Ch. 4, equations (19) and (20))

$$Q(r,\theta) = Q(r)C_{l+2}^{-3/2}(\theta), \quad (137)$$

and

$$B(r,\theta) = \frac{B(r)}{\sin\theta}\frac{dC_{l+2}^{-3/2}}{d\theta} = 3B(r)C_{l+1}^{-1/2}(\theta), \quad (138)$$

where, in writing the alternative form for $B(r,\theta)$, we have made use of the recurrence relation,

$$\frac{1}{\sin\theta}\frac{dC_n^{\nu}}{d\theta} = -2\nu C_{n-1}^{\nu+1}. \quad (139)$$

By making use of the equations satisfied by the Gegenbauer functions $C_{l+2}^{-3/2}$ and $C_{l+1}^{-1/2}$, we find that the substitutions (137) and (138) in equations (135) and (136) yield the radial equations (cf. Ch. 4, equation (23))

$$\Delta\frac{d}{dr}\left(\frac{\Delta}{r^4}\frac{dQ}{dr}\right)-\mu^2\frac{\Delta}{r^4}Q+\sigma^2Q = -\frac{4Q_*\mu^2}{r^3}\Delta e^{\nu}B \quad (140)$$

and

$$[e^{2\nu}(re^{\nu}B)_{,r}]_{,r}-(\mu^2+2)\frac{e^{\nu}}{r}B+\left(\sigma^2re^{-\nu}-\frac{4Q_*^2}{r^3}e^{\nu}\right)B = -Q_*\frac{Q}{r^4}, \quad (141)$$

where

$$\mu^2 = 2n = (l-1)(l+2). \quad (142)$$

Changing to the variable r_* (defined in equation (42)) and further letting

$$Q(r) = rH_2^{(-)} \qquad \text{and} \qquad re^{\nu}B = -\frac{H_1^{(-)}}{2\mu}, \quad (143)$$

we find that $H_2^{(-)}$ and $H_1^{(-)}$ satisfy the pair of coupled equations,

$$\Lambda^2 H_2^{(-)} = \frac{\Delta}{r^5}\left\{\left[(\mu^2+2)r - 3M + 4\frac{Q_*^2}{r}\right]H_2^{(-)} - 3MH_2^{(-)} + 2Q_*\mu H_1^{(-)}\right\}, \tag{144}$$

$$\Lambda^2 H_1^{(-)} = \frac{\Delta}{r^5}\left\{\left[(\mu^2+2)r - 3M + 4\frac{Q_*^2}{r}\right]H_1^{(-)} + 2Q_*\mu H_2^{(-)} + 3MH_1^{(-)}\right\}, \tag{145}$$

where, in accordance with our standard usage,

$$\Lambda^2 = \frac{d^2}{dr_*^2} + \sigma^2. \tag{146}$$

Equations (144) and (145) can be decoupled by the substitutions

$$Z_1^{(-)} = +q_1 H_1^{(-)} + (-q_1 q_2)^{1/2} H_2^{(-)} \tag{147}$$

and

$$Z_2^{(-)} = -(-q_1 q_2)^{1/2} H_1^{(-)} + q_1 H_2^{(-)}, \tag{148}$$

where

$$q_1 = 3M + \surd(9M^2 + 4Q_*^2\mu^2) \quad \text{and} \quad q_2 = 3M - \surd(9M^2 + 4Q_*^2\mu^2). \tag{149}$$

We find that $Z_1^{(-)}$ and $Z_2^{(-)}$ satisfy the one-dimensional Schrödinger-type wave-equations,

$$\Lambda^2 Z_i^{(-)} = V_i^{(-)} Z_i^{(-)} \quad (i = 1, 2), \tag{150}$$

where

$$V_i^{(-)} = \frac{\Delta}{r^5}\left[(\mu^2+2)r - q_j\left(1 + \frac{q_i}{\mu^2 r}\right)\right] \quad (i, j = 1, 2; i \neq j) \tag{151}$$

$$q_1 + q_2 = 6M, \qquad \text{and} \qquad -q_1 q_2 = 4Q_*^2\mu^2. \tag{152}$$

The reduction of the equations governing the axial perturbations to the pair of one-dimensional wave equations (150) was first accomplished by Moncrief and by Zerilli (though by very different procedures).

It will be observed that in the limit $Q_* = 0$, when $q_1 = 6M$ and $q_2 = 0$, the equation governing $Z_2^{(-)}$ reduces to the Regge–Wheeler equation (Ch. 4, equations (27) and (28)).

(*d*) *Polar perturbations*

Polar perturbations are characterized by non-vanishing increments in the metric functions ν, μ_2, μ_3, and ψ. Linearizing the expressions for R_{02}, R_{03}, R_{23}, R_{11}, and G_{22} given in Chapter 4 (equations (4)) and substituting from equations (127) for the perturbations in these quantities, we find that, in place of Ch. 4, equations (31)–(35), we now have

$$(\delta\psi+\delta\mu_3)_{,r}+\left(\frac{1}{r}-\nu_{,r}\right)(\delta\psi+\delta\mu_3)-\frac{2}{r}\delta\mu_2=-\delta R_{02}=0, \tag{153}$$

$$[(\delta\psi+\delta\mu_2)_{,\theta}+(\delta\psi-\delta\mu_3)\cot\theta]_{,0}=-e^{\nu+\mu_3}\delta R_{03}=-2Q_*\frac{e^{\nu}}{r}F_{23}, \tag{154}$$

$$\begin{aligned}(\delta\psi+\delta\nu)_{,r,\theta}+(\delta\psi-\delta\mu_3)_{,r}\cot\theta+\left(\nu_{,r}-\frac{1}{r}\right)\delta\nu_{,\theta}&\\ -\left(\nu_{,r}+\frac{1}{r}\right)\delta\mu_{2,\theta}=-e^{\mu_2+\mu_3}\delta R_{23}&=-2Q_*\frac{e^{-\nu}}{r}F_{03},\end{aligned} \tag{155}$$

$$\begin{aligned}e^{2\nu}\left[\frac{2}{r}\delta\nu_{,r}+\left(\frac{1}{r}+\nu_{,r}\right)(\delta\psi+\delta\mu_3)_{,r}-2\left(\frac{1}{r^2}+2\frac{\nu_{,r}}{r}\right)\delta\mu_2\right]&\\ +\frac{1}{r^2}[(\delta\psi+\delta\nu)_{,\theta,\theta}+(2\delta\psi+\delta\nu-\delta\mu_3)_{,\theta}\cot\theta+2\delta\mu_3]&\\ -e^{-2\nu}(\delta\psi+\delta\mu_3)_{,0,0}=\delta G_{22}=\delta R_{22}&=2\frac{Q_*}{r^2}\delta F_{02}\end{aligned} \tag{156}$$

and

$$\begin{aligned}e^{2\nu}\left[\delta\psi_{,r,r}+2\left(\frac{1}{r}+\nu_{,r}\right)\delta\psi_{,r}+\frac{1}{r}(\delta\psi+\delta\nu+\delta\mu_3-\delta\mu_2)_{,r}-2\left(\frac{1}{r}+2\nu_{,r}\right)\frac{\delta\mu_2}{r}\right]&\\ +\frac{1}{r^2}[\delta\psi_{,\theta,\theta}+\delta\psi_{,\theta}\cot\theta+(\delta\psi+\delta\nu-\delta\mu_3+\delta\mu_2)_{,\theta}\cot\theta+2\delta\mu_3]&\\ -e^{-2\nu}\delta\psi_{,0,0}=-\delta R_{11}&=2\frac{Q_*}{r^2}\delta F_{02}.\end{aligned} \tag{157}$$

These equations must be considered along with equations (123)–(125) obtained from the linearization of Maxwell's equations.

The variables r and θ in equations (123)–(125) and (153)–(157) can be separated by the substitutions

$$\left.\begin{aligned}\delta\nu=N(r)P_l(\theta);\quad \delta\mu_2&=L(r)P_l(\theta);\\ \delta\mu_3=[T(r)P_l&+V(r)P_{l,\theta,\theta}],\\ \delta\psi=[T(r)P_l&+V(r)P_{l,\theta}\cot\theta],\end{aligned}\right\} \tag{158}$$

$$\delta F_{02}=\frac{r^2e^{2\nu}}{2Q_*}B_{02}(r)P_l,\quad F_{03}=-\frac{re^{\nu}}{2Q_*}B_{03}(r)P_{l,\theta}, \tag{159}$$

and

$$F_{23}=-i\sigma\frac{re^{-\nu}}{2Q_*}B_{23}(r)P_{l,\theta}, \tag{160}$$

if proper use is made of the relations in Chapter 4, equations (40)–(42). We thus obtain the following equations for the various radial functions we have introduced:

$$\left[\frac{\mathrm{d}}{\mathrm{d}r}+\left(\frac{1}{r}-\nu_{,r}\right)\right][2T-l(l+1)V]-\frac{2}{r}L=0, \tag{161}$$

$$(T-V+L)=B_{23}, \tag{162}$$

$$(T-V+N)_{,r}-\left(\frac{1}{r}-\nu_{,r}\right)N-\left(\frac{1}{r}+\nu_{,r}\right)L=B_{03}, \tag{163}$$

$$\frac{2}{r}N_{,r}+\left(\frac{1}{r}+\nu_{,r}\right)[2T-l(l+1)V]-\frac{2}{r}\left(\frac{1}{r}+2\nu_{,r}\right)L$$
$$-\frac{l(l+1)}{r^2}e^{-2\nu}N-\frac{2n}{r^2}e^{-2\nu}T+\sigma^2e^{-4\nu}[2T-l(l+1)V]=B_{02} \tag{164}$$

$$B_{03}=\frac{1}{r^2}(r^2B_{23})_{,r}=B_{23,r}+\frac{2}{r}B_{23}, \tag{165}$$

$$r^4e^{2\nu}B_{02}=2Q_*^2[2T-l(l+1)V]-l(l+1)r^2B_{23}, \tag{166}$$

and

$$(r^2e^{2\nu}B_{03})_{,r}+r^2e^{2\nu}B_{02}+\sigma^2r^2e^{-2\nu}B_{23}=2Q_*^2\frac{N+L}{r^2}. \tag{167}$$

We observe that two of these equations (namely, equations (162) and (166)) are algebraic.

We shall find it convenient to write

$$X=nV=\tfrac{1}{2}(l-1)(l+2)V, \tag{168}$$

in which case (by virtue of equation (162))

$$2T-l(l+1)V=-2(L+X-B_{23}). \tag{169}$$

By making use of this last relation, we can rewrite equation (161) in the form

$$(L+X-B_{23})_{,r}=-\left(\frac{1}{r}-\nu_{,r}\right)(L+X-B_{23})-\frac{1}{r}L. \tag{170}$$

Similarly, by combining equations (162), (163) and (165)

$$N_{,r}-L_{,r}=\left(\frac{1}{r}-\nu_{,r}\right)N+\left(\frac{1}{r}+\nu_{,r}\right)L+\frac{2}{r}B_{23}. \tag{171}$$

Finally, we may note that by separating the variables in equation (157) by the substitutions (158), we are led to the equation

$$V_{,r,r}+2\left(\frac{1}{r}+\nu_{,r}\right)V_{,r}+\frac{e^{-2\nu}}{r^2}(N+L)+\sigma^2e^{-4\nu}V=0. \tag{172}$$

We observe that this is the same equation that we obtained in the context of the Schwarzschild perturbations (Ch. 4, equation (51)).

Equations (164), (170), and (171) provide three linear first-order equations for the three functions L, N, and $V(=X/n)$. By suitably combining these equations, we can express each of them as linear combinations of L, N, V, B_{23}, and B_{03}. Thus, (cf. Ch. 4, equations (52)–(55))

$$N_{,r} = aN + bL + c(X - B_{23}), \tag{173}$$

$$L_{,r} = \left(a - \frac{1}{r} + \nu_{,r}\right)N + \left(b - \frac{1}{r} - \nu_{,r}\right)L + c(X - B_{23}) - \frac{2}{r}B_{23}, \tag{174}$$

$$X_{,r} = -\left(a - \frac{1}{r} + \nu_{,r}\right)N - \left(b + \frac{1}{r} - 2\nu_{,r}\right)L - \left(c + \frac{1}{r} - \nu_{,r}\right)(X - B_{23}) + B_{03}, \tag{175}$$

where

$$a = \frac{n+1}{r}e^{-2\nu}, \tag{176}$$

$$\begin{aligned} b &= -\frac{1}{r} - \frac{n}{r}e^{-2\nu} + \nu_{,r} + r\nu_{,r}^2 + \sigma^2 r e^{-4\nu} - 2\frac{Q_*^2}{r^3}e^{-2\nu} \\ &= -\frac{1}{r} - \frac{n}{r}e^{-2\nu} + \frac{M}{r^2}e^{-2\nu} + \frac{M^2}{r^3}e^{-4\nu} + \sigma^2 r e^{-4\nu} - \frac{Q_*^2}{r^3}(1 + 2e^{2\nu})e^{-4\nu}, \end{aligned} \tag{177}$$

$$\begin{aligned} c &= -\frac{1}{r} + \frac{e^{-2\nu}}{r} + r\nu_{,r}^2 + \sigma^2 r e^{-4\nu} - 2\frac{Q_*^2}{r^3}e^{-2\nu} \\ &= -\frac{1}{r} + \frac{e^{-2\nu}}{r} + \frac{M^2}{r^3}e^{-4\nu} + \sigma^2 r e^{-4\nu} - \frac{Q_*^2}{r^3}(1 + e^{2\nu})e^{-4\nu}. \end{aligned} \tag{178}$$

The following alternative form of equation (175) may be noted:

$$\begin{aligned} X_{,r} = &-\frac{1}{r^2}\left(nr + 3M - 2\frac{Q_*^2}{r}\right)e^{-2\nu}N \\ &+ \left[-\frac{1}{r}e^{-2\nu} + \frac{M}{r^2}e^{-2\nu} - \frac{1}{r^3}(M^2 - Q_*^2)e^{-4\nu} - \sigma^2 r e^{-4\nu}\right](L + X - B_{23}) \\ &+ aL + B_{03}. \end{aligned} \tag{179}$$

It is a remarkable fact that the system of equations of order five, represented by equations (165)–(166) and (173)–(175), can be reduced to two independent equations of the second order. Thus, it can be directly verified that the functions $H_2^{(+)}$ and $H_1^{(+)}$, defined in the manner (cf. Ch. 4, equation (59))

$$H_2^{(+)} = \frac{r}{n}X - \frac{r^2}{\varpi}(L + X - B_{23}), \tag{180}$$

and

$$H_1^{(+)} = -\frac{1}{Q_*\mu}\left[r^2 B_{23} + 2Q_*^2\frac{r}{\varpi}(L+X-B_{23})\right]$$

$$= -\frac{1}{Q_*\mu}\left[r^2 B_{23} + 2\frac{Q_*^2}{r}\left(\frac{r}{n}X - H_2^{(+)}\right)\right], \tag{181}$$

where

$$\varpi = nr + 3M - 2Q_*^2/r, \tag{182}$$

satisfy the pair of equations (cf. equations (144) and (145)),

$$\Lambda^2 H_2^{(+)} = \frac{\Delta}{r^5}[UH_2^{(+)} + W(-3MH_2^{(+)} + 2Q_*\mu H_1^{(+)})], \tag{183}$$

$$\Lambda^2 H_1^{(+)} = \frac{\Delta}{r^5}[UH_1^{(+)} + W(+2Q_*\mu H_2^{(+)} + 3MH_1^{(+)})], \tag{184}$$

where

$$U = (2nr+3M)W + (\varpi - nr - M) - \frac{2n\Delta}{\varpi}$$

and

$$W = \frac{\Delta}{r\varpi^2}(2nr+3M) + \frac{1}{\varpi}(nr+M). \tag{185}$$

Equations (183) and (184) can be decoupled by the same substitutions (147) and (148) that were used to decouple equations (144) and (145) obtained in the context of the axial perturbations. Thus, the functions

$$Z_1^{(+)} = +q_1 H_1^{(+)} + (-q_1 q_2)^{1/2} H_2^{(+)} \tag{186}$$

and

$$Z_2^{(+)} = -(-q_1 q_2)^{1/2} H_1^{(+)} + q_1 H_2^{(+)}, \tag{187}$$

with the same definitions of q_1 and q_2, satisfy the one-dimensional wave-equations

$$\Lambda^2 Z_i^{(+)} = V_i^{(+)} Z_i^{(+)} \quad (i = 1, 2), \tag{188}$$

where

$$V_1^{(+)} = \frac{\Delta}{r^5}[U + \tfrac{1}{2}(q_1 - q_2)W] \quad \text{and} \quad V_2^{(+)} = \frac{\Delta}{r^5}[U - \tfrac{1}{2}(q_1 - q_2)W]. \tag{189}$$

These decoupled equations were first derived by Moncrief and Zerilli (though by very different methods).

Also, it can be verified that in the limit $Q_* = 0$, the equation governing $Z_2^{(+)}$ reduces to Zerilli's equation (Ch. 4, equations (62) and (63)).

(i) The completion of the solution

From our discussion in Chapter 4 (§25) of the reduction of the equations governing the polar perturbations of the Schwarzschild black-hole to Zerilli's equation, it is clear that the present reducibility of a system of equations of order five to the pair of equations (183) and (184) must be the result of the system allowing a special solution. By applying an extension of the algorism described in Chapter 4, §25, to the more general case, when the reduction of a linear system of differential equations to more than one second-order equation has been accomplished, Xanthopoulos discovered that the present system of equations (165), (167), and (173)–(175) allows the special solution (cf. Ch. 4, equation (126))

$$\left.\begin{aligned} N^{(0)} &= r^{-2}e^{\nu}\left[M-\frac{r}{\Delta}(M^2-Q_*^2+\sigma^2r^4)-2\frac{Q_*^2}{r}\right], \\ L^{(0)} &= r^{-3}e^{\nu}(3Mr-4Q_*^2),\ X^{(0)} = ne^{\nu}r^{-1}, \\ B_{23}^{(0)} &= -2Q_*^2r^{-3}e^{\nu}, \quad \text{and} \quad B_{03}^{(0)} = 2Q_*^2r^{-6}e^{-\nu}(2Q_*^2+r^2-3Mr). \end{aligned}\right\} \tag{190}$$

The completion of the solution for the remaining radial functions with the aid of the special solution (190) is relatively straightforward. Xanthopoulos finds

$$\begin{aligned} N = N^{(0)}\Phi + 2n\frac{e^{2\nu}}{\varpi}H_2^{(+)} - \frac{e^{2\nu}}{\varpi}(nrH_2^{(+)}+Q_*\mu H_1^{(+)})_{,r} \\ + \frac{1}{r\varpi^2}[e^{2\nu}(\varpi-2nr-3M)-(n+1)\varpi]\,(nrH_2^{(+)}+Q_*\mu H_1^{(+)}), \end{aligned} \tag{191}$$

$$L = L^{(0)}\Phi - \frac{1}{r^2}(nrH_2^{(+)}+Q_*\mu H_1^{(+)}), \tag{192}$$

$$X = X^{(0)}\Phi + \frac{n}{r}H_2^{(+)}, \tag{193}$$

$$B_{23} = B_{23}^{(0)}\Phi - \frac{Q_*\mu}{r^2}H_1^{(+)}, \tag{194}$$

$$B_{03} = B_{03}^{(0)}\Phi - \frac{Q_*\mu}{r^2}H_{1,r}^{(+)} - 2\frac{Q_*^2}{r^4\varpi}(nrH_2^{(+)}+Q_*\mu H_1^{(+)}), \tag{195}$$

where

$$\Phi = \int (nrH_2^{(+)}+Q_*\mu H_1^{(+)})\frac{e^{-\nu}}{\varpi r}dr. \tag{196}$$

It may be noted here that in the limit $Q_* = 0$, Φ as defined in equation (196) differs from Φ as defined in Chapter 4, equation (70) by a factor $e^{-\nu}$.

43. The relations between $V_i^{(+)}$ and $V_i^{(-)}$ and $Z_i^{(+)}$ and $Z_i^{(-)}$

As in the case of the Schwarzschild perturbations, the potentials $V_i^{(\pm)}$ $(i = 1, 2)$, associated with the polar and the axial perturbations, are related in a

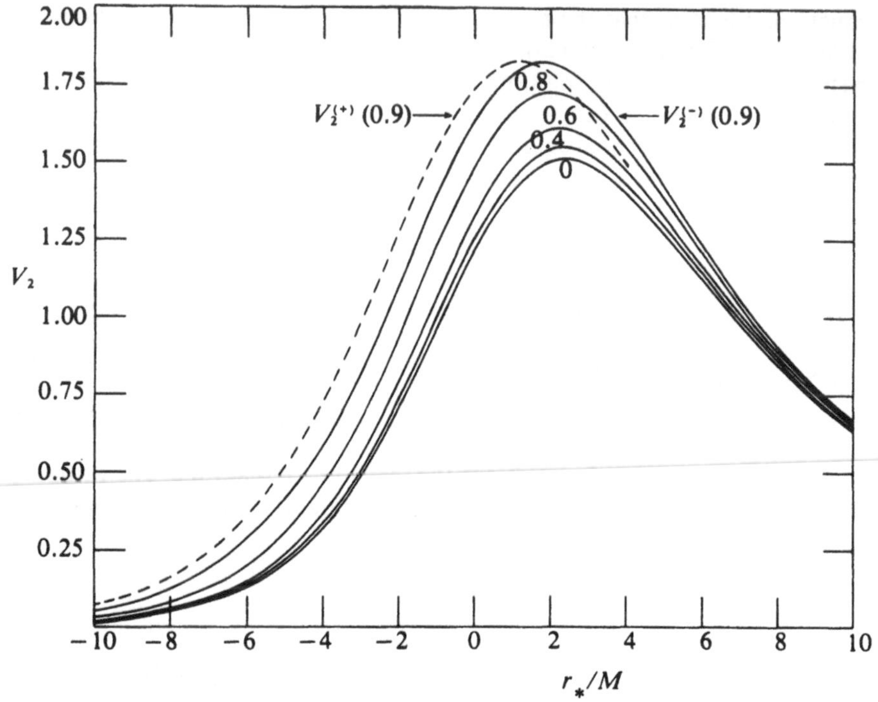

FIG. 19. The potential barriers $V_2^{(\pm)}$ surrounding the Reissner–Nordström black-hole: the full-line curves are appropriate for axial perturbations for $Q_* = 0, 0.4, 0.6, 0.8$, and 0.9. The curves are labelled by the values of Q_* to which they belong. The potential barriers for the polar perturbations differ very little from those for the axial perturbations: cf. the neighbouring full-line and dashed-line curves belonging to the axial and polar perturbations for $Q_* = 0.9$.

manner which guarantees the equality of the reflexion and the transmission coefficients determined by the equations governing $Z_i^{(\pm)}$. Thus, it can be verified (and as we shall later deduce in §45), the potentials are, in fact, given by

$$V_i^{(\pm)} = \pm\beta_i \frac{\mathrm{d}f_i}{\mathrm{d}r_*} + \beta_i^2 f_i^2 + \kappa f_i, \tag{197}$$

where

$$\kappa = \mu^2(\mu^2+2),\ \beta_i = q_j, \quad \text{and} \quad f_i = \frac{\Delta}{r^3(\mu^2 r + q_j)} \quad (i, j = 1, 2; i \neq j), \tag{198}$$

and the q_j's have the same meanings as hitherto. The solutions, $Z_i^{(+)}$ and $Z_i^{(-)}$ of the respective equations are, therefore, related in the manner (cf. Ch. 4,

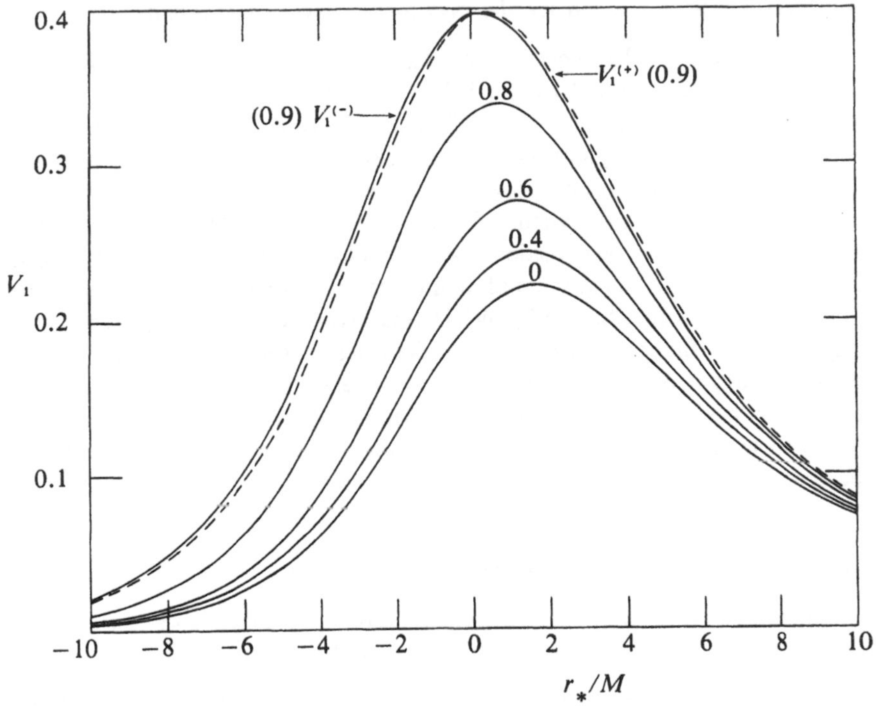

FIG. 20. The potential barriers $V_1^{(\pm)}$surrounding the Reissner–Nordström black-hole: the full-line curves are appropriate for axial perturbations for $Q_* = 0, 0.4, 0.6, 0.8$, and 0.9. The curves are labelled by the values of Q_* to which they belong. The potential barriers for the polar perturbations cannot be distinguished from those for the axial perturbations in the scale of the diagrams: cf. the neighbouring full-line and dashed-line curves belonging to the axial and polar perturbations for $Q_* = 0.9$.

equations (152) and (153))

$$[\mu^2(\mu^2+2)\pm 2i\sigma q_j]Z_i^{(\pm)} = \left[\mu^2(\mu^2+2)+\frac{2q_j^2\Delta}{r^3(\mu^2 r+q_j)}\right]Z_i^{(\mp)} \pm 2q_j\frac{\mathrm{d}Z_i^{(\mp)}}{\mathrm{d}r_*} \quad (i,j=1,2;\, i\neq j). \tag{199}$$

It is the existence of this relation which guarantees the equality of the reflexion and the transmission coefficients determined by the wave equations governing $Z_i^{(+)}$ and $Z_i^{(-)}$. We shall return in §47 to a consideration of the implications of the solutions of these equations to the problem of reflexion and transmission of gravitational and electromagnetic waves simultaneously incident on the Reissner–Nordström black-hole. Meantime, we exhibit in Figs. 19 and 20 the potentials $V_i^{(\pm)}$ for a range of values of Q_*.

44. Perturbations treated via the Newman–Penrose formalism

The solution for the Weyl scalars, Ψ_0 and Ψ_4, via the Newman–Penrose formalism, in the context of the Schwarzschild perturbations, was readily possible only because four of the Bianchi identities and two of the Ricci identities (listed in Chapter 4, equations (232) and (233)) are already linearized in the sense that they are homogeneous in the quantities which vanish in the background geometry (by virtue of its type-D character). This fact meant that these equations could be considered without a knowledge of the perturbations in the basis vectors—a problem of an altogether different order of magnitude (as we shall find in Chapter 9). While this favoured situation persists with regard to the Bianchi and the Ricci identities in the Reissner–Nordström geometry (again, by virtue of its type-D character) it is not present in Maxwell's equations (see equations (200)–(203) below)—they include terms in the Maxwell scalar ϕ_1, which do not vanish in the background. But equations can be derived which are already linearized in the same sense that the Bianchi and the Ricci identities are. We begin by deriving these equations.

(a) *Maxwell's equations which are already linearized*

Equations governing the Maxwell scalars, ϕ_0, ϕ_1, and ϕ_2 in the Newman–Penrose formalism are (Ch. 1, equations (330)–(333))

$$(D-2\rho)\phi_1-(\delta^*+\pi-2\alpha)\phi_0+\kappa\phi_2=0, \tag{200}$$

$$(\delta-2\tau)\phi_1-(\triangle+\mu-2\gamma)\phi_0+\sigma\phi_2=0, \tag{201}$$

$$(\delta^*+2\pi)\phi_1-(D-\rho+2\varepsilon)\phi_2-\lambda\phi_0=0, \tag{202}$$

$$(\triangle+2\mu)\phi_1-(\delta-\tau+2\beta)\phi_2-\nu\phi_0=0. \tag{203}$$

Since $\phi_1(=Q_*/2r^2)$ is non-vanishing in the background geometry, the linearized version of these equations will involve the perturbations in the basis vectors (acting as directional derivatives on ϕ_1) as well as of the spin coefficients ρ, τ, π, and μ (which are also non-vanishing in the background). To rectify this situation we proceed as follows.

Applying the operator $(\delta-2\tau-\alpha^*-\beta+\pi^*)$ to equation (200) and the operator $(D-\varepsilon+\varepsilon^*-2\rho-\rho^*)$ to equation (201) and subtracting one from the other of the resulting equations, we obtain

$$\begin{aligned}
&[(\delta-2\tau-\alpha^*-\beta+\pi^*)(\delta^*+\pi-2\alpha)\\
&-(D-\varepsilon+\varepsilon^*-2\rho-\rho^*)(\triangle+\mu-2\gamma)]\phi_0\\
&\qquad=[(\delta-2\tau-\alpha^*-\beta+\pi^*)\kappa-(D-\varepsilon+\varepsilon^*-2\rho-\rho^*)\sigma]\phi_2\\
&\qquad\quad+\kappa\delta\phi_2-\sigma D\phi_2+(\delta D-D\delta)\phi_1-(2\delta\rho)\phi_1+(2D\tau)\phi_1\\
&\qquad\quad-(\alpha^*+\beta-\pi^*)D\phi_1+2\rho(\alpha^*+\beta-\pi^*)\phi_1\\
&\qquad\quad-(-\varepsilon+\varepsilon^*-\rho^*)\delta\phi_1+2\tau(-\varepsilon+\varepsilon^*-\rho^*)\phi_1.
\end{aligned} \tag{204}$$

On the right-hand side of this equation, we replace the commutator $(\delta D - D\delta)$ operating on ϕ_1, by its value given in Chapter 1, equation (304) and also substitute for $D\tau$ and $\delta\rho$ from the Ricci identities (Ch. 1, equation (310), (c) and (k), respectively). Considerable simplification results and we are left with

$$\begin{aligned}
&[(\delta - 2\tau - \alpha^* - \beta + \pi^*)(\delta^* + \pi - 2\alpha) \\
&\quad - (D - \varepsilon + \varepsilon^* - 2\rho - \rho^*)(\triangle + \mu - 2\gamma)]\phi_0 \\
&= [(\delta - 2\tau - \alpha^* - \beta + \pi^*)\kappa - (D - \varepsilon + \varepsilon^* - 2\rho - \rho^*)\sigma]\phi_2 + \kappa\delta\phi_2 - \sigma D\phi_2 \\
&\quad + 2\phi_1[(\triangle - 3\gamma - \gamma^* - \mu + \mu^*)\kappa - (\delta^* - 3\alpha + \beta^* - \tau^* - \pi)\sigma] \\
&\quad + 4\phi_1\Psi_1 + \kappa\triangle\phi_1 - \sigma\delta^*\phi_1. \qquad (205)
\end{aligned}$$

Similarly, by the application of the operators $-(\triangle + \mu^* - \gamma^* + \gamma + 2\mu)$ and $+(\delta^* - \tau^* + \alpha + \beta^* + 2\pi)$ on equations (202) and (203), addition, and subsequent simplification in like manner, we obtain the further equation

$$\begin{aligned}
&[(\delta^* - \tau^* + \alpha + \beta^* + 2\pi)(\delta - \tau + 2\beta) \\
&\quad - (\triangle + \mu^* - \gamma^* + \gamma + 2\mu)(D - \rho + 2\varepsilon)]\phi_2 \\
&= [(\triangle + \mu^* - \gamma^* + \gamma + 2\mu)\lambda - (\delta^* - \tau^* + \alpha + \beta^* + 2\pi)\nu]\phi_0 \\
&\quad + \lambda\triangle\phi_0 - \nu\delta^*\phi_0 \\
&\quad + 2\phi_1[-(D + 3\varepsilon + \varepsilon^* + \rho - \rho^*)\nu + (\delta + \pi^* + \tau - \alpha^* + 3\beta)\lambda] \\
&\quad + 4\phi_1\Psi_3 - \nu D\phi_1 + \lambda\delta\phi_1. \qquad (206)
\end{aligned}$$

It will be observed that equations (205) and (206) are already 'linearized' in the sense we have been using that term. Indeed, the terms in ϕ_2 in equation (205) and in ϕ_0 in equation (206) are quantities of the second order of smallness and can be ignored in a linear perturbation theory. Also, the terms $\triangle\phi_1$ and $\delta^*\phi_1$ (in equation (205)) and $D\phi_1$ and $\delta\phi_1$ (in equation (206)) can be replaced, respectively, by $-2\mu\phi_1$, $-2\pi\phi_1$, $2\rho\phi_1$, and $2\tau\phi_1$ as required by equations (200)–(203) for the background. We thus obtain the following equations appropriate for a linear perturbation theory:

$$\begin{aligned}
&[(\delta - 2\tau - \alpha^* - \beta + \pi^*)(\delta^* + \pi - 2\alpha) \\
&\quad - (D - \varepsilon + \varepsilon^* - 2\rho - \rho^*)(\triangle + \mu - 2\gamma)]\phi_0 \\
&= 2\phi_1[(\triangle - 3\gamma - \gamma^* - 2\mu + \mu^*)\kappa - (\delta^* - 3\alpha + \beta^* - \tau^* - 2\pi)\sigma + 2\Psi_1] \qquad (207)
\end{aligned}$$

and

$$\begin{aligned}
&[(\delta^* - \tau^* + \alpha + \beta^* + 2\pi)(\delta - \tau + 2\beta) \\
&\quad - (\triangle + \mu^* - \gamma^* + \gamma + 2\mu)(D - \rho + 2\varepsilon)]\phi_2 \\
&= 2\phi_1[(\delta + \pi^* + 2\tau - \alpha^* + 3\beta)\lambda - (D + 3\varepsilon + \varepsilon^* + 2\rho - \rho^*)\nu + 2\Psi_3]. \qquad (208)
\end{aligned}$$

(b) The 'phantom' gauge

It is clear that it would not be a simple matter to consider equations (207) and (208), as they stand, along with the other equations derived from the Bianchi and the Ricci identities. A gauge that would manifestly simplify the equations considerably is one in which

$$\phi_0 = \phi_2 = 0. \tag{209}$$

For, in this gauge, equations (207) and (208) simply become equations relating the spin coefficients, κ, σ, λ, and ν, with the Weyl scalars Ψ_1 and Ψ_3; thus, we shall have

$$(\Delta - 3\gamma - \gamma^* - 2\mu + \mu^*)\kappa - (\delta^* - 3\alpha + \beta^* - \tau^* - 2\pi)\sigma + 2\Psi_1 = 0 \tag{210}$$

and

$$(\delta + \pi^* + 2\tau - \alpha^* + 3\beta)\lambda - (D + 3\varepsilon + \varepsilon^* + 2\rho - \rho^*)\nu + 2\Psi_3 = 0. \tag{211}$$

It is a remarkable fact that equations (210) and (211) survive even in the limit $\phi_1 = 0$; and in the limit $\phi_1 = 0$, they may be said to describe a '*phantom gauge*'. Indeed, we shall find that it is, in fact, *the* phantom gauge to which we were led in Chapter 4, §29(*b*) (see equations (217) and (218) below).

It follows from our discussion of the effect of tetrad rotations on the various scalars and spin coefficients in Chapter 1 (§8(*g*)) that a gauge in which ϕ_0 and ϕ_2 vanish can be chosen. An infinitesimal rotation of class II, for example, will change the Weyl and the Maxwell scalars, to the first order in *b*, according to the scheme (cf. Ch. 1, equations (345) and (346))

$$\Psi_0 \to \Psi_0 + 4b\Psi_1, \quad \Psi_1 \to \Psi_1 + 3b\Psi_2, \quad \Psi_2 \to \Psi_2 + 2b\Psi_3, \quad \Psi_3 \to \Psi_3 + b\Psi_4,$$
$$\Psi_4 \to \Psi_4; \quad \phi_0 \to \phi_0 + 2b\phi_1, \quad \phi_1 \to \phi_1 + b\phi_2, \quad \text{and} \quad \phi_2 \to \phi_2. \tag{212}$$

Since Ψ_0, Ψ_1, Ψ_3, Ψ_4, ϕ_0, and ϕ_2 vanish in the unperturbed Reissner–Nordström space-time, it is clear that Ψ_0, Ψ_2, Ψ_3, Ψ_4, ϕ_1, and ϕ_2 are unaffected to the first order by the infinitesimal rotation. But Ψ_1 and ϕ_0 are not: they are affected to the first order in *b* since Ψ_2 and ϕ_1 are non-vanishing in the background. It may, however, be noted, parenthetically, that while a gauge, in which Ψ_1 and ϕ_1 vanish simultaneously, cannot be chosen, the combination

$$2\Psi_1\phi_1 - 3\phi_0\Psi_2 \quad \text{is invariant to the first order} \tag{213}$$

for the infinitesimal rotations.

In like manner, a gauge in which ϕ_2 or Ψ_3 is zero can be chosen.

Returning to equations (210) and (211) and substituting for the non-vanishing spin-coefficients their values given in equations (112), we obtain

$$\frac{\Delta}{2r^2}\left(\mathscr{D}_2^\dagger - \frac{5}{r}\right)\kappa + \frac{1}{r\sqrt{2}}\mathscr{L}_2\sigma = 2\Psi_1 \tag{214}$$

and

$$\frac{1}{r\sqrt{2}}\mathscr{L}_2^\dagger\lambda-\left(\mathscr{D}_0-\frac{1}{r}\right)\nu=-2\Psi_3, \tag{215}$$

where the operators $\mathscr{D}_n$, $\mathscr{D}_n^\dagger$, $\mathscr{L}_n$, and $\mathscr{L}_n^\dagger$ have the same meanings as in Chapter 4, equations (228) with a difference only in the definition of Δ. In terms of the variables (cf. Ch. 4, equations (236))

$$\left.\begin{aligned}\kappa=(r^2\sqrt{2})k,\quad \sigma=sr,\quad \lambda=2l/r,\quad \nu=n\sqrt{2}/r^2,\\ \Psi_1=\Phi_1/r\sqrt{2},\qquad \text{and}\qquad \Psi_3=\Phi_3\sqrt{2}/r^3,\end{aligned}\right\} \tag{216}$$

equations (214) and (215) take the simpler forms,

$$\Delta\left(\mathscr{D}_2^\dagger-\frac{3}{r}\right)k+\mathscr{L}_2 s=2\frac{\Phi_1}{r} \tag{217}$$

and

$$\left(\mathscr{D}_0-\frac{3}{r}\right)n-\mathscr{L}_2^\dagger l=2\frac{\Phi_3}{r}. \tag{218}$$

It will be observed that equations (217) and (218) are exactly those that were introduced in Chapter 4 (equations (273) and (276)) to rectify the 'truncated symmetry' of Chapter 4, equations (237)–(242); they represent in fact the phantom gauge.

(c) The basic equations

The Ricci and the Bianchi identities we shall consider are the same as the ones we considered in the context of the Schwarzschild perturbations, namely, Chapter 1, equations (310), (b) and (j) and Chapter 1, equations (321), (a), (e), (d), and (h). The Ricci identities are the same in the present context as well: they are the third in each of the two groups of three equations (Ch. 4, equations (232) and (233)). But in the first pair of equations, which represent the Bianchi identities, we must include the 'Ricci terms' listed in Chapter 1, equation (339). In the chosen gauge, $\phi_0=\phi_2=0$, these additional terms are, respectively,

$$-4\kappa\phi_1\phi_1^*,\ \text{(a)};\quad 4\sigma\phi_1\phi_1^*,\ \text{(e)};\quad -4\lambda\phi_1\phi_1^*,\ \text{(d)};\quad \text{and}\quad 4\nu\phi_1\phi_1^*\ \text{(h)}. \tag{219}$$

By the inclusion of these terms, the right-hand sides of the first pair of equations in Chapter 4, equations (232) and (233), become:

$$\left.\begin{aligned}+\kappa(3\Psi_2-4\phi_1\phi_1^*)&=-\kappa(3Mr-2Q_*^2)r^{-4},\\ +\sigma(3\Psi_2+4\phi_1\phi_1^*)&=-\sigma(3Mr-4Q_*^2)r^{-4},\\ -\lambda(3\Psi_2+4\phi_1\phi_1^*)&=+\lambda(3Mr-4Q_*^2)r^{-4},\\ -\nu(3\Psi_2-4\phi_1\phi_1^*)&=+\nu(3Mr-2Q_*^2)r^{-4},\end{aligned}\right\} \tag{220}$$

on substituting for Ψ_2 and ϕ_1 their values given in equations (114) and (115), namely,

$$\Psi_2 = -(Mr - Q_*^2)r^{-4} \quad \text{and} \quad \phi_1 = Q_*/2r^2. \tag{221}$$

With the terms (220) on the right-hand sides, our present basic set of equations (including equations (217) and (218)) are (cf. Ch. 4, equations (237)–(242))

$$\mathscr{L}_2\Phi_0 - \left(\mathscr{D}_0 + \frac{3}{r}\right)\Phi_1 = -2k\left(3M - 2\frac{Q_*^2}{r}\right), \tag{222}$$

$$\Delta\left(\mathscr{D}_{\frac{1}{2}}^\dagger - \frac{3}{r}\right)\Phi_0 + \mathscr{L}_{-1}^\dagger\Phi_1 = +2s\left(3M - 4\frac{Q_*^2}{r}\right), \tag{223}$$

$$\left(\mathscr{D}_0 + \frac{3}{r}\right)s - \mathscr{L}_{-1}^\dagger k = \frac{\Phi_0}{r}, \tag{224}$$

$$\Delta\left(\mathscr{D}_{\frac{1}{2}}^\dagger - \frac{3}{r}\right)k + \mathscr{L}_2 s = 2\frac{\Phi_1}{r} \tag{225}$$

and

$$\left(\mathscr{D}_0 - \frac{3}{r}\right)\Phi_4 - \mathscr{L}_{-1}\Phi_3 = +2l\left(3M - 4\frac{Q_*^2}{r}\right), \tag{226}$$

$$\mathscr{L}_{\frac{1}{2}}^\dagger\Phi_4 + \Delta\left(\mathscr{D}_{-1}^\dagger + \frac{3}{r}\right)\Phi_3 = 2n\left(3M - 2\frac{Q_*^2}{r}\right) \tag{227}$$

$$\Delta\left(\mathscr{D}_{-1}^\dagger + \frac{3}{r}\right)l + \mathscr{L}_{-1}n = \frac{\Phi_4}{r}, \tag{228}$$

$$\left(\mathscr{D}_0 - \frac{3}{r}\right)n - \mathscr{L}_{\frac{1}{2}}^\dagger l = \frac{2}{r}\Phi_3, \tag{229}$$

where

$$\Phi_0 = \Psi_0 \quad \text{and} \quad \Phi_4 = \Psi_4 r^4, \tag{230}$$

and the remaining variables have already been defined in equation (216).

(*d*) *The separation of the variables and the decoupling and reduction of the equations*

Considering the first group of equations (222)–(225), we find that the variables can be separated by the substitutions (cf. Ch. 4, equations (247), (278), and (283)),

$$\left.\begin{aligned} \Phi_0(r,\theta) &= R_{+2}(r)S_{+2}(\theta); \quad & \Phi_1(r,\theta) &= R_{+1}(r)S_{+1}(\theta); \\ k(r,\theta) &= k(r)S_{+1}(\theta); \quad & s(r,\theta) &= s(r)S_{+2}(\theta), \end{aligned}\right\} \tag{231}$$

where the angular functions $S_{+2}(\theta)$ and $S_{+1}(\theta)$ are the normalized proper solutions of the equations

$$\mathscr{L}^{\dagger}_{-1}\mathscr{L}_2 S_{+2} = -\mu^2 S_{+2} \qquad \text{and} \qquad \mathscr{L}_2\mathscr{L}^{\dagger}_{-1}S_{+1} = -\mu^2 S_{+1}, \tag{232}$$

where

$$\mu^2 = 2n = (l-1)(l+2). \tag{233}$$

Besides, the functions $S_{+2}(\theta)$ and $S_{+1}(\theta)$ are related in the manner (cf. Ch. 4, equations (281) and (282))

$$\mathscr{L}_2 S_{+2} = \mu S_{+1} \qquad \text{and} \qquad \mathscr{L}^{\dagger}_{-1}S_{+1} = -\mu S_{+2}. \tag{234}$$

The substitutions (231) effect a separation of the variables in equations (222)–(225) by virtue of the relations (234) among the angular functions; and we obtain the following coupled system of equations for the radial functions we have defined:

$$\mu R_{+2} - \left(\mathscr{D}_0 + \frac{3}{r}\right)R_{+1} = -2k\left(3M - 2\frac{Q_*^2}{r}\right), \tag{235}$$

$$\Delta\left(\mathscr{D}_2^{\dagger} - \frac{3}{r}\right)R_{+2} - \mu R_{+1} = +2s\left(3M - 4\frac{Q_*^2}{r}\right), \tag{236}$$

$$\left(\mathscr{D}_0 + \frac{3}{r}\right)s + \mu k = \frac{R_{+2}}{r}, \tag{237}$$

$$\Delta\left(\mathscr{D}_2^{\dagger} - \frac{3}{r}\right)k + \mu s = 2\frac{R_{+1}}{r}. \tag{238}$$

The system of equations (235)–(238) can be decoupled to provide a pair of independent equations of the second order by considering the functions

$$\left.\begin{aligned} F_{+1} &= R_{+2} + q_1 k/\mu; \quad G_{+1} = R_{+1} + q_1 s/\mu, \\ F_{+2} &= R_{+2} + q_2 k/\mu; \quad G_{+2} = R_{+1} + q_2 s/\mu, \end{aligned}\right\} \tag{239}$$

where q_1 and q_2 have the same meanings as in §42 (equation (149)). Thus, by adding equation (236) to q_1/μ times equation (238), we obtain

$$\begin{aligned} \Delta\left(\mathscr{D}_2^{\dagger} - \frac{3}{r}\right)F_{+1} &= \mu R_{+1} + 2s\left(3M + \frac{q_1 q_2}{\mu^2 r}\right) - q_1 s + 2\frac{q_1}{\mu r}R_{+1} \\ &= \mu R_{+1} + q_2 s + \frac{2q_1}{\mu r}\left(R_{+1} + \frac{q_2}{\mu}s\right) \\ &= \mu\left(1 + \frac{2q_1}{\mu^2 r}\right)\left(R_{+1} + \frac{q_2}{\mu}s\right) = \mu\left(1 + \frac{2q_1}{\mu^2 r}\right)G_{+2}, \end{aligned} \tag{240}$$

where, in the reductions, we have made use of the relations (152) among q_1 and q_2. The relation obtained by interchanging the indices 1 and 2 in equation (240) is manifestly also true. Both these relations can be expressed in the single equation

$$\Delta\left(\mathscr{D}_2^\dagger-\frac{3}{r}\right)F_{+i}=\mu\left(1+\frac{2q_i}{\mu^2 r}\right)G_{+j}\quad (i,j=1,2;\ i\neq j). \tag{241}$$

The convention that i and j take the values 1 and 2 but i ≠ j will be adhered to, strictly, in the rest of this chapter; it will not be restated on every occasion.

By combining equations (235) and (237) in a similar fashion, we obtain the equation

$$\left(\mathscr{D}_0+\frac{3}{r}\right)G_{+i}=\mu\left(1+\frac{q_i}{\mu^2 r}\right)F_{+j}. \tag{242}$$

Now letting

$$F_{+i}=\frac{r^3}{\Delta^2}Y_{+i}\qquad\text{and}\qquad G_{+i}=\frac{1}{r^3}X_{+i}, \tag{243}$$

we find that equations (241) and (242) become

$$\mathscr{D}_0^\dagger Y_{+i}=\mu\frac{\Delta}{r^6}\left(1+\frac{2q_i}{\mu^2 r}\right)X_{+j} \tag{244}$$

and

$$\mathscr{D}_0 X_{+j}=\mu\frac{r^6}{\Delta^2}\left(1+\frac{q_j}{\mu^2 r}\right)Y_{+i}. \tag{245}$$

Since

$$\frac{\Delta}{r^2}\mathscr{D}_0=\frac{\mathrm{d}}{\mathrm{d}r_*}+i\sigma=\Lambda_+\qquad\text{and}\qquad \frac{\Delta}{r^2}\mathscr{D}_0^\dagger=\frac{\mathrm{d}}{\mathrm{d}r_*}-i\sigma=\Lambda_-, \tag{246}$$

we can rewrite equations (244) and (245) in the forms

$$\Lambda_- Y_{+i}=\mu\frac{\Delta^2}{r^8}\left(1+\frac{2q_i}{\mu^2 r}\right)X_{+j} \tag{247}$$

and

$$\Lambda_+ X_{+j}=\mu\frac{r^4}{\Delta}\left(1+\frac{q_j}{\mu^2 r}\right)Y_{+i}. \tag{248}$$

Finally, eliminating the X's in favour of the Y's, we obtain the pair of basic equations,

$$\Lambda^2 Y_{+i}+P_i\Lambda_- Y_{+i}-Q_i Y_{+i}=0\quad (i=1,2), \tag{249}$$

where

$$P_i=\frac{\mathrm{d}}{\mathrm{d}r_*}\lg\left(\frac{r^8}{D_i}\right),\quad D_i=\Delta^2\left(1+\frac{2q_i}{\mu^2 r}\right), \tag{250}$$

and

$$Q_i=\mu^2\frac{\Delta}{r^4}\left(1+\frac{2q_i}{\mu^2 r}\right)\left(1+\frac{q_j}{\mu^2 r}\right). \tag{251}$$

By a similar sequence of reductions, we obtain from the second group of equations, (226)–(229), the conjugate pair of equations,

$$\Lambda^2 Y_{-i} + P_i \Lambda_+ Y_{-i} - Q_i Y_{-i} = 0, \tag{252}$$

by the substitutions

$$\left.\begin{aligned} \Phi_4(r,\theta) &= R_{-2}(r)S_{-2}(\theta), & \Phi_3(r,\theta) &= R_{-1}(r)S_{-1}(\theta), \\ n(r,\theta) &= n(r)S_{-1}(\theta), & l(r,\theta) &= l(r)S_{-2}(\theta), \end{aligned}\right\} \tag{253}$$

$$F_{-i} = r^3 Y_{-i}, \qquad \text{and} \qquad G_{-i} = \frac{\Delta}{r^3} X_{-i}. \tag{254}$$

The angular functions, $S_{-1}(\theta)$ and $S_{-2}(\theta)$, are now defined by the 'adjoint equations'

$$\mathscr{L}_{-1}\mathscr{L}_2^\dagger S_{-2} = -\mu^2 S_{-2} \qquad \text{and} \qquad \mathscr{L}_2^\dagger \mathscr{L}_{-1} S_{-1} = -\mu^2 S_{-1}; \tag{255}$$

and they are related in the manner

$$\mathscr{L}_2^\dagger S_{-2} = -\mu S_{-1} \qquad \text{and} \qquad \mathscr{L}_{-1} S_{-1} = +\mu S_{-2}. \tag{256}$$

Thus, the Newman–Penrose equations, governing the Weyl and the Maxwell scalars describing the perturbations of the Reissner–Nordström black-hole, have been reduced to the same standard equation that we considered in Chapter 4 in the context of the Schwarzschild black-hole.

45. The transformation theory

The pair of equations, (249) and (252), to which the Newman–Penrose equations were reduced in §44, differ from the equations considered in §30 (Ch. 4, equation (284) and its complex conjugate) only in one inconsequential respect: Δ^2, in the definition of P_i, is replaced by D_i. The transformation theory developed in §30 will, therefore, apply in the present context with only very minor modifications.

The problem to which we now address ourselves is the transformation of equations (249) and (252) to a pair of one-dimensional wave equations of the form

$$\Lambda^2 Z_i = V_i Z_i \tag{257}$$

for some suitably defined potential.

We shall, in the first instance, restrict our consideration to the transformation of equation (249). Also for the sake of convenience, we shall write Y_i in place of Y_{+i}. The analysis appropriate to equation (252) will proceed along exactly the same lines with the only difference that the sign of σ should be changed and, in particular, Λ_-, wherever it occurs, must be replaced by Λ_+, and conversely.

As in §30, we start with a substitution of the form (cf. Ch. 4, equation (287))

$$Y_i = f_i V_i Z_i + (W_i + 2i\sigma f_i)\Lambda_+ Z_i, \tag{258}$$

and obtain the equations (cf. Ch. 4, equations (290)–(292))

$$\Lambda_- Y_i = -\frac{D_i}{r^8}\beta_i Z_i + R_i \Lambda_+ Z_i, \tag{259}$$

$$-\frac{D_i}{r^8}\beta_i = \frac{\mathrm{d}}{\mathrm{d}r_*}(f_i V_i) + W_i V_i, \tag{260}$$

and

$$R_i = f_i V_i + \frac{\mathrm{d}}{\mathrm{d}r_*}(W_i + 2i\sigma f_i). \tag{261}$$

The compatibility of equations (249) and (257), then, leads to the equations (Ch. 4, equations (298)–(300))

$$-\frac{D_i}{r^8}\frac{\mathrm{d}\beta_i}{\mathrm{d}r_*} = (Q_i f_i - R_i)V_i, \tag{262}$$

$$\frac{\mathrm{d}}{\mathrm{d}r_*}\left(\frac{r^8}{D_i}R_i\right) = \frac{r^8}{D_i}[Q_i W_i + 2i\sigma(Q_i f_i - R_i)] + \beta_i, \tag{263}$$

and the integral

$$\frac{r^8}{D_i}R_i f_i V_i + \beta_i(W_i + 2i\sigma f_i) = K_i = \text{constant}. \tag{264}$$

(a) The admissibility of dual transformations

We shall now verify that, as in the case of the Schwarzschild perturbations, equations (260)–(264) are compatible with

$$\beta_i = \text{constant} \qquad \text{and} \qquad f_i = 1, \tag{265}$$

and allow dual transformations with two values for β_i of opposite signs.

As in §30 (Ch. 4, equations (307) and (311)) the condition for existence of dual transformations is that

$$F_i = r^8\frac{Q_i}{D_i} = \frac{r^3}{\Delta}(\mu^2 r + q_j) \quad (i,\ j = 1,\ 2;\ i \neq j) \tag{266}$$

satisfies the differential equation

$$\frac{1}{F_i}\left(\frac{\mathrm{d}F_i}{\mathrm{d}r_*}\right)^2 - \frac{\mathrm{d}^2 F_i}{\mathrm{d}r_*^2} + \frac{D_i}{r^8}F_i^2 = \frac{\beta_i^2}{F_i} + \kappa_i \tag{267}$$

for some suitably chosen constants,

$$\beta_i^2 \qquad \text{and} \qquad \kappa_i = K_i - 2i\sigma\beta_i. \tag{268}$$

We find that F_i, as defined in equation (266), does satisfy equation (267) with

$$\beta_i^2 = q_j^2 \quad (i, j = 1, 2; i \neq j) \qquad \text{and} \qquad \kappa_i = \kappa = \mu^2(\mu^2+2). \tag{269}$$

The potentials, $V_i^{(\pm)}$, associated with the dual transformations belonging to $\pm q_j$ are (cf. Ch. 4, equation (317))

$$V_i^{(\pm)} = \pm q_j \frac{\mathrm{d}f_i}{\mathrm{d}r_*} + q_j^2 f_i^2 + \kappa f_i \quad (i, j = 1, 2; i \neq j), \tag{270}$$

where

$$f_i = \frac{1}{F_i} = \frac{\Delta}{r^3(\mu^2 r + q_j)} \quad (i, j = 1, 2; i \neq j). \tag{271}$$

Comparison with equation (197) shows that the potentials (270) are in fact the same that occur in the wave equations which determine the axial and the polar perturbations.

Also, we may note that (cf. Ch. 4, equation (312))

$$W_i^{(\pm)} = -\frac{\mathrm{d}}{\mathrm{d}r_*}\lg f_i \mp q_j f_i, \tag{272}$$

or, explicitly,

$$W_i^{(-)} = W^{(-)} = \frac{2}{r^2}\left(r - 3M + 2\frac{Q_*^2}{r}\right) \tag{273}$$

and

$$W_i^{(+)} = W^{(-)} - 2q_j \frac{\Delta}{r^3(\mu^2 r + q_j)} \quad (i, j = 1, 2; i \neq j). \tag{274}$$

The explicit forms of the associated transformations are (cf. Ch. 4, equations (318) and (319))

$$Y_i = V_i^{(\pm)} Z_i^{(\pm)} + (W_i^{(\pm)} + 2i\sigma)\Lambda_+ Z_i^{(\pm)}, \tag{275}$$

$$\Lambda_- Y_i = \mp \frac{D_i}{r^8} q_j Z_i^{(\pm)} + Q_i \Lambda_+ Z_i^{(\pm)}, \tag{276}$$

$$K_i^{(\pm)} Z_i^{(\pm)} = \frac{r^8}{D_i} Q_i Y_i - \frac{r^8}{D_i}(W_i^{(\pm)} + 2i\sigma)\Lambda_- Y_i, \tag{277}$$

$$K_i^{(\pm)} \Lambda_+ Z_i^{(\pm)} = \pm q_j Y_i + \frac{r^8}{D_i} V_i^{(\pm)} \Lambda_- Y_i, \tag{278}$$

where it may be recalled that

$$Q_i = \mu^2 \frac{\Delta}{r^4}\left(1 + \frac{2q_i}{\mu^2 r}\right)\left(1 + \frac{q_j}{\mu^2 r}\right) = \mu^2 \frac{D_i}{r^4 \Delta}\left(1 + \frac{q_j}{\mu^2 r}\right), \tag{279}$$

and

$$K_i^{(\pm)} = \mu^2(\mu^2+2) \pm 2i\sigma q_j. \tag{280}$$

Making use of equation (247), we obtain from equations (276) and (277) the further relations

$$\mu X_j = \mp q_j Z_i^{(\pm)} + \mu^2 \frac{r^4}{\Delta}\left(1+\frac{q_j}{\mu^2 r}\right)\Lambda_+ Z_i^{(\pm)}, \tag{281}$$

$$K_i^{(\pm)} Z_i^{(\pm)} = \mu^2 \frac{r^4}{\Delta}\left(1+\frac{q_j}{\mu^2 r}\right) Y_i - (W_i^{(\pm)} + 2i\sigma)\mu X_j. \tag{282}$$

As we have stated earlier, in the foregoing equations, we have written Y_i and X_j for Y_{+i} and X_{+j}. To obtain the corresponding equations relating Y_{-i} and X_{-j} to $Z_i^{(\pm)}$ we need only to change the sign of σ and write Λ_+ wherever Λ_- occurs, and conversely.

(b) The asymptotic behaviours of $Y_{\pm i}$ and $X_{\pm j}$

The short range character of the potentials, $V_i^{(\pm)}$, ensure that the solutions $Z_i^{(\pm)}$ have the asymptotic behaviours, $e^{\pm i\sigma r_*}$ both for $r_* \to +\infty$ and for $r_* \to -\infty$ (cf. §47 below). We can use this knowledge (as we did in a similar context in §32) to deduce the associated asymptotic behaviours of $Y_{\pm i}$ and $X_{\pm j}$ with the aid of equations (275)–(278), (281), and (282). In making these deductions, we need only to know that

$$V_i^{(\pm)},\ Q_i,\ f_i, \text{ and } D_i \quad \text{all tend to} \quad 0 \text{ for } r_* \to \pm\infty, \tag{283}$$

and that

$$\left.\begin{aligned} &W_i^{(\pm)} \to 0 \quad \text{for} \quad r_* \to +\infty \\ \text{and}\quad &W_i^{(\pm)} \to -\frac{2}{r_+^3}(Mr_+ - Q_*^2) \quad \text{for} \quad r_* \to -\infty. \end{aligned}\right\} \tag{284}$$

Notice that the limiting value of $W_i^{(\pm)}$ at the horizon is independent of i and of the distinguishing superscript $(\pm)$.

To avoid the ambiguity of "$\pm$" used both as superscripts (to distinguish the axial from the polar in $Z_i^{(\pm)}$) and as subscripts (to distinguish Y_{+i} from Y_{-i} which satisfies the complex-conjugate equation) we shall explicitly write out the relations appropriate for $Z_i^{(+)}$ and suppress also the distinguishing superscript. We find:

$r_* \to +\infty$:

$$\begin{aligned} &Z_i \to e^{+i\sigma r_*}, \quad Y_{+i} \to -4\sigma^2 e^{+i\sigma r_*}; \quad X_{+j} \to 2i\sigma\mu r^2 e^{+i\sigma r_*}, \\ &\qquad\qquad Y_{-i} \to -\frac{K_i^*}{4\sigma^2}\frac{e^{+i\sigma r_*}}{r^4}; \quad X_{-j} \to \frac{K_i^*}{2i\sigma\mu} e^{+i\sigma r_*}, \\ &Z_i \to e^{-i\sigma r_*}, \quad Y_{+i} \to -\frac{K_i}{4\sigma^2}\frac{e^{-i\sigma r_*}}{r^4}; \quad X_{+j} \to -\frac{K_i}{2i\sigma\mu} e^{-i\sigma r_*}, \\ &\qquad\qquad Y_{-i} \to -4\sigma^2 e^{-i\sigma r_*}; \quad X_{-j} \to -2i\sigma\mu r^2 e^{-i\sigma r_*}, \end{aligned} \tag{285}$$

and

$r_* \to -\infty$:

$$
\begin{aligned}
Z_i \to e^{+i\sigma r_*}, \quad & Y_{+i} \to 4i\sigma[i\sigma-(Mr_+-Q_*^2)/r_+^3]e^{+i\sigma r_*},\\
& Y_{-i} \to \frac{\Delta^2}{r_+^8}\frac{K_i^*(1+2q_i/\mu^2 r_+)e^{+i\sigma r_*}}{4[i\sigma+(Mr_+-Q_*^2)/r_+^3][i\sigma+2(r_+-M)/r_+^2]},\\
& X_{+j} \to 2i\sigma\mu\frac{r_+^4}{\Delta}\left(1+\frac{q_j}{\mu^2 r_+}\right)e^{+i\sigma r_*},\\
& X_{-j} \to \frac{K_i^* e^{+i\sigma r_*}}{2\mu[i\sigma+(Mr_+-Q_*^2)/r_+^3]};\\
Z_i \to e^{-i\sigma r_*}, \quad & Y_{+i} \to \frac{\Delta^2}{r_+^8}\frac{K_i(1+2q_i/\mu^2 r_+)e^{-i\sigma r_*}}{4[i\sigma-(Mr_+-Q_*^2)/r_+^3][i\sigma-2(r_+-M)/r_+^2]},\\
& Y_{-i} \to 4i\sigma[i\sigma+(Mr_+-Q_*^2)/r_+^3]e^{-i\sigma r_*},\\
& X_{+j} \to -\frac{K_i e^{-i\sigma r_*}}{2\mu[i\sigma-(Mr_+-Q_*^2)/r_+^3]},\\
& X_{-j} \to -2i\sigma\mu\frac{r_+^4}{\Delta}\left(1+\frac{q_j}{\mu^2 r_+}\right)e^{-i\sigma r_*}.
\end{aligned}
\tag{286}
$$

To obtain the relations appropriate for $Z_i^{(-)}$, we need, in accordance with equations (280), only to replace K_i by K_i^*, and conversely.

46. A direct evaluation of the Weyl and the Maxwell scalars in terms of the metric perturbations

As we have stated, one of the principal objectives in the study of the perturbations of the Reissner–Nordström black-hole is to ascertain how incident electromagnetic and gravitational waves will be scattered and absorbed by the black hole. To apply the perturbation theory developed in the preceding sections towards this end, it is necessary that we first relate the functions defined in the theory with the amplitudes of the waves of the two kinds. The functions, $Z_i^{(\pm)}$, must, in some direct way, specify the amplitudes of the incident, the reflected, and the transmitted waves since, in the limit $Q_* \to 0$, the equations governing $Z_2^{(\pm)}$ reduce to the Zerilli and the Regge–Wheeler equations. In general, we may expect that the required amplitudes are some linear combinations of the functions $Z_1^{(+)}$ and $Z_2^{(+)}$ and, similarly, of $Z_1^{(-)}$ and $Z_2^{(-)}$. The question is: what linear combination? The answer to this question via the solution of the Newman–Penrose equations in §44 is not also straightforward (as it was in Chapter 4, §32) since they have been solved in a special gauge; and, in consequence, the functions $R_{\pm 2}$ and $R_{\pm 1}$, in terms of which the solutions have been expressed, do not have simple physical interpretations. For these reasons, we shall evaluate the Weyl scalars Ψ_0 and

Ψ_4 and the Maxwell scalars ϕ_0 and ϕ_2, *ab initio*, from our knowledge of the Riemann and the Maxwell tensors.

Since the Schwarzschild and the Reissner–Nordström metrices differ only in the definition of the horizon function Δ $(=r^2 e^{2\nu}$ in both cases), much of the analysis of §31 can be carried over. In particular, the expression for Ψ_0 given in Chapter 4, equations (334), (339), and (350) are valid, as written, in the present context (cf. Chapter 1, equations (293) and (294)).

Considering the axial part, Im Ψ_0 of Ψ_0, we start with Chapter 4, equation (339):

$$-\operatorname{Im}\Psi_0=\left\{\frac{1}{2}\frac{d}{dr}\left(\frac{Q}{\Delta}\right)+\frac{Q}{r\Delta}+\frac{1}{2}i\sigma\frac{r^2}{\Delta^2}Q+\frac{1}{2}\frac{Q_{,r}}{\Delta}\right.$$
$$\left.+\frac{r^2}{2i\sigma\Delta}\left[\frac{d}{dr}\left(\frac{\Delta}{r^4}\frac{dQ}{dr}\right)+2\frac{\Delta}{r^4}\left(\frac{1}{r}-\nu_{,r}\right)\frac{dQ}{dr}\right]\right\}\frac{C_{l+2}^{-3/2}}{\sin^2\theta}. \tag{287}$$

On the right-hand side of this equation, we replace the first term in the square brackets by

$$\frac{d}{dr}\left(\frac{\Delta}{r^4}\frac{dQ}{dr}\right)=\frac{1}{\Delta}\left(\frac{\mu^2\Delta}{r^4}Q-\sigma^2 Q+2\frac{Q_*\mu}{r^4}\Delta H_1^{(-)}\right). \tag{288}$$

(This is equation (140) in which B has been replaced by $H_1^{(-)}$ in accordance with its definition in equation (143).) Making the replacement, we find after some rearrangement of the terms

$$-i\sigma\operatorname{Im}\Psi_0=\left\{i\sigma\left[\frac{1}{2}\frac{d}{dr}\left(\frac{Q}{\Delta}\right)+\frac{Q}{r\Delta}+\frac{1}{2\Delta}\frac{dQ}{dr}\right]-\sigma^2\frac{r^2}{\Delta^2}Q\right.$$
$$\left.+\frac{\mu^2}{2r^2\Delta}Q+\frac{1}{r^2}\left(\frac{1}{r}-\nu_{,r}\right)\frac{dQ}{dr}+\frac{Q_*\mu}{r^2\Delta}H_1^{(-)}\right\}\frac{C_{l+2}^{-3/2}}{\sin^2\theta}, \tag{289}$$

where

$$\frac{1}{r}-\nu_{,r}=\frac{1}{r\Delta}(r^2-3Mr+2Q_*^2). \tag{290}$$

Now letting

$$Q=rH_2^{(-)} \tag{291}$$

as in equation (143), we find, after some further simplifications, that the terms in curly brackets on the right-hand side of equation (289) become

$$i\sigma\left[\frac{r}{\Delta}\frac{dH_2^{(-)}}{dr}+\frac{1}{\Delta^2}(r^2-3Mr+2Q_*^2)H_2^{(-)}\right]-\sigma^2\frac{r^3}{\Delta^2}H_2^{(-)}$$
$$+\frac{\mu^2}{2r\Delta}H_2^{(-)}+\frac{Q_*\mu}{r^2\Delta}H_1^{(-)}+\frac{1}{r^3\Delta}(r^2-3Mr+2Q_*^2)\left(r\frac{dH_2^{(-)}}{dr}+H_2^{(-)}\right). \tag{292}$$

Replacing the derivatives with respect to r by derivatives with respect to r_* and recalling that (cf. equation (273))

$$W^{(-)} = \frac{2}{r^3}(r^2 - 3Mr + 2Q_*^2), \tag{293}$$

we regroup the terms in (292) in the manner

$$\frac{r^3}{2\Delta^2}(W^{(-)} + 2i\sigma)\frac{\mathrm{d}H_2^{(-)}}{\mathrm{d}r_*} + \frac{Q_*\mu}{r^2\Delta}H_1^{(-)}$$
$$+\left[\left(\frac{\mu^2}{2r\Delta} + \frac{r^2 - 3Mr + 2Q_*^2}{r^3\Delta}\right) + i\sigma\frac{r^3}{2\Delta^2}(W^{(-)} + 2i\sigma)\right]H_2^{(-)}. \tag{294}$$

Returning to equation (289), we can now write

$$-i\sigma \operatorname{Im}\Psi_0 = \frac{r^3}{2\Delta^2}\Bigg\{(W^{(-)} + 2i\sigma)\Lambda_+ H_2^{(-)}$$
$$+\frac{\Delta}{r^5}\left[(\mu^2 + 2)r - 6M + 4\frac{Q_*^2}{r}\right]H_2^{(-)} + \frac{\Delta}{r^5}(2Q_*\mu)H_1^{(-)}\Bigg\}\frac{C_{l+2}^{-3/2}}{\sin^2\theta}; \tag{295}$$

or, in view of equation (144) satisfied by $H_2^{(-)}$,

$$-i\sigma \operatorname{Im}\Psi_0 = \frac{r^3}{2\Delta^2}\{(W^{(-)} + 2i\sigma)\Lambda_+ + \Lambda^2\}H_2^{(-)}\frac{C_{l+2}^{-3/2}}{\sin^2\theta}. \tag{296}$$

We shall now express $H_2^{(-)}$ as a linear combination of $Z_1^{(-)}$ and $Z_2^{(+)}$ by solving equations (147) and (148). Rewriting these equations in the forms

$$\left.\begin{aligned} Z_1^{(-)} &= [q_1(q_1 - q_2)]^{1/2}(H_1^{(-)}\cos\psi + H_2^{(-)}\sin\psi), \\ \text{and}\qquad Z_2^{(-)} &= [q_1(q_1 - q_2)]^{1/2}(H_2^{(-)}\cos\psi - H_1^{(-)}\sin\psi), \end{aligned}\right\} \tag{297}$$

where

$$\sin 2\psi = \frac{2(-q_1q_2)^{1/2}}{q_1 - q_2} = \frac{2Q_*\mu}{\sqrt{(9M^2 + 4Q_*^2\mu^2)}}, \tag{298}$$

we have the solutions

$$\left.\begin{aligned} [q_1(q_1 - q_2)]^{1/2}H_1^{(-)} &= Z_1^{(-)}\cos\psi - Z_2^{(-)}\sin\psi, \\ \text{and}\qquad [q_1(q_1 - q_2)]^{1/2}H_2^{(-)} &= Z_1^{(-)}\sin\psi + Z_2^{(-)}\cos\psi. \end{aligned}\right\} \tag{299}$$

Inserting for $H_2^{(-)}$ from this last equation in equation (296), we obtain

$$-i\sigma[q_1(q_1 - q_2)]^{1/2}\operatorname{Im}\Psi_0 = \frac{r^3}{2\Delta^2}\{(W^{(-)} + 2i\sigma)\Lambda_+ + \Lambda^2\}$$
$$\times (Z_2^{(-)}\cos\psi + Z_1^{(-)}\sin\psi) \times \frac{C_{l+2}^{-3/2}}{\sin^2\theta}, \tag{300}$$

or, by virtue of the equations satisfied by $Z_i^{(-)}$, we have (cf. Ch. 4, equation (345))

$$-i\sigma[q_1(q_1-q_2)]^{1/2}\,\mathrm{Im}\,\Psi_0=\frac{r^3}{2\Delta^2}\{[V_2^{(-)}Z_2^{(-)}+(W^{(-)}+2i\sigma)\Lambda_+Z_2^{(-)}]\cos\psi$$

$$+[V_1^{(-)}Z_1^{(-)}+(W^{(-)}+2i\sigma)\Lambda_+Z_1^{(-)}]\sin\psi\}\frac{C_{l+2}^{-3/2}}{\sin^2\theta}. \quad (301)$$

Finally, by making use of equation (275) of the transformation theory, we obtain the important relation

$$-2i\sigma[q_1(q_1-q_2)]^{1/2}\,\mathrm{Im}\,\Psi_0=\frac{r^3}{\Delta^2}(Y_{+2}\cos\psi+Y_{+1}\sin\psi)\frac{C_{l+2}^{-3/2}}{\sin^2\theta}$$

$$=(F_{+2}\cos\psi+F_{+1}\sin\psi)\frac{C_{l+2}^{-3/2}}{\sin^2\theta}. \quad (302)$$

Similarly, we shall find

$$-2i\sigma[q_1(q_1-q_2)]^{1/2}\,\mathrm{Im}\,r^4\Psi_4=\tfrac{1}{4}r^3(Y_{-2}\cos\psi+Y_{-1}\sin\psi)\frac{C_{l+2}^{-3/2}}{\sin^2\theta}$$

$$=\tfrac{1}{4}(F_{-2}\cos\psi+F_{-1}\sin\psi)\frac{C_{l+2}^{-3/2}}{\sin^2\theta}. \quad (303)$$

An analogous calculation for $\mathrm{Re}\,\Psi_0$ can be carried out; but it is hardly necessary since it is clear on general grounds that we shall obtain the same relation except that the factor $2i\sigma$ on the left-hand side of equation (302) will be missing (cf. Ch. 4, equations (345) and (353)).

(*a*) *The Maxwell scalars ϕ_0 and ϕ_2*

By definition,

$$\phi_0=F_{(p)(q)}l^{(p)}m^{(q)}, \quad (304)$$

where we have enclosed the indices in parentheses to emphasize that they signify the components in the tetrad frame in which equations (117) and (118) are written. The basis null-vectors in the tetrad frame are given in Chapter 4, equation (333); and by contraction with these vectors, we obtain

$$\phi_0=\frac{1}{\sqrt{2}}e^{-\nu}[i(F_{01}+F_{21})+(F_{03}+F_{23})]. \quad (305)$$

We observe that ϕ_0 is again decomposed into an axial and a polar part.

Considering first the axial part,

$$\mathrm{Im}\,\phi_0=\frac{e^{-\nu}}{\sqrt{2}}(F_{01}+F_{21})=\frac{e^{-2\nu}}{(r\sin\theta)\sqrt{2}}(re^{\nu}F_{01}\sin\theta+re^{\nu}F_{21}\sin\theta), \quad (306)$$

we write the term in F_{21}, by making use of equation (120), in the manner

$$i\sigma r e^{\nu} F_{21} \sin\theta = e^{2\nu}(re^{\nu}F_{01}\sin\theta)_{,r} = (re^{\nu}F_{01}\sin\theta)_{,r_*}. \tag{307}$$

Thus,

$$\begin{aligned}\operatorname{Im}\phi_0 &= \frac{r}{(i\sigma\Delta\sin\theta)\sqrt{2}}\Lambda_+(re^{\nu}F_{01}\sin\theta)\\ &= \frac{r}{(i\sigma\Delta\sin\theta)\sqrt{2}}\Lambda_+(re^{\nu}B),\end{aligned} \tag{308}$$

where, in accordance with equation (131), we have written B in place of $F_{01}\sin\theta$. Now substituting for $re^{\nu}B$ from equations (138) and (143), we obtain

$$-(2\sqrt{2})i\sigma\mu\operatorname{Im}\phi_0 = \frac{r}{\Delta}\Lambda_+ H_1^{(-)}\frac{1}{\sin^2\theta}\frac{dC_{l+2}^{-3/2}}{d\theta}, \tag{309}$$

where, by the recurrence relations satisfied by the Gegenbauer functions,

$$\frac{1}{\sin^2\theta}\frac{dC_{l+2}^{-3/2}}{d\theta} = -\frac{3}{\mu^2+2}P_{l,\theta}. \tag{310}$$

Now substituting for $H_1^{(-)}$ from equation (299), we have

$$\frac{2}{3}i\sigma\mu(\mu^2+2)[2q_1(q_1-q_2)]^{1/2}\operatorname{Im}\phi_0 = \frac{r}{\Delta}\Lambda_+(Z_1^{(-)}\cos\psi - Z_2^{(-)}\sin\psi)P_{l,\theta}, \tag{311}$$

where it may be noted that by equation (281)

$$\frac{r}{\Delta}\Lambda_+ Z_i^{(-)} = \frac{1}{r^2(\mu^2 r+q_j)}(\mu X_j - q_j Z_i^{(-)}). \tag{312}$$

Similarly, by considering the expression

$$\phi_2 = F_{(p)(q)}\bar{m}^{(p)}n^{(q)} = \frac{e^{\nu}}{2\sqrt{2}}[i(F_{01}+F_{21})+(F_{03}+F_{23})], \tag{313}$$

we find

$$\frac{2}{3}i\sigma\mu(\mu^2+2)[2q_1(q_1-q_2)]^{1/2}\operatorname{Im}\phi_2 = \frac{1}{2r}\Lambda_-(Z_1^{(-)}\cos\psi - Z_2^{(-)}\sin\psi)P_{l,\theta}. \tag{314}$$

where

$$\frac{1}{r}\Lambda_- Z_i^{(-)} = \frac{\Delta}{r^4(\mu^2 r+q_j)}(\mu X_{-j} - q_j Z_i^{(-)}). \tag{315}$$

Considering next the polar part, $\operatorname{Re}\phi_0$, of ϕ_0 and substituting for F_{03} and F_{23} from equations (159) and (165), we find

$$\operatorname{Re}\phi_0 = -\frac{r}{2\Delta Q_*\sqrt{2}}\Lambda_+(r^2B_{23})P_{l,\theta}. \tag{316}$$

On the other hand, by equations (181), (190), and (193)

$$r^2 B_{23} = -Q_*\mu H_1^{(+)} - \frac{2Q_*^2}{r}\left(\frac{r}{n}X - H_2^{(+)}\right)$$
$$= -Q_*\mu H_1^{(+)} - \frac{2Q_*^2}{r}e^{\nu}\Phi. \tag{317}$$

Therefore,

$$(2\sqrt{2})\operatorname{Re}\phi_0 = \mu\frac{r}{\Delta}\Lambda_+\left(H_1^{(+)} + \frac{2Q_*}{\mu r}e^{\nu}\Phi\right)P_{l,\theta}. \tag{318}$$

We may again replace the term in $H_1^{(+)}$ in the foregoing equation by

$$\frac{r}{\Delta}\Lambda_+ H_1^{(+)} = \frac{r}{\Delta}[q_1(q_1-q_2)]^{-1/2}\Lambda_+(Z_1^{(+)}\cos\psi - Z_2^{(+)}\sin\psi), \tag{319}$$

where by equation (281)

$$\frac{r}{\Delta}\Lambda_+ Z_i^{(+)} = \frac{1}{r^2(\mu^2 r + q_j)}(\mu X_j + q_j Z_i^{(+)}). \tag{320}$$

47. The problem of reflexion and transmission; the scattering matrix

We now turn to the fundamental problem to which the theory of the perturbations of black holes is addressed: the manner of their interaction with incident waves of different sorts.

First, we recall that the axial and the polar perturbations have each been reduced to a pair of one-dimensional wave equations with real positive potentials which have an inverse-square behaviour for $r_* \to +\infty$ and an exponentially decreasing behaviour for $r_* \to -\infty$. On account of this short-range character of the potentials, the problem with respect to each of the equations is one of barrier penetration: i.e., we have to seek solutions of the equations which satisfy the standard boundary conditions

$$\left.\begin{aligned} Z_i^{(\pm)} &\to e^{+i\sigma r_*} + R_i^{(\pm)}(\sigma)e^{-i\sigma r_*} \quad (r_* \to +\infty), \\ &\to \qquad T_i^{(\pm)}(\sigma)e^{+i\sigma r_*} \quad (r_* \to -\infty). \end{aligned}\right\} \tag{321}$$

As we have shown in §43, the forms (197), which the potentials, $V_i^{(\pm)}$, have, enable us to express, as in equation (199), the solution $Z_i^{(+)}$ as a linear combination of $Z_i^{(-)}$ and its derivative, and conversely; and from equation (199) it follows from the analysis of §27 that (cf. Ch. 4, equations (168) and (169))

$$R_i^{(+)}(\sigma) = R_i^{(-)}(\sigma)e^{i\delta_i} \quad \text{and} \quad T_i^{(+)}(\sigma) = T_i^{(-)}(\sigma), \tag{322}$$

where

$$e^{i\delta_i} = \frac{\mu^2(\mu^2+2) - 2i\sigma q_j}{\mu^2(\mu^2+2) + 2i\sigma q_j} \quad (i, j = 1, 2;\ i \neq j). \tag{323}$$

In view of these very simple relations between the amplitudes of the reflected and the transmitted waves belonging to the two classes of perturbations, we shall, in the rest of this section, suppress the distinguishing superscripts with the understanding that the analysis which follows applies, equally, to $Z_i^{(+)}$ and $Z_i^{(-)}$.

Now, with respect to scattering by each of the two potentials V_i ($i = 1, 2$; suppressing the distinguishing superscript!) we can define a scattering matrix (Ch. 4, §28)

$$\boldsymbol{S}_i = \begin{vmatrix} T_i(\sigma) & \tilde{R}_i(\sigma) \\ R_i(\sigma) & T_i(\sigma) \end{vmatrix} \quad (i = 1, 2), \tag{324}$$

whose elements (representing the reflexion and the transmission amplitudes) satisfy the unitarity requirements (Ch. 4, equations (181)–(186))

$$R_i(\sigma)T_i^*(\sigma) + T_i(\sigma)\tilde{R}_i^*(\sigma) = 0, \quad |R_i(\sigma)| = |\tilde{R}_i(\sigma)|, \tag{325}$$

and

$$|R_i(\sigma)|^2 + |T_i(\sigma)|^2 = 1. \tag{326}$$

(We are here denoting by $\tilde{R}(\sigma)$ what was denoted by $R_2(\sigma)$ in §28.) By virtue of the relations (325) and (326),

$$\boldsymbol{S}_i(\sigma)\tilde{\boldsymbol{S}}_i(\sigma) = 1, \tag{327}$$

where $\tilde{\boldsymbol{S}}_i$ denotes the conjugate transposed-matrix of $\boldsymbol{S}_i$.

The scattering matrix has an important property with respect to 'time-reversibility': If $\overleftarrow{Z}_i^{(r)}$ and $\vec{Z}_i^{(l)}$ are the amplitudes of the waves incident on the barrier from the right and from the left (see Fig. 21), then the amplitudes, $\overleftarrow{Z}_i^{(l)}$ and $\vec{Z}_i^{(r)}$, of the waves reflected to the left and to the right, are given by

$$\boldsymbol{S}_i \begin{vmatrix} \overleftarrow{Z}_i^{(r)} \\ \vec{Z}_i^{(l)} \end{vmatrix} = \begin{vmatrix} \overleftarrow{Z}_i^{(l)} \\ \vec{Z}_i^{(r)} \end{vmatrix}, \tag{328}$$

so that by unitarity,

$$[\overleftarrow{Z}_i^{(r)}, \vec{Z}_i^{(l)}] = [\overleftarrow{Z}_i^{(l)}, \vec{Z}_i^{(r)}]\, \tilde{\boldsymbol{S}}_i\,. \tag{329}$$

Our problem is now to relate the scattering matrices $\boldsymbol{S}_1$ and $\boldsymbol{S}_2$ to a scattering matrix (of order four) which will describe, similarly, the reflexion and transmission of electromagnetic and gravitational waves, simultaneously, incident on the Reissner–Nordström black-hole. To this end, we must relate the amplitudes of the waves of the two sorts with the functions Z_1 and Z_2. For this purpose consider the boundary conditions satisfied by the solutions for Y_{+i} and Y_{-i} derived from the solutions for Z_i satisfying the boundary conditions

$$\left.\begin{aligned} Z_i &\to Z_i^{(\infty)} e^{+i\sigma r_*} + Z_i^{(\infty)} R_i(\sigma) e^{-i\sigma r_*} \quad (r_* \to +\infty) \\ &\to \qquad\qquad Z_i^{(\infty)} T_i(\sigma) e^{+i\sigma r_*} \quad (r_* \to -\infty). \end{aligned}\right\} \tag{330}$$

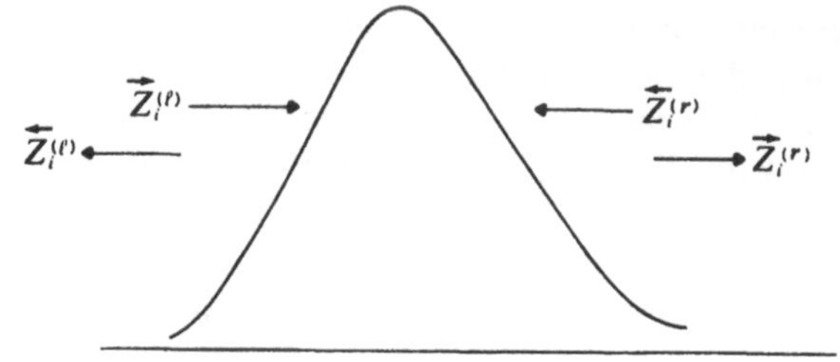

FIG. 21. Incoming and outgoing waves from $\pm\infty$ are reflected by the potential barrier to outgoing and ingoing waves. The scattering matrix (equation (328)) relates the amplitudes of the waves incident on the barrier to the amplitudes of the waves reflected by the barrier.

Writing the asymptotic behaviours of the solutions for $Y_{\pm i}$, derived from the solution Z_i, in the forms

$$\left.\begin{aligned} \frac{r^3}{\Delta^2} Y_{+i} &\rightarrow Y_{+i}^{(\text{in})} \frac{e^{+i\sigma r_*}}{r} + Y_{+i}^{(\text{ref})} \frac{e^{-i\sigma r_*}}{r^5} \qquad (r_* \rightarrow +\infty), \\ &\rightarrow Y_{+i}^{(\text{tr})} \frac{e^{+i\sigma r_*}}{\Delta^2} \qquad (r_* \rightarrow -\infty), \end{aligned}\right\} \tag{331}$$

$$\left.\begin{aligned} r^3 Y_{-i} &\rightarrow Y_{-i}^{(\text{in})} \frac{e^{+i\sigma r_*}}{r} + Y_{-i}^{(\text{ref})} r^3 e^{-i\sigma r_*} \qquad (r_* \rightarrow +\infty), \\ &\rightarrow Y_{-i}^{(\text{tr})} \Delta^2 e^{+i\sigma r_*} \qquad (r_* \rightarrow -\infty), \end{aligned}\right\} \tag{332}$$

we conclude, from the correspondences in the asymptotic behaviours listed in equations (285) and (286), that

$$\left.\begin{aligned} Y_{+i}^{(\text{in})} &= -4\sigma^2 Z_i^{(\infty)}; \qquad Y_{-i}^{(\text{in})} = -\frac{K_i^*}{4\sigma^2} Z_i^{(\infty)}, \\ Y_{+i}^{(\text{ref})} &= -\frac{K_i}{4\sigma^2} Z_i^{(\infty)} R_i(\sigma); Y_{-i}^{(\text{ref})} = -4\sigma^2 Z_i^{(\infty)} R_i(\sigma); \end{aligned}\right\} \tag{333}$$

$$\left.\begin{aligned} Y_{+i}^{(\text{tr})} &= 4i\sigma[i\sigma - (Mr_+ - Q_*^2)/r_+^3]\,(r_+)^3 Z_i^{(\infty)} T_i(\sigma), \\ Y_{-i}^{(\text{tr})} &= \frac{1}{(r_+)^5}\left(1 + \frac{2q_i}{\mu^2 r_+}\right) \frac{K_i^* Z_i^{(\infty)} T_i(\sigma)}{4[i\sigma + (Mr_+ - Q_*^2)/r_+^3]\,[i\sigma + 2(r_+ - M)/r_+^2]}. \end{aligned}\right\} \tag{334}$$

Considering first the fluxes of the incident and of the reflected gravitational waves, we recall that they are determined by the leading r^{-1}-terms in the asymptotic behaviours of Ψ_0 and Ψ_4 representing the incoming and the outgoing waves, respectively. From the expressions (302) and (303) for Ψ_0 and

Ψ_4, we infer (cf. Ch. 4, equations (368) and (369))

$$\frac{d^2 E^{(\text{in})}_{\text{gr.}}}{dt\, d\Omega} = \frac{(S_{+2})^2}{2\pi} \frac{1}{2\sigma^2} |Y^{(\text{in})}_{+2} \cos\psi + Y^{(\text{in})}_{+1} \sin\psi|^2 \tag{335}$$

and

$$\frac{d^2 E^{(\text{ref})}_{\text{gr.}}}{dt\, d\Omega} = \frac{(S_{-2})^2}{2\pi} \frac{1}{2\sigma^2} |Y^{(\text{ref})}_{-2} \cos\psi + Y^{(\text{ref})}_{-1} \sin\psi|^2. \tag{336}$$

From the expressions for $Y^{(\text{in})}_{+i}$ and $Y^{(\text{ref})}_{-i}$ given in equations (333), it follows that the reflexion coefficient for the incident gravitational waves is given by

$$\mathbb{R}_{\text{gr.}} = \frac{|Z^{(\infty)}_{+2} R_2(\sigma) \cos\psi + Z^{(\infty)}_{+1} R_1(\sigma) \sin\psi|^2}{|Z^{(\infty)}_{+2} \cos\psi + Z^{(\infty)}_{+1} \sin\psi|^2}. \tag{337}$$

From this expression it is manifest that the amplitude H_2 of gravitational waves is, apart from a constant of proportionality, given by

$$H_2 = Z_2 \cos\psi + Z_1 \sin\psi, \tag{338}$$

i.e., precisely the same quantity which satisfied the coupled equations (144) and (145) and (183) and (184) before their decoupling to yield the equations for Z_1 and Z_2 (cf. equations (299)).

(a) *The energy-momentum tensor of the Maxwell field and the flux of electromagnetic energy*

The energy-momentum tensor of a Maxwell field, written out explicitly in terms of the scalars ϕ_0, ϕ_1, and ϕ_2, is

$$\begin{aligned} 4\pi T_{ij} = \{&\phi_0 \phi_0^* n_i n_j + \phi_2 \phi_2^* l_i l_j + 2\phi_1 \phi_1^* [l_{(i} n_{j)} + m_{(i} \bar{m}_{j)}] \\ &- 4\phi_0^* \phi_1 n_{(i} m_{j)} - 4\phi_1^* \phi_2 l_{(i} m_{j)} + 2\phi_2 \phi_0^* m_i m_j\} \\ &+ \text{complex-conjugate terms.} \end{aligned} \tag{339}$$

In the Reissner–Nordström background, the only non-vanishing scalar is $\phi_1\ (= Q_*/2r^2)$. On the assumption that the perturbations in the field, represented by ϕ_0, ϕ_2, and $\delta\phi_1$, vanish at least as rapidly as $1/r$, one verifies that the evaluation of the perturbed T_{ij} does not require a knowledge of the perturbation in the basis vectors; and ignoring the terms which fall off more rapidly than r^{-2}, we have

$$\begin{aligned} 4\pi T_{ij} = \{&\phi_0 \phi_0^* n_i n_j + \phi_2 \phi_2^* l_i l_j + 2\delta\phi_1 \delta\phi_1^* [l_{(i} n_{j)} + m_{(i} \bar{m}_{j)}] \\ &- 4\phi_0^* \delta\phi_1 n_{(i} m_{j)} - 4\delta\phi_1^* \phi_2 l_{(i} m_{j)} + 2\phi_2 \phi_0^* m_i m_j\} \\ &+ \text{complex-conjugate terms} + O(r^{-3}) \quad (r \to \infty). \end{aligned} \tag{340}$$

With the basis vectors given in equation (108), we find from the foregoing equation that the flux of energy, per unit time and per unit solid angle, is given by

$$\frac{d^2E_{\text{ele}}}{dt\,d\Omega} = \underset{r\to\infty}{\text{limit}}\,(r^2T^r{}_t) = \underset{r\to\infty}{\text{limit}}\frac{r^2}{4\pi}(-\tfrac{1}{4}|\phi_0|^2 + |\phi_2|^2). \tag{341}$$

From the expressions for ϕ_0 and ϕ_2 given in equations (311) and (314), it follows that, apart from the same constant of proportionality, for incoming waves, $Z_i^{(\infty)}\exp(+i\sigma r_*)$,

$$\frac{d^2E_{\text{ele}}^{(\text{in})}}{dt\,d\Omega} = \frac{1}{4\pi}(P_{l,\theta})^2\,(16\sigma^4)|Z_1^{(\infty)}\cos\psi - Z_2^{(\infty)}\sin\psi|^2; \tag{342}$$

and for the reflected waves, $Z_i^{(\infty)}R_i(\sigma)\exp(-i\sigma r_*)$,

$$\frac{d^2E_{\text{ele}}^{(\text{ref})}}{dt\,d\Omega} = \frac{1}{4\pi}(P_{l,\theta})^2\,(16\sigma^4)|Z_1^{(\infty)}R_1(\sigma)\cos\psi - Z_2^{(\infty)}R_2(\sigma)\sin\psi|^2. \tag{343}$$

Accordingly, the reflexion coefficient, for the incident electromagnetic waves, is given by

$$\mathbb{R}_{\text{ele}} = \frac{|Z_1^{(\infty)}R_1(\sigma)\cos\psi - Z_2^{(\infty)}R_2(\sigma)\sin\psi|^2}{|Z_1^{(\infty)}\cos\psi - Z_2^{(\infty)}\sin\psi|^2}. \tag{344}$$

From this expression it is manifest that the amplitude, H_1, of the electromagnetic wave is, apart from a constant of proportionality, given by

$$H_1 = Z_1\cos\psi - Z_2\sin\psi, \tag{345}$$

i.e., precisely the same quantity which satisfied the coupled equations (183) and (184) before their decoupling to yield the equations for Z_1 and Z_2.

One can arrive at the same results (344) and (345) by reducing the problem to one in which the asymptotic behaviours of the solutions for $X_{\pm j}$ play the same role as the behaviours of the solutions for $Y_{\pm i}$ do in the treatment of the gravitational waves. But the arguments are somewhat less direct.

(*b*) *The scattering matrix*

We shall now show how the general process of scattering and absorption of electromagnetic and gravitational waves by the black hole can be described by a symmetric unitary scattering-matrix of order four exhibiting time-reversibility.

As we have seen, the amplitudes, H_1 and H_2 of the electromagnetic and gravitational (wave-like) disturbances (of some specified frequency) are related to the functions Z_1 and Z_2 by

$$\left.\begin{aligned} H_1 &= Z_1\cos\psi - Z_2\sin\psi \\ \text{and}\qquad H_2 &= Z_2\cos\psi + Z_1\sin\psi, \end{aligned}\right\} \tag{346}$$

where it may be recalled that superscripts, $(\pm)$, distinguishing the axial and the polar perturbations have been suppressed. Relations, inverse to (346), are

$$\left.\begin{aligned} Z_1 &= H_1 \cos\psi + H_2 \sin\psi \\ Z_2 &= H_2 \cos\psi - H_1 \sin\psi. \end{aligned}\right\} \quad \text{and} \tag{347}$$

Now suppose that a pure electromagnetic wave of amplitude $\overleftarrow{H}_1^{(r)}$ is incident on the black hole from the right. Then $\overleftarrow{H}_2^{(r)} = 0$ and

$$\overleftarrow{Z}_1^{(r)} = \overleftarrow{H}_1^{(r)} \cos\psi \quad \text{and} \quad \overleftarrow{Z}_2^{(r)} = -\overleftarrow{H}_1^{(r)} \sin\psi. \tag{348}$$

Each of these incident Z -fields will give rise to reflected and transmitted amplitudes given by

$$\vec{Z}_i^{(r)} = \overleftarrow{Z}_i^{(r)} R_i(\sigma) \quad \text{and} \quad \overleftarrow{Z}_i^{(l)} = \overleftarrow{Z}_i^{(r)} T_i(\sigma); \tag{349}$$

and these amplitudes, recombined in accordance with equations (346), will give reflected and transmitted amplitudes in both electromagnetic and gravitational waves. Thus,

$$\left.\begin{aligned} \vec{H}_1^{(r)} &= \overleftarrow{H}_1^{(r)} (R_1 \cos^2\psi + R_2 \sin^2\psi), && \text{(a)} \\ \overleftarrow{H}_1^{(l)} &= \overleftarrow{H}_1^{(r)} (T_1 \cos^2\psi + T_2 \sin^2\psi), && \text{(b)} \\ \vec{H}_2^{(r)} &= \overleftarrow{H}_1^{(r)} (R_1 - R_2) \sin\psi \cos\psi, && \text{(c)} \\ \overleftarrow{H}_2^{(l)} &= \overleftarrow{H}_1^{(r)} (T_1 - T_2) \sin\psi \cos\psi. && \text{(d)} \end{aligned}\right\} \tag{350}$$

If on the other hand, a pure gravitational wave of amplitude $\overleftarrow{H}_2^{(r)}$ is incident on the black hole from the right, then $\overleftarrow{H}_1^{(r)} = 0$ and

$$\overleftarrow{Z}_1^{(r)} = \overleftarrow{H}_2^{(r)} \sin\psi \quad \text{and} \quad \overleftarrow{Z}_2^{(r)} = \overleftarrow{H}_2^{(r)} \cos\psi. \tag{351}$$

The reflexion and transmission of these incident Z-fields, when recombined, will give rise (in analogous fashion) to the following reflected and transmitted amplitudes in electromagnetic and gravitational waves:

$$\left.\begin{aligned} \vec{H}_1^{(r)} &= \overleftarrow{H}_2^{(r)} (R_1 - R_2) \sin\psi \cos\psi, && \text{(a)} \\ \overleftarrow{H}_1^{(l)} &= \overleftarrow{H}_2^{(r)} (T_1 - T_2) \sin\psi \cos\psi, && \text{(b)} \\ \vec{H}_2^{(r)} &= \overleftarrow{H}_2^{(r)} (R_1 \sin^2\psi + R_2 \cos^2\psi), && \text{(c)} \\ \overleftarrow{H}_2^{(l)} &= \overleftarrow{H}_2^{(r)} (T_1 \sin^2\psi + T_2 \cos^2\psi). && \text{(d)} \end{aligned}\right\} \tag{352}$$

Comparison of equations (350) (c), (d) and (352) (a), (b) shows that the conversion of incident energy of one kind, into reflected and transmitted energy of the other kind, takes place entirely symmetrically with respect to the two kinds—a consequence, clearly, of time-reversibility.

Turning next to the situation described in Fig. 22, the scattering matrix which will describe the transformation

$$(\overleftarrow{H}_1^{(r)}, \vec{H}_1^{(l)}; \overleftarrow{H}_2^{(r)}, \vec{H}_2^{(l)}) \rightarrow (\overleftarrow{H}_1^{(l)}, \vec{H}_1^{(r)}; \overleftarrow{H}_2^{(l)}, \vec{H}_2^{(r)}) \tag{353}$$

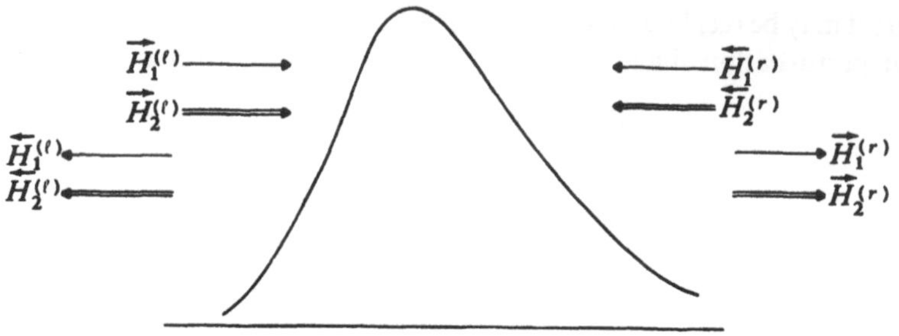

FIG. 22. In the case of the Reissner–Nordström black-hole, electromagnetic and gravitational waves simultaneously incident on the potential barrier from opposite ends, from $\pm\infty$, are reflected to $\mp\infty$; and a scattering matrix of order 4 is required to describe this process.

can be readily written down with the aid of equations (350) and (351). Thus,

$$S\begin{vmatrix}\overleftarrow{H}_1^{(r)}\\ \vec{H}_1^{(l)}\\ \overleftarrow{H}_2^{(r)}\\ \vec{H}_2^{(l)}\end{vmatrix} = \begin{vmatrix}T_{11} & \tilde{R}_{11} & T_{21} & \tilde{R}_{21}\\ R_{11} & T_{11} & R_{21} & T_{21}\\ T_{12} & \tilde{R}_{12} & T_{22} & \tilde{R}_{22}\\ R_{12} & T_{12} & R_{22} & T_{22}\end{vmatrix}\begin{vmatrix}\overleftarrow{H}_1^{(r)}\\ \vec{H}_1^{(l)}\\ \overleftarrow{H}_2^{(r)}\\ \vec{H}_2^{(l)}\end{vmatrix} = \begin{vmatrix}\overleftarrow{H}_1^{(l)}\\ \vec{H}_1^{(r)}\\ \overleftarrow{H}_2^{(l)}\\ \vec{H}_2^{(r)}\end{vmatrix}, \tag{354}$$

where

$$\begin{aligned}
R_{11} &= R_1\cos^2\psi + R_2\sin^2\psi; \quad R_{22} = R_1\sin^2\psi + R_2\cos^2\psi,\\
T_{11} &= T_1\cos^2\psi + T_2\sin^2\psi; \quad T_{22} = T_1\sin^2\psi + T_2\cos^2\psi,\\
R_{12} &= R_{21} = (R_1 - R_2)\sin\psi\cos\psi,\\
T_{12} &= T_{21} = (T_1 - T_2)\sin\psi\cos\psi.
\end{aligned} \tag{355}$$

It can be readily verified that the scattering matrix S defined in equation (354) is, by virtue of the unitarity of S_1 and S_2, unitary itself:

$$S\tilde{S} = 1. \tag{356}$$

In particular, the identities

and

$$\left.\begin{aligned}
|T_{11}|^2 + |R_{11}|^2 + |T_{12}|^2 + |R_{21}|^2 &= 1\\
|T_{11}|^2 + |R_{11}|^2 &= |T_{22}|^2 + |R_{22}|^2
\end{aligned}\right\} \tag{357}$$

required by the conservation of energy are satisfied.

In Fig. 23 the dependence of the conversion factor $\mathbb{C} = |R_{12}^{(\pm)}|^2$ on Q_* and σ is illustrated for the case $l = 2$.

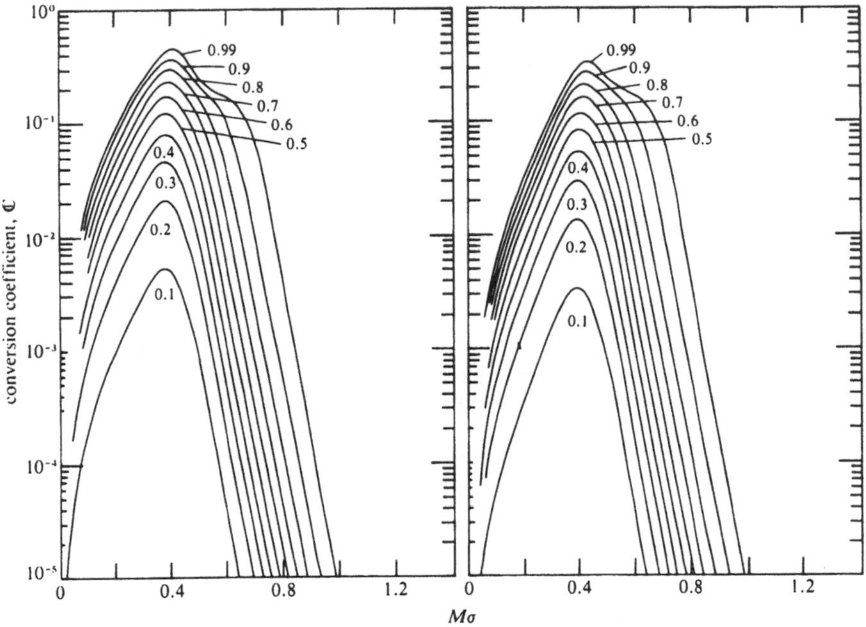

FIG. 23. The conversion coefficient $\mathbb{C}^{(+)}$ (the figures on the left) and $\mathbb{C}^{(-)}$(the figures on the right) for the axial and polar perturbations belonging to $l = 2$. Each curve is labelled by the value Q_* to which it belongs.

48. The quasi-normal modes of the Reissner-Nordström black-hole

The Reissner–Nordström black-hole can be characterized by quasi-normal modes satisfying the same boundary conditions and with the same meanings as described in Chapter 4, §35 in the context of the Schwarzschild black-hole. There is, however, one important difference: quasi-normal modes can be defined with respect to each of the equations governing $Z_i^{(\pm)}$ $(i = 1, 2)$. Since the potentials, $V_i^{(\pm)}$, are real and positive (in the space external to the event horizon) the imaginary parts of the complex characteristic frequencies of the quasi-normal modes will be positive. Also, in view of the relation (199) between the solutions belonging to axial and polar perturbations, the characteristic frequencies will be the same for $Z_i^{(+)}$ and $Z_i^{(-)}$. It should also be noticed that there is no quasi-normal mode which is purely electromagnetic or purely gravitational: any quasi-normal mode of oscillation will be accompanied by the emission of both electromagnetic and gravitational radiation in accordance with equation (346).

In Table V, we list the complex characteristic frequencies σ_1 and σ_2

TABLE V

The complex characteristic frequencies belonging to the quasi-normal modes of Z_1 and Z_2

(σ is expressed in the unit M^{-1})

	Z_1				Z_2			
Q_*	$l=1$	$l=2$	$l=3$	$l=4$	$l=1$	$l=2$	$l=3$	$l=4$
0	0.24828 +0.09250*i*	0.45760 +0.09500*i*	0.65690 +0.09562*i*	0.85310 +0.09586*i*	0.11252 +0.10040*i*	0.37367 +0.08896*i*	0.59944 +0.09270*i*	0.80918 +0.09416*i*
0.2	0.25150 +0.09291*i*	0.46296 +0.09537*i*	0.66437 +0.09597*i*	0.86260 +0.09621*i*	0.11320 +0.10061*i*	0.37475 +0.08907*i*	0.60103 +0.09279*i*	0.81134 +0.09425*i*
0.4	0.26194 +0.09416*i*	0.47993 +0.09644*i*	0.68728 +0.09697*i*	0.89100 +0.09716*i*	0.11537 +0.10123*i*	0.37844 +0.08940*i*	0.60705 +0.09306*i*	0.82020 +0.09453*i*
0.6	0.28276 +0.09619*i*	0.51201 +0.09802*i*	0.72919 +0.09837*i*	0.94192 +0.09845*i*	0.11957 +0.10206*i*	0.38622 +0.08981*i*	0.62066 0.9341*i*	0.84056 +0.09493*i*
0.8	0.32349 +0.09827*i*	0.57013 +0.09907*i*	0.80284 +0.09911*i*	1.03039 +0.09903*i*	1.12712 +0.10198*i*	0.40122 +0.08964*i*	0.64755 +0.09312*i*	0.88057 +0.09467*i*
0.9	0.36082 +0.09744*i*	0.61939 +0.09758*i*	0.86375 +0.09752*i*	1.10286 +0.09742*i*	0.13276 +0.09980*i*	0.41357 +0.08833*i*	0.67002 +0.09164*i*	0.91396 +0.09314*i*
0.95	0.38927 +0.09442*i*	0.65476 +0.09460*i*	0.90668 +0.09469*i*	1.1535 +0.09467*i*	0.13540 +0.09675*i*	0.42169 +0.08666*i*	0.68519 +0.08978*i*	0.93664 +0.09117*i*
0.999	0.43031 +0.08388*i*	0.70310 +0.08627*i*	0.96434 +0.08726*i*	1.22097 +0.08773*i*	0.13416 +0.09666*i*	0.43113 +0.08354*i*	0.70387 +0.08608*i*	0.96510 +0.08712*i*

(belonging to $Z_1^{(\pm)}$ and $Z_2^{(\pm)}$) of the quasi-normal modes for a range of values of Q_* and l.

49. Considerations relative to the stability of the Reissner-Nordström space-time

The considerations, relative to the stability of the Schwarzschild black-hole to external perturbations, in Chapter 4, §34 apply, quite literally, to the Reissner–Nordström black-hole since the only fact relevant to those considerations was that the potential barriers, external to the event horizon, are real and positive; and stability follows from this fact. But quite different circumstances prevail in the interval, $r_- < r < r_+$, between the two horizons. The altered circumstances are the following.

While the equations governing $Z_i^{(\pm)}$ remain formally unaltered, the potential barriers, $V_i^{(\pm)}$, are negative in the interval, $r_- < r < r_+$, and in the associated range of r_*, namely $+\infty > r_* > -\infty$; they are in fact *potential*

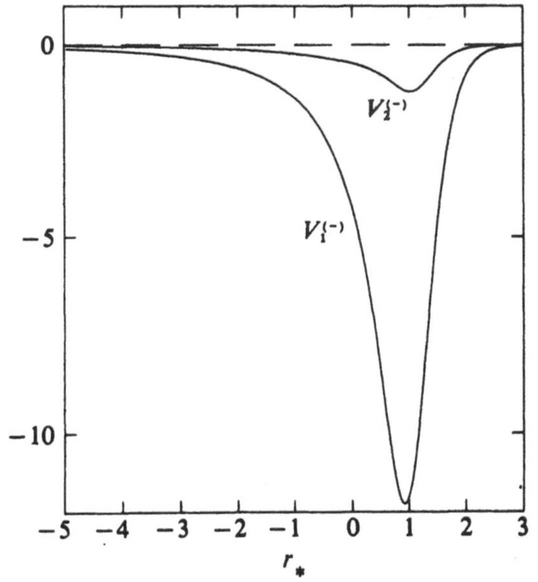

FIG. 24. The potential wells, $V_1^{(-)}$ and $V_2^{(-)}$ for $Q_* = 0.75$, governing the dispersion of perturbations in the region between the two horizons.

wells rather than potential barriers (see Fig. 24). Thus, the equation now governing $Z_i^{(-)}$ is, for example (cf. equation (151)),

$$\frac{d^2 Z_i^{(-)}}{dr_*^2} + \sigma^2 Z_i^{(-)} = -\frac{|\Delta|}{r^5}\left[(\mu^2+2)r - q_j + \frac{4Q_*^2}{r}\right] Z_i^{(-)}$$

$$(i, j = 1, 2;\ i \neq j; \quad \text{and} \quad r_- < r < r_+,\ +\infty > r_* > -\infty), \tag{358}$$

where

$$r_* = r + \frac{1}{2\kappa_+}\lg|r_+ - r| - \frac{1}{2\kappa_-}\lg|r - r_-|, \tag{359}$$

$$\kappa_+ = \frac{r_+ - r_-}{2r_+^2}, \quad \text{and} \quad \kappa_- = \frac{r_+ - r_-}{2r_-^2}. \tag{360}$$

In view of the relation (199) between the solutions, belonging to axial and polar perturbations, it will, again, suffice to restrict our consideration to equation (358); and for convenience, we shall suppress the distinguishing superscript.

An important consequence of the fact that we are now concerned with a short-range one-dimensional potential-well, is that equation (358) will allow a *finite number* of discrete, non-degenerate, bound states:

$$\sigma = \pm i\sigma_j \quad [j = 1, 2; n = 1, 2, \ldots, m \text{ (say)}]. \tag{361}$$

Besides these bound states, we continue to have the standard problem of reflexion and transmission associated with the continuous spectrum including all real values of σ. However, in the present context, the relevant solutions must satisfy boundary conditions different from the ones we have used hitherto. The boundary conditions we must now impose are

$$\begin{aligned} Z(r_*) &\to A(\sigma)e^{-i\sigma r_*} + B(\sigma)e^{+i\sigma r_*} \quad (r \to r_- + 0;\ r_* \to +\infty) \\ &\to e^{-i\sigma r_*} \qquad\qquad\qquad\qquad\quad (r \to r_+ - 0;\ r_* \to -\infty). \end{aligned} \tag{362}$$

The reason for these altered boundary conditions is that, by virtue of the light-cone structure of the region between the two horizons, one can have only waves entering the region, by crossing the event horizon at r_+ (from the 'outside world'), and none leaving it, by crossing the event horizon in the reverse direction.

The coefficients $A(\sigma)$ and $B(\sigma)$ in equation (362) are related to the reflexion and the transmission amplitudes, as we have defined them hitherto and in Chapter 4 (equations (175) and (176))

$$A(\sigma) = \frac{1}{T^*(\sigma)} = \frac{1}{T_1(-\sigma)} \quad B(\sigma) = \frac{R^*(\sigma)}{T^*(\sigma)} = \frac{R_1(-\sigma)}{T_1(-\sigma)}, \tag{363}$$

so that

$$|A(\sigma)|^2 - |B(\sigma)|^2 = 1. \tag{364}$$

In Table VI we have listed the *amplification factors*, $|A(\sigma)|^2$, for $Q_*^2 = 0.75\,M^2$. We observe that $|A(\sigma)|^2$, and, therefore, also $|B(\sigma)|^2$, tend to finite limits as $\sigma \to 0$. This fact has its origin in the existence of bound states of zero energy in the potential wells, V_1 and V_2. (See Bibliographical Notes at the end of the chapter.)

In analyzing the radiation arriving at the Cauchy horizon at r_-, we must distinguish the edges EC' and EF in the Penrose diagram (Fig. 14). For this reason, we restore the time-dependence, $e^{i\sigma t}$, of the solutions; and remembering that in the interval, $r_- < r < r_+$ (cf. equations (51)),

$$u = r_* + t \quad \text{and} \quad v = r_* - t, \tag{365}$$

we write, in place of equation (362),

$$Z(r_*, t) \to e^{-i\sigma v} + [A(\sigma) - 1]e^{-i\sigma v} + B(\sigma)e^{+i\sigma u}. \tag{366}$$

If we now suppose that the flux of radiation emerging from $D'C'$ is $\hat{Z}(v)$, then

$$Z(\sigma) = \frac{1}{2\pi}\int_{-\infty}^{+\infty} \hat{Z}(v)e^{i\sigma v}\,dv. \tag{367}$$

TABLE VI

Amplification factors appropriate for the potential $V_1^{(\pm)}$ and $V_2^{(\pm)}$ for $Q_*^2 = 0.75\,M^2$

σ	for V_1	for V_2	σ	for V_1	for V_2
0.009375	2.7746	2.7747	0.50	1.2631	1.2653
0.01875	2.7652	2.7655	0.55	1.2165	1.2169
0.0375	2.7287	2.7299	0.65	1.1494	1.1471
0.05	2.6923	2.6944	0.70	1.1248	1.1220
0.075	2.5962	2.6005	0.75	1.1044	1.1016
0.10	2.4785	2.4853	0.80	1.0877	1.0849
0.15	2.2201	2.2312	0.85	1.0741	1.0710
0.20	1.9781	1.9914	0.90	1.0628	1.0596
0.25	1.7764	1.7898	0.95	1.0534	1.0502
0.30	1.6168	1.6286	0.100	1.0453	1.0423
0.35	1.4928	1.5022	0.105	1.0384	1.0358
0.40	1.3967	1.4035	0.110	1.0325	1.0304
0.45	1.3219	1.3262	0.115	1.0276	1.0257

This flux disperses in the domain between the two horizons and at the Cauchy horizon it is determined by

$$Z(r_*, t) \to X(v) + Y(u) \quad (v \to \infty;\ u \to \infty), \tag{368}$$

where

$$X(v) = \int_{-\infty}^{+\infty} Z(\sigma)[A(\sigma) - 1]e^{-i\sigma v}\,d\sigma, \tag{369}$$

and

$$Y(u) = \int_{-\infty}^{+\infty} Z(\sigma)B(\sigma)e^{i\sigma u}\,d\sigma. \tag{370}$$

However, our interest is *not* in $X(v)$ or $Y(u)$, *per se*, but rather in quantities related to them.

As we have explained in §38 (p. 214), we are primarily interested in the radiation an observer receives at the instant of his (or her) crossing the Cauchy horizon. To evaluate this quantity, we consider a freely falling observer following a radial geodesic. The four-velocity, U, of the observer is given by equations (65) for $L = 0$; thus,

$$U^t = \frac{r^2}{\Delta}E, \quad U^{r_*} = \frac{r^2}{\Delta}\left(E^2 - \frac{\Delta}{r^2}\right)^{1/2}, \quad \text{and} \quad U^\theta = U^\varphi = 0, \tag{371}$$

where, consistently with the time-like character of the coordinate r in the interval $r_- < r < r_+$, we have chosen the positive square-root in the expression for U^{r_*}. Also, it should be noted that we are allowed to assign negative values for E since the coordinate t is space-like in the same interval.

With the prevalent radiation-field expressed in terms of $Z(r_*, t)$, a measure of the flux of radiation, $\mathscr{F}$, received by the freely falling observer is given by

$$\mathscr{F} = U^j Z_{,j} = \frac{r^2}{\Delta}\left[EZ_{,t} + \left(E^2 - \frac{\Delta}{r^2}\right)^{1/2} Z_{,r_*}\right]. \tag{372}$$

We have seen that as we approach the Cauchy horizon (cf. equations (369) and (370)),

$$Z(r_*, t) \to X(t - r_*) + Y(t + r_*). \tag{373}$$

Accordingly,

$$Z_{,t} \to X_{,-v} + Y_{,u} \quad \text{and} \quad Z_{,r_*} \to -X_{,-v} + Y_{,u}; \tag{374}$$

and the expression (372) for $\mathscr{F}$ becomes

$$\mathscr{F} \to \frac{r^2}{\Delta}\left\{X_{,-v}\left[E - \left(E^2 - \frac{\Delta}{r^2}\right)^{1/2}\right] + Y_{,u}\left[E + \left(E^2 - \frac{\Delta}{r^2}\right)^{1/2}\right]\right\}; \tag{375}$$

On EF, v remains finite while $u \to \infty$; therefore,

$$r_* \to +t, \quad u \to 2r_* \to -\frac{1}{\kappa_-}\lg|r - r_-| \quad \text{as} \quad r \to r_- \text{ on } EF. \tag{376}$$

Also for $E > 0$, the term in $X_{,-v}$ remains finite while the term in $Y_{,u}$ has a divergent factor (namely, $1/\Delta$). Hence,

$$\mathscr{F}_{EF} \to -\frac{2r_-^2}{r_+ - r_-} EY_{,u} e^{\kappa_- u} \quad (u \to \infty \text{ on } EF). \tag{377}$$

On EC', u remains finite while $v \to \infty$; therefore,

$$r_* \to -t, \quad v \to 2r_* \to -\frac{1}{\kappa_-}\lg|r - r_-| \text{ as } \quad r \to r_- \text{ on } EC'. \tag{378}$$

And for $E < 0$, the term in $Y_{,u}$ remains finite while the term in $X_{,-v}$ has the divergent factor. Hence,

$$\mathscr{F}_{EC'} \to +\frac{2r_-^2}{r_+ - r_-}|E| X_{,-v} e^{\kappa_- v} \quad (v \to \infty \text{ on } EC'). \tag{379}$$

We conclude from equations (377) and (399) that the divergence, or otherwise, of the received fluxes on the Cauchy horizon, at EF and EC', depend on

$$Y_{,u} = \int_{-\infty}^{+\infty} i\sigma \frac{R_1(-\sigma)}{T_1(-\sigma)} Z(\sigma) e^{i\sigma u}\,d\sigma \tag{380}$$

and

$$X_{,-v} = \int_{-\infty}^{+\infty} i\sigma \left[\frac{1}{T_1(-\sigma)} - 1\right] Z(\sigma) e^{-i\sigma v}\,d\sigma, \tag{381}$$

where we have substituted for $A(\sigma)$ and $B(\sigma)$ from equation (363). In particular, if we wish to evaluate the infinite integrals, as is naturally suggested, by contour integration, closing the contour appropriately in the upper half-plane (for determining the behaviour of $Y_{,u}$ for $u \to \infty$) and in the lower half-plane (for determining the behaviour of $X_{,-v}$ for $v \to \infty$), then we need to specify the domains of analyticity of $A(\sigma)$ and $B(\sigma)$, as defined in equations (363).

Returning then to the definitions of $A(\sigma)$ and $B(\sigma)$, we can, in accordance with Chapter 4, equations (180), write

$$B(\sigma) = \frac{R_1(-\sigma)}{T_1(-\sigma)} = \frac{1}{2i\sigma}[\,f_2(x, -\sigma), f_1(x, +\sigma)\,] \tag{382}$$

and

$$A(\sigma) = \frac{1}{T_1(-\sigma)} = \frac{1}{2i\sigma}[\,f_1(x, -\sigma), f_2(x, -\sigma)], \tag{383}$$

where, for convenience, we have written x in place of r_* and $f_1(x, \pm\sigma)$, and $f_2(x, +\sigma)$ are solutions of the one-dimensional wave equations which satisfy the boundary conditions

$$\left.\begin{aligned} f_1(x, \pm\sigma) &\to e^{\mp i\sigma x} \qquad x \to +\infty, \\ \text{and} \qquad f_2(x, \pm\sigma) &\to e^{\pm i\sigma x} \qquad x \to -\infty. \end{aligned}\right\} \tag{384}$$

Also, as we have shown in Chapter 4, equations (194) and (198), $f_2(x, -\sigma)$ satisfies the integral equation

$$f_2(x, -\sigma) = e^{-i\sigma x} + \int_{-\infty}^{x} \frac{\sin\sigma(x-x')}{\sigma} V(x') f_2(x', -\sigma)\, dx'. \tag{385}$$

The corresponding integral equation satisfied by $f_1(x, \pm\sigma)$ is

$$f_1(x, \mp\sigma) = e^{\pm i\sigma x} - \int_{x}^{\infty} \frac{\sin\sigma(x-x')}{\sigma} V(x') f_1(x', \mp\sigma)\, dx'. \tag{386}$$

Adapting a more general investigation of Hartle and Wilkins to the simpler circumstances of our present problem, we can determine the domains of analyticity of the functions, $f_2(x, -\sigma)$ and $f_1(x, \pm\sigma)$, in the complex σ-plane, by solving the Volterra integral-equations (385) and (386) by successive iterations. (As is known, iterations of Volterra-type integral equations normally provide solutions whose uniform convergence is guaranteed.) Thus, considering equation (385), we may express its solution as a series in the form

$$f_2(x, -\sigma) = e^{-i\sigma x} + \sum_{n=1}^{\infty} f_2^{(n)}(x, -\sigma), \tag{387}$$

where

$$f_2^{(n)}(x, -\sigma) = \int_{-\infty}^{x} dx_1 \frac{\sin\sigma(x-x_1)}{\sigma} V(x_1) f_2^{(n-1)}(x_1, -\sigma). \tag{388}$$

By this last recurrence relation,

$$f_2^{(n)}(x, -\sigma)$$
$$= \int_{-\infty}^{x_0} dx_1 \int_{-\infty}^{x_1} dx_2 \ldots \int_{-\infty}^{x_{n-1}} dx_n \prod_{i=1}^{n} \frac{\sin\sigma(x_{i-1}-x_i)}{\sigma} V(x_i) e^{-i\sigma x_n} \tag{389}$$

where $x_0 = x$; or, after some rearrangement,

$$f_2^{(n)}(x, -\sigma) = \frac{e^{-i\sigma x}}{(2i\sigma)^n} \int_{-\infty}^{x_0} dx_1 \ldots \int_{-\infty}^{x_{n-1}} dx_n \prod_{i=1}^{n} \left\{ [e^{2i\sigma(x_{i-1}-x_i)} - 1] V(x_i) \right\}. \tag{390}$$

Since

$$V(x) \to \text{constant } e^{2\kappa_+ x} \quad (x \to -\infty), \tag{391}$$

it is manifest that each of the multiplicands in (390) tends to zero, exponentially, for $x \to -\infty$ for all

$$\text{Im}\,\sigma > -\kappa_+. \tag{392}$$

Consequently, $f_2^{(n)}(x, -\sigma)$ exists for each n for $\text{Im}\,\sigma > -\kappa_+$. And it can be shown (as Hartle and Wilkins have shown in their more general context) that the series (387) converges uniformly for all σ with $\text{Im}\,\sigma > -\kappa_+$. *The domain of analyticity of $f_2(x, -\sigma)$, therefore, includes the entire upper half-plane and the infinite strip of width κ_+ in the lower half-plane.**

It can be further shown that there is a sequence of poles along the imaginary axis at $\sigma = -im\kappa_+ (m = 1, 2 \ldots)$. While we shall not show this entirely rigorously, we shall indicate the origin of the singularities. In view of the asymptotic behaviour (391) for $V(x)$, we may, compatible with this behaviour, expect a representation of $V(x)$ for $x < 0$, in the manner of a Laplace transform, by

$$V(x) = \int_{2\kappa_+}^{\infty} d\mu\, \mathscr{V}(\mu) e^{\mu x}, \tag{393}$$

where $\mathscr{V}(\mu)$ includes δ-functions at various locations, i.e., $\mathscr{V}(\mu)$ is a distribution in the technical sense. With the foregoing representation for $V(x)$, the first iterate, $f_2^{(1)}(x, -\sigma)$, of the solution for $f_2(x, -\sigma)$, becomes (cf. equation (389))

$$f_2^{(1)}(x, -\sigma) = \frac{e^{-i\sigma x}}{2i\sigma} \int_{-\infty}^{x} dx_1 \left[e^{2i\sigma(x-x_1)} - 1 \right] \int_{2\kappa_+}^{\infty} d\mu\, \mathscr{V}(\mu) e^{\mu x_1}; \tag{394}$$

* I am grateful to Dr. Roza Trautman for pointing out to me that a result equivalent to the one stated is contained in M. J. Ablowitz, D. J. Kaup, A. C. Newell, and H. Segur, *Studies in Applied Mathematics*, 53, 249 (1974).

or, inverting the order of the integrations, we have

$$f_2^{(1)}(x,\ -\sigma) = \frac{e^{-i\sigma x}}{2i\sigma}\int_{2\kappa_+}^{\infty} d\mu\,\mathscr{V}(\mu)e^{\mu x}\int_{-\infty}^{x} dx_1\,[e^{2i\sigma(x-x_1)}-1]\,e^{\mu(x_1-x)} \tag{395}$$

After effecting the integration over x_1, we are left with

$$f_2^{(1)}(x,\ -\sigma) = e^{-i\sigma x}\int_{2\kappa_+}^{\infty} d\mu\frac{\mathscr{V}(\mu)}{\mu(\mu-2i\sigma)}e^{\mu x}. \tag{396}$$

From this last expression, it is evident that $f_2^{(1)}(x,\ -\sigma)$ has singularities along the negative imaginary axis beginning at $\text{Im}\sigma = -\kappa_+$. Similar results obtain for the higher iterates. While these arguments do not conclusively establish the existence of cuts and singularities, for the sum $\sum f_2^{(n)}(x,\ -\sigma)$, beginning at $\text{Im}\,\sigma = -\kappa_+$, they can be refined to show that this is the case; but we shall not attempt to do so.

Turning now to the integral equation (386) satisfied by $f_1(x,\ \pm\sigma)$, we can show, in the same manner as in the context of $f_2(x,\ -\sigma)$, by solving the equation by iterations, that by virtue, now, of the asymptotic behaviour,

$$V(x) \to \text{constant}\, e^{-2\kappa_- x} \quad (x \to +\infty), \tag{397}$$

the domain of analyticity of $f_1(x,\ -\sigma)$ *includes the entire upper half-plane and an infinite strip of width* κ_- *in the lower half-plane; and the domain of analyticity of* $f_1(x,\ +\sigma)$ *includes the entire lower half-plane and an infinite strip of width* κ_- *in the upper half-plane.*

In view of the domains of analyticity of $f_2(x,\ -\infty)$ and $f_1(x,\ +\sigma)$, it follows that as far as $\sigma B(\sigma)$ is concerned, its domain of analyticity, being the Wronskian of $f_2(x,\ -\sigma)$ and $f_1(x,\ +\sigma)$, certainly includes the strip of width κ_- in the upper half-plane. (It can, in fact, be shown that $f_1(x,\ \sigma)$ is analytic in the entire complex σ-plane except for singularities along the imaginary axis at $\sigma = im\kappa_-(m = 1,2,\ldots)$.) In any event, by evaluating the integral $Y_{,u}$, by closing the contour along the real axis by extending it into the upper half-plane, we can conclude that, so long as $Z(\sigma)$ is analytic in the upper half-plane, that $Y_{,u}$ must fall off at least as rapidly as $e^{-\kappa_- u}$ for $u \to \infty$, i.e.,

$$Y_{,u} \prec e^{-\kappa_- u}, \qquad u \to \infty. \tag{398}$$

It now follows from equation (377) that *the flux* $\mathscr{F}_{EF}$ *at the Cauchy horizon at EF is necessarily bounded.*

Turning next to $\sigma A(\sigma)$, which appears in the integrand of $X_{,-v}$, we observe that, by virtue of what we have shown with respect to $f_2(x,\ -\sigma)$ and $f_1(x,\ -\sigma)$, its domain of analyticity in the lower half-plane includes the strip of width κ_+; and the function has cuts or singularities along the entire negative imaginary axis beginning at $-\kappa_+ i$. Therefore, by evaluating the integral $X_{,-v}$, by a contour in the lower half-plane avoiding the cut along the negative imaginary axis which starts at $\operatorname{Im}\sigma = -\kappa_+$, we shall find, on the assumption that $Z(\sigma)$ is regular in the domain of integration, that

$$X_{,-v} \to \text{constant}\, e^{-\kappa_+ v} \qquad \text{for } v \to \infty. \tag{399}$$

Consequently, by equation (379),

$$\mathscr{F}_{EC'} \to \text{constant}\; e^{(\kappa_- - \kappa_+)v} \qquad (v \to \infty). \tag{400}$$

Since $\kappa_- > \kappa_+$, it follows that *the flux, $\mathscr{F}_{EC'}$, diverges on the Cauchy horizon at EC'*. This establishes that an observer, attempting to cross the Cauchy horizon, by following a time-like geodesic, and to emancipate himself (or herself) from the past, will experience the impact of an infinite flux of radiation at the instant of the crossing even if the perturbation crossing the event horizon is of compact support.

50. Some general observations on the static black-hole solutions

With our consideration of the Schwarzschild and the Reissner–Nordström space-times completed, it is opportune to bring to a focus their similarities and their differences—both equally striking.

As for the similarities: both are of Petrov type-D; both have an event horizon which bounds an inner region incommunicable to an asymptotically flat world outside; in both cases, the event horizon is surrounded by potential barriers which determine the reflective power of the black hole to radiation incident on them; and the potential barriers all belong to a very special class which ensures the equality of their reflective power to waves of axial and of polar types; also, the perturbation equations that follow from the Newman–Penrose formalism allow, in both cases, dual transformations—a fact which, in turn, is related to the special class to which the potential barriers belong. But the variations bring their own surprises: the description of the manner of the interaction of the Reissner–Nordström black-hole, to the simultaneous incidence of electromagnetic and gravitational waves, requires a scattering matrix of order 4; but it is symmetric and unitary! Similarly, the decoupling of the perturbation equations of the Reissner–Nordström black-hole in the Newman–Penrose formalism requires the choice of a special gauge which appeared, first, as a phantom in the treatment of the Schwarzschild perturbations. And so one variation follows another.

And among the differences, the principal one—from which all others derive—is the fact that the Reissner–Nordström space-time has an inner Cauchy horizon which makes its maximal analytic extension reveal the possibility, in principle, of exploring other worlds and emancipating oneself from one's past. But the metamorphosis of a coordinate, space-like outside the event horizon, into one which is time-like as the event horizon is crossed, brings its own danger to such prospects: radiation crossing the event horizon gets amplified, as it progresses towards the Cauchy horizon to an extent that impact with it may prove fatal. And this last passage provides a startling coda to the theme and variations.

BIBLIOGRAPHICAL NOTES

H. Reissner (1874–1967) and G. Nordström (1881–1923) independently derived the solution which has come to be known as the Reissner–Nordström solution:

1. H. Reissner, *Ann. d. Physik*, 50, 106–20 (1916).
2. G. Nordström, *Proc. Kon. Ned. Akad. Wet.*, 20, 1238–45 (1918).

§37. The derivation of the Reissner–Nordström solution is patterned after the derivation of the Schwarzschild solution in §17.

§38. Our discussion of the nature of the Reissner–Nordström space-time follows closely:

3. M. Simpson and R. Penrose, *Internat. J. Theoret. Phys.*, 7, 183–97 (1973).

See also:

4. J. C. Graves and D. R. Brill, *Phys. Rev.*, 120, 1507–13 (1960).
5. S. W. Hawking and G. F. R. Ellis, *The Large-Scale Structure of Space-Time*, 156–9, Cambridge, England, 1973.

§40. For a complete bibliography of papers on the geodesics in the Reissner–Nordström space-time, see:

6. N. A. Sharp, *Gen. Relativity and Gravitation*, 10, 659–70 (1979).

Special reference may be made to:

7. A. R. Prasanna and R. K. Varma, *Pramana*, 8, 229–44 (1977)

which is complete on the formal side. But in none of the papers on the subject are the orbits properly classified and arranged; and they do not point out the similarities and the differences with the orbits in the Schwarzschild space-time.

Fig. 16 is taken from:

8. A. Armenti, Jr., *Nuovo Cim.*, 25B, 442–8 (1973).

The geodesics illustrated in Figs. 15, 17, and 18 were calculated and drawn by Mr. Garret Toomey to whom I am greatly indebted.

The motion of charged particles, with special emphasis on the extreme case $Q_*^2 = M^2$, is considered in:

9. R. Ruffini in *Black Holes*, 497–508, edited by C. DeWitt and B. S. DeWitt, Gordon and Breach Science Publishers, New York, 1973.
10. G. Denardo and R. Ruffini, ibid., R33–44.

The primary emphasis in these papers is on the possibility of extracting charge from a Reissner–Nordström black-hole by encounters with particles of opposite charge.

§41. The fact that the spin-coefficients and the principal null-directions are the same for the Reissner–Nordström black-hole as they are for the Schwarzschild black-hole, except for a redefinition of the 'horizon function' Δ, is pointed out in:

11. S. K. Bose, *J. Math. Phys.*, 16, 772–5 (1975).
12. D. M. Chitre, *Phys. Rev.* D, 13, 2713–19 (1976).
13. C. H. Lee, *J. Math. Phys.*, 17, 1226–35 (1976).

§42. The analysis in this section is based on:

14. S. Chandrasekhar and B. C. Xanthopoulos, *Proc. Roy. Soc.* (*London*) A, 367, 1–14 (1979).

15. B. C. Xanthopoulos, *Proc. Roy. Soc.* (*London*) A, 378, 73–88 (1981).
For alternative treatments using different methods see:
16. V. Moncrief, *Phys. Rev.* D, 9, 2707–9 (1974).
17. ———, ibid., 10, 1057–9 (1974).
18. ———, ibid., 12, 1526–37 (1975).
19. F. J. Zerilli, ibid., 9, 860–8 (1974).

§43. The relations in this section are given in:
20. S. Chandrasekhar, *Proc. Roy. Soc.* (*London*) A, 369, 425–33 (1980).

§44. The analysis in this section follows:
21. S. Chandrasekhar, *Proc. Roy. Soc.* (*London*) A, 365, 453–65 (1979).
See also:
22. A. W. C.-Lun, *Nuovo Cim. Letters*, 10, 681–4 (1976).
23. J.Bicak, *Czech. J. Phys.* B, 29, 945–80 (1979).
24. R. M. Wald, *Proc. Roy. Soc.* (*London*) A, 369, 67–81 (1979).

§45. See reference 21.

§§46–47. Much of the discussions in these sections is new. I am grateful to Dr. R. Sorkin who suggested the introduction of the scattering matrix.

The conversion of gravitational energy into electromagnetic energy in the low-frequency limit has been considered in great detail by:
25. R. A. Matzner, *Phys. Rev.* D, 14, 3274–80 (1976).
See also:
26. D. W. Olson and W. G. Unruh, *Phys. Rev. Letters*, 33, 1116–8 (1974).

§48. The numerical results given in Table V are taken from:
27. D. Gunter, *Phil. Trans. Roy. Soc.* (*London*), 296, 457–526 (1980) and 301, 705–9 (1981).

§49. The importance of examining the development of perturbations on the Cauchy horizon was first recognized and explored by Penrose (see reference 3). The subject has been discussed extensively since that time; and the principal papers are those of:
28. J. M. McNamara, *Proc. Roy. Soc.* (*London*) A, 364, 121–34 (1978).
29. ———, ibid., 358, 499–517 (1978).
30. Y. Gürsel, V. D. Sandberg, I. D. Novikov, and A. A. Starobinsky, *Phys. Rev.* D, 19, 413–20 (1979).
31. Y. Gürsel, I. D. Novikov, V. D. Sandberg, and A. A. Starobinsky, ibid., 20, 1260–70 (1979).
32. R. A. Matzner, N. Zamorano, and V. D. Sandberg, ibid., 19, 2821–6 (1979).
I should add that I find the use of the word 'stability' in these contexts unfortunate: it does not seem to relate to what one is actually investigating, namely, how waves from the outside world, entering the domain between the two horizons via the event horizon, develop as they approach the Cauchy horizon and affect a freely-falling observer.
I am indebted to Dr. S. Detweiler for the amplification factors listed in Table VI.
The existence of bound states of zero-energy in the potential wells V_1 and V_2 is proved in paper 31.
The treatment in this section departs in some essential respects from those in papers 28–32. The principal arguments are derived from:
33. J. B. Hartle and D. C. Wilkins, *Commun. Math. Phys.*, 38, 47–63 (1974).
For a fuller treatment of the finer issues, side-stepped in the account, see:
34. S. Chandrasekhar and J. B. Hartle, *Proc. Roy. Soc.* (*London*) A 384, 301–15 (1982).

6

THE KERR METRIC

51. Introduction

In this chapter, we begin the study of the Kerr solution. It represents, as we have stated in the Prologue, the unique solution which the general theory of relativity provides for the description of all black holes that can occur in the astronomical universe by the gravitational collapse of stellar masses; and it is the only instance of a physical theory providing an exact description of a macroscopic object. In large measure, the preceding chapters are but preludes to the study we now begin.

It has been stated that "there is no constructive analytic derivation of the [Kerr] metric that is adequate in its physical ideas, and even a check of this solution of Einstein's equations involves cumbersome calculations" (Landau and Lifshitz). Contrary to this statement, we shall find that, once the basic equations have been properly written and reduced, the derivation of the Kerr metric is really very simple and proceeds with an adequate base of physical and mathematical motivations.

In this chapter, besides the derivation of the Kerr metric and the establishment of its uniqueness, we shall give a description of the space-time in a Newman–Penrose formalism which makes its type-D character manifest.

52. Equations governing vacuum space-times which are stationary and axisymmetric

As we have shown in Chapter 2 (§11), a metric appropriate to stationary axisymmetric space-times can be written in the form

$$ds^2 = e^{2\nu}(dt)^2 - e^{2\psi}(d\varphi - \omega\, dt)^2 - e^{2\mu_2}(dx^2)^2 - e^{2\mu_3}(dx^3)^2, \qquad (1)$$

where ν, ψ, ω, μ_2, and μ_3 are functions of x^2 and x^3 with the freedom to impose a coordinate condition on μ_2 and μ_3.

We can now write down the components of the Riemann tensor, in a tetrad-frame with the basis one-forms,

$$\omega^0 = e^{\nu}\, dt,\ \omega^1 = e^{\psi}(d\varphi - \omega dt),\ \omega^2 = e^{\mu_2}\, dx^2,\ \text{ and } \ \omega^3 = e^{\mu_3}\, dx^3, \qquad (2)$$

by suitably specializing the expressions listed in Chapter 2, equations (75). We find that the non-vanishing components of the Riemann tensor are

$$
\begin{aligned}
R_{1212} &= +e^{-\psi-\mu_2}(e^{\psi-\mu_2}\psi_{,2})_{,2}+e^{-2\mu_3}\psi_{,3}\mu_{2,3}+\tfrac{1}{4}e^{2\psi-2\mu_2-2\nu}\omega_{,2}{}^2,\\
R_{1313} &= +e^{-\psi-\mu_3}(e^{\psi-\mu_3}\psi_{,3})_{,3}+e^{-2\mu_2}\psi_{,2}\mu_{3,2}+\tfrac{1}{4}e^{2\psi-2\mu_3-2\nu}\omega_{,3}{}^2,\\
R_{1010} &= -e^{-2\mu_2}\psi_{,2}\nu_{,2}-e^{-2\mu_3}\psi_{,3}\nu_{,3}-\tfrac{1}{4}e^{2\psi-2\nu}(e^{-2\mu_2}\omega_{,2}{}^2+e^{-2\mu_3}\omega_{,3}{}^2),\\
R_{2323} &= +e^{-\mu_2-\mu_3}[(e^{\mu_2-\mu_3}\mu_{2,3})_{,3}+(e^{\mu_3-\mu_2}\mu_{3,2})_{,2}],\\
R_{2020} &= -e^{-\mu_2-\nu}(e^{\nu-\mu_2}\nu_{,2})_{,2}-e^{-2\mu_3}\nu_{,3}\mu_{2,3}+\tfrac{3}{4}e^{2\psi-2\mu_2-2\nu}\omega_{,2}{}^2,\\
R_{3030} &= -e^{-\mu_3-\nu}(e^{\nu-\mu_3}\nu_{,3})_{,3}-e^{-2\mu_2}\nu_{,2}\mu_{3,2}+\tfrac{3}{4}e^{2\psi-2\mu_3-2\nu}\omega_{,3}{}^2,\\
R_{1213} &= +e^{-\psi-\mu_3}(e^{\psi-\mu_2}\psi_{,2})_{,3}-e^{-\mu_2-\mu_3}\psi_{,3}\mu_{3,2}+\tfrac{1}{4}e^{2\psi-\mu_2-\mu_3-2\nu}\omega_{,2}\omega_{,3},\\
R_{1220} &= +e^{\psi-2\mu_2-\nu}\omega_{,2}(\psi_{,2}-\tfrac{1}{2}\nu_{,2})+\tfrac{1}{2}e^{-\mu_2-\nu}(e^{\psi-\mu_2}\omega_{,2})_{,2}\\
&\quad+\tfrac{1}{2}e^{\psi-2\mu_3-\nu}\omega_{,3}\mu_{2,3},\\
R_{1330} &= +e^{\psi-2\mu_3-\nu}\omega_{,3}(\psi_{,3}-\tfrac{1}{2}\nu_{,3})+\tfrac{1}{2}e^{-\mu_3-\nu}(e^{\psi-\mu_3}\omega_{,3})_{,3}\\
&\quad+\tfrac{1}{2}e^{\psi-2\mu_2-\nu}\omega_{,2}\mu_{3,2},\\
R_{3002} &= +e^{-\nu-\mu_3}(e^{\nu-\mu_2}\nu_{,2})_{,3}-e^{-\mu_2-\mu_3}\nu_{,3}\mu_{3,2}-\tfrac{3}{4}e^{2\psi-\mu_2-\mu_3-2\nu}\omega_{,2}\omega_{,3},\\
R_{1230} &= +\tfrac{1}{2}e^{-\mu_3-\nu}(e^{\psi-\mu_2}\omega_{,2})_{,3}+\tfrac{1}{2}e^{\psi-\mu_2-\mu_3-\nu}\omega_{,3}(2\psi_{,2}-\mu_{3,2}-\nu_{,2}),\\
R_{1302} &= -\tfrac{1}{2}e^{-\mu_2-\nu}(e^{\psi-\mu_3}\omega_{,3})_2-\tfrac{1}{2}e^{\psi-\mu_2-\mu_3-\nu}\omega_{,2}(2\psi_{,3}-\mu_{2,3}-\nu_{,3}),\\
R_{1023} &= -\tfrac{1}{2}e^{-\mu_2-\mu_3}[(e^{\psi-\nu}\omega_{,3})_{,2}-(e^{\psi-\nu}\omega_{,2})_{,3}]. \qquad (3)
\end{aligned}
$$

The components which vanish are

$$
\left.\begin{aligned}
R_{1210}=R_{1310}=R_{1223}=R_{1332}=R_{1002}=R_{1003}\\
=R_{2330}=R_{3220}=0.
\end{aligned}\right\} \qquad (4)
$$

The components of the Ricci and the Einstein tensors are given by certain linear combinations of the components of the Riemann tensor (specified in Ch. 2, equations (75)); and by setting them equal to zero, we obtain the equations governing stationary axisymmetric vacuum space-times. For our present purposes it will suffice to consider the following equations.

$$
\begin{aligned}
&e^{-2\mu_2}[\nu_{,2;2}+\nu_{,2}(\psi+\nu-\mu_2+\mu_3)_{,2}]+e^{-2\mu_3}[\nu_{,3,3}+\nu_{,3}(\psi+\nu+\mu_2-\mu_3)_{,3}]\\
&\quad=+\tfrac{1}{2}e^{2\psi-2\nu}[e^{-2\mu_2}(\omega_{,2})^2+e^{-2\mu_3}(\omega_{,3})^2] \qquad (R_{00}=0), \quad (5)
\end{aligned}
$$

$$
\begin{aligned}
&e^{-2\mu_2}[\psi_{,2,2}+\psi_{,2}(\psi+\nu-\mu_2+\mu_3)_{,2}]\\
&\qquad+e^{-2\mu_3}[\psi_{,3,3}+\psi_{,3}(\psi+\nu+\mu_2-\mu_3)_{,3}]\\
&\quad=-\tfrac{1}{2}e^{2\psi-2\nu}[e^{-2\mu_2}(\omega_{,2})^2+e^{-2\mu_3}(\omega_{,3})^2] \qquad (R_{11}=0), \quad (6)
\end{aligned}
$$

$$
(e^{3\psi-\nu-\mu_2+\mu_3}\omega_{,2})_{,2}+(e^{3\psi-\nu+\mu_2-\mu_3}\omega_{,3})_{,3}=0 \qquad (R_{01}=0), \quad (7)
$$

$$
\begin{aligned}
&(\psi+\nu)_{,2,3}-(\psi+\nu)_{,2}\mu_{2,3}-(\psi+\nu)_3\mu_{3,2}+\psi_{,2}\psi_{,3}+\nu_{,2}\nu_{,3}\\
&\quad=\tfrac{1}{2}e^{2\psi-2\nu}\omega_{,2}\omega_{,3} \qquad (R_{23}=0), \quad (8)
\end{aligned}
$$

$$
\begin{aligned}
&e^{-2\mu_3}[(\psi+\nu)_{,3,3}+(\psi+\nu)_{,3}(\nu-\mu_3)_{,3}+\psi_{,3}\psi_{,3}]\\
&\qquad+e^{-2\mu_2}[\nu_{,2}(\psi+\mu_3)_{,2}+\psi_{,2}\mu_{3,2}]\\
&\quad=-\tfrac{1}{4}e^{2\psi-2\nu}[e^{-2\mu_2}(\omega_{,2})^2-e^{-2\mu_3}(\omega_{,3})^2] \qquad (G_{22}=0), \quad (9)
\end{aligned}
$$

$$e^{-2\mu_2}[(\psi+\nu)_{,2,2}+(\psi+\nu)_{,2}(\nu-\mu_2)_{,2}+\psi_{,2}\psi_{,2}]$$
$$+e^{-2\mu_3}[\nu_{,3}(\psi+\mu_2)_{,3}+\psi_{,3}\mu_{2,3}]$$
$$=+\tfrac{1}{4}e^{2\psi-2\nu}[e^{-2\mu_2}(\omega_{,2})^2-e^{-2\mu_3}(\omega_{,3})^2] \qquad (G_{33}=0). \quad (10)$$

Letting

$$\beta=\psi+\nu, \tag{11}$$

we can rewrite equations (5) and (6) in the forms

$$(e^{\beta+\mu_3-\mu_2}\nu_{,2})_{,2}+(e^{\beta+\mu_2-\mu_3}\nu_{,3})_{,3}$$
$$=+\tfrac{1}{2}e^{3\psi-\nu}[e^{\mu_3-\mu_2}(\omega_{,2})^2+e^{\mu_2-\mu_3}(\omega_{,3})^2], \tag{12}$$

$$(e^{\beta+\mu_3-\mu_2}\psi_{,2})_{,2}+(e^{\beta+\mu_2-\mu_3}\psi_{,3})_{,3}$$
$$=-\tfrac{1}{2}e^{3\psi-\nu}[e^{\mu_3-\mu_2}(\omega_{,2})^2+e^{\mu_2-\mu_3}(\omega_{,3})^2]. \tag{13}$$

The sum and the difference of these equations give

$$[e^{\mu_3-\mu_2}(e^{\beta})_{,2}]_{,2}+[e^{\mu_2-\mu_3}(e^{\beta})_{,3}]_{,3}=0, \tag{14}$$

$$[e^{\beta+\mu_3-\mu_2}(\psi-\nu)_{,2}]_{,2}+[e^{\beta+\mu_2-\mu_3}(\psi-\nu)_{,3}]_{,3}$$
$$=-e^{3\psi-\nu}[e^{\mu_3-\mu_2}(\omega_{,2})^2+e^{\mu_2-\mu_3}(\omega_{,3})^2]. \tag{15}$$

The addition of equations (9) and (10) yields the same equation (14), while subtraction gives

$$4e^{\mu_3-\mu_2}(\beta_{,2}\mu_{3,2}+\psi_{,2}\nu_{,2})-4e^{\mu_2-\mu_3}(\beta_{,3}\mu_{2,3}+\psi_{,3}\nu_{,3})$$
$$=2e^{-\beta}\{[e^{\mu_3-\mu_2}(e^{\beta})_{,2}]_{,2}-[e^{\mu_2-\mu_3}(e^{\beta})_{,3}]_{,3}\}$$
$$-e^{2\psi-2\nu}[e^{\mu_3-\mu_2}(\omega_{,2})^2-e^{\mu_2-\mu_3}(\omega_{,3})^2]: \tag{16}$$

An examination of the foregoing equations suggests that we consider them in the following sequence.

First, exercising the gauge-freedom, we have to impose a coordinate condition on μ_2 and μ_3, we may specify

$$e^{2(\mu_3-\mu_2)}=\Delta(x^2,x^3) \tag{17}$$

in some convenient manner. Equation (14) then becomes an equation for $\beta=\psi+\nu$. Its solution presents no difficulty. Indeed, we shall show presently that e^{β} can be 'absorbed' as one of the coordinates (see §(*b*) below). Equations (7) and (15) then become a pair of coupled equations for $\psi-\nu$ and ω; and their solution is the central problem in the theory of stationary axisymmetric solutions of Einstein's equation.

The significance of the foregoing grouping of the metric functions, suggested by the structure of the equations, becomes clear when we write the metric (1) in the form

$$ds^2=e^{\beta}\left[\chi(dt)^2-\frac{1}{\chi}(d\varphi-\omega dt)^2\right]-\frac{e^{\mu_2+\mu_3}}{\sqrt{\Delta}}[(dx^2)^2+\Delta(dx^3)^2], \tag{18}$$

where

$$\Delta = e^{2(\mu_3-\mu_2)}, \quad \beta = \psi + \nu, \quad \text{and} \quad \chi = e^{-\psi+\nu}. \tag{19}$$

As we have stated, Δ is at our disposal; β can be solved for, independently of the others; the central problem is to solve for χ and ω; and the solution for $\mu_2+\mu_3$ presents no difficulty of principle once χ and ω are known: it follows by simple quadratures as we shall show in detail in §52, below.

It is important to observe that $-e^{2\beta}$ is the determinant of the metric of the two-dimensional space spanned by the vectors $\partial/\partial t$ and $\partial/\partial\varphi$.

One feature of the equations governing χ and ω is deserving of notice. Rewriting equation (7) in the form

$$\begin{aligned}(e^{3\psi-\nu-\mu_2+\mu_3}\omega\omega_{,2})_{,2}+(e^{3\psi-\nu+\mu_2-\mu_3}\omega\omega_{,3})_{,3}\\ = +e^{3\psi-\nu}[e^{\mu_3-\mu_2}(\omega_{,2})^2+e^{\mu_2-\mu_3}(\omega_{,3})^2],\end{aligned} \tag{20}$$

we can combine it with equation (15) to give

$$\begin{aligned}\{e^{3\psi-\nu-\mu_2+\mu_3}[(\omega^2)_{,2}+2e^{-2\psi+2\nu}(\psi-\nu)_{,2}]\}_{,2}\\ +\{e^{3\psi-\nu+\mu_2-\mu_3}[(\omega^2)_{,3}+2e^{-2\psi+2\nu}(\psi-\nu)_{,3}]\}_{,3}=0,\end{aligned} \tag{21}$$

or, alternatively,

$$[e^{3\psi-\nu-\mu_2+\mu_3}(\chi^2-\omega^2)_{,2}]_{,2}+[e^{3\psi-\nu+\mu_2-\mu_3}(\chi^2-\omega^2)_{,3}]_{,3}=0. \tag{22}$$

Comparison of equations (7) and (21) shows that ω and $\chi^2-\omega^2$, formally, satisfy the same equation.

(*a*) *Conjugate metrics*

An important feature of the equations governing stationary axisymmetric space-times is that starting from a solution represented by, say, (χ, ω), one can obtain other solutions. An example is provided by subjecting the metric (18) to the transformation

$$t \to +i\varphi \quad \text{and} \quad \varphi \to -it. \tag{23}$$

Then

$$\begin{aligned}\chi(\mathrm{d}t)^2-\frac{1}{\chi}(\mathrm{d}\varphi-\omega\,\mathrm{d}t)^2 &\to \frac{1}{\chi}(\mathrm{d}t)^2+\frac{2\omega}{\chi}\mathrm{d}t\,\mathrm{d}\varphi-\frac{\chi^2-\omega^2}{\chi}(\mathrm{d}\varphi)^2\\ &= \left[\tilde\chi(\mathrm{d}t)^2-\frac{1}{\tilde\chi}(\mathrm{d}\varphi-\tilde\omega\,\mathrm{d}t)^2\right],\end{aligned} \tag{24}$$

where

$$\tilde\omega = \frac{\omega}{\chi^2-\omega^2} \quad \text{and} \quad \tilde\chi = \frac{\chi}{\chi^2-\omega^2}. \tag{25}$$

None of the other metric coefficients are affected by this transformation. We conclude that *if* (χ, ω) *represent solutions of the equations, then* $(\tilde\chi, \tilde\omega)$, *also, represent solutions of the same equations.* In §53 we shall explicitly verify that this is the case: it is related to the fact that ω and $\chi^2 - \omega^2$ satisfy the same equation.

We shall call (χ, ω) and $(\tilde{\chi}, \tilde{\omega})$ *conjugate solutions* and the transformation (23), which yields these solutions, *conjugation.*

(b) *The Papapetrou transformation*

With the choice of a gauge,

$$\mu_2 = \mu_3 = \mu \qquad \text{and} \qquad \Delta = 1, \tag{26}$$

the metric (18) becomes

$$ds^2 = e^{\beta}\left[\chi(dt)^2 - \frac{1}{\chi}(d\varphi - \omega dt)^2\right] - e^{2\mu}[(dx^2)^2 + (dx^3)^2], \tag{27}$$

while the equation satisfied by e^{β} becomes

$$(e^{\beta})_{,\alpha,\alpha} = 0 \quad (\alpha = 2, 3). \tag{28}$$

We may now take advantage of the fact, that e^{β} satisfies the two-dimensional Laplace's equation, to consider e^{β} as one of the coordinates and seek a coordinate transformation

$$(x^2, x^3) \rightarrow (\rho, z), \tag{29}$$

such that

$$e^{2\mu}[(dx^2)^2 + (dx^3)^2] \rightarrow f(\rho, z)[(d\rho)^2 + (dz)^2]. \tag{30}$$

It is clear that such a coordinate transformation can be made if (cf. Ch. 2, §11)

$$\left.\begin{aligned} &(\rho_{,2})^2 + (z_{,2})^2 = (\rho_{,3})^2 + (z_{,3})^2, \\ \text{and} \qquad &\rho_{,2}\rho_{,3} + z_{,2}z_{,3} = 0. \end{aligned}\right\} \tag{31}$$

These conditions can be satisfied by setting

$$\rho_{,2} = +z_{,3} \qquad \text{and} \qquad \rho_{,3} = -z_{,2}, \tag{32}$$

equations which we can certainly satisfy by virtue of the equation

$$\rho_{,\alpha,\alpha} = 0 \quad (\rho = e^{\beta}) \tag{33}$$

satisfied by ρ.

With $e^{\beta} = \rho$ chosen as one of the coordinates, the metric (27) takes the form

$$ds^2 = \rho\left[\chi(dt)^2 - \frac{1}{\chi}(d\varphi - \omega dt)^2\right] - e^{2\mu}[(d\rho)^2 + (dz)^2], \tag{34}$$

where χ, ω, and μ are functions of ρ and z. This is the *Papapetrou form* of the metric.

53. The choice of gauge and the reduction of the equations to standard forms

The particular choice of gauge we shall presently make implies no loss of generality; and, indeed, it is not strictly necessary to make at this stage. But it does provide a meaningful physical motivation for the form of the metric we shall seek.

It is convenient to choose the polar angle θ (with respect to the axis of symmetry) as the spatial coordinate x^3. (Later in this chapter, we shall replace θ by $\cos\theta$ as one of the independent variables.)

We shall suppose that the metric allows an *event horizon*, which we shall define, in the present context, *as a smooth two-dimensional, null surface, which is spanned by the tangent vectors, $\partial/\partial t$ and $\partial/\partial\varphi$*—the "Killing vectors" of the space-time.

In conformity with the assumed stationarity and axisymmetry of the space-time, let the equation of the event horizon be

$$N(x^2, x^3) = 0. \tag{35}$$

The condition that it be null is

$$g^{ij}N_{,i}N_{,j} = 0. \tag{36}$$

For the chosen form of the metric, equation (36) gives

$$e^{2(\mu_3-\mu_2)}(N_{,r})^2 + (N_{,\theta})^2 = 0, \tag{37}$$

where we have written $x^2 = r$. Exercising the gauge freedom we have, we shall suppose that

$$e^{2(\mu_3-\mu_2)} = \Delta(r), \tag{38}$$

where $\Delta(r)$ is some function of r which we leave unspecified for the present. From equation (37) it now follows that the equation of the null surface is, in fact, given by

$$\Delta(r) = 0. \tag{39}$$

The second condition that the null-surface be spanned by $\partial/\partial t$ and $\partial/\partial\varphi$ requires that the determinant of the metric of the subspace, (t, φ), vanish on $\Delta(r) = 0$:

$$e^{2\beta} = 0 \quad \text{on} \quad \Delta(r) = 0. \tag{40}$$

Since we have left $\Delta(r)$ unspecified, we may, without loss of generality, suppose that

$$e^{\beta} = \Delta^{1/2} f(r, \theta), \tag{41}$$

where $f(r,\theta)$ is some function of r and θ which is regular on $\Delta(r) = 0$ and on the axis, $\theta = 0$. We shall suppose, instead, that e^{β} has the somewhat more restricted form

$$e^{\beta} = \Delta^{1/2} f(\theta), \tag{42}$$

i.e., e^{β} is separable in the variables r and θ.

With $e^{\mu_3 - \mu_2}$ and e^{β} given by equations (38) and (42), equation (14) for e^{β} becomes

$$[\Delta^{1/2} (\Delta^{1/2})_{,r}]_{,r} + \frac{1}{f} f_{,\theta,\theta} = 0. \tag{43}$$

A solution of this equation, compatible with the requirements of regularity on the axis and convexity of the horizon, is determined by

$$\Delta_{,r,r} = 2 \qquad \text{and} \qquad f = \sin\theta; \tag{44}$$

and the solution for Δ that is appropriate is

$$\Delta = r^2 - 2Mr + a^2, \tag{45}$$

where M and a are constants.* (We shall later show that these constants signify the mass and the angular momentum per unit mass of the black hole.)

Thus, with a choice of gauge that is consistent with the existence of an event horizon, we have the solution

$$e^{\mu_3 - \mu_2} = \Delta^{1/2} \qquad \text{and} \qquad e^{\beta} = \Delta^{1/2} \sin\theta, \tag{46}$$

where Δ is given by equation (45).

The fact that the choice of the solutions for e^{β} and $e^{\mu_3 - \mu_2}$ we have made implies no loss of generality becomes apparent when it is noted that they are consistent with the Papapetrou form of the metric for

$$\rho = e^{\beta} = \Delta^{1/2} \sin\theta \qquad \text{and} \qquad z = (r - M)\cos\theta. \tag{47}$$

With the solutions for $e^{\mu_3 - \mu_2}$ and e^{β} given in equations (46) and by using

$$\mu = \cos\theta, \tag{48}$$

(instead of θ) as the variable indicated by the index '3', we can bring equations (15) and (7), respectively to the forms

$$[\Delta(\psi - \nu)_{,2}]_{,2} + [\delta(\psi - \nu)_{,3}]_{,3} = -e^{2(\psi - \nu)}[\Delta(\omega_{,2})^2 + \delta(\omega_{,3})^2], \tag{49}$$

and

$$[\Delta e^{2(\psi - \nu)} \omega_{,2}]_{,2} + [\delta e^{2(\psi - \nu)} \omega_{,3}]_{,3} = 0, \tag{50}$$

* It should be noted that, while M and a are here introduced as constants of integration, Δ and $(M^2 - a^2)$ are invariant to the transformation (due to Xanthopolous),

$$M \to M' = (M^2 - a^2)^{1/2}/p, \; a \to a' = q(M^2 - a^2)^{1/2}/p, \quad \text{and} \quad r - M \to r' - M',$$

where p and q are constants and $p^2 + q^2 = 1$.

where we have written

$$\delta = 1 - \mu^2 = \sin^2\theta. \tag{51}$$

Letting χ have the meaning given in equation (19), we can rewrite equations (49) and (50) in the forms

$$\left(\frac{\Delta}{\chi}\chi_{,2}\right)_{,2} + \left(\frac{\delta}{\chi}\chi_{,3}\right)_{,3} = \frac{1}{\chi^2}[\Delta(\omega_{,2})^2 + \delta(\omega_{,3})^2], \tag{52}$$

$$\left(\frac{\Delta}{\chi^2}\omega_{,2}\right)_{,2} + \left(\frac{\delta}{\chi^2}\omega_{,3}\right)_{,3} = 0; \tag{53}$$

or, alternatively,

$$\chi[(\Delta\chi_{,2})_{,2} + (\delta\chi_{,3})_{,3}] = \Delta[(\chi_{,2})^2 + (\omega_{,2})^2] + \delta[(\chi_{,3})^2 + (\omega_{,3})^2], \tag{54}$$

$$\chi[(\Delta\omega_{,2})_{,2} + (\delta\omega_{,3})_{,3}] = 2\Delta\chi_{,2}\omega_{,2} + 2\delta\chi_{,3}\omega_{,3}. \tag{55}$$

Now letting

$$X = \chi + \omega \quad \text{and} \quad Y = \chi - \omega, \tag{56}$$

we obtain the pair of symmetric equations

$$\tfrac{1}{2}(X+Y)[(\Delta X_{,2})_{,2} + (\delta X_{,3})_{,3}] = \Delta(X_{,2})^2 + \delta(X_{,3})^2, \tag{57}$$

$$\tfrac{1}{2}(X+Y)[(\Delta Y_{,2})_{,2} + (\delta Y_{,3})_{,3}] = \Delta(Y_{,2})^2 + \delta(Y_{,3})^2. \tag{58}$$

Equations (8) and (16) (which determine $\mu_2 + \mu_3$), after some elementary reductions, become

$$-\frac{\mu}{\delta}(\mu_3 + \mu_2)_{,2} + \frac{r-M}{\Delta}(\mu_3 + \mu_2)_{,3} = \frac{2}{(X+Y)^2}(X_{,2}Y_{,3} + Y_{,2}X_{,3}) \tag{59}$$

and

$$2(r-M)(\mu_3+\mu_2)_{,2} + 2\mu(\mu_3+\mu_2)_{,3}$$
$$= \frac{4}{(X+Y)^2}(\Delta X_{,2}Y_{,2} - \delta X_{,3}Y_{,3}) - 3\frac{(r-M)^2 - \Delta}{\Delta} + \frac{\mu^2+\delta}{\delta}. \tag{60}$$

It is now manifest that once equations (57) and (58) have been solved for X and Y, the solution for $(\mu_2 + \mu_3)$ follows from equations (59) and (60) by simple quadratures.

Equations (57)–(60) can be written in more symmetrical forms by changing r to the new variable

$$\left.\begin{aligned} \eta &= (r-M)/(M^2-a^2)^{1/2} \\ \text{when} \qquad \Delta &= (M^2-a^2)(\eta^2-1). \end{aligned}\right\} \tag{61}$$

Thus, we find

$$\tfrac{1}{2}(X+Y)\{[(\eta^2-1)X_{,\eta}]_{,\eta}+[(1-\mu^2)X_{,\mu}]_{,\mu}\}$$
$$=(\eta^2-1)(X_{,\eta})^2+(1-\mu^2)(X_{,\mu})^2, \tag{62}$$

$$\tfrac{1}{2}(X+Y)\{[(\eta^2-1)Y_{,\eta}]_{,\eta}+[(1-\mu^2)Y_{,\mu}]_{,\mu}\}$$
$$=(\eta^2-1)(Y_{,\eta})^2+(1-\mu^2)(Y_{,\mu})^2, \tag{63}$$

$$-\frac{\mu}{1-\mu^2}(\mu_3+\mu_2)_{,\eta}+\frac{\eta}{\eta^2-1}(\mu_3+\mu_2)_{,\mu}=\frac{2}{(X+Y)^2}(X_{,\eta}Y_{,\mu}+Y_{,\eta}X_{,\mu}), \tag{64}$$

$$2\eta(\mu_3+\mu_2)_{,\eta}+2\mu(\mu_3+\mu_2)_{,\mu}$$
$$=\frac{4}{(X+Y)^2}[(\eta^2-1)X_{,\eta}Y_{,\eta}-(1-\mu^2)X_{,\mu}Y_{,\mu}]-\frac{3}{\eta^2-1}+\frac{1}{1-\mu^2}. \tag{65}$$

A convenient form of equations (62) and (63) which enables one to find by inspection some special solutions is obtained by the transformation

$$X=\frac{1+F}{1-F}\quad\text{and}\quad Y=\frac{1+G}{1-G}. \tag{66}$$

We find

$$(1-FG)\{[(\eta^2-1)F_{,\eta}]_{,\eta}+[(1-\mu^2)F_{,\mu}]_{,\mu}\}$$
$$=-2G[(\eta^2-1)(F_{,\eta})^2+(1-\mu^2)(F_{,\mu})^2], \tag{67}$$

and

$$(1-FG)\{[(\eta^2-1)G_{,\eta}]_{,\eta}+[(1-\mu^2)G_{,\mu}]_{,\mu}\}$$
$$=-2F[(\eta^2-1)(G_{,\eta})^2+(1-\mu^2)(G_{,\mu})^2]. \tag{68}$$

The metric functions χ and ω are related to F and G by

$$\chi=\frac{1-FG}{(1-F)(1-G)}\quad\text{and}\quad\omega=\frac{F-G}{(1-F)(1-G)}. \tag{69}$$

We may parenthetically note that equations (67) and (68) allow the simple solution

$$F=-p\eta-q\mu\quad\text{and}\quad G=-p\eta+q\mu, \tag{70}$$

where p and q are real constants restricted by the condition

$$p^2-q^2=1. \tag{70$'$}$$

(*a*) *Some properties of the equations governing X and Y*

Let (X, Y) represent a solution of equations (62) and (63). Then the following are also solutions.

(i) (Y, X) also represents a solution. For this solution χ remains unchanged while ω changes sign—a trivial change equivalent to letting $\varphi\to-\varphi$.

(ii) $(X+c, Y-c)$, where c is an arbitrary constant, also represents a solution. For this solution, χ again remains unchanged while $\omega \to \omega + 2c$—again, a trivial change equivalent to letting $\varphi \to \varphi - 2ct$.

(iii) (X^{-1}, Y^{-1}) and (Y^{-1}, X^{-1}) also represent solutions. This fact follows from the invariance of equations (67) and (68) to simultaneous changes in the signs of F and G (when X and Y are replaced by their reciprocals). To ascertain the significance of these new solutions, let the metric functions derived from the solution (Y^{-1}, X^{-1}) be distinguished from the functions derived from the solution (X, Y) by a tilde. Then

$$\left.\begin{aligned} 2\tilde{\chi} &= \left(\frac{1}{Y}+\frac{1}{X}\right) = \frac{X+Y}{XY} = \frac{2\chi}{\chi^2-\omega^2} \\ \text{and}\qquad 2\tilde{\omega} &= \left(\frac{1}{Y}-\frac{1}{X}\right) = \frac{X-Y}{XY} = \frac{2\omega}{\chi^2-\omega^2}. \end{aligned}\right\} \tag{71}$$

Comparison with equation (25) shows that the solutions derived from (Y^{-1}, X^{-1}) are the *conjugates* of the solutions derived from (X, Y). Therefore, *the transformation* $(X, Y) \to (Y^{-1}, X^{-1})$ *is equivalent to conjugation.*

(iv) By combining the results of (i), (ii), and (iii), we infer that if (X, Y) represents a solution, then so does

$$[X/(1+cX),\ Y/(1-cY)]$$

where c is an arbitrary constant. But by these transformations we do not obtain any really new solutions beyond conjugation.

(*b*) *Alternative forms of the equations*

Returning to equations (7), (12), and (13), we find that, with our choice of gauge and the solution for e^{β}, they become

$$\left(e^{4\psi}\frac{\omega_{,r}}{\sin\theta}\right)_{,r} + \left(e^{4\psi}\frac{\omega_{,\theta}}{\Delta\sin\theta}\right)_{,\theta} = 0, \tag{72}$$

$$[(\Delta\sin\theta)\nu_{,r}]_{,r} + [(\sin\theta)\nu_{,\theta}]_{,\theta} = +\frac{e^{4\psi}}{2\Delta\sin\theta}[\Delta(\omega_{,r})^2 + (\omega_{,\theta})^2], \tag{73}$$

$$[(\Delta\sin\theta)\psi_{,r}]_{,r} + [(\sin\theta)\psi_{,\theta}]_{,\theta} = -\frac{e^{4\psi}}{2\Delta\sin\theta}[\Delta(\omega_{,r})^2 + (\omega_{,\theta})^2]. \tag{74}$$

It is apparent from equation (72) that ω can be derived from a 'potential' Y in the manner

$$e^{4\psi}\frac{\omega_{,r}}{\sin\theta} = Y_{,\theta} \qquad\text{and}\qquad e^{4\psi}\frac{\omega_{,\theta}}{\Delta\sin\theta} = -Y_{,r}, \tag{75}$$

so that

$$\omega_{,r} = +e^{-4\psi} Y_{,\theta} \sin\theta \qquad \text{and} \qquad \omega_{,\theta} = -e^{-4\psi} Y_{,r} \Delta \sin\theta. \tag{76}$$

The integrability condition of this last equation is

$$(e^{-4\psi}\Delta Y_{,r})_{,r} + \frac{1}{\sin\theta}(e^{-4\psi} Y_{,\theta}\sin\theta)_{,\theta} = 0; \tag{77}$$

and equation (74), expressed in terms of Y becomes

$$(\Delta\psi_{,r})_{,r} + \frac{1}{\sin\theta}(\psi_{,\theta}\sin\theta)_{,\theta} = -\tfrac{1}{2}e^{-4\psi}[(Y_{,\theta})^2 + \Delta(Y_{,r})^2]. \tag{78}$$

Now letting

$$X = e^{2\psi}, \tag{79}$$

and reverting to the variables η and μ, we have the equations

$$\left[(\eta^2-1)\frac{X_{,\eta}}{X}\right]_{,\eta} + \left[(1-\mu^2)\frac{X_{,\mu}}{X}\right]_{,\mu} = -\frac{1}{X^2}[(\eta^2-1)(Y_{,\eta})^2 + (1-\mu^2)(Y_{,\mu})^2], \tag{80}$$

$$\left[(\eta^2-1)\frac{Y_{,\eta}}{X^2}\right]_{,\eta} + \left[(1-\mu^2)\frac{Y_{,\mu}}{X^2}\right]_{,\mu} = 0. \tag{81}^*$$

The behaviours of the solutions of equations (80) and (81) for $\eta \to \infty$, compatible with the requirements of asymptotic flatness of the space-time, can be obtained as follows.

For our present purposes, it will suffice to state that asymptotic flatness requires that

$$e^{2\nu} \to 1 - \frac{2M^*}{r} + O(r^{-2}) \quad \text{and} \quad \omega \to \frac{2J}{r^3} + O(r^{-4}) \quad \text{as} \quad r \to \infty, \tag{82}$$

where M^* (not necessarily the same as M introduced in our definition of Δ—see footnote on p. 279) denotes the mass and J the angular momentum of the source. The foregoing behaviours are seen to be consistent with equation (73); and, as we shall see presently, they are also consistent with equations (80) and (81).

In terms of the variable η, the required behaviours are

$$e^{2\nu} \to 1 - \frac{2M^*}{(M^2-a^2)^{1/2}\eta} + O(\eta^{-2}) \quad \text{and} \quad \omega \to \frac{J}{(M^2-a^2)^{3/2}\eta^3} + O(\eta^{-4}). \tag{83}$$

* Since there will be no occasion to use equations (62) and (63) and (80) and (81) at the same time, the use of X and Y with different meanings, in the two pairs of equations, is not likely to cause any confusion.

Since

$$e^{2\psi+2\nu} = Xe^{2\nu} = \Delta(1-\mu^2) = (M^2-a^2)(\eta^2-1)(1-\mu^2), \tag{84}$$

we conclude that

$$X \to (M^2-a^2)(1-\mu^2)\eta^2\left[1+\frac{2M^*}{(M^2-a^2)^{1/2}\eta}\right]+O(1). \tag{85}$$

From equation (81) it now follows that consistent with this behaviour of X,

$$Y \to 2J\mu(3-\mu^2)+O(\eta^{-1}); \tag{86}$$

and we verify that the corresponding behaviour of ω derived with the aid of equations (75) and (85) is consistent with the requirement that $\omega \to 2Jr^{-3}$ as $r \to \infty$.

(*c*) *The Ernst equation*

A still another form of the basic equations, which has played a central role in the investigations related to the finding of all stationary axisymmetric solutions of Einstein's equations for the vacuum, is due to Ernst. Besides, his equation provides the shortest and the simplest route to the Kerr metric.

First, we observe that equation (53) allows ω to be derived from a potential Φ in the manner:

$$\Phi_{,2} = \frac{\delta}{\chi^2}\omega_{,3} \quad \text{and} \quad \Phi_{,3} = -\frac{\Delta}{\chi^2}\omega_{,2}. \tag{87}$$

The potential Φ will be governed by the equation

$$\left(\frac{\chi^2}{\delta}\Phi_{,2}\right)_{,2}+\left(\frac{\chi^2}{\Delta}\Phi_{,3}\right)_{,3} = 0; \tag{88}$$

and, expressed in terms of Φ, equation (52) becomes

$$[\Delta(\lg\chi)_{,2}]_{,2}+[\delta(\lg\chi)_{,3}]_{,3} = \frac{\chi^2}{\Delta}(\Phi_{,3})^2+\frac{\chi^2}{\delta}(\Phi_{,2})^2. \tag{89}$$

Letting

$$\Psi = \sqrt{(\Delta\delta)}/\chi, \tag{90}$$

we find that equations (88) and (89) can be reduced to the forms (cf. equations (54) and (55))

$$\Psi[(\Delta\Psi_{,2})_{,2}+(\delta\Psi_{,3})_{,3}] = \Delta[(\Psi_{,2})^2-(\Phi_{,2})^2]+\delta[(\Psi_{,3})^2-(\Phi_{,3})^2], \tag{91}$$

$$\Psi[(\Delta\Phi_{,2})_{,2}+(\delta\Phi_{,3})_{,3}] = 2\Delta\Psi_{,2}\Phi_{,2}+2\delta\Psi_{,3}\Phi_{,3}. \tag{92}$$

Now expressing Ψ and Φ as the real and the imaginary parts of a complex function,

$$Z = \Psi + i\Phi, \tag{93}$$

we can combine equations (91) and (92) into the single equation

$$\mathrm{Re}\,(Z)\,[(\Delta Z_{,2})_{,2} + (\delta Z_{,3})_{,3}] = \Delta(Z_{,2})^2 + \delta(Z_{,3})^2. \tag{94}$$

From the close similarity of equation (94) with the pair of equations (57) and (58) governing X and Y, it follows that properties, analogous to those enumerated in §(a) in the context of equations (57) and (58), must also exist in the context of equation (94). Thus, if Z is a solution, then so is Z^{-1}; and, more generally, $Z/(1+icZ)$ is also a solution where c is an arbitrary real constant. The transformation, $Z \to Z/(1+icZ)$ is equivalent to what is generally called '*Ehler's transformation*'. Ehler's transformation, unlike the corresponding transformation allowed by the equations governing X and Y, does provide significantly new solutions derived from the circumstance that an integration is involved in passing from Z to the metric functions.

By the transformation,

$$Z = -\frac{1+E}{1-E}, \tag{95}$$

(analogous to the transformation (66)), equation (94) becomes

$$(1-EE^*)\,[(\Delta E_{,2})_{,2} + (\delta E_{,3})_{,3}] = -2E^*[\Delta(E_{,2})^2 + \delta(E_{,3})^2]. \tag{96}$$

This is Ernst's equation.

We have seen that associated with a solution (χ, ω), we have the conjugate solution,

$$\tilde{\chi} = \frac{\chi}{\chi^2 - \omega^2} \qquad \text{and} \qquad \tilde{\omega} = \frac{\omega}{\chi^2 - \omega^2}. \tag{97}$$

Accordingly, we may derive a conjugate Ernst-equation. Thus, by defining

$$\tilde{\Psi} = \frac{\sqrt{(\Delta\delta)}}{\tilde{\chi}} = e^{\psi+\nu}\frac{\chi^2-\omega^2}{\chi} = e^{2\nu} - \omega^2 e^{2\psi}; \tag{98}$$

$$\tilde{\Phi}_{,2} = \frac{\delta}{\tilde{\chi}^2}\tilde{\omega}_{,3} = \frac{\tilde{\Psi}^2}{\Delta}\tilde{\omega}_{,3}; \; \tilde{\Phi}_{,3} = -\frac{\Delta}{\tilde{\chi}^2}\tilde{\omega}_{,2} = -\frac{\tilde{\Psi}^2}{\delta}\tilde{\omega}_{,2}; \tag{99}$$

and

$$\tilde{Z} = \tilde{\Psi} + i\tilde{\Phi} = -\frac{1+\tilde{E}}{1-\tilde{E}}, \tag{100}$$

we shall have

$$(1-\tilde{E}\tilde{E}^*)\,[(\Delta\tilde{E}_{,2})_{,2} + (\delta\tilde{E}_{,3})_{,3}] = -2\tilde{E}^*[\Delta(\tilde{E}_{,2})^2 + \delta(\tilde{E}_{,3})^2]; \tag{101}$$

or, reverting to the variables η and μ, we have

$$\begin{aligned}(1-\tilde{E}\tilde{E}^*)\,&\{[(\eta^2-1)\tilde{E}_{,\eta}]_{,\eta} + [(1-\mu^2)\tilde{E}_{,\mu}]_{,\mu}\}\\ &= -2\tilde{E}^*[(\eta^2-1)\,(\tilde{E}_{,\eta})^2 + (1-\mu^2)\,(\tilde{E}_{,\mu})^2].\end{aligned} \tag{102}$$

In terms of $\tilde{E}$, the metric functions $\tilde{\Psi}$ and $\tilde{\Phi}$ are given by

$$\tilde{\Psi} = \operatorname{Re}\tilde{Z} = -\frac{1-\tilde{E}\tilde{E}^*}{|1-\tilde{E}|^2} \quad \text{and} \quad \tilde{\Phi} = \operatorname{Im}\tilde{Z} = \frac{i(\tilde{E}-\tilde{E}^*)}{|1-\tilde{E}|^2}. \tag{103}$$

54. The derivation of the Kerr metric

It is a simple matter to verify that Ernst's equation (102) allows the elementary solution (analogous to the solution (70) of equations (67) and (68))

$$\tilde{E} = -p\eta - iq\mu, \tag{104}$$

where

$$p^2 + q^2 = 1 \tag{105}$$

and p and q are real constants. (We have chosen to write $-p$ and $-q$ in the solution (104) for later convenience.) The corresponding solution for $\tilde{Z}$ is

$$\tilde{Z} = \tilde{\Psi} + i\tilde{\Phi} = -\frac{1-p\eta-iq\mu}{1+p\eta+iq\mu}. \tag{106}$$

Separating the real and the imaginary parts of $\tilde{Z}$, we have

$$\left.\begin{aligned} \tilde{\Psi} &= \frac{p^2\eta^2+q^2\mu^2-1}{1+2p\eta+p^2\eta^2+q^2\mu^2} = \frac{p^2(\eta^2-1)-q^2(1-\mu^2)}{(p\eta+1)^2+q^2\mu^2}, \\ \text{and} \qquad \tilde{\Phi} &= \frac{2q\mu}{(p\eta+1)^2+q^2\mu^2}, \end{aligned}\right\} \tag{107}$$

or, reverting to the variable r,

$$\tilde{\Psi} = \frac{\Delta - [q^2(M^2-a^2)/p^2]\delta}{[(r-M)+(M^2-a^2)^{1/2}/p]^2 + [q^2(M^2-a^2)/p^2]\mu^2}, \tag{108}$$

and

$$\tilde{\Phi} = \frac{2[q(M^2-a^2)/p]\mu}{[(r-M)+(M^2-a^2)^{1/2}/p]^2 + [q^2(M^2-a^2)/p^2]\mu^2}. \tag{109}$$

With the choices,

$$p = (M^2-a^2)^{1/2}/M \qquad \text{and} \qquad q = a/M, \tag{110}$$

consistent with the condition (105), the solutions for $\tilde{\Psi}$ and $\tilde{\Phi}$ simplify considerably to give

$$\tilde{\Psi} = \frac{1}{\rho^2}(\Delta - a^2\delta) \qquad \text{and} \qquad \tilde{\Phi} = \frac{2aM\mu}{\rho^2}, \tag{111}$$

where

$$\rho^2 = r^2 + a^2\mu^2 = r^2 + a^2\cos^2\theta. \tag{112}$$

Now, by equations (99),

$$\left.\begin{aligned}\tilde{\Phi}_{,2} &= -\frac{4aMr\mu}{\rho^4} = \frac{\tilde{\Psi}^2}{\Delta}\tilde{\omega}_{,3} = \frac{(\Delta - a^2\delta)^2}{\rho^4\Delta}\tilde{\omega}_{,3}, \\ \text{and}\qquad & \\ \tilde{\Phi}_{,3} &= \frac{2aM}{\rho^4}(r^2 - a^2\mu^2) = -\frac{\tilde{\Psi}^2}{\delta}\tilde{\omega}_{,2} = -\frac{(\Delta - a^2\delta)^2}{\rho^4\delta}\tilde{\omega}_{,2}.\end{aligned}\right\} \tag{113}$$

Accordingly,

$$\tilde{\omega}_{,3} = -\frac{4aMr\mu\Delta}{(\Delta - a^2\delta)^2} \quad \text{and} \quad \tilde{\omega}_{,2} = -\frac{2aM(r^2 - a^2\mu^2)\delta}{(\Delta - a^2\delta)^2}; \tag{114}$$

and the solution for $\tilde{\omega}$ is

$$\tilde{\omega} = \frac{\omega}{\chi^2 - \omega^2} = \frac{2aMr\delta}{\Delta - a^2\delta}. \tag{115}$$

We also have (cf. equation (98))

$$\tilde{\Psi} = e^{2\psi}(\chi^2 - \omega^2) = e^{2\nu} - \omega^2 e^{2\psi} = \frac{1}{\rho^2}(\Delta - a^2\delta). \tag{116}$$

Making use of this last equation, we can write

$$\omega = \frac{2aMr\delta}{\Delta - a^2\delta}(\chi^2 - \omega^2) = \frac{2aMr\delta}{\rho^2}e^{-2\psi}. \tag{117}$$

We can now combine equations (116) and (117) to give

$$\frac{\Delta - a^2\delta}{\rho^2}e^{2\psi} = e^{2\beta} - \omega^2 e^{4\psi} = \frac{\delta}{\rho^4}(\Delta\rho^4 - 4a^2M^2r^2\delta), \tag{118}$$

where we have made use of our knowledge of $e^{2\beta}$ ($= \Delta\delta$). The solutions for ω and $e^{2\psi}$, which follow from equations (117) and (118), take simple forms by virtue of the following readily verifiable identities:

$$[(r^2 + a^2) \mp a\sqrt{(\Delta\delta)}](\sqrt{\Delta} \pm a\sqrt{\delta}) = \rho^2\sqrt{\Delta} \pm 2aMr\sqrt{\delta}, \tag{119}$$

and

$$\Sigma^2(\Delta - a^2\delta) = \rho^4\Delta - 4a^2M^2r^2\delta, \tag{120}$$

where

$$\Sigma^2 = (r^2 + a^2)^2 - a^2\Delta\delta. \tag{121}$$

Thus, making use of the identity (120), we find from equation (118) that

$$e^{2\psi} = \frac{\delta\Sigma^2}{\rho^2}. \tag{122}$$

From equation (117) it now follows that

$$\omega = \frac{2aMr}{\Sigma^2}. \tag{123}$$

Also,

$$e^{2\nu} = e^{2\beta-2\psi} = \frac{\rho^2\Delta}{\Sigma^2}, \tag{124}$$

and

$$\chi = e^{-\psi+\nu} = \frac{\rho^2\sqrt{\Delta}}{\Sigma^2\sqrt{\delta}}. \tag{125}$$

Next, making use of the identity (119), we find from equations (123) and (125) that

$$X = \chi+\omega = \frac{\sqrt{\Delta}+a\sqrt{\delta}}{[(r^2+a^2)+a\sqrt{(\Delta\delta)}]\sqrt{\delta}} \tag{126}$$

and

$$Y = \chi-\omega = \frac{\sqrt{\Delta}-a\sqrt{\delta}}{[(r^2+a^2)-a\sqrt{(\Delta\delta)}]\sqrt{\delta}}. \tag{127}$$

Finally, to complete the solution, we turn to equations (59) and (60). The reduction of these equations is facilitated by the following formulae giving the derivatives of X and Y:

$$\left.\begin{aligned}
X_{,2} &= \frac{1}{[(r^2+a^2)+a\sqrt{(\Delta\delta)}]^2\sqrt{(\Delta\delta)}}[\rho^2(r-M)-2r(\sqrt{\Delta}+a\sqrt{\delta})\sqrt{\Delta}],\\
X_{,3} &= \frac{\mu\sqrt{\Delta}}{[(r^2+a^2)+a\sqrt{(\Delta\delta)}]^2\delta^{3/2}}[(r^2+a^2)+a^2\delta+2a\sqrt{(\Delta\delta)}],\\
Y_{,2} &= \frac{1}{[(r^2+a^2)-a\sqrt{(\Delta\delta)}]^2\sqrt{(\Delta\delta)}}[\rho^2(r-M)-2r(\sqrt{\Delta}-a\sqrt{\delta})\sqrt{\Delta}],\\
Y_{,3} &= \frac{\mu\sqrt{\Delta}}{[(r^2+a^2)-a\sqrt{(\Delta\delta)}]^2\delta^{3/2}}[(r^2+a^2)+a^2\delta-2a\sqrt{(\Delta\delta)}].
\end{aligned}\right\} \tag{128}$$

Making use of the foregoing formulae, we find that equations (59) and (60) become

$$-\frac{\mu}{\delta}(\mu_3+\mu_2)_{,2}+\frac{r-M}{\Delta}(\mu_3+\mu_2)_{,3} = \frac{\mu}{\rho^2\Delta\delta}[(r-M)(\rho^2+2a^2\delta)-2r\Delta] \tag{129}$$

and

$$2(r-M)(\mu_3+\mu_2)_{,2}+2\mu(\mu_3+\mu_2)_{,3} = 4-\frac{2(r-M)^2}{\Delta}-\frac{4rM}{\rho^2}. \tag{130}$$

We readily verify that the solution of these equations is given by

$$e^{\mu_3+\mu_2} = \frac{\rho^2}{\sqrt{\Delta}}; \tag{131}$$

and since $e^{\mu_3-\mu_2} = \sqrt{\Delta}$ (by our choice of gauge), the solutions for e^{μ_3} and e^{μ_2}, separately, are

$$e^{2\mu_2} = \rho^2/\Delta \quad \text{and} \quad e^{2\mu_3} = \rho^2. \tag{132}$$

We have now completed the solution for all the metric coefficients; and the metric is

$$ds^2 = \rho^2\frac{\Delta}{\Sigma^2}(dt)^2 - \frac{\Sigma^2}{\rho^2}\left(d\varphi - \frac{2aMr}{\Sigma^2}dt\right)^2 \sin^2\theta - \frac{\rho^2}{\Delta}(dr)^2 - \rho^2(d\theta)^2. \tag{133}$$

This is the Kerr metric. The covariant and the contravariant forms of it are:

$$(g_{ij}) = \begin{vmatrix} 1-2Mr/\rho^2 & 0 & 0 & 2aMr\sin^2\theta/\rho^2 \\ 0 & -\rho^2/\Delta & 0 & 0 \\ 0 & 0 & -\rho^2 & 0 \\ 2aMr\sin^2\theta/\rho^2 & 0 & 0 & -[(r^2+a^2)+2a^2Mr\sin^2\theta/\rho^2]\sin^2\theta \end{vmatrix} \tag{134}$$

$$(g^{ij}) = \begin{vmatrix} \Sigma^2/\rho^2\Delta & 0 & 0 & 2aMr/\rho^2\Delta \\ 0 & -\Delta/\rho^2 & 0 & 0 \\ 0 & 0 & -1/\rho^2 & 0 \\ 2aMr/\rho^2\Delta & 0 & 0 & -(\Delta-a^2\sin^2\theta)/\rho^2\Delta\sin^2\theta \end{vmatrix}. \tag{135}$$

We observe that when $a = 0$, the Kerr metric (133) reduces to the Schwarzschild metric in its standard form. Also, from the asymptotic behaviours,

$$\left.\begin{aligned} &e^{2\nu} \to 1 - \frac{2M}{r} + O(r^{-2}),\ e^{2\psi} \to r^2\sin^2\theta + O(r),\ \omega \to \frac{2aM}{r^3} + O(r^{-4}), \\ &e^{-2\mu_2} \to 1 - \frac{2M}{r} + O(r^{-2}), \quad \text{and} \quad e^{2\mu_3} \to r^2 + O(r) \quad (r \to \infty), \end{aligned}\right\} \tag{136}$$

of the various metric coefficients, it is further clear that the metric approaches the Schwarzschild metric for $r \to \infty$. We conclude that the Kerr metric is asymptotically flat and that the parameter M is to be identified with the mass of the black hole. Again, from the interpretation of ω as representing the dragging of the inertial frame, we conclude from its asymptotic behaviour ($\omega \to 2aMr^{-3}$) that a is to be interpreted as the angular momentum per unit mass of black hole.

Finally, we may note the following algebraic relations among the metric functions which we shall find useful:

$$a - (r^2+a^2)\omega = a\frac{\Delta\rho^2}{\Sigma^2} \quad \text{and} \quad 1 - a\omega\sin^2\theta = (r^2+a^2)\frac{\rho^2}{\Sigma^2}. \tag{137}$$

(a) The tetrad components of the Riemann tensor

There is, in principle, no difficulty in inserting the various metric coefficients of the Kerr solution in the expressions listed in equation (3) and obtaining the components of the Riemann tensor. The required reductions can be simplified by first noting that, besides the components listed in equation (4) which vanish and the cyclic identity,

$$R_{1230}+R_{1302}+R_{1023}=0, \tag{138}$$

we have the following relations which are consequences of the vanishing of the Ricci tensor:

$$\left.\begin{aligned} R_{1213} &= -R_{3002};\quad R_{1330}=-R_{1220};\quad R_{0202}=-R_{1313};\\ R_{0303} &= -R_{1212};\quad R_{2323}=-R_{1010}=R_{0202}+R_{0303}. \end{aligned}\right\} \tag{139}$$

The reductions can be further facilitated by having at hand the following list of the derivatives of the various metric functions:

$$\begin{aligned}
\mu_{2,\theta} &= -\frac{a^2\sin\theta\cos\theta}{\rho^2}=\mu_{3,\theta};\quad \mu_{3,2}=\frac{r}{\rho^2};\\
(\mu_2+\mu_3)_{,2} &= \frac{2r}{\rho^2}-\frac{r-M}{\Delta},\quad (\mu_2+\mu_3)_{,\theta}=-\frac{2a^2\sin\theta\cos\theta}{\rho^2};\\
\psi_{,2} &= \frac{1}{\Sigma^2}[2r(r^2+a^2)-a^2(r-M)\sin^2\theta]-\frac{r}{\rho^2}=-\nu_{,2}+\frac{r-M}{\Delta},\\
\psi_{,\theta} &= \cot\theta+a^2\left(\frac{1}{\rho^2}-\frac{\Delta}{\Sigma^2}\right)\sin\theta\cos\theta=-\nu_{,3}+\cot\theta;\\
\psi_{,2,\theta} &= 2a^2\left\{-\frac{r-M}{\Sigma^2}+\frac{\Delta}{\Sigma^4}[2r(r^2+a^2)-a^2(r-M)\sin^2\theta]-\frac{r}{\rho^4}\right\}\sin\theta\cos\theta\\
&= -\nu_{,2,\theta},\\
\psi_{,\theta,\theta} &= -\operatorname{cosec}^2\theta-\frac{a^2\Delta\cos 2\theta}{\Sigma^2}-\frac{2a^4\Delta^2}{\Sigma^4}\sin^2\theta\cos^2\theta\\
&\qquad +\frac{a^2\cos 2\theta}{\rho^2}+\frac{2a^4}{\rho^4}\sin^2\theta\cos^2\theta=-\nu_{,\theta,\theta}-\operatorname{cosec}^2\theta;\\
\omega_{,\theta} &= +\frac{4a^3Mr\Delta}{\Sigma^4}\sin\theta\cos\theta,\\
\omega_{,2} &= -\frac{2aM}{\Sigma^4}[(r^2+a^2)(3r^2-a^2)-a^2(r^2-a^2)\sin^2\theta].
\end{aligned} \tag{140}$$

We find that the reductions involved in deriving the following formulae are not unduly excessive:

$$R_{1023} = -\frac{\sqrt{(\Delta\delta)}}{4\rho^2\chi^2}(X_{,3}Y_{,2} - X_{,2}Y_{,3})$$

$$= -\frac{aM\cos\theta}{\rho^6}(3r^2 - a^2\cos^2\theta) = -(R_{1230} + R_{1302}),$$

$$R_{1230} - R_{1302} = \frac{\Sigma^2\sin\theta}{2\rho^4}\bigg[-\omega_{,r}(\mu_2+\mu_3)_{,\theta} - \omega_{,\theta}(\mu_2+\mu_3)_{,r}$$

$$+ e^{-4\psi}\left(e^{4\psi}\frac{\omega_{,r}}{\sin\theta}\right)_{,\theta}\sin\theta + e^{-4\psi}\left(e^{4\psi}\frac{\omega_{,\theta}}{\Delta\sin\theta}\right)_{,r}\Delta\sin\theta\bigg]$$

$$= -3\frac{aM\cos\theta}{\rho^6}(3r^2 - a^2\cos^2\theta)\frac{1}{\Sigma^2}[(r^2+a^2)^2 + a^2\Delta\sin^2\theta],$$

$$R_{2323} = \frac{\sqrt{\Delta}}{\rho^2}\left[\left(\frac{\mu_{2,3}}{\sqrt{\Delta}}\right)_{,3} + (\mu_{3,2}\sqrt{\Delta})_{,2}\right]$$

$$= \frac{Mr}{\rho^6}(r^2 - 3a^2\cos^2\theta) = R_{0202} + R_{0303} = -R_{1010},$$

$$2R_{1330} = e^{-2\psi-\mu_3}(e^{3\psi-\nu-\mu_3}\omega_{,3})_{,3} + e^{\psi-2\mu_2-\nu}\omega_{,2}\mu_{3,2}$$

$$= -2\frac{Mr}{\rho^6}(r^2 - 3a^2\cos^2\theta)\frac{3a\sqrt{\Delta}}{\Sigma^2}(r^2+a^2)\sin\theta,$$

$$2R_{1213} = \frac{\sqrt{\Delta}}{\rho^2}[(3\psi+\nu)_{,3,2} + 3\psi_{,3}\psi_{,2} + \nu_{,3}\nu_{,2} - (3\psi+\nu)_{,2}\mu_{2,3}$$

$$-(3\psi+\nu)_{,3}\mu_{3,2}]$$

$$= -2\frac{aM\cos\theta}{\rho^6}(3r^2 - a^2\cos^2\theta)\frac{3a\sqrt{\Delta}}{\Sigma^2}(r^2+a^2)\sin\theta,$$

$$3R_{1313} - R_{3030} = e^{-2\mu_3}[3\psi_{,3,3} + \nu_{,3,3} + 3(\psi_{,3})^2 + (\nu_{,3})^2 - (3\psi+\nu)_{,3}\mu_{3,3}]$$

$$+ e^{-2\mu_2}(3\psi+\nu)_{,2}\mu_{3,2}$$

$$= -\frac{Mr}{\rho^6}(r^2 - 3a^2\cos^2\theta)\frac{1}{\Sigma^2}[5(r^2+a^2)^2 + a^2\Delta\sin^2\theta]. \quad (141)$$

And, finally, with the aid of the relations (138) and (139), we obtain

$$R_{1023} = -\frac{aM\cos\theta}{\rho^6}(3r^2 - a^2\cos^2\theta),$$

$$R_{1230} = -\frac{aM\cos\theta}{\rho^6}(3r^2 - a^2\cos^2\theta)\frac{1}{\Sigma^2}[(r^2+a^{2)2} + 2a^2\Delta\sin^2\theta],$$

$$R_{1302} = +\frac{aM\cos\theta}{\rho^6}(3r^2 - a^2\cos^2\theta)\frac{1}{\Sigma^2}[2(r^2+a^2)^2 + a^2\Delta\sin^2\theta],$$

$$
\begin{aligned}
-R_{3002} = R_{1213} &= -\frac{aM\cos\theta}{\rho^6}(3r^2 - a^2\cos^2\theta)\frac{3a\sqrt{\Delta}}{\Sigma^2}(r^2+a^2)\sin\theta,\\
-R_{1220} = R_{1330} &= -\frac{Mr}{\rho^6}(r^2 - 3a^2\cos^2\theta)\frac{3a\sqrt{\Delta}}{\Sigma^2}(r^2+a^2)\sin\theta,\\
-R_{1010} = R_{2323} &= +\frac{Mr}{\rho^6}(r^2 - 3a^2\cos^2\theta) = R_{0202}+R_{0303},\\
-R_{1313} = R_{0202} &= +\frac{Mr}{\rho^6}(r^2 - 3a^2\cos^2\theta)\frac{1}{\Sigma^2}[2(r^2+a^2)^2 + a^2\Delta\sin^2\theta],\\
-R_{1212} = R_{0303} &= -\frac{Mr}{\rho^6}(r^2 - 3a^2\cos^2\theta)\frac{1}{\Sigma^2}[(r^2+a^2)^2 + 2a^2\Delta\sin^2\theta].
\end{aligned}
\tag{142}
$$

We observe that these non-vanishing *components of the Riemann tensor become singular only for* $\theta = \pi/2$ *and* $r = 0$; *this, then, is the only singularity of the Kerr space-time.*

55. The uniqueness of the Kerr metric; the theorems of Robinson and Carter

We have seen that the Kerr metric, which endows a stationary, axisymmetric, asymptotically flat space-time with a smooth convex event-horizon, is characterized by just two parameters—the mass M and the angular momentum $J(=aM)$. The uniqueness of the Kerr metric for the description of the black holes of nature follows from the theorem of Robinson which states: *stationary axisymmetric solutions of Einstein's equation for the vacuum, which have a smooth convex event-horizon, are asymptotically flat, and are non-singular outside of the horizon, are uniquely specified by the two parameters, the mass and the angular momentum, and these two parameters only.*

The proof of Robinson's theorem is based on an identity (see equation (156) below) derived from the equations (cf. equations (80) and (81))

$$
E = 0 \qquad \text{and} \qquad F = 0, \tag{143}
$$

where

$$
\begin{aligned}
E(X,Y) = &\left[(\eta^2-1)\frac{X_{,\eta}}{X}\right]_{,\eta} + (\eta^2-1)\frac{Y_{,\eta}^2}{X^2}\\
&+\left[(1-\mu^2)\frac{X_{,\mu}}{X}\right]_{,\mu} + (1-\mu^2)\frac{Y_{,\mu}^2}{X^2},
\end{aligned}
\tag{144}
$$

and

$$
F(X,Y) = \left[(\eta^2-1)\frac{Y_{,\eta}}{X^2}\right]_{,\eta} + \left[(1-\mu^2)\frac{Y_{,\mu}}{X^2}\right]_{,\mu}. \tag{145}
$$

We recall that in deriving equations (80) and (81), we exercised the gauge freedom at our disposal to specify the event-horizon as occurring at $\eta = 1$. Besides, the expressions (144) and (145) for the operators E and F make it manifest that if (X, Y) represents a solution of the equations then so does (cX, cY) where c is an arbitrary real constant. On these accounts, there is clearly no loss of generality in supposing that the solutions for X in which we are interested all have at infinity the same asymptotic behaviour (cf. equation (85)),

$$X \to (1-\mu^2)\eta^2 + O(\eta) \qquad (\eta \to \infty). \tag{146}$$

This restricton to solutions having the behaviour (146) is clearly equivalent to considering solutions for the same mass.

The associated asymptotic behaviour of Y is (cf. equation (86)),

$$Y \to 2J\mu(3-\mu^2) + O(\eta^{-1}) \qquad (\eta \to \infty), \tag{147}$$

where J denotes the angular momentum.

Besides the behaviours (146) and (147) required by asymptotic flatness, we shall also require that the solutions X and Y are smooth and regular on the axis ($\mu = \pm 1$), and on the horizon ($\eta = 1$), and are non-singular outside ($\eta > 1$). It would appear that, compatible with these requirements, one may very well have several solutions for the same assigned J. But Robinson's theorem is precisely to the effect that one can have no more than one solution for any given J. The proof consists in showing that if the contrary were true and we should have two solution-pairs, (X_1, Y_1) and (X_2, Y_2), belonging to the same value of J, then they must coincide.

We suppose then that there exist two solution-pairs, (X_1, Y_1) and (X_2, Y_2), belonging to the same value of J and, following Robinson, consider the functional

$$\begin{aligned} \mathrm{R} = \int_1^\infty \int_{-1}^{+1} \Bigg[& \frac{X_1}{X_2}(Y_2 - Y_1)F(X_1, Y_1) + \frac{X_2}{X_1}(Y_1 - Y_2)F(X_2, Y_2) \\ & + \frac{1}{2X_1X_2}[(Y_2-Y_1)^2 + (X_2^2 - X_1^2)]E(X_1, Y_1) \\ & + \frac{1}{2X_1X_2}[(Y_2-Y_1)^2 + (X_1^2 - X_2^2)]E(X_2, Y_2)\Bigg] \mathrm{d}\eta\, \mathrm{d}\mu. \end{aligned} \tag{148}$$

By the equations satisfied by (X_1, Y_1) and (X_2, Y_2), R must identically vanish. But we shall now show that, by virtue solely of the definitions of the operators E and F and of the boundary conditions imposed on the solutions, R can be reduced to an integral with a positive-definite integrand. The reduction crucially depends on two elementary algebraic identities which we shall state as Robinson's lemmas.

ROBINSON'S LEMMAS:

$$\text{I.}\quad \frac{1}{2X_1^3X_2^3}[X_2(Y_2-Y_1)Y_{1,\eta}+X_1(X_1X_{2,\eta}-X_2X_{1,\eta})]^2$$
$$+\frac{1}{2X_2^3X_1^3}[X_1(Y_2-Y_1)Y_{2,\eta}+X_2(X_1X_{2,\eta}-X_2X_{1,\eta})]^2$$
$$=\frac{1}{2X_1^3X_2}(Y_2-Y_1)^2(Y_{1,\eta})^2+\frac{1}{2X_1^3X_2^3}(X_1^2+X_2^2)(X_1X_{2,\eta}-X_2X_{1,\eta})^2$$
$$+\frac{1}{2X_1X_2^3}(Y_2-Y_1)^2(Y_{2,\eta})^2$$
$$+\frac{1}{X_1^2X_2^2}(Y_2-Y_1)(Y_2+Y_1)_{,\eta}(X_1X_{2,\eta}-X_2X_{1,\eta}), \tag{149}$$

$$\text{II.}\quad \frac{1}{4X_1^3X_2^3}[(X_2+X_1)(X_1Y_{2,\eta}-X_2Y_{1,\eta})-(Y_2-Y_1)(X_1X_{2,\eta}+X_2X_{1,\eta})]^2$$
$$+\frac{1}{4X_1^3X_2^3}[(X_2-X_1)(X_1Y_{2,\eta}+X_2Y_{1,\eta})$$
$$-(Y_2-Y_1)(X_1X_{2,\eta}+X_2X_{1,\eta})]^2$$
$$=\frac{1}{2X_1^3X_2^3}(Y_2-Y_1)^2(X_1X_{2,\eta}+X_2X_{1,\eta})^2$$
$$-\frac{1}{X_1^2X_2^2}(Y_2-Y_1)(Y_2-Y_1)_{,\eta}(X_1X_{2,\eta}+X_2X_{1,\eta})$$
$$+\frac{1}{2X_1^3X_2^3}\{(X_1^2+X_2^2)[X_1^2(Y_{2,\eta})^2+X_2^2(Y_{1,\eta})^2]-4X_1^2X_2^2Y_{1,\eta}Y_{2,\eta}\}, \tag{150}$$

and two further identities in which the derivatives with respect to η are replaced by derivatives with respect to μ.

In view of the entire symmetry of the terms in E and F (and, therefore, also in R) involving the derivatives with respect to η—the η-terms—and the derivatives with respect to μ—the μ-terms—except for (η^2-1) replacing (μ^2-1), we shall explicitly indicate the manner of the reductions only with respect to the η-terms. Analogous reductions apply to the μ-terms.

Considering R, we integrate by parts the terms which occur with F as a factor and the terms which occur with the quantity in square brackets in the definition of E (cf. equation (144)) as a factor. The integrated parts arising from the η-terms are

$$\frac{\eta^2-1}{X_1X_2}\bigg\{(Y_1-Y_2)_{,\eta}(Y_2-Y_1)+\frac{1}{2X_1X_2}[(Y_2-Y_1)^2(X_2X_{1,\eta}+X_1X_{2,\eta})$$
$$+(X_2^2-X_1^2)(X_2X_{1,\eta}-X_1X_{2,\eta})]\bigg\}$$

$$= \frac{1}{2}(\eta^2-1)\left\{-\frac{1}{X_1X_2}[(Y_2-Y_1)]^2_{,\eta}+\frac{1}{X_1^2X_2^2}(Y_2-Y_1)^2(X_2X_{1,\eta}+X_1X_{2,\eta})\right.$$
$$\left.+\frac{1}{X_1^2X_2^2}(X_2^2-X_1^2)(X_2X_{1,\eta}-X_1X_{2,\eta})\right\}$$
$$= -\frac{1}{2}(\eta^2-1)\left[\frac{(Y_2-Y_1)^2+(X_2-X_1)^2}{X_1X_2}\right]_{,\eta}. \tag{151}$$

Combined with a similar term arising from the integration by parts of the μ-terms, we have

$$-\frac{1}{2}\int_{-1}^{+1}\mathrm{d}\mu\left\{(\eta^2-1)\left[\frac{(Y_2-Y_1)^2+(X_2-X_1)^2}{X_1X_2}\right]_{,\eta}\right\}_{\eta=1}^{\eta=+\infty}$$
$$-\frac{1}{2}\int_{1}^{\infty}\mathrm{d}\eta\left\{(1-\mu^2)\left[\frac{(Y_2-Y_1)^2+(X_2-X_1)^2}{X_1X_2}\right]_{,\mu}\right\}_{\mu=-1}^{\mu=+1} \tag{152}$$

By the required smoothness of the functions X and Y on the horizon and on the axis, the terms in (152) evaluated on the axis ($\mu = \pm 1$) and on the horizon ($\eta = 1$) do not survive. And by the required asymptotic behaviours (146) and (147) of the solutions at infinity,

$$\left[\frac{(Y_2-Y_1)^2+(X_2-X_1)^2}{X_1X_2}\right]_{,\eta} \quad \text{is} \quad O(\eta^{-3}). \tag{153*}$$

There is, therefore, no contribution to the surface integral from this term as well; and it vanishes altogether.

The η-terms of the integrand of the volume integral which remains, after the integration by parts, are apart from a factor (η^2-1),

$$\frac{1}{2X_1^3X_2}[(Y_2-Y_1)^2+X_2^2-X_1^2](Y_{1,\eta})^2$$
$$+\frac{1}{2X_2^3X_1}[(Y_2-Y_1)^2+X_1^2-X_2^2](Y_{2,\eta})^2$$
$$-\frac{Y_{1,\eta}}{X_1^2}\left[\frac{X_1}{X_2}(Y_2-Y_1)\right]_{,\eta}-\frac{X_{1,\eta}}{X_1}\left\{\frac{1}{2X_1X_2}[(Y_2-Y_1)^2+X_2^2-X_1^2]\right\}_{,\eta}$$
$$+\frac{Y_{2,\eta}}{X_2^2}\left[\frac{X_2}{X_1}(Y_2-Y_1)\right]_{,\eta}-\frac{X_{2,\eta}}{X_2}\left\{\frac{1}{2X_1X_2}[(Y_2-Y_1)^2+X_1^2-X_2^2]\right\}_{,\eta}$$

* Actually, this is of $O(\eta^{-5})$ if account is taken of the term of $O(\eta^{-1})$ in equation (146) (cf. equation (85)).

$$
\begin{aligned}
&= \frac{1}{2X_1^3X_2}(Y_2-Y_1)^2(Y_{1,\eta})^2+\frac{1}{2X_2^3X_1}(Y_2-Y_1)^2(Y_{2,\eta})^2\\
&\quad+\frac{1}{X_1X_2}[(Y_2-Y_1)_{,\eta}]^2\\
&\quad+\frac{1}{X_1^2X_2^2}(Y_2-Y_1)(Y_2+Y_1)_{,\eta}(X_1X_{2,\eta}-X_2X_{1,\eta})\\
&\quad-\frac{1}{X_1^2X_2^2}(Y_2-Y_1)(Y_2-Y_1)_{,\eta}(X_2X_{1,\eta}+X_1X_{2,\eta})\\
&\quad-\frac{1}{X_1^2X_2^2}(X_1X_{1,\eta}-X_2X_{2,\eta})(X_1X_{2,\eta}-X_2X_{1,\eta})\\
&\quad+\frac{1}{2X_1^3X_2^3}(X_2^2-X_1^2)\{X_2^2[(X_{1,\eta})^2+(Y_{1,\eta})^2]-X_1^2[(X_{2,\eta})^2+(Y_{2,\eta})^2]\}\\
&\quad+\frac{1}{2X_1^3X_2^3}(Y_2-Y_1)^2(X_1X_{2,\eta}+X_2X_{1,\eta})^2. \qquad (154)
\end{aligned}
$$

We observe that some terms on the right-hand side of (154) are common with those included on the right-hand side of Robinson's Lemma I. Eliminating these terms, we find, after some rearrangements,

$$
\begin{aligned}
&\frac{1}{2X_1^3X_2^3}[X_2(Y_2-Y_1)Y_{1,\eta}+X_1(X_1X_{2,\eta}-X_2X_{1,\eta})]^2\\
&\quad+\frac{1}{2X_1^3X_2^3}[X_1(Y_2-Y_1)Y_{2,\eta}+X_2(X_1X_{2,\eta}-X_2X_{1,\eta})]^2\\
&\quad-\frac{1}{2X_1^3X_2^3}(X_1^2+X_2^2)[X_1^2(X_{2,\eta})^2+X_2^2(X_{1,\eta})^2-2X_1X_2X_{1,\eta}X_{2,\eta}]\\
&\quad+\frac{1}{X_1X_2}[(Y_{1,\eta})^2+(Y_{2,\eta})^2-2Y_{1,\eta}Y_{2,\eta}]\\
&\quad-\frac{1}{X_1^2X_2^2}(Y_2-Y_1)(Y_2-Y_1)_{,\eta}(X_2X_{1,\eta}+X_1X_{2,\eta})\\
&\quad-\frac{1}{X_1^2X_2^2}\{(X_1^2+X_2^2)X_{1,\eta}X_{2,\eta}-X_1X_2[(X_{1,\eta})^2+(X_{2,\eta})^2]\}\\
&\quad+\frac{1}{2X_1^3X_2^3}(Y_2-Y_1)^2(X_2X_{1,\eta}+X_1X_{2,\eta})^2\\
&\quad+\frac{1}{2X_1^3X_2^3}(X_2^2-X_1^2)\{X_2^2[(X_{1,\eta})^2+(Y_{1,\eta})^2]-X_1^2[(X_{2,\eta})^2+(Y_{2,\eta})^2]\}.
\end{aligned}
\qquad (155)
$$

We find that many of the terms in lines 3, 4, 6, and 8 of the foregoing expression cancel; and the surviving terms are precisely those included on the right-hand side of Robinson's Lemma II. Therefore, the η-terms of the integrand, apart from the suppressed factor (η^2-1), are the positive-definite expressions in Robinson's Lemmas. The μ-terms will give similar contributions to the integrand. Thus, the result of the reduction is

$$\begin{aligned}\int_1^\infty \int_{-1}^{+1} \frac{d\eta\, d\mu}{4X_1^3 X_2^3} \Bigg[\!\!\Bigg[(\eta^2-1)\Big\{ & 2[X_2(Y_2-Y_1)Y_{1,\eta}+X_1(X_1X_{2,\eta}-X_2X_{1,\eta})]^2 \\ & +2[X_1(Y_2-Y_1)Y_{2,\eta}+X_2(X_1X_{2,\eta}-X_2X_{1,\eta})]^2 \\ & +[(X_2+X_1)(X_1Y_{2,\eta}-X_2Y_{1,\eta})-(Y_2-Y_1)(X_1X_{2,\eta}+X_2X_{1,\eta})]^2 \\ & +[(X_2-X_1)(X_2Y_{1,\eta}+X_1Y_{2,\eta})-(Y_2-Y_1)(X_1X_{2,\eta}+X_2X_{1,\eta})]^2\Big\} \\ & +(1-\mu^2)\Big\{2[X_2(Y_2-Y_1)Y_{1,\mu}+X_1(X_1X_{2,\mu}-X_2X_{1,\mu})]^2 \\ & +2[X_1(Y_2-Y_1)Y_{2,\mu}+X_2(X_1X_{2,\mu}-X_2X_{1,\mu})]^2 \\ & +[(X_2+X_1)(X_1Y_{2,\mu}-X_2Y_{1,\mu})-(Y_2-Y_1)(X_1X_{2,\mu}+X_2X_{1,\mu})]^2 \\ & +[(X_2-X_1)(X_2Y_{1,\mu}+X_1Y_{2,\mu}) \\ & -\ (Y_2-Y_1)(X_1X_{2,\mu}+X_2X_{1,\mu})]^2\Big\}\Bigg]\!\!\Bigg] \\ =0. \qquad & \qquad\qquad (156)\end{aligned}$$

It follows from equation (156) that each of the eight positive-definite expressions in the integrand must vanish identically. Accordingly,

$$\begin{aligned}(Y_2-Y_1)(X_1X_{2,\alpha}+X_2X_{1,\alpha}) &= (X_2+X_1)(X_1Y_{2,\alpha}-X_2Y_{1,\alpha}) \\ &= (X_2-X_1)(X_2Y_{1,\alpha}+X_1Y_{2,\alpha}),\end{aligned} \qquad (157)$$

and

$$\left.\begin{aligned}X_2(Y_2-Y_1)Y_{1,\alpha} &= -X_1(X_1X_{2,\alpha}-X_2X_{1,\alpha}), \\ X_1(Y_2-Y_1)Y_{2,\alpha} &= -X_2(X_1X_{2,\alpha}-X_2X_{1,\alpha}).\end{aligned}\right\}(\alpha=\eta,\mu). \qquad (158)$$

The equalities of the two expressions on the right-hand side of equations (157) require that

$$X_1^2 Y_{2,\alpha} = X_2^2 Y_{1,\alpha}. \qquad (159)$$

Using this result, we find that the equality of the expressions on the first line of equations (157) gives

$$X_1X_2(Y_2-Y_1)_{,\alpha}-(Y_2-Y_1)(X_1X_{2,\alpha}+X_2X_{1,\alpha})=0, \qquad (160)$$

or, alternatively,

$$\left(\frac{Y_2-Y_1}{X_1X_2}\right)_{,\alpha}=0 \quad (\alpha=\eta,\mu). \qquad (161)$$

Since Y_1 and Y_2 asymptotically approach equality for $\eta \to \infty$, it follows from equation (161) that

$$Y_2 \equiv Y_1. \tag{162}$$

In view of this last equality, either of the equations (158) implies that

$$(X_1/X_2)_{,\alpha} = 0 \quad (\alpha = \eta, \mu). \tag{163}$$

Again, from the equality of X_1 and X_2 for $\eta \to \infty$, we conclude that

$$X_1 \equiv X_2; \tag{164}$$

and the uniqueness of the solution for assigned M (by virtue of the assumed normalization of the solutions) and J follows.

Since the Kerr metric does satisfy the required boundary conditions of Robinson's theorem and provides a solution for a given M and J ($< M^2$), *it follows that it is the unique solution for the given M and J.* In other words, Kerr's discovery provides proof for the *existence* of a solution satisfying the requirements of Robinson's theorem!

Two special cases of Robinson's identity (156) are of some interest.

First, we consider the case of two 'neighbouring solutions', (X, Y) and $(X + \delta X, Y + \delta Y)$, i.e., we consider *quasi-stationary axisymmetric perturbations which preserve the mass and the angular momentum.* By considering δX and δY as small quantities of the first order (as behoves perturbations!) we obtain, by linearizing equation (156), the identity

$$\begin{aligned}\int_1^\infty \int_{-1}^{+1} \frac{d\eta\, d\mu}{X^4} \{ (\eta^2 - 1)[(X_{,\eta}\delta X - Y_{,\eta}\delta Y - X\delta X_{,\eta})^2 \\ + (X_{,\eta}\delta Y + Y_{,\eta}\delta X - X\delta Y_{,\eta})^2 + (X_{,\eta}\delta Y - Y_{,\eta}\delta X)^2] \\ + (1 - \mu^2)[(X_{,\mu}\delta X - Y_{,\mu}\delta Y - X\delta X_{,\mu})^2 \\ + (X_{,\mu}\delta Y + Y_{,\mu}\delta X - X\delta Y_{,\mu})^2 \\ + (X_{,\mu}\delta Y - Y_{,\mu}\delta X)^2]\} = 0,\end{aligned} \tag{165}$$

which is equivalent to one first derived by Carter. From equation (165) it follows, as in the proof of Robinson's theorem, that

$$\delta X \equiv 0 \quad \text{and} \quad \delta Y \equiv 0. \tag{166}$$

In other words, the Kerr solution cannot be subject to quasi-stationary axisymmetric perturbations which leave the mass and the angular momentum unchanged; or, as one says, *there can be no point of bifurcation along the Kerr sequence.* This is *Carter's theorem* proved prior to Robinson's *tour de force.*

Next, we consider the case of non-rotating black-holes when $Y = 0$.

Robinson's identity then reduces to

$$\int_1^\infty \int_{-1}^{+1} \left[(\eta^2-1)(X_1 X_{2,\eta} - X_2 X_{1,\eta})^2 + (1-\mu^2)(X_1 X_{2,\mu} - X_2 X_{1,\mu})^2 \right] \times \frac{X_1^2 + X_2^2}{X_1^3 X_2^3}\, \mathrm{d}\eta\, \mathrm{d}\mu = 0, \quad (167)$$

where X_1 and X_2 are two solutions belonging to the same mass. From equation (167), we conclude that

$$X_1 \equiv X_2. \quad (168)$$

In other words, under the assumption of axisymmetry, non-rotating black-holes are uniquely specified by their mass. This result, without the assumption of axisymmetry, is *Israel's theorem* which states that *the Schwarzschild metric is the unique solution for representing non-rotating static black-holes.*

56. The description of the Kerr space-time in a Newman–Penrose formalism

For a description of the Kerr space-time in a Newman–Penrose formalism, we first need to construct a null tetrad-frame. In Chapter 7 (§63(*b*)) we shall show that in the Kerr space-time we have a simple class of null geodesics given by the tangent vectors

$$\frac{\mathrm{d}t}{\mathrm{d}\tau} = \frac{r^2+a^2}{\Delta}E, \quad \frac{\mathrm{d}r}{\mathrm{d}\tau} = \pm E, \quad \frac{\mathrm{d}\theta}{\mathrm{d}\tau} = 0, \quad \text{and} \quad \frac{\mathrm{d}\varphi}{\mathrm{d}\tau} = \frac{a}{\Delta}E, \quad (169)$$

where E is a constant. Defining the real null-vectors $\boldsymbol{l}$ and $\boldsymbol{n}$ of the Newman–Penrose formalism in terms of these geodesics and adjoining to them a complex null-vector $\boldsymbol{m}$, orthogonal to them, we obtain the basis

$$\left.\begin{aligned} l^i &= \frac{1}{\Delta}(r^2+a^2,\ +\Delta, 0, a), \\ n^i &= \frac{1}{2\rho^2}(r^2+a^2,\ -\Delta, 0, a), \\ m^i &= \frac{1}{\bar\rho\sqrt{2}}(ia\sin\theta, 0, 1, i\operatorname{cosec}\theta), \end{aligned}\right\} \quad (170)$$

where

$$\bar\rho = r + ia\cos\theta \qquad \text{and} \qquad \bar\rho^* = r - ia\cos\theta. \quad (171)$$

[It is unfortunate that the letter "ρ" is used in so many different contexts: as a spin coefficient, as $\rho^2 = r^2 + a^2\cos^2\theta$, and now as $\bar\rho$ and $\bar\rho^*$. Normally, there will be no ambiguity as to which meaning is intended; but in case of overlapping usage, we shall denote the spin coefficient ρ by $\tilde\rho$ to distinguish it from $\bar\rho$ and $\bar\rho^*$. Also, ρ^2 will always signify $\rho^2 = r^2 + a^2\cos^2\theta$.]

The vector $\boldsymbol{l}$ is affinely parametrized while $\boldsymbol{n}$ and $\boldsymbol{m}$ are not: they have been arranged to satisfy the required normalization conditions

$$\boldsymbol{l}\cdot\boldsymbol{n} = 1 \quad \text{and} \quad \boldsymbol{m}\cdot\bar{\boldsymbol{m}} = -1. \tag{172}$$

The covariant form of the basis vectors is

$$\left.\begin{aligned} l_i &= \frac{1}{\Delta}(\Delta, -\rho^2, 0, -a\Delta\sin^2\theta), \\ n_i &= \frac{1}{2\rho^2}(\Delta, +\rho^2, 0, -a\Delta\sin^2\theta), \\ m_i &= \frac{1}{\bar{\rho}\sqrt{2}}(ia\sin\theta, 0, -\rho^2, -i(r^2+a^2)\sin\theta). \end{aligned}\right\} \tag{173}$$

The spin coefficients (as defined in Ch. 1, equations (286)) with respect to the chosen basis are most conveniently evaluated via the λ-symbols (Ch. 1, equation (265)). We find that the non-vanishing λ-symbols are:

$$\begin{aligned} \lambda_{122} &= -\frac{1}{\rho^4}[(r-M)\rho^2 - r\Delta]; & \lambda_{134} &= -\frac{2ia\cos\theta}{\rho^2} \\ \lambda_{132} &= +\sqrt{2}\frac{iar\sin\theta}{\rho^2\bar{\rho}}; & \lambda_{324} &= -\frac{ia\Delta\cos\theta}{\rho^4} \\ \lambda_{213} &= -\sqrt{2}\frac{a^2\sin\theta\cos\theta}{\rho^2\bar{\rho}}; & \lambda_{334} &= +\frac{(ia+r\cos\theta)\operatorname{cosec}\theta}{\bar{\rho}^2\sqrt{2}} \\ \lambda_{243} &= -\frac{\Delta}{2\rho^2\bar{\rho}}; & \lambda_{341} &= -\frac{1}{\bar{\rho}}; \end{aligned} \tag{174}$$

and the spin-coefficients derived with the aid of these λ-symbols are:

$$\left.\begin{gathered} \kappa = \sigma = \lambda = \nu = \varepsilon = 0 \\ \tilde{\rho} = -\frac{1}{\bar{\rho}^*}; \quad \beta = \frac{\cot\theta}{\bar{\rho}2\sqrt{2}}; \quad \pi = \frac{ia\sin\theta}{(\bar{\rho}^*)^2\sqrt{2}}; \quad \tau = -\frac{ia\sin\theta}{\rho^2\sqrt{2}} \\ \mu = -\frac{\Delta}{2\rho^2\bar{\rho}^*}; \quad \gamma = \mu + \frac{r-M}{2\rho^2}; \quad \alpha = \pi - \beta^*. \end{gathered}\right\} \tag{175}$$

The fact that the spin-coefficients, κ, σ, λ, and ν vanish, shows that the congruence of the null geodesics, $\boldsymbol{l}$ and $\boldsymbol{n}$, are shear free. And from the shear-free character of these congruences, we can conclude, on the basis of the Goldberg–Sachs theorem, that *the Kerr space-time is of Petrov type*-D.

The Goldberg–Sachs theorem also allows us to conclude that in the chosen basis, the Weyl scalars Ψ_0, Ψ_1, Ψ_3, and Ψ_4 vanish. These conclusions can be directly verified by contracting the non-vanishing components of the Riemann tensor listed in equations (142) with the vectors $\boldsymbol{l}$, $\boldsymbol{n}$, $\boldsymbol{m}$, and $\bar{\boldsymbol{m}}$ in accordance with the definitions of these scalars.

Since the components of the Riemann tensor listed in equations (142) are in the tetrad-frame, it is convenient to express the basis vectors also in the same frame by subjecting them to the transformation represented by

$$(e^{(a)}{}_i) = \begin{vmatrix} e^{\nu} & 0 & 0 & 0 \\ -\omega e^{\psi} & e^{\psi} & 0 & 0 \\ 0 & 0 & e^{\mu_2} & 0 \\ 0 & 0 & 0 & e^{\mu_3} \end{vmatrix}. \tag{176}$$

We find

$$\begin{aligned} l^{(p)} &= \left\{\frac{e^{\nu}}{\Delta}(r^2+a^2), \quad a\frac{e^{\psi}\rho^2}{\Sigma^2}, \quad e^{\mu_2}, \quad 0\right\}, \\ n^{(p)} &= \left\{\frac{e^{\nu}}{2\rho^2}(r^2+a^2), \quad a\frac{e^{\psi}\Delta}{2\Sigma^2}, \quad -\frac{\Delta}{2\rho^2}e^{\mu_2}, \quad 0\right\}, \\ m^{(p)} &= \left\{i\frac{e^{\nu}}{\bar{\rho}\sqrt{2}}a\sin\theta, \quad i\frac{e^{\psi}\rho^2}{\Sigma^2\sin\theta}\frac{(r^2+a^2)}{\bar{\rho}\sqrt{2}}, \quad 0 \quad \frac{e^{\mu_3}}{\bar{\rho}\sqrt{2}}\right\}, \end{aligned} \tag{177}$$

where, in obtaining the vectors in these forms, use has been made of the relations (137).

Using the foregoing representation of the basis vectors and making use of equations (4) and (139), we find, for example, that

$$\begin{aligned} -\Psi_0 &= R_{pqrs}l^p m^q l^r m^s \\ &= R_{0101}[(l^0m^1-l^1m^0)^2-(l^2m^3)^2]+R_{0303}[(l^0m^3)^2-(l^2m^1)^2] \\ &\quad + R_{2020}[(l^2m^0)^2-(l^1m^3)^2]+2R_{1330}[(l^2)^2m^0m^1-(m^3)^2l^0l^1] \\ &\quad + 2R_{3002}[l^2l^0m^0m^3+l^2l^1m^1m^3]+2R_{2301}l^2m^3(l^0m^1-l^1m^0) \\ &\quad + 2R_{2103}l^2l^0m^1m^3+2R_{2013}l^2l^1m^0m^3 \\ &= \frac{\rho^4}{\Delta\bar{\rho}^2}\left\{-R_{0101}+\frac{(r^2+a^2)^2}{\Sigma^2}R_{0303}-\frac{a^2\Delta\sin^2\theta}{\Sigma^2}R_{0202}\right. \\ &\qquad \left. -\left[\frac{2a\sqrt{\Delta}}{\Sigma^2}(r^2+a^2)\sin\theta\right]R_{1330}\right\} \\ &\quad - \frac{i\rho^4}{\Delta\bar{\rho}^2}\left\{-R_{2301}+\frac{(r^2+a^2)^2}{\Sigma^2}R_{2103}+\frac{a^2\Delta\sin^2\theta}{\Sigma^2}R_{2013}\right. \\ &\qquad \left. -\left[\frac{2a\sqrt{\Delta}}{\Sigma^2}(r^2+a^2)\sin\theta\right]R_{3002}\right\}. \end{aligned} \tag{178}$$

Now substituting for the components of the Reimann tensor from equations (142), we find that, as required,

$$\Psi_0 = 0. \tag{179}$$

The vanishing of the Weyl scalars Ψ_1, Ψ_3, Ψ_4 can be similarly verified.

The value of the non-vanishing Weyl scalar, Ψ_2, can be found by contracting the Reimann tensor, with the vectors of the null basis, in accordance with its definition. Thus, we find

$$\begin{aligned}
\Psi_2 &= R_{pqrs} l^p m^q n^r \bar{m}^s \\
&= R_{0101}[(l^0 m^1 - l^1 m^0)(n^0 \bar{m}^1 - n^1 \bar{m}^0) - l^2 n^2 |m^3|^2] \\
&\quad + R_{0303}[l^0 n^0 |m^3|^2 - l^2 n^2 |m^1|^2] + R_{0202}[l^2 n^2 |m^0|^2 - l^1 n^1 |m^3|^2] \\
&\quad + R_{1330}[l^2 n^2 (m^1 \bar{m}^0 + \bar{m}^1 m^0) - (l^1 n^0 + l^0 n^1)|m^3|^2] \\
&\quad + R_{3002}[l^2 n^0 m^0 \bar{m}^3 + l^0 n^2 \bar{m}^0 m^3 + l^2 n^1 m^1 \bar{m}^3 + l^1 n^2 \bar{m}^1 m^3] \\
&\quad + R_{2301}[l^2 m^3 (n^0 \bar{m}^1 - n^1 \bar{m}^0) + n^2 \bar{m}^3 (l^0 m^1 - l^1 m^0)] \\
&\quad + R_{2103}[l^2 n^0 m^1 \bar{m}^3 + l^0 n^2 \bar{m}^1 m^3] + R_{2013}[l^1 n^2 m^3 \bar{m}^0 + l^2 n^1 \bar{m}^3 m^0] \\
&= \frac{1}{2}\left\{ + R_{0101} + \frac{(r^2+a^2)^2}{\Sigma^2} R_{0303} - \frac{a^2 \Delta \sin^2\theta}{\Sigma^2} R_{0202} \right. \\
&\qquad \left. - \left[\frac{2a\sqrt{\Delta}}{\Sigma^2}(r^2+a^2)\sin\theta\right] R_{1330} \right\} \\
&\quad + \frac{1}{2} i \left\{ - R_{2301} + \frac{(r^2+a^2)}{\Sigma^2} R_{2103} + \frac{a^2 \Delta \sin^2\theta}{\Sigma^2} R_{2013} \right. \\
&\qquad \left. + \left[\frac{2a\sqrt{\Delta}}{\Sigma^2}(r^2+a^2)\sin\theta\right] R_{3002} \right\} \\
&= -\frac{Mr}{\rho^6}(r^2 - 3a^2\cos^2\theta) - i\frac{aM\cos\theta}{\rho^6}(3r^2 - a^2\cos^2\theta) \\
&= -\frac{M}{(\bar{\rho}^*)^3}.
\end{aligned} \tag{180}$$

With the specification of the null basis, the spin-coefficients, and the Weyl scalars, we have completed the description of the Kerr space-time in a Newman–Penrose formalism which exemplifies its special algebraic character.

57. The Kerr–Schild form of the metric

The derivation of the Kerr metric via the route described in §§52–54 is not the one which was followed by Kerr in his original investigation leading to his discovery. He was seeking, instead, solutions of Einstein's equation for forms of the metric appropriate to algebraically special space-times. The form of the metric he chose for his special consideration (now known as the *Kerr–Schild form*) is

$$g_{ij} = \eta_{ij} + l_i l_j, \tag{181}$$

where η_{ij} is a flat metric and $\boldsymbol{l}$ is a null vector with respect to η_{ij}. The Kerr metric belongs to this class. Indeed, it will appear that it is in the Kerr–Schild form that the nature of the Kerr space-time becomes most transparent. On this account, we shall preface our consideration of the nature of the Kerr space-time showing why a space-time with the metric of the form (181) is necessarily of Petrov type-II. The following manner of presentation is due to Xanthopoulos.

We begin with a few preliminaries.

Consider a manifold endowed with two metrics g_{ij} and g'_{ij}. Let ∇_i and ∇'_i denote the operators defining covariant differentiation with respect to the Christoffel connections derived from each of the two metrics. If ξ_j denotes a covariant vector defined on the manifold, then we should have a relation of the form

$$\nabla'_i \xi_j = \nabla_i \xi_j - C^m_{ij} \xi_m, \tag{182}$$

where C^m_{ij} has all the properties of a symmetric connection. Thus, we should have, for example,

$$\nabla'_i T^k_j = \nabla_i T^k_j - C^m_{ij} T^k_m + C^k_{im} T^m_j, \tag{183}$$

where T^k_j is a tensor of type (1, 1). With the aid of these formulae, we can relate the Riemann tensors, defined with respect to the two metrics, via the Ricci identity. Thus,

$$\begin{aligned} R'_{ijk}{}^m \xi_m &= 2\nabla'_{[i} \nabla'_{j]} \xi_k \\ &= 2\nabla_{[i}(\nabla_{j]} \xi_k - C^m_{j]k} \xi_m) - 2C^n_{[ij]}(\nabla_n \xi_k - C^m_{nk} \xi_m) \\ &\qquad - 2C^n_{k[i}(\nabla_{j]} \xi_n - C^m_{j]n} \xi_m). \end{aligned} \tag{184}$$

On further simplification (making use of the symmetry of C^m_{ij} in the indices i and j), we obtain the relation

$$R'_{ijk}{}^m = R_{ijk}{}^m - 2\nabla_{[i} C^m_{j]k} + 2C^n_{k[i} C^m_{j]n}. \tag{185}$$

The corresponding relation between the Ricci tensors follows by contraction; thus,

$$R'_{ik} = R_{ik} - 2\nabla_{[i} C^m_{m]k} + 2C^n_{k[i} C^m_{m]n}. \tag{186}$$

In considering the Kerr–Schild form of the metric (181), we shall identify the unprimed and the primed metrics in the foregoing formulae with

$$\eta_{ij} \qquad \text{and} \qquad g_{ij} = \eta_{ij} + l_i l_j, \tag{187}$$

where, as we have stated, l_i is a null vector with respect to η_{ij}, i.e.,

$$l^i = \eta^{ij} l_j \qquad \text{and} \qquad l^i l_i = 0. \tag{188}$$

Now, with the *definition*,

$$g^{ij} = \eta^{ij} - l^i l^j, \tag{189}$$

we verify that

$$g^{ij}g_{jk} = (\eta^{ij} - l^i l^j)(\eta_{jk} + l_j l_k) = \delta^i_k. \tag{190}$$

In other words, g^{ij} as defined is, indeed, the contravariant form of g_{ij}. Moreover,

$$g^{ij}l_j = (\eta^{ij} - l^i l^j)l_j = l_i. \tag{191}$$

Accordingly, the index of the null vector can be raised or lowered by $\boldsymbol{g}$ as well as by $\boldsymbol{\eta}$. It follows that $\boldsymbol{l}$ is null, also, with respect to $\boldsymbol{g}$. (However, it is to be understood that, in general, the raising and the lowering of the indices is effected only by $\boldsymbol{\eta}$.)

We shall now obtain an explicit formula for the connection C^m_{ij}. Since ∇'_i is defined in terms of the Christoffel connection derived from g_{ij},

$$\nabla'_i g_{jk} = 0. \tag{192}$$

But we can also evaluate this quantity with the aid of equation (183). Thus,

$$\nabla'_i g_{jk} = \nabla_i g_{jk} - C^m_{ij} g_{mk} - C^m_{ik} g_{jm} = 0. \tag{193}$$

On the other hand,

$$\nabla_i g_{jk} = \nabla_i(\eta_{jk} + l_j l_k) = \nabla_i(l_j l_k). \tag{194}$$

Therefore,

$$\nabla_i(l_j l_k) = C^m_{ij} g_{mk} + C^m_{ik} g_{jm}. \tag{195}$$

From this last relation we obtain (in the same manner as one obtains the Christoffel connection from the vanishing of the covariant derivative of the metric tensor):

$$C^m_{ij} g_{mk} = \tfrac{1}{2}[\nabla_i(l_j l_k) + \nabla_j(l_k l_i) - \nabla_k(l_i l_j)]. \tag{196}$$

By contracting this last relation with g^{kn}, we obtain (by virtue of the relation (190))

$$C^n_{ij} = \tfrac{1}{2}(\eta^{kn} - l^k l^n)[\nabla_i(l_j l_k) + \nabla_j(l_i l_k) - \nabla_k(l_i l_j); \tag{197}$$

or, after some simplification (remembering that $\boldsymbol{l}$ is a null vector) we find

$$C^n_{ij} = \tfrac{1}{2}[\nabla_i(l_j l^n) + \nabla_j(l_i l^n) - \nabla^n(l_i l_j) + l^n l^k \nabla_k(l_i l_j)]. \tag{198}$$

Contracting equation (198) with respect to the indices n and i, we obtain

$$C^n_{jn} = 0. \tag{199}$$

Similarly, we find

$$l_n C^n_{ij} = -\tfrac{1}{2} l_n \nabla^n(l_i l_j) = -\tfrac{1}{2}(l_i l_n \nabla^n l_j + l_j l_n \nabla^n l_i) \tag{200}$$

and

$$l^i C^n_{ij} = \tfrac{1}{2}(l^n l^i \nabla_i l_j + l_j l^i \nabla_i l^n). \tag{201}$$

Defining

$$X_j = l^i \nabla_i l_j, \tag{202}$$

we can rewrite the relations (200) and (201) in the forms

$$l_n C^n_{ij} = -\tfrac{1}{2}(l_i X_j + l_j X_i)$$

and

$$l^i C^n_{ij} = +\tfrac{1}{2}(l^n X_j + l_j X^n). \tag{203}$$

Besides, we must also have

$$l^j X_j = 0. \tag{204}$$

We now assume that g_{ij} represents a solution of Einstein's equation for the vacuum. Then the Ricci tensor R'_{ij} must vanish. But R_{ij} also vanishes by virtue of our assumption that η_{ij} is the metric of a flat space-time. Therefore, by equation (186),

$$\begin{aligned} 0 &= -2\nabla_{[i}C^m_{m]k} + 2C^n_{k[i}C^m_{m]n} \\ &= -\nabla_i C^m_{mk} + \nabla_m C^m_{ik} + C^n_{ki}C^m_{mn} - C^n_{km}C^m_{in} \\ &= \nabla_m C^m_{ik} - C^n_{km}C^m_{in}, \end{aligned} \tag{205}$$

where, in the reductions, we have made use of equation (199). Now contracting equation (205) with $l^i l^k$ and making use of the relations (203) and (204), we obtain

$$\begin{aligned} 0 &= -l^k[\nabla_m(l^i C^m_{ik}) - C^m_{ik}\nabla_m l^i] + (l^k C^n_{km})(l^i C^m_{in}) \\ &= -\tfrac{1}{2}l^k\nabla_m(l^m X_k + X^m l_k) + \tfrac{1}{2}(l^m X^i + l^i X^m)\nabla_m l_i \\ &\quad + \tfrac{1}{4}(l^n X_m + l_m X^n)(l^m X_n + l_n X^m) \\ &= -\tfrac{1}{2}l^k l^m \nabla_m X_k + \tfrac{1}{2}X_i l^m \nabla_m l^i \\ &= +\tfrac{1}{2}X_k l^m \nabla_m l^k + \tfrac{1}{2}X_i X^i = X_i X^i. \end{aligned} \tag{206}$$

Therefore, X is a null vector; and since it is also orthogonal to the null vector $\boldsymbol{l}$, it must be a multiple of $\boldsymbol{l}$. Accordingly, we may write

$$X_j = \phi l_j, \tag{207}$$

where ϕ is some function of proportionality. Now, by the definition (202) of X_j,

$$X_j = l^i \nabla_i l_j = \phi l_j. \tag{208}$$

Therefore, *l defines a null geodesic congruence*. And finally, by combining equations (203) and (207), we obtain the relations

$$l_n C^n_{ij} = -\phi l_i l_j \qquad \text{and} \qquad l^i C^n_{ij} = +\phi l^n l_j. \tag{209}$$

We now return to equation (185) and contract it with $l^j l_m$. Remembering that $R_{ijk}{}^m$ vanishes (since the metric η_{ij}, by assumption, is that of a flat space-time), we have

$$l^j l_m R'_{ijk}{}^m = l^j l_m(\nabla_j C^m_{ik} - \nabla_i C^m_{jk} + C^n_{ki}C^m_{jn} - C^n_{kj}C^m_{in}). \tag{210}$$

We readily verify that all the terms on the right-hand side, except the first, vanish by virtue of the relations (209). We are thus left with

$$\begin{aligned} l^j l_m R'_{ijk}{}^m &= l^j l_m \nabla_j C^m_{ik} = l^j [\nabla_j (l_m C^m_{ik}) - C^m_{ik} \nabla_j l_m] \\ &= -l^j \nabla_j (\phi l_i l_k) - C^m_{ik} \phi l_m \\ &= -(l^j \nabla_j \phi + \phi^2) l_i l_k, \end{aligned} \tag{211}$$

where, in the reductions, we have again made use of the relations (209). Since the index of l can be raised or lowered by either of the metrics η or g, we can rewrite the relation we have arrived at in the form

$$R'_{ijkm} l^j l^m = H l_i l_k \quad \text{where} \quad H = -(l^j \nabla_j \phi + \phi^2), \tag{212}$$

where it may be recalled that l represents a null geodesic congruence.

From equation (212), it clearly follows that

$$R'_{ijm[k} l_{n]} l^j l^m = -H l_i l_{[k} l_{n]} = 0. \tag{213}$$

By Chapter 1, equation (385), this last relation is precisely the one which ensures that the space-time is of Petrov type-II and that l is a principal null direction.

(a) *Casting the Kerr metric in the Kerr-Schild form*

We start with the congruence of null vectors l defined in equation (169). With the choice $E = 1$, we have

$$dt = \frac{r^2 + a^2}{\Delta} d\lambda, \quad dr = d\lambda, \quad d\theta = 0, \quad \text{and} \quad d\varphi = \frac{a}{\Delta} d\lambda. \tag{214}$$

We now introduce, in place of t and φ, the new variables

$$du = dt - \frac{r^2 + a^2}{\Delta} dr \qquad \text{and} \qquad d\tilde{\varphi} = d\varphi - \frac{a}{\Delta} dr. \tag{215}$$

In these new coordinates, the null geodesic is given by

$$l^i = (0, 1, 0, 0). \tag{216}$$

For expressing the Kerr metric in the new coordinates, it is convenient to rewrite it, first, in the form

$$\begin{aligned} ds^2 = \frac{\Delta}{\rho^2} [dt - (a \sin^2\theta) d\varphi]^2 - \frac{\sin^2\theta}{\rho^2} [(r^2 + a^2) d\varphi - a\, dt]^2 \\ - \frac{\rho^2}{\Delta} (dr)^2 - \rho^2 (d\theta)^2. \end{aligned} \tag{217}$$

A direct substitution of the relations (215) now gives

$$\mathrm{d}s^2 = \frac{\Delta}{\rho^2}(\mathrm{d}u - a\,\mathrm{d}\tilde{\varphi}\sin^2\theta)^2 - \frac{\sin^2\theta}{\rho^2}[(r^2+a^2)\mathrm{d}\tilde{\varphi} - a\,\mathrm{d}u]^2 + 2\mathrm{d}r(\mathrm{d}u - a\,\mathrm{d}\tilde{\varphi}\sin^2\theta) - \rho^2(\mathrm{d}\theta)^2. \quad (218)$$

The corresponding explicit form of the metric is

$$(g_{ij}) = \begin{vmatrix} (u) & (r) & (\theta) & (\tilde{\varphi}) \\ 1-2Mr/\rho^2 & 1 & 0 & 2aMr\sin^2\theta/\rho^2 \\ 1 & 0 & 0 & -a\sin^2\theta \\ 0 & 0 & -\rho^2 & 0 \\ 2aMr\sin^2\theta/\rho^2 & -a\sin^2\theta & 0 & -\Sigma^2\sin^2\theta/\rho^2 \end{vmatrix}. \quad (219)$$

By a further regrouping of the terms, we find that the metric (218) can be brought to the form

$$\mathrm{d}s^2 = [(\mathrm{d}u+\mathrm{d}r)^2 - (\mathrm{d}r)^2 - \rho^2(\mathrm{d}\theta)^2 - (r^2+a^2)(\mathrm{d}\tilde{\varphi})^2\sin^2\theta - 2a\,\mathrm{d}r\,\mathrm{d}\tilde{\varphi}\sin^2\theta] - \frac{2Mr}{\rho^2}(\mathrm{d}u - a\,\mathrm{d}\tilde{\varphi}\sin^2\theta)^2; \quad (220)$$

or, in view of the covariant form of $\boldsymbol{l}$ being

$$l_i = (1, 0, 0, -a\sin^2\theta), \quad (221)$$

we can also write

$$\mathrm{d}s^2 = [(\mathrm{d}x^0)^2 - (\mathrm{d}r)^2 - \rho^2(\mathrm{d}\theta)^2 - (r^2+a^2)(\mathrm{d}\tilde{\varphi})^2\sin^2\theta - 2a\,\mathrm{d}r\,\mathrm{d}\tilde{\varphi}\sin^2\theta] - \frac{2Mr}{\rho^2}l_i l_j\,\mathrm{d}x^i\,\mathrm{d}x^j, \quad (222)$$

where

$$\mathrm{d}x^0 = \mathrm{d}u + \mathrm{d}r. \quad (223)$$

We shall presently show that the part of the metric in square brackets in equation (222) represents flat space-time. We have thus reduced the Kerr metric to the Kerr–Schild form; and its algebraic speciality is now a part of its definition.

We can now verify that the substitutions,

$$\left.\begin{aligned} x = (r\cos\tilde{\varphi} + a\sin\tilde{\varphi})\sin\theta, \quad y = (r\sin\tilde{\varphi} - a\cos\tilde{\varphi})\sin\theta, \\ x^2+y^2 = (r^2+a^2)\sin^2\theta \quad \text{and} \quad z = r\cos\theta, \end{aligned}\right\} \quad (224)$$

bring the part of the metric in square brackets in equation (222) to the form of the manifestly flat metric,

$$(\mathrm{d}x^0)^2 - (\mathrm{d}x)^2 - (\mathrm{d}y)^2 - (\mathrm{d}z)^2. \quad (225)$$

Explicitly, the entire metric in these new coordinates is

$$ds^2 = (dx^0)^2 - (dx)^2 - (dy)^2 - (dz)^2$$

$$-\frac{2Mr^3}{r^4 + a^2 z^2}\left\{dx^0 - \frac{1}{r^2 + a^2}[r(x\,dx + y\,dy) + a(x\,dy - y\,dx)] - \frac{1}{r}z\,dz\right\}^2, \tag{226}$$

where, according to the substitutions (224), r^2 is defined, implicitly, in terms of x, y, and z by

$$r^4 - r^2(x^2 + y^2 + z^2 - a^2) - a^2 z^2 = 0. \tag{227}$$

The metric (226) is clearly analytic everywhere except at

$$x^2 + y^2 + z^2 = a^2 \quad \text{and} \quad z = 0; \tag{228}$$

in other words, it has a *ring singularity* in the (x, y)-plane. We shall consider the nature of this singularity in greater detail in §58 below. Here, we shall only note that it is precisely in the form (226) that Kerr originally presented his solution.

58. The nature of the Kerr space-time

There are two principal features of the Kerr space-time which require to be examined and clarified: the first relates to the nature of the null surfaces and the second relates to the nature of the singularity at $r = 0$ and $\theta = \pi/2$.

Concerning the first, we know that the null surfaces occur at the zeros of

$$\Delta(r) = r^2 - 2Mr + a^2 = 0, \tag{229}$$

i.e., at

$$r = r_+ = M + \sqrt{(M^2 - a^2)} \quad \text{and} \quad r = r_- = M - \sqrt{(M^2 - a^2)}. \tag{230}$$

These zeros are real, positive, and distinct so long as $a^2 < M^2$, an inequality which we shall assume obtains. Since

$$\Delta > 0 \text{ for } r > r_+ \text{ and } r < r_- \quad \text{and} \quad \Delta < 0 \text{ for } r_- < r < r_+, \tag{231}$$

it is clear that, as in the case of the Reissner–Nordström space-time, we must distinguish the three regions:

$$A\colon r < r_-; \quad B\colon r_- < r < r_+; \quad \text{and} \quad C\colon r > r_+. \tag{232}$$

(We shall presently see that the region A can be extended to *all negative values of* r even as the region C extends to all positive values of r.) We shall find, again as in the Reissner–Nordström geometry, that while the surface at $r = r_+$ represents an event horizon, the surface at $r = r_-$ represents a Cauchy horizon.

Concerning the singularity in the Kerr space-time, we have already pointed out that the non-vanishing components of the Riemann tensor (listed in equations (142)) diverge only for $r = 0$ and $\theta = \pi/2$; and that *this is the only singularity there is*. Since the divergence at $r = 0$ occurs only for $\theta = \pi/2$, it is clear that its nature cannot be the same as the singularity at $r = 0$ of the Schwarzschild and the Reissner–Nordström space-times. To understand the real nature of the singularity of the Kerr space-time, we must, first, eliminate the inherent ambiguity in the coordinate system, (r, θ, φ), at $r = 0$. This ambiguity was abolished in §57(a) by the choice of the 'Cartesian' coordinate system (x, y, z). In the metric (226), written in these coordinates, r^2 is implicitly defined in terms of x, y, and z by equation (227), while

$$(x^2 + y^2) = (r^2 + a^2)\sin^2\theta. \tag{233}$$

The surfaces of constant r are confocal ellipsoids whose principal axes coincide with the coordinate axes. These ellipsoids degenerate, for $r = 0$, to the *disc*,

$$x^2 + y^2 \leqslant a^2,\ z = 0. \tag{234}$$

The point, $(r = 0, \theta = \pi/2)$, corresponds, then, to the *ring*,

$$x^2 + y^2 = a^2,\ z = 0; \tag{235}$$

and the *singularity along this ring is the only singularity of the Kerr space-time.*

The points, interior to the ring (235), correspond to $r = 0$ and $\pi/2 > \theta > 0$ and $0 < \tilde{\varphi} < 2\pi$. The surfaces of constant θ cross the disc and pass to the other side. On these accounts, we are entitled to *extend the domain of r to all negative values*. In other words, we analytically continue the function r, defined by equation (227), to all negative values. We accomplish this continuation by attaching another chart (x', y', z') to the disc (234) and by identifying a point on the top side of the disc to a point, with the same x and y coordinates, on the bottom side of the disc in the (x', y', z')-chart. Similarly, we identify a point on the bottom side of the disc in the (x, y, z)-chart to a point on the top side of the disc in the (x', y', z')-chart. (See Fig. 25.) The enlargement of the original manifold in this manner is analogous to the enlargement of the complex plane to the Riemann surfaces for the representation of analytic functions with singularities.

With the manifold enlarged, in the region A, distinguished in (232), r is allowed to take all negative values.

The metric (226), extended to the enlarged manifold, has the same form in the (x', y', z')-chart as in the (x, y, z)-chart except that r is now assigned negative values. Therefore, in the (x', y', z')-chart, the terms in the curly brackets in (226) occur with the factor, $+2Mr^3/(r^4 + a^2z^2)$. For large negative values of r, the space is again asymptotically flat but with a negative mass for the source.

Turning to the elimination of the coordinate singularities at r_+ and r_-, we first realize that, on account of the principal null congruences, $\boldsymbol{l}$ and $\boldsymbol{n}$, not being hyper-surface orthogonal, we cannot make them the basis for deriving non-singular coordinates as we were able to do in the context of the Schwarzschild and the Reissner–Nordström space-times. It is, therefore, useful

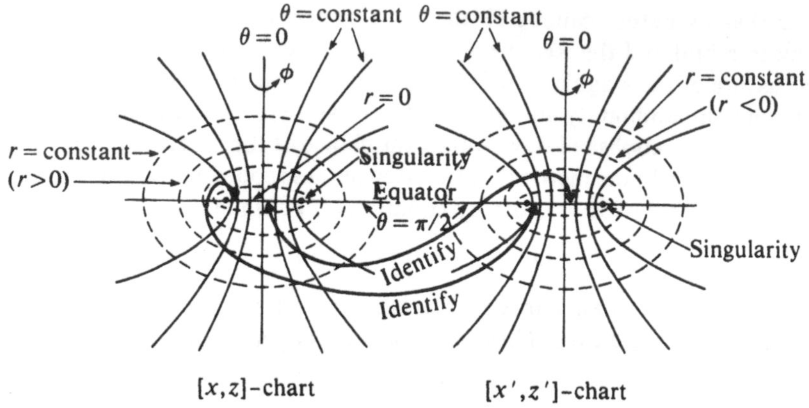

FIG. 25. The ring singularity of the Kerr metric in the equatorial plane. The nature of the singularity becomes manifest when the metric is written in "Cartesian coordinates" in the form (226) when the curves of constant r become confocal spheroids and the curves of constant θ become confocal hyperbolae. The analytic extension to all negative values of r is accomplished by attaching to the disc ($x^2+y^2 \leqslant a^2$; $z=0$), another disc in a chart (x', y', z'), by the identifications described in the text and illustrated in the figure. The figure shows the sections $y=0$ and $y'=0$ of these planes.

to restrict ourselves, in the first instance, to the two-dimensional (r, t)-manifold along the symmetry axis, $\theta = 0$, to clarify that the nature of the horizons is essentially the same as in the Riessner–Nordström space-time and that the manifold can be completed with the same basic ideas.

Along the symmetry axis, $\theta = 0$, the Kerr metric degenerates to the form

$$ds^2 = \frac{\Delta}{r^2+a^2}\left(dt - \frac{r^2+a^2}{\Delta}dr\right)\left(dt + \frac{r^2+a^2}{\Delta}dr\right). \tag{236}$$

By defining the null coordinates,

$$u = t - r_* \quad \text{and} \quad v = t + r_*, \tag{237}$$

where

$$r_* = \int \frac{r^2+a^2}{\Delta} dr = r + \frac{r_+^2+a^2}{r_+-r_-}\lg|r-r_+| - \frac{r_-^2+a^2}{r_+-r_-}\lg|r-r_-|, \tag{238}$$

we can write

$$ds^2 = \frac{\Delta}{r^2+a^2} du\, dv. \tag{239}$$

The substitutions (237) are the proper ones for the regions A and C (cf. Ch. 5, equations (46) and (48)). In the region B, Δ is negative and the proper substitutions are

$$u = r_* + t \quad \text{and} \quad v = r_* - t, \tag{240}$$

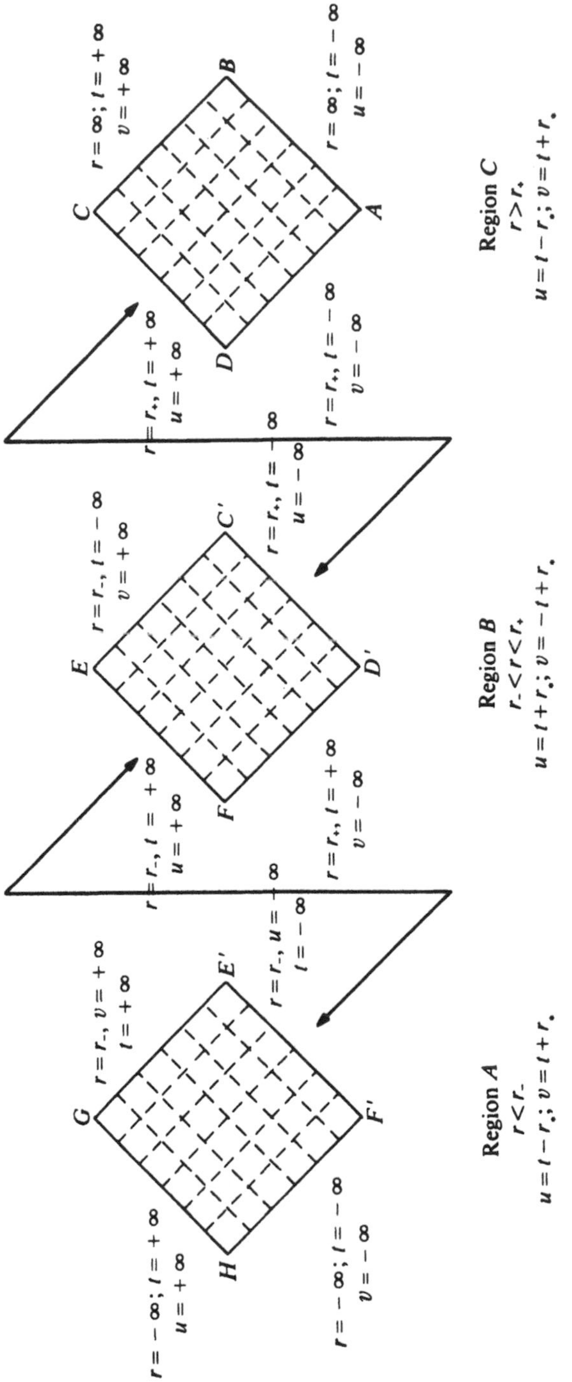

FIG. 26. The different regions in the Kerr space-time along the axis $\theta = 0$. For explanation see text.

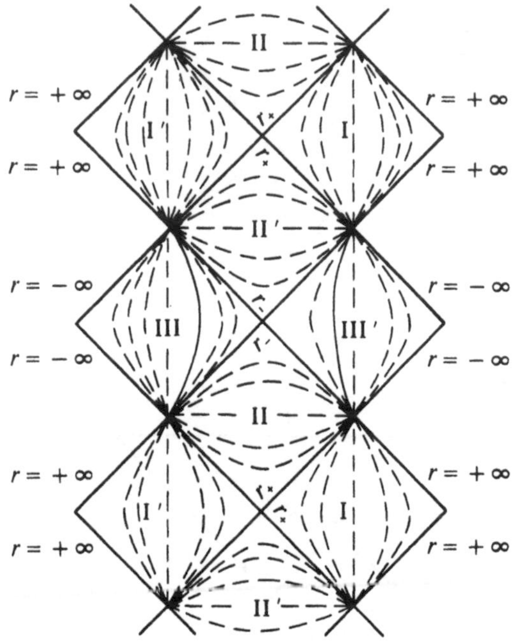

FIG. 27. The maximum analytic extension of the Kerr space-time for $\theta = 0$ is achieved by piecing together the blocks in Fig. 26 in the manner described for the Reissner–Nordström space-time in Fig. 14.

when the metric takes the form

$$\mathrm{d}s^2 = \frac{|\Delta|}{r^2 + a^2}\,\mathrm{d}u\,\mathrm{d}v. \tag{241}$$

By the foregoing choices, the parts of the manifold in the three regions can be represented by the 'blocks' in Fig. 26. The edges of the blocks are to be identified as indicated. Also, to the regions A, B, and C, there correspond further regions, A', B', and C', obtained by applying the transformation $u \to -u$ and $v \to -v$ which reverses the light-cone structure. A maximal analytic extension of the (r, t)-manifold may then be obtained by piecing together copies of the six blocks in an analytic fashion so that an overlapping edge is covered by a $u(r)$ or a $v(r)$-chart (with the exception of the corners of the blocks). The resulting 'ladder' is shown in Fig. 27; it can be extended, indefinitely, in both directions.

An analytic representation of the maximally extended space-time on the axis can be obtained, exactly as in the context of the Reissner–Nordström space-time, as follows.

In regions A and C, we set

$$\tan U = -e^{-\alpha u} \quad \text{and} \quad \tan V = +e^{+\alpha v}, \tag{242}$$

while in the region B, we set

$$\tan U = +e^{+\alpha u} \quad \text{and} \quad \tan V = +e^{+\alpha v}, \tag{243}$$

where

$$\alpha = \frac{r_+ - r_-}{2(r_+^2 + a^2)}, \tag{244}$$

and u and v, in the substitutions (242) and (243), have, respectively, the same meanings as in equations (237) and (240). With the foregoing substitutions, the metric takes the 'universal' form

$$ds^2 = -\frac{4}{\alpha^2}|\Delta| \operatorname{cosec} 2U \operatorname{cosec} 2V \, dU \, dV, \tag{245}$$

where r is defined implicitly in terms of U and V by

$$\begin{aligned} \tan U \tan V &= -e^{2\alpha r}|r - r_+||r - r_-|^{-\beta} \quad (r > r_+ \text{ and } r < r_-) \\ &= +e^{2\alpha r}|r - r_+||r - r_-|^{-\beta} \quad (r_- < r < r_+), \end{aligned} \tag{246}$$

where

$$\beta = \frac{r_-^2 + a^2}{r_+^2 + a^2} = \frac{r_-}{r_+}. \tag{247}$$

The delineation of the coordinate axes by null geodesics in Figs. 26 and 27 enables us to visualize, pictorially, the nature of the underlying space-time. Thus, the partition at $r \doteq r_+$ is an event horizon and the partition at $r = r_-$ is a Cauchy horizon in the senses we have made clear in the earlier chapters.

To eliminate the coordinate singularities at r_+ and r_- in general in a unified manner, and to achieve, at the same time, a maximal analytic extension of the entire manifold is more difficult. The difficulty arises from the fact that both t and φ have singular behaviours as the horizons are approached along null (or, time-like) geodesics (see Chapter 7, §61). On this account, a simple change of coordinates, which will be satisfactory at both horizons simultaneously, does not appear practicable. We shall, therefore, be content in treating the event and the Cauchy horizons separately.

We consider first the substitutions appropriate for crossing the event horizon smoothly.

We start with the metric in the form (cf. equation (217))

$$ds^2 = \frac{\Delta}{\rho^2}(dt - a\,d\varphi \sin^2\theta)^2 - \frac{\sin^2\theta}{\rho^2}[(r^2 + a^2)\,d\varphi - a\,dt]^2 - \frac{\rho^2}{\Delta}(dr)^2 - \rho^2(d\theta)^2; \tag{248}$$

and write, as before,

$$u = t - r_* \quad \text{and} \quad v = t + r_*, \tag{249}$$

where r_* is the same variable as defined in equation (238). In addition, we shall now introduce, in place of φ,

$$\tilde{\varphi}_+ = \varphi - at/2Mr_+, \tag{250}$$

where we have distinguished the new variable by a subscript, +, to emphasize that we are presently concerned with a smooth crossing of the event horizon. (The introduction of a new angle variable, depending on the horizon we are crossing, is an essential feature of considerations pertaining to the Kerr space-time.)

With the change of variables defined,

$$\begin{aligned} dt - a\,d\varphi \sin^2\theta &= \frac{\rho_+^2}{r_+^2 + a^2} dt - a\,d\tilde{\varphi}_+ \sin^2\theta \\ &= \frac{1}{2}\frac{\rho_+^2}{r_+^2 + a^2}(du + dv) - a\,d\tilde{\varphi}_+ \sin^2\theta, \end{aligned} \tag{251}$$

where

$$\rho_+^2 = r_+^2 + a^2 \cos^2\theta; \tag{252}$$

similarly,

$$(r^2 + a^2)\,d\varphi - a\,dt = (r^2 + a^2)\,d\tilde{\varphi}_+ + \frac{1}{2} a \frac{r^2 - r_+^2}{r_+^2 + a^2}(du + dv). \tag{253}$$

We also have

$$dr = \frac{1}{2}\frac{\Delta}{r^2 + a^2}(dv - du). \tag{254}$$

Inserting the expressions (251), (253), and (254) in equation (248), we obtain after some further simplifications,

$$\begin{aligned} ds^2 = {} & \frac{\Delta}{2\rho^2}\left[\frac{\rho_+^4}{(r_+^2 + a^2)^2} + \frac{\rho^4}{(r^2 + a^2)^2}\right] du\,dv - \rho^2 (d\theta)^2 \\ & - \frac{\sin^2\theta}{\rho^2}\left[(r^2 + a^2)\,d\tilde{\varphi}_+ + \frac{1}{2} a \frac{r^2 - r_+^2}{r_+^2 + a^2}(du + dv)\right]^2 \\ & - \frac{a^2 \Delta \sin^2\theta}{4\rho^2}\left(\frac{\rho_+^2}{r_+^2 + a^2} + \frac{\rho^2}{r^2 + a^2}\right)\frac{r^2 - r_+^2}{(r_+^2 + a^2)(r^2 + a^2)}[(du)^2 + (dv)^2] \\ & + \frac{a\Delta \sin^2\theta}{\rho^2}\left[a \sin^2\theta\,d\tilde{\varphi}_+ - \frac{\rho_+^2}{r_+^2 + a^2}(du + dv)\right] d\tilde{\varphi}_+. \end{aligned} \tag{255}$$

To include the four adjoining regions, I, II, I′, and II′ in Fig. 27 in a single coordinate neighbourhood, we can define, as before

$$\left.\begin{aligned}\tan U_+ &= -e^{-\alpha_+ u} = -e^{-\alpha_+ t}e^{+\alpha_+ r_*}\\ \tan V_+ &= +e^{+\alpha_+ v} = +e^{+\alpha_+ t}e^{+\alpha_+ r_*},\end{aligned}\right\} \tag{256}$$

where

$$\alpha_+ = \frac{r_+ - r_-}{2(r_+^2 + a^2)}. \tag{257}$$

Then

$$du = -\frac{2}{\alpha_+}\operatorname{cosec} 2U_+ \, dU_+ \quad \text{and} \quad dv = +\frac{2}{\alpha_+}\operatorname{cosec} 2V_+ \, dV_+. \tag{258}$$

By substituting these expressions for du and dv in equation (255), we shall obtain the metric in the required form, where r is to be understood as defined implicitly in terms of U_+ and V_+ by the equation

$$\tan U_+ \tan V_+ = -e^{2\alpha_+ r}|r - r_+||r - r_-|^{-\beta_+}, \tag{259}$$

where

$$\beta_+ = r_-/r_+. \tag{260}$$

Considering next the substitutions, appropriate for crossing the Cauchy horizon at $r = r_-$ smoothly, we write

$$u = r_* + t \qquad \text{and} \qquad v = r_* - t \tag{261}$$

and

$$\tilde{\varphi}_- = \varphi - at/2Mr_-. \tag{262}$$

The resulting form of the metric can be obtained from equation (255) by replacing "+" by "−" and writing du in place of dv and −dv in place of du. And the transformations analogous to those given in equations (256)–(260) are

$$\begin{aligned}&\tan U_- = e^{+\alpha_- u} = e^{+\alpha_-(r_* + t)}, \quad \tan V_- = e^{+\alpha_- v} = e^{+\alpha_-(r_* - t)},\\ &du = \frac{2}{\alpha_-}\operatorname{cosec} 2U_- \, dU_-, \quad dv = \frac{2}{\alpha_-}\operatorname{cosec} 2V_- \, dV_-,\end{aligned} \tag{263}$$

where

$$\alpha_- = \frac{r_- - r_+}{2(r_-^2 + a^2)}; \tag{264}$$

and r is defined implicitly in terms U_- and V_- by

$$\tan U_- \tan V_- = e^{2\alpha_- r}|r - r_-||r - r_+|^{-\beta_-}, \tag{265}$$

where

$$\beta_- = r_+/r_-. \tag{266}$$

It is an easy matter to verify that the metric, as we have defined, in terms (U_+, V_+) and (U_-, V_-), is analytic everywhere except at the curvature singularity where ρ^2 vanishes.

In Chapter 7 we shall see that with the maximal extension of the manifold, all time-like and null geodesics are of infinite affine length (both when extended to the past and when extended to the future) except those which terminate at the singularity; in other words, the extended space-time is *geodesically complete.*

(a) *The ergosphere*

Besides the principal difference in the nature of the singularities in the Kerr and in the Schwarzschild and in the Reissner–Nordström space-times, there is a further feature which distinguishes the Kerr space-time from the other two. The distinction arises from the fact that unlike in the Schwarzschild and in the Reissner–Nordström space-times, the event horizon does not coincide with the surface

$$g_{tt} = 1 - 2Mr/\rho^2 = 0 \tag{267}$$

or

$$r^2 - 2Mr + a^2\cos^2\theta = \Delta - a^2\sin^2\theta = 0. \tag{268}$$

The surface,

$$r = r_e(\theta) = M + \sqrt{(M^2 - a^2\cos^2\theta)}, \tag{269}$$

external to the event horizon, on which g_{tt} vanishes, is called the *ergosphere*. It coincides with the event horizon only at the poles $\theta = 0$ and $\theta = \pi$.

The ergosphere is a *stationary limit surface* in the sense that it is the inner boundary of the region in which a particle with a word-line $dt = 0$ is time-like. This fact becomes apparent when we inquire, for example, into the implications of the obvious requirement

$$|v^{(\varphi)}| \leqslant 1, \tag{270}$$

where

$$v^{(\varphi)} = e^{\psi - \nu}(\Omega - \omega) \tag{271}$$

is the φ-component of the 3-velocity in the ortho-normal tetrad-frame considered in §52. In the context of the Kerr metric, the inequality (270) gives (cf. equation (125))

$$|\Omega - \omega| \leqslant e^{\nu - \psi} = \frac{\rho^2\sqrt{\Delta}}{\Sigma^2\sqrt{\delta}}. \tag{272}$$

It follows that

$$\Omega_{\substack{\max\\ \min}} = \omega \pm \frac{\rho^2\sqrt{\Delta}}{\Sigma^2\sqrt{\delta}} = \frac{2aMr\sqrt{\delta} \pm \rho^2\sqrt{\Delta}}{\Sigma^2\sqrt{\delta}}, \tag{273}$$

or, by virtue of the identities (119)–(121),

$$\Omega_{\substack{\max\\ \min}} = \pm \frac{\sqrt{\Delta} \pm a\sqrt{\delta}}{[(r^2 + a^2) \pm a\sqrt{(\Delta\delta)}]\sqrt{\delta}}. \tag{274}$$

In particular, we have

$$\Omega_{\min} = -\frac{\sqrt{\Delta} - a\sqrt{\delta}}{[(r^2+a^2) - a\sqrt{(\Delta\delta)}]\sqrt{\delta}} = -\frac{\Delta - a^2\delta}{[\rho^2\sqrt{\Delta} + 2aMr\sqrt{\delta}]\sqrt{\delta}}, \tag{275}$$

on making use, once again, of the identities (119)–(121). Since

$$\Delta - a^2\delta < 0 \quad \text{inside the ergosphere,} \tag{276}$$

we conclude that

$$\Omega_{\min} > 0 \quad \text{inside the ergosphere} \quad \text{and} = 0 \quad \text{on} \quad r = r_e(\theta). \tag{277}$$

This result is in accord with our earlier statement concerning the stationary limit-surface derived from the vanishing of g_{tt} on it.

The fact that there can be no stationary observer inside the ergosphere does not, of course, preclude the existence of time-like trajectories in the region, $r_+ < r < r_e(\theta)$, which can escape to infinity. This part of space is, therefore, communicable to the space outside. We shall see in §65 in Chapter 7 that the existence of this finite region of space between the event horizon and the stationary limit-surface has important consequences: it allows, for example, processes—the Penrose processes—which will result in the extraction of the rotational energy of the black hole.

Finally, we may note here, for future reference, that the surface area of the event horizon is given by

$$\begin{aligned}\text{Surface area} &= \int_0^\pi \int_0^{2\pi} (g_{\theta\theta} g_{\varphi\varphi})^{1/2}_{r=r_+} \, d\theta d\varphi \\ &= \int_0^\pi \int_0^{2\pi} \left[\frac{\sin^2\theta}{\rho^2}\{(r^2+a^2)^2 - a^2\Delta\sin^2\theta\}\rho^2\right]^{1/2}_{r=r_+} d\theta d\varphi \\ &= 4\pi(r_+^2 + a^2) = 8\pi M r_+ = 8\pi M[M + \sqrt{(M^2-a^2)}].\end{aligned} \tag{278}$$

BIBLIOGRAPHICAL NOTES

Kerr's discovery was announced very briefly in a short letter (dated 26 July 1963):

1. R. P. KERR, *Phys. Rev. Letters*, 11, 237–8 (September 1, 1963) and reported at a meeting (December 16–18, 1963) a few months later:
2. R. P. KERR in *Quasistellar Sources and Gravitational Collapse*, 99–103, edited by I. Robinson and E. Schuking, University of Chicago Press, Chicago, 1965.

In spite of its brevity, the original announcement is surprisingly complete in enumerating the essential features of the solution. But the announcement gave not even an outline of the derivation; it was provided only some two years later in:

3. R. P. KERR and A. SCHILD in *Comitato Nazionale per le Manifestazioni Celebrative Del IV Centenario della Nascita di Galileo Galilei, Atti del Convegno Sulla Relatività Generale: Problemi Dell'Energia E Onde Gravitazionali*, 1–12, edited by G. Barbèra, Florence, 1965.

4. ——— in *Proceedings of Symposia in Applied Mathematics*, 17, 199–209, American Math. Soc., 1965.

However, the authors warn the reader (in reference 3) that "the calculations giving these results are by no means simple."

§52. The literature on stationary axisymmetric space-times is voluminous; the extent of it can be judged from:

5. W. Kinnersley in *General Relativity and Gravitation*, 109–35, edited by G. Shaviv and J. Rosen, John Wiley & Sons, Inc., New York, 1975.

Since the foregoing report was written, the advances in the subject have been spectacular. But here we are concerned with the general theory only to the extent that we have an adequate base for a simple derivation of the Kerr metric.

While the equations in this section have been derived as special cases of the ones given in Chapter 2, they are contained in essentially the same forms in:

6. S. Chandrasekhar and J. L. Friedman, *Astrophys. J.*, 175, 379–405 (1972).

What we have called the Papapetrou transformation is described in:

7. A. Papapetrou, *Ann. d. Physik*, (6) 12, 309–15 (1953).
8. ———, *Ann. d'l Institut Henri Poincaré Section* A, 4, 83–105 (1966).

§53. The treatment in this section derives largely from:

9. S. Chandrasekhar, *Proc. Roy. Soc.* (*London*) A, 358, 406–20 (1978).

The Ernst equation is derived in:

10. F. J. Ernst, *Phys. Rev.*, 167, 1175–8 (1968).
11. ———, ibid., 168, 1415–7 (1968).

See also:

12. B. C. Xanthopoulos, *Proc. Roy. Soc.* (*London*) A, **365**, 381–411, (1979).

§55. The principal papers dealing with the uniqueness of the Kerr space-time are those of:

13. B. Carter, *Phys. Rev. Letters*, 26, 331–3 (1972).
14. D. C. Robinson, ibid., 34, 905–6 (1975).

See also:

15. B. Carter in *Black Holes*, edited by C. DeWitt and B. S. DeWitt, Gordon and Breach Science Publishers, New York, 1973.
16. ——— in *General Relativity—An Einstein Centenary Survey*, chapter 6, 294–369, edited by S. W. Hawking and W. Israel, Cambridge, England, 1979.

Israel's theorem (derived in this section as a special case of Robinson's theorem and assuming somewhat more restrictive conditions than in Israel's original statement) is proved in:

17. W. Israel, *Phys. Rev.*, 164, 1776–9 (1968).

§56. An important reference in this context is:

18. W. Kinnersley, *J. Math. Phys.*, 10, 1195–1203 (1969).

§57. The 'Kerr–Schild' space-times were introduced in references 3 and 4.

The principal ideas in the presentation in this section (due to Xanthopoulos) are derived from:

19. B. C. Xanthopoulos, *J. Math. Phys.*, 19, 1607–9 (1978).

§58. The basic papers dealing with the nature of the Kerr space-time are those of:

20. B. Carter, *Phys. Rev.*, 141, 1242–7 (1966).
21. R. H. Boyer and R. W. Lindquist, *J. Math. Phys.*, 8, 265–81 (1967).
22. B. Carter, *Phys. Rev.*, 174, 1559–70 (1968).

The coordinates in which we have derived and written the Kerr metric are often called the 'Boyer–Lindquist' coordinates; they are introduced in reference 21.

7

THE GEODESICS IN THE KERR SPACE-TIME

59. Introduction

This chapter is devoted to the study of the geodesics in the Kerr space-time. We are devoting an entire chapter to this study for several reasons: apart from the fact that the delineation of the geodesics exhibits the essential features of the space-time (as for example, the relevance of extending the manifold to negative values of the radial coordinate r), the separability of the Hamilton–Jacobi equation (discovered by Carter) was the first of the many properties which have endowed the Kerr metric with an aura of the miraculous. Besides, the study discloses the possibility (discovered by Penrose) of extracting energy from the Kerr black-hole, thus revealing the existence of physical processes one had not contemplated before. At the same time the study of the geodesics in the Kerr space-time directed attention to certain quite unexpected properties of space-times of Petrov type-D in general. We begin, in fact, with the establishment of one of these properties.

60. Theorems on the integrals of geodesic motion in type-D space-times

We have shown in Chapter 1 (§9(*c*)) that in a space-time of Petrov type-D, the principal null congruences are geodesic and shear-free; and that in a null tetrad-basis ($\boldsymbol{l}, \boldsymbol{n}, \boldsymbol{m}, \bar{\boldsymbol{m}}$), in which $\boldsymbol{l}$ and $\boldsymbol{n}$ are the principal null-directions, the spin coefficients, κ, σ, λ, and ν, and the Weyl scalars, Ψ_0, Ψ_1, Ψ_3, and Ψ_4, vanish:

$$\kappa = \sigma = \lambda = \nu = 0, \tag{1}$$

and

$$\Psi_0 = \Psi_1 = \Psi_3 = \Psi_4 = 0. \tag{2}$$

And if, as we shall suppose, that $\boldsymbol{l}$ is affinely parametrized, then the spin-coefficient ε is also zero:

$$\varepsilon = 0. \tag{3}$$

Finally, the Bianchi identities (Chapter 1, equations 321)) provide for the non-vanishing Weyl-scalar Ψ_2 the equations

$$\left.\begin{aligned} D \lg \Psi_2 &= +3\rho; & \delta \lg \Psi_2 &= +3\tau, \\ \triangle \lg \Psi_2 &= -3\mu; & \delta^* \lg \Psi_2 &= -3\pi, \end{aligned}\right\} \tag{4}$$

where D, Δ, δ, and δ^* are the directional derivatives along $\boldsymbol{l}$, $\boldsymbol{n}$, $\boldsymbol{m}$, and $\bar{\boldsymbol{m}}$, respectively.

The metric tensor, g_{ij}, in a coordinate basis, expressed in terms of $\boldsymbol{l}$, $\boldsymbol{n}$, $\boldsymbol{m}$, and $\bar{\boldsymbol{m}}$, is given by (cf. Ch. 1, equation(246))

$$g_{ij} = l_i n_j + l_j n_i - m_i \bar{m}_j - m_j \bar{m}_i; \tag{5}$$

and the covariant derivatives of $\boldsymbol{l}$, $\boldsymbol{n}$, and $\boldsymbol{m}$ expressed in terms of the spin coefficients are (cf. Ch. 1, equation (357))

$$\begin{aligned} l_{j;i} = &+(\gamma+\gamma^*)l_j l_i - (\alpha^*+\beta)l_j \bar{m}_i - (\alpha+\beta^*)l_j m_i \\ &-\tau \bar{m}_j l_i - \tau^* m_j l_i + \rho \bar{m}_j m_i + \rho^* m_j \bar{m}_i; \end{aligned} \tag{6}$$

$$\begin{aligned} n_{j;i} = &-(\gamma+\gamma^*)n_j l_i + (\alpha^*+\beta)n_j \bar{m}_i + (\alpha+\beta^*)n_j m_i \\ &+\pi^* \bar{m}_j n_i + \pi m_j n_i - \mu m_j \bar{m}_i - \mu^* \bar{m}_j m_i; \end{aligned} \tag{7}$$

$$\begin{aligned} m_{j;i} = &+(\gamma-\gamma^*)m_j l_i + (\alpha^*-\beta)m_j \bar{m}_i + (\beta^*-\alpha)m_j m_i \\ &+\pi^* l_j n_i - \mu^* l_j m_i - \tau n_j l_i + \rho n_j m_i. \end{aligned} \tag{8}$$

(The foregoing relations are applicable only to type-D space-times, since the relations (1) and (3) have been assumed.)

If $\boldsymbol{k}$ and $\boldsymbol{f}$ denote any two vectors, then, according to equation (5),

$$\begin{aligned} \boldsymbol{k}\cdot\boldsymbol{f} = g_{ij}k^i f^j = (\boldsymbol{k}\cdot\boldsymbol{l})(\boldsymbol{f}\cdot\boldsymbol{n}) + (\boldsymbol{k}\cdot\boldsymbol{n})(\boldsymbol{f}\cdot\boldsymbol{l}) \\ -(\boldsymbol{k}\cdot\boldsymbol{m})(\boldsymbol{f}\cdot\bar{\boldsymbol{m}}) - (\boldsymbol{k}\cdot\bar{\boldsymbol{m}})(\boldsymbol{f}\cdot\boldsymbol{m}), \end{aligned} \tag{9}$$

$$|\boldsymbol{k}|^2 = 2[(\boldsymbol{k}\cdot\boldsymbol{l})(\boldsymbol{k}\cdot\boldsymbol{n}) - (\boldsymbol{k}\cdot\boldsymbol{m})(\boldsymbol{k}\cdot\bar{\boldsymbol{m}})] \tag{10}$$

and

$$k^i\nabla_i = (\boldsymbol{k}\cdot\boldsymbol{n})D + (\boldsymbol{k}\cdot\boldsymbol{l})\Delta - (\boldsymbol{k}\cdot\bar{\boldsymbol{m}})\delta - (\boldsymbol{k}\cdot\boldsymbol{m})\delta^*. \tag{11}$$

We now state the principal theorem of the subject due to Walker and Penrose.

THEOREM 1. *If* $\boldsymbol{k}$ *is a null geodesic, affinely parametrized, and* $\boldsymbol{f}$ *is a vector orthogonal to* $\boldsymbol{k}$ *and parallelly propagated along it, then, in a type-D space-time, the quantity*

$$\begin{aligned} K_s &= k^i f^j (l_i n_j - l_j n_i - m_i \bar{m}_j + \bar{m}_i m_j)\Psi_2^{-1/3} \\ &= k^i f^j (2 l_i n_j - 2 m_i \bar{m}_j - g_{ij})\Psi_2^{-1/3} \\ &= 2[(\boldsymbol{k}\cdot\boldsymbol{l})(\boldsymbol{f}\cdot\boldsymbol{n}) - (\boldsymbol{k}\cdot\boldsymbol{m})(\boldsymbol{f}\cdot\bar{\boldsymbol{m}})]\Psi_2^{-1/3}, \end{aligned} \tag{12}$$

is conserved along the geodesic, i.e.,

$$k^i\nabla_i K_s = 0. \tag{13}$$

A more common formulation of the theorem is that '*a type-D space-time admits a conformal Killing tensor*'.

Proof. The assumptions of the theorem that k is a null geodesic, affinely parametrized, and that f is orthogonal to k and parallelly propagated along it require that

$$k^i\nabla_i k_j = 0, \quad k^i\nabla_i f_j = 0, \tag{14}$$

$$(\boldsymbol{k}\cdot\boldsymbol{l})(\boldsymbol{k}\cdot\boldsymbol{n})-(\boldsymbol{k}\cdot\boldsymbol{m})(\boldsymbol{k}\cdot\bar{\boldsymbol{m}})=0, \tag{15}$$

and

$$(\boldsymbol{k}\cdot\boldsymbol{l})(\boldsymbol{f}\cdot\boldsymbol{n})+(\boldsymbol{k}\cdot\boldsymbol{n})(\boldsymbol{f}\cdot\boldsymbol{l})-(\boldsymbol{k}\cdot\boldsymbol{m})(\boldsymbol{f}\cdot\bar{\boldsymbol{m}})-(\boldsymbol{k}\cdot\bar{\boldsymbol{m}})(\boldsymbol{f}\cdot\boldsymbol{m})=0. \tag{16}$$

Since

$$\begin{aligned}
k^i\nabla_i\{[(\boldsymbol{k}\cdot\boldsymbol{l})(\boldsymbol{f}\cdot\boldsymbol{n})-(\boldsymbol{k}\cdot\boldsymbol{m})(\boldsymbol{f}\cdot\bar{\boldsymbol{m}})]\Psi_2^{-1/3}\}
&= \Psi_2^{-1/3}\{k^i\nabla_i[(\boldsymbol{k}\cdot\boldsymbol{l})(\boldsymbol{f}\cdot\boldsymbol{n})-(\boldsymbol{k}\cdot\boldsymbol{m})(\boldsymbol{f}\cdot\bar{\boldsymbol{m}})]\\
&\quad+[(\boldsymbol{k}\cdot\boldsymbol{l})(\boldsymbol{f}\cdot\boldsymbol{n})-(\boldsymbol{k}\cdot\boldsymbol{m})(\boldsymbol{f}\cdot\bar{\boldsymbol{m}})]k^i\nabla_i\lg\Psi_2^{-1/3}\},
\end{aligned} \tag{17}$$

the theorem requires us to show that

$$\begin{aligned}
k^i\nabla_i[(\boldsymbol{k}\cdot\boldsymbol{l})(\boldsymbol{f}\cdot\boldsymbol{n})-(\boldsymbol{k}\cdot\boldsymbol{m})(\boldsymbol{f}\cdot\bar{\boldsymbol{m}})]
&= -[(\boldsymbol{k}\cdot\boldsymbol{l})(\boldsymbol{f}\cdot\boldsymbol{n})-(\boldsymbol{k}\cdot\boldsymbol{m})(\boldsymbol{f}\cdot\bar{\boldsymbol{m}})]k^i\nabla_i\lg\Psi_2^{-1/3}.
\end{aligned} \tag{18}$$

Expanding the left-hand side of equation (18) and making use of equations (14), we obtain

$$\begin{aligned}
k^i\nabla_i[(\boldsymbol{k}\cdot\boldsymbol{l})(\boldsymbol{f}\cdot\boldsymbol{n})-(\boldsymbol{k}\cdot\boldsymbol{m})(\boldsymbol{f}\cdot\bar{\boldsymbol{m}})]
&= (\boldsymbol{f}\cdot\boldsymbol{n})k^ik^jl_{j;i}+(\boldsymbol{k}\cdot\boldsymbol{l})k^if^jn_{j;i}\\
&\quad-(\boldsymbol{f}\cdot\bar{\boldsymbol{m}})k^ik^jm_{j;i}-(\boldsymbol{k}\cdot\boldsymbol{m})k^if^j\bar{m}_{j;i}.
\end{aligned} \tag{19}$$

Now substituting for $l_{j;i}$, etc., from equations (6)–(8) and simplifying, we find that we are left with

$$\begin{aligned}
k^i\nabla_i[(\boldsymbol{k}\cdot\boldsymbol{l})(\boldsymbol{f}\cdot\boldsymbol{n})-(\boldsymbol{k}\cdot\boldsymbol{m})(\boldsymbol{f}\cdot\bar{\boldsymbol{m}})]
&= \rho[+(\boldsymbol{f}\cdot\boldsymbol{n})(\boldsymbol{k}\cdot\bar{\boldsymbol{m}})(\boldsymbol{k}\cdot\boldsymbol{m})-(\boldsymbol{f}\cdot\bar{\boldsymbol{m}})(\boldsymbol{k}\cdot\boldsymbol{n})(\boldsymbol{k}\cdot\boldsymbol{m})]\\
&\quad+\tau[-(\boldsymbol{f}\cdot\boldsymbol{n})(\boldsymbol{k}\cdot\bar{\boldsymbol{m}})(\boldsymbol{k}\cdot\boldsymbol{l})+(\boldsymbol{f}\cdot\bar{\boldsymbol{m}})(\boldsymbol{k}\cdot\boldsymbol{n})(\boldsymbol{k}\cdot\boldsymbol{l})]\\
&\quad+\mu[-(\boldsymbol{k}\cdot\boldsymbol{l})(\boldsymbol{k}\cdot\bar{\boldsymbol{m}})(\boldsymbol{f}\cdot\boldsymbol{m})+(\boldsymbol{k}\cdot\boldsymbol{m})(\boldsymbol{k}\cdot\bar{\boldsymbol{m}})(\boldsymbol{f}\cdot\boldsymbol{l})]\\
&\quad+\pi[+(\boldsymbol{k}\cdot\boldsymbol{l})(\boldsymbol{k}\cdot\boldsymbol{n})(\boldsymbol{f}\cdot\boldsymbol{m})-(\boldsymbol{k}\cdot\boldsymbol{m})(\boldsymbol{k}\cdot\boldsymbol{n})(\boldsymbol{f}\cdot\boldsymbol{l})].
\end{aligned} \tag{20}$$

Considering next the right-hand side of equation (18) and making use of equations (4) and (11), we find

$$\begin{aligned}
-[(\boldsymbol{k}\cdot\boldsymbol{l})(\boldsymbol{f}\cdot\boldsymbol{n})-(\boldsymbol{k}\cdot\boldsymbol{m})(\boldsymbol{f}\cdot\bar{\boldsymbol{m}})]k^i\nabla_i\lg\Psi_2^{-1/3}
&= \tfrac{1}{3}[(\boldsymbol{k}\cdot\boldsymbol{l})(\boldsymbol{f}\cdot\boldsymbol{n})-(\boldsymbol{k}\cdot\boldsymbol{m})(\boldsymbol{f}\cdot\bar{\boldsymbol{m}})]\\
&\quad\times[(\boldsymbol{k}\cdot\boldsymbol{n})D+(\boldsymbol{k}\cdot\boldsymbol{l})\triangle-(\boldsymbol{k}\cdot\bar{\boldsymbol{m}})\delta-(\boldsymbol{k}\cdot\boldsymbol{m})\delta^*]\lg\Psi_2\\
&= [(\boldsymbol{k}\cdot\boldsymbol{l})(\boldsymbol{f}\cdot\boldsymbol{n})-(\boldsymbol{k}\cdot\boldsymbol{m})(\boldsymbol{f}\cdot\bar{\boldsymbol{m}})]\\
&\quad\times[+\rho(\boldsymbol{k}\cdot\boldsymbol{n})-\mu(\boldsymbol{k}\cdot\boldsymbol{l})-\tau(\boldsymbol{k}\cdot\bar{\boldsymbol{m}})+\pi(\boldsymbol{k}\cdot\boldsymbol{m})].
\end{aligned} \tag{21}$$

The equality (18), which the theorem asserts, requires that the terms on the right-hand side of equation (20) agree with the term in the last two lines of equation (21). We verify that this is, indeed, the case. Thus, considering the terms which occur with the spin coefficient μ as a factor on the right-hand side of equation (20), we have, by virtue of equations (15) and (16),

$$\begin{aligned}-(\boldsymbol{k}\cdot\boldsymbol{l})(\boldsymbol{k}\cdot\bar{\boldsymbol{m}})(\boldsymbol{f}\cdot\boldsymbol{m})+(\boldsymbol{k}\cdot\boldsymbol{m})(\boldsymbol{k}\cdot\bar{\boldsymbol{m}})(\boldsymbol{f}\cdot\boldsymbol{l})\\ =-(\boldsymbol{k}\cdot\boldsymbol{l})[(\boldsymbol{k}\cdot\boldsymbol{l})(\boldsymbol{f}\cdot\boldsymbol{n})+(\boldsymbol{k}\cdot\boldsymbol{n})(\boldsymbol{f}\cdot\boldsymbol{l})-(\boldsymbol{k}\cdot\boldsymbol{m})(\boldsymbol{f}\cdot\bar{\boldsymbol{m}})]\\ +(\boldsymbol{k}\cdot\boldsymbol{m})(\boldsymbol{k}\cdot\bar{\boldsymbol{m}})(\boldsymbol{f}\cdot\boldsymbol{l})\\ =-(\boldsymbol{k}\cdot\boldsymbol{l})[(\boldsymbol{k}\cdot\boldsymbol{l})(\boldsymbol{f}\cdot\boldsymbol{n})-(\boldsymbol{k}\cdot\boldsymbol{m})(\boldsymbol{f}\cdot\bar{\boldsymbol{m}})];\end{aligned}\tag{22}$$

and this expression agrees with the coefficient of μ in equation (21). The equality of the factors of the other spin coefficients in equations (20) and (21) follows in similar fashion by virtue of the same equations (15) and (16). The conservation of K_s, along a null geodesic, affinely parametrized, is thus established.

THEOREM 2. *If* $\boldsymbol{k}$ *is an affinely parametrized geodesic in a type-D space-time, then*

$$\begin{aligned}K&=2|\Psi_2|^{-2/3}(\boldsymbol{k}\cdot\boldsymbol{l})(\boldsymbol{k}\cdot\boldsymbol{n})-Q|\boldsymbol{k}|^2\\ &=2|\Psi_2|^{-2/3}(\boldsymbol{k}\cdot\boldsymbol{m})(\boldsymbol{k}\cdot\bar{\boldsymbol{m}})-(Q-|\Psi_2|^{-2/3})|\boldsymbol{k}|^2\end{aligned}\tag{23}$$

is conserved along $\boldsymbol{k}$ *if, and only if, a scalar* Q *exists which satisfies the equations*

$$DQ=D|\Psi_2|^{-2/3},\quad \triangle Q=\triangle|\Psi_2|^{-2/3},\quad \textit{and}\quad \delta Q=\delta^*Q=0.\tag{24}$$

Proof. First we observe that the equivalence of the two alternative forms of K in the enunciation of the theorem follows from equation (10).

The conservation of K for geodesic motion along $\boldsymbol{k}$ requires that

$$2k^i\nabla_i[(\boldsymbol{k}\cdot\boldsymbol{l})(\boldsymbol{k}\cdot\boldsymbol{n})|\Psi_2|^{-2/3}]=|\boldsymbol{k}|^2k^i\nabla_iQ\tag{25}$$

since $|\boldsymbol{k}|^2$ is conserved. By equation (11), the right-hand side of equation (25) can be rewritten in the form

$$\begin{aligned}|\boldsymbol{k}|^2k^i\nabla_iQ=|\boldsymbol{k}|^2[(\boldsymbol{k}\cdot\boldsymbol{n})DQ+(\boldsymbol{k}\cdot\boldsymbol{l})\triangle Q\\ -(\boldsymbol{k}\cdot\bar{\boldsymbol{m}})\delta Q-(\boldsymbol{k}\cdot\boldsymbol{m})\delta^*Q].\end{aligned}\tag{26}$$

Considering the left-hand side of equation (25), we have

$$\begin{aligned}&2k^i\nabla_i[(\boldsymbol{k}\cdot\boldsymbol{l})(\boldsymbol{k}\cdot\boldsymbol{n})|\Psi_2|^{-2/3}]\\ &\quad=2|\Psi_2|^{-2/3}[k^i\nabla_i(\boldsymbol{k}\cdot\boldsymbol{l})(\boldsymbol{k}\cdot\boldsymbol{n})+(\boldsymbol{k}\cdot\boldsymbol{l})(\boldsymbol{k}\cdot\boldsymbol{n})k^i\nabla_i\lg|\Psi_2|^{-2/3}].\end{aligned}\tag{27}$$

On the other hand, by virtue of equations (4) and the geodesic character of $\boldsymbol{k}$,

$$k^i\nabla_i \lg|\Psi_2|^{-2/3}$$
$$= -\tfrac{1}{3}[(\boldsymbol{k}\cdot\boldsymbol{n})D+(\boldsymbol{k}\cdot\boldsymbol{l})\triangle-(\boldsymbol{k}\cdot\bar{\boldsymbol{m}})\delta-(\boldsymbol{k}\cdot\boldsymbol{m})\delta^*](\lg\Psi_2+\lg\Psi_2^*)$$
$$= -[(\boldsymbol{k}\cdot\boldsymbol{n})(\rho+\rho^*)-(\boldsymbol{k}\cdot\boldsymbol{l})(\mu+\mu^*)-(\boldsymbol{k}\cdot\bar{\boldsymbol{m}})(\tau-\pi^*)-(\boldsymbol{k}\cdot\boldsymbol{m})(\tau^*-\pi)], \tag{28}$$

and

$$k^i\nabla_i(\boldsymbol{k}\cdot\boldsymbol{l})(\boldsymbol{k}\cdot\boldsymbol{n}) = (\boldsymbol{k}\cdot\boldsymbol{n})k^ik^jl_{j;i}+(\boldsymbol{k}\cdot\boldsymbol{l})k^ik^jn_{j;i}. \tag{29}$$

Now inserting equations (28) and (29) in equation (27) and substituting for $l_{j;i}$ and $n_{j;i}$ from equations (6) and (7), we find that considerable simplification results and we are left with

$$2k^i\nabla_i[(\boldsymbol{k}\cdot\boldsymbol{l})(\boldsymbol{k}\cdot\boldsymbol{n})|\Psi_2|^{-2/3}]$$
$$= 2|\Psi_2|^{-2/3}\{[(\boldsymbol{k}\cdot\boldsymbol{n})(\rho+\rho^*)-(\boldsymbol{k}\cdot\boldsymbol{l})(\mu+\mu^*)]$$
$$\times[(\boldsymbol{k}\cdot\boldsymbol{m})(\boldsymbol{k}\cdot\bar{\boldsymbol{m}})-(\boldsymbol{k}\cdot\boldsymbol{l})(\boldsymbol{k}\cdot\boldsymbol{n})]\}$$
$$= |\Psi_2|^{-2/3}|\boldsymbol{k}|^2[(\boldsymbol{k}\cdot\boldsymbol{l})(\mu+\mu^*)-(\boldsymbol{k}\cdot\boldsymbol{n})(\rho+\rho^*)]$$
$$= |\boldsymbol{k}|^2[(\boldsymbol{k}\cdot\boldsymbol{n})D|\Psi_2|^{-2/3}+(\boldsymbol{k}\cdot\boldsymbol{l})\triangle|\Psi_2|^{-2/3}], \tag{30}$$

where, in the last step of the reductions, we have made use, once again, of equations (4). A comparison of the final result of the foregoing reduction with the right-hand side of equation (26) leads to the requirements (24) stated in the theorem.

COROLLARY 1. *A null geodesic,* **k**, *in any type-D space-time allows the integral of motion*

$$K_0 = 2|\Psi_2|^{-2/3}(\boldsymbol{k}\cdot\boldsymbol{m})(\boldsymbol{k}\cdot\bar{\boldsymbol{m}}) = 2|\Psi_2|^{-2/3}(\boldsymbol{k}\cdot\boldsymbol{l})(\boldsymbol{k}\cdot\boldsymbol{n}). \tag{31}$$

This follows trivially from the theorem stated, since for a null geodesic $|\boldsymbol{k}|^2 = 0$ and none of conditions (24) are required for its validity.

COROLLARY 2. *The constant K_0 of Corollary 1 is, apart from a multiplicative factor, the square of the absolute value of the complex constant K_s of Theorem 1.*

Proof. By equation (12),

$$K_sK_s^* = 4|\Psi_2|^{-2/3}[(\boldsymbol{k}\cdot\boldsymbol{l})(\boldsymbol{f}\cdot\boldsymbol{n})-(\boldsymbol{k}\cdot\boldsymbol{m})(\boldsymbol{f}\cdot\bar{\boldsymbol{m}})]$$
$$\times[(\boldsymbol{k}\cdot\boldsymbol{l})(\boldsymbol{f}\cdot\boldsymbol{n})-(\boldsymbol{k}\cdot\bar{\boldsymbol{m}})(\boldsymbol{f}\cdot\boldsymbol{m})]; \tag{32}$$

or, alternatively, by virtue of equation (16),

$$|K_s|^2 = -4|\Psi_2|^{-2/3}[(\boldsymbol{k}\cdot\boldsymbol{l})(\boldsymbol{f}\cdot\boldsymbol{n})-(\boldsymbol{k}\cdot\boldsymbol{m})(\boldsymbol{f}\cdot\bar{\boldsymbol{m}})]$$
$$\times[(\boldsymbol{k}\cdot\boldsymbol{n})(\boldsymbol{f}\cdot\boldsymbol{l})-(\boldsymbol{k}\cdot\boldsymbol{m})(\boldsymbol{f}\cdot\bar{\boldsymbol{m}})]. \tag{33}$$

Expanding and simplifying by repeated use of the identities (15) and (16), we obtain

$$\begin{aligned}|K_s|^2 &= -4|\Psi_2|^{-2/3}\{(\boldsymbol{k}\cdot\boldsymbol{l})(\boldsymbol{k}\cdot\boldsymbol{n})(\boldsymbol{f}\cdot\boldsymbol{n})(\boldsymbol{f}\cdot\boldsymbol{l})+(\boldsymbol{k}\cdot\boldsymbol{m})^2(\boldsymbol{f}\cdot\bar{\boldsymbol{m}})^2 \\ &\qquad -(\boldsymbol{k}\cdot\boldsymbol{m})(\boldsymbol{f}\cdot\bar{\boldsymbol{m}})[(\boldsymbol{k}\cdot\boldsymbol{n})(\boldsymbol{f}\cdot\boldsymbol{l})+(\boldsymbol{k}\cdot\boldsymbol{l})(\boldsymbol{f}\cdot\boldsymbol{n})]\} \\ &= -4|\Psi_2|^{-2/3}\{(\boldsymbol{k}\cdot\boldsymbol{m})(\boldsymbol{k}\cdot\bar{\boldsymbol{m}})(\boldsymbol{f}\cdot\boldsymbol{n})(\boldsymbol{f}\cdot\boldsymbol{l})+(\boldsymbol{k}\cdot\boldsymbol{m})^2(\boldsymbol{f}\cdot\bar{\boldsymbol{m}})^2 \\ &\qquad -(\boldsymbol{k}\cdot\boldsymbol{m})(\boldsymbol{f}\cdot\bar{\boldsymbol{m}})[(\boldsymbol{k}\cdot\boldsymbol{m})(\boldsymbol{f}\cdot\bar{\boldsymbol{m}})+(\boldsymbol{k}\cdot\bar{\boldsymbol{m}})(\boldsymbol{f}\cdot\boldsymbol{m})]\} \\ &= -4|\Psi_2|^{-2/3}(\boldsymbol{k}\cdot\boldsymbol{m})(\boldsymbol{k}\cdot\bar{\boldsymbol{m}})[(\boldsymbol{f}\cdot\boldsymbol{n})(\boldsymbol{f}\cdot\boldsymbol{l})-(\boldsymbol{f}\cdot\boldsymbol{m})(\boldsymbol{f}\cdot\bar{\boldsymbol{m}})] \\ &= -4|\Psi_2|^{-2/3}(\boldsymbol{k}\cdot\boldsymbol{m})(\boldsymbol{k}\cdot\bar{\boldsymbol{m}})|\boldsymbol{f}|^2. \end{aligned} \tag{34}$$

Since $\boldsymbol{f}$ is parallelly propagated, $|\boldsymbol{f}|^2$ remains constant along the geodesic $\boldsymbol{k}$ and the corollary stated follows.

COROLLARY 3. *The Kerr metric allows a conserved quantity of the form enunciated in Theorem 2.*

In the coordinate system adopted in Chapter 6 (cf. Ch. 6, equation (180))

$$|\Psi_2|^{-2/3} = M^{-2/3}\rho^2 = M^{-2/3}(r^2+a^2\cos^2\theta). \tag{35}$$

Accordingly, the requirements of the theorem are met by the choice

$$Q = M^{-2/3}r^2. \tag{36}$$

Suppressing the factor $M^{-2/3}$, we conclude that *in Kerr geometry*

$$\left.\begin{aligned} K &= 2\rho^2(\boldsymbol{k}\cdot\boldsymbol{l})(\boldsymbol{k}\cdot\boldsymbol{n})-r^2|\boldsymbol{k}|^2 \\ \textit{and}\qquad & \\ K &= 2\rho^2(\boldsymbol{k}\cdot\boldsymbol{m})(\boldsymbol{k}\cdot\bar{\boldsymbol{m}})+a^2|\boldsymbol{k}|^2\cos^2\theta \end{aligned}\right\} \tag{37}$$

are integrals of geodesic motion.

The two integrals (37) may be considered as *independent* since their equivalence derives from the constancy of $|\boldsymbol{k}|^2$ which in common parlance is equated with the conservation of the rest mass.

THEOREM 3. *The necessary and sufficient conditions for an integral of geodesic motion, of the form considered in Theorem 2 to exist, for a type-D space-time, are that the spin coefficients ρ, τ, μ, and π are related in the manner*

$$\frac{\rho}{\rho^*} = \frac{\mu}{\mu^*} = \frac{\tau}{\pi^*} = \frac{\pi}{\tau^*}. \tag{38}$$

Proof. The conditions obtained in Theorem 2 for the existence of an integral of the form (23), for geodesic motions in a type-D space-time, are that a scalar Q exists with the properties

$$DQ = Df, \quad \triangle Q = \triangle f \quad \text{and} \quad \delta Q = \delta^* Q = 0, \tag{39}$$

where, for convenience, we have written

$$f = |\Psi_2|^{-2/3}. \tag{40}$$

In terms of f, the Bianchi identities (4) governing Ψ_2 become

$$\left.\begin{aligned} Df &= -(\rho+\rho^*)f; & \delta f &= (\pi^*-\tau)f, \\ \triangle f &= +(\mu+\mu^* f; & \delta^* f &= (\pi-\tau^*)f. \end{aligned}\right\} \tag{41}$$

The proof of the theorem consists in applying the available commutation relations (Ch. 1, §8(*c*)) to Q and tracing the consequences of the requirements (39) consistently with equations (41). For a type-D space-time, compatible with the relations (1) and (3), the commutation relations are

$$\left.\begin{aligned} \delta^*\delta - \delta\delta^* &= (\mu^*-\mu)D + (\rho^*-\rho)\triangle + (\alpha-\beta^*)\delta + (\beta-\alpha^*)\delta^*, \\ \triangle D - D\triangle &= (\gamma+\gamma^*)D - (\tau^*+\pi)\delta - (\tau+\pi^*)\delta^*, \\ \delta D - D\delta &= (\alpha^*+\beta-\pi^*)D - \rho^*\delta, \\ \delta\triangle - \triangle\delta &= (\tau-\alpha^*-\beta)\triangle + (\mu-\gamma-\gamma^*)\delta. \end{aligned}\right\} \tag{42}$$

Applying these relations to Q and making use of the requirements (39), we obtain

$$(\mu^*-\mu)Df + (\rho^*-\rho)\triangle f = 0, \tag{43}$$

$$(\triangle D - D\triangle)f = (\gamma+\gamma^*)Df, \tag{44}$$

$$\delta Df = (\alpha^*+\beta-\pi^*)Df, \tag{45}$$

and

$$\delta\triangle f = (\tau-\alpha^*-\beta)\triangle f. \tag{46}$$

We now make use of the relations (41). Equation (43) directly yields the relation

$$(\mu^*-\mu)(\rho+\rho^*) = (\rho^*-\rho)(\mu+\mu^*), \tag{47}$$

which on simplification gives

$$\rho\mu^* = \rho^*\mu \qquad \text{or} \qquad \rho/\rho^* = \mu/\mu^*. \tag{48}$$

Next, evaluating the left-hand side of equation (44), in accordance with equation (42), we obtain

$$(\tau^*+\pi)\delta f + (\tau+\pi^*)\delta^* f = 0, \tag{49}$$

or, by virtue of the relations (41),

$$(\tau^*+\pi)(\tau-\pi^*) + (\tau+\pi^*)(\tau^*-\pi) = 0. \tag{50}$$

On simplification, equation (50) gives

$$\tau\tau^* - \pi\pi^* = 0, \qquad \text{or} \qquad \tau/\pi^* = \pi/\tau^*. \tag{51}$$

Tracing the consequences of equations (45) and (46) is somewhat more involved. First, with the aid of equations (41), equations (45) and (46) can be reduced to give

$$\delta \lg(\rho+\rho^*) = (\alpha^*+\beta-2\pi^*+\tau) = \delta \lg \rho + \delta \lg(1+\rho^*/\rho), \qquad (52)$$

$$\delta \lg(\mu+\mu^*) = (2\tau-\pi^*-\alpha^*-\beta) = \delta \lg \mu + \delta \lg(1+\mu^*/\mu). \qquad (53)$$

Since $\rho/\rho^* = \mu/\mu^*$ by equation (48), we obtain by subtracting equation (53) from equation (52)

$$\delta \lg \rho - \delta \lg \mu = 2\alpha^* + 2\beta - \pi^* - \tau, \qquad (54)$$

or

$$\delta\rho - \rho\delta \lg \mu = (2\alpha^* + 2\beta - \pi^* - \tau)\rho. \qquad (55)$$

On the other hand, by the Ricci identity, Chapter 1, equation (310(k)),

$$\delta\rho = \rho(\alpha^*+\beta+\tau) - \tau\rho^*. \qquad (56)$$

Eliminating $\delta\rho$ from equation (55) with the aid of this last relation, we obtain

$$-\delta\mu = (\alpha^*+\beta-\pi^*-2\tau)\mu + \tau\mu^*, \qquad (57)$$

where we have, once again, made use of equation (48). But by Ricci identity, Chapter 1, equation (310(m)),

$$-\delta\mu^* = (\alpha^*+\beta+\pi^*)\mu^* - \pi^*\mu. \qquad (58)$$

By adding equations (57) and (58), we obtain

$$-\delta(\mu+\mu^*) = (\alpha^*+\beta-2\pi^*-2\tau)\mu + (\alpha^*+\beta+\tau+\pi^*)\mu^*. \qquad (59)$$

But we also have the relation (cf. equation (53))

$$-\delta(\mu+\mu^*) = -(2\tau-\pi^*-\alpha^*-\beta)(\mu+\mu^*). \qquad (60)$$

From a comparison of equations (59) and (60), we find that

$$\tau\mu^* - \pi^*\mu = 0, \qquad \text{or} \qquad \tau/\pi^* = \mu/\mu^*. \qquad (61)$$

Combining equations (48), (51), and (61), we obtain the relations (35) which were to be established.

It appears that the relations (35) are, in fact, satisfied by most of the type-D metrics.

61. The geodesics in the equatorial plane

It is clear that the geodesics in the equatorial plane can be delineated in very much the same way as we did in the Schwarzschild and in the Reissner–Nordström space-times: the energy and the angular-momentum integrals will suffice to reduce the problem to one of quadratures. But two essential differences must be kept in mind. First, a distinction should be made between *direct* and *retrograde* orbits whose rotations about the axis of symmetry are in the same sense or in the opposite sense to that of the black hole. And, second, the coordinate φ, like the coordinate t, is not a 'good' co-ordinate for describing what 'really' happens with respect to a co-moving observer: a trajectory approaching the horizon (at r_+ or r_-) will spiral round the black hole an infinite number of times even as it will take an infinite co-ordinate time t to cross the horizon; and neither will be the experience of the co-moving observer.

The Lagrangian appropriate to motions in the equatorial plane (for which $\dot{\theta} = 0$ and $\theta =$ a constant $= \pi/2$) is

$$2\mathscr{L} = \left(1 - \frac{2M}{r}\right)\dot{t}^2 + \frac{4aM}{r}\dot{t}\dot{\varphi} - \frac{r^2}{\Delta}\dot{r}^2 - \left[(r^2 + a^2) + \frac{2a^2M}{r}\right]\dot{\varphi}^2; \tag{62}$$

and we deduce from it that the generalized momenta are given by

$$p_t = \left(1 - \frac{2M}{r}\right)\dot{t} + \frac{2aM}{r}\dot{\varphi} = E = \text{constant} \tag{63}$$

$$-p_\varphi = -\frac{2aM}{r}\dot{t} + \left[(r^2 + a^2) + \frac{2a^2M}{r}\right]\dot{\varphi} = L = \text{constant} \tag{64}$$

and

$$-p_r = \frac{r^2}{\Delta}\dot{r}, \tag{65}$$

where we have used superior dots to denote differentiation with respect to an affine parameter τ. (The constancy of p_t and p_φ follows from the independence of the Lagrangian on t and φ which, in turn, is a manifestation of the stationary and the axisymmetric character of the Kerr geometry.)

The Hamiltonian is given by

$$\begin{aligned}\mathscr{H} &= p_t\dot{t} + p_\varphi\dot{\varphi} + p_r\dot{r} - \mathscr{L}\\ &= \frac{1}{2}\left(1 - \frac{2M}{r}\right)\dot{t}^2 + \frac{2aM}{r}\dot{t}\dot{\varphi} - \frac{r^2}{2\Delta}\dot{r}^2 - \frac{1}{2}\left(r^2 + a^2 + \frac{2a^2M}{r}\right)\dot{\varphi}^2;\end{aligned} \tag{66}$$

and from the independence of the Hamiltonian on t, we deduce that

$$\begin{aligned}2\mathscr{H} &= \left[\left(1 - \frac{2M}{r}\right)\dot{t} + \frac{2aM}{r}\dot{\varphi}\right]\dot{t} - \left[\left(r^2 + a^2 + \frac{2a^2M}{r}\right)\dot{\varphi} - \frac{2aM}{r}\dot{t}\right]\dot{\varphi} - \frac{r^2}{\Delta}\dot{r}^2\\ &= E\dot{t} - L\dot{\varphi} - \frac{r^2}{\Delta}\dot{r}^2 = \delta_1 = \text{constant}\end{aligned} \tag{67}$$

We may, without loss of generality, set

$$\left.\begin{aligned}\delta_1 &= 1 \quad \text{for time-like geodesics}\\ &= 0 \quad \text{for null geodesics.}\end{aligned}\right\} \tag{68}$$

(Setting $\delta_1 = 1$ for time-like geodesics requires E to be interpreted as the specific energy or the energy per unit mass.)

Solving equations (63) and (64) for $\dot{\varphi}$ and $\dot{t}$, we obtain

$$\dot{\varphi} = \frac{1}{\Delta}\left[\left(1 - \frac{2M}{r}\right)L + \frac{2aM}{r}E\right], \tag{69}$$

$$\dot{t} = \frac{1}{\Delta}\left[\left(r^2 + a^2 + \frac{2a^2M}{r}\right)E - \frac{2aM}{r}L\right]; \tag{70}$$

and inserting these solutions in the second line of equation (67), we obtain the radial equation

$$r^2\dot{r}^2 = r^2E^2 + \frac{2M}{r}(aE - L)^2 + (a^2E^2 - L^2) - \delta_1\Delta. \tag{71}$$

(a) The null geodesics

As we have noted, $\delta_1 = 0$ for null geodesics and the radial equation (71) becomes

$$\dot{r}^2 = E^2 + \frac{2M}{r^3}(L - aE)^2 - \frac{1}{r^2}(L^2 - a^2E^2). \tag{72}$$

In our further considerations, it will be more convenient to distinguish the geodesics by the impact parameter

$$D = L/E \tag{73}$$

rather than by L.

First, we observe that geodesics with the impact parameter

$$D = a, \quad \text{when} \quad L = aE, \tag{74}$$

play, in the present context, the same role as the radial geodesics in the Schwarzschild and in the Reissner–Nordström geometry. Thus, in this case, equations (69), (70), and (72) reduce to

$$\dot{r} = \pm E, \; \dot{t} = (r^2 + a^2)E/\Delta, \quad \text{and} \quad \dot{\varphi} = aE/\Delta. \tag{75}$$

The radial coordinate is described uniformly with respect to the affine parameter while the equations governing t and φ are

$$\frac{dt}{dr} = \pm\frac{r^2 + a^2}{\Delta} \quad \text{and} \quad \frac{d\varphi}{dr} = \pm\frac{a}{\Delta}. \tag{76}$$

The solutions of these equations are (cf. Ch. 6, equation (238))

$$\left.\begin{aligned} \pm t &= r + \frac{r_+^2 + a^2}{r_+ - r_-}\lg\left(\frac{r}{r_+} - 1\right) - \frac{r_-^2 + a^2}{r_+ - r_-}\lg\left(\frac{r}{r_-} - 1\right), \\ \pm\varphi &= \frac{a}{r_+ - r_-}\lg\left(\frac{r}{r_+} - 1\right) - \frac{a}{r_+ - r_-}\lg\left(\frac{r}{r_-} - 1\right). \end{aligned}\right\} \tag{77}$$

These solutions exhibit the characteristic behaviours of t and φ of tending to $\pm\infty$ as the horizons at r_+ and r_- are approached—a fact to which we have already drawn attention.

As we shall see later, the null geodesics described by the equations (76) are members of the principal null congruences that are confined to the equatorial plane.

In general it is clear that we must distinguish, as in Schwarzschild's geometry, orbits with impact parameters greater than or less than a certain critical value D_c (which will in turn be different for the direct and for the retrograde orbits). For $D = D_c$, the geodesic equations will allow an unstable circular orbit of radius r_c (say). For $D > D_c$, we shall have orbits of two kinds: those of the *first kind* which, arriving from infinity, have perihelion distances greater than r_c; and those of the *second kind* which, having aphelion distances less than r_c, terminate at the singularity at $r = 0$ (and $\theta = \pi/2$!). For $D = D_c$, the orbits of the two kinds coalesce: they both spiral, indefinitely, about the same unstable circular orbit at $r = r_c$. For $D < D_c$, there are only orbits of one kind: arriving from infinity, they cross both horizons and terminate at the singularity.

The equations determining the radius r_c of the unstable circular 'photon-orbit' are (cf. equation (72))

$$E^2 + \frac{2M}{r_c^3}(L-aE)^2 - \frac{1}{r_c^2}(L^2 - a^2E^2) = 0 \tag{78}$$

and

$$-\frac{6M}{r_c^4}(L-aE)^2 + \frac{2}{r_c^3}(L^2 - a^2E^2) = 0. \tag{79}$$

From equation (79), we conclude that

$$r_c = 3M\frac{L-aE}{L+aE} = 3M\frac{D_c - a}{D_c + a}. \tag{80}$$

Inserting this last relation in equation (78), we find

$$E^2 = \frac{1}{27M^2}\frac{(L+aE)^3}{L-aE} = \frac{E^2}{27M^2}\frac{(D_c+a)^3}{D_c - a}, \tag{81}$$

or

$$(D_c + a)^3 = 27M^2(D_c - a). \tag{82*}$$

Letting

$$y = D_c + a, \tag{83}$$

* It is a readily verifiable consequence of equations (80) and (82) that $D_c^2 \doteq 3r_c^2 + a^2$—a simple generalization of a result which obtains in the Schwarzschild limit, $a = 0$.

we obtain the cubic equation

$$y^3 - 27M^2 y + 54aM^2 = 0. \tag{84}$$

We must now distinguish $a > 0$ and $a < 0$ corresponding to the direct and the retrograde orbits. For $a > 0$,

$$\left.\begin{aligned} y &= -6M\cos(\vartheta + 120°) \quad \text{where} \quad \cos 3\vartheta = a/M, \\ D_c &= y - a \quad \text{and} \quad r_c = 3M(1 - 2a/y); \end{aligned}\right\} \tag{85}$$

and for $-a = |a| > 0$,

$$\left.\begin{aligned} y &= 6M\cos\vartheta \quad \text{where} \quad \cos 3\vartheta = |a|/M, \\ D_c &= y + |a| \quad \text{and} \quad r_c = 3M(1 + 2|a|/y). \end{aligned}\right\} \tag{86}$$

It can be directly verified that the solution for r_c, expressed directly in terms of ϑ, is given by

$$r_c = 4M\cos^2\vartheta = 2M\left\{1 + \cos\left[\frac{2}{3}\cos^{-1}\left(\pm\frac{a}{M}\right)\right]\right\}, \tag{87}$$

where the upper sign applies to retrograde orbits and the lower sign to direct orbits.

From equations (85) and (86), we find

$$\text{for } a = 0: \quad D_c = (3\sqrt{3})M \quad \text{and} \quad r_c = 3M, \tag{88}$$

$$\left.\begin{aligned} &\text{for } a = M: \quad D_c = 2M \quad \text{and} \quad r_c = M \text{ (for direct orbits)}, \\ &\text{for } a = -M: \quad D_c = 7M \quad \text{and} \quad r_c = 4M \text{ (for retrograde orbits)}. \end{aligned}\right\} \tag{89}$$

Turning to the equations governing the orbits when the impact parameter has the critical value D_c, and the expression on the right-hand side of equation (72) allows a double root, we find that the equation can be reduced to the form

$$\dot{u}^2 = ME^2(D_c - a)^2 u^4 (u - u_c)^2 (2u + u_c), \tag{90}$$

where

$$u = \frac{1}{r} \quad \text{and} \quad u_c = \frac{1}{r_c} = \frac{D_c + a}{3M(D_c - a)}. \tag{91}$$

Equation (90) can be integrated directly to give

$$\begin{aligned}[E(D_c - a)\sqrt{M}]\tau &= \pm\int\frac{du}{u^2(u - u_c)(2u + u_c)^{1/2}} \\ &= \pm\frac{1}{u_c^2}\left\{\frac{1}{u}(2u + u_c)^{1/2} + \frac{1}{\sqrt{(3u_c)}}\lg\left|\frac{\sqrt{(2u + u_c)} - \sqrt{(3u_c)}}{\sqrt{(2u + u_c)} + \sqrt{(3u_c)}}\right|.\right\}\end{aligned} \tag{92}$$

But if we wish to exhibit the orbit in the equatorial plane, we may combine it with the equation

$$\dot{\varphi} = \frac{Eu^2}{3(a^2u^2 - 2Mu + 1)u_c}[3D_cu_c - 2(D_c + a)u] \tag{93}$$

(which follows directly from equation (69)) to obtain

$$\frac{\mathrm{d}u}{\mathrm{d}\varphi} = \frac{D_c + a}{\sqrt{M}}(a^2u^2 - 2Mu + 1)\frac{(u - u_c)(2u + u_c)^{1/2}}{3D_cu_c - 2(D_c + a)u}, \tag{94}$$

or

$$\varphi = \pm\frac{\sqrt{M}}{a^2(D_c + a)}\int\frac{3D_cu_c - 2(D_c + a)u}{(u - u_+)(u - u_-)(u - u_c)(2u + u_c)^{1/2}}\mathrm{d}u, \tag{95}$$

where $u_\pm = 1/r_\pm$. The integral on the right-hand side of equation (95) is elementary and can be evaluated explicitly. The explicit expression, involving, as it does, partial fractions, is not simple and we shall not write it down.

In Fig. 28, orbits derived from the solutions (92) and (95) are illustrated. They exhibit the features we have already described. The nature of the orbits, in general, can be readily visualized from the orbits with the critical impact parameters illustrated.

(b) The time-like geodesics

For time-like geodesics, equations (69) and (70) for $\dot{\varphi}$ and $\dot{t}$ remain unchanged; but equation (72) is replaced by

$$r^2\dot{r}^2 = -\Delta + r^2E^2 + \frac{2M}{r}(L - aE)^2 - (L^2 - a^2E^2), \tag{96}$$

where E is now to be interpreted as the energy per unit mass of the particle describing the trajectory.

(i) *The special case, $L = aE$.*

Time-like geodesics with $L = aE$, like the null geodesics with $D = a$, are of interest in that their behaviour as they cross the horizons is characteristic of the orbits in general.

When $L = aE$, equation (96) becomes

$$r^2\dot{r}^2 = (E^2 - 1)r^2 + 2Mr - a^2, \tag{97}$$

while the equations for $\dot{\varphi}$ and $\dot{t}$ are the same as for the null geodesics (cf. equations (75)):

$$\dot{\varphi} = aE/\Delta \qquad \text{and} \qquad \dot{t} = E(r^2 + a^2)/\Delta. \tag{98}$$

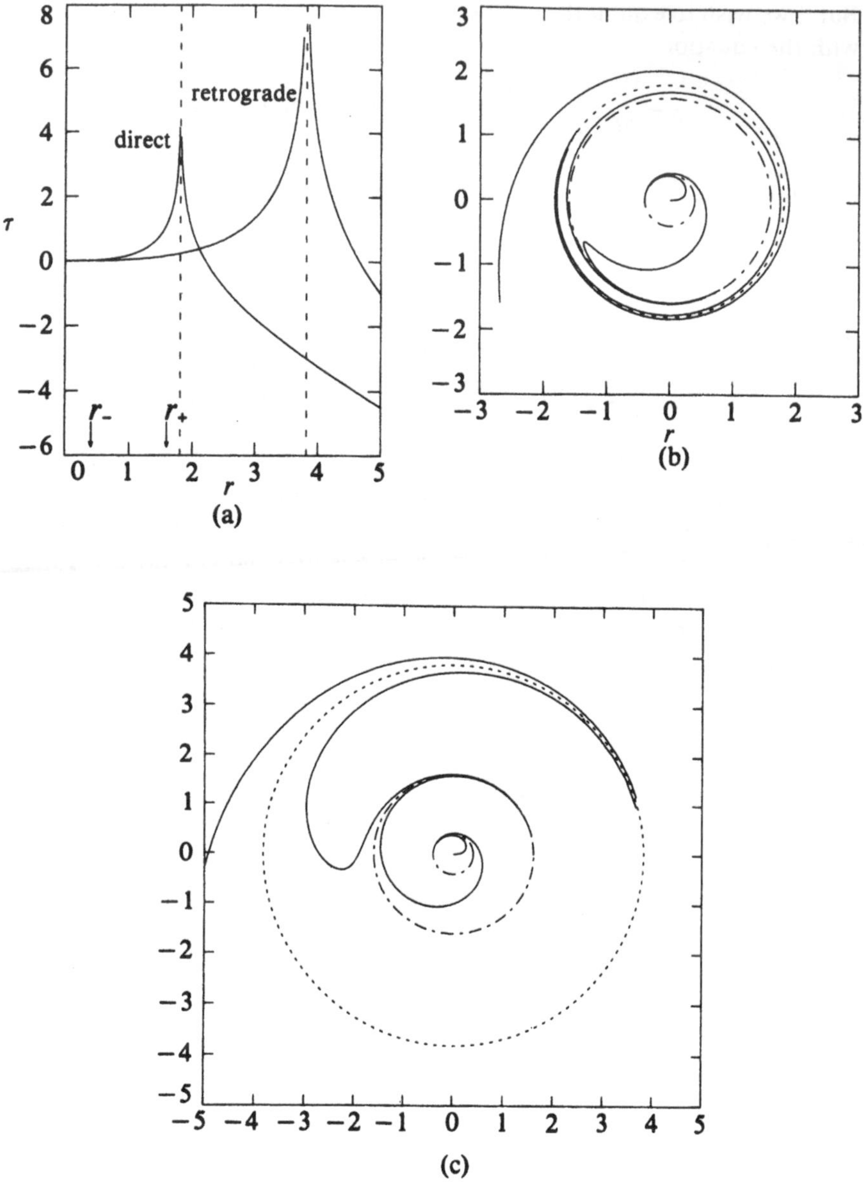

FIG. 28. The critical null-geodesics (direct and retrograde) in the equatorial plane of a Kerr black-hole with a parameter $a = 0.8$. The variation of the radial coordinate r with respect to an affine parameter τ is exhibited in Fig. (*a*) while the corresponding variations with φ of the same orbits are illustrated in Figs. (*b*) and (*c*). The radii of the unstable circular orbits and the impact parameters to which these geodesics correspond are $r_c = 1.811\,M$ and $D_c = 3.237\,M$ for the direct orbit and $r_c = 3.819\,M$ and $D_c = 6.662\,M$ for the retrograde orbit. The unit of length along the coordinate axes is M. The two horizons are shown by the dashed circles in Figs. (*b*) and (*c*).

Equation (97) on integration gives

$$\tau = \frac{1}{E^2-1}\left[\!\!\left[[(E^2-1)r^2+2Mr-a^2]^{1/2} - \frac{M}{\sqrt{(E^2-1)}}\lg\left\{[(E^2-1)r^2+2Mr-a^2]^{1/2}+r\sqrt{(E^2-1)} + \frac{M}{\sqrt{(E^2-1)}}\right\}\right]\!\!\right] \quad (\text{for } E^2>1) \tag{99}$$

and

$$\tau = -\frac{1}{1-E^2}\left\{[-(1-E^2)r^2+2Mr-a^2]^{1/2} + \frac{1}{\sqrt{(1-E^2)}}\sin^{-1}\frac{M-(1-E^2)r}{[M^2-a^2(1-E^2)]^{1/2}}\right\} \quad (\text{for } E^2<1). \tag{100}$$

(The solutions, as written, are valid only for $a^2 < M^2$.)

Alternatively, we may combine equations governing $\dot{r}$ and $\dot{\varphi}$ in the manner

$$\frac{du}{d\varphi} = \pm\frac{a}{E}(u-u_+)(u-u_-)[(E^2-1)+2Mu-a^2u^2]^{1/2}, \tag{101}$$

where $u = 1/r$ and $u_\pm = 1/r_\pm$, and obtain the solution

$$\varphi = \frac{1}{a(u_+-u_-)}\left[\!\!\left[\lg\{2E[E^2\xi_+^2+2(M-a^2u_+)\xi_+-a^2]^{1/2}+2E^2\xi_+ + 2(M-a^2u_+)\} - \lg\{2E[E^2\xi_-^2+2(M-a^2u_-)\xi_- - a^2]^{1/2} + 2E^2\xi_- + 2(M-a^2u_-)\}\right]\!\!\right], \tag{102}$$

where

$$\xi_\pm = (u-u_\pm)^{-1}. \tag{103}$$

An example of an orbit derived with the aid of equation (102) is illustrated in Fig. 29.

(ii) *The circular and associated orbits*

We now turn to a consideration of the radial equation (96) in general. With the reciprocal radius $u(=1/r)$ as the independent variable, the equation takes the form

$$u^{-4}\dot{u}^2 = -(a^2u^2-2Mu+1)+E^2+2M(L-aE)^2u^3-(L^2-a^2E^2)u^2. \tag{104}$$

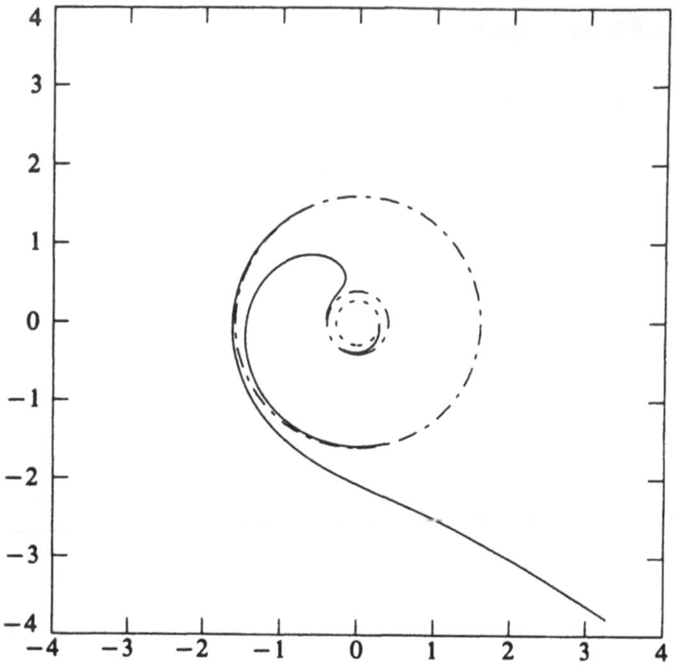

FIG. 29. An example of an unbound time-like geodesic with $L \doteq aE$, described in the equatorial of a Kerr black-hole with $a = 0.8$. For the orbit illustrated, $E = 1.4$. The unit of length along the coordinate axes is M; and the two horizons are shown by dashed circles.

As in the Schwarzschild and the Reissner-Nordström geometries, the circular orbits play an important role in the classification of the orbits. Besides, they are useful in providing simple examples of orbits which exhibit the essential features at the same time; and this is, after all, the reason for studying the geodesics.

We seek then the values of L and E which a circular orbit at some assigned radius, $r = 1/u$, will have. When L and E have these values, the cubic polynomial on the right-hand side of equation (104) will have a double root. The conditions for the occurrence of a double root are

$$-(a^2u^2 - 2Mu + 1) + E^2 + 2Mx^2u^3 - (x^2 + 2aEx)u^2 = 0 \tag{105}$$

and

$$-(a^2u - M) + 3Mx^2u^2 - (x^2 + 2aEx)u = 0, \tag{106}$$

where we have written

$$x = L - aE. \tag{107}$$

Equations (105) and (106) can be combined to give

$$E^2 = (1 - Mu) + Mx^2u^3 \tag{108}$$

and

$$2axEu = x^2(3Mu-1)u-(a^2u-M). \tag{109}$$

By eliminating E between these equations, we obtain the following quadratic equation for x:

$$x^4u^2[(3Mu-1)^2-4a^2Mu^3]$$
$$-2x^2u[(3Mu-1)(a^2u-M)-2a^2u(Mu-1)]+(a^2u-M)^2=0. \tag{110}$$

Quite remarkably, the discriminant "$\frac{1}{4}(b^2-4ac)$" of this equation is

$$4a^2M\Delta_u^2u \qquad \text{where} \qquad \Delta_u = a^2u^2-2Mu+1; \tag{111}$$

and the solution of equation (110) takes a particularly simple form by writing

$$(3Mu-1)^2-4a^2Mu^3 = Q_+Q_-, \tag{112}$$

where

$$Q_\pm = 1-3Mu\pm 2a\sqrt{(Mu^3)}. \tag{113}$$

Thus, we find

$$x^2u^2 = \frac{Q_\pm\Delta_u - Q_+Q_-}{Q_+Q_-} = \frac{1}{Q_\mp}(\Delta_u - Q_\mp). \tag{114}$$

On the other hand (as we may verify),

$$\Delta_u - Q_\mp = u(a\sqrt{u}\pm\sqrt{M})^2. \tag{115}$$

The solution for x thus takes the simple form

$$x = -\frac{a\sqrt{u}\pm\sqrt{M}}{\sqrt{(uQ_\mp)}}. \tag{116}$$

It will appear presently that the upper sign in the foregoing equations applies to retrograde orbits, while the lower sign applies to direct orbits. This convention will be adhered to consistently in this section.

Inserting the solution (116) for x in equation (108), we find

$$E = \frac{1}{\sqrt{Q_\mp}}[1-2Mu\mp a\sqrt{(Mu^3)}]; \tag{117}$$

and the value of L to be associated with this value of E is

$$L = aE+x = \mp\frac{\sqrt{M}}{\sqrt{(uQ_\mp)}}[a^2u^2+1\pm 2a\sqrt{(Mu^3)}]. \tag{118}$$

As we have explained, and as the manner of derivation makes it explicit, E and L given by equations (117) and (118) are the energy and the angular momentum, per unit mass, of a particle describing a circular orbit of reciprocal radius u. The angular velocity, Ω, and the rotational velocity, $v^{(\varphi)}$, follow from these equations and equations (69) and (70); thus,

$$\Omega = \frac{d\varphi}{dt} = \frac{L - 2Mux}{(r^2 + a^2)E - 2aMxu} = \frac{(L - 2Mux)u^2}{(1 + a^2u^2)E - 2aMux^3} \tag{119}$$

and

$$v^{(\varphi)} = e^{\psi - \nu}(\Omega - \omega) = \left(r^2 + a^2 + \frac{2Ma^2}{r}\right)\frac{\Omega}{\sqrt{\Delta}} - \frac{2aM}{r\sqrt{\Delta}}. \tag{120}$$

We find*

$$\Omega = \frac{\mp\sqrt{(Mu^3)}}{1 \mp a\sqrt{(Mu^3)}}, \tag{121}$$

and

$$v^{(\varphi)} = \frac{\mp\sqrt{(Mu)}}{[1 \mp a\sqrt{(Mu^3)}]\sqrt{\Delta_u}}[1 + a^2u^2 \pm 2a\sqrt{(Mu^3)}]. \tag{122}$$

We may parenthetically note here that we can recover from equation (117) the condition for the occurrence of the unstable circular null geodesic by considering the limit $E \to \infty$, when

$$Q_{\mp} = 1 - 3Mu \mp 2a\sqrt{(Mu^3)} = 0, \tag{123}$$

or, equivalently,

$$r^{3/2} - 3Mr^{1/2} \mp 2a\sqrt{M} = 0. \tag{124}$$

We can directly verify that the solution of this cubic equation for $\sqrt{r}$ agrees with that given in equation (87).

Let L and E have the values appropriate to a circular orbit of some assigned reciprocal radius u_c, i.e., L and E have the values given by equations (117) and (118) for $u = u_c$. To avoid any ambiguity, we shall distinguish the quantities (L, E, x, etc.) evaluated for a particular u_c by a subscript c. With L_c and E_c chosen in this manner, the cubic polynomial on the right-hand side of equation (104) will allow a double root at $u = u_c$; and we find that, in consequence, the equation reduces to

$$u^{-4}\dot{u}^2 = 2M(L_c - aE_c)^2(u - u_c)^2\left[u + 2u_c - \frac{L_c^2 - a^2E_c^2 + a^2}{2M(L_c - aE_c)^2}\right]. \tag{125}$$

For L_c and E_c given by equations (117) and (118) (for $u = u_c$), we find

$$\frac{L_c^2 - a^2E_c^2 + a^2}{2Mx_c^2} = [1 + 3a^2u_c^2 \pm 4a\sqrt{(Mu_c^3)}]\frac{1}{2(a\sqrt{u_c} \pm \sqrt{M})^2}, \tag{126}$$

* We may note here the following relations which were found useful in the reductions:

$$L - 2Mux = \frac{\mp\sqrt{M}}{\sqrt{(uQ_{\mp})}}\Delta_u,$$

$$(1 + a^2u^2)E - 2aMxu^3 = \frac{\Delta_u}{\sqrt{Q_{\mp}}}[1 \mp a\sqrt{(Mu^3)}].$$

and
$$u_c - \frac{L_c^2 - a^2 E_c^2 + a^2}{2Mx_c^2} = -\frac{\Delta_{u_c}}{2(a\sqrt{u_c} \pm M)^2}. \tag{127}$$

We may accordingly rewrite equation (125) in the simpler form
$$\dot{u}^2 = 2Mx_c^2 u^4 (u-u_c)^2 (u-u_*), \tag{128}$$
where
$$u_* = -u_c + \frac{\Delta_{u_c}}{2(a\sqrt{u_c} \pm \sqrt{M})^2}. \tag{129}$$

It is clear that u_* defines the reciprocal radius of the orbit of the second kind associated with a stable circular orbit of reciprocal radius u_c. The appropriate solution of equation (128) is
$$\tau = \frac{1}{x_c\sqrt{(2M)}} \int \frac{du}{u^2(u-u_c)(u-u_*)^{1/2}}. \tag{130}$$

Alternatively, we may combine equation (128) with
$$\dot{\varphi} = \frac{u^2}{\Delta_u}(L_c - 2Mx_c u), \tag{131}$$
to obtain the trajectory
$$\varphi = \frac{1}{x_c a^2 \sqrt{(2M)}} \int \frac{(L_c - 2Mx_c u)}{(u-u_+)(u-u_-)(u-u_c)(u-u_*)^{1/2}} du. \tag{132}$$

Orbits of the second kind derived with the aid of equation (130) are illustrated in Fig. 30.

It is clear that the condition for the instability of the circular orbit is $u_* = u_c$, or by equation (129)
$$4u_c(a\sqrt{u_c} \pm \sqrt{M})^2 = \Delta_{u_c} = a^2u_c^2 - 2Mu_c + 1. \tag{133}$$

Expanding this equation and suppressing (as unnecessary) the subscript c, we obtain the equation
$$3a^2u^2 + 6Mu \pm 8a\sqrt{(Mu^3)} - 1 = 0; \tag{134}$$

or, reverting to the variable r, we have
$$r^2 - 6Mr \mp 8a\sqrt{(Mr)} - 3a^2 = 0. \tag{135}$$

This biquadratic equation for $\sqrt{r}$ can be reduced by standard methods to the quadratic equation
$$r - 2q\sqrt{r} - 2M\frac{\cosh\vartheta}{\cosh 3\vartheta} \mp 2a\frac{\sqrt{M}}{q} - M = 0, \tag{136}$$

where $\quad \vartheta = \tfrac{1}{3}\tanh^{-1}\dfrac{|a|}{M} \quad$ and $\quad q^2 = 4M\dfrac{\cosh\vartheta}{\cosh 3\vartheta}\sinh^2\vartheta.$

(137)

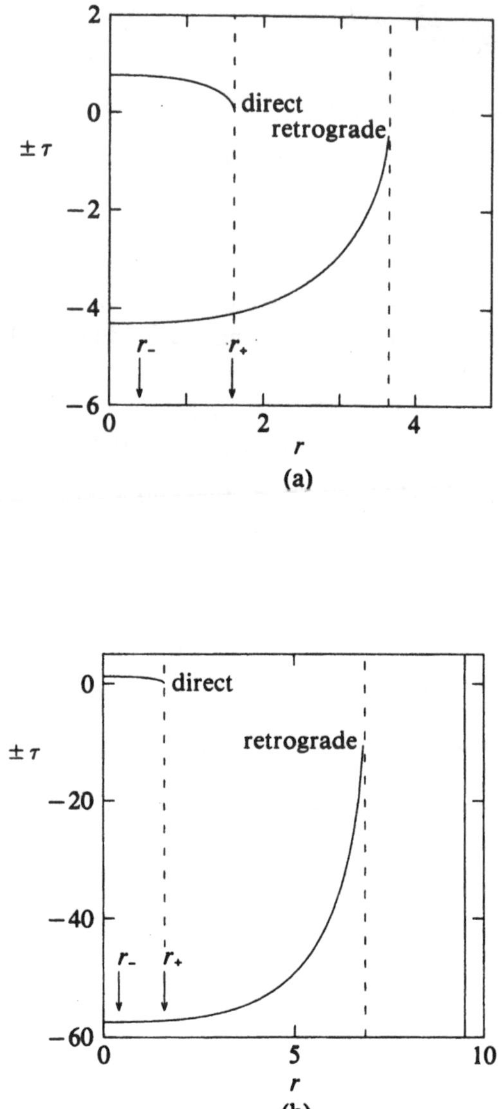

FIG. 30. Examples of orbits of the second kind associated with stable circular orbits with a radii, 20 M and 9.5 M. The parameters of the orbits considered are:

r_c	E	L		
20	0.9780	4.7222	Direct	FIG. (*a*)
	0.8911	−4.9981	Retrograde	
9.5	0.9503	3.4153	Direct	FIG. (*b*)
	0.9605	−4.1243	Retrograde	

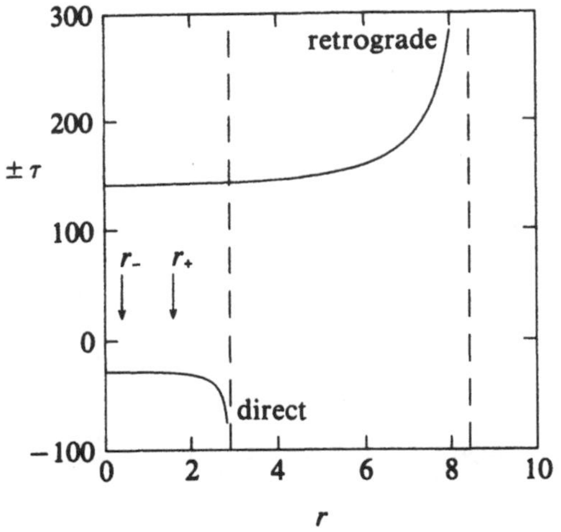

FIG. 31. Marginally stable circular orbits, retrograde and direct, in a Kerr space-time with $a = 0.8$.

The geodesic described by a particle when its energy and angular momentum are those appropriate to an unstable circular orbit can be derived from equations (130) and (132) by simply letting u_c and u_* coincide. An example of such an orbit is illustrated in Fig. 31.

The circular orbits one formally obtains from equations (117) and (118) for radii less than r given by equation (135) is the time-like analogues, for $E^2 > 1$, of the unstable null geodesic considered in § (a): indeed, as we have verified, when $E^2 \to \infty$, we are led to the same limiting case.

Besides the limiting case $E^2 \to \infty$, the case of the 'marginally bound' orbit with $E^2 = 1$ is of some interest: it corresponds to the case of a particle, at rest at infinity, falling towards the black hole. By equations (108) and (116), the radius of the marginally bound circular orbit is given by

$$1 = x^2u^2 = \frac{u}{Q_\mp}(a\sqrt{u} \pm \sqrt{M})^2, \tag{138}$$

or

$$\begin{aligned} Q_\mp &= 1 - 3Mu \mp 2a\sqrt{(Mu^3)} \\ &= u[a^2u + M \pm 2a\sqrt{(Mu)}]. \end{aligned} \tag{139}$$

This last equation simplifies to the form

$$[au \pm 2\sqrt{(Mu)}]^2 = 1; \tag{140}$$

and we derive from this equation

$$r = 2M \pm a + 2\sqrt{(M^2 \pm aM)}. \tag{141}$$

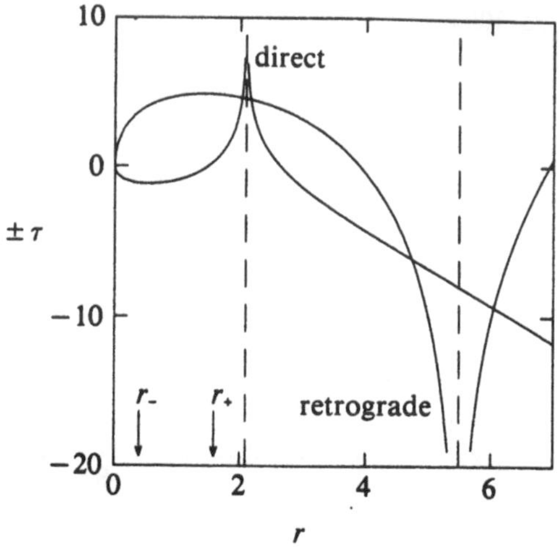

FIG. 32. Examples of critical marginally bound orbits (direct and retrograde) with $E^2 = 1$ falling from rest at infinity. (In Figures 30–32, by a choice of convention, the affine parameter varies in opposite senses along the direct and the retrograde orbits.)

The circular orbit of this radius is the envelope of the trajectories of particles which are at rest at infinity. An example of a marginally bound trajectory is illustrated in Fig. 32.

In the foregoing discussion of the null and the time-like geodesics, we have encountered limiting circular orbits of three kinds: the unstable photon-orbit, the last stable time-like orbit, and the marginally bound orbit. The dependence of the radii of these orbits on a/M is exhibited in Fig. 33. We observe that the radii of the direct orbits, of all three kinds, tend to M as $a \to M$. A more careful consideration of how this limit is approached leads to behaviours shown in Table VII.

The coincidence of the three limiting radii for the direct orbits as $a \to M$ is, in fact, 'illusory' since the proper distances between them tend to non-zero limits. Thus, for

$$a = M(1-\delta) \quad \text{and} \quad r_+ = M[1+\sqrt{(2\delta)}] + O(\delta^{3/2}), \tag{142}$$

the proper distance between $r = M(1+\varepsilon)$ and r_+ is given by

$$\int_{r_+ = M[1+\sqrt{(2\delta)}]}^{r = M(1+\varepsilon)} \frac{r\,dr}{\sqrt{\Delta}} \to M \lg \frac{\sqrt{(\varepsilon^2 - 2\delta)} + \varepsilon}{\sqrt{(2\delta)}}. \tag{143}$$

With the ε's given in Table VII, we find

$$\left.\begin{aligned}
(r_{\text{ph}} - r_+)_{\text{proper distance}} &\to \tfrac{1}{2}M \lg 3 \qquad [\varepsilon = \sqrt{(8\delta/3)}],\\
(r_{\text{m.b.}} - r_+)_{\text{proper distance}} &\to M \lg(1+\sqrt{2}) \qquad [\varepsilon = 2\sqrt{\delta}],\\
(r_{\text{st.}} - r_+)_{\text{proper distance}} &\to M \lg(2^{5/6}\delta^{-1/6}) \qquad [\varepsilon = \sqrt[3]{(4\delta)}].
\end{aligned}\right\} \tag{144}$$

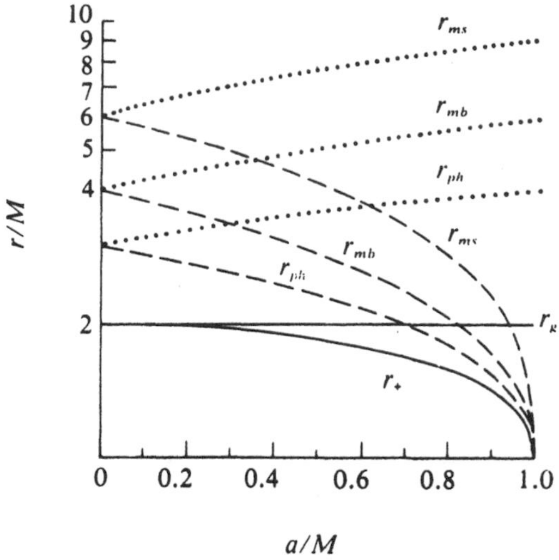

FIG. 33. Radii of circular equatorial orbits around a Kerr black-hole as functions of the parameter a. Dashed and dotted curves (for direct and retrograde orbits) refer to the innermost stable (ms), innermost bound (mb), and photon (ph) orbits. Solid curves indicate the event horizon (r_+) and the equatorial boundary of the ergosphere which always occurs at $2\,M$.

The fact that these proper distances tend to non-zero limits as $a \to M$ is a manifestation of the ambiguity in the chosen radial coordinate as $a \to M$. A related fact is that the energy of the last stable circular orbit does not tend to infinity for $a^2 \to M^2 - 0$ as one might have expected from the coincidence of its radius with that of the horizon. Thus, for $a = M$, the energy of a direct circular orbit with a reciprocal radius u is given by (cf. equation (117))

$$E_{\text{direct}}(a = M) = \frac{1 - 2\,Mu + \sqrt{(Mu)^3}}{[1 - 3\,Mu + 2\sqrt{(Mu)^3}]^{1/2}}; \tag{145}$$

TABLE VII

The radii of the limiting circular orbits for $a = 0$ and $a = \pm M$

	$a = 0$	Direct $a = M(1-\delta)$ $(\delta \to 0)$	Retrograde $a = -M$
r_{photon}	$3\,M$	$M[1 + \sqrt{(8\delta/3)}]$	$4\,M$
$r_{\text{mar. bound}}$	$4\,M$	$M[1 + 2\sqrt{\delta}]$	$(3 + 2\sqrt{2})M$
$r_{\text{stability}}$	$6\,M$	$M[1 + \sqrt[3]{(4\delta)}]$	$9\,M$

and this expression, for $u \to M^{-1} \mp 0$, tends to the limits

$$E_{\text{direct}}(a = M) \to \pm 3^{-1/2} \quad (u \to M^{-1} \mp 0), \tag{146}$$

exhibiting a discontinuity.

The limit $E = 3^{-1/2}$ for $r \to M + 0$, when $a = M$, is important: it gives the maximum energy (per unit mass) which a stable circular orbit can have in Kerr geometry for $a^2 < M^2$.

We shall return to the implications of the discontinuity exhibited by E in §66.

62. The general equations of geodesic motion and the separability of the Hamilton–Jacobi equation

Quite generally, geodesic motion in a stationary axisymmetric space-time will allow two integrals of motion: the energy and the angular momentum about the axis of symmetry. Besides, the norm of the four-velocity will also be conserved by virtue of its parallel propagation. These three conservation laws will not, in general, suffice to reduce the problem of solving the equations of geodesic motion to one involving quadratures only. That, nevertheless, such a reduction is in fact possible was discovered by Carter who demonstrated explicitly the separability of the Hamilton–Jacobi equation and deduced the existence of a further conserved quantity. Subsequent to Carter's discovery, Walker and Penrose showed that any type-D space-time has a conformal Killing-tensor—and, more generally, the conserved complex quantity, K_s, of Theorem 1 of §60 for null geodesics—and, further, that in Kerr geometry we have, in addition, the integral K established in Theorem 2, Corollary 3. In this section, we shall consider the reduction of equations of geodesic motion, in general, which follows from these discoveries.

Starting from the general Lagrangian,

$$2\mathscr{L} = \left(1 - \frac{2Mr}{\rho^2}\right)\dot{t}^2 + \frac{4aMr\sin^2\theta}{\rho^2}\dot{t}\dot{\varphi} - \frac{\rho^2}{\Delta}\dot{r}^2 - \rho^2\dot{\theta}^2$$
$$- \left(r^2 + a^2 + \frac{2a^2Mr}{\rho^2}\sin^2\theta\right)(\sin^2\theta)\dot{\varphi}^2, \tag{147}$$

we deduce the energy and the angular-momentum integrals:

$$+p_t = \left(1 - \frac{2Mr}{\rho^2}\right)\dot{t} + \frac{2aMr\sin^2\theta}{\rho^2}\dot{\varphi} = E = \text{constant} \tag{148}$$

and

$$-p_\varphi = -\frac{2aMr\sin^2\theta}{\rho^2}\dot{t} + \left(r^2 + a^2 + \frac{2a^2Mr}{\rho^2}\sin^2\theta\right)(\sin^2\theta)\dot{\varphi}$$
$$= L_z = \text{constant}. \tag{149}$$

Besides, we have the integrals (cf. equations (37))

$$2\rho^2(\boldsymbol{k}\cdot\boldsymbol{l})(\boldsymbol{k}\cdot\boldsymbol{n})-r^2|\boldsymbol{k}|^2 = K \tag{150}$$

and

$$2\rho^2(\boldsymbol{k}\cdot\boldsymbol{m})(\boldsymbol{k}\cdot\bar{\boldsymbol{m}})+(a^2\cos^2\theta)|\boldsymbol{k}|^2 = K, \tag{151}$$

where $\boldsymbol{l}$, $\boldsymbol{n}$, $\boldsymbol{m}$, and $\bar{\boldsymbol{m}}$ are the basis null vectors. We may consider the integrals (150) and (151) as independent since it presupposes the conservation of $|\boldsymbol{k}|^2$. And, as usual, we shall set

$$\left.\begin{aligned}|\boldsymbol{k}|^2 = \delta_1 &= 1 \qquad \text{for time-like geodesics}\\ &= 0 \qquad \text{for null geodesics.}\end{aligned}\right\} \tag{152}$$

In the present context,

$$k^i = (\dot{t}, \dot{r}, \dot{\theta}, \dot{\varphi}). \tag{153}$$

Making use of our present knowledge of the null basis-vectors (Ch. 6, equation (173)), we find that in Kerr geometry the integrals (150) and (151) have the explicit forms

$$\frac{1}{\Delta}[\Delta\dot{t}-(a\Delta\sin^2\theta)\dot{\varphi}]^2-\frac{\rho^4}{\Delta}\dot{r}^2-\delta_1 r^2 = K \tag{154}$$

and

$$[(a\sin\theta)\dot{t}-(r^2+a^2)(\sin\theta)\dot{\varphi}]^2+\rho^4\dot{\theta}^2+\delta_1 a^2\cos^2\theta = K. \tag{155}$$

On the other hand, by equations (148) and (149))

$$(a\sin\theta)\dot{t}-[(r^2+a^2)\sin\theta]\dot{\varphi} = aE\sin\theta-L_z\operatorname{cosec}\theta \tag{156}$$

and

$$\Delta\dot{t}-(a\Delta\sin^2\theta)\dot{\varphi} = (r^2+a^2)E-aL_z. \tag{157}$$

Inserting these expressions in equations (154) and (155), we obtain the equations

$$\frac{1}{\Delta}[(r^2+a^2)E-aL_z]^2-\frac{\rho^4}{\Delta}\dot{r}^2-\delta_1 r^2 = K, \tag{158}$$

and

$$(aE\sin\theta-L_z\operatorname{cosec}\theta)^2+\rho^4\dot{\theta}^2+\delta_1 a^2\cos^2\theta = K; \tag{159}$$

or, alternatively,

$$\rho^4\dot{r}^2 = [(r^2+a^2)E-aL_z]^2-\Delta(\delta_1 r^2+K) \tag{160}$$

and

$$\rho^4\dot{\theta}^2 = -(aE\sin\theta-L_z\operatorname{cosec}\theta)^2+(-\delta_1 a^2\cos^2\theta+K). \tag{161}$$

Finally, equations (156) and (157) provide the complementary pair of equations

$$\rho^2\dot{t} = \frac{1}{\Delta}(\Sigma^2E-2aMrL_z) \tag{162}$$

and

$$\rho^2\dot{\varphi} = \frac{1}{\Delta}[2aMrE + (\rho^2 - 2Mr)L_z \operatorname{cosec}^2\theta]. \tag{163}$$

It is now clear that the problem of solving the equations of geodesic motion has been reduced to one of quadratures.

An alternative form of equation (161) which we shall find useful is

$$\rho^4\dot{\theta}^2 = [K - (L_z - aE)^2] - [a^2(\delta_1 - E^2) + L_z^2 \operatorname{cosec}^2\theta]\cos^2\theta. \tag{164}$$

(a) *The separability of the Hamilton–Jacobi equation and an alternative derivation of the basic equations*

As we have stated, the existence of a fourth quantity that is conserved along a geodesic was first discovered by Carter by explicitly demonstrating the separability of the Hamilton–Jacobi equation. At the time, it was wholly unexpected; and it suggested that the other equations of mathematical physics might be similarly separable. Indeed, they were all eventually separated as we shall see in Chapters 8, 9, and 10. As the first of the chain of remarkable properties that characterize Kerr geometry, it is useful to follow Carter's demonstration of the separability of the Hamilton–Jacobi equation.

The Hamilton–Jacobi equation governing geodesic motion in a space-time with the metric tensor g^{ij} is given by

$$2\frac{\partial S}{\partial \tau} = g^{ij}\frac{\partial S}{\partial x^i}\frac{\partial S}{\partial x^j}, \tag{165}$$

where S denotes Hamilton's principal function. With g^{ij} for the Kerr geometry given in Chapter 6, equation (135), equation (165) becomes

$$2\frac{\partial S}{\partial \tau} = \frac{\Sigma^2}{\rho^2\Delta}\left(\frac{\partial S}{\partial t}\right)^2 + \frac{4aMr}{\rho^2\Delta}\frac{\partial S}{\partial t}\frac{\partial S}{\partial \varphi} - \frac{\Delta - a^2\sin^2\theta}{\rho^2\Delta\sin^2\theta}\left(\frac{\partial S}{\partial \varphi}\right)^2 - \frac{\Delta}{\rho^2}\left(\frac{\partial S}{\partial r}\right)^2 - \frac{1}{\rho^2}\left(\frac{\partial S}{\partial \theta}\right)^2. \tag{166}$$

It is convenient to rewrite this equation in the alternative form

$$2\frac{\partial S}{\partial \tau} = \frac{1}{\rho^2\Delta}\left[(r^2+a^2)\frac{\partial S}{\partial t} + a\frac{\partial S}{\partial \varphi}\right]^2 - \frac{1}{\rho^2\sin^2\theta}\left[(a\sin^2\theta)\frac{\partial S}{\partial t} + \frac{\partial S}{\partial \varphi}\right]^2 - \frac{\Delta}{\rho^2}\left(\frac{\partial S}{\partial r}\right)^2 - \frac{1}{\rho^2}\left(\frac{\partial S}{\partial \theta}\right)^2. \tag{167}$$

Assuming that the variables can be separated, we seek a solution of equation (167) of the form

$$S = \tfrac{1}{2}\delta_1\tau - Et + L_z\varphi + S_r(r) + S_\theta(\theta), \tag{168}$$

where, as the notation indicates, S_r and S_θ are functions only of the variable specified. For the chosen form of S, equation (167) becomes

$$\delta_1\rho^2 = \frac{1}{\Delta}[(r^2+a^2)E - aL_z]^2 - \frac{1}{\sin^2\theta}(aE\sin^2\theta - L_z)^2 - \Delta\left(\frac{dS_r}{dr}\right)^2 - \left(\frac{dS_\theta}{d\theta}\right)^2. \tag{169}$$

With the aid of the identity

$$(aE\sin^2\theta - L_z)^2\operatorname{cosec}^2\theta = (L_z^2\operatorname{cosec}^2\theta - a^2E^2)\cos^2\theta + (L_z - aE)^2, \tag{170}$$

we can rewrite equation (169) in the form

$$\left\{\Delta\left(\frac{dS_r}{dr}\right)^2 - \frac{1}{\Delta}[(r^2+a^2)E - aL_z]^2 + (L_z - aE)^2 + \delta_1 r^2\right\} + \left\{\left(\frac{dS_\theta}{d\theta}\right)^2 + (L_z^2\operatorname{cosec}^2\theta - a^2E^2)\cos^2\theta + \delta_1 a^2\cos^2\theta\right\} = 0. \tag{171}$$

The separability of the equation is now manifest and we infer that

$$\Delta\left(\frac{dS_r}{dr}\right)^2 = \frac{1}{\Delta}[(r^2+a^2)E - aL_z]^2 - [\mathcal{Q} + (L_z - aE)^2 + \delta_1 r^2] \tag{172}$$

and

$$\left(\frac{dS_\theta}{d\theta}\right)^2 = \mathcal{Q} - (L_z^2\operatorname{cosec}^2\theta - a^2E^2 + \delta_1 a^2)\cos^2\theta, \tag{173}$$

where $\mathcal{Q}$ is a separation constant. With the abbreviations

$$R = [(r^2+a^2)E - aL_z]^2 - \Delta[\mathcal{Q} + (L_z - aE)^2 + \delta_1 r^2] \tag{174}$$

and

$$\Theta = \mathcal{Q} - [a^2(\delta_1 - E^2) + L_z^2\operatorname{cosec}^2\theta]\cos^2\theta, \tag{175}$$

the solution for S is

$$S = \tfrac{1}{2}\delta_1\tau - Et + L_z\varphi + \int^r \frac{\sqrt{R(r)}}{\Delta}\,dr + \int^\theta d\theta\,\sqrt{\Theta(\theta)}. \tag{176}$$

The basic equations governing the motion can be deduced from the solution (176) for the principal function S by the standard procedure of setting to zero the partial derivatives of S with respect to the different constants of the motion—$\mathcal{Q}$, δ_1, E, and L_z in this instance. Thus, we find that

$$\frac{\partial S}{\partial \mathcal{Q}} = \frac{1}{2}\int \frac{1}{\Delta\sqrt{R}}\frac{\partial R}{\partial \mathcal{Q}}\,dr + \frac{1}{2}\int \frac{1}{\sqrt{\Theta}}\frac{\partial \Theta}{\partial \mathcal{Q}}\,d\theta = 0 \tag{177}$$

leads to the equation

$$\int^r \frac{dr}{\sqrt{R}} = \int^\theta \frac{d\theta}{\sqrt{\Theta}}. \tag{178}$$

Similarly, we find

$$\tau = \int^r \frac{r^2}{\sqrt{R}}\,dr + a^2 \int^\theta \frac{\cos^2\theta}{\sqrt{\Theta}}\,d\theta, \tag{179}$$

$$\begin{aligned} t &= \frac{1}{2}\int^r \frac{1}{\Delta\sqrt{R}}\frac{\partial R}{\partial E}\,dr + \frac{1}{2}\int^\theta \frac{1}{\sqrt{\Theta}}\frac{\partial \Theta}{\partial E}\,d\theta \\ &= \tau E + 2M\int^r r[r^2E - a(L_z - aE)]\frac{dr}{\Delta\sqrt{R}}, \end{aligned} \tag{180}$$

and

$$\begin{aligned} \varphi &= -\frac{1}{2}\int^r \frac{1}{\Delta\sqrt{R}}\frac{\partial R}{\partial L_z}\,dr - \frac{1}{2}\int^\theta \frac{1}{\sqrt{\Theta}}\frac{\partial \Theta}{\partial L_z}\,d\theta \\ &= a\int^r [(r^2+a^2)E - aL_z]\frac{dr}{\Delta\sqrt{R}} + \int^\theta (L_z \operatorname{cosec}^2\theta - aE)\frac{d\theta}{\sqrt{\Theta}}, \end{aligned} \tag{181}$$

where simplifications have been effected with the aid of equation (178).

It can now be verified that equations (178)–(181) are entirely equivalent to the set (160)–(163) with the identification

$$\mathcal{Q} = K - (L_z - aE)^2. \tag{182}$$

In particular, with this identification, the right-hand sides of equations (160) and (164) agree with the present definitions of R and Θ in equations (174) and (175); and the relation (178) is an immediate consequence of equations (160) and (164). Our basic equations, then, are

$$\rho^4\dot{r}^2 = R; \quad \rho^4\dot{\theta}^2 = \Theta, \tag{183}$$

$$\left.\begin{aligned} \rho^2\dot{\varphi} &= \frac{1}{\Delta}[2aMrE + (\rho^2 - 2Mr)L_z \operatorname{cosec}^2\theta], \\ \text{and} \qquad \rho^2\dot{t} &= \frac{1}{\Delta}(\Sigma^2E - 2aMrL_z), \end{aligned}\right\} \tag{184}$$

where

$$R = E^2r^4 + (a^2E^2 - L_z^2 - \mathcal{Q})r^2 + 2Mr[\mathcal{Q} + (L_z - aE)^2] - a^2\mathcal{Q} - \delta_1 r^2\Delta \tag{185}$$

and

$$\Theta = \mathcal{Q} + (a^2E^2 - L_z^2\operatorname{cosec}^2\theta)\cos^2\theta - \delta_1 a^2\cos^2\theta. \tag{186}$$

It is convenient to assemble in one place the various formulae giving the tensor and the tetrad components of the four momentum. We have

$$\left.\begin{aligned} -p_r &= \frac{\rho^2}{\Delta}p^r = \frac{\sqrt{R}}{\Delta}, \quad -p_\theta = \rho^2 p^\theta = \sqrt{\Theta}, \\ p_\varphi &= \frac{2aMr\sin^2\theta}{\rho^2}p^t - \left(r^2+a^2+\frac{2a^2Mr}{\rho^2}\sin^2\theta\right)(\sin^2\theta)p^\varphi = -L_z, \\ p_t &= \left(1-\frac{2Mr}{\rho^2}\right)p^t + \frac{2aMr\sin^2\theta}{\rho^2}p^\varphi = E, \end{aligned}\right\} \quad (187)$$

and

$$\left.\begin{aligned} p^{(t)} &= p_{(t)} = e^{-\nu}(E-\omega L_z) = e^{+\nu}p^t, \\ p^{(r)} &= -p_{(r)} = -e^{-\mu_2}p_r = +e^{+\mu_2}p^r, \\ p^{(\theta)} &= -p_{(\theta)} = -e^{-\mu_3}p_\theta = e^{+\mu_3}p^\theta, \\ \text{and}\quad p^{(\varphi)} &= -p_{(\varphi)} = e^{\psi}p^\varphi = -e^{-\psi}p_\varphi = e^{-\psi}L_z. \end{aligned}\right\} \quad (188)$$

63. The null geodesics

In our considerations of the general non-planar orbits, we shall concentrate on delineating the projection of the orbits on to the (r, θ) plane: the variations of t and φ along the orbits do not reveal any special features that have not already been displayed by the planar orbits on the equatorial plane.

For the null geodesics $\delta_1 = 0$, and it is convenient to minimize the number of parameters by letting

$$\xi = L_z/E \qquad \text{and} \qquad \eta = \mathcal{Q}/E^2, \quad (189)$$

and writing R and Θ in place of R/E^2 and Θ/E^2:

$$R = r^4 + (a^2 - \xi^2 - \eta)r^2 + 2M[\eta + (\xi - a)^2]r - a^2\eta \quad (190)$$

and

$$\Theta = \eta + a^2\cos^2\theta - \xi^2\cot^2\theta. \quad (191)$$

The two parameters ξ and η replace the single impact parameter, D, by which we distinguished the null geodesics in the equatorial plane. The parameters ξ and η are in fact related very simply to the 'celestial coordinates' α and β of the image as seen by an observer at infinity who receives the light ray. Making use of the expressions (188), we readily verify that

$$\left.\begin{aligned} \alpha &= \left(\frac{rp^{(\varphi)}}{p^{(t)}}\right)_{r\to\infty} = \xi\operatorname{cosec}\theta_0 \\ \text{and}\quad \beta &= \left(\frac{rp^{(\theta)}}{p^{(t)}}\right)_{r\to\infty} = (\eta + a^2\cos^2\theta_0 - \xi^2\cot^2\theta_0)^{1/2} = -p_{\theta_0}. \end{aligned}\right\} \quad (192)$$

where θ_0 is the angular coordinate of the observer at infinity. Precisely, α is the apparent perpendicular distance of the image from the axis of symmetry and β is the apparent perpendicular distance of the image from its projection on the equatorial plane.

The equation governing the projection of the orbit in the (r, θ)-plane is

$$\int_{r_i}^{r} \frac{dr}{\sqrt{R}} = \int_{\theta_i}^{\theta} \frac{d\theta}{\sqrt{\Theta}}, \tag{193}$$

where r_i and θ_i are certain assigned initial values of r and θ. Also, it should be noted that while the signs of $\sqrt{R}$ and $\sqrt{\Theta}$ can be chosen independently, they must be adhered to once the choice has been made.

(a) *The θ-motion*

We shall first consider the θ-motion, as specified by the integral over $\Theta^{-1/2}$ in equation (193), since it already gives some measure of insight into the character of the orbits. Thus, the fundamental requirement, that Θ be not negative, restricts the constant ξ and η by the inequality

$$\eta + (a - \xi)^2 \geqslant 0, \tag{194}$$

which follows from Θ written in the alternative form:

$$\Theta = \eta + (a - \xi)^2 - (a \sin\theta - \xi \operatorname{cosec}\theta)^2. \tag{195}$$

We shall find it convenient to replace θ by $\cos\theta = \mu$ as the variable of integration and consider, instead, the integral

$$I_\mu = \int \frac{d\mu}{\sqrt{\Theta_\mu}} \quad (\mu = \cos\theta), \tag{196}$$

where

$$\Theta_\mu = \eta - (\xi^2 + \eta - a^2)\mu^2 - a^2\mu^4. \tag{197}$$

Since

$$\left.\begin{aligned} &\Theta_\mu = \eta \quad \text{when} \quad \mu = 0 \quad \text{and} \quad \theta \leqslant \pi/2 \\ \text{and} \quad &\Theta_\mu = -\xi^2 \leqslant 0 \quad \text{when} \quad \mu = 1 \quad \text{and} \quad \theta = 0, \end{aligned}\right\} \tag{198}$$

it follows that we must distinguish the cases $\eta > 0$ and $\eta < 0$. When $\eta > 0$, the allowed range of μ^2, for $\Theta_\mu \geqslant 0$, will be between $\mu^2 = 0$ and a certain maximum $\mu^2_{\max}$: $0 \leqslant \mu^2 \leqslant \mu^2_{\max}$; and when $\eta < 0$, μ^2 will be restricted to an interval $0 < \mu_1^2 \leqslant \mu^2 \leqslant \mu_2^2 < 1$ (allowing the possibility that no such interval exists). In other words, when $\eta > 0$, the orbits will intersect the equatorial plane and oscillate symmetrically about it; and when $\eta < 0$, the orbits cannot intersect the equatorial plane and must be confined to cones which exclude both the axis ($\theta = 0$) and the orthogonal direction ($\theta = \pi/2$). The case $\eta = 0$ should be considered separately.

Explicit expressions for the integral I_μ, appropriate for the cases $\eta > 0$, $\eta = 0$, and $\eta < 0$, can be readily written down:

(i) $\eta > 0$: $$\Theta_\mu = a^2(\mu_-^2 + \mu^2)(\mu_+^2 - \mu^2) \quad (0 \leqslant \mu^2 \leqslant \mu_+^2), \tag{199}$$

where

$$\left.\begin{aligned} \mu_+^2 &= \frac{1}{2a^2}\{[(\xi^2+\eta-a^2)^2+4a^2\eta]^{1/2}-(\xi^2+\eta-a^2)\},\\ \mu_-^2 &= \frac{1}{2a^2}\{[(\xi^2+\eta-a^2)^2+4a^2\eta]^{1/2}+(\xi^2+\eta-a^2)\};\end{aligned}\right\} \tag{200}$$

and

$$\left.\begin{aligned} &\int_\mu^{\mu_+}\frac{d\mu}{\sqrt{\Theta_\mu}} = \frac{1}{a(\mu_+^2+\mu_-^2)^{1/2}}F(\psi, k),\\ &\text{where}\\ &k^2 = \mu_+^2/(\mu_+^2+\mu_-^2) \quad \text{and} \quad \cos\psi = \mu/\mu_+.\end{aligned}\right\} \tag{201}$$

(ii) $\eta = 0$: $$\left.\begin{aligned} &\Theta_\mu = (a^2-\xi^2)\mu^2 - a^2\mu^4 \quad (0 \leqslant \mu^2 \leqslant \mu_{\max}^2),\\ &\text{where}\\ &\mu_{\max}^2 = 1 - \xi^2/a^2 \quad (|\xi|^2 \leqslant |a|^2),\end{aligned}\right\} \tag{202}$$

and

$$\int_\mu^{\mu_{\max}}\frac{d\mu}{\sqrt{\Theta_\mu}} = -\frac{1}{\sqrt{(a^2-\xi^2)}}\operatorname{sech}^{-1}\frac{\mu}{\mu_{\max}}. \tag{203}$$

(iii) $\eta < 0$: $$\Theta_\mu = a^2(\mu^2 - \mu_-^2)(\mu_+^2 - \mu^2) \quad (\mu_-^2 \leqslant \mu^2 \leqslant \mu_+^2), \tag{204}$$

where

$$\mu_\pm^2 = \frac{1}{2a^2}\{(|\eta| + a^2 - \xi^2) \pm [(|\eta| + a^2 - \xi^2)^2 - 4a^2|\eta|]^{1/2}\} \tag{205}$$

and

$$\left.\begin{aligned} &\int_{\mu_-}^{\mu}\frac{d\mu}{\sqrt{\Theta_\mu}} = \frac{1}{a\mu_+}F(\psi, k) \quad \text{where} \quad k^2 = (\mu_+^2 - \mu_-^2)/\mu_+^2,\\ &|\eta| \leqslant a^2, \quad 0 \leqslant |\xi| \leqslant |a| - \sqrt{|\eta|}, \quad \text{and} \quad \sin\psi = \left[\frac{\mu_+^2(\mu^2-\mu_-^2)}{\mu^2(\mu_+^2-\mu_-^2)}\right]^{1/2}.\end{aligned}\right\} \tag{206}$$

In equations (201) and (206), $F(\psi, k)$ denotes the Jacobian elliptic integral of the first kind.

(b) *The principal null-congruences*

From equations (194) and (195) it is clear that the case

$$\eta + (a - \xi)^2 = 0 \tag{207}$$

is distinguished by the fact that it is compatible with the requirement $\Theta_\mu \geqslant 0$ only for

$$\theta = \theta_0 = \text{a constant}, \qquad \text{and} \qquad \xi = a\sin^2\theta_0, \tag{208}$$

when $\Theta_\mu = 0$ for $\theta = \theta_0$. Equation (207) now requires that

$$\eta = -a^2\cos^4\theta_0. \tag{209}$$

Inserting the foregoing values of ξ and η in equation (190), we find that

$$R = (r^2 + a^2\cos^2\theta_0)^2 E^2, \tag{210}$$

where we have restored the suppressed factor E^2. We now conclude from equation (183) that in this case

$$\dot{r} = \pm E. \tag{211}$$

We similarly find from equations (184) that

$$\dot{\varphi} = aE/\Delta \qquad \text{and} \qquad \dot{t} = (r^2 + a^2)E/\Delta. \tag{212}$$

Accordingly, these special null geodesics are defined by

$$\frac{dt}{d\tau} = \frac{r^2 + a^2}{\Delta}E, \quad \frac{dr}{d\tau} = \pm E, \quad \frac{d\theta}{d\tau} = 0, \quad \text{and} \quad \frac{d\varphi}{d\tau} = \frac{a}{\Delta}E. \tag{213}$$

These equations in fact define the shear-free null-congruences which we have used for constructing a null basis for a description of the Kerr space-time, in a Newman–Penrose formalism, adapted to its type-D character (cf. Ch. 6, §56).

(c) *The r-motion*

We now turn to a consideration of the radial motion as specified by the integral over $R^{-1/2}$ in equation (193).

As in the case of the planar orbits considered in §61, we can distinguish between the orbits which, arriving from infinity, either cross or do not cross the event horizon. In the planar case, the distinction was one of whether the impact parameter, D, was less than or greater than a certain critical value D_c, the orbit with the impact parameter D_c, itself, representing an unstable circular orbit. In the same way, in the non-planar case we are presently considering, the distinction must be one of whether the constants of the motion, ξ and η, are on one side or the other of a critical locus, (ξ_s, η_s), in the (ξ, η)-plane, with the orbits, with the constants of the motion on the locus itself, representing unstable orbits with r = constant. The locus, (ξ_s, η_s), will play the same role in the present context as the impact parameter, D_c, in the earlier context.

The equations determining the unstable orbits of constant radius are

$$R = r^4 + (a^2 - \xi^2 - \eta)r^2 + 2M[\eta + (\xi - a)^2]r - a^2\eta = 0 \tag{214}$$

and

$$\partial R/\partial r = 4r^3 + 2(a^2 - \xi^2 - \eta)r + 2M[\eta + (\xi - a)^2] = 0. \tag{215}$$

Equations (214) and (215) can be combined to give

$$3r^4 + r^2 a^2 - \eta(r^2 - a^2) = r^2 \xi^2 \tag{216}$$

and

$$r^4 - a^2 Mr + \eta(a^2 - Mr) = Mr(\xi^2 - 2a\xi). \tag{217}$$

These equations can be solved for ξ and η. Thus, eliminating η between them, we obtain

$$a^2(r-M)\xi^2 - 2aM(r^2 - a^2)\xi - (r^2 + a^2)[r(r^2 + a^2) - M(3r^2 - a^2)] = 0. \tag{218}$$

Quite remarkably, we find that the discriminant, "$\frac{1}{4}(b^2 - ac)$" of this equation is

$$a^2 r^2 \Delta^2; \tag{219}$$

and we find that the solution of equation (218) is given by

$$\xi = \frac{1}{a(r-M)}[M(r^2 - a^2) \pm r\Delta]. \tag{220}$$

If we choose the upper sign in the solution (220),

$$\xi = (r^2 + a^2)/a; \tag{221}$$

and from equation (216), we find that the corresponding solution for η is

$$\eta = -r^4/a^2. \tag{222}$$

With ξ and η given by equations (221) and (222),

$$\eta + (a - \xi)^2 = 0. \tag{223}$$

But this relation is inconsistent with our present requirements: as we have seen in §(*b*), it requires θ to be a constant. We must, therefore, choose the lower sign in equation (220). In this manner, we find

$$\xi = \frac{1}{a(r-M)}[M(r^2 - a^2) - r\Delta] \tag{224}$$

and

$$\eta = \frac{r^3}{a^2(r-M)^2}[4a^2 M - r(r-3M)^2] = \frac{r^3}{a^2(r-M)^2}[4M\Delta - r(r-M)^2]. \tag{225}$$

These equations determine, parametrically, the critical locus (ξ_s, η_s). (See Fig. 34.)

We have seen in §(*a*), that the character of the orbits depends, crucially, on

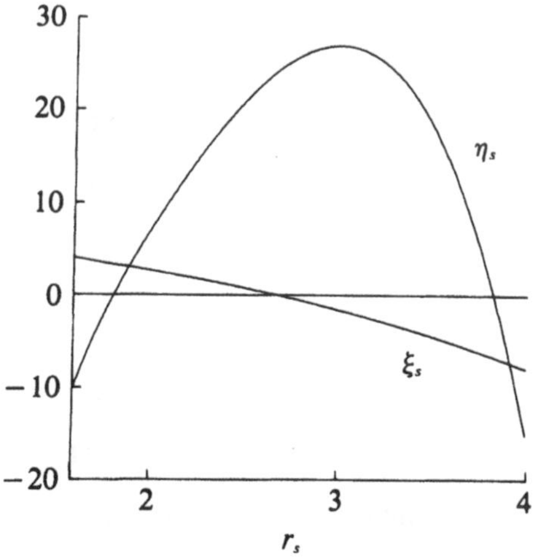

FIG. 34. The locus (ξ_s, η_s) determining the constants of the motion for three-dimensional orbits of constant radius described around a Kerr black-hole with $a = 0.8$. The unit of length along the abscissa is M.

whether η is greater than, equal to, or less than zero. We must, accordingly, distinguish these three cases.

By equation (225), $\eta = 0$ when

$$4a^2 M = r(r-3M)^2 \tag{226}$$

or, equivalently, when

$$\pm 2a\sqrt{M} = r^{3/2} - 3Mr^{1/2}. \tag{227}$$

But this equation is the same equation (124) which determines the radii of the unstable circular photon-orbits, direct and retrograde, in the equatorial plane. Let the radii of these unstable photon-orbits be $r_{\text{ph}}^{(+)}$ and $r_{\text{ph}}^{(-)}$. Then, it is clear that

$$\left.\begin{aligned} &\eta > 0 \quad \text{for} \quad r_{\text{ph}}^{(+)} < r < r_{\text{ph}}^{(-)}, \\ &\eta = 0 \quad \text{for} \quad r = r_{\text{ph}}^{(+)} \quad \text{and} \quad r = r_{\text{ph}}^{(-)}, \\ \text{and} \qquad &\eta < 0 \quad \text{for} \quad r_+ < r < r_{\text{ph}}^{(+)} \quad \text{and} \quad r > r_{\text{ph}}^{(-)}. \end{aligned}\right\} \tag{228}$$

The solutions for the θ-motion appropriate to the three cases (228) have been given in §(*a*). It follows from these solutions that *orbits of constant radius with* $\eta < 0$ *are not allowed* since for $\mu_{\pm}^2$, given by equations (205), to be positive it is necessary that $0 \leqslant \xi \leqslant a - \sqrt{|\eta|}$ and $|\eta| \leqslant a^2$; and these conditions (as one can verify) cannot be met by ξ and η given by equations (224) and (225).

It is also important to observe that for $\eta = 0$, the orbits lie entirely in the equatorial plane. For when $\eta = 0$, it follows from equation (216) that $\xi^2 = 3r^2 + a^2$; and this relation is the same as that noted in the footnote to equation (82) since the present meaning of $\xi\,(= L_z/a)$ and the former meaning of $D\,(= L/a)$ are the same.

Returning to the consideration of the r-motion, let the constants of the motion, ξ and η, have the values ξ_s and η_s appropriate to an orbit of constant radius r_s, i.e., ξ_s and η_s have the values given by equations (224) and (225) for $r = r_s$. For this choice of the constants of the motion, r_s is a double root of $R = 0$; and it can be shown that R reduces to

$$R = (r - r_s)^2 (r^2 + 2rr_s - a^2\eta_s/r_s^2), \tag{229}$$

and the integral we have to consider is

$$\int \frac{dr}{\sqrt{R}} = \int \frac{dr}{(r - r_s)(r^2 + 2rr_s - a^2\eta_s/r_s^2)^{1/2}}. \tag{230}$$

From this equation, it is apparent that the trajectory consists of two parts: a part exterior to $r = r_s$ and a part interior to $r = r_s$. Since $\eta_s > 0$, it will terminate before $r = 0$; for the larger of the two roots of the equation

$$r^2 + 2rr_s - a^2\eta_s/r_s^2 = 0, \tag{231}$$

namely,

$$-r_s + (r_s^2 + a^2\eta_s/r_s^2)^{1/2} = -r_s + 2\sqrt{(M\Delta r_s)}/(r_s - M), \tag{232}$$

is positive and, as can be shown, is less than r_s provided $a^2 < M^2$.

Now, letting

$$x = (r - r_s)^{-1}, \tag{233}$$

and evaluating the integral (230), we obtain the equation determining the projection of the orbit on the $(r, \theta = \cos^{-1}\mu)$-plane. Thus,

$$\int \frac{d\mu}{\sqrt{\Theta_\mu}} = \pm \frac{1}{\sqrt{c}} \lg\{2[c(1 + 4r_s x + cx^2)]^{1/2} + 2cx + 4r_s\}, \tag{234}$$

where

$$c = 3r_s^2 - a^2\eta_s/r_s^2 \quad (> 0). \tag{235}$$

The integral over μ on the right-hand side of the equation is given by equation (201).

Examples of trajectories derived from equation (234) are illustrated in Fig. 35.

While we have considered only the orbits with the constants of the motion on the critical locus, (ξ_s, η_s), it is clear how they must look when the constants of the motion are one or the other side of the locus: the two parts of the trajectory which approaches the sphere at $r = r_s$ from one side, or the other, will either separate into two disjoint parts (which we have designated as orbits

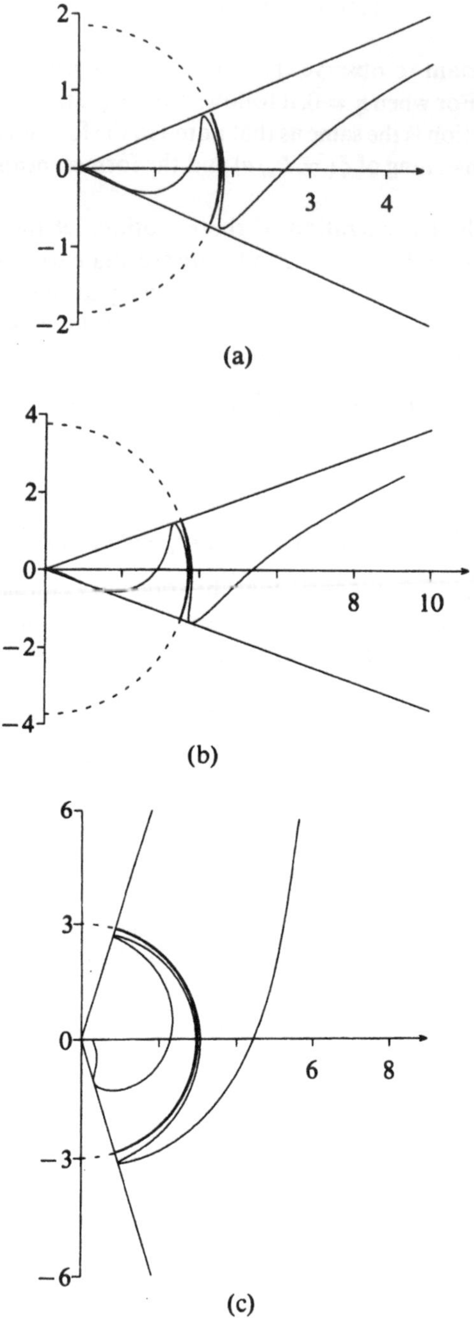

FIG. 35. Critical null-geodesics in the (r, θ)-plane for the three pairs of values: (*a*), $r_s = 1.85$, $\eta_s = 1.6524$; (*b*), $r_s = 3.75$, $\eta_s = 4.9087$; and (*c*), $r_s = 3.0$ and $\eta_s = 27.0$. These orbits do not reach the singularity. The unit of length along the coordinate axes is M; and the chosen Kerr parameter a is 0.8.

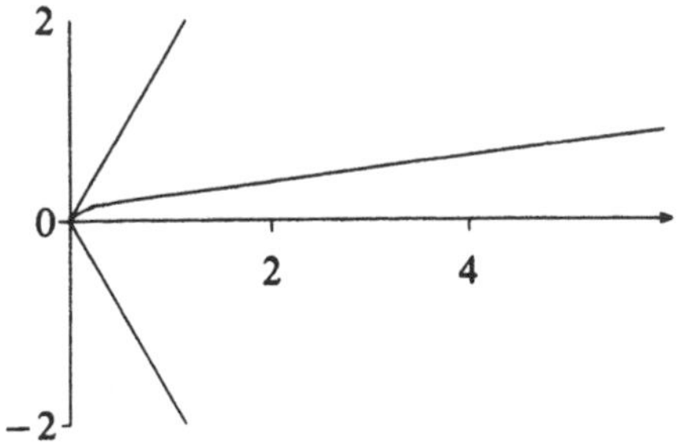

FIG. 36. An example of a null geodesic with $\eta = 0$ in the (r, θ)-plane. These are the only orbits, not on the equatorial plane, which terminate at the singularity. The unit of length along the coordinate axes is M; and the chosen Kerr parameter a is 0.8.

of the first and the second kind) or join together as a single contiguous trajectory.

The case $\eta = 0$ is of some interest. As we have seen, the orbits of constant radius, $r = r_{ph}^{(\pm)}$, with $\eta = 0$, lie entirely in the equatorial plane. Also, all orbits in the equatorial plane are characterized by the Carter constant being zero. But *not all orbits with $\eta = 0$ lie in the equatorial plane.* For, when $\eta = 0$, the r-motion is, quite generally, determined by (cf. equation (214))

$$\int \frac{dr}{\sqrt{R}} = \int \frac{dr}{\{r[r^3 + (a^2 - \xi^2)r + 2M(a-\xi)^2]\}^{1/2}}, \tag{236}$$

while the θ-motion is determined by equation (203). The integral (236) can be reduced to a Jacobian elliptic integral; and the trajectory in the (r, θ)-plane is expressible in the form

$$\left.\begin{aligned}
&\frac{1}{\sqrt{(a^2-\xi^2)}}\operatorname{sech}^{-1}\frac{\mu}{\mu_{max}} = \pm\frac{1}{\sqrt{(AB)}}F(\psi, k),\\
&\text{where}\\
&\mu_{max} = (1-\xi^2/a^2)^{1/2}, \quad k^2 = \frac{(A+B)^2 - r_1^2}{4AB}, \quad \cos\psi = \frac{(A-B)r + r_1 A}{(A+B)r + r_1 A},\\
&A^2 = r_1^2 + (a^2-\xi^2); \quad B^2 = 3r_1^2 + (a^2-\xi^2),\\
&r_1 = -[\tfrac{4}{3}(a^2-\xi^2)]^{1/2}\sinh\vartheta \quad\text{and}\quad \sinh 3\vartheta = \left[27M^2\frac{a-\xi}{(a+\xi)^3}\right]^{1/3}.
\end{aligned}\right\} \tag{237}$$

It is clear from the various definitions that these trajectories exist only for $\xi \leqslant a$. For orbits in the equatorial plane, the restriction $\xi \leqslant a$ is equivalent to the restriction $D \leqslant a$ (cf. the definitions of D and ξ in equations (73) and (189)). *The existence of three-dimensional orbits, with $\eta = 0$ and $\xi \leqslant a$, is the*

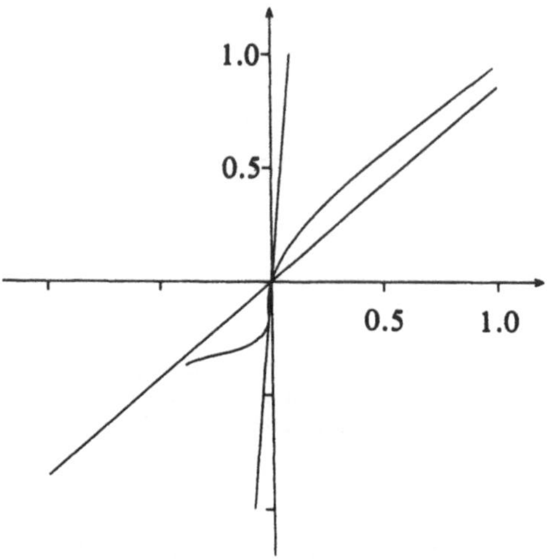

(a)

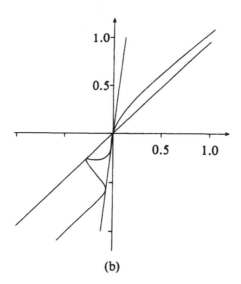

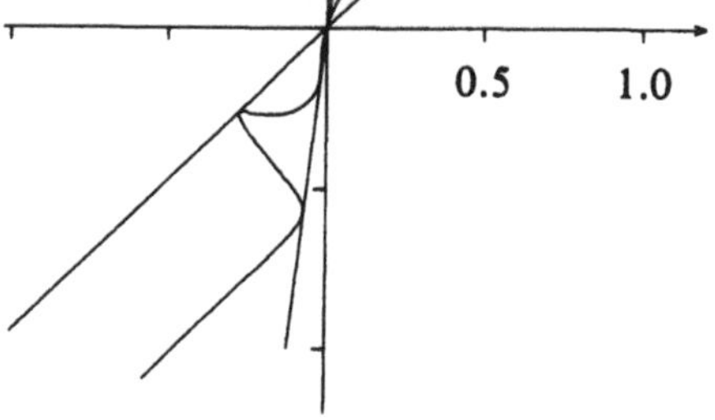

(b)

FIG. 37. Two examples of orbits with negative η which extend into the domain of negative r: (*a*), $\eta = -0.3$ and $\xi = -0.05$; and (*b*), $\eta = -0.3$ and $\xi = -0.07$. These orbits are confined between two angles, θ_{max} and θ_{min}. While in the (r, θ)-plane these orbits pass through zero, they do not in fact meet the singularity: in the Kerr–Schild coordinates of Fig. 25 they smoothly pass through the disc avoiding the singularity. (The chosen Kerr parameter a is 0.8.)

manifestation of the instability of the orbits in the equatorial plane with impact parameters $D < a$. It is also clear that all these trajectories terminate at the singularity and do not have a natural extension into the domain of negative r (see Fig. 36 where examples of trajectories derived from equation (237) are shown).

Finally, we shall consider the class of orbits with $\eta < 0$. These orbits are of interest since they explore regions of negative r in a natural way and make the enlargement of the manifold to include these regions physically necessary. But the conditions for their occurrence are stringent: $|\eta| < a^2$ and $0 < |\xi| \leqslant |a| - \sqrt{|\eta|}$. But in view of the interest in these orbits, the r- and the θ-motions were numerically integrated by Garret Toomey and the results of his integrations are shown in Fig. 37.

(*d*) *The case* $a = M$

While it has been our judgement (contrary to the commonly held view) not to give any special attention to a consideration of 'what happens' when $a = M$, we shall briefly note the limiting forms which the formulae of the preceding section take when $a = M$. (We have already noticed in §61(*b*), in the discussion following equation (142), the inappropriateness of the coordinate r for considerations involving matters close to the event horizon when $a \to M$.)

When $a = M$, equations (224) and (225) become

$$\xi = \frac{1}{M}(-r^2 + 2Mr + M^2); \quad \eta = \frac{r^3}{M^2}(4M - r), \tag{238}$$

and we can obtain the locus $\eta(\xi)$ explicitly. Thus, solving for r in terms of ξ, we obtain

$$r(\xi) = M + (2M^2 - M\xi)^{1/2}; \tag{239}$$

and inserting this value in the expression for η, we obtain

$$\eta(\xi) = \frac{1}{M^2}[M + (2M^2 - M\xi)^{1/2}]^3[3M - (2M^2 - M\xi)^{1/2}], \tag{240}$$

where it may be recalled that

$$\alpha = +\xi \operatorname{cosec}\theta_0 \quad \text{and} \quad \beta = (\eta + a^2\cos^2\theta_0 - \xi^2\cot^2\theta_0)^{1/2}. \tag{241}$$

The apparent shape of the black hole, for an observer at infinity, will be determined by the critical locus, (ξ, η), projected on the 'celestial sphere' even as the unstable circular photon-orbit of radius $3M$ (with the impact parameter, $3M\sqrt{3}$) determines the shape of the Schwarzschild black-hole. In the present instance, we shall simply have to transcribe the locus (ξ, η) into a closed curve in the (α, β)-plane. Thus, for an observer on the equatorial plane, viewing the

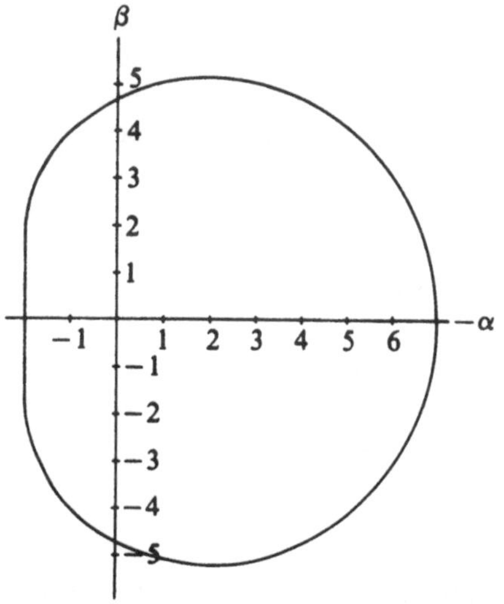

FIG. 38. The apparent shape of an extreme ($a = M$) Kerr black-hole as seen by a distant observer in the equatorial plane, if the black hole is in front of a source of illumination with an angular size larger than that of the black hole. The unit of length along the coordinate axes α and β (defined in equation (241) is M.

black hole from infinity, the apparent shape will be determined by

$$(\alpha, \pm\beta) = [\xi, \sqrt{\eta}(\xi)]. \tag{242}$$

This locus is exhibited in Fig. 38.

(e) *The propagation of the direction of polarization along a null geodesic*

It is customary in general relativity to describe the propagation of light in the language of geometrical optics. The basic notion underlying this description is that of *light rays* defined as trajectories orthogonal to the spreading wave-fronts. Precisely, a light ray is a curve in space-time whose tangent vector along any point of it is in the direction of the average Poynting-vector. These curves, often described as the trajectories of photons, are the null geodesics of general relativity. In our considerations of the null geodesics in this and in the earlier chapters, we have adhered to this picture.

Another characteristic of light, which has acquired some interest in astrophysics, in the context exactly of black holes, is that of *polarization*. The reason for this interest is that the radiation emerging from accretion discs surrounding black holes is expected to be polarized. And the question arises as to how one may relate the state of polarization of the radiation received by an observer at infinity with the state of polarization as it emerges from the accretion disc. The question is simply answered. The state of polarization is described by the direction of the electric vector in the plane transverse to the direction of propagation, i.e., of the wave-vector $\boldsymbol{k}$; and the electric vector, orthogonal to $\boldsymbol{k}$, is parallelly propagated along the light ray, i.e., along the null geodesic described by $\boldsymbol{k}$. Formulated in this fashion, it is clear that in Kerr geometry the question is answered directly by identifying the vector $\boldsymbol{f}$, in the Walker–Penrose theorem of §60, with the electric vector and taking advantage of the conservation of the complex quantity, K_s, defined in equation (12), along the null geodesic.

In the Kerr space-time (cf. Ch. 6, equation 180))

$$\Psi_2 = -M(r - ia\cos\theta)^{-3}; \tag{243}$$

and the Walker–Penrose conserved quantity is

$$[(\boldsymbol{k}\cdot\boldsymbol{l})(\boldsymbol{f}\cdot\boldsymbol{n}) - (\boldsymbol{k}\cdot\boldsymbol{n})(\boldsymbol{f}\cdot\boldsymbol{l}) - (\boldsymbol{k}\cdot\boldsymbol{m})(\boldsymbol{f}\cdot\bar{\boldsymbol{m}}) + (\boldsymbol{k}\cdot\bar{\boldsymbol{m}})(\boldsymbol{f}\cdot\boldsymbol{m})](r - ia\cos\theta) = K_2 + iK_1, \quad \text{(say)}. \tag{244}$$

With the known expressions for the basis vectors given in Chapter 6, equation (173), we find

$$(\boldsymbol{k}\cdot\boldsymbol{l})(\boldsymbol{f}\cdot\boldsymbol{n}) - (\boldsymbol{k}\cdot\boldsymbol{n})(\boldsymbol{f}\cdot\boldsymbol{l}) = (k^t f^r - k^r f^t) + a(k^r f^\varphi - k^\varphi f^r)\sin^2\theta = A \quad \text{(say)}, \tag{245}$$

and

$$(\boldsymbol{k}\cdot\boldsymbol{m})(\boldsymbol{f}\cdot\bar{\boldsymbol{m}}) - (\boldsymbol{k}\cdot\bar{\boldsymbol{m}})(\boldsymbol{f}\cdot\boldsymbol{m}) = i[(r^2+a^2)(k^\varphi f^\theta - k^\theta f^\varphi) - a(k^t f^\theta - k^\theta f^t)]\sin\theta = iB \quad \text{(say)}. \tag{246}$$

With these definitions, the conservation law, expressed by equation (244), after multiplication by $(r + ia\cos\theta)$, gives

$$\rho^2(A + iB) = (r + ia\cos\theta)(K_2 + iK_1); \tag{247}$$

and equating the real and the imaginary parts, we obtain

$$A = \frac{1}{\rho^2}(K_2 r - aK_1\cos\theta) \tag{248}$$

and

$$B = \frac{1}{\rho^2}(K_1 r + aK_2\cos\theta). \tag{249}$$

Besides, the orthogonality of $\boldsymbol{k}$ and $\boldsymbol{f}$ requires

$$\left(1 - \frac{2Mr}{\rho^2}\right)k^t f^t + \frac{2aMr\sin^2\theta}{\rho^2}(k^t f^\varphi + k^\varphi f^t) - \rho^2 k^\theta f^\theta - \frac{\rho^2}{\Delta}k^r f^r - \left(r^2 + a^2 + \frac{2a^2 Mr}{\rho^2}\sin^2\theta\right)k^\varphi f^\varphi \sin^2\theta = 0. \tag{250}$$

And finally, the geodesic equations (183) and (184) give

$$\rho^2 k^r = \sqrt{R}, \quad \rho^2 k^\theta = \sqrt{\Theta},$$

$$\rho^2 k^t = \frac{1}{\Delta}[(r^2+a^2)^2 - a^2\Delta\sin^2\theta - 2aMr\xi],$$

and

$$\rho^2 k^\varphi = \frac{1}{\Delta}[2aMr + (\rho^2 - 2Mr)\xi\,\mathrm{cosec}^2\,\theta]. \tag{251}$$

With the components of $\boldsymbol{k}$ given by these equations, equations (248)–(250) provide three equations for determining the components of f; and the three equations will just suffice since f is determinate only *modulo* the null vector $\boldsymbol{k}$.

The principal interest in the foregoing equations is to relate the direction of f, given by some physical theory, at some point in the neighbourhood of the black hole to the direction of f at infinity. Accordingly, it will suffice to consider the asymptotic form of the solution of equations (248)–(250).

The asymptotic form of $\boldsymbol{k}$, as one may deduce from equations (251), is given by

$$k^r \to 1, \quad k^t \to 1, \quad k^\theta \to \frac{\beta}{r^2}, \qquad \text{and} \qquad k^\varphi \to \frac{\xi}{r^2}\mathrm{cosec}^2\,\theta_0, \tag{252}$$

where (cf. equation (192))

$$\beta^2 = \eta + a^2\cos^2\theta_0 - \xi^2\cot^2\theta_0. \tag{253}$$

Since f is determinate only *modulo* $\boldsymbol{k}$, there is no loss of generality in setting

$$f^t = 0. \tag{254}$$

The orthogonality condition (250), now, gives

$$-r^2 k^\theta f^\theta - k^r f^r - (r^2\sin^2\theta_0)k^\varphi f^\varphi = 0 \quad (r\to\infty), \tag{255}$$

or, by equations (252),

$$f^r = -\beta f^\theta - \xi f^\varphi. \tag{256}$$

Similarly, the asymptotic forms of equations (248) and (249) give

$$f^r + (a\sin^2\theta_0)f^\varphi = K_2 r^{-1} \tag{257}$$

and

$$[r^2(k^\varphi f^\theta - k^\theta f^\varphi) - af^\theta]\sin\theta_0 = K_1 r^{-1}. \tag{258}$$

Eliminating f^r from equation (257) with the aid of equation (256) and substituting for k^θ and k^φ from equations (252), we obtain

$$-\beta f^\theta - (\xi - a\sin^2\theta_0)f^\varphi = K_2 r^{-1} \tag{259}$$

and

$$(\xi\,\mathrm{cosec}\,\theta_0 - a\sin\theta_0)f^\theta - (\beta\sin\theta_0)f^\varphi = K_1 r^{-1}, \tag{260}$$

or, letting

$$\gamma = \xi \operatorname{cosec}\theta_0 - a\sin\theta_0, \tag{261}$$

we have

$$\left.\begin{aligned} \beta f^{\theta} + (\gamma\sin\theta_0)f^{\varphi} &= -K_2 r^{-1} \\ \text{and} \qquad \gamma f^{\theta} - (\beta\sin\theta_0)f^{\varphi} &= +K_1 r^{-1}. \end{aligned}\right\} \tag{262}$$

Solving these equations for f^{θ} and f^{φ}, we find

$$\left.\begin{aligned} f^{\varphi}\sin\theta_0 &= -\frac{1}{r(\beta^2+\gamma^2)}(\beta K_1 + \gamma K_2) \\ \text{and} \qquad f^{\theta} &= -\frac{1}{r(\beta^2+\gamma^2)}(\beta K_2 - \gamma K_1). \end{aligned}\right\} \tag{263}$$

If we now denote by $\mathscr{E}^{\theta}$ and $\mathscr{E}^{\varphi}$ the components of the electric vector, $\boldsymbol{\mathscr{E}}$, in the transverse plane which determines the state of polarization at infinity, then we may identify

$$\mathscr{E}^{\varphi} = -(r\sin\theta_0)f^{\varphi} \qquad \text{and} \qquad \mathscr{E}^{\theta} = -rf^{\theta}, \tag{264}$$

and equations (263) give

$$\mathscr{E}^{\varphi} = \frac{1}{\beta^2+\gamma^2}(\beta K_1 + \gamma K_2) \qquad \text{and} \qquad \mathscr{E}^{\theta} = \frac{1}{\beta^2+\gamma^2}(\beta K_2 - \gamma K_1). \tag{265}$$

We can clearly arrange that

$$K_1^2 + K_2^2 = 1. \tag{266}$$

Then, $(\mathscr{E}^{\varphi}, \mathscr{E}^{\theta})$ is a unit vector in the transverse plane which determines the state of polarization: the Stokes parameters (normalized to unit intensity), commonly used to describe the state of polarization, can be readily expressed in terms of $(\mathscr{E}^{\varphi}, \mathscr{E}^{\theta})$.

64. The time-like geodesics

For time-like geodesics, we can put $\delta_1 = 1$ in equations (185) and (186); and the equation determining the projection of the geodesic on the (r, θ)-plane is, again,

$$\int^{r} \frac{dr}{\sqrt{R}} = \int^{\theta} \frac{d\theta}{\sqrt{\Theta}}, \tag{267}$$

but now with the definitions

$$R = r^4 + (a^2 - \xi^2 - \eta)r^2 + 2M[\eta + (\xi - a)^2]r - a^2\eta - r^2\Delta/E^2 \tag{268}$$

and

$$\Theta = \eta + a^2(1 - E^{-2})\cos^2\theta - \xi^2\cot^2\theta, \tag{269}$$

where $\xi = L_z/E$, $\eta = \mathscr{Q}/E^2$, and E is the energy per unit mass. Also, we have written R and Θ in place of R/E^2 and Θ/E^2.

(a) *The θ-motion*

Considering first the θ-motion specified by the integral over $\Theta^{-1/2}$, we replace, as before, θ by $\mu = \cos\theta$ as the variable of integration. The right-hand side of equation (267) then becomes

$$\int^{\mu} \frac{d\mu}{\sqrt{\Theta_\mu}}, \tag{270}$$

where

$$\Theta_\mu = \eta - [\xi^2 + \eta - a^2(1 - E^{-2})]\mu^2 - a^2(1 - E^{-2})\mu^4. \tag{271}$$

It is clear that we must distinguish three cases: the *bound orbits* with $E^2 < 1$, the *marginally bound orbits* with $E^2 = 1$, and the *unbound orbits* with $E^2 > 1$.

For *bound orbits* with $E^2 < 1$, it is convenient to replace a^2 by

$$\alpha^2 = a^2(E^{-2} - 1) > 0 \tag{272}$$

and write Θ_μ in the form

$$\Theta_\mu = \eta - (\xi^2 + \eta + \alpha^2)\mu^2 + \alpha^2\mu^4. \tag{273}$$

From this expression for Θ_μ and the requirements $\Theta_\mu \geqslant 0$ and $0 \leqslant \mu^2 \leqslant 1$, we conclude that negative values of η are not allowed; and that the range of μ^2 is between zero and the smaller of the two roots of the equation

$$\eta - (\xi^2 + \eta + \alpha^2)\mu^2 + \alpha^2\mu^4 = 0 \quad (\eta > 0). \tag{274}$$

The integral (270) can now be reduced to an elliptic integral of the first kind in the usual manner. We find

$$\int_0^{\mu} \frac{d\mu}{\sqrt{\Theta_\mu}} = \frac{1}{\alpha}\int_0^{\mu} \frac{d\mu}{\sqrt{[(\mu_+^2 - \mu^2)(\mu_-^2 - \mu^2)]}} = \frac{1}{\alpha\mu_+}F(\psi, k), \tag{275}$$

where

$$\mu_\pm^2 = \frac{1}{2\alpha^2}\{(\xi^2 + \eta + \alpha^2) \pm [(\xi^2 + \eta + \alpha^2)^2 - 4\alpha^2\eta]^{1/2}\}, \tag{276}$$

$$k = \mu_-/\mu_+ \qquad \text{and} \qquad \sin\psi = \mu/\mu_-. \tag{277}$$

For *marginally bound orbits* with $E^2 = 1$,

$$\Theta_\mu = \eta - (\xi^2 + \eta)\mu^2. \tag{278}$$

Again, negative values of η are not allowed and

$$\int^{\mu} \frac{d\mu}{\sqrt{\Theta_\mu}} = -\frac{1}{\sqrt{(\xi^2 + \eta)}}\cos^{-1}\left[\left(\frac{\xi^2 + \eta}{\eta}\right)^{1/2}\mu\right]. \tag{279}$$

For *unbound orbits* with $E^2 > 1$, the discussion for the case of the null geodesics in §63(*a*) applies with the replacement of a^2 by

$$\alpha^2 = a^2(1 - E^{-2}) > 0. \tag{280}$$

In summary then, the bound and the marginally bound orbits must necessarily cross the equatorial plane and oscillate about it. The unbound orbits are very much like the null geodesics: orbits with $\eta > 0$ which intersect the equatorial plane, orbits with $\eta < 0$ which do not intersect the equatorial plane, are confined to cones, and extend into the regions of negative r, and orbits with $\eta = 0$.

(*b*) *The r-motion*

Turning next to the r-motion specified by the integral over $R^{-1/2}$, we consider in accordance with our standard procedure, the special orbits which provide not only a basis for a classification of the orbits but also simple integrable cases which sufficiently illustrate the nature of the orbits to be expected. In the present instance, the special orbits are those for which the radial coordinate remains constant. The conditions for their occurrence are

$$R = 0 \qquad \text{and} \qquad \partial R/\partial r = 0. \tag{281}$$

The constraints on the constants of the motion, ξ, η, and E^2, implied by the conditions (281), can be derived exactly as in §63(*c*) except that we must now allow for the additional term, $-r^2\Delta/E^2$, in the definition of R. Following, then, the same steps, we now find that equations (216) and (217) are replaced by

$$3r^4 + r^2a^2 - \eta(r^2 - a^2) - \frac{r^2}{E^2}(3r^2 - 4Mr + a^2) = r^2\xi^2 \tag{282}$$

and

$$r^4 - a^2Mr + \eta(a^2 - Mr) - \frac{r^3}{E^2}(r - M) = rM(\xi^2 - 2a\xi); \tag{283}$$

and, as before, we solve these equations for ξ and η.

Eliminating η from equations (282) and (283) we obtain (in place of equation (218))

$$a^2(r - M)\xi^2 - 2aM(r^2 - a^2)\xi - \left\{(r^2 + a^2)[r(r^2 + a^2) - M(3r^2 - a^2)] - \frac{r}{E^2}\Delta^2\right\} = 0; \tag{284}$$

and the solution of this equation is (cf. equation (220))

$$\xi = \frac{1}{a(r - M)}\left\{M(r^2 - a^2) \pm r\Delta\left[1 - \frac{1}{E^2}\left(1 - \frac{M}{r}\right)\right]^{1/2}\right\}; \tag{285}$$

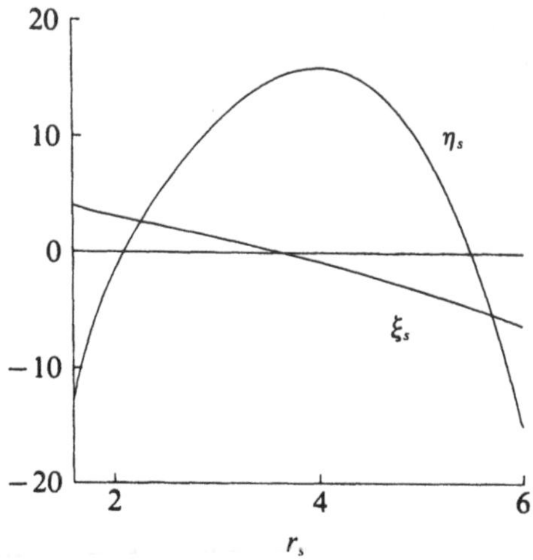

FIG. 39. The (ξ_s, η_s) locus defined by equations (285) and (286) for marginally-bound time-like orbits with $E^2 = 1.0$. The unit of length along the abscissa is M; and the chosen Kerr parameter a is 0.8.

the corresponding solution for η is (cf. equation (225)) (see Fig. 39)

$$\eta a^2(r-M) = \frac{r^3}{r-M}[4a^2M - r(r-3M)^2] + \frac{r^2}{E^2}[r(r-2M)^2 - a^2M] - \frac{2r^3M}{r-M}\Delta\left\{1 \pm \left[1 - \frac{1}{E^2}\left(1 - \frac{M}{r}\right)\right]^{1/2}\right\}. \tag{286}$$

Returning to the consideration of the r-motions, let the constants of the motion, ξ and η, have the values ξ_s and η_s appropriate to a particle with energy E, per unit mass, describing an orbit of constant radius r_s, i.e., ξ_s and η_s have the values given by equations (285) and (286) for $r = r_s$ and the assigned E^2. (It is of course necessary that for the chosen value of r_s and E^2, η_s is positive for bound and marginally bound orbits.) For this choice of the constants of the motion, r_s is a double root of $R = 0$ and, in consequence, R can be reduced to the form

$$R = (r-r_s)^2\left\{r^2\left(1 - \frac{1}{E^2}\right) + 2rr_s\left[1 - \frac{1}{E^2}\left(1 - \frac{M}{r_s}\right)\right] - \frac{a^2}{r_s^2}\eta_s\right\}. \tag{287}$$

Now, letting

$$x = (r-r_s)^{-1}, \tag{288}$$

the integral which we must equate with one or the other of the expressions for

$$\int\frac{\mathrm{d}\mu}{\sqrt{\Theta_\mu}} \tag{289}$$

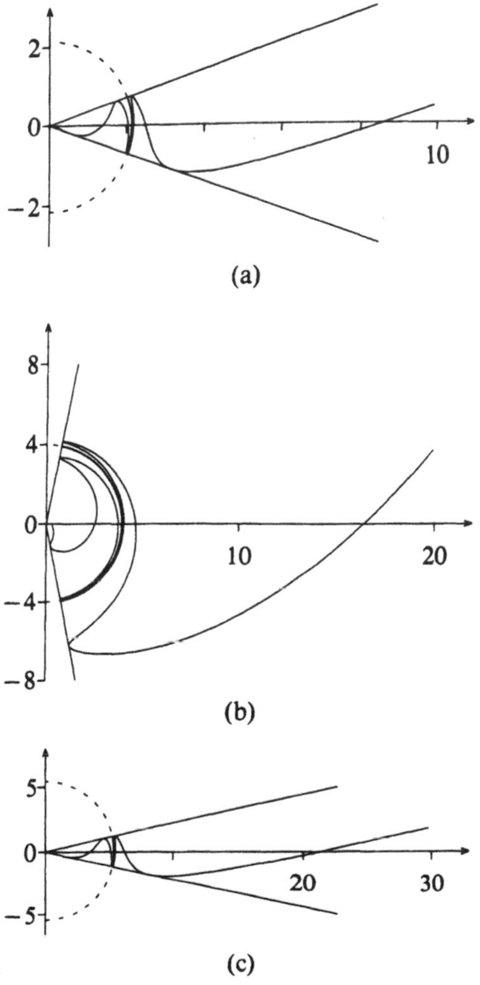

FIG. 40. Marginally-bound, critical, time-like geodesics in the (r, θ)-plane described by equations (279) and (292) for three values of r_s: (*a*), 2.154; (*b*), 4.0; and (*c*), 5.44. The unit of length along the coordinate axes is M; and the chosen Kerr parameter a is 0.8.

given in §(*a*) above is

$$\int \frac{\mathrm{d}r}{\sqrt{R}} = -\int \mathrm{d}x \left\{ \left(1 - \frac{1}{E^2}\right) + \left[4r_s\left(1 - \frac{1}{E^2}\right) + \frac{2M}{E^2}\right]x + \left[3r_s^2\left(1 - \frac{1}{E^2}\right) - \frac{a^2\eta_s}{r_s^2} + \frac{2Mr_s}{E^2}\right]x^2 \right\}^{-1/2} \tag{290}$$

—an elementary integral whose values for the different cases, $E^2 > 1$, $E^2 = 1$, and $E^2 < 1$, can be readily written down. Thus, with the definition

$$F(x) = \alpha + \beta x + \gamma x^2, \tag{291}$$

where α, β, and γ are the coefficients of the quadratic in x which appears in the integral (290), we have

$$\left.\begin{aligned} \int \frac{dr}{\sqrt{R}} &= \pm\frac{1}{\sqrt{\gamma}} \lg[2\sqrt{(\gamma F)} + 2\gamma x + \beta] \quad (\gamma > 0), \\ &= \pm\frac{1}{\sqrt{\gamma}} \sinh^{-1}\frac{2\gamma x + \beta}{\sqrt{D}} \quad (\gamma > 0,\ D > 0), \\ &= \mp\frac{1}{\sqrt{-\gamma}} \sin^{-1}\frac{2\gamma x + \beta}{\sqrt{-D}} \quad (\gamma < 0,\ D < 0), \end{aligned}\right\} \tag{292}$$

where

$$D = -4\left\{\left[r_s\left(1 - \frac{1}{E^2}\right) + \frac{M}{E^2}\right]^2 + \frac{\alpha^2 \eta_s}{r_s^2}\left(1 - \frac{1}{E^2}\right)\right\}. \tag{293}$$

Examples of orbits derived from the equations we have derived are illustrated in Fig. 40.

As in the case of the null geodesics, orbits of constant radius r, with $\eta = 0$, are necessarily confined to the equatorial plane. This follows from the fact that equations (282) and (283), which, for $\eta = 0$, become

$$3r^2 + a^2 - \frac{1}{E^2}(3r^2 - 4Mr + a^2) = \xi^2$$

and

$$r^3 - \frac{r^2}{E^2}(r - M) = M(\xi - a)^2, \tag{294}$$

are the same as equations (108) and (109) which describe the circular orbits in the equatorial plane.

65. The Penrose process

In the last chapter (§58(*a*)) we have drawn attention to the significance of the surface on which g_{tt} vanishes and the fact that, in Kerr geometry, this surface does not coincide with the event horizon except at the poles. In the toroidal space between the two surfaces, i.e., in the ergosphere, the Killing vector $\partial/\partial t$ becomes space-like and, likewise, the conserved component, p_t, of the four-momentum. The energy of a particle in this region of space, as perceived by an observer at infinity, can be negative. This last fact has important consequences: it makes possible, as Penrose first showed, physical processes which, in effect, extract energy and angular momentum from the black hole. In this section, we shall consider the nature of these processes and the limits on the energy that can be extracted.

For a consideration of the Penrose processes, it is convenient to have limits on the energy which a particle, at a specified location, can have. From equations (183) and (185) it follows that the limit is set by

$$[r(r^2+a^2)+2a^2M]E^2-(4aML_z)E \\ -L_z^2(r-2M)-(\delta_1 r+\mathscr{Q}/r)\Delta=0, \tag{295}$$

since this equation presupposes that there is no contribution to E from the kinetic energy derived from $\dot{r}^2$. Solving equation (295) for E and L_z, separately, we obtain

$$E=\frac{1}{[r(r^2+a^2)+2a^2M]}\Big[\!\Big[2aML_z\pm\Delta^{1/2}\{r^2L_z^2 \\ +[r(r^2+a^2)+2a^2M](\delta_1 r+\mathscr{Q}r^{-1})\}^{1/2}\Big]\!\Big] \tag{296}$$

and

$$L_z=\frac{1}{r-2M}\{-2aME\pm\Delta^{1/2}[r^2E^2-(r-2M)(\delta_1 r+\mathscr{Q}r^{-1})]^{1/2}\}, \tag{297}$$

where, in deriving these solutions, we have made use of the identity

$$\Delta r^2-4a^2M^2=[r^2(r^2+a^2)+2a^2Mr]\left(1-\frac{2M}{r}\right). \tag{298}$$

The circumstances under which E, as perceived by an observer at infinity, can be negative can be inferred from equation (296). First, it is important to observe that a particle of unit mass, at rest at infinity, must, in accordance with our conventions, be assigned an energy $E=1$; and to be consistent with this requirement, we must, in the present context, choose the positive sign on the right-hand side of equation (296). With this choice of the sign, it is clearly necessary that for $E<0$,

$$L_z<0, \tag{299}$$

and

$$4a^2M^2L_z^2>\Delta\{r^2L_z^2+[r^2(r^2+a^2)+2a^2Mr](\delta_1+\mathscr{Q}/r^2)\}. \tag{300}$$

With the aid of the identity (298), this inequality can be brought to the form

$$[r^2(r^2+a^2)+2a^2Mr]\left\{\left(1-\frac{2M}{r}\right)L_z^2+\Delta\left(\delta_1+\frac{\mathscr{Q}}{r^2}\right)\right\}<0. \tag{301}$$

It follows that

$$\left.\begin{aligned} &E < 0, \quad \text{if and only if} \quad L_z < 0 \\ \text{and} \\ &(r-2M) < -\frac{\Delta r}{L_z^2}\left(\delta_1 + \frac{\mathscr{Q}}{r^2}\right). \end{aligned}\right\} \tag{302}$$

We conclude that only counter-rotating particles can have negative energy; and, on the equatorial plane, it is further necessary that $r < 2M$, i.e., the particle be inside the ergosphere.

(*a*) *The original Penrose process*

The process originally conceived by Penrose to illustrate the manner in which energy may be extracted from the Kerr black-hole is the following.

A particle, at rest at infinity, arrives by a geodesic in the equatorial plane, at a point $r(< 2M)$ where it has a turning point (so that $\dot{r} = 0$). At r, it decays into two photons, one of which crosses the event horizon and is 'lost' while the other escapes to infinity. We arrange that the photon which crosses the event horizon has negative energy and the photon which escapes to infinity has an energy in excess of the particle which arrived from infinity.

Let

$$E^{(0)} = 1, \quad L_z^{(0)}; \quad E^{(1)}, L_z^{(1)}; \qquad \text{and} \qquad E^{(2)}, \quad L_z^{(2)}, \tag{303}$$

denote the energies and the angular momenta of the particle arriving from infinity and of the photons which cross the event horizon and escape to infinity, respectively.

Since the particle from infinity arrives at r by a time-like geodesic and has a turning point at r, its angular momentum, $L_z^{(0)}$, can be inferred from equation (297) by setting $\delta_1 = 1$, $E = 1$, and $\mathscr{Q} = 0$. Thus,

$$L_z^{(0)} = \frac{1}{r-2M}[-2aM + \sqrt{(2Mr\Delta)}] = \alpha^{(0)} \quad \text{(say)}. \tag{304}$$

Similarly, by setting $\delta_1 = \mathscr{Q} = 0$ and choosing, respectively, the negative and the positive sign in equation (297), we can obtain the relations between the energies and the angular momenta of the photon which crosses the event horizon and the photon which escapes to infinity. We find

$$L_z^{(1)} = \frac{1}{r-2M}(-2aM - r\sqrt{\Delta})E^{(1)} = \alpha^{(1)}E^{(1)} \quad \text{(say)} \tag{305}$$

and

$$L_z^{(2)} = \frac{1}{r-2M}(-2aM + r\sqrt{\Delta})E^{(2)} = \alpha^{(2)}E^{(2)} \quad \text{(say)}. \tag{306}$$

The conservation of energy and angular momentum now requires that

$$E^{(1)} + E^{(2)} = E^{(0)} = 1 \tag{307}$$

and

$$L_z^{(1)} + L_z^{(2)} = \alpha^{(1)}E^{(1)} + \alpha^{(2)}E^{(2)} = L_z^{(0)} = \alpha^{(0)}. \tag{308}$$

Solving these equations, we find

$$E^{(1)} = \frac{\alpha^{(0)} - \alpha^{(2)}}{\alpha^{(1)} - \alpha^{(2)}} \quad \text{and} \quad E^{(2)} = \frac{\alpha^{(1)} - \alpha^{(0)}}{\alpha^{(1)} - \alpha^{(2)}}, \tag{309}$$

or, substituting for $\alpha^{(0)}$, $\alpha^{(1)}$, and $\alpha^{(2)}$ from equations (304)–(306), we find

$$E^{(1)} = -\frac{1}{2}\left(\sqrt{\frac{2M}{r}} - 1\right) \quad \text{and} \quad E^{(2)} = +\frac{1}{2}\left(\sqrt{\frac{2M}{r}} + 1\right). \tag{310}$$

The photon escaping to infinity has, indeed, an energy in excess of $E^{(0)} = 1$ so long as $r < 2M$ (as we have postulated). The energy, ΔE, that has been gained is

$$\Delta E = \frac{1}{2}\left(\sqrt{\frac{2M}{r}} - 1\right) = -E^{(1)}. \tag{311}$$

It is apparent from equation (311), that by the process we have described, the maximum gain in energy that can be achieved is when the particle, arriving from infinity, has a turning point at the event horizon. Therefore

$$\Delta E \leqslant \frac{1}{2}\left(\sqrt{\frac{2M}{r_+}} - 1\right). \tag{312}$$

Since the minimum value of r_+ is M (when $a^2 = M^2$)

$$\Delta E \leqslant \tfrac{1}{2}(\sqrt{2} - 1) = 0.207. \tag{313}$$

An alternative form of the inequality (312), which we shall find instructive, is to express it in terms of the 'irreducible mass'

$$M_{\text{irr}} = \tfrac{1}{2}(r_+^2 + a^2)^{1/2} = (\tfrac{1}{2}Mr_+)^{1/2}. \tag{314}$$

(We shall explain in §(d) below why this mass is called 'irreducible'.) By virtue of this definition, the inequality (312) can be written in the form

$$\Delta E \leqslant \frac{1}{2}\left(\frac{M}{M_{\text{irr}}} - 1\right). \tag{315}$$

A more general inequality than the foregoing can be derived directly from equation (296). This equation requires

$$[r(r^2 + a^2) + 2a^2M]E - 2aML_z \geqslant 0. \tag{316}$$

In particular,

$$[r_+(r_+^2 + a^2) + 2a^2M]E - 2aML_z \geqslant 0. \tag{317}$$

But

$$r_+(r_+^2 + a^2) + 2a^2M = 2M(r_+^2 + a^2) = 4M^2r_+. \tag{318}$$

Therefore,

$$2Mr_+E - aL_z \geqslant 0. \tag{319}$$

We shall return to this inequality in §(*d*) below.

(*b*) *The Wald inequality*

There are limits to the energy that can be extracted by the Penrose process. We shall establish two inequalities (due to Wald and to Bardeen, Press, and Teukolsky) which indicate the nature and the origin of the limitations.

In the spirit of the process considered in §(*a*), we suppose that a particle, with a four-velocity U^i and specific energy E, breaks up into fragments. Let ε be the specific energy and u^i the four-velocity of one of the fragments. We seek limits on ε.

Choose an orthonormal tetrad-frame, $e_{(a)}{}^i$, in which U^i coincides with $e_{(0)}{}^i$ and the remaining (space-like) basis vectors are $e_{(\alpha)}{}^i$ ($\alpha = 1, 2, 3$):

$$e_{(0)}{}^i = U^i \quad \text{and} \quad e_{(\alpha)}{}^i \quad (\alpha = 1, 2, 3). \tag{320}$$

In this frame,

$$u^i = \gamma(U^i + v^{(\alpha)}e_{(\alpha)}{}^i), \tag{321}$$

where $v^{(\alpha)}$ are the spatial components of the three-velocity of the fragment

$$\gamma = (1 - |v|^2)^{-1/2}, \quad \text{and} \quad |v|^2 = v^{(\alpha)}v_{(\alpha)}. \tag{322}$$

We now suppose that the space-time allows a time-like Killing vector $\xi_i = \partial/\partial x^0$. Let its representation in the chosen tetrad-frame be

$$\xi_i = \xi_{(0)}U_i + \xi_{(\alpha)}e^{(\alpha)}{}_i \quad \text{and} \quad \xi^i = \xi^{(0)}U^i + \xi^{(\alpha)}e_{(\alpha)}{}^i. \tag{323}$$

From this representation of ξ_i, it follows that

$$E = \xi_i U^i = \xi_{(0)} = \xi^i U_i = \xi^{(0)}, \tag{324}$$

and

$$g_{00} = \xi^i\xi_i = \xi_{(0)}{}^2 - \xi_{(\alpha)}\xi^{(\alpha)} = E^2 - |\xi|^2. \tag{325}$$

We have thus the relation

$$|\xi|^2 = \xi_{(\alpha)}\xi^{(\alpha)} = E^2 - g_{00}. \tag{326}$$

From the representation (321) of the four-velocity of the fragment, we obtain

$$\varepsilon = \xi_i u^i = \gamma(\xi_{(0)} + v^{(\alpha)}\xi_{(\alpha)}) = \gamma(E + |v||\xi|\cos\vartheta), \tag{327}$$

where ϑ is the angle between the three-dimensional vectors $v^{(\alpha)}$ and $\xi_{(\alpha)}$. Making use of the relation (326), we can rewrite equation (327) in the form

$$\varepsilon = \gamma E + \gamma|v|(E^2 - g_{00})^{1/2}\cos\vartheta\,. \tag{328}$$

This equation clearly implies the inequality

$$\gamma E-\gamma|v|(E^2-g_{00})^{1/2}\leqslant \varepsilon\leqslant \gamma E+\gamma|v|(E^2-g_{00})^{1/2}. \tag{329}$$

This is Wald's inequality.

In Kerr geometry,

$$g_{00}=1-2Mr/\rho^2<0 \text{ in the ergosphere;} \tag{330}$$

and its lower bound is -1 which it attains at the event horizon for $a=M$ and $\theta=\pi/2$. Hence in Kerr geometry, we must under all conditions have

$$\gamma E-\gamma|v|(E^2+1)^{1/2}\leqslant \varepsilon\leqslant \gamma E+\gamma|v|(E^2+1)^{1/2}. \tag{331}$$

We have seen that the maximum energy that a particle describing a stable circular orbit can have is (cf. equation (146))

$$E_{\max}=3^{-1/2}. \tag{332}$$

For ε to be negative, it is therefore necessary that

$$|v|>\frac{E}{\sqrt{(E^2+1)}}=\tfrac{1}{2}. \tag{333}$$

In other words, the fragments must have relativistic energies before any extraction of energy by a Penrose process becomes possible.

It is of interest to compare the inequality (331) with what one would have in the framework of special relativity without the advantage of an ergosphere, namely,

$$\gamma E-\gamma|v|(E^2-1)^{1/2}\leqslant \varepsilon\leqslant \gamma E+\gamma|v|(E^2-1)^{1/2}. \tag{334}$$

It is clear that no very great gain can be achieved over that possible by more conventional processes in the framework of special relativity.

(c) *The Bardeen–Press–Teukolsky inequality*

Again, in the spirit of the process considered in §(*a*), let two particles with specific energies, E_+ and E_-, following two orbits, collide at some point. We ask for the lower bound on the magnitude of the relative three-velocity, $|w|$, between them.

Choose an orthonormal tetrad-frame,

$$e_{(0)}{}^i=U^i \qquad \text{and} \qquad e_{(\alpha)}{}^i \quad (\alpha=1,2,3), \tag{335}$$

in which the two orbits cross with equal and opposite three-velocities, $+v^{(\alpha)}$ and $-v^{(\alpha)}$ so that

$$|w|=\frac{2|v|}{1+|v|^2} \qquad \text{where} \qquad |v|^2=v^{(\alpha)}v_{(\alpha)}. \tag{336}$$

The representations of the four-velocities, $u_+{}^i$ and $u_-{}^i$ of the two particles in the chosen tetrad-frame, at the instant of collision, are

$$\left.\begin{aligned} u_+{}^i &= \gamma(U^i + v^{(\alpha)} e_{(\alpha)}{}^i) \\ \text{and} \qquad u_-{}^i &= \gamma(U^i - v^{(\alpha)} e_{(\alpha)}{}^i), \end{aligned}\right\} \tag{337}$$

where

$$\gamma = (1-|v|^2)^{-1/2}. \tag{338}$$

We now suppose that the space-time allows a time-like Killing vector $\xi_i(=\partial/\partial x^0)$. Let its representation in the chosen tetrad-frame be

$$\left.\begin{aligned} \xi^i &= \xi^{(0)} U^i + \xi^{(\alpha)} e_{(\alpha)}{}^i \\ \text{and} \qquad \xi_i &= \xi_{(0)} U_i + \xi_{(\beta)} e^{(\beta)}{}_i \quad (\xi^{(0)} = \xi_{(0)}). \end{aligned}\right\} \tag{339}$$

Then, by definition,

$$g_{00} = \xi^i \xi_i = \xi^{(0)} \xi_{(0)} - \xi_{(\alpha)} \xi^{(\alpha)} = \xi_{(0)}^2 - |\xi|^2, \tag{340}$$

so that

$$|\xi|^2 = \xi_{(0)}^2 - g_{00}. \tag{341}$$

The specific energies of the two particles at the instant of collision are given by

$$\left.\begin{aligned} E_+ &= \xi_i u_+{}^i = \gamma(\xi_{(0)} + v^{(\alpha)} \xi_{(\alpha)}) = \gamma(\xi_{(0)} + |v||\xi|\cos\vartheta) \\ \text{and} \qquad E_- &= \xi_i u_-{}^i = \gamma(\xi_{(0)} - v^{(\alpha)} \xi_{(\alpha)}) = \gamma(\xi_{(0)} - |v||\xi|\cos\vartheta), \end{aligned}\right\} \tag{342}$$

where ϑ is the angle between the 3-vectors $\xi^{(\alpha)}$ and $v_{(\alpha)}$. From the foregoing equations, it follows that

$$\gamma\xi_{(0)} = \tfrac{1}{2}(E_+ + E_-) \qquad \text{and} \qquad \gamma|v||\xi|\cos\vartheta = \tfrac{1}{2}(E_+ - E_-). \tag{343}$$

Accordingly,

$$\begin{aligned} (E_+ - E_-)^2 &= 4\gamma^2 |v|^2 |\xi|^2 \cos^2\vartheta \\ &= |v|^2 (4\gamma^2 \xi_{(0)}^2 - 4\gamma^2 g_{00}) \cos^2\vartheta \\ &= |v|^2 [(E_+ + E_-)^2 - 4\gamma^2 g_{00}] \cos^2\vartheta , \end{aligned} \tag{344}$$

where we have made use of the relations (341) and (343). Equation (344) clearly implies the inequality

$$(E_+ - E_-)^2 \leqslant |v|^2 [(E_+ + E_-)^2 - 4\gamma^2 g_{00}]; \tag{345}$$

or, remembering the definition of γ, we have

$$|v|^2 \left[(E_+ + E_-)^2 - \frac{4}{1-|v|^2} g_{00} \right] \geqslant (E_+ - E_-)^2, \tag{346}$$

or, alternatively,

$$-|v|^4(E_+ + E_-)^2 + 2|v|^2(E_+^2 + E_-^2 - 2g_{00}) - (E_+ - E_-)^2 \geqslant 0. \quad (347)$$

With the equality sign, the roots of equation (347) are

$$\frac{1}{(E_+ + E_-)^2}[\sqrt{(E_+^2 - g_{00})} \pm \sqrt{(E_-^2 - g_{00})}]^2. \quad (348)$$

We conclude that

$$|v| \geqslant \frac{1}{E_+ + E_-}|\sqrt{(E_+^2 - g_{00})} - \sqrt{(E_-^2 - g_{00})}|; \quad (349)$$

and the required lower bound on $|w|$ follows from equation (336); and the resulting inequality is that of Bardeen, Press, and Teukolsky.

If the particle with the energy E_+ is describing a stable circular orbit in the equatorial plane of a Kerr black-hole, then its maximum energy (as we have seen) is $3^{-1/2}$. With the most favourable values, $g_{00} = -1$ and $E_- = 0$, the inequality (349) gives

$$|v| > [\sqrt{(1+\tfrac{1}{3})} - 1]\sqrt{3} = 2 - \sqrt{3}; \quad (350)$$

and the corresponding inequality for $|w|$ is

$$|w| \geqslant \tfrac{1}{2}, \quad (351)$$

in agreement with the result (333) derived from Wald's inequality.

The principal conclusion to be drawn from the Wald and the Bardeen–Press–Teukolsky inequalities is that to achieve substantial energy extraction by the Penrose process, one must first accelerate the particle fragments to more than half the speed of light by hydrodynamical or other forces. The specific example considered in §(*a*) confirms this conclusion.

(*d*) *The reversible extraction of energy*

We return to the inequality (319) derived in §(*a*). The inequality becomes an equality only if the process considered takes place at the event horizon. In general, the gain in energy δM ($= E$) and the gain in the angular momentum δJ ($= L_z$), resulting from a particle with negative energy, $-E$, and an angular momentum, $-L_z$, arriving at the event horizon, are subject to the inequality

$$2r_+ M\delta M \geqslant a\delta J. \quad (352)$$

If we suppose that the process takes place 'adiabatically' so that the black hole evolves along the Kerr sequence, then

$$\delta J = \delta(aM) = M\delta a + a\delta M, \quad (353)$$

and the inequality (352) gives

$$(2Mr_+ - a^2)\delta M = r_+^2 \,\delta M \geqslant Ma\,\delta a. \tag{354}$$

Now, by the definition of the irreducible mass, M_{irr}, in equation (314),

$$M_{\text{irr}}^2 = \tfrac{1}{2}Mr_+ = \tfrac{1}{2}M[M + \sqrt{(M^2 - a^2)}]. \tag{355}$$

From this equation it follows that

$$\delta M_{\text{irr}}^2 = \frac{1}{2\sqrt{(M^2 - a^2)}}(r_+^2\,\delta M - Ma\,\delta a). \tag{356}$$

Therefore, the inequality (354) is equivalent to the restriction

$$\delta M_{\text{irr}}^2 \geqslant 0. \tag{357}$$

In other words, *by no continuous process can the irreducible mass of a black hole be decreased*—a result which justifies the designation. Since (cf. Ch. 6, equation (278))

$$\text{Surface area} = 4\pi(r_+^2 + a^2) = 16\pi M_{\text{irr}}^2, \tag{358}$$

we may also say that *by no continuous process can the surface area of a black hole be decreased.* The more general assertion, that "*no interaction, whatsoever, among black holes can result in a decrease of their total surface area*" is the *area theorem* of Hawking. The considerations we have set out, however, apply only to infinitesimal processes involving a single Kerr black-hole.

Since the best that can be achieved is to keep the irreducible mass unchanged, processes in which it remains constant are said to be *reversible.* It may also be noted that by virtue of the definition (355), we have the relation

$$M^2 = M_{\text{irr}}^2 + J^2/4M_{\text{irr}}^2. \tag{359}$$

Since M_{irr} *is* irreducible, we may interpret the second term, $J^2/4M_{\text{irr}}^2$, as the contribution of the rotational kinetic energy to the square of the inertial mass of the black hole; and that it is this rotational energy that is being extracted by the Penrose processes.

A further result of some general interest follows from considering the effect of Penrose processes on the extreme Kerr black-hole with $a = M$ and $r_+ = M$. In this case, the inequality (354) gives

$$M\,\delta M \geqslant a\,\delta a \quad \text{or} \quad \delta(M^2 - a^2) \geqslant 0. \tag{360}$$

In other words, the black hole evolves *down* the Kerr sequence to lower values of a^2/M^2 and away from equality. Thus, by infinitesimal processes of the kind we have been considering, a naked singularity with $a^2 > M^2$ cannot be created. The question whether a naked singularity can be created by non-linear 'explosive' processes is an open one. The conjecture, that *no naked singularity can ever be created when there are none*, is the *cosmic-censorship hypothesis* of Penrose.

66. Geodesics for $a^2 > M^2$

Because our primary interest is in black holes, we have restricted our consideration to the Kerr space-times with $a^2 \leqslant M^2$ since, only then, does the space-time allow an event horizon and represent a black hole. If $a^2 > M^2$, the horizons cease to exist and only the ring singularity remains. What remains, in fact, is a naked singularity. Naked singularities are believed not to exist in nature: the hypothesis of cosmic censorship forbids their coming into being in the natural course of events. Nevertheless, considerable interest attaches to knowing the sort of things space-times with naked singularities are and whether there are any essential differences in the manifestations of space-times with naked singularities and space-times with singularities concealed behind event horizons. A contrastive study of the geodesics in Kerr space-times with $a^2 < M^2$ and $a^2 > M^2$ may contribute a measure of understanding to what these differences may be.

(*a*) *The null geodesics*

First, we observe that there are some essential features of the null geodesics which depend on whether $a^2 < M^2$ or $a^2 > M^2$. Thus, we have seen in §61(*a*) that when $a^2 < M^2$, the null geodesics in the equatorial plane are distinguished by the critical impact parameters $D_c^{\mp}$ (the upper sign referring to retrograde orbits and the lower sign to direct orbits): orbits arriving from infinity with impact parameters $D > D_c^{\mp}$ recede once again to infinity after perihelion passages while orbits with impact parameters $D < D_c^{\mp}$ terminate at the singularity. But when $a^2 > M^2$, such a distinction exists only for retrograde orbits: *all direct orbits, arriving from infinity with $D > a$, recede again into infinity*. For the distinction depends upon the existence, or otherwise, of unstable circular photon-orbits. The radii of these unstable circular photon-orbits are determined by the real positive roots, $\sqrt{r}$, of the cubic equation (cf. equation (124))

$$r^{3/2}Q_{\mp} = r^{3/2} - 3Mr^{1/2} \mp 2aM^{1/2} = 0; \tag{361}$$

and for $a^2 > M^2$, this equation with the positive sign (corresponding to direct orbits) does not allow any real positive root. However, with the negative sign (corresponding to retrograde orbits) equation (361) allows the root (cf. equation (87))

$$\left.\begin{aligned} &\sqrt{r_{\text{ph.}}^{(-)}} = (2\sqrt{M})\cosh\vartheta \quad \text{or} \quad r_{\text{ph.}}^{(-)} = 4M\cosh^2\vartheta \\ \text{where} \quad &\cosh 3\vartheta = |a|/M. \end{aligned}\right\} \tag{362}$$

The corresponding impact parameter, associated with this retrograde, unstable, circular orbit is

$$D_c^{(-)} = M(6\cosh\vartheta + \cosh 3\vartheta). \tag{363}$$

The discussion of the non-planar orbits of constant radius (r_s) in §63(a) applies equally for $a^2 > M^2$. In particular, orbits with $\eta > 0$ (for $r_s < r_{\text{ph}}^{(-)}$) cross the equatorial plane and avoid the singularity while the orbits with $\eta < 0$ (which includes retrograde orbits with $r_s > r_{\text{ph}}^{(-)}$ and all direct orbits) are confined to cones (not including the directions $\theta = 0$ and $\theta = \pi/2$) and have turning points for some negative r. (The case $\eta = 0$ is possible only for the retrograde orbit with $r_s = r_{\text{ph}}^{(-)}$.) However, in the context of the orbits with $\eta < 0$, there is one important difference. When $a^2 < M^2$, the orbits which have turning points in the domain of negative r cannot return to the world which they left: the interposition of the horizons will necessitate their migration to other worlds. But when $a^2 > M^2$, there are no interposing horizons; and the trajectories which have turning points in the domain of negative r can return to the realm of positive r and recount their 'wondrous tales of travel' (cf. §(c) below).

(b) *The time-like geodesics*

Some interesting features which bear on the transition across the 'barrier' $a^2 = M^2$ emerge (as first shown by de Felice) from a consideration of the expression (cf. equation (117))

$$E^{\mp} = \frac{1}{\sqrt{Q_{\mp}}}[1 - 2Mu \mp a\sqrt{(Mu)^3}], \tag{364}$$

which gives the energies of the direct (+) and the retrograde (−) circular orbits with the assigned reciprocal radius $u(=1/r)$. In the present context, we are interested only in the direct orbits: the indirect orbits are restricted by the requirement $Q_- \geqslant 0$, i.e., by the radius of the retrograde circular photon-orbit (which, as we have seen, exists for all a^2) while there is no such restriction for the direct orbits.

Restricting ourselves, then, to the direct orbits, we have already noted that $E^{(+)}$ for $a = M$ has a discontinuity at $r = M \pm 0$; thus (cf. equation (146))

$$E^{(+)} \to \pm 3^{-1/2} \quad (r \to M \pm 0). \tag{365}$$

While the discontinuity is a consequence of the ambiguity of the chosen radial coordinate r at $r = M$ for $a^2 = M^2$, we are here concerned with the behaviour of $E^{(+)}$ for $r < M$. By equation (364), $E^{(+)}$ will have a zero, if the equation

$$r^{3/2} - 2Mr^{1/2} + aM^{1/2} = 0 \tag{366}$$

allows a real positive root for $\sqrt{r}$. For $a = M$ equation (366) allows two such roots:

$$\sqrt{r} = \sqrt{M} \qquad \text{and} \qquad \sqrt{r} = \tfrac{1}{2}(\sqrt{5}-1)\sqrt{M}, \tag{367}$$

or

$$r = M \qquad \text{and} \qquad r = \tfrac{1}{2}(3-\sqrt{5})M = 0.38197\,M. \tag{368}$$

Therefore, $E^{(+)}$ is negative in the interval

$$\tfrac{1}{2}(3-\sqrt{5})M < r < M \quad (E^{(+)} < 0). \tag{369}$$

In other words, stable circular orbits of negative energy exist for $a^2 = M^2$ in the interval (369). Such stable circular orbits of negative energy continue to exist for $a^2 > M^2$ but only in the limited range

$$(32/27)^2 M^2 > a^2 > M^2. \tag{370}$$

This follows from the fact that equation (366), for $a^2/M^2 = 32/27$, namely,

$$r^{3/2} - 2Mr^{1/2} + (32/27)^{1/2} M^{3/2} = 0, \tag{371}$$

allows two coincident roots at

$$\sqrt{r} = \sqrt{(2M/3)} \qquad \text{or} \qquad r = \tfrac{2}{3}M; \tag{372}$$

and no real positive root for $|a| > (32/27)^{1/2}M = 1.088\,M$.

The discussion of the non-planar time-like geodesics in §64 applies equally for $a^2 > M^2$. And again, the orbits associated with $\eta < 0$ are of special interest since they return to the domain of positive r after visiting the domain of negative r—a matter to which we return presently in §(*c*) below.

(*c*) *Violation of causality*

One of the important features of space-times with singularities is that one may have domains in which causality may be violated in the sense that one may have closed time-like curves which will permit one's future to influence one's past—a feature one would describe as unphysical. In the Kerr space-time, extended to allow r to be negative, one can certainly violate causality in the domain in which $g_{\varphi\varphi} > 0$ and φ become time-like.

The boundary of the domain in which φ is time-like is defined by the equation

$$\begin{aligned}\Sigma^2 &= (r^2+a^2)^2 - a^2\Delta\sin^2\theta \\ &= (r^2+a^2)(r^2+a^2\cos^2\theta) + 2Ma^2 r\sin^2\theta = 0\end{aligned} \tag{373}$$

—an equation which can be satisfied only if r were allowed to be negative. Letting $x = -r$, to avoid ambiguity, we can rewrite equation (373) in the form

$$\frac{(x^2+a^2)^2}{a^2(x^2+a^2+2Mx)} = \sin^2\theta \quad (x = -r). \tag{374}$$

This equation clearly requires that

$$\frac{(x^2+a^2)^2}{a^2(x^2+a^2+2Mx)} \leqslant 1; \tag{375}$$

and this inequality bounds the range of x:

$$0 \leqslant x \leqslant x_{max} \tag{376}$$

where

$$x_{max} = \frac{2a}{\sqrt{3}}\sinh\left(\frac{1}{3}\sinh^{-1}\frac{3M}{a}\sqrt{3}\right), \tag{377}$$

is the positive root of the equation

$$x^3 + a^2x - 2Ma^2 = 0. \tag{378}$$

In particular,

$$0 \leqslant x \leqslant M \quad \text{for} \quad a^2 = M^2$$

and

$$0 \leqslant x \leqslant 1.2878M \quad \text{for} \quad a^2 = 3M^2. \tag{379}$$

Similarly θ is also bounded:

$$\sin^2\theta_{min} \leqslant \sin^2\theta \leqslant 1, \tag{380}$$

where

$$\sin^2\theta_{min} = \min\frac{(x^2+a^2)^2}{a^2(x^2+a^2+2Mx)}. \tag{381}$$

The minimum of the expression on the right-hand side occurs at $x = zM$ where z is the positive root of the equation

$$(z+1)^3 - (3-a^2/M^2)(z+1) + 2(1-a^2/M^2) = 0. \tag{382}$$

We find, for example, that

$$z = \sqrt{2}-1, \quad \sin^2\theta_{min} = 4(3-2\sqrt{2}), \quad \theta_{min} = 55.75° \quad (a = M)$$

and

$$z = \sqrt[3]{4}-1, \quad \sin^2\theta_{min} = 0.9080, \quad \theta_{min} = 65.25° \quad (a = M\sqrt{3}). \tag{383}$$

Thus, equation (374), which defines the boundary of the region in which causality can be violated, provides rather severe bounds on the ranges of $-r$ and θ.

While regions in which φ is time-like exist for all values of $a^2 > 0$, for $a^2 < M^2$, these regions are incommunicable to the space outside and we can afford to take a nonchalant view. But we cannot afford to be indifferent when $a^2 > M^2$ and no event horizon interposes and the region in which φ is time-like *is* communicable to the space outside. A question of some interest in this connection is whether causality (in some sense) can be violated by travelling along time-like (or null) geodesics. If such violations occur, they can only be along unbound geodesics with $\eta < 0$ and which have turning points in the domain of negative r. Since (cf. equations (179) and (180))

$$t = \int^r \frac{r^2}{\sqrt{R}}\,\mathrm{d}r + a^2 \int^\theta \frac{\cos^2\theta}{\sqrt{\Theta}}\,\mathrm{d}\theta + 2M \int^r r[r^2 - a(\xi - a)]\frac{\mathrm{d}r}{\Delta\sqrt{R}}, \qquad (384)$$

It follows that for violation of causality, the negative contribution by the last integral derived from the part of the geodesic in the domain of negative r (which we can maximize by arranging that $a(\xi - a)$ is negative and the turning point occurs in the region in which φ is time-like) suffices to compensate for the positive contribution by all three integrals from the parts of the geodesic in the domain of positive r. The results of the calculation are inconclusive.

BIBLIOGRAPHICAL NOTES

It was Carter's discovery of the separability of the Hamilton–Jacobi equation that made possible a complete analytical discussion of the geodesics in the Kerr space-time:

1. B. CARTER, *Phys. Rev.*, 174, 1559–71 (1968).

See also:

2. B. CARTER, *Commun. Math. Phys.*, 10, 280–310 (1968).

§60. In presenting the subject, we have inverted the historical order, and preferred to begin with an account of the integral for null-geodesic motion (allowed by any type-D space-time) discovered by Walker and Penrose:

3. M. WALKER and R. PENROSE, *Commun. Math. Phys.*, 18, 265–74 (1970).

Walker and Penrose also derived the integral (equivalent to Carter's) for time-like geodesic motion for the particular case of the Kerr metric.

The treatment in the text differs from the one adopted by Walker and Penrose and avoids the machinery of the spinor formalism. It is more direct and perhaps 'pedestrian'. However, it has the advantage of yielding simple necessary and sufficient conditions, in terms of the spin-coefficients, for a type-D space-time to allow Carter-type integrals for time-like geodesic motions.

The proof of Theorem 1 was devised in collaboration with B. Xanthopoulos.

It appears that the conditions equivalent to those established in Theorem 3 are included in the following papers by Hauser and Malhiot (though the author has been unable to unravel their notation):

4. I. HAUSER and R. J. MALHIOT, *J. Math. Phys.*, 16, 150–2 (1975).
5. ———, ibid., 1625–9 (1975).
6. ———, ibid., 17, 1306–12 (1976).

§61. The geodesics in the equatorial plane of the Kerr black-hole have been discussed extensively in the literature. For a complete listing see:

7. N. A. SHARP, *General Relativity and Gravitation*, 10, 659–70, (1979).

In writing the account the author has consulted the following papers:

8. F. DE FELICE, *Il Nuovo Cim.*, 57 B, 351–88 (1968).
9. J. M. BARDEEN, W. H. PRESS, and S. A. TEUKOLSKY, *Astrophys. J.*, 178, 347–69 (1972).
10. ———, ibid., 161, 103–9 (1970).

11. ———, in *Black Holes*, 241–89, edited by C. DeWitt and B. S. DeWitt, Gordon and Breach Science Publishers, New York, 1973.

The common procedure in treating the geodesics in the Kerr space-time is to start with the general equations of motion (as they are derived in §62(a)) and then specialize them appropriately for the orbits in the equatorial plane. For the delineation of the geodesics in the equatorial plane we do not, of course, require the general equations of motion: they can be treated, *ab initio*, as they are in the text.

The illustrations of the geodesics in this and in the following sections were prepared by Mr. Garret Toomey to whom the author is again indebted.

§62. For readable accounts see:

12. C. W. Misner, K. S. Thorne, and J. W. Wheeler, *Gravitation*, W. H. Freeman and Co., San Francisco, 1970.
13. J. Stewart and M. Walker, *Springer Tracts in Modern Physics*, 69, 69–115 (1973).

§§63 and 64. Among the many papers devoted to the geodesics in the Kerr space-time, the author found the following the most useful:

14. J. M. Bardeen in *Black Holes*, 215–39, edited by C. DeWitt and B. S. DeWitt, Gordon and Breach Science Publishers, New York, 1973.
15. F. de Felice and M. Calvani, *Il Nuovo Cim.*, 10 B, 447–58 (1972).
16. M. Calvani and F. de Felice, *General Relativity and Gravitation*, 9, 889–902 (1978).

See also:

17. R. H. Boyer and R. W. Lindquist, *J. Math. Phys.*, 8, 265–81 (1967).
18. D. C. Wilkins, *Phys. Rev.* D, 5, 814–22 (1972).
19. C. T. Cunningham and J. M. Bardeen, *Astrophys. J.*, 183, 237–64, (1973).
20. ———, ibid., 202, 788–802 (1975).
21. ———, ibid., 208, 534–49 (1976).

The first application of the complex integral of Walker and Penrose to the problem of tracing the changing direction of polarizations along null geodesics is due to Stark and Connors:

22. P. A. Connors and R. F. Stark, *Nature*, 269, 128–9 (1977).

See also:

23. R. F. Stark and P. A. Connors, *Nature*, 266, 429–30 (1977).
24. P. A. Connors, T. Piran, and R. F. Stark, *Astrophys. J.*, 235, 224–44 (1980).

The treatment in §63(e) is an amplification of that sketched in reference 22.

§65. The existence of physical processes which, in effect, extract the rotational energy of a Kerr black-hole, was first demonstrated by:

25. R. Penrose and R. M. Floyd, *Nature Phys. Sci.*, 229, 177–9 (1971).

In this paper the 'area theorem' is also stated, albeit tentatively.

More detailed considerations relative to the process conceived by Penrose are those of:

26. D. Christodoulou, 'Investigations in gravitational collapse and the physics of black holes', Ph.D. dissertation, Princeton University, Princeton, N. J., 1971; also, *Phys. Rev. Lett.*, 25, 1596–7 (1970).
27. ——— and R. Ruffini, *Phys. Rev.* D, 4, 3552–5 (1971).

The notion of 'irreducible mass' is introduced in reference 26.

Limits to the energy that may be extracted by Penrose processes were derived by Bardeen, Press, and Teukolsky in reference 9 and also by Wald:

28. R. M. Wald, *Astrophys. J.*, 191, 231–3 (1974).
29. ———, *Ann. Phys.*, 82, 548–56 (1974).

See also:

30. T. PIRAN and J. SHAHAM, *Phys. Rev.* D, 16, 1615—35 (1977).

§66. The geodesics in the Kerr space-time for $a^2 > M^2$ and the possibility of causality violations in these space-times have been considered most persistently by de Felice and his associates:

31. F. DE FELICE, *Astron. Astrophys.*, 34, 15–19 (1974).
32. ———, ibid., 45, 65–8 (1975).
33. ———, M. CALVANI, and L. NOBILI, *Il Nuovo Cim.*, 26 B, 1–15 (1975).
34. M. CALVANI, F. DE FELICE, B. MUCHOTRZEB, and F. SALMISTRARO, *General Relativity and Gravitation*, 9, 155–63 (1977).
35. F. DE FELICE, *Nature*, 273, 429–31 (1978).
36. ——— and M. CALVANI, *General Relativity and Gravitation*, 10, 335–43 (1979).

In the text we have attempted to summarize the principal results in the foregoing papers.

8

ELECTROMAGNETIC WAVES IN KERR GEOMETRY

67. Introduction

As our studies in the contexts of the Schwarzschild and the Reissner–Nordström black-holes have shown, our understanding of the space-times is deepened and enriched by analyzing the manner of their reaction to external perturbations. Since external perturbations can, in general, be represented by incident waves of different sorts, their study, in essence, reduces to one of the propagation of waves of the different sorts in the space-times of the black holes and of how, in particular, they are scattered and absorbed. While the problem, as thus formulated, has a physical base, its solution reveals an analytic richness of the space-times themselves for which one is unprepared. In the case of the Kerr space-time, this richness becomes manifest almost at the outset.

Since the space-time of the Kerr black-hole is stationary and axisymmetric, one naturally expects to express a general perturbation as a superposition of waves of different frequencies (σ^{+})* and of different periods ($2m\pi$, $m = 0, 1, 2, \ldots$) in φ. In other words, one expects to analyze the perturbation as a superposition of different *modes* with a time- and a φ-dependence given by

$$e^{i(\sigma^{+}t + m\varphi)}, \tag{1}$$

where m is an integer, positive, negative, or zero. But one does not expect that the dependence of the amplitudes of these waves (with the foregoing t- and φ-dependence) on the remaining two variables, r and θ, can be separated as well. But Teukolsky showed in 1971 that this further separation can, in fact, be achieved if the equations of the massless particles—the photons, the gravitons, and the two-component neutrinos—are written in a Newman–Penrose formalism with the basis vectors chosen as in Chapter 6, §56, And as it turned out, this unexpected separability of the basic equations of mathematical physics was a sort of 'open Sesame' for unlocking further doors.

In this chapter, we begin our study of the perturbations of the Kerr black-hole with Maxwell's equations and the propagation of electromagnetic waves. The theory is sufficiently simple that a complete and a self-contained account can be given; and it provides a prototype for the study of other fields.

* In this and the following chapter, we shall denote the frequency by σ^{+} (instead of the customary σ) in order to distinguish it from the spin-coefficient σ.

68. Definitions and lemmas

Consistent with the formulation of the perturbation problem as we have sketched in §67 and with our treatment of the same problem in the contexts of the Schwarzschild and the Reissner-Nordström black-holes, we shall suppose that the perturbations can be analyzed into modes having a time- and a φ-dependence specified in equation (1). The common factor (1), in all quantities describing the perturbation, will be suppressed; and the symbols representing them will be their amplitudes.

The basis vectors $(\boldsymbol{l}, \boldsymbol{n}, \boldsymbol{m}, \bar{\boldsymbol{m}})$ given in Chapter 6, equations (170)–(173), when applied as tangent vectors to the functions with a t- and a φ-dependence specified in equation (1), become the derivative operators

$$\left.\begin{aligned} \boldsymbol{l} &= D = \mathscr{D}_0, \quad \boldsymbol{n} = \triangle = -\frac{\Delta}{2\rho^2}\mathscr{D}_0^\dagger \\ \boldsymbol{m} &= \delta = \frac{1}{\bar{\rho}\sqrt{2}}\mathscr{L}_0^\dagger, \quad \text{and} \quad \bar{\boldsymbol{m}} = \delta^* = \frac{1}{\bar{\rho}^*\sqrt{2}}\mathscr{L}_0, \end{aligned}\right\} \tag{2}$$

where

$$\left.\begin{aligned} \mathscr{D}_n &= \partial_r + \frac{iK}{\Delta} + 2n\frac{r-M}{\Delta}, \quad & \mathscr{D}_n^\dagger &= \partial_r - \frac{iK}{\Delta} + 2n\frac{r-M}{\Delta}, \\ \mathscr{L}_n &= \partial_\theta + Q + n\cot\theta, & \mathscr{L}_n^\dagger &= \partial_\theta - Q + n\cot\theta, \end{aligned}\right\} \tag{3}$$

and

$$K = (r^2 + a^2)\sigma + am, \quad Q = a\sigma\sin\theta + m\,\text{cosec}\,\theta, \tag{4}$$

$$\bar{\rho} = r + ia\cos\theta, \quad \bar{\rho}^* = r - ia\cos\theta, \quad \rho^2 = r^2 + a^2\cos^2\theta. \tag{5}$$

It will be noticed that while $\mathscr{D}_n$ and $\mathscr{D}_n^\dagger$ are purely radial operators, $\mathscr{L}_n$ and $\mathscr{L}_n^\dagger$ are purely angular operators. Also, it is clear that when applied to 'background' quantities (independent of t and φ) the operators $\mathscr{D}$ and $\mathscr{L}$ reduce to ∂_r and ∂_θ, respectively.

The differential operators we have defined satisfy a number of elementary identities which we shall have occasion to use frequently in our subsequent analysis in this and in the following chapters. On this account, we shall collect them here as a series of lemmas (cf. Ch. 4, equations (227)–(229)).

LEMMA 1.

$$\left.\begin{aligned} \mathscr{L}_n(\theta) &= -\mathscr{L}_n^\dagger(\pi - \theta), \quad \mathscr{D}_n^\dagger = (\mathscr{D}_n)^*, \\ (\sin\theta)\mathscr{L}_{n+1} &= \mathscr{L}_n\sin\theta, \quad (\sin\theta)\mathscr{L}_{n+1}^\dagger = \mathscr{L}_n^\dagger\sin\theta, \\ \Delta\mathscr{D}_{n+1} &= \mathscr{D}_n\Delta, \quad \Delta\mathscr{D}_{n+1}^\dagger = \mathscr{D}_n^\dagger\Delta. \end{aligned}\right\} \tag{6}$$

LEMMA 2.

$$\left(\mathscr{D}+\frac{m}{\bar{\rho}^*}\right)\left(\mathscr{L}+\frac{ima\sin\theta}{\bar{\rho}^*}\right)=\left(\mathscr{L}+\frac{ima\sin\theta}{\bar{\rho}^*}\right)\left(\mathscr{D}+\frac{m}{\bar{\rho}^*}\right), \tag{7}$$

where $\mathscr{D}$ can be any $\mathscr{D}_n$ or $\mathscr{D}_n^\dagger$, and $\mathscr{L}$ any $\mathscr{L}_n$ or $\mathscr{L}_n^\dagger$ and m is a constant (generally an integer, positive or negative).

LEMMA 3.

$$\mathscr{L}_{n+1}\mathscr{L}_{n+2}\ldots\mathscr{L}_{n+m}(f\cos\theta)=(\cos\theta)\mathscr{L}_{n+1}\ldots\mathscr{L}_{n+m}f$$
$$-(m\sin\theta)\mathscr{L}_{n+2}\ldots\mathscr{L}_{n+m}f, \tag{8}$$

where f is any smooth function of θ and the $\mathscr{L}_n$'s can be replaced by $\mathscr{L}_n^\dagger$'s.

LEMMA 4. *If $f(\theta)$ and $g(\theta)$ are any two smooth functions of θ, in the interval, $0\leqslant\theta\leqslant\pi$, then*

$$\int_0^\pi g(\mathscr{L}_nf)\sin\theta\,d\theta=-\int_0^\pi f(\mathscr{L}^\dagger_{-n+1}g)\sin\theta\,d\theta. \tag{9}$$

And finally, we may note two elementary identities which play crucial roles in many simplifications:

$$Q_{,\theta}+Q\cot\theta=2a\sigma\cos\theta \quad\text{and}\quad K-aQ\sin\theta=\rho^2\sigma. \tag{10}$$

69. Maxwell's equations: their reduction and their separability

Maxwell's equations, appropriate to Kerr geometry, can be obtained by inserting in Chapter 1, equations (330)–(333), the spin coefficients listed in Chapter 6, equations (175) and the directional derivatives defined in equations (2). They are

$$\begin{aligned}
\frac{1}{\bar{\rho}^*\sqrt{2}}\left(\mathscr{L}_1-\frac{ia\sin\theta}{\bar{\rho}^*}\right)\phi_0&=+\left(\mathscr{D}_0+\frac{2}{\bar{\rho}^*}\right)\phi_1,\\
\frac{1}{\bar{\rho}^*\sqrt{2}}\left(\mathscr{L}_0+\frac{2ia\sin\theta}{\bar{\rho}^*}\right)\phi_1&=+\left(\mathscr{D}_0+\frac{1}{\bar{\rho}^*}\right)\phi_2,\\
\frac{1}{\bar{\rho}\sqrt{2}}\left(\mathscr{L}_1^\dagger+\frac{ia\sin\theta}{\bar{\rho}^*}\right)\phi_2&=-\frac{\Delta}{2\rho^2}\left(\mathscr{D}_0^\dagger+\frac{2}{\bar{\rho}^*}\right)\phi_1,\\
\frac{1}{\bar{\rho}\sqrt{2}}\left(\mathscr{L}_0^\dagger+\frac{2ia\sin\theta}{\bar{\rho}^*}\right)\phi_1&=-\frac{\Delta}{2\rho^2}\left(\mathscr{D}_1^\dagger-\frac{1}{\bar{\rho}^*}\right)\phi_0.
\end{aligned} \tag{11}$$

These equations take simpler and more symmetric forms when they are written in terms of the variables

$$\Phi_0=\phi_0,\ \Phi_1=\phi_1\bar{\rho}^*\sqrt{2},\quad\text{and}\quad\Phi_2=2\phi_2(\bar{\rho}^*)^2. \tag{12}$$

We find

$$\left(\mathscr{L}_1-\frac{ia\sin\theta}{\bar\rho^*}\right)\Phi_0=\left(\mathscr{D}_0+\frac{1}{\bar\rho^*}\right)\Phi_1,\tag{13}$$

$$\left(\mathscr{L}_0+\frac{ia\sin\theta}{\bar\rho^*}\right)\Phi_1=\left(\mathscr{D}_0-\frac{1}{\bar\rho^*}\right)\Phi_2,\tag{14}$$

$$\left(\mathscr{L}_1^\dagger-\frac{ia\sin\theta}{\bar\rho^*}\right)\Phi_2=-\Delta\left(\mathscr{D}_0^\dagger+\frac{1}{\bar\rho^*}\right)\Phi_1,\tag{15}$$

$$\left(\mathscr{L}_0^\dagger+\frac{ia\sin\theta}{\bar\rho^*}\right)\Phi_1=-\Delta\left(\mathscr{D}_1^\dagger-\frac{1}{\bar\rho^*}\right)\Phi_0.\tag{16}$$

(a) The reduction and the separability of the equations for Φ_0 and Φ_2

It is evident that the commutativity of the operators $(\mathscr{D}_0+1/\bar\rho^*)$ and $(\mathscr{L}_0^\dagger+ia\sin\theta/\bar\rho^*)$ (by Lemma 2 of §68) enables us to eliminate Φ_1 from equations (13) and (16) and obtain a decoupled equation for Φ_0. We thus obtain

$$\left[\left(\mathscr{L}_0^\dagger+\frac{ia\sin\theta}{\bar\rho^*}\right)\left(\mathscr{L}_1-\frac{ia\sin\theta}{\bar\rho^*}\right)+\Delta\left(\mathscr{D}_1+\frac{1}{\bar\rho^*}\right)\left(\mathscr{D}_1^\dagger-\frac{1}{\bar\rho^*}\right)\right]\Phi_0=0.\tag{17}$$

Similarly, the commutativity of the operators $(\mathscr{L}_0+ia\sin\theta/\bar\rho^*)$ and $\Delta(\mathscr{D}_0^\dagger+1/\bar\rho^*)$ enables us to eliminate Φ_1 from equations (14) and (15) and obtain a decoupled equation for Φ_2. Thus,

$$\left[\left(\mathscr{L}_0+\frac{ia\sin\theta}{\bar\rho^*}\right)\left(\mathscr{L}_1^\dagger-\frac{ia\sin\theta}{\bar\rho^*}\right)+\Delta\left(\mathscr{D}_0^\dagger+\frac{1}{\bar\rho^*}\right)\left(\mathscr{D}_0-\frac{1}{\bar\rho^*}\right)\right]\Phi_2=0.\tag{18}$$

Equation (17) can be simplified with the aid of the readily verifiable identities,

$$\left.\begin{aligned}\Delta\left(\mathscr{D}_1+\frac{1}{\bar\rho^*}\right)\left(\mathscr{D}_1^\dagger-\frac{1}{\bar\rho^*}\right)&=\Delta\mathscr{D}_1\mathscr{D}_1^\dagger-\frac{2iK}{\bar\rho^*},\\ \left(\mathscr{L}_0^\dagger+\frac{ia\sin\theta}{\bar\rho^*}\right)\left(\mathscr{L}_1-\frac{ia\sin\theta}{\bar\rho^*}\right)&=\mathscr{L}_0^\dagger\mathscr{L}_1+\frac{2ia\sin\theta}{\bar\rho^*}Q;\end{aligned}\right\}\tag{19}$$

and the further relation

$$K-aQ\sin\theta=\rho^2\sigma.\tag{20}$$

We thus obtain

$$[\Delta\mathscr{D}_1\mathscr{D}_1^\dagger+\mathscr{L}_0^\dagger\mathscr{L}_1-2i\sigma(r+ia\cos\theta)]\Phi_0=0.\tag{21}$$

A similar reduction of equation (18) yields

$$[\Delta \mathscr{D}_0^\dagger \mathscr{D}_0 + \mathscr{L}_0 \mathscr{L}_1^\dagger + 2i\sigma(r + ia\cos\theta)]\,\Phi_2 = 0. \tag{22}$$

Equations (21) and (22) are clearly separable. Thus, with the substitutions

$$\Phi_0 = R_{+1}(r)S_{+1}(\theta) \qquad \text{and} \qquad \Phi_2 = R_{-1}(r)S_{-1}(\theta), \tag{23}$$

where $R_{\pm 1}(r)$ and $S_{\pm 1}(\theta)$ are, respectively, functions of r and θ only, equations (21) and (22) separate to give the two pairs of equations,

$$(\Delta \mathscr{D}_1 \mathscr{D}_1^\dagger - 2i\sigma r)R_{+1} = \lambda R_{+1}, \tag{24}$$

$$(\mathscr{L}_0^\dagger \mathscr{L}_1 + 2a\sigma\cos\theta)S_{+1} = -\lambda S_{+1}; \tag{25}$$

and

$$(\Delta \mathscr{D}_0^\dagger \mathscr{D}_0 + 2i\sigma r)R_{-1} = \lambda R_{-1}, \tag{26}$$

$$(\mathscr{L}_0 \mathscr{L}_1^\dagger - 2a\sigma\cos\theta)S_{-1} = -\lambda S_{-1}, \tag{27}$$

where λ is a separation constant. (We are using λ in place of our customary λ to distinguish it from the spin-coefficient λ.)

It will be noticed that we have not distinguished the separation constants that derive from equations (21) and (22). The reason is the same as in the earlier context of Chapter 4, equations (248) and (250), which we have explained.

Equations equivalent to (24)–(27) were first derived by Teukolsky. We shall call them Teukolsky's equations.

Also, we may note that a comparison of equation (26) with equation (24), rewritten in the form

$$(\Delta \mathscr{D}_0 \mathscr{D}_0^\dagger - 2i\sigma r)\Delta R_{+1} = \lambda \Delta R_{+1} \tag{28}$$

shows that R_{-1} and ΔR_{+1} satisfy complex-conjugate equations even as $S_{+1}(\theta)$ and $S_{-1}(\theta)$ satisfy a pair of equations, one of which can be obtained from the other by replacing θ by $\pi - \theta$.

70. The Teukolsky–Starobinsky identities

The decoupling of equations (13)–(16) to provide a pair of independent separable equations for Φ_0 and Φ_2 solves the problem only partially: for, apart from the fact the solution for Φ_1 is yet to be found, the relative normalization of the solutions Φ_0 and Φ_2 also remains to be resolved. The resolution of the latter problem is, in some ways, the more fundamental: the solution to *any* linear-perturbation problem must be determinate apart from a single constant of proportionality. In the present context, the determinacy requires that the solution for Φ_2 (for example) can have no arbitrariness—not even to the extent of a constant of proportionality—if the amplitude of the solution for Φ_0 has been specified, and conversely. With Φ_0 and Φ_2 expressed as in equations (23), the solutions for $S_{+1}(\theta)$ and $S_{-1}(\theta)$ can be made unambiguous by requiring that they both be normalized to unity:

$$\int_0^\pi S_{+1}^2 \sin\theta \, d\theta = \int_0^\pi S_{-1}^2 \sin\theta \, d\theta = 1. \tag{29}$$

But this will leave the relative normalization of the radical functions R_{+1} and R_{-1} unresolved.

The completion of the solution of Maxwell's equations requires, as we shall see, a careful examination of equations (24)–(28) and the relationships among the solutions R_{+1} and R_{-1}, and S_{+1} and S_{-1}, which the same equations imply. Such an examination by Starobinsky and Teukolsky led to some very remarkable identities which are basic to the entire theory. We shall give an account of these identities (though not in the forms in which they were originally formulated).

THEOREM 1. *$\Delta\mathscr{D}_0\mathscr{D}_0 R_{-1}$ is a constant multiple of ΔR_{+1} and $\Delta\mathscr{D}_0^\dagger\mathscr{D}_0^\dagger\Delta R_{+1}$ is a constant multiple of R_{-1}.*

Proof. Applying the operator $\mathscr{D}_0\mathscr{D}_0$ to equation (26) satisfied by R_{-1}, we find, successively,

$$\begin{aligned}
\lambda\mathscr{D}_0\mathscr{D}_0 R_{-1} &= \mathscr{D}_0\mathscr{D}_0(\Delta\mathscr{D}_0^\dagger\mathscr{D}_0 R_{-1}) + 2i\sigma\mathscr{D}_0\mathscr{D}_0(rR_{-1}) \\
&= \mathscr{D}_0\mathscr{D}_0(\Delta\mathscr{D}_0 - 2iK)\mathscr{D}_0 R_{-1} + 2i\sigma r\mathscr{D}_0\mathscr{D}_0 R_{-1} + 4i\sigma\mathscr{D}_0 R_{-1} \\
&= \mathscr{D}_0\Delta\mathscr{D}_1\mathscr{D}_0\mathscr{D}_0 R_{-1} - 2i\mathscr{D}_0(K\mathscr{D}_0\mathscr{D}_0 R_{-1} + 2r\sigma\mathscr{D}_0 R_{-1}) \\
&\qquad + 2i\sigma r\mathscr{D}_0\mathscr{D}_0 R_{-1} + 4i\sigma\mathscr{D}_0 R_{-1} \\
&= \mathscr{D}_0(\Delta\mathscr{D}_1^\dagger + 2iK)\mathscr{D}_0\mathscr{D}_0 R_{-1} - 2i\mathscr{D}_0(K\mathscr{D}_0\mathscr{D}_0 R_{-1}) - 2i\sigma r\mathscr{D}_0\mathscr{D}_0 R_{-1} \\
&= (\Delta\mathscr{D}_1\mathscr{D}_1^\dagger - 2i\sigma r)\mathscr{D}_0\mathscr{D}_0 R_{-1}.
\end{aligned} \tag{30}$$

Therefore, $\mathscr{D}_0\mathscr{D}_0 R_{-1}$ satisfies the same equation as R_{+1} and the first part of the theorem stated follows. At the same time, the result of the foregoing reductions is equivalent to establishing the validity of the identity

$$\mathscr{D}_0\mathscr{D}_0(\Delta\mathscr{D}_0^\dagger\mathscr{D}_0 + 2i\sigma r) = (\Delta\mathscr{D}_1\mathscr{D}_1^\dagger - 2i\sigma r)\mathscr{D}_0\mathscr{D}_0. \tag{31}$$

The complex conjugate of this identity, namely,

$$\mathscr{D}_0^\dagger\mathscr{D}_0^\dagger(\Delta\mathscr{D}_0\mathscr{D}_0^\dagger - 2i\sigma r) = (\Delta\mathscr{D}_1^\dagger\mathscr{D}_1 + 2i\sigma r)\mathscr{D}_0^\dagger\mathscr{D}_0^\dagger, \tag{32}$$

establishes the second part of the theorem.

COROLLARY. *By a suitable choice of the relative normalization of the functions ΔR_{+1} and R_{-1} we can arrange that*

$$\Delta\mathscr{D}_0\mathscr{D}_0 R_{-1} = \mathscr{C}\Delta R_{+1} \tag{33}$$

and

$$\Delta\mathscr{D}_0^\dagger\mathscr{D}_0^\dagger\Delta R_{+1} = \mathscr{C}^* R_{-1}, \tag{34}$$

where $\mathscr{C}$ is a constant (which can be complex).*

We can clearly arrange for the relative normalization as prescribed since ΔR_{+1} and R_{-1} satisfy complex-conjugate equations.

THEOREM 2. *If the relative normalization of the functions ΔR_{+1} and R_{-1} is so arranged that equations (33) and (34) are valid, then the square of the modulus of $\mathscr{C}$ is given by*

$$|\mathscr{C}|^2 = \lambda^2 - 4\alpha^2\sigma^{+2} \qquad \textit{where} \qquad \alpha^2 = a^2 + (am/\sigma^+). \tag{35}$$

Proof. Applying the operator $\Delta\mathscr{D}_0^\dagger\mathscr{D}_0^\dagger$ to equation (33), we obtain

$$\Delta\mathscr{D}_0^\dagger\mathscr{D}_0^\dagger\Delta\mathscr{D}_0\mathscr{D}_0 R_{-1} = \mathscr{C}\Delta\mathscr{D}_0^\dagger\mathscr{D}_0^\dagger\Delta R_{+1}, \tag{36}$$

or, by virtue of equation (34),

$$\Delta\mathscr{D}_0^\dagger\mathscr{D}_0^\dagger\Delta\mathscr{D}_0\mathscr{D}_0 R_{-1} = |\mathscr{C}|^2 R_{-1}. \tag{37}$$

We have, therefore, the identity,

$$\Delta\mathscr{D}_0^\dagger\mathscr{D}_0^\dagger\Delta\mathscr{D}_0\mathscr{D}_0 \equiv |\mathscr{C}|^2 \quad \text{mod} \quad \Delta\mathscr{D}_0^\dagger\mathscr{D}_0 + 2i\sigma^+ r - \lambda = 0; \tag{38}$$

and the stated value of $|\mathscr{C}|^2$ must follow from a direct evaluation of the identity. Considering then the left-hand side of equation (38), we have

$$\begin{aligned}\Delta\mathscr{D}_0^\dagger\mathscr{D}_0^\dagger\Delta\mathscr{D}_0\mathscr{D}_0 &= \Delta\mathscr{D}_0^\dagger(\mathscr{D}_0\Delta\mathscr{D}_0^\dagger + 4ir\sigma^+)\mathscr{D}_0\\ &= \Delta\mathscr{D}_0^\dagger\mathscr{D}_0(\Delta\mathscr{D}_0^\dagger\mathscr{D}_0) + 4ir\sigma^+\Delta\mathscr{D}_0^\dagger\mathscr{D}_0 + 4i\sigma^+\Delta\mathscr{D}_0\\ &\equiv \Delta\mathscr{D}_0^\dagger\mathscr{D}_0(\lambda - 2i\sigma^+ r) + 4ir\sigma^+\Delta\mathscr{D}_0^\dagger\mathscr{D}_0 + 4i\sigma^+\Delta\mathscr{D}_0.\end{aligned} \tag{39}$$

On the other hand,

$$\begin{aligned}\Delta\mathscr{D}_0^\dagger\mathscr{D}_0 r &= \Delta\mathscr{D}_0^\dagger(r\mathscr{D}_0 + 1) = r\Delta\mathscr{D}_0^\dagger\mathscr{D}_0 + \Delta\mathscr{D}_0 + \Delta\mathscr{D}_0^\dagger\\ &= r\Delta\mathscr{D}_0^\dagger\mathscr{D}_0 + 2\Delta\mathscr{D}_0 - 2iK.\end{aligned} \tag{40}$$

Making use of this last relation, we can further reduce the result (39) in the manner

$$\begin{aligned}\Delta\mathscr{D}_0^\dagger\mathscr{D}_0^\dagger\Delta\mathscr{D}_0\mathscr{D}_0 &\equiv (\lambda + 2ir\sigma^+)\Delta\mathscr{D}_0^\dagger\mathscr{D}_0 - 4\sigma^+K\\ &= (\lambda + 2ir\sigma^+)(\lambda - 2ir\sigma^+) - 4\sigma^+[(r^2 + a^2)\sigma^+ + am]\\ &= \lambda^2 - 4a^2\sigma^{+2} - 4a\sigma^+ m = |\mathscr{C}|^2,\end{aligned} \tag{41}$$

and establish the required identity.

* We shall show in §71 that $\mathscr{C}$ is in fact a real constant.

THEOREM 3. *$\mathscr{L}_0\mathscr{L}_1 S_{+1}$ is a constant multiple of S_{-1}*
and *$\mathscr{L}_0^\dagger\mathscr{L}_1^\dagger S_{-1}$ is a constant multiple of S_{+1}.*

Proof. Applying the operator $\mathscr{L}_0\mathscr{L}_1$ to equation (25) satisfied by S_{+1}, we obtain

$$\begin{aligned}-\lambda\mathscr{L}_0\mathscr{L}_1 S_{+1} &= \mathscr{L}_0\mathscr{L}_1(\mathscr{L}_0^\dagger\mathscr{L}_1 + 2a\sigma\cos\theta)S_{+1}\\ &= \mathscr{L}_0\mathscr{L}_1\mathscr{L}_0^\dagger\mathscr{L}_1 S_{+1} + 2a\sigma(\cos\theta)\mathscr{L}_0\mathscr{L}_1 S_{+1}\\ &\qquad -4a\sigma(\sin\theta)\mathscr{L}_1 S_{+1};\end{aligned} \tag{42}$$

and by successive reductions, we find

$$\begin{aligned}\mathscr{L}_0\mathscr{L}_1\mathscr{L}_0^\dagger\mathscr{L}_1 &= \mathscr{L}_0\mathscr{L}_1(\mathscr{L}_0 - 2Q)\mathscr{L}_1\\ &= \mathscr{L}_0(\mathscr{L}_1^\dagger + 2Q)\mathscr{L}_0\mathscr{L}_1 - 2\mathscr{L}_0\mathscr{L}_1 Q\mathscr{L}_1\\ &= \mathscr{L}_0\mathscr{L}_1^\dagger\mathscr{L}_0\mathscr{L}_1 + 2\mathscr{L}_0(Q\mathscr{L}_0\mathscr{L}_1) - 2\mathscr{L}_0(\mathscr{L}_0 Q + Q\cot\theta)\mathscr{L}_1\\ &= \mathscr{L}_0\mathscr{L}_1^\dagger\mathscr{L}_0\mathscr{L}_1 + 2\mathscr{L}_0(Q\mathscr{L}_0\mathscr{L}_1) - 2\mathscr{L}_0(a\sigma\cos\theta + m\cot\theta\operatorname{cosec}\theta)\mathscr{L}_1\\ &\qquad - 2\mathscr{L}_0[Q\mathscr{L}_0\mathscr{L}_1 + (a\sigma\cos\theta - m\cot\theta\operatorname{cosec}\theta)\mathscr{L}_1]\\ &= \mathscr{L}_0\mathscr{L}_1^\dagger\mathscr{L}_0\mathscr{L}_1 - 4a\sigma\mathscr{L}_0(\cos\theta)\mathscr{L}_1\\ &= \mathscr{L}_0\mathscr{L}_1^\dagger\mathscr{L}_0\mathscr{L}_1 - 4a\sigma(\cos\theta)\mathscr{L}_0\mathscr{L}_1 + 4a\sigma(\sin\theta)\mathscr{L}_1.\end{aligned} \tag{43}$$

Combining the results of the reductions (42) and (43), we obtain

$$(\mathscr{L}_0\mathscr{L}_1^\dagger - 2a\sigma\cos\theta)\mathscr{L}_0\mathscr{L}_1 S_{+1} = -\lambda\mathscr{L}_0\mathscr{L}_1 S_{+1}. \tag{44}$$

Therefore, $\mathscr{L}_0\mathscr{L}_1 S_{+1}$ satisfies the same equation (27) as S_{-1} and the first part of the theorem stated follows. At the same time, the result of the foregoing reductions is equivalent to establishing the identity

$$\mathscr{L}_0\mathscr{L}_1(\mathscr{L}_0^\dagger\mathscr{L}_1 + 2a\sigma\cos\theta) = (\mathscr{L}_0\mathscr{L}_1^\dagger - 2a\sigma\cos\theta)\mathscr{L}_0\mathscr{L}_1. \tag{45}$$

The *adjoint* of this relation (obtained by replacing θ by $\pi - \theta$), namely,

$$\mathscr{L}_0^\dagger\mathscr{L}_1^\dagger(\mathscr{L}_0\mathscr{L}_1^\dagger - 2a\sigma\cos\theta) = (\mathscr{L}_0^\dagger\mathscr{L}_1 + 2a\sigma\cos\theta)\mathscr{L}_0^\dagger\mathscr{L}_1^\dagger, \tag{46}$$

establishes the second part of the theorem.

Theorem 3 is clearly equivalent to establishing the existence of relations of the form

$$\mathscr{L}_0\mathscr{L}_1 S_{+1} = D_1 S_{-1} \quad\text{and}\quad \mathscr{L}_0^\dagger\mathscr{L}_1^\dagger S_{-1} = D_2 S_{+1}, \tag{47}$$

where D_1 and D_2 are two real constants.

THEOREM 4. *If $S_{+1}(\theta)$ and $S_{-1}(\theta)$ are both normalized to unity, then in the relations* (47), *$D_1 = D_2$, and we shall have*

$$\mathscr{L}_0\mathscr{L}_1 S_{+1} = DS_{-1} \quad\text{and}\quad \mathscr{L}_0^\dagger\mathscr{L}_1^\dagger S_{-1} = DS_{+1}. \tag{48}$$

Proof. The proof follows from two successive applications of Lemma 4 to the normalization integral (29). Thus,

$$\begin{aligned} D_1^2 &= D_1^2 \int_0^\pi S_{-1}^2 \sin\theta \mathrm{d}\theta \\ &= \int_0^\pi (\mathscr{L}_0 \mathscr{L}_1 S_{+1})(\mathscr{L}_0 \mathscr{L}_1 S_{+1}) \sin\theta \mathrm{d}\theta \\ &= \int_0^\pi (\mathscr{L}_0^\dagger \mathscr{L}_1^\dagger \mathscr{L}_0 \mathscr{L}_1 S_{+1}) S_{+1} \sin\theta \mathrm{d}\theta. \end{aligned} \tag{49}$$

On the other hand,

$$\mathscr{L}_0^\dagger \mathscr{L}_1^\dagger \mathscr{L}_0 \mathscr{L}_1 S_{+1} = D_1 \mathscr{L}_0^\dagger \mathscr{L}_1^\dagger S_{-1} = D_1 D_2 S_{+1}; \tag{50}$$

and equation (49) gives

$$D_1^2 = D_1 D_2 \int_0^\pi S_{+1}^2 \sin\theta \mathrm{d}\theta = D_1 D_2, \tag{51}$$

since S_{+1} has also been assumed to be normalized to unity. The equality of D_1 and D_2 follows from this last relation.

THEOREM 5. *The constant D of Theorem 4 has the value*

$$D^2 = \lambda\!\!\!^{-2} - 4\alpha^2 \sigma^{+2} = |\mathscr{C}|^2. \tag{52}$$

Proof. It is clear from equation (50) and the equality of D_1 and D_2 ($= D$) that what is required is to establish by direct evaluation that

$$\mathscr{L}_0^\dagger \mathscr{L}_1^\dagger \mathscr{L}_0 \mathscr{L}_1 = \lambda\!\!\!^{-2} - 4\alpha^2 \sigma^{+2} \quad \text{mod} \quad \mathscr{L}_0^\dagger \mathscr{L}_1 + 2a\sigma^{+} \cos\theta + \lambda\!\!\!^{-} = 0. \tag{53}$$

First, we observe that

$$\mathscr{L}_1^\dagger \mathscr{L}_0 = (\mathscr{L}_1 - 2Q)(\mathscr{L}_0^\dagger + 2Q) = \mathscr{L}_1 \mathscr{L}_0^\dagger + 4a\sigma^{+} \cos\theta; \tag{54}$$

and that, therefore,

$$\begin{aligned} \mathscr{L}_0^\dagger \mathscr{L}_1^\dagger \mathscr{L}_0 \mathscr{L}_1 &= \mathscr{L}_0^\dagger (\mathscr{L}_1 \mathscr{L}_0^\dagger + 4a\sigma^{+} \cos\theta) \mathscr{L}_1 \\ &\equiv \mathscr{L}_0^\dagger \mathscr{L}_1 (-\lambda\!\!\!^{-} - 2a\sigma^{+} \cos\theta) + 4a\sigma^{+} (\cos\theta) \mathscr{L}_0^\dagger \mathscr{L}_1 \\ &\qquad - 4a\sigma^{+} (\sin\theta) \mathscr{L}_1. \end{aligned} \tag{55}$$

On the other hand, by Lemma 3 of §68,

$$\begin{aligned} \mathscr{L}_0^\dagger \mathscr{L}_1 \cos\theta &= (\cos\theta) \mathscr{L}_0^\dagger \mathscr{L}_1 - (\sin\theta)(\mathscr{L}_1 + \mathscr{L}_1^\dagger) \\ &= (\cos\theta) \mathscr{L}_0^\dagger \mathscr{L}_1 - 2(\sin\theta) \mathscr{L}_1 + 2Q \sin\theta. \end{aligned} \tag{56}$$

Combining the results of the reductions (55) and (56), we obtain the desired result:

$$\mathscr{L}_0^\dagger \mathscr{L}_1^\dagger \mathscr{L}_0 \mathscr{L}_1 = -(\lambda + 2a\sigma\cos\theta)\mathscr{L}_0^\dagger \mathscr{L}_1 - 4a\sigma Q\sin\theta$$

$$\equiv (\lambda + 2a\sigma\cos\theta)(\lambda - 2a\sigma\cos\theta) - 4a\sigma(a\sigma\sin\theta + m\operatorname{cosec}\theta)\sin\theta$$

$$= \lambda^2 - 4a^2\sigma^2 - 4a\sigma m = \lambda^2 - 4\alpha^2\sigma^2 = D^2 = |\mathscr{C}|^2. \tag{57}$$

We shall call $\mathscr{C}$ and D the *Starobinsky constants*.

Returning to equations (33) and (34), we shall denote ΔR_{+1} and R_{-1} by P_{+1} and P_{-1} when their relative normalization is compatible with these equations. Thus, we shall write

$$\Delta \mathscr{D}_0 \mathscr{D}_0 P_{-1} = \mathscr{C} P_{+1} \qquad \text{and} \qquad \Delta \mathscr{D}_0^\dagger \mathscr{D}_0^\dagger P_{+1} = \mathscr{C}^* P_{-1}. \tag{58}$$

The first of these equations can be rewritten in the form

$$\mathscr{C} P_{+1} = \Delta \mathscr{D}_0 \mathscr{D}_0 P_{-1} = \Delta\left(\mathscr{D}_0^\dagger + \frac{2iK}{\Delta}\right)\mathscr{D}_0 P_{-1}$$

$$= \Delta \mathscr{D}_0^\dagger \mathscr{D}_0 P_{-1} + 2iK\mathscr{D}_0 P_{-1}; \tag{59}$$

or, by virtue of the equation satisfied by P_{-1},

$$\mathscr{C} P_{+1} = (\lambda - 2i\sigma r)P_{-1} + 2iK\mathscr{D}_0 P_{-1}. \tag{60}$$

Similarly, we find from the second of the equations (58), that

$$\mathscr{C}^* P_{-1} = (\lambda + 2i\sigma r)P_{+1} - 2iK\mathscr{D}_0^\dagger P_{+1}. \tag{61}$$

Equations (60) and (61) enable us to express the derivatives of P_{+1} and P_{-1} in terms of functions themselves; thus,

$$\frac{dP_{+1}}{dr} = +\frac{iK}{\Delta}P_{+1} - \frac{i}{2K}[(\lambda + 2i\sigma r)P_{+1} - \mathscr{C}^* P_{-1}]$$

and (62)

$$\frac{dP_{-1}}{dr} = -\frac{iK}{\Delta}P_{-1} + \frac{i}{2K}[(\lambda - 2i\sigma r)P_{-1} - \mathscr{C} P_{+1}],$$

where, it should be noted that we do not as yet know the real and the imaginary parts of $\mathscr{C}$ separately, though we do know its absolute value. This lacuna in our information will be rectified in §71.

We find from equations (48), in similar fashion, that when S_{+1} and S_{-1} are both normalized to unity, then

$$\mathscr{L}_1^\dagger S_{-1} = -\frac{1}{2Q}[\lambda - 2a\sigma\cos\theta)S_{-1} + DS_{+1}]$$

and (63)

$$\mathscr{L}_1 S_{+1} = +\frac{1}{2Q}[(\lambda + 2a\sigma\cos\theta)S_{+1} + DS_{-1}].$$

These equations clearly enable us to express the derivatives of S_{+1} and S_{-1} in terms of the functions themselves.

71. The completion of the solution

To complete the solution of Maxwell's equations, beyond separating the variables of Φ_0 and Φ_2, it is necessary to determine their relative normalization. If we choose the radial functions ΔR_{+1} and R_{-1} to be P_{+1} and P_{-1} (consistently with equations (58)), and let S_{+1} and S_{-1} be normalized to unity (consistently with equations (48)), then what remains to be ascertained is the numerical factor by which we must multiply $\Phi_2 = P_{-1}S_{-1}$, for example, if $\Delta\Phi_0$ is chosen to be $P_{+1}S_{+1}$. This factor can be ascertained only from an equation which directly relates Φ_0 and Φ_2. Such an equation can be obtained by eliminating Φ_1 from equations (13) and (14) with the aid of Lemma 2 of §68. Thus, by applying the operator $(\mathscr{L}_0 + ia\sin\theta/\bar{\rho}^*)$ to equation (13) and $(\mathscr{D}_0 + 1/\bar{\rho}^*)$ to equation (14) and adding, we obtain

$$\left(\mathscr{L}_0 + \frac{ia\sin\theta}{\bar{\rho}^*}\right)\left(\mathscr{L}_1 - \frac{ia\sin\theta}{\bar{\rho}^*}\right)\Phi_0 = \left(\mathscr{D}_0 + \frac{1}{\bar{\rho}^*}\right)\left(\mathscr{D}_0 - \frac{1}{\bar{\rho}^*}\right)\Phi_2. \tag{64}$$

On simplification, this equation yields

$$\mathscr{L}_0\mathscr{L}_1\Phi_0 = \mathscr{D}_0\mathscr{D}_0\Phi_2. \tag{65}$$

With the solutions for Φ_0 and Φ_2 given in equations (23), equation (65) requires that

$$(\mathscr{L}_0\mathscr{L}_1 S_{+1})/S_{-1} = (\Delta\mathscr{D}_0\mathscr{D}_0 R_{-1})/\Delta R_{+1}. \tag{66}$$

If we now suppose that both S_{+1} and S_{-1} are normalized to unity, then by equation (48)

$$\Delta\mathscr{D}_0\mathscr{D}_0 R_{-1} = D\Delta R_{+1}. \tag{67}$$

This last relation is consistent with the identification

$$R_{-1} = P_{-1} \qquad \text{and} \qquad \Delta R_{+1} = P_{+1}, \tag{68}$$

if

$$\mathscr{C} = \mathscr{C}^* = D = (\lambda\!\!\!^{-2} - 4\alpha^2\sigma^{+2})^{1/2}. \tag{69}$$

By equation (52), this last condition requires that $\mathscr{C}$ is real and that we need not distinguish between $\mathscr{C}$ and $\mathscr{C}^*$ (in equations (60)–(62), for example).

The solutions for ϕ_0 and ϕ_2 have now become determinate and in accordance with equations (12) and (23), we may write

$$\Delta\phi_0 = P_{+1}S_{+1} \qquad \text{and} \qquad \phi_2 = \frac{1}{2(\bar{\rho}^*)^2}P_{-1}S_{-1}. \tag{70}$$

Thus, the relation (65) has not only resolved the relative normalization of the solutions for ϕ_0 and ϕ_2, it has also determined the constant $\mathscr{C}$ without any ambiguity.

(a) *The solution for* ϕ_1

We shall now complete the solution of Maxwell's equations by determining the remaining scalar Φ_1.

First, we define the functions

$$\left.\begin{aligned} g_{+1}(r) &= \frac{1}{\mathscr{C}}(r\mathscr{D}_0 P_{-1} - P_{-1}); \quad g_{-1}(r) = \frac{1}{\mathscr{C}}(r\mathscr{D}_0^\dagger P_{+1} - P_{+1}), \\ f_{+1}(\theta) &= \frac{1}{\mathscr{C}}[(\cos\theta)\mathscr{L}_1^\dagger S_{-1} + (\sin\theta)S_{-1}], \\ f_{-1}(\theta) &= \frac{1}{\mathscr{C}}[(\cos\theta)\mathscr{L}_1 S_{+1} + (\sin\theta)S_{+1}]. \end{aligned}\right\} \tag{71}$$

It can be directly verified with the aid of the Teukolsky–Starobinsky identities that the functions $g_\pm(r)$ and $f_\pm(\theta)$ satisfy the differential equations

$$\Delta\mathscr{D}_0 g_{+1} = rP_{+1}; \quad \Delta\mathscr{D}_0^\dagger g_{-1} = rP_{-1},$$

and

$$\mathscr{L}_0^\dagger f_{+1} = S_{+1}\cos\theta; \quad \mathscr{L}_0 f_{-1} = S_{-1}\cos\theta. \tag{72}$$

Now writing equation (13) in the form

$$\Delta\mathscr{D}_0(\bar{\rho}^*\Phi_1) = (\bar{\rho}^*\mathscr{L}_1 - ia\sin\theta)\Delta\Phi_0, \tag{73}$$

and substituting for $\Delta\Phi_0$ its solution $P_{+1}S_{+1}$, we can rewrite it, successively, with the aid of equations (71) and of the identities established in §70, in the manner

$$\begin{aligned} \Delta\mathscr{D}_0(\bar{\rho}^*\Phi_1) &= (rP_{+1})\mathscr{L}_1 S_{+1} - iaP_{+1}[(\cos\theta)\mathscr{L}_1 S_{+1} + (\sin\theta)S_{+1}] \\ &= (\Delta\mathscr{D}_0 g_{+1})\mathscr{L}_1 S_{+1} - ia\mathscr{C}P_{+1}f_{-1} \\ &= (\Delta\mathscr{D}_0 g_{+1})\mathscr{L}_1 S_{+1} - ia(\Delta\mathscr{D}_0\mathscr{D}_0 P_{-1})f_{-1}. \end{aligned} \tag{74}$$

We thus obtain the relation,

$$\mathscr{D}_0(\bar{\rho}^*\Phi_1) = \mathscr{D}_0(g_{+1}\mathscr{L}_1 S_{+1} - iaf_{-1}\mathscr{D}_0 P_{-1}). \tag{75}$$

In similar fashion, we find from equation (14)

$$\begin{aligned} \mathscr{L}_0(\bar{\rho}^*\Phi_1) &= [(r - ia\cos\theta)\mathscr{D}_0 - 1]P_{-1}S_{-1} \\ &= (r\mathscr{D}_0 P_{-1} - P_{-1})S_{-1} - ia(\mathscr{D}_0 P_{-1})(\cos\theta)S_{-1} \\ &= g_{+1}\mathscr{L}_0\mathscr{L}_1 S_{+1} - ia(\mathscr{D}_0 P_{-1})\mathscr{L}_0 f_{-1}; \end{aligned} \tag{76}$$

or, equivalently,

$$\mathscr{L}_0(\bar{\rho}^*\Phi_1) = \mathscr{L}_0(g_{+1}\mathscr{L}_1 S_{+1} - iaf_{-1}\mathscr{D}_0 P_{-1}). \tag{77}$$

From a comparison of equations (75) and (77), we conclude that the required solution for Φ_1 is given by

$$\bar{\rho}^*\Phi_1 = g_{+1}(r)\mathscr{L}_1 S_{+1}(\theta) - iaf_{-1}(\theta)\mathscr{D}_0 P_{-1}(r). \qquad (78)^*$$

By treating equations (15) and (16) in analogous fashion, we find that we have the following alternative form for the solution of Φ_1:

$$-\bar{\rho}^*\Phi_1 = g_{-1}(r)\mathscr{L}_1^\dagger S_{-1}(\theta) - iaf_{+1}(\theta)\mathscr{D}_0^\dagger P_{+1}(r). \qquad (79)$$

A comparison of the solutions (78) and (79) leads to the interesting identity

$$g_{+1}\mathscr{L}_1 S_{+1} + g_{-1}\mathscr{L}_1^\dagger S_{-1} = ia(f_{-1}\mathscr{D}_0 P_{-1} + f_{+1}\mathscr{D}_0^\dagger P_{+1}). \qquad (80)$$

Postponing the verification of this identity, we observe that combining the solutions (78) and (79), we can write the solution for ϕ_1 more symmetrically in the form

$$\phi_1 = \frac{\Phi_1}{\bar{\rho}^*\sqrt{2}} = \frac{\sqrt{2}}{4(\bar{\rho}^*)^2}[(g_{+1}\mathscr{L}_1 S_{+1} - g_{-1}\mathscr{L}_1^\dagger S_{-1}) - ia(f_{-1}\mathscr{D}_0 P_{-1} - f_{+1}\mathscr{D}_0^\dagger P_{+1})]. \qquad (81)$$

(b) The verification of the identity (80)

By making use of equations (60), (61), and (63), we can rewrite the expression for $g_{\pm 1}(r)$ and $f_{\pm 1}(\theta)$ as linear combinations of the functions $P_{\pm 1}(r)$ and $S_{\pm}(\theta)$. We find

$$\left.\begin{aligned}
g_{+1} &= \frac{1}{2\mathscr{C}K}[(ir\lambda - 2\alpha^2\sigma)P_{-1} - ir\mathscr{C}P_{+1}],\\
g_{-1} &= \frac{1}{2\mathscr{C}K}[-(ir\lambda + 2\alpha^2\sigma)P_{+1} + ir\mathscr{C}P_{-1}],\\
f_{+1} &= \frac{1}{2\mathscr{C}Q}[(-\lambda\cos\theta + 2\alpha^2\sigma/a)S_{-1} - \mathscr{C}S_{+1}\cos\theta],\\
f_{-1} &= \frac{1}{2\mathscr{C}Q}[(+\lambda\cos\theta + 2\alpha^2\sigma/a)S_{+1} + \mathscr{C}S_{-1}\cos\theta].
\end{aligned}\right\} \qquad (82)$$

With the aid of these expressions, the identity (80) can be readily verified.

* Strictly, we should have added to the particular solution (78) a solution P (say) of the corresponding homogeneous equations $\mathscr{D}_0(P) = \mathscr{L}_0(P) = 0$. But the solution for P, namely,

$$P = \text{constant}\exp[-i\sigma(r_* + ia\cos\theta)]\cdot\cot^m(\theta/2),$$

where r_* is defined by the equation (see equation (100) below)

$$dr_* = (r^2 + \alpha^2)dr/\Delta \quad (\alpha^2 = a^2 + am/\sigma),$$

is singular at $\theta = 0$ and $\theta = \pi/2$; and on this account we have not included it in the solution for $\bar{\rho}^*\Phi_1$.

(c) *The solution for the vector potential*

We shall now show how an explicit solution for the vector potential, A, can be obtained from the solutions for the Maxwell scalars ϕ_0, ϕ_1, and ϕ_2 we have found.

We start with the expression

$$F_{ij} = \partial_j A_i - \partial_i A_j, \tag{83}$$

and express ϕ_0 and ϕ_2 in terms of A. Thus,

$$\begin{aligned}\phi_0 &= F_{ij} l^i m^j = l^i m^j (\partial_j A_i - \partial_i A_j) \\ &= l^i \delta A_i - m^j D A_j = \frac{1}{\bar{\rho}\sqrt{2}} l^i \mathscr{L}_0^\dagger A_i - m^j \mathscr{D}_0 A_j \\ &= \frac{1}{\bar{\rho}\sqrt{2}} \mathscr{L}_0^\dagger \left(\frac{r^2+a^2}{\Delta} A_t + A_r + \frac{a}{\Delta} A_\varphi \right) \\ &\quad - \frac{1}{\bar{\rho}\sqrt{2}} \mathscr{D}_0 (iaA_t \sin\theta + A_\theta + iA_\varphi \operatorname{cosec}\theta),\end{aligned} \tag{84}$$

and

$$\begin{aligned}\phi_2 &= \bar{m}^i n^j (\partial_j A_i - \partial_i A_j) = \bar{m}^i \vartriangle A_i - n^j \delta^* A_j \\ &= -\frac{\bar{m}^i \Delta}{2\rho^2} \mathscr{D}_0^\dagger A_i - \frac{n^j}{\bar{\rho}^* \sqrt{2}} \mathscr{L}_0 A_j \\ &= -\frac{1}{\rho^2 \bar{\rho}^* 2\sqrt{2}} \{ \Delta \mathscr{D}_0^\dagger (-iaA_t \sin\theta + A_\theta - iA_\varphi \operatorname{cosec}\theta) \\ &\qquad + \mathscr{L}_0 [-\Delta A_r + (r^2+a^2) A_t + aA_\varphi] \}.\end{aligned} \tag{85}$$

Now, letting

$$\left.\begin{aligned}\Delta F_{+1} &= (r^2+a^2)A_t + \Delta A_r + aA_\varphi, \\ \Delta F_{-1} &= (r^2+a^2)A_t - \Delta A_r + aA_\varphi, \\ G_{+1} &= iaA_t \sin\theta + A_\theta + iA_\varphi \operatorname{cosec}\theta, \\ \text{and}\qquad G_{-1} &= -iaA_t \sin\theta + A_\theta - iA_\varphi \operatorname{cosec}\theta,\end{aligned}\right\} \tag{86}$$

and making use of the solutions for ϕ_0 and ϕ_2 given in equations (70) and of the definitions (72), we can rewrite equations (84) and (85) in the forms

$$\begin{aligned}\frac{1}{\sqrt{2}} (\mathscr{L}_0^\dagger \Delta F_{+1} - \Delta \mathscr{D}_0 G_{+1}) &= (r + ia\cos\theta) P_{+1} S_{+1} \\ &= S_{+1} \Delta \mathscr{D}_0 g_{+1} + iaP_{+1} \mathscr{L}_0^\dagger f_{+1},\end{aligned} \tag{87}$$

and

$$-\frac{1}{\sqrt{2}}(\Delta\mathscr{D}_0^\dagger G_{-1}+\mathscr{L}_0\Delta F_{-1})=(r+ia\cos\theta)P_{-1}S_{-1}$$
$$=S_{-1}\Delta\mathscr{D}_0^\dagger g_{-1}+iaP_{-1}\mathscr{L}_0 f_{-1}. \quad (88)$$

These equations are readily solved for $F_{\pm1}$ and $G_{\pm1}$. We find

$$\left.\begin{aligned}\Delta F_{+1}&=(iaP_{+1}f_{+1}+\Delta\mathscr{D}_0 H_{+1})\surd 2,\\ \Delta F_{-1}&=(-iaP_{-1}f_{-1}-\Delta\mathscr{D}_0^\dagger H_{-1})\surd 2,\\ G_{+1}&=(-g_{+1}S_{+1}+\mathscr{L}_0^\dagger H_{+1})\surd 2,\\ \text{and}\quad G_{-1}&=(-g_{-1}S_{-1}+\mathscr{L}_0 H_{-1})\surd 2,\end{aligned}\right\} \quad (89)$$

where H_{+1} and H_{-1}, as introduced here, are arbitrary functions but which, as we shall show presently, are not independent.

With the foregoing solutions for $F_{\pm1}$ and $G_{\pm1}$, we can solve equations (86) explicitly for the components of the vector potential. We find

$$\Delta A_r=\tfrac{1}{2}\Delta(F_{+1}-F_{-1})$$
$$=\frac{ia}{\sqrt{2}}(P_{+1}f_{+1}+P_{-1}f_{-1})+\frac{1}{\sqrt{2}}(\Delta\mathscr{D}_0 H_{+1}+\Delta\mathscr{D}_0^\dagger H_{-1}), \quad (90)$$

$$A_\theta=\tfrac{1}{2}(G_{+1}+G_{-1})$$
$$=-\frac{1}{\sqrt{2}}(g_{+1}S_{+1}+g_{-1}S_{-1})+\frac{1}{\sqrt{2}}(\mathscr{L}_0^\dagger H_{+1}+\mathscr{L}_0 H_{-1}), \quad (91)$$

$$\rho^2 A_t=\tfrac{1}{2}[\Delta(F_{+1}+F_{-1})+ia(G_{+1}-G_{-1})\sin\theta]$$
$$=\frac{ia}{\sqrt{2}}[(P_{+1}f_{+1}-P_{-1}f_{-1})-(g_{+1}S_{+1}-g_{-1}S_{-1})\sin\theta]$$
$$+\frac{1}{\sqrt{2}}[(\Delta\mathscr{D}_0 H_{+1}-\Delta\mathscr{D}_0^\dagger H_{-1})+ia(\mathscr{L}_0^\dagger H_{+1}-\mathscr{L}_0 H_{-1})\sin\theta], \quad (92)$$

and

$$\rho^2 A_\varphi=-\tfrac{1}{2}[a\Delta(F_{+1}+F_{-1})\sin^2\theta+i(r^2+a^2)(G_{+1}-G_{-1})\sin\theta]$$
$$=-\frac{i}{\sqrt{2}}[a^2(P_{+1}f_{+1}-P_{-1}f_{-1})\sin^2\theta$$
$$-(r^2+a^2)(g_{+1}S_{+1}-g_{-1}S_{-1})\sin\theta]$$
$$-\frac{1}{\sqrt{2}}[a(\Delta\mathscr{D}_0 H_{+1}-\Delta\mathscr{D}_0^\dagger H_{-1})\sin^2\theta$$
$$+i(r^2+a^2)(\mathscr{L}_0^\dagger H_{+}-\mathscr{L}_0 H_{-1})\sin\theta]. \quad (93)$$

There is, as we have stated, a restriction on the choice of the functions H_{+1} and H_{-1}: it follows from evaluating ϕ_1 in terms A and comparing it with the solution (81) we have already found. Thus, evaluating ϕ_1 in terms of the vector potential, we find

$$\phi_1 = \tfrac{1}{2}(l^i n^j + \bar{m}^i m^j)F_{ij} = \tfrac{1}{2}(l^i n^j + \bar{m}^i m^j)(\partial_j A_i - \partial_i A_j)$$

$$= -\frac{\Delta}{2\rho^2} l^i \mathscr{D}_0^\dagger A_i - n^j \mathscr{D}_0 A_j + \frac{\bar{m}^i}{\bar{\rho}\sqrt{2}} \mathscr{L}_0^\dagger A_i - \frac{m^j}{\bar{\rho}^*\sqrt{2}} \mathscr{L}_0 A_j$$

$$= \frac{1}{2\rho^2}[iKA_r - (r^2+a^2)A_{t,r} - aA_{\varphi,r} - QA_\theta - i(aA_{t,\theta}\sin^2\theta + A_{\varphi,\theta})\operatorname{cosec}\theta]. \quad (94)$$

Now substituting for the components of A from equations (90)–(93) and equating the resulting expression with the solution (81) for ϕ_1, we find, after some considerable simplifications, that we are left with

$$\mathscr{D}_0^\dagger \frac{\Delta \mathscr{D}_0 H_{+1}}{(\bar{\rho}^*)^2} + \mathscr{L}_1 \frac{\mathscr{L}_0^\dagger H_{+1}}{(\bar{\rho}^*)^2} - \mathscr{D}_0 \frac{\Delta \mathscr{D}_0^\dagger H_{-1}}{(\bar{\rho}^*)^2} - \mathscr{L}_1^\dagger \frac{\mathscr{L}_0 H_{-1}}{(\bar{\rho}^*)^2} = 0. \quad (95)$$

This equation defines the freedom we have in the choice of gauge for the vector potential. It is reminiscent of the Coulomb gauge.

72. The transformation of Teukolsky's equations to a standard form

We shall consider in place of equations (26) and (28) the more general equations

$$[\Delta \mathscr{D}_{1-|s|} \mathscr{D}_0^\dagger - 2(2|s|-1)i\sigma r]P_{+|s|} = \lambda P_{+|s|} \quad (96)$$

and

$$[\Delta \mathscr{D}_{1-|s|}^\dagger \mathscr{D}_0 + 2(2|s|-1)i\sigma r]P_{-|s|} = \lambda P_{-|s|}, \quad (97)^*$$

applicable to massless fields of spin $|s|$. Thus, for $|s| = 1$, the equations reduce to equations (26) and (28) appropriate to photons of spin 1; and we shall find in Chapters 9 and 10 that for $|s| = 2$ and $|s| = 1/2$ they are the equations governing the propagation of gravitational waves and of the two-component neutrinos, respectively.

Since $P_{+|s|}$ and $P_{-|s|}$ satisfy complex–conjugate equations, it will suffice to consider the equation for $P_{+|s|}$ only. Also, for convenience we shall write simply s in place of $|s|$ with the understanding that it is to be considered positive and allowed the values 2, 1, and 1/2. The equation we shall consider, then, is

$$[\Delta \mathscr{D}_{1-s} \mathscr{D}_0^\dagger - 2(2s-1)i\sigma r]P_{+s} = \lambda P_{+s}. \quad (98)$$

* While $P_{+|s|}$ and $P_{-|s|}$, in general, stand for $\Delta^s R_{+s}$ and R_{-s}, we are *not* here assuming that the functions have been relatively normalized in any particular way.

In this equation m explicitly occurs through

$$K = (r^2 + a^2)\sigma + am \tag{99}$$

in the operators $\mathscr{D}$ and $\mathscr{D}^\dagger$. We shall now show how an *explicit* reference to m can be eliminated by a suitable change of variables.

First, we introduce in place of r a new independent variable r_* defined by

$$\frac{\mathrm{d}}{\mathrm{d}r_*} = \frac{\Delta}{\varpi^2}\frac{\mathrm{d}}{\mathrm{d}r}, \tag{100}$$

where

$$\varpi^2 = r^2 + \alpha^2 \qquad \text{and} \qquad \alpha^2 = a^2 + (am/\sigma). \tag{101}$$

In view of the admissibility of negative values of α^2 (when m is negative and $\sigma \to 0$), it is clear that the $r_*(r)$-relation, that follows from equation (100), can under certain circumstances become double-valued. We shall have to consider, in due course, how this contingency, when it arises, should be met; but, meantime, we shall consider only the formal consequences of this change of variable.

One immediate consequence of the change in the independent variable to r_* is the resulting simplicity of the operators $\mathscr{D}_0$ and $\mathscr{D}_0^\dagger$:

$$\mathscr{D}_0 = \frac{\varpi^2}{\Delta}\Lambda_+ \qquad \text{and} \qquad \mathscr{D}_0^\dagger = \frac{\varpi^2}{\Delta}\Lambda_-, \tag{102}$$

where, in conformity with our standard usage,

$$\Lambda_\pm = \frac{\mathrm{d}}{\mathrm{d}r_*} \pm i\sigma. \tag{103}$$

The simplicity arises from the fact that by virtue of the definition (101)

$$K = \varpi^2\sigma. \tag{104}$$

Second, in addition to the change in the independent variable, we shall also change the dependent variable. We shall let

$$Y = |\varpi^2|^{-s+1/2}P_{+s}. \tag{105}$$

For integral values of s, this transformation is singular if ϖ^2 should change sign in the range of r of interest—a contingency to which we shall return presently.

With the change of variables made, equation (98) becomes

$$\Delta^{s-1}\varpi^2\left\{\Lambda_+\left[\frac{\varpi^2}{\Delta^s}\Lambda_-(|\varpi^2|^{s-1/2}Y)\right]\right\}$$
$$-2(2s-1)i\sigma r|\varpi^2|^{s-1/2}Y - \lambda|\varpi^2|^{s-1/2}Y = 0. \tag{106}$$

On expanding this equation, we find

$$\frac{|\varpi^2|^{s+3/2}}{\Delta}\left\{\Lambda^2 Y+\left[\frac{d}{dr_*}\lg\frac{|\varpi^2|^{2s}}{\Delta^s}\right]\Lambda_- Y+2i\sigma^+(2s-1)r\frac{\Delta}{\varpi^4}Y\right\}$$
$$+\Delta^{s-1}\varpi^2\left[\frac{d}{dr_*}\left(\frac{\varpi^2}{\Delta^s}\frac{d}{dr_*}|\varpi^2|^{s-1/2}\right)\right]Y-[2(2s-1)i\sigma^+ r|\varpi^2|^{s-1/2}$$
$$+\lambda|\varpi^2|^{s-1/2}]Y=0. \quad (107)$$

With the definitions

$$P=\frac{d}{dr_*}\lg\frac{|\varpi^2|^{2s}}{\Delta^s}=\frac{2s}{\varpi^4}[2r\Delta-\varpi^2(r-M)], \quad (108)$$

and

$$Q=\lambda\frac{\Delta}{\varpi^4}-\frac{\Delta^{s+1}}{|\varpi^2|^{s+3/2}}\frac{d}{dr}\left(\frac{1}{\Delta^{s-1}}\frac{d}{dr}|\varpi^2|^{s-1/2}\right)$$
$$=\frac{\Delta}{\varpi^4}\left\{\lambda-(2s-1)\left[\frac{\Delta-2(s-1)r(r-M)}{\varpi^2}+(2s-3)\frac{r^2\Delta}{\varpi^4}\right]\right\}, \quad (109)$$

equation (107) can be reduced to the form

$$\Lambda^2 Y+P\Lambda_- Y-QY=0 \quad (110)$$

—a form which we have encountered already in our treatment of the perturbations of the Schwarzschild black-hole (Ch. 4, equation (284)).

(a) The $r_*(r)$-relation

Integrating equation (100), we obtain the relation

$$r_*=r+\frac{2Mr_++(am/\sigma^+)}{r_+-r_-}\lg\left(\frac{r}{r_+}-1\right)-\frac{2Mr_-+(am/\sigma^+)}{r_+-r_-}\lg\left(\frac{r}{r_-}-1\right) \quad (r>r_+). \quad (111)$$

When considering the external perturbations of the Kerr black-hole we are indifferent to what 'happens' inside the event horizon at $r=r_+$. We need, therefore, be concerned only with the $r_*(r)$-relation for $r>r_+$.

It is now apparent from equation (111) that the $r_*(r)$-relation is single-valued for $r>r_+$ only so long as

$$r_+^2+\alpha^2=2Mr_+-a^2+\alpha^2=2Mr_++(am/\sigma^+)>0. \quad (112)$$

When this inequality obtains,

$$r_*\to+\infty \text{ when } r\to\infty \text{ and } r_*\to-\infty \text{ when } r\to r_++0, \quad (113)$$

and the $r_*(r)$-relation is a monotonic one. Letting

$$\sigma_s^+=-am/2Mr_+ \quad (\text{for } m \text{ negative}), \quad (114)$$

and remembering that our convention with respect to σ^+ is that it is to be positive, we conclude that *so long as* $\sigma^+ > \sigma_s$, *the* $r_*(r)$-*relation is single-valued outside the event horizon and the range of* r_* *of interest is the entire interval,* $(-\infty, +\infty)$. *But if* $0 < \sigma^+ < \sigma^+_s$ *and* $r_+^2 + \alpha^2 < 0$, *the* $r_*(r)$-*relation is double-valued:* $r_* \to +\infty$ *both when* $r \to \infty$ *and when* $r \to r_+ + 0$. In the latter case, in the neighbourhood of $r = |\alpha|$, the $r_*(r)$-relation has the behaviour

$$r_* = r_*(|\alpha|) + \frac{|\alpha|}{\Delta_{|\alpha|}}(r - |\alpha|)^2 + O((r - |\alpha|)^3). \tag{115}$$

Also, when $0 < \sigma^+ < \sigma^+_s$, the functions P and Q in equation (110) become singular at $r = |\alpha|$. Accordingly, in these cases, the equation must be considered separately in the two branches of the $r_*(r)$-relation, namely for $\infty > r > |\alpha|$ and $|\alpha| > r > r_+$.

We shall show in §§ 74 and 79 that *in the interval,* $0 < \sigma^+ < \sigma^+_s$, *the reflexion coefficient for incident waves of integral spins exceeds unity.* This is the phenomenon of *super-radiance*: it is the analogue, in the domain of wave propagation, of the Penrose process in the domain of particle dynamics. While we shall consider the origins of this super-radiance in the framework of equations (96) and (97) in § 75, we may yet draw attention here to the similarity of the present inequality,

$$2Mr_+\sigma^+ < -am, \tag{116}$$

(equivalent to $\sigma^+ < \sigma^+_s$) with the inequality, Chapter 7, (352): they are the *same* if we identify $h\sigma^+$ with δM and hm with δJ, where h is Planck's constant!

Finally, it should be noted that α^2 becomes negative already before σ^+_s:

$$\alpha^2 \leqslant 0 \text{ for } \sigma^+ \leqslant \sigma^+_c = -m/a; \tag{117}$$

and $\alpha^2 = 0$ when σ^+ is '*co-rotational*'. *In the interval,* $\sigma^+_s < \sigma^+ < \sigma^+_c$, *while* α^2 *is negative,* $r_+^2 + \alpha^2 > 0$ *and the* $r_*(r)$-*relation continues to be single-valued.*

73. A general transformation theory and the reduction to a one-dimensional wave-equation

The form (110) to which the general Teukolsky equation (98) was reduced in §72 is identical to the one to which the equations of the Newman–Penrose formalism, in the context of the perturbations of the Schwarzschild black-hole, were reduced in § 20. And as in that context, we shall seek a transformation which will bring equation (110) to a one-dimensional wave-equation of the form

$$\Lambda^2 Z = VZ, \tag{118}$$

where V is a potential function to be determined.

The theory we shall now outline differs from the one presented in § 30 only in the one respect derived from the different definitions of P (cf. equations

(108) and Ch. 4, (286)). It is, however, convenient to have the basic formulae written in the forms we shall need.

We assume, then, that Y is related to Z in the manner (cf. Ch. 4, equations (287) and (288))

$$Y = f\Lambda_+\Lambda_+ Z + W\Lambda_+ Z, \tag{119}$$

where f and W are certain functions of r_* to be determined. Equation (119) can be written alternatively in the form

$$Y = f V Z + T\Lambda_+ Z, \tag{120}$$

where

$$T = W + 2i\sigma f. \tag{121}$$

Applying the operator Λ_- to equation (119) and making use of the fact Z has been assumed to satisfy equation (118), we find that we can write (cf. Ch. 4, equations (290)–(292))

$$\Lambda_- Y = -\frac{\Delta^s}{\varpi^{4s}}\beta Z + R\Lambda_+ Z, \tag{122}$$

where

$$-\frac{\Delta^s}{\varpi^{4s}}\beta(r_*) = \frac{\mathrm{d}}{\mathrm{d}r_*} f V + W V, \tag{123}$$

and

$$R = f V + \frac{\mathrm{d}T}{\mathrm{d}r_*}. \tag{124}$$

We must now require that Y defined by equation (119) does satisfy equation (110) by virtue of the equation satisfied by Z. By the process of elimination described in Chapter 4, § 30, we obtain the equations (cf. Ch. 4, equations (298) and (299))

$$RV - \frac{\Delta^s}{\varpi^{4s}}\frac{\mathrm{d}\beta}{\mathrm{d}r_*} = Q f V \tag{125}$$

and

$$\frac{\mathrm{d}}{\mathrm{d}r_*}\left(\frac{\varpi^{4s}}{\Delta^s}R\right) = \frac{\varpi^{4s}}{\Delta^s}\left(QT - 2i\sigma R\right) + \beta. \tag{126}$$

It can now be verified that equations (122)–(126) allow the integral

$$\frac{\varpi^{4s}}{\Delta^s} R f V + \beta T = K = \text{constant}. \tag{127}$$

This integral enables us to write the inverse of equations (120) and (122) in the forms

$$\frac{\Delta^s}{\varpi^{4s}} K Z = RY - T\Lambda_- Y \tag{128}$$

and

$$K\Lambda_+ Z = \beta Y + \frac{\varpi^{4s}}{\Delta^s} f V \Lambda_- Y. \tag{129}$$

It can now be verified that equations (120), (122), (128), and (129), by virtue of equations (123)–(126), are necessary and sufficient conditions to ensure that equation (110) implies equation (118), and conversely.

Since equations (122)–(126) allow the integral (127), it will suffice to consider the following equations:

$$R - f V = \frac{dT}{dr_*}, \tag{130}$$

$$\frac{d}{dr_*}\left(\frac{\varpi^{4s}}{\Delta^s} R\right) = \frac{\varpi^{4s}}{\Delta^s}(QT - 2i\sigma^+ R) + \beta, \tag{131}$$

$$R\left(R - \frac{dT}{dr_*}\right) + \frac{\Delta^s}{\varpi^{4s}} \beta T = \frac{\Delta^s}{\varpi^{4s}} K, \tag{132}$$

and

$$(R - Q f) V = \frac{\Delta^s}{\varpi^{4s}} \frac{d\beta}{dr_*}, \tag{133}$$

where it may be noted that equation (132) is an alternative form of the integral (127) in which $f V$ has been replaced by $R - T_{,r_*}$ in accordance with equation (130).

Equations (130)–(133) provide four equations for the five functions f, β, R, T, and V. There is, accordingly, considerable latitude in seeking useful solutions of the equations. But for the particular functions Q which follow from equation (109) for $s = 1/2$, 1, and 2, it will appear that we can write down explicit solutions of the equations. The transformation of equation (110) to the form of a one-dimensional wave-equation can be effected and explicit expressions for the potential V can be found.

Assuming, then, that equation (110) can be transformed into a one-dimensional wave-equation of the form (118), let Z_1 and Z_2 denote two independent solutions. Their Wronskian will clearly be a constant (cf. Ch. 4, equation (354), *et seq.*); thus,

$$[Z_1, Z_2]_{r_*} = \text{constant}, \tag{134}$$

where the subscript r_* signifies that the derivatives in evaluating the Wronskian are with respect to r_*. Some care should be exercized in interpreting (134) when σ^+ is in the super-radiant interval. For in that case, the potential V will be found to be singular (even as the functions P and Q in equation (110) are) at $r = |\alpha| > r_+$. In these cases, as we have indicated earlier, the equation must be considered, separately, in the two branches of the $r_*(r)$-

relation. The Wronskian of two independent solutions will take constant values in the two branches separately; but they need not be the same. On the other hand, if the two solutions, Z_1 and Z_2 considered, correspond to two distinct solutions of Teukolsky's equations, then the Wronskian, $[Z_1, Z_2]_{r_*}$, in the two branches must be related; and the relation between them can be found as follows.

Let Y_1 and Y_2 correspond to the two solutions Z_1 and Z_2; then,

$$[Y_1, Y_2]_{r_*} = Y_1\Lambda_- Y_2 - Y_2\Lambda_- Y_1. \tag{135}$$

Now substituting for Y and $\Lambda_- Y$ in accordance with equations (120) and (122), we find

$$[Y_1, Y_2]_{r_*} = \left(R \not{f} V + \frac{\Delta^s}{\varpi^{4s}}\beta T\right)(Z_1\Lambda_+ Z_2 - Z_2\Lambda_+ Z_1); \tag{136}$$

or, making use of the integral (127), we have

$$[Y_1, Y_2]_{r_*} = -K\frac{\Delta^s}{\varpi^{4s}}[Z_1, Z_2]_{r_*}. \tag{137}$$

But by equation (105)

$$\begin{aligned}[Y_1, Y_2]_{r_*} &= -\frac{1}{|\varpi^2|^{2s-1}}[P_s(1), P_s(2)]_{r_*} \\ &= -\frac{\Delta}{\varpi^2|\varpi^2|^{2s-1}}[P_s(1), P_s(2)]_{r},\end{aligned} \tag{138}$$

where $P_s(1)$ and $P_s(2)$ are the independent solutions of the Teukolsky equation from which the solutions Y_1 and Y_2 are derived. By combining equations (137) and (138), we obtain

$$K[Z_1, Z_2]_{r_*} = -\Delta^{1-s}\frac{\varpi^{4s-2}}{|\varpi^2|^{2s-1}}[P_s(1), P_s(2)]_r. \tag{139}$$

Since the solutions of Teukolsky's equations can have no singularities in the interval, $r_+ < r < \infty$, we conclude from equation (139) that

$$\{[Z_1, Z_2]_{r_*}\}_{r<|\alpha|} = (-1)^{2s-1}\{[Z_1, Z_2]_{r_*}\}_{r>|\alpha|}, \tag{140}$$

in case $\alpha^2 < 0$ and $r_+ < |\alpha|$, a situation which will occur in the super-radiant interval, $0 < \sigma^+ < \sigma^+_s$. Therefore, *for* $s = 1$ *and* 2, *only the sign of the Wronskian,* $[Z_1, Z_2]_{r_*}$, *changes as we cross the singularity at* $r = |\alpha|(> r_+)$ *while for* $s = 1/2$, *the Wronskian,* $[Z_1, Z_2]_{r_*}$, *retains its value.*

Our considerations, so far, have been restricted to equation (110) governing P_{+s}. We now inquire how these considerations will be affected when they are applied to the complex-conjugate equation governing P_{-s}. When the occasion to use both equations arises, we shall find it convenient to distinguish the corresponding solutions Y, of the complex-conjugate equations, by $Y^{(+\sigma)}$ and $Y^{(-\sigma)}$. Thus, we shall write

$$\Lambda^2 Y^{(\pm\sigma)} + P\Lambda_{\mp} Y^{(\pm\sigma)} - QY^{(\pm\sigma)} = 0 \tag{141}$$

for the equations governing them. And when we seek to transform the equation governing $Y^{(-\sigma)}$ to a one-dimensional wave-equation for a function $Z^{(-\sigma)}$ (in contrast to $Z^{(+\sigma)}$) by a sequence of transformations similar to those adopted in the context of the equation governing $Y^{(+\sigma)}$, we find that the same equations apply if we reverse the sign of σ wherever it occurs *explicitly*; and this remark applies in particular to equations (130)–(133). It should, however, be noted that α^2 remains unchanged in both sets of equations: for, 'complex-conjugation' requires that the signs of σ and m are simultaneously reversed: for example, $\mathscr{D}_n$ and $\mathscr{D}_n^\dagger$, as defined, are complex conjugates only for this *simultaneous* reversal in the signs of σ and m.

74. Potential barriers for incident electromagnetic waves

The considerations of the two preceding sections have ignored or side-stepped several questions which arise from the singular nature of the underlying transformations: the double-valuedness of the $r_*(r)$-relation in the super-radiant interval, $0 < \sigma < \sigma_s$, and the associated singularities in the derived potentials at $r = |\alpha|\,(> r_+)$. We shall clarify these questions in the context of the electromagnetic perturbations ($s = 1$) we are presently considering. Fortunately, in this case, the solutions of the various equations are sufficiently simple that the different aspects of the problem can be isolated and resolved.

For $s = 1$, equations (109) and (130)–(133) give

$$Q = \frac{\Delta}{\varpi^4}\left(\lambda - \alpha^2 \frac{\Delta}{\varpi^4}\right) \tag{142}$$

and

$$R - fV = \frac{dT}{dr_*}, \tag{143}$$

$$\frac{\Delta}{\varpi^4}\frac{d}{dr_*}\left(\frac{\varpi^4}{\Delta}R\right) = QT - 2i\sigma R + \frac{\Delta}{\varpi^4}\beta, \tag{144}$$

$$R\left(R - \frac{dT}{dr_*}\right) + \frac{\Delta}{\varpi^4}\beta T = \frac{\Delta}{\varpi^4}K, \tag{145}$$

$$RV - QfV = \frac{\Delta}{\varpi^4}\frac{d\beta}{dr_*}. \tag{146}$$

We shall now show that these equations allow solutions compatible with the assumptions

$$T = \text{a constant} \qquad \text{and} \qquad R = q\frac{\Delta}{\varpi^4}, \tag{147}$$

where q is a further constant. With these assumptions, equation (144) gives

$$\beta = 2i\sigma^+ q - T\left(\lambda - \alpha^2\frac{\Delta}{\varpi^4}\right). \tag{148}$$

Inserting this solution for β and the assumed form for R in equation (145), we find

$$q^2\frac{\Delta}{\varpi^4} + T\left[2i\sigma^+ q - T\left(\lambda - \alpha^2\frac{\Delta}{\varpi^4}\right)\right] = K. \tag{149}$$

This equation clearly requires that

$$q^2 = -T^2\alpha^2 \qquad \text{or} \qquad q = \pm iT\alpha \tag{150}$$

and

$$K = 2i\sigma^+ qT - T^2\lambda = -T^2(\lambda \pm 2\sigma^+\alpha) = \text{a constant}, \tag{151}$$

as required. Finally, equation (146) gives

$$\pm i(V - Q) = \alpha\frac{d}{dr_*}\left(\frac{\Delta}{\varpi^4}\right), \tag{152}$$

or,

$$\begin{aligned} V &= Q \mp i\alpha\frac{\Delta}{\varpi^2}\frac{d}{dr}\left(\frac{\Delta}{\varpi^4}\right) \\ &= \frac{\Delta}{\varpi^4}\left[\lambda - \alpha^2\frac{\Delta}{\varpi^4} \mp i\alpha\varpi^2\frac{d}{dr}\left(\frac{\Delta}{\varpi^4}\right)\right]. \end{aligned} \tag{153}$$

Thus, for Q given by equation (142), we do have a solution of equations (143)–(146) compatible with the assumptions (147). We also notice that the solution for V is independent of the choice of the constant T; that we obtain two distinct solutions by choosing one or the other sign in equation (153); and finally that V is complex for $\sigma^+ > \sigma^+_c \; (= -m/a)$.

Since T appears as a simple scaling factor in the solutions for R, β, and K, we may, without loss of generality, suppose that

$$T = 2i\sigma^+, \tag{154}$$

in which case,

$$\left.\begin{aligned} &R = \mp 2\sigma^+\alpha\frac{\Delta}{\varpi^4}, \quad K = 4\sigma^{+2}(\lambda \pm 2\sigma^+\alpha), \\ \text{and} \\ &\beta = -2i\sigma^+\left(\lambda - \alpha^2\frac{\Delta}{\varpi^4} \pm 2\sigma^+\alpha\right); \end{aligned}\right\} \tag{155}$$

and the equations relating the solutions of equations (110) and (118) (for the case $s = 1$) become

$$Y = \mp 2\sigma^{+}\alpha \frac{\Delta}{\varpi^4} Z + 2i\sigma^{+}\Lambda_+ Z, \tag{156}$$

$$\Lambda_- Y = 2i\sigma^{+}\frac{\Delta}{\varpi^4}\left(\lambda - \alpha^2\frac{\Delta}{\varpi^4} \pm 2\sigma^{+}\alpha\right)Z \mp 2\sigma^{+}\alpha\frac{\Delta}{\varpi^4}\Lambda_+ Z, \tag{157}$$

$$KZ = \mp 2\sigma^{+}\alpha Y - 2i\sigma^{+}\frac{\varpi^4}{\Delta}\Lambda_- Y, \tag{158}$$

$$K\Lambda_+ Z = -2i\sigma^{+}\left(\lambda - \alpha^2\frac{\Delta}{\varpi^4} \pm 2\sigma^{+}\alpha\right)Y \mp 2\sigma^{+}\alpha\Lambda_- Y. \tag{159}$$

The two solutions, which we shall distinguish by Z_+ and Z_- and which follow from equations (156)–(159) by choosing the upper or the lower sign, are simply related (even as the solutions, belonging to the axial and the polar perturbations of the Schwarzschild and the Reissner–Nordstrom black-holes, are related). Thus, substituting in the relation,

$$K_+ Z_+ = -2\sigma^{+}\alpha Y - 2i\sigma^{+}\frac{\varpi^4}{\Delta}\Lambda_- Y, \tag{160}$$

appropriate for Z_+, the expressions for Y and $\Lambda_- Y$ relating them to Z_-, we obtain

$$\begin{aligned} K_+ Z_+ = & -2\sigma^{+}\alpha\left(+2\sigma^{+}\alpha\frac{\Delta}{\omega^4}Z_- + 2i\sigma^{+}\Lambda_+ Z_-\right) \\ & -2i\sigma^{+}\left[2i\sigma^{+}\left(\lambda - \alpha^2\frac{\Delta}{\varpi^4} - 2\sigma^{+}\alpha\right)Z_- + 2\sigma^{+}\alpha\Lambda_+ Z_-\right]. \end{aligned} \tag{161}$$

On simplification, equation (161) yields

$$K_+ Z_+ = 4\sigma^{+2}\left(\lambda - 2\alpha^2\frac{\Delta}{\varpi^4} - 2\sigma^{+}\alpha\right)Z_- - 8i\sigma^{+2}\alpha\Lambda_+ Z_-. \tag{162}$$

Therefore, given any solution Z_-, belonging to the potential V_-, we can derive a solution Z_+, belonging to the potential V_+. In particular, it follows from equation (162) that

$$K_+ Z_+ \to K_- Z_- - 8i\alpha\sigma^{+2}\Lambda_+ Z_- \quad (r \to \infty \quad \text{and} \quad r \to r_+ + 0). \tag{163}$$

(a) The distinction between $Z^{(+\sigma^{+})}$ *and* $Z^{(-\sigma^{+})}$

Distinguishing Y and the function satisfying the complex-conjugate equation by $Y^{(+\sigma^{+})}$ and $Y^{(-\sigma^{+})}$ (as we have in equation (141)) and the functions satisfying the associated one-dimensional wave-equations by $Z^{(+\sigma^+)}$ and $Z^{(-\sigma^+)}$, we find by substitutions, analogous to (154) and (155), in equations (143)–(146) with the signs of σ^+ reversed, that *$Z^{(+\sigma^+)}$ and $Z^{(-\sigma^+)}$ satisfy wave equations with the same potential* (153):

$$\Lambda^2 Z^{(\pm\sigma)} = V Z^{(\pm\sigma^+)}. \tag{164}$$

Since V is complex for $\sigma^+ > \sigma^+_c = -m/a$, it follows that in these cases *$Z^{(+\sigma^+)}$ and $Z^{(-\sigma^+)}$, unlike $Y^{(+\sigma^+)}$ and $Y^{(-\sigma^+)}$, do not satisfy complex-conjugate equations.*

If we select for $Z^{(+\sigma^+)}$ the solution belonging to the potential

$$V_+ = \frac{\Delta}{\varpi^4}\left[\lambda - \alpha^2\frac{\Delta}{\varpi^4} - i\alpha\varpi^2\frac{d}{dr}\left(\frac{\Delta}{\varpi^4}\right)\right], \tag{165}$$

(i.e., the solution designated by Z_+ in equations (160)–(163)), then we have the relations

$$T^{(\pm\sigma^+)} = \pm 2i\sigma^+, \quad R^{(\pm\sigma^+)} = \mp 2\sigma^+\alpha\frac{\Delta}{\varpi^4},$$
$$\beta^{(\pm\sigma^+)} = \mp 2i\sigma^+\left(\lambda - \alpha^2\frac{\Delta}{\varpi^4} \pm 2\sigma^+\alpha\right); \quad K^{(\pm\sigma^+)} = 4\sigma^{+2}(\lambda \pm 2\sigma^+\alpha); \tag{166}$$

and the equations relating $Z^{(+\sigma^+)}$ and $Z^{(-\sigma^+)}$ are:

$$Y^{(\pm\sigma^+)} = \mp 2\sigma^+\alpha\frac{\Delta}{\varpi^4}Z^{(\pm\sigma^+)} \pm 2i\sigma^+\Lambda_\pm Z^{(\pm\sigma^+)}, \tag{167}$$

$$\Lambda_\mp Y^{(\pm\sigma^+)} = \pm 2i\sigma^+\frac{\Delta}{\varpi^4}\left(\lambda - \alpha^2\frac{\Delta}{\varpi^4} \pm 2\sigma^+\alpha\right)Z^{(\pm\sigma^+)} \mp 2\sigma^+\alpha\frac{\Delta}{\varpi^4}\Lambda_\pm Z^{(\pm\sigma^+)}, \tag{168}$$

$$K^{(\pm\sigma^+)}Z^{(\pm\sigma^+)} = \mp 2\sigma^+\alpha Y^{(\pm\sigma^+)} \mp 2i\sigma^+\frac{\varpi^4}{\Delta}\Lambda_\mp Y^{(\pm\sigma^+)}, \tag{169}$$

$$K^{(\pm\sigma^+)}\Lambda_\pm Z^{(\pm\sigma^+)} = \mp 2i\sigma^+\left(\lambda - \alpha^2\frac{\Delta}{\varpi^4} \pm 2\sigma^+\alpha\right)Y^{(\pm\sigma^+)} \mp 2\sigma^+\alpha\Lambda_\mp Y^{(\pm\sigma^+)}. \tag{170}$$

(Note that with the choice of the upper signs in both cases, equations (156)–(159) and (167)–(170) are the same, as they should be by definition.)

From equations (167)–(170), it can be shown, by a procedure analogous to that used earlier to relate the solutions, distinguished by Z_+ and Z_- in equations (160)–(163), that $Z^{(+\sigma^+)}$ is related to the complex conjugate of a solution $Z^{(-\sigma^+)}$ by

$$K^{(+\sigma^+)}Z^{(+\sigma^+)} = 4\sigma^{+2}\left(\lambda - 2\alpha^2\frac{\Delta}{\varpi^4} - 2\sigma^+\alpha\right)[Z^{(-\sigma^+)}]^* - 8i\alpha\sigma^{+2}\Lambda_+[Z^{(-\sigma^+)}]^*. \tag{171}$$

In particular,

$$K^{(\pm\sigma)}Z^{(\pm\sigma)} \to K^{(\mp\sigma)}[Z^{(\mp\sigma)}]^* - 8i\alpha\sigma^2\Lambda_{\pm}[Z^{(\mp\sigma)}]^* \quad (r \to \infty, \text{ and } r \to r_+ + 0). \tag{172}$$

Finally, we may note the following relation (cf. equation (69)):

$$K^{(+\sigma)}K^{(-\sigma)} = 16\sigma^4(\lambda^2 - 4\sigma^2\alpha^2) = 16\sigma^4\mathscr{C}^2. \tag{173}$$

(*b*) *The asymptotic behaviour of the solutions*

The potential V given by equation (153) vanishes on the event horizon, exponentially in r_*, as $r \to r_+ + 0$ and falls off like r^{-2} as $r \to \infty$. Therefore, so long as σ is outside of the super-radiant interval, the potential is of short range: its integral over r_* is finite and, *indeed*, always real. In the super-radiant interval, the potential has a singularity at $r = |\alpha| > r_+$ (see §75(*c*) below); but its behaviour at infinity and at the horizon is unaffected. Hence, in all cases, the solutions of the wave equations have the asymptotic behaviours

$$Z \to e^{\pm i\sigma r_*} \qquad (r \to \infty \quad \text{and} \quad r \to r_+ + 0). \tag{174}$$

By inserting these behaviours of Z in equations (156)–(159) and (167)–(170), we can deduce the corresponding behaviours of Y (and, therefore, also of the Teukolsky functions). Besides, by making use of equations (163) and (172), we can also relate the asymptotic behaviours of the solutions belonging to the two potentials and the solutions $Z^{(+\sigma)}$ and $Z^{(-\sigma)}$ belonging to the same potential.

Thus, from the relation (163) it follows that the solutions for Z_-, which have the asymptotic behaviours

$$Z_-^{(+\sigma)} \to e^{-i\sigma r_*} \quad \text{and} \quad Z_-^{(+\sigma)} \to e^{+i\sigma r_*} \quad (r \to \infty \quad \text{and} \quad r \to r_+ + 0) \tag{175}$$

lead to solutions for $Z_+^{(+\sigma)}$ which have, respectively, the asymptotic behaviours

$$Z_+^{(+\sigma)} \to \frac{\lambda - 2\sigma\alpha}{\lambda + 2\sigma\alpha}e^{-i\sigma r_*} \quad \text{and} \quad Z_+^{(+\sigma)} \to e^{+i\sigma r_*} \quad (r \to \infty \quad \text{and} \quad r \to r_+ + 0). \tag{176}$$

Similarly, we conclude from the relation (172) that solutions for $Z_+^{(-\sigma)}$ which have the asymptotic behaviours,

$$Z_+^{(-\sigma)} \to e^{+i\sigma r_*} \quad \text{and} \quad Z_+^{(-\sigma)} \to e^{-i\sigma r_*} \quad (r \to \infty \quad \text{and} \quad r \to r_+ + 0) \tag{177}$$

lead to solutions for $Z_+^{(+\sigma)}$ which have, respectively, the asymptotic behaviours

$$Z_+^{(+\sigma)} \to \frac{\lambda - 2\alpha\sigma}{\lambda + 2\alpha\sigma}e^{-i\sigma r_*} \quad \text{and} \quad Z_+^{(+\sigma)} \to e^{+i\sigma r_*} \quad (r \to \infty \quad \text{and} \quad r \to r_+ + 0). \tag{178}$$

TABLE VIII

Asymptotic behaviours of $Z^{(\pm\sigma)}$ and $Y^{(\pm\sigma)}$ compatible with the behaviours of $Z_{+}^{(+\sigma)}$ belonging to V_+

	$V_+ = \frac{\Delta}{\varpi^4}\left[\lambda - \alpha^2\frac{\Delta}{\varpi^4} - i\alpha\varpi^2\frac{d}{dr}\left(\frac{\Delta}{\varpi^4}\right)\right]$		$V_- = \frac{\Delta}{\varpi^4}\left[\lambda - \alpha^2\frac{\Delta}{\varpi^4} + i\alpha\varpi^2\frac{d}{dr}\left(\frac{\Delta}{\varpi^4}\right)\right]$		
$Z^{(+\sigma)}$	$e^{+i\sigma r_*}$	$e^{-i\sigma r_*}$	$e^{+i\sigma r_*}$	$\frac{\lambda+2\sigma\alpha}{\lambda-2\sigma\alpha}e^{-i\sigma r_*}$	$r_* \to \pm\infty$
$Z^{(-\sigma)}$	$e^{-i\sigma r_*}$	$\frac{\lambda+2\sigma\alpha}{\lambda-2\sigma\alpha}e^{+i\sigma r_*}$	$e^{-i\sigma r_*}$	$e^{+i\sigma r_*}$	$r_* \to \pm\infty$
$Y^{(+\sigma)}$	$-4\sigma^2 e^{+i\sigma r_*}$	$-\frac{\lambda+2\sigma\alpha}{r^2}e^{-i\sigma r_*}$	$-4\sigma^2 e^{+i\sigma r_*}$	$-\frac{\lambda+2\sigma\alpha}{r^2}e^{-i\sigma r_*}$	$r \to +\infty$
$Y^{(+\sigma)}$	$-4\sigma^2 e^{+i\sigma r_*}$	$-\frac{(\lambda+2\sigma\alpha)\Delta e^{-i\sigma r_*}}{\varpi_+^2[\varpi_+^2+i(r_+-M)/\sigma]}$	$-4\sigma^2 e^{+i\sigma r_*}$	$-\frac{(\lambda+2\sigma\alpha)\Delta e^{-i\sigma r_*}}{\varpi_+^2[\varpi_+^2+i(r_+-M)/\sigma]}$	$r \to r_+ + 0$
$Y^{(-\sigma)}$	$-4\sigma^2 e^{-i\sigma r_*}$	$-\frac{\lambda+2\sigma\alpha}{r^2}e^{+i\sigma r_*}$	$-4\sigma^2 e^{-i\sigma r_*}$	$-\frac{\lambda+2\sigma\alpha}{r^2}e^{+i\sigma r_*}$	$r \to +\infty$
$Y^{(-\sigma)}$	$-4\sigma^2 e^{-i\sigma r_*}$	$-\frac{(\lambda+2\sigma\alpha)\Delta e^{+i\sigma r_*}}{\varpi_+^2[\varpi_+^2-i(r_+-M)/\sigma]}$	$-4\sigma^2 e^{-i\sigma r_*}$	$-\frac{(\lambda+2\sigma\alpha)\Delta e^{+i\sigma r_*}}{\varpi_+^2[\varpi_+^2-i(r_+-M)/\sigma]}$	$r \to r_+ + 0$

Note: $\varpi_+^2 = r_+^2 + a^2 + (am/\sigma) = 2Mr_+(1-\sigma_s/\sigma)$.

In Table VIII we summarize the asymptotic behaviours of $Z^{(\pm\sigma)}$ and $Y^{(\pm\sigma)}$ compatible with the asymptotic behaviours $e^{\pm i\sigma r_*}$ for $Z_+^{(+\sigma)}$ belonging to V_+.

One important aspect of the asymptotic behaviours $Y^{(\pm\sigma)}$ listed in Table VIII requires emphasis. It is that, while in the analysis leading to the relations (156)–(159) and (167)–(170), no prior assumptions were made with respect to their relative normalization, a relative normalization has in effect, been defined in Table VIII, since the behaviours of $Y^{(+\sigma)}$ and $Y^{(-\sigma)}$, as listed, have been deduced consistently with the behaviours $e^{+i\sigma r_*}$ and $e^{-i\sigma r_*}$ (for $r \to \infty$ and $r \to r_+ + 0$) of $Z_+^{(+\sigma)}$ *belonging to* V_+.

75. The problem of reflexion and transmission

With the completion of the formal solution to the problem of reducing Teukolsky's equations (for $s = 1$) to the form of one-dimensional wave-equations, we now turn to the problem of how to use these equations for determining the reflexion and the transmission coefficients for incident electromagnetic waves. But one reservation requires to be made at the outset: while the one-dimensional wave-equations to which we have reduced the basic equations naturally suggest the consideration of the associated problem of barrier penetration, it remains to be established that the reflexion and the transmission coefficients determined with their aid are the physically relevant coefficients. Postponing to §76 the consideration of this and related questions of physical interpretation, we shall continue, for the present, the formal problems suggested by the equations themselves.

In considering the problem of reflexion and transmission by the potential barrier defined in equation (153), we shall distinguish the three cases, $\sigma > \sigma_c$ $(= -a/m)$ when $\alpha^2 > 0$; $\sigma_s < \sigma < \sigma_c$ when $\alpha^2 < 0$ but $r_+ > |\alpha|$; and $0 < \sigma < \sigma_s$ when $\alpha^2 < 0$ and $r_+ < |\alpha|$.

(a) *The case* $\sigma > \sigma_c$ $(= -m/a)$ *and* $\alpha^2 > 0$

Since $\alpha^2 > 0$, the $r_*(r)$-relation is single-valued in the range of r of interest and the potentials we have designated by $V_\pm$ in Table VIII are bounded and of short range. The wave equation (164) will, therefore, admit solutions satisfying the boundary conditions

$$\left.\begin{aligned} Z_+^{(+\sigma)} &\to e^{+i\sigma r_*} + A_+^{(+\sigma)} e^{-\sigma r_*} & (r_* \to +\infty), \\ &\to \qquad B_+^{(+\sigma)} e^{+i\sigma r_*} & (r_* \to -\infty), \end{aligned}\right\} \tag{179}$$

and

$$\left.\begin{aligned} Z_+^{(-\sigma)} &\to e^{-i\sigma r_*} + A_+^{(-\sigma)} e^{+i\sigma r_*} & (r_* \to +\infty), \\ &\to \qquad B_+^{(-\sigma)} e^{-i\sigma r_*} & (r_* \to -\infty). \end{aligned}\right\} \tag{180}$$

In writing the foregoing behaviours, we have further distinguished the solutions by subscripts "+" to indicate that we are, in this instance, considering solutions belonging to V_+.

Since $Z_+^{(+\sigma^t)}$ and $Z_+^{(-\sigma^t)}$ both satisfy wave equations with the same potential V_+, the reflexion and transmission coefficients defined in the manner

$$\mathbb{R} = A_+^{(+\sigma^t)} A_+^{(-\sigma^t)} \quad \text{and} \quad \mathbb{T} = B_+^{(+\sigma^t)} B_+^{(-\sigma^t)} \tag{181}$$

will satisfy the *conservation law*

$$\mathbb{R} + \mathbb{T} = 1. \tag{182}$$

On the other hand, by the relations between the asymptotic behaviours of the solutions $Z_+^{(+\sigma^t)}$ and $Z_+^{(-\sigma^t)}$ listed in Table VIII,

$$A_+^{(-\sigma^t)} = \frac{\lambda + 2\alpha\sigma^t}{\lambda - 2\alpha\sigma^t}[A_+^{(+\sigma^t)}]^* \quad \text{and} \quad B_+^{(-\sigma^t)} = [B_+^{(+\sigma^t)}]^*. \tag{183}$$

The expressions for $\mathbb{R}$ and $\mathbb{T}$, given in equations (181), can, therefore, be rewritten in the forms

$$\mathbb{R} = \frac{\lambda + 2\alpha\sigma^t}{\lambda - 2\alpha\sigma^t}|A_+^{(+\sigma^t)}|^2 \quad \text{and} \quad \mathbb{T} = |B_+^{(+\sigma^t)}|^2. \tag{184}$$

These alternative forms for $\mathbb{R}$ and $\mathbb{T}$ show that, as defined in equation (181), they are indeed real. However, the importance of the expressions (184) for $\mathbb{R}$ and $\mathbb{T}$ consists in showing that for the purposes of evaluating these coefficients, it is not necessary to integrate the wave equation twice in order to obtain the solutions with the different asymptotic behaviours (179) and (180): it will suffice to integrate the equation only once, for example, appropriate to the boundary conditions (179).

The expressions for $\mathbb{R}$ and $\mathbb{T}$ given in equations (184) apply for solutions belonging to the potential V_+. If we had considered, instead, solutions belonging to V_-, we should have found

$$\mathbb{R} = \frac{\lambda - 2\alpha\sigma^t}{\lambda + 2\alpha\sigma^t}|A_-^{(+\sigma^t)}|^2 \quad \text{and} \quad \mathbb{T} = |B_-^{(+\sigma^t)}|^2. \tag{185}$$

On the other hand, according to the relations listed in Table VIII,

$$A_+^{(+\sigma^t)} = \frac{\lambda - 2\alpha\sigma^t}{\lambda + 2\alpha\sigma^t}A_-^{(+\sigma^t)} \quad \text{and} \quad B_+^{(+\sigma^t)} = B_-^{(+\sigma^t)}. \tag{186}$$

Therefore, equations (184) and (185) define the same reflexion and transmission coefficients.

The foregoing discussion clarifies how the standard methods of treating the penetration of real one-dimensional potential-barriers have to be modified when the potentials are complex and the solutions, satisfying complex-conjugate boundary-conditions, are not, themselves, complex-conjugate functions.

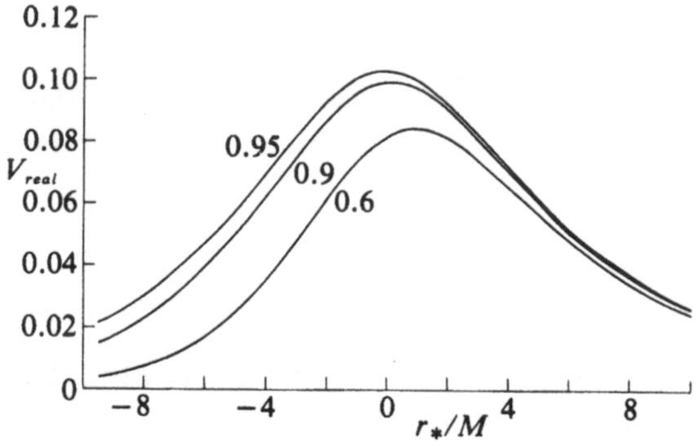

(a)

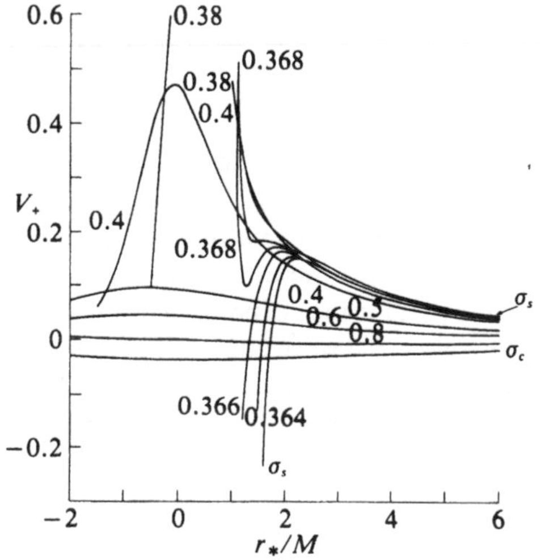

(c)

(b) The case $\sigma^+_s < \sigma^+ < \sigma^+_c$

Now $\alpha^2 < 0$ but $r_+ > |\alpha|$. Therefore, the $r_*(r)$-relation continues to be single-valued in the range of r of interest. On the other hand, since α is now imaginary, the potential (153) now becomes real:

$$V_{\pm} = \frac{\Delta}{\varpi^4}\left[\lambda + |\alpha|^2 \frac{\Delta}{\varpi^4} \pm |\alpha|\,\varpi^2 \frac{d}{dr}\left(\frac{\Delta}{\varpi^4}\right)\right], \tag{187}$$

where

$$\varpi^2 = r^2 - |\alpha|^2 \quad \text{and} \quad |\alpha|^2 = -\alpha^2. \tag{188}$$

The potentials, in addition, being bounded and of short range, the theory of the penetration of one-dimensional potential-barriers, familiar in elementary

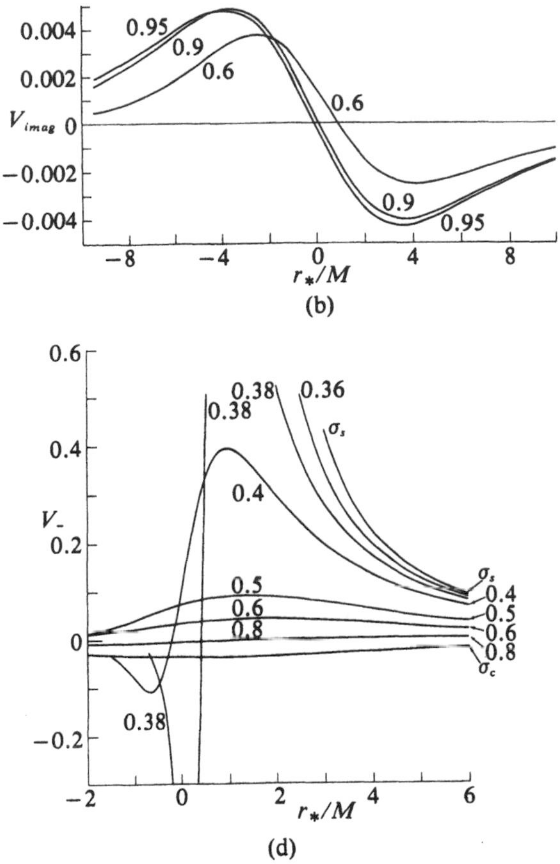

FIG. 41. The potential barriers surrounding a Kerr black-hole ($a = 0.95$) for the incidence of electromagnetic waves. (*a*), (*b*): The real (*a*) and the imaginary (*b*) parts of the complex potentials belonging to $l = 1$ and $m = -1$. The curves are labelled by the values of $a\sigma$ to which they belong. (*c*), (*d*): The family of real potentials, V_+ and V_-, in the interval $\sigma^+_c \leqslant \sigma^+ \leqslant \sigma^+_s$. The curves are labelled by the values of σ to which they belong.

quantum theory, now becomes applicable. It should, however, be noted that the potentials V_+ and V_- both yield the same reflexion and transmission coefficients. The reason is that the corresponding amplitudes of the reflected and the transmitted waves are related in the manner (cf. equation (186))

$$A_+ = \frac{\bar{\lambda} - 2i|\alpha|\sigma^+}{\bar{\lambda} + 2i|\alpha|\sigma^+} A_- \qquad \text{and} \qquad B_+ = B_-; \tag{189}$$

and, therefore,

$$|A_+|^2 = |A_-|^2 \quad \text{and} \quad |B_+|^2 = |B_-|^2. \tag{190}$$

While the potentials given by equation (187) are bounded so long as $\sigma^+ > \sigma^+_s$, they become singular at the horizon as $\sigma^+ \to \sigma^+_s + 0$: the barrier presented to the incoming waves, accordingly, becomes increasingly difficult to tunnel as σ^+ approaches σ^+_s. (See Fig. 41 in which a family of potentials in the interval, $\sigma^+_s < \sigma^+ \leqslant \sigma^+_c$, is illustrated.) We might on this account expect that

$$\mathbb{R} \to 1 \quad \text{and} \quad \mathbb{T} \to 0 \quad \text{as} \quad \sigma^+ \to \sigma^+_s + 0. \tag{191}$$

This argument, while it is consistent with the expectation that *super-radiance*, with an accompanying reflexion coefficient $\mathbb{R} > 1$, begins at $\sigma^+ = \sigma^+_s$, it is not rigorous as will become manifest in Chapter 10, §105(*a*). But in the following section (*c*), we shall show that, in fact, $\mathbb{R} > 1$ for $\sigma^+ < \sigma^+_s$; and continuity will require that $\mathbb{R} = 1$ for $\sigma^+ = \sigma^+_s$.

(*c*) *The case* $0 \leqslant \sigma^+ < \sigma^+_s$

As we have seen, when $\sigma^+ < \sigma^+_s$ and $r_+^2 - |\alpha|^2 < 0$, the $r_*(r)$-relation attains a minimum value at $r = |\alpha|$ and tends to $+\infty$ both when $r \to \infty$ and when $r \to r_+ + 0$. Therefore, we must consider separately the solutions along the two branches of the $r_*(r)$-relation which, starting at $r = |\alpha|$, progress either towards $r \to \infty$ or towards $r \to r_+ + 0$. This requirement of separate consideration of the solutions along the two branches is further compelled by the fact that the potentials given by equation (187) have singularities at $r = |\alpha|$. Thus, by rewriting the expression for $V_\pm$ in the form

$$V_\pm = \frac{\Delta}{\varpi^4}\left[\lambda + \frac{\Delta}{\varpi^4}|\alpha|(|\alpha| \mp 4r) \mp 2|\alpha|\frac{r-M}{\varpi^2}\right], \tag{192}$$

we find that in the neighbourhood of $r = |\alpha|$, $V_\pm$ has the behaviour

$$V_\pm = \pm(-4 \pm 1)\frac{\Delta^2_{|\alpha|}}{16|\alpha|^2}\frac{1}{(r-|\alpha|)^4}. \tag{193}$$

Postponing for the present the manner in which the singularity in the potential at $r = |\alpha|$ is to be taken into account in the solution of the wave equation, we observe that the boundary conditions with respect to which the wave equation has to be solved are the same as hitherto, namely,

$$\left.\begin{aligned} Z &\to e^{+i\sigma^+ r_*} + A_\pm e^{-i\sigma^+ r_*} \quad (r_* \to \infty;\quad r \to \infty) \\ &\to \qquad B_\pm e^{+i\sigma^+ r_*} \quad (r_* \to \infty;\quad r \to r_+ + 0), \end{aligned}\right\} \tag{194}$$

where the subscript "$\pm$" distinguishes the solutions belonging to V_+ or V_-. A question may be raised here as to why the wave, as one approaches the horizon along the branch, $r_* \to +\infty$ and $r \to r_+ + 0$, should be assumed to have the same $e^{+i\sigma^+ r_*}$-dependence as when the horizon was approached for $r_* \to -\infty$ and $r \to r_+ + 0$ when $\sigma^+ > \sigma^+_s$. We shall consider this question in §76 below; meantime we shall continue the discussion on the basis of the boundary conditions (194).

Even with the boundary conditions as assumed, there is one important additional consideration that must be taken into account: it is that, according to equation (146), the sign of the Wronskian, $[Z, Z^*]_{r_*}$, must be reversed as we cross the singularity at $r = |\alpha|$. Therefore, with the usual definitions,

$$\mathbb{R} = |A_\pm|^2 \quad \text{and} \quad \mathbb{T} = |B_\pm|^2, \tag{195}$$

we shall now obtain the *conservation law*

$$\mathbb{R} - \mathbb{T} = 1. \tag{196}$$

In other words, $\mathbb{R} > 1$ in the interval $0 < \sigma^+ < \sigma^+_s$. This is the phenomenon of *super-radiance*.

It remains to clarify how a solution of the wave equation satisfying the boundary conditions (194) can be obtained, duly allowing for the singularity (193) in the potentials at $r = |\alpha|$. For this purpose, we must examine the behaviour of Z near the singularity. A straightforward calculation shows that in the neighbourhood of $r = |\alpha|$, Z allows two independent solutions with the behaviours

$$\left.\begin{aligned} Z_+ &\sim |(r-|\alpha|)|^{3/2} \quad \text{and} \quad |(r-|\alpha|)|^{1/2} \quad \text{for} \quad V_+, \\ Z_- &\sim |(r-|\alpha|)|^{5/2} \quad \text{and} \quad |(r-|\alpha|)|^{-1/2} \quad \text{for} \quad V_-; \end{aligned}\right\} \tag{197}$$

and a general solution for Z in the neighbourhood of $|\alpha|$ will be a linear combination of these.

A method of solving the wave equation satisfying the boundary conditions (194) is the following. We start with a solution for Z_+ (for the potential V_+, say) with the behaviour

$$Z_+ \to e^{+i\sigma r_*} \quad \text{and} \quad r_* \to \infty \text{ along the branch } r \to r_+ + 0; \tag{198}$$

and continue the integration forward from the horizon (but backward in r_*). As we approach the singularity at $r = |\alpha|$, from the left (in r), the solution will tend to a determinate linear combination of the solutions that obtain here. Let the linear combination be (cf. equation (197))

$$Z_+ \to C_1(|\alpha| - r)^{3/2} + C_2(|\alpha| - r)^{1/2} \text{ as } r \to |\alpha| - 0, \tag{199}$$

where C_1 and C_2 are certain constants that will be determined by the integration. The requirement that the Wronskian, $[Z_+, Z_+^*]_{r_*}$, reverses its sign at $r = |\alpha|$ implies that as $r \to |\alpha| + 0$, the appropriate linear combination is

$$Z_+ \to iC_1(r - |\alpha|)^{3/2} - iC_2(r - |\alpha|)^{1/2} \quad \text{as} \quad r \to |\alpha| + 0. \tag{200}$$

With this form for Z_+, we can continue the integration forward (in r and in r_*) beyond $r = |\alpha|$ along the branch $r \to +\infty$. By such forward integration, we shall eventually find that as $r \to \infty$, the solution tends towards a limiting behaviour of the form

$$Z_+ \to C_{\text{inc}}\, e^{+i\sigma^+ r_*} + C_{\text{ref}}\, e^{-i\sigma^+ r_*} \quad (r_* \to \infty,\ r \to \infty), \tag{201}$$

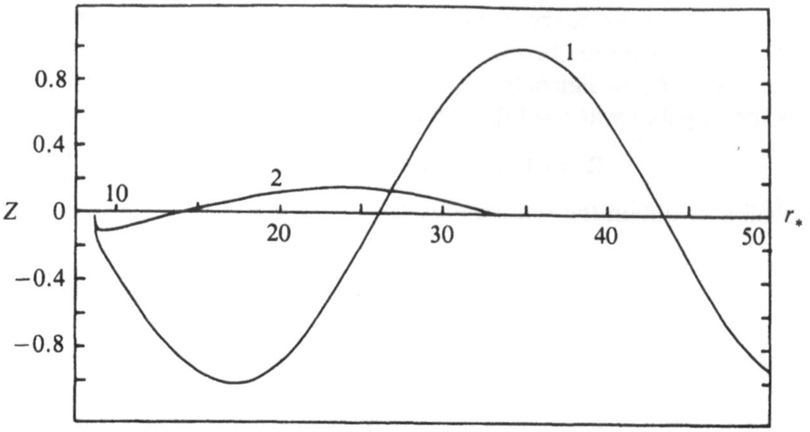

FIG. 42. Illustrating a solution of the wave equation for Z representing a standing electromagnetic wave ($l = -m = 1$) with a super-radiant frequency ($\sigma^+ = \frac{1}{2}\sigma^+_s$) in the field of a Kerr black-hole ($a = 0.95$). The part of the curve labelled '1' represents the *real* amplitude of the wave for $\infty > r > |\alpha|$ while the part of the curve labelled '2' represents the *imaginary* amplitude of the wave—imaginary on account of equation (200)— for $|\alpha| > r > r_+$. The turning point occurs at $r_* = 8.707$.

where C_{inc} and C_{ref} are certain constants that will be determined by the integration. Since the solution we started with corresponds to a transmitted wave of unit amplitude approaching the horizon, it is apparent that the required reflexion and transmission coefficients will be given by

$$\mathbb{R} = |C_{\text{ref}}|^2/|C_{\text{inc}}|^2 \quad \text{and} \quad \mathbb{T} = |C_{\text{inc}}|^{-2}. \tag{202}$$

The coefficients $\mathbb{R}$ and $\mathbb{T}$ derived in this fashion will satisfy the conservation law (196).

TABLE IX
Reflexion coefficients for electromagnetic waves incident on a Kerr black-hole with $a = 0.95$
($l = 1, m = -1$)

σ^+	σ^+/σ^+_s	$\mathbb{R}$	σ^+	σ^+/σ^+_s	$\mathbb{R}$
0.325000	0.8979	1.02428	0.405593	1.1205	0.70998
0.345000	0.9531	1.01919	0.415593	1.1481	0.56810
0.350000	0.9669	1.01565	0.425593	1.1758	0.41943
0.365593	1.0100	0.99241	0.435593	1.2034	0.28686
0.375593	1.0376	0.96100	0.445593	1.2310	0.18435
0.385593	1.0653	0.90807	0.455593	1.2586	0.11332
0.395593	1.0929	0.82563			

In Fig. 42 we illustrate a solution for Z (a standing wave in this instance) obtained by a direct integration of the equation for Z_+, appropriate for a value of $\sigma = \frac{1}{2}\sigma_s$.

And finally in Table IX we list the reflexion and transmission coefficients for electromagnetic waves of various frequencies incident on a Kerr black-hole with $a = 0.95$.

76. Further amplifications and physical interpretation

As we stated at the outset in §75, it remains to be established that the reflexion and the transmission coefficients, deduced from the one-dimensional wave-equations satisfied by $Z^{(\pm\sigma)}$, are, indeed, the relevant physical quantities which describe the interaction of the Kerr black-hole with incident electromagnetic waves. But before we can enter into a meaningful discussion of these questions of physical interpretation, it is necessary to show how these same reflexion and transmission coefficients can be deduced from solutions of Teukolsky's equations satisfying suitable boundary conditions.

First, we recall that (cf. equation (12))

$$\left.\begin{aligned} \phi_0 &= \Phi_0 = R_{+1}S_{+1}e^{i(\sigma t + m\varphi)}, \\ \text{and} \qquad 2(\bar{\rho}^*)^2\phi_2 &= \Phi_2 = R_{-1}S_{-1}e^{i(\sigma t + m\varphi)}, \end{aligned}\right\} \tag{203}$$

where we have restored the time- and the φ-dependent factors. The Teukolsky functions R_{+1} and R_{-1} are, in turn, related to $Y^{(+\sigma)}$ and $Y^{(-\sigma)}$ by (cf. equation (105))

$$R_{+1} = \frac{|\varpi^2|^{1/2}}{\Delta}Y^{(+\sigma)} \quad \text{and} \quad R_{-1} = |\varpi^2|^{1/2}Y^{(-\sigma)}, \tag{204}$$

where

$$\varpi^2 = r^2 + a^2 + (am/\sigma). \tag{205}$$

To avoid ambiguities, we shall explicitly restrict ourselves to the case $\sigma > \sigma_s$ (when $\alpha^2 > 0$) and to the solutions $Z^{(\pm\sigma)}$ belonging to the potential V_+ and satisfying the boundary conditions (179) and (180). These restrictions do not imply any essential loss of generality: the modifications necessary to allow for $\alpha^2 < 0$ and for solutions $Z^{(\pm\sigma)}$ belonging to V_- are minor and mostly *pro forma*.

From the behaviours listed in Table VIII, we deduce that the boundary conditions satisfied by the solutions $Y^{(\pm\sigma)}$, which follow from equation (167), when the solutions for $Z_+^{(\pm\sigma)}$ satisfy the boundary conditions (179) and (180), are

$$\begin{aligned} Y^{(+\sigma)} &\rightarrow -4\sigma^2 e^{+i\sigma r_*} - (\lambda + 2\sigma\alpha)A^{(+\sigma)}\frac{e^{-i\sigma r_*}}{r^2} \qquad (r \rightarrow \infty), \\ &\rightarrow \qquad -4\sigma^2 B^{(+\sigma)}e^{+i\sigma r_*} \qquad (r \rightarrow r_+ + 0), \end{aligned} \tag{206}$$

and

$$\begin{aligned} Y^{(-\sigma^+)} &\to -(\lambda+2\sigma^+\alpha)\frac{e^{+i\sigma^+ r_*}}{r^2} - 4\sigma^{+2}A^{(-\sigma^+)}e^{-i\sigma^+ r_*} \quad (r\to\infty), \\ &\to -\frac{(\lambda+2\sigma^+\alpha)B^{(-\sigma^+)}\Delta}{\varpi_+^2[\varpi_+^2 - i(r_+-M)/\sigma^+]}e^{+i\sigma^+ r_*} \quad (r\to r_+ + 0), \end{aligned} \tag{207}$$

where it will be recalled that (cf. equation (183))

$$[A^{(+\sigma^+)}]^* = \frac{\lambda - 2\alpha\sigma^+}{\lambda + 2\alpha\sigma^+}A^{(-\sigma^+)}; \quad [B^{(+\sigma^+)}]^* = B^{(-\sigma^+)}. \tag{208}$$

From the relations (204), it now follows that the corresponding solutions of Teukolsky's radial equations satisfy the boundary conditions

$$\begin{aligned} R_{+1} &\to R_{+1}^{(\text{inc})}\frac{e^{+i\sigma^+ r_*}}{r} + R_{+1}^{(\text{ref})}\frac{e^{-i\sigma^+ r_*}}{r^3} \quad (r\to\infty), \\ &\to R_{+1}^{(\text{trans})}\frac{e^{+i\sigma^+ r_*}}{\Delta} \quad (r\to r_+ + 0), \end{aligned} \tag{209}$$

and

$$\begin{aligned} R_{-1} &\to R_{-1}^{(\text{inc})}\frac{e^{+i\sigma^+ r_*}}{r} + R_{-1}^{(\text{ref})}re^{-i\sigma^+ r_*} \quad (r\to\infty), \\ &\to R_{-1}^{(\text{trans})}\Delta e^{+i\sigma^+ r_*} \quad (r\to r_+ + 0), \end{aligned} \tag{210}$$

where, by comparison with equations (206) and (207),

$$\begin{aligned} &R_{+1}^{(\text{inc})} = -4\sigma^{+2}; && R_{-1}^{(\text{inc})} = -(\lambda+2\sigma^+\alpha), \\ &R_{+1}^{(\text{ref})} = -(\lambda+2\sigma^+\alpha)A^{(+\sigma^+)}; && R_{-1}^{(\text{ref})} = -4\sigma^{+2}A^{(-\sigma^+)}, \\ &R_{+1}^{(\text{trans})} = -4\sigma^{+2}(\varpi_+^2)^{1/2}B^{(+\sigma^+)}; && R_{-1}^{(\text{trans})} = -[(\lambda+2\sigma^+\alpha)(\varpi_+^2)^{1/2}/H^*]B^{(-\sigma^+)}. \end{aligned} \tag{211}$$

In equation (211), we have used the abbreviations

$$\varpi_+^2 = r_+^2 + \alpha^2 = 2Mr_+(1-\sigma_s^+/\sigma^+) \tag{212}$$

and

$$\begin{aligned} H &= \varpi_+^2[\varpi_+^2 + i(r_+-M)/\sigma^+] \\ &= \frac{4M^2r_+^2}{\sigma^{+2}}(\sigma^+-\sigma_s^+)[(\sigma^+-\sigma_s^+) + i\sqrt{(M^2-a^2)}/2Mr_+]. \end{aligned} \tag{213}$$

It now follows from the relations (211) that

$$\left|\frac{R_{+1}^{(\text{ref})}}{R_{+1}^{(\text{inc})}}\right|^2 = \frac{(\lambda+2\sigma^+\alpha)^2}{16\sigma^{+4}}|A^{(+\sigma^+)}|^2 \quad \text{and} \quad \left|\frac{R_{-1}^{(\text{ref})}}{R_{-1}^{(\text{inc})}}\right|^2 = \frac{16\sigma^{+4}}{(\lambda+2\alpha\sigma^+)^2}|A^{(-\sigma^+)}|^2 \tag{214}$$

or, by virtue of the relation between $A^{(+\sigma^+)}$ and $A^{(-\sigma^+)}$ given in equations (208),

$$\left|\frac{R_{+1}^{(\text{ref})}}{R_{+1}^{(\text{inc})}}\right| = \frac{\lambda^2 - 4\alpha^2\sigma^{+2}}{16\sigma^{+4}} A^{(+\sigma^+)} A^{(-\sigma^+)} \quad \text{and} \quad \left|\frac{R_{-1}^{(\text{ref})}}{R_{-1}^{(\text{inc})}}\right|^2 = \frac{16\sigma^{+4}}{\lambda^2 - 4\alpha^2\sigma^{+2}} A^{(+\sigma^+)} A^{(-\sigma^+)}. \tag{215}$$

By equation (181), the reflexion coefficient, $\mathbb{R}$, is given by either of the two formulae

$$\mathbb{R} = \frac{16\sigma^{+4}}{\lambda^2 - 4\alpha^2\sigma^{+2}} \left|\frac{R_{+1}^{(\text{ref})}}{R_{+1}^{(\text{inc})}}\right|^2 \quad \text{and} \quad \mathbb{R} = \frac{\lambda^2 - 4\alpha^2\sigma^{+2}}{16\sigma^{+4}} \left|\frac{R_{-1}^{(\text{ref})}}{R_{-1}^{(\text{inc})}}\right|^2. \tag{216}$$

Similarly, we find for the transmission coefficient $\mathbb{T}$, the two formulae

$$\mathbb{T} = |B^{(+\sigma^+)}|^2 = \frac{1}{\varpi_+^2} \left|\frac{R_{+1}^{(\text{trans})}}{R_{+1}^{(\text{inc})}}\right|^2 = \frac{\sigma^+}{2Mr_+(\sigma^+ - \sigma^+_s)} \left|\frac{R_{+1}^{(\text{trans})}}{R_{+1}^{(\text{inc})}}\right|^2 \tag{217}$$

and

$$\mathbb{T} = |B^{(-\sigma^+)}|^2 = \frac{|H|^2}{\varpi_+^2} \left|\frac{R_{-1}^{(\text{trans})}}{R_{-1}^{(\text{inc})}}\right|^2$$
$$= \frac{8M^3 r_+^3}{\sigma^{+3}} (\sigma^+ - \sigma^+_s)\left[(\sigma^+ - \sigma^+_s)^2 + (M^2 - a^2)/4M^2 r_+^2\right] \left|\frac{R_{-1}^{(\text{trans})}}{R_{-1}^{(\text{inc})}}\right|^2. \tag{218}$$

The conservation law (182) now requires that

$$\frac{4\sigma^{+4}}{\lambda^2 - 4\alpha^2\sigma^{+2}} |R_{+1}^{(\text{ref})}|^2 + \frac{\sigma^+}{8Mr_+(\sigma^+ - \sigma^+_s)} |R_{+1}^{(\text{trans})}|^2 = \tfrac{1}{4} |R_{+1}^{(\text{inc})}|^2 \tag{219}$$

and

$$\frac{\lambda^2 - 4\alpha^2\sigma^{+2}}{16\sigma^{+4}} |R_{-1}^{(\text{ref})}|^2 + \frac{8M^3 r_+^3}{\sigma^{+3}} (\sigma^+ - \sigma^+_s)\left[(\sigma^+ - \sigma^+_s)^2 + \frac{M^2 - a^2}{4M^2 r_+^2}\right] |R_{-1}^{(\text{trans})}|^2$$
$$= |R_{-1}^{(\text{inc})}|^2 \tag{220}$$

—relations which, in fact, are equivalent to the Wronskian identity (139) if appropriate use is made of the Teukolsky–Starobinsky identities.

While we have explicitly restricted ourselves to $\alpha^2 > 0$, it can be readily verified that equations (216) and (217) are generally valid without the restriction. And we observe that in agreement with what was found in §75(*c*),

$$\mathbb{T} < 0 \qquad \text{for} \qquad \sigma^+ < \sigma^+_s, \tag{221}$$

i.e., we have super-radiance as predicted.

(*a*) *Implications of unitarity*

It is clear that we should obtain the same reflexion and transmission coefficients, $\mathbb{R}$ and $\mathbb{T}$, by seeking solutions for $Z^{(\pm\sigma^+)}$ which satisfy the boundary conditions

$$\begin{aligned} Z^{(+\sigma^+)} &\to C^{(+\sigma^+)} e^{-i\sigma^+ r_*} + e^{+i\sigma^+ r_*} && (r \to r_+ + 0), \\ &\to D^{(+\sigma^+)} e^{-i\sigma^+ r_*} && (r \to \infty), \end{aligned} \tag{222}$$

and

$$\begin{aligned} Z^{(-\sigma^+)} &\to C^{(-\sigma^+)} e^{+i\sigma^+ r_*} + e^{-i\sigma^+ r_*} && (r \to r_+ + 0), \\ &\to D^{(-\sigma^+)} e^{+i\sigma r_*} && (r \to \infty), \end{aligned} \tag{223}$$

instead of the boundary conditions (179) and (180).

In place of the relations (183), we shall now have

$$\frac{[C^{(+\sigma^+)}]^*}{C^{(-\sigma^+)}} = \frac{[D^{(+\sigma^+)}]^*}{D^{(-\sigma^+)}} = \frac{\lambda - 2\alpha\sigma^+}{\lambda + 2\alpha\sigma^+}; \tag{224}$$

and the reflexion and the transmission coefficients will be given by

$$\mathbb{R} = \frac{1}{C^{(+\sigma^+)} C^{(-\sigma^+)}} = \frac{\lambda \mp 2\alpha\sigma^+}{\lambda \pm 2\alpha\sigma^+} \frac{1}{|C^{(\pm\sigma^+)}|^2}, \tag{225}$$

and

$$\mathbb{T} = \left| \frac{D^{(\pm\sigma^+)}}{C^{(\pm\sigma^+)}} \right|^2. \tag{226}$$

Starting with the solutions for $Z^{(\pm\sigma^+)}$ satisfying the boundary conditions (219) and (220), we can now deduce (as in the case we have already considered) that the associated solutions of Teukolsky's equations satisfy the boundary conditions

$$\begin{aligned} R_{+1} &\to R_{+1}^{(\mathrm{inc})} e^{-i\sigma^+ r_*} + R_{+1}^{(\mathrm{ref})} \frac{e^{+\sigma^+ r_*}}{\Delta} && (r \to r_+ + 0), \\ &\to R_{+1}^{(\mathrm{trans})} \frac{e^{-i\sigma^+ r_*}}{r^3} && (r \to \infty); \end{aligned} \tag{227}$$

and

$$\begin{aligned} R_{-1} &\to R_{-1}^{(\mathrm{inc})} e^{-i\sigma^+ r_*} + R_{-1}^{(\mathrm{ref})} \Delta e^{+i\sigma^+ r_*} && (r \to r_+ + 0), \\ &\to R_{-1}^{(\mathrm{trans})} r e^{-i\sigma^+ r_*} && (r \to \infty), \end{aligned} \tag{228}$$

where

$$\begin{aligned} R_{+1}^{(\mathrm{inc})} &= -[(\lambda + 2\alpha\sigma^+)(\varpi_+^2)^{1/2}/H] C^{(+\sigma^+)}; & R_{-1}^{(\mathrm{inc})} &= -4\sigma^{+2}(\varpi_+^2)^{1/2} C^{(-\sigma^+)}, \\ R_{+1}^{(\mathrm{ref})} &= -4\sigma^{+2}(\varpi_+^2)^{1/2}; & R_{-1}^{(\mathrm{ref})} &= -[(\lambda + 2\alpha\sigma^+)(\varpi_+^2)^{1/2}/H^*], \\ R_{+1}^{(\mathrm{trans})} &= -(\lambda + 2\sigma^+\alpha) D^{(+\sigma^+)}; & R_{-1}^{(\mathrm{trans})} &= -4\sigma^{+2} D^{(-\sigma^+)}. \end{aligned} \tag{229}$$

From the foregoing equations, we find that the reflexion and the transmission coefficients are now given by

$$\mathbb{R} = \frac{\lambda^2 - 4\alpha^2\sigma^{+2}}{16\sigma^{+4}|H|^2}\left|\frac{R_{+1}^{(\text{ref})}}{R_{+1}^{(\text{inc})}}\right|^2 = \frac{16\sigma^{+4}|H|^2}{\lambda^2 - 4\alpha^2\sigma^{+2}}\left|\frac{R_{-1}^{(\text{ref})}}{R_{-1}^{(\text{inc})}}\right|^2 \tag{230}$$

and

$$\mathbb{T} = \frac{\varpi_+^2}{|H|^2}\left|\frac{R_{+1}^{(\text{trans})}}{R_{+1}^{(\text{inc})}}\right| = \frac{1}{\varpi_+^2}\left|\frac{R_{-1}^{(\text{trans})}}{R_{-1}^{(\text{inc})}}\right]^2. \tag{231}$$

Finally, we may draw attention to the fact that we have been able to deduce the asymptotic behaviours of the solutions, R_{+1} and R_{-1}, of Teukolsky's equations from the known asymptotic behaviours, $e^{\pm i\sigma^+ r_*}$, of the solutions, $Z^{(\pm\sigma^+)}$, of the one-dimensional wave-equations. These behaviours, which can be read off from the first lines of equations (209), (210), (227), and (228), can also be obtained directly from Teukolsky's equations themselves; but the passage from these behaviours to the expressions (216)–(218) and (230)–(231) for the reflexion and the transmission coefficients is not a direct nor an easy one (see §(b) below).

(b) *A direct evaluation of the flux of radiation at infinity and at the event horizon*

The notion of reflexion and transmission of waves is so natural and immediate when one is presented—as we are—with a problem in the penetration of potential barriers with an associated conservation law, it is difficult to avoid the assumption that the reflexion and the transmission coefficients one obtains have their natural physical interpretations. Our discussion hitherto (since §75) has been predicated on this assumption. We shall now show that it is, in fact, justified.

To justify the physical meanings we have attributed to the reflexion and the transmission coefficients we have derived, we must go back to the expression for the energy-momentum tensor (of the Maxwell field in the present instance) and establish the following facts. *First*, we must show that for $r \to \infty$,

$$\operatorname*{limit}_{r\to\infty} r^2 T^r{}_t \quad \text{exists;} \tag{232}$$

i.e., the component $T^r{}_t$ of the energy-momentum tensor has a $1/r^2$-behaviour at large distances, appropriate for wave propagation; and, further, that the flux of energy at infinity (per unit time and per unit solid angle), when evaluated for the outgoing (i.e., the reflected) and the ingoing (i.e., the incident) waves in terms of its proper definition,

$$\frac{d^2E}{dt\,d\Omega} = \operatorname*{limit}_{r\to\infty} r^2 T^r{}_t, \tag{233}$$

leads to the same reflexion coefficient, $\mathbb{R}$, that we have derived. And *second*, we must show that, at the horizon, the quantity

$$T_i^{\ j}\xi^i(t)\,\mathrm{d}\Sigma_j, \tag{234}$$

where $\xi(t)$ is the time-like Killing-vector that the background space-time has and $\mathrm{d}\Sigma_j$ is the 3-surface element of the horizon normal to the inward radial direction, when properly evaluated leads to the same transmission coefficient, $\mathbb{T}$, that we have derived.

In terms of the Maxwell scalars, the energy-momentum tensor, in a Newman–Penrose formalism, is given by (cf. Ch. 5, equation (339))

$$\begin{aligned}4\pi T_{ij} = \{&\phi_0\phi_0{}^* n_i n_j + \phi_2\phi_2{}^* l_i l_j + 2\phi_1\phi_1{}^*[l_{(i}n_{j)} + m_{(i}\bar{m}_{j)}] \\ &-4\phi_0{}^*\phi_1 n_{(i}m_{j)} - 4\phi_1{}^*\phi_2 l_{(i}m_{j)} + 2\phi_2\phi_0{}^* m_i m_j\} \\ &+\text{complex conjugates.}\end{aligned} \tag{235}$$

With the basis vectors chosen as in Chapter 6, equations (170) and (173), we find that the quantity we have to consider for a proper evaluation of $\mathbb{R}$ is

$$\underset{r\to\infty}{\text{limit}}\ \frac{r^2}{2\pi}(-\tfrac{1}{4}\phi_0\phi_0{}^* + \phi_2\phi_2{}^*). \tag{236}$$

For ϕ_0 and ϕ_2, we now have the solutions (cf. equations (70))

$$\phi_0 = R_{+1}S_{+1} \qquad \text{and} \qquad \phi_2 = \frac{1}{2(\bar{\rho}^*)^2}R_{-1}S_{-1}, \tag{237}$$

where the angular functions S_{+1} and S_{-1} are normalized to unity and the radial functions, R_{+1} and R_{-1} are relatively normalized so as to be compatible with equations (33) and (34)—a full knowledge which is a prerequisite to the use of equation (236). With the known solutions for ϕ_0 and ϕ_2, the flux of energy at infinity is given by

$$\left(\frac{\mathrm{d}^2E}{\mathrm{d}t\,\mathrm{d}\Omega}\right)_\infty = \underset{r\to\infty}{\text{limit}}\ \frac{r^2}{2\pi}\left\{-\frac{1}{4}S^2_{+1}\,[R^{(+\sigma)}_{+1}R^{(-\sigma)}_{+1}] + \frac{1}{4r^4}S^2_{-1}\,[R^{(+\sigma)}_{-1}R^{(-\sigma)}_{-1}]\right\}. \tag{238}$$

Equations (209) and (210) show that Teukolsky's equations do allow solutions for which the limit does (238) exist. Indeed, with the definitions of the coefficients $R^{(\text{inc})}_{+}$ and $R^{(\text{ref})}_{-1}$ in equations (209) and (210)

$$\left(\frac{\mathrm{d}^2E^{(\text{inc})}}{\mathrm{d}t\,\mathrm{d}\Omega}\right)_\infty = \frac{S^2_{+1}}{2\pi}\frac{1}{4}[R^{(\text{inc}),+\sigma)}_{+1}R^{(\text{inc},-\sigma)}_{+1}]$$

and (239)

$$\left(\frac{\mathrm{d}^2E^{(\text{ref})}}{\mathrm{d}t\,\mathrm{d}\Omega}\right)_\infty = \frac{S^2_{-1}}{2\pi}\frac{1}{4}[R^{(\text{ref},+\sigma)}_{-1}R^{(\text{ref},-\sigma)}_{-1}].$$

Inserting for the coefficients, $R^{(\text{inc},+\sigma)}_{+1}$ etc., listed in equations (211), we find from the foregoing equations, after integration over the angles, that the reflexion coefficient is given by

$$\mathbb{R} = A^{(+\sigma)} A^{(-\sigma)}, \tag{240}$$

in agreement with the expression for the reflexion coefficient we derived earlier (cf. equation (181)) from considerations pertaining to the one-dimensional wave-equations satisfied by $Z^{(\pm\sigma)}$.

To evaluate the flow of energy (234) across the event horizon, the energy-momentum tensor expressed in the basis we have used hitherto (namely that defined in Chapter 6, equations (170) and (171)) is unsuitable since the vector, $\boldsymbol{l}$, is singular on the horizon (where Δ vanishes). To obtain a singularity-free basis, we follow Hawking and Hartle in subjecting the basis, Chapter 6, (170), to a rotation of class III (cf. Ch. 1 §8 (g)) with "A" $= 2(r^2+a^2)/\Delta$. By this rotation, the vectors, $\boldsymbol{m}$ and $\bar{\boldsymbol{m}}$, are unaffected while $\boldsymbol{l}$ and $\boldsymbol{n}$ are changed to

$$\boldsymbol{l} \to \frac{\Delta}{2(r^2+a^2)}\boldsymbol{l} = \left[\frac{1}{2}, \frac{\Delta}{2(r^2+a^2)}, 0, \frac{a}{2(r^2+a^2)}\right]$$

and

$$\boldsymbol{n} \to \frac{2(r^2+a^2)}{\Delta}\boldsymbol{n} = \left[\frac{(r^2+a^2)^2}{\rho^2\Delta}, -\frac{(r^2+a^2)}{\rho^2}, 0, a\frac{(r^2+a^2)}{\rho^2\Delta}\right]. \tag{241}$$

In addition, we shall go over to a Kerr–Schild frame (cf. Ch. 6, §57 (a)) by the transformation

$$\mathrm{d}v = \mathrm{d}t + \frac{r^2+a^2}{\Delta}\,\mathrm{d}r \qquad \text{and} \qquad d\tilde{\varphi} = \mathrm{d}\varphi + \frac{a}{\Delta}\mathrm{d}r. \tag{242}$$

The choice of this frame is appropriate to the consideration of inward-directed radiation (as our consideration, presently, will be) in contrast to the choice made in Chapter 6, equations (215) which is appropriate to outward-directed radiation. The null vectors, $\boldsymbol{l}^{(HH)}$ and $\boldsymbol{n}^{(HH)}$, of the new basis are

$$\boldsymbol{l}^{(HH)} = \left[1, \frac{\Delta}{2(r^2+a^2)}, 0, \frac{a}{r^2+a^2}\right]$$

and

$$\boldsymbol{n}^{(HH)} = \left[0, -\frac{r^2+a^2}{\rho^2}, 0, 0\right]. \tag{243}$$

We observe that by this choice, the basis is well-behaved on the future event-horizon, i.e., for in-falling observers.

The element of 3-surface of the horizon, normal to the inward radial direction, is given by

$$\mathrm{d}\Sigma_j = l_j^{(HH)}\, 2Mr_+ \sin\theta\, \mathrm{d}\theta\, \mathrm{d}\tilde{\varphi}\, \mathrm{d}v. \tag{244}$$

This expression follows from the metric in the Kerr–Schild frame given in

Chapter 6, equation (219) and the evaluation in Chapter 6, equation (272) (which is applicable to the present context). Since the Jacobian of the transformation $\partial(\tilde{\varphi}, v)/\partial(\varphi, t) = 1$, we can also write

$$d\Sigma_j = l_j^{(HH)} 2Mr_+ \sin\theta \, d\theta \, d\varphi \, dt. \tag{245}$$

Since the Kerr space-time is stationary and axisymmetric, it has a time-like Killing-vector, $\boldsymbol{\xi}(t)$ $(= \partial/\partial t = \partial/\partial v)$, and an axial Killing-vector, $\boldsymbol{\xi}(\varphi)$ $(= \partial/\partial\varphi = \partial/\partial\tilde{\varphi})$. Therefore, any field with a well-defined energy-momentum tensor will allow us to define the flux vectors,

$$T_i^{\,j}\xi^i(t) \qquad \text{and} \qquad T_i^{\,j}\xi^i(\varphi), \tag{246}$$

associated with the conservation of energy and the conservation of angular momentum. And the flow of energy and of angular momentum across an element of 2-surface, formed by the intersection of an element of the horizon and two surfaces of constant v separated by dv, will be given by

$$dE = T_i^{\,j}\xi^i(t)\,d\Sigma_j \qquad dL_z = T_i^{\,j}\xi^i(\varphi)\,d\Sigma_j. \tag{247}$$

Inserting for $d\Sigma_j$ from equation (245), we obtain for the flow of energy and of angular momentum into the black hole, across the event horizon, the expressions

$$\left(\frac{d^2 E}{dt\,d\Omega}\right)_{r_+} = 2Mr_+ T_i^{\,j}\xi^i(t)\,l_j^{(HH)} \tag{248}$$

and

$$\left(\frac{d^2 L_z}{dt\,d\Omega}\right)_{r_+} = 2Mr_+ T_i^{\,j}\xi^i(\varphi)\,l_j^{(HH)}. \tag{249}$$

From equations (243), it is clear that on the horizon (where $\Delta = 0$)

$$\boldsymbol{l}^{(HH)} = \boldsymbol{\xi}(t) + \frac{a}{2Mr_+}\boldsymbol{\xi}(\varphi). \tag{250}$$

Accordingly, we obtain from equations (248) and (249)

$$\left(\frac{d^2 E}{dt\,d\Omega}\right)_{r_+} + \frac{a}{2Mr_+}\left(\frac{d^2 L_z}{dt\,d\Omega}\right)_{r_+} = 2Mr_+ T_{ij} l^{(HH)i} l^{(HH)j}. \tag{251}$$

On the other hand, for the chosen t- and φ-dependences of the perturbations,

$$\frac{\partial}{\partial t} = i\sigma^+ \qquad \text{and} \qquad \frac{\partial}{\partial\varphi} = im; \tag{252}$$

and it follows from a comparison of the expressions for dE and dL_z, given in equations (247), that

$$dL_z = (m/\sigma^+)\,dE. \tag{253}$$

In view of this relation, equation (251) gives

$$\left(1+\frac{am}{2Mr_+\sigma^+}\right)\left(\frac{\mathrm{d}^2E}{\mathrm{d}t\,\mathrm{d}\Omega}\right)_{r_+}=2Mr_+T_{ij}l^{(HH)i}l^{(HH)j}, \tag{254}$$

or, remembering that $\sigma^+_s = -am/2Mr_+$, we have

$$\left(\frac{\mathrm{d}^2E}{\mathrm{d}t\,\mathrm{d}\Omega}\right)_{r_+}=\frac{2Mr_+\sigma^+}{\sigma^+-\sigma^+_s}T_{ij}l^{(HH)i}l^{(HH)j}. \tag{255}$$

With T_{ij} given by equation (235),

$$T_{ij}l^{(HH)i}l^{(HH)j}=\frac{1}{2\pi}\phi_0^{(HH)}\phi_0^{*(HH)}. \tag{256}$$

The Maxwell scalar ϕ_0, in the basis in which the solutions (237) were obtained, is related to $\phi_0^{(HH)}$, evaluated in the basis defined in equation (243), by

$$\phi_0^{(HH)}=\frac{\Delta}{2(r^2+a^2)}\phi_0. \tag{257}$$

Therefore, equation (255) now gives

$$\left(\frac{\mathrm{d}^2E}{\mathrm{d}t\,\mathrm{d}\Omega}\right)_{r_+}=\frac{\sigma^+}{8Mr_+(\sigma^+-\sigma^+_s)}\frac{\Delta^2}{2\pi}\phi_0\phi_0^*, \tag{258}$$

or, inserting for ϕ_0 its solution, we have

$$\left(\frac{\mathrm{d}^2E}{\mathrm{d}t\,\mathrm{d}\Omega}\right)_{r_+}=\frac{S^2_{+1}}{2\pi}\frac{\sigma^+}{8Mr_+(\sigma^+-\sigma^+_s)}[\Delta^2R^{(+\sigma^+)}_{+1}R^{(-\sigma^+)}_{+1}]_{r_+}. \tag{259}$$

From the behaviour of R_{+1} at the horizon given in equation (209) it follows that

$$\left(\frac{\mathrm{d}^2E^{(\mathrm{trans})}}{\mathrm{d}t\,\mathrm{d}\Omega}\right)_{r_+}=\frac{S^2_{+1}}{2\pi}\frac{\sigma^+}{8Mr_+(\sigma^+-\sigma^+_s)}[R^{(\mathrm{trans},+\sigma^+)}_{+1}R^{(\mathrm{trans},-\sigma^+)}_{+1}]. \tag{260}$$

It is now manifest that, together with the expression for $(\mathrm{d}E^{(\mathrm{inc})}/\mathrm{d}t\,\mathrm{d}\Omega)_\infty$ given in equations (239), equation (260) yields a transmission coefficient, $\mathbb{T}$, in agreement with what we derived earlier (cf. equations (217) and (219)).

The justification is now complete that the reflexion and the transmission coefficients derived in §75 have the physical meanings that were attributed to them.

(c) *Further amplifications*

There are some questions concerning the proper boundary conditions for ϕ_0 at the horizon which require some clarification.

The solution for R_{+1}, selected in equation (209), as appropriate for ingoing waves at the horizon, has the behaviour

$$R_{+1} \to \Delta^{-1} e^{+i\sigma^{+} r_*} \qquad (r \to r_+ + 0). \tag{261}$$

It is accordingly singular; and the question occurs as to why it is acceptable. It *is* acceptable because the singularity derives, solely, from the ill-behaviour, at the horizon, of the basis in which the solution for ϕ_0 has been derived. It is, in fact, for this reason that we changed to the singularity-free Hawking–Hartle basis (245) in our considerations of the preceding section. As we have seen in equation (257), in the Hawking–Hartle basis, the Maxwell scalar ϕ_0 gets multiplied by the factor $\Delta/2(r^2+a^2)$ ($=\Delta/4Mr_+$ at the horizon); and this factor removes the singularity in (261).

A second question concerns the representation of the ingoing waves at the horizon by $e^{+i\sigma^{+} r_*}$. It will be recalled that the variable r_* has implicitly in it a dependence on σ^{+}. To make the dependence on σ^{+} explicit, it is convenient to transform to the more conventional variable, $r_*^{(c)}$, defined by (cf. equation (100))

$$\frac{d}{dr_*^{(c)}} = \frac{\Delta}{r^2+a^2}\frac{d}{dr}, \tag{262}$$

or, explicitly,

$$r_*^{(c)} = r + \frac{2Mr_+}{r_+ - r_-}\lg\left(\frac{r}{r_+}-1\right) - \frac{2Mr_+}{r_+ - r_-}\lg\left(\frac{r}{r_-}-1\right) \quad (r > r_+). \tag{263}$$

This $r_*^{(c)}(r)$-relation—in contrast to the $r^*(r)$-relation—is always single-valued (which is the reason for its common preference):

$$r_*^{(c)} \to +\infty \text{ for } r \to \infty \text{ and } \quad r_*^{(c)} \to -\infty \text{ for } r \to r_+ + 0. \tag{264}$$

From a comparison of equations (111) and (263), it follows that, while $r_* \to r_*^{(c)}$ for $r \to \infty$,

$$r_* \to \left(1 + \frac{am}{2Mr_+\sigma^{+}}\right) r_*^{(c)} = \left(1 - \frac{\sigma^{+}_s}{\sigma^{+}}\right) r_*^{(c)} \qquad (r \to r_+ + 0). \tag{265}$$

Accordingly, the selected representation for the ingoing waves at the horizon is

$$\exp[i(\sigma^{+} - \sigma^{+}_s) r_*^{(c)} + i\sigma^{+} t] = \exp[i(kr_*^{(c)} + \sigma^{+} t)], \tag{266}$$

where

$$k = \sigma^{+} - \sigma^{+}_s. \tag{267}$$

Therefore, the *group velocity* of the waves at the horizon is

$$v_{\text{group}} = -d\sigma^{+}/dk = -1, \tag{268}$$

while the *phase velocity* is given by

$$v_{\text{phase}} = -\sigma^{+}/k = -(1 - \sigma^{+}_s/\sigma^{+})^{-1}. \tag{269}$$

Thus, to an observer in any local frame, the waves will appear as progressing towards the horizon (with the velocity of light!); but this will not always be the case for an observer at infinity for whom the phase velocity is the relevant one. To an observer at infinity, the waves represented by (266) will appear as progressing towards the black hole only so long as $\sigma^+ > \sigma^+_s$; but they will appear as *emerging from the black hole when σ^+ is in the super-radiant interval*—a premonition of super-radiance yet to come. In any event, the only physical requirement for ingoing waves is that their group velocity be negative—a requirement that is met by the representation (266).

We conclude that the behaviour (261) of the solution for R_{+1}, representing ingoing waves at the horizon, is entirely consistent with all the physical requirements. The remarkable fact is that the choice of the variable r_* and the analysis flowing from it seem to have anticipated it all.

77. Some general observations on the theory

The solution of Maxwell's equation, which we have completed, gives us a first glimpse of the analytic richness of the Kerr space-time. We shall briefly recapitulate some salient features.

In some ways, the most remarkable fact is the separability of the equations when written in the Newman–Penrose formalism. It reveals at once the appropriateness of writing the equations in a frame which grasps the algebraic character of the space-time in a fundamental way. It is equally noteworthy that the separation is effected, not in terms of any of the known transcendental functions of classical analysis, but in terms of new functions, newly defined. In particular the angular functions involve the frequency, σ^+, as a parameter. This is manifest when the equation governing $S_{+1}(\theta)$ is written in its expanded form,

$$\frac{1}{\sin\theta}\frac{d}{d\theta}\left(\sin\theta\frac{dS_{+1}}{d\theta}\right) + \left(\lambda - 2a\sigma^+m - a^2\sigma^{+2}\sin^2\theta\right.$$
$$\left. + 2a\sigma^+\cos\theta - \frac{m^2 + 1 + 2m\cos\theta}{\sin^2\theta}\right)S_{+1} = 0. \quad (270)$$

(The equation governing $S_{-1}(\theta)$ is obtained by simply reversing the signs of the terms in $\cos\theta$ in equation (230).) Only when a (or σ^+) is zero does $S_{+1}(\theta)$ reduce to a classical transcendental function—the Gegenbauer functions or the 'spin-weighted' harmonics as they are commonly called. The occurence of $a\sigma^+$ as a parameter in the angular functions means that no discussion pertaining to stability can be carried out in terms of the solutions of the radial equations only.

The separability of the equations is not the only cause for surprise. The occurrence of a pair of functions (which we have distinguished by the subscripts $+1$ and -1) related via the Teukolsky–Starobinsky identities is shrouded in mystery. To be specific, if the angular functions are both normalized to unity, then $S_{+1}(\theta) \equiv S_{-1}(\pi - \theta)$. But why should this 'discrete symmetry'—perhaps 'understandable'—be incorporated in the Teukolsky–Starobinsky identities? The same question occurs with respect to the radial functions, ΔR_{+1} and R_{-1}, which satisfy complex-conjugate equations. An important and, as it emerges, also a necessary consequence of the Teukolsky–Starobinsky identities is that the derivatives of R_{+1} and R_{-1} and of S_{+1} and S_{-1} can be expressed as linear combinations of R_{+1} and R_{-1} and of S_{+1} and S_{-1}, respectively. And most mysterious of all is the identity of the Starobinsky constants for the radial and the angular functions.

Questions of a different sort occur when we turn to the transformation theory of §73 and its implications. Is there, for example, any significance to the fact that $Y^{(\pm\sigma)}$ allows 'dual' transformations leading to a pair of potentials yielding the same reflexion and transmission coefficients? In the context of the Schwarzschild and the Reissner–Nordström space-times the existence of dual transformations was related to physically distinct classes of perturbations (which we distinguished as axial and polar). Is there any corresponding physical distinction in the present context?

On the level of details, there are other facets to which we may refer. For example, the elementary theory of barrier penetration, with which we are familiar in elementary quantum theory, now emerges with novel aspects: how complex potentials *can* represent conservative scattering and how real potentials with singularities provide the key to the phenomenon of super-radiance.

We shall find, when we come to studying the gravitational perturbations in the next chapter, that all the features we have described, and much besides, will emerge in an incredibly more rococo setting.

BIBLIOGRAPHICAL NOTES

The decoupling and the separability of the Newman–Penrose equations governing the Maxwell and the Weyl scalars were discovered by Teukolsky:

1. S. A. Teukolsky, *Phys. Rev. Lett.*, 29, 1114–8 (1972).

Teukolsky's discovery made the subsequent developments possible. For a more detailed account of the discovery and further developments, see:

2. S. A. Teukolsky, *Astrophys. J.*, 185, 635–49 (1973).

The application of the decoupled and separated equations to the problem of the reflexion, absorption, and amplification of incident electromagnetic waves by the Kerr black-hole was further considered by:

3. A. A. Starobinsky and S. M. Churilov, *Zh. Exp. i. Teoret. Fiz.*, 65, 3–8 (1973); translated in *Soviet Phys. JETP*, 38, 1–5 (1973).
4. W. H. Press and S. A. Teukolsky, *Astrophys. J.*, 185, 649–73 (1973).

What we have called the 'Teukolsky–Starobinsky identities' are given in papers 3 and 4. A related paper of interest is that of:

5. J. D. Bekenstein, *Phys. Rev. D*, 7, 949–53 (1973).

The present chapter is almost exclusively based on the following papers:

6. S. CHANDRASEKHAR, *Proc. Roy. Soc.* (*London*) A, 348, 39–55 (1976).
7. ———, ibid., 349, 1–8 (1976).
8. ———, ibid., 358, 421–37 (see Appendix A, 434–7) (1978).

See also:

9. S. CHANDRASEKHAR and S. DETWEILER, *Proc. Roy. Soc.* (*London*) A, 352, 325–38 (see Appendix, 335–8) (1977).

The transformation theory as described in §74 is essentially the same as first developed in:

10. S. CHANDRASEKHAR and S. DETWEILER, *Proc. Roy. Soc.* (*London*) A, 345, 145–67 (1975).

For a review of the subject matter (from a somewhat different point of view) see:

11. S. CHANDRASEKHAR in *General Relativity—An Einstein Centenary Survey* (§§7.5 and 7.81–7.84), edited by S. W. Hawking and W. Israel, Cambridge, England, 1979.

§§73–75. The treatment in these sections is based largely on papers 6, 7. For an alternative treatment see:

12. S. DETWEILER, *Proc. Roy. Soc.* (*London*), 349, 217–30 (1976).

§76 (*b*), (*c*). The discussion in this section follows closely that of Teukolsky and Press in papers 2 and 4. See also:

13. S. W. HAWKING and J. B. HARTLE, *Commun. Math. Phys.*, 27, 283–90 (1972).

For alternative treatments of Maxwell's equations in Kerr geometry see:

14. J. M. COHEN and L. S. KEGELES, *Phys. Rev. D*, 10, 1070–84 (1974).
15. P. L. CHRZANOWSKI, *Phys. Rev. D*, 11, 2042–62 (1975).
16. ———, ibid., 13, 806–18 (1976).

This chapter provides the occasion to explain why the author has chosen to work with the original Newman–Penrose formalism in preference to the later more symmetric version by:

17. R. GEROCH, A. HELD, and R. PENROSE, *J. Math. Phys.*, 14, 874–81 (1973).

The reason is that had we adopted this later version, we should have had to replace the basis, Chapter 5, equation (170), to the more symmetric one,

$$l^i = \frac{1}{(2\rho^2\Delta)^{1/2}}(r^2+a^2, \Delta, 0, a),$$

$$n^i = \frac{1}{(2\rho^2\Delta)^{1/2}}(r^2+a^2, \Delta, 0, -a).$$

and

$$m^i = \frac{1}{(2\rho^2)^{1/2}}(ia\sin\theta, 0, 1, i\,\mathrm{cosec}\,\theta),$$

where $\rho^2 = r^2 + a^2\cos^2\theta$. This latter basis can be obtained by subjecting the former one to a rotation of class III (Chapter 1, equation (347)). On this account, neither of the present vectors $\boldsymbol{l}$ and $\boldsymbol{n}$ are affinely parametrized. And moreover, the fact that this basis requires us to work with $(r^2+a^2\cos^2\theta)^{1/2}$ rather than with $(r \pm ia\cos\theta)$ introduces algebraic inelegance in the formal developments. Besides, it will become abundantly clear in Chapter 9 that the essential complexities of the problem lie much deeper and cannot be eliminated by an early fair appearance.

9

THE GRAVITATIONAL PERTURBATIONS OF THE KERR BLACK-HOLE

78. Introduction

This chapter is devoted to the theory of the gravitational perturbations of the Kerr black-hole. The subject is one of considerable complexity; and, in spite of the length of this chapter, the account, in large parts, is hardly more than an outline.

We begin with a brief statement of the problem.

In the description of the Kerr space-time in a Newman–Penrose formalism, in Chapter 6, §56, the Weyl scalars, Ψ_0, Ψ_1, Ψ_3, and Ψ_4, and the spin-coefficients, κ, σ, λ, ν, and ε, vanish in the chosen null-basis $(\boldsymbol{l}, \boldsymbol{n}, \boldsymbol{m}, \bar{\boldsymbol{m}})$. The vanishing of these quantities reflects the type-D character of the space-time, the shear-free geodesic character of the principal null-directions, $\boldsymbol{l}$ and $\boldsymbol{n}$, and the affine parametrization of null geodesic $\boldsymbol{l}$. The spin-coefficients, $\tilde{\rho}$, τ, μ, π, α, β, and γ, and the Weyl scalar, Ψ_2, do not vanish; their values are given in Chapter 6, equations (175) and (180).

When the Kerr black-hole is gravitationally perturbed, by the incidence of gravitational waves, for example, the quantities which vanish in the stationary state will, in general, cease to vanish and will become quantities of the first order of smallness; and the quantities which are finite in the stationary state will, likewise, experience first-order changes. A gravitational perturbation will, therefore, be described by

$$\Psi_0, \Psi_1, \Psi_3, \Psi_4, \kappa, \sigma, \lambda, \quad \text{and} \quad \nu, \tag{1}$$

and

$$\left.\begin{array}{c}\Psi_2^{(1)}, \tilde{\rho}^{(1)}, \tau^{(1)}, \mu^{(1)}, \pi^{(1)}, \alpha^{(1)}, \beta^{(1)}, \gamma^{(1)}, \varepsilon^{(1)} \\ \boldsymbol{l}^{(1)}, \boldsymbol{n}^{(1)}, \boldsymbol{m}^{(1)}, \quad \text{and} \quad \bar{\boldsymbol{m}}^{(1)},\end{array}\right\} \tag{2}$$

where the superscripts "(1)" in the second set of quantities distinguish them from their unperturbed values in the stationary state. The quantities in the first set are not so distinguished since they vanish in the stationary state. (It will be noticed that ε is not included in the first set since its vanishing in the stationary state was 'contrived': it is not required by the algebraic character of the space-time.)

The problem now is to determine the two sets of quantities (1) and (2). We shall find that the problem very naturally divides itself into two parts: a part pertaining to the first set of quantities (considered in §§79–81) and a part pertaining to the second set of quantities (considered in §§82–95). And yet, the theory is all of a piece.

79. The reduction and the decoupling of the equations governing the Weyl scalars Ψ_0, Ψ_1, Ψ_3, and Ψ_4

As we have already noticed in Chapter 4 (§29(*a*)) there are among the Newman–Penrose equations a set of six equations which are linear and homogeneous in the quantities Ψ_0, Ψ_1, Ψ_3, Ψ_4, κ, σ, λ, and ν which vanish in the background geometry. Thus, four of the Bianchi identities (Ch. 1, equations (321, a, d, e, and h)) and two of the Ricci identities (Ch. 1, equations (310, b and j)) are

$$\left.\begin{aligned}(\delta^* - 4\alpha + \pi)\Psi_0 - (D - 2\varepsilon - 4\tilde{\rho})\Psi_1 &= 3\kappa\Psi_2,\\ (\triangle - 4\gamma + \mu)\Psi_0 - (\delta - 4\tau - 2\beta)\Psi_1 &= 3\sigma\Psi_2,\\ (D - \tilde{\rho} - \tilde{\rho}^* - 3\varepsilon + \varepsilon^*)\sigma - (\delta - \tau + \pi^* - \alpha^* - 3\beta)\kappa &= \Psi_0;\end{aligned}\right\} \tag{3}$$

and

$$\left.\begin{aligned}(D + 4\varepsilon - \tilde{\rho})\Psi_4 - (\delta^* + 4\pi + 2\alpha)\Psi_3 &= -3\lambda\Psi_2,\\ (\delta + 4\beta - \tau)\Psi_4 - (\triangle + 2\gamma + 4\mu)\Psi_3 &= -3\nu\Psi_2,\\ (\triangle + \mu + \mu^* + 3\gamma - \gamma^*)\lambda - (\delta^* + 3\alpha + \beta^* + \pi - \tau^*)\nu &= -\Psi_4.\end{aligned}\right\} \tag{4}$$

These are the same equations that were considered in Chapter 4, §29 in the context of the perturbations of the Schwarzschild black-hole.

Equations (3) and (4) are already linearized in the sense that, since Ψ_0, Ψ_1, κ, and σ in equations (3) and Ψ_4, Ψ_3, λ, and ν in equations (4) are to be considered as quantities of the first order of smallness, we can replace all the other quantities (including the derivative operators) which occur in the equations by their unperturbed values. An important additional feature of the equations is that the two sets of three equations, governing Ψ_0, Ψ_1, κ, and σ and Ψ_4, Ψ_3, λ, and ν, are entirely decoupled. It follows that the solutions for each of the two sets of quantities are independent.

We shall assume, as we have in the past, that the perturbations, to which the various quantities are subjected, have at t- and a φ-dependence given by

$$e^{i\sigma t + im\varphi}, \tag{5}$$

where σ is a constant (which we shall consider mostly as real and positive) and m is an integer (positive, negative, or zero). The basis vectors, $\boldsymbol{l}$, $\boldsymbol{n}$, $\boldsymbol{m}$, and $\bar{\boldsymbol{m}}$ (given in Ch. 6, equations (170)–(173)), when applied as tangent vectors to functions with the foregoing t- and φ-dependence, become the derivative operators defined in Chapter 8, equations (2). We shall, in fact, carry over the notation and the definitions of Chapter 8, §68. In particular, we shall suppress the factor (5) in all the quantities expressing the perturbations so that the symbols representing them are their amplitudes.

As we have stated, we can replace the various spin-coefficients (with the exception of κ, σ, λ, and ν) and Ψ_2 in equations (3) and (4) by their values given in Chapter 6, equations (175) and (180); and the derivative operators by the ones defined in Chapter 8, equations (2)–(5). We find that the resulting equations take simple and symmetrical forms if we write them in terms of the variables

$$\left.\begin{aligned}&\Phi_0 = \Psi_0, \quad \Phi_1 = \Psi_1\bar{\rho}^*\sqrt{2}, \quad k = \kappa/(\bar{\rho}^*)^2\sqrt{2}, \quad \text{and} \quad s = \sigma\bar{\rho}/(\bar{\rho}^*)^2,\\ &\Phi_4 = \Psi_4(\bar{\rho}^*)^4, \quad \Phi_3 = \Psi_3(\bar{\rho}^*)^3/\sqrt{2}, \quad l = \lambda\bar{\rho}^*/2, \quad \text{and} \quad n = \nu\rho^2/\sqrt{2}.\end{aligned}\right\} \tag{6}$$

After the various replacements and substitutions, equations (3) and (4) reduce to

$$\left(\mathscr{L}_2 - \frac{3ia\sin\theta}{\bar{\rho}^*}\right)\Phi_0 - \left(\mathscr{D}_0 + \frac{3}{\bar{\rho}^*}\right)\Phi_1 = -6Mk, \tag{7}$$

$$\Delta\left(\mathscr{D}_2^\dagger - \frac{3}{\bar{\rho}^*}\right)\Phi_0 + \left(\mathscr{L}_{-1}^\dagger + \frac{3ia\sin\theta}{\bar{\rho}^*}\right)\Phi_1 = +6Ms, \tag{8}$$

$$\left(\mathscr{D}_0 + \frac{3}{\bar{\rho}^*}\right)s - \left(\mathscr{L}_{-1}^\dagger + \frac{3ia\sin\theta}{\bar{\rho}^*}\right)k = +\frac{\bar{\rho}}{(\bar{\rho}^*)^2}\Phi_0, \tag{9}$$

and

$$\left(\mathscr{D}_0 - \frac{3}{\bar{\rho}^*}\right)\Phi_4 - \left(\mathscr{L}_{-1} + \frac{3ia\sin\theta}{\bar{\rho}^*}\right)\Phi_3 = +6Ml, \tag{10}$$

$$\left(\mathscr{L}_2^\dagger - \frac{3ia\sin\theta}{\bar{\rho}^*}\right)\Phi_4 + \Delta\left(\mathscr{D}_{-1}^\dagger + \frac{3}{\bar{\rho}^*}\right)\Phi_3 = +6Mn, \tag{11}$$

$$\Delta\left(\mathscr{D}_{-1}^\dagger + \frac{3}{\bar{\rho}^*}\right)l + \left(\mathscr{L}_{-1} + \frac{3ia\sin\theta}{\bar{\rho}^*}\right)n = +\frac{\bar{\rho}}{(\bar{\rho}^*)^2}\Phi_4. \tag{12}$$

The decoupling of the equations, in the two sets of equations (7)–(9) and (10)–(12), can be accomplished as in the context of Chapter 8, equations (13)–(16). Thus we can eliminate Φ_1 from equations (7) and (8) by operating equation (7) by $(\mathscr{L}_{-1}^\dagger + 3ia\sin\theta/\bar{\rho}^*)$ and equation (8) by $(\mathscr{D}_0 + 3/\bar{\rho}^*)$ and adding. The right-hand side of the resulting equation, apart from a factor $6M$, is precisely the quantity which occurs on the left-hand side of equation (9). We thus obtain the decoupled equation

$$\left[\left(\mathscr{L}_{-1}^\dagger + \frac{3ia\sin\theta}{\bar{\rho}^*}\right)\left(\mathscr{L}_2 - \frac{3ia\sin\theta}{\bar{\rho}^*}\right) + \left(\mathscr{D}_0 + \frac{3}{\bar{\rho}^*}\right)\Delta\left(\mathscr{D}_2^\dagger - \frac{3}{\bar{\rho}^*}\right)\right]\Phi_0$$

$$= 6M\frac{\bar{\rho}}{(\bar{\rho}^*)^2}\Phi_0. \tag{13}$$

In similar fashion, we obtain from equations (10)–(12) the decoupled equation

$$\left[\left(\mathscr{L}_{-1}+\frac{3ia\sin\theta}{\bar{\rho}^*}\right)\left(\mathscr{L}_2^\dagger-\frac{3ia\sin\theta}{\bar{\rho}^*}\right)+\Delta\left(\mathscr{D}_{-1}^\dagger+\frac{3}{\bar{\rho}^*}\right)\left(\mathscr{D}_0-\frac{3}{\bar{\rho}^*}\right)\right]\Phi_4$$

$$=6M\frac{\bar{\rho}}{(\bar{\rho}^*)^2}\Phi_4. \quad (14)$$

Considering the two groups of terms on the left-hand side of equation (13), we find, on simplification,

$$\Delta\left(\mathscr{D}_1+\frac{3}{\bar{\rho}^*}\right)\left(\mathscr{D}_2^\dagger-\frac{3}{\bar{\rho}^*}\right)=\Delta\mathscr{D}_1\mathscr{D}_2^\dagger+\frac{6}{\bar{\rho}^*}[-iK+(r-M)]-\frac{6\Delta}{(\bar{\rho}^*)^2}, \quad (15)$$

and

$$\left(\mathscr{L}_{-1}^\dagger+\frac{3ia\sin\theta}{\bar{\rho}^*}\right)\left(\mathscr{L}_2-\frac{3ia\sin\theta}{\bar{\rho}^*}\right)=\mathscr{L}_{-1}^\dagger\mathscr{L}_2+\frac{6ia\sin\theta}{\bar{\rho}^*}(Q+\cot\theta)$$

$$+\frac{6a^2\sin^2\theta}{(\bar{\rho}^*)^2}. \quad (16)$$

Combining the results of this reduction and making further use of the elementary identities in Chapter 8, equation (10), we find that equation (13) reduces to

$$[\Delta\mathscr{D}_1\mathscr{D}_2^\dagger+\mathscr{L}_{-1}^\dagger\mathscr{L}_2-6i\sigma(r+ia\cos\theta)]\Phi_0=0. \quad (17)$$

Similarly, equation (14) reduces to

$$[\Delta\mathscr{D}_{-1}^\dagger\mathscr{D}_0+\mathscr{L}_{-1}\mathscr{L}_2^\dagger+6i\sigma(r+ia\cos\theta)]\Phi_4=0. \quad (18)$$

Equations (17) and (18) clearly allow a separation of the variables. Thus by the substitutions

$$\Phi_0=R_{+2}(r)S_{+2}(\theta) \quad \text{and} \quad \Phi_4=R_{-2}(r)S_{-2}(\theta), \quad (19)$$

where $R_{\pm 2}(r)$ and $S_{\pm 2}(\theta)$ are functions, respectively, of r and θ only, we obtain the two pairs of equations,

$$(\Delta\mathscr{D}_1\mathscr{D}_2^\dagger-6i\sigma r)R_{+2}=\lambda R_{+2}, \quad (20)$$

$$(\mathscr{L}_{-1}^\dagger\mathscr{L}_2+6a\sigma\cos\theta)S_{+2}=-\lambda S_{+2}, \quad (21)$$

and

$$(\Delta\mathscr{D}_{-1}^\dagger\mathscr{D}_0+6i\sigma r)R_{-2}=\lambda R_{-2}, \quad (22)$$

$$(\mathscr{L}_{-1}\mathscr{L}_2^\dagger-6a\sigma\cos\theta)S_{-2}=-\lambda S_{-2}. \quad (23)$$

It will be noticed (again!) that we have not distinguished the separation constants that derive from equations (17) and (18) and for the same reasons as in the earlier contexts.

Equations equivalent to (20)–(23) were first derived by Teukolsky; and we shall call them Teukolsky's equations.

We observe that equation (20) can be rewritten in the form

$$(\Delta \mathscr{D}_{-1} \mathscr{D}_0^\dagger - 6i\sigma r)\Delta^2 R_{+2} = \lambda \Delta^2 R_{+2}. \tag{24}$$

Therefore, $\Delta^2 R_{+2}$ and R_{-2} (like ΔR_{+1} and R_- in Chapter 8) satisfy complex-conjugate equations. We also notice that equations (22) and (24) agree with Chapter 8, equations (96) and (97) for $|s| = 2$, i.e., equations appropriate for gravitons with spin 2.

80. The choice of gauge and the solutions for the spin-coefficients κ, σ, λ, and ν

Quite generally, the linearized equations governing the perturbations must be consistent with the freedom we have in the choice of the tetrad frame and in the choice of the coordinates. Precisely, we have six degrees of freedom to make infinitesimal rotations of the local tetrad-frame and four degrees of freedom to make infinitesimal coordinate transformations: altogether, a total of ten degrees of *gauge freedom*. We can exercise these available degrees of freedom as convenience and occasion may dictate.

As we have explained in Chapter 4, §29(*b*), in a linear perturbation-theory, Ψ_0 and Ψ_4 are gauge invariant while Ψ_1 and Ψ_3 are not. Consequently, we may choose a gauge (i.e., subject the tetrad null-basis to an infinitesimal rotation) in which Ψ_1 and Ψ_3 vanish without affecting Ψ_0 and Ψ_4.

If a choice of gauge in which Ψ_1 and Ψ_3 vanish is made, the corresponding solutions for k, s, l, and n can be directly read off from equations (7), (8), (10), and (11). Reverting to the original variables κ, σ, λ, and ν, we have

$$\kappa = -\frac{\sqrt{2}}{6M}(\bar{\rho}^*)^2 R_{+2}\left(\mathscr{L}_2 - \frac{3ia\sin\theta}{\bar{\rho}^*}\right)S_{+2}, \tag{25}$$

$$\sigma = +\frac{1}{6M}\frac{(\bar{\rho}^*)^2}{\bar{\rho}}S_{+2}\Delta\left(\mathscr{D}_2^\dagger - \frac{3}{\bar{\rho}^*}\right)R_{+2}, \tag{26}$$

$$\lambda = +\frac{1}{6M}\frac{2}{\bar{\rho}^*}S_{-2}\left(\mathscr{D}_0 - \frac{3}{\bar{\rho}^*}\right)R_{-2}, \tag{27}$$

$$\nu = +\frac{\sqrt{2}}{6M}\frac{1}{\rho^2}R_{-2}\left(\mathscr{L}_2^\dagger - \frac{3ia\sin\theta}{\bar{\rho}^*}\right)S_{-2}, \tag{28}$$

where it should be noted that we have yet to specify the relative normalization of the functions $\Phi_0\,(= R_{+2}S_{+2})$ and $\Phi_4\,(= R_{-2}S_{-2})$; and to this extent there is an ambiguity in the solutions for κ and σ relative to the solutions for λ and ν.

(*a*) *The phantom gauge*

Other choices of gauge, besides the one in which Ψ_1 and Ψ_3 vanish, can be made; in particular one which rectifies the strangely truncated appearances of equations (7)–(9) and (10)–(12) and restores to them a manifest symmetry they lack. The lack of symmetry consists of the following.

Considering equations (7)–(9), for example, we have seen how the first pair of equations permits the elimination of Φ_1, while the last equation is exactly the 'right' relation between k and s to obtain a decoupled equation for Φ_0. The first pair of equations equally permits the elimination of Φ_0 by virtue of the commutativity of the operators $(\mathscr{L}_2 - 3ia\sin\theta/\bar{\rho}^*)$ and $\Delta(\mathscr{D}_2^\dagger - 3/\bar{\rho}^*)$; but we do not have a 'right relation' between k and s to obtain a decoupled equation. But exercising the freedom we have to subject the local perturbed tetrad-frame to an infinitesimal rotation, we can rectify the situation by supplying (ad hoc?) the needed relation. Thus, with the additional equation

$$\Delta\left(\mathscr{D}_2^\dagger - \frac{3}{\bar{\rho}^*}\right)k + \left(\mathscr{L}_2 - \frac{3ia\sin\theta}{\bar{\rho}^*}\right)s = 2\frac{\bar{\rho}}{(\bar{\rho}^*)^2}\Phi_1, \tag{29}$$

we can complete the elimination to obtain

$$\left[\Delta\left(\mathscr{D}_2^\dagger - \frac{3}{\bar{\rho}^*}\right)\left(\mathscr{D}_0 + \frac{3}{\bar{\rho}^*}\right) + \left(\mathscr{L}_2 - \frac{3ia\sin\theta}{\bar{\rho}^*}\right)\left(\mathscr{L}_{-1}^\dagger + \frac{3ia\sin\theta}{\bar{\rho}^*}\right)\right]\Phi_1$$
$$= 12M\frac{\bar{\rho}}{(\bar{\rho}^*)^2}\Phi_1. \tag{30}$$

On expanding this equation, we obtain

$$[\Delta\mathscr{D}_2^\dagger\mathscr{D}_0 + \mathscr{L}_2\mathscr{L}_{-1}^\dagger - 6i\sigma(r + ia\cos\theta)]\Phi_1 = 0. \tag{31}$$

This equation is clearly separable: by the substitution

$$\Phi_1 = R_{+1}(r)S_{+1}(\theta) \tag{32}$$

—we justify, presently, the designation of these functions with the subscripts "+1" —we obtain the pair of equations

$$(\Delta\mathscr{D}_2^\dagger\mathscr{D}_0 - 6i\sigma r)R_{+1} = (\lambda^{(1)} - 2)R_{+1}, \tag{33}$$

and

$$(\mathscr{L}_2\mathscr{L}_{-1}^\dagger + 6a\sigma\cos\theta)S_{+1} = -(\lambda^{(1)} - 2)S_{+1}. \tag{34}$$

By making use of the readily verifiable identities

$$\left.\begin{aligned} &\Delta\mathscr{D}_2^\dagger\mathscr{D}_0 = \Delta\mathscr{D}_1\mathscr{D}_1^\dagger + 4ir\sigma - 2, \\ \text{and}\quad &\mathscr{L}_2\mathscr{L}_{-1}^\dagger = \mathscr{L}_0^\dagger\mathscr{L}_1 - 4a\sigma\cos\theta + 2, \end{aligned}\right\} \tag{35}$$

we can rewrite equations (33) and (34) in the forms

$$(\Delta\mathscr{D}_1\mathscr{D}_1^\dagger - 2i\sigma r)R_{+1} = \lambda^{(1)}R_{+1}, \tag{36}$$

and

$$(\mathscr{L}_0^\dagger\mathscr{L}_1 + 2a\sigma\cos\theta)S_{+1} = -\lambda^{(1)}S_{+1} \tag{37}$$

But these are the same equations we derived in Chapter 8, §69 for the Maxwell scalar ϕ_0 (Ch. 8, equations (24) and (25)).

Similarly, by supplementing equations (10)–(12) with the equation

$$\left(\mathscr{D}_0 - \frac{3}{\bar{\rho}^*}\right)n - \left(\mathscr{L}_2^\dagger - \frac{3ia\sin\theta}{\bar{\rho}^*}\right)l = 2\frac{\bar{\rho}}{(\bar{\rho}^*)^2}\Phi_3, \tag{38}$$

we can obtain a decoupled equation for Φ_3, which on separation of the variables, enables us to express Φ_3 as $R_{-1}(r)S_{-1}(\theta)$ where $R_{-1}(r)$ and $S_{-1}(\theta)$ satisfy the pair of equations Chapter 8, (26) and (27) for the Maxwell scalar ϕ_2.

Thus, in a gauge which restores to equations (7)–(12) the symmetry they lack, the Weyl scalars Ψ_1 and Ψ_3 satisfy the same equations as the Maxwell scalars ϕ_0 and ϕ_2 describing the propagation of electromagnetic waves in Kerr geometry. We shall show in Chapter 11, §111, that in a gauge which would appear to be most suitable for treating the perturbations of the Kerr–Newman charged black-hole, we recover equations (29) and (38) which consideration of symmetry has required us to postulate in the present context.

81. The Teukolsky–Starobinsky identities

Identities analogous to those established in Chapter 8, §70, obtain among the functions $\Delta^2 R_{+2}$ and R_{-2} and S_{+2} and S_{-2}. The proofs of these identities (in contrast to those established in §70) require much elaborate algebraic manipulations. But in view of the very crucial roles they play in the entire theory, we shall provide some of the principal steps in the proofs beyond a simple enunciation of them.

THEOREM 1.

$\Delta^2\mathscr{D}_0\mathscr{D}_0\mathscr{D}_0\mathscr{D}_0 R_{-2}$ *is a constant multiple of* $\Delta^2 R_{+2}$

and

$\Delta^2\mathscr{D}_0^\dagger\mathscr{D}_0^\dagger\mathscr{D}_0^\dagger\mathscr{D}_0^\dagger\Delta^2 R_{+2}$ *is a constant multiple of* R_{-2}.

Proof: In view of equations (22) and (24) governing R_{-2} and $\Delta^2 R_{+2}$, the theorem is equivalent to asserting the identity

$$\Delta^2\mathscr{D}_0\mathscr{D}_0\mathscr{D}_0\mathscr{D}_0(\Delta\mathscr{D}_{-1}^\dagger\mathscr{D}_0 + 6i\sigma r) = (\Delta\mathscr{D}_{-1}\mathscr{D}_0^\dagger - 6i\sigma r)\Delta^2\mathscr{D}_0\mathscr{D}_0\mathscr{D}_0\mathscr{D}_0 \tag{39}$$

and its complex conjugate.

By considering the left-hand side of the equation (39), we can establish the identity by successive reductions, the principal steps of which are

$$\begin{aligned}
&\mathscr{D}_0\mathscr{D}_0\mathscr{D}_0\mathscr{D}_0(\Delta\mathscr{D}^\dagger_{-1}\mathscr{D}_0+6i\sigma r)\\
&=(\mathscr{D}_0\mathscr{D}_0\mathscr{D}_0\mathscr{D}_0\Delta\mathscr{D}^\dagger_{-1}+6i\sigma r\mathscr{D}_0\mathscr{D}_0\mathscr{D}_0+24i\sigma\mathscr{D}_0\mathscr{D}_0)\mathscr{D}_0\\
&=[\mathscr{D}_0\mathscr{D}_0\mathscr{D}_0\mathscr{D}_0\Delta-2(iK+r-M)\mathscr{D}_0\mathscr{D}_0\mathscr{D}_0-10i\sigma r\mathscr{D}_0\mathscr{D}_0\\
&\quad-8\mathscr{D}_0\mathscr{D}_0]\mathscr{D}_0\mathscr{D}_0\\
&=[\Delta\mathscr{D}_0\mathscr{D}_0+6(r-M)\mathscr{D}_0-2iK\mathscr{D}_0-10i\sigma r+4]\mathscr{D}_0\mathscr{D}_0\mathscr{D}_0\mathscr{D}_0\\
&=(\Delta\mathscr{D}_1\mathscr{D}^\dagger_2-6i\sigma r)\mathscr{D}_0\mathscr{D}_0\mathscr{D}_0\mathscr{D}_0.
\end{aligned}\tag{40}$$

COROLLARY. *By a suitable choice of the relative normalization of the functions Δ^2R_{+2} and R_{-2}, we can arrange that*

$$\Delta^2\mathscr{D}_0\mathscr{D}_0\mathscr{D}_0\mathscr{D}_0R_{-2}=\mathscr{C}\Delta^2R_{+2}\tag{41}$$

and

$$\Delta^2\mathscr{D}^\dagger_0\mathscr{D}^\dagger_0\mathscr{D}^\dagger_0\mathscr{D}^\dagger_0\Delta^2R_{+2}=\mathscr{C}^*R_{-2},\tag{42}$$

where $\mathscr{C}$ is some complex constant.

We can arrange the relative normalization in the way prescribed since Δ^2R_{+2} and R_{-2} satisfy complex-conjugate equations.

THEOREM 2. *If the relative normalization of the functions Δ^2R_{+2} and R_{-2} is so arranged that equations* (41) *and* (42) *are valid, then the square of the modulus of the constant $\mathscr{C}$ is*

$$|\mathscr{C}|^2=\lambda^2(\lambda+2)^2-8\sigma^2\lambda[\alpha^2(5\lambda+6)-12a^2]+144\sigma^4\alpha^4+144\sigma^2M^2,\tag{43}$$

where

$$\alpha^2=a^2+am/\sigma.\tag{44}$$

Proof: A manifest consequence of equations (41) and (42) is

$$\Delta^2\mathscr{D}^\dagger_0\mathscr{D}^\dagger_0\mathscr{D}^\dagger_0\mathscr{D}^\dagger_0\Delta^2\mathscr{D}_0\mathscr{D}_0\mathscr{D}_0\mathscr{D}_0R_{-2}=|\mathscr{C}|^2R_{-2}.\tag{45}$$

We have therefore the identity

$$\Delta^2\mathscr{D}^\dagger_0\mathscr{D}^\dagger_0\mathscr{D}^\dagger_0\mathscr{D}^\dagger_0\Delta^2\mathscr{D}_0\mathscr{D}_0\mathscr{D}_0\mathscr{D}_0=|\mathscr{C}|^2 \bmod \Delta\mathscr{D}^\dagger_{-1}\mathscr{D}_0+6i\sigma r=\lambda,\tag{46}$$

and the value (43) of $|\mathscr{C}|^2$ must follow from a direct evaluation of the identity; yet the evaluation of the identity is not an easy task.

We first establish, modulo $\Delta\mathscr{D}^\dagger_{-1}\mathscr{D}_0+6i\sigma r=\lambda$, the following identities:

$$\Delta\mathscr{D}_0\mathscr{D}_0\equiv2(iK+r-M)\mathscr{D}_0+(\lambda-6i\sigma r),\tag{47}$$

$$\begin{aligned}\Delta^2\mathscr{D}_0\mathscr{D}_0\mathscr{D}_0\equiv{}&[4iK(iK+r-M)+(\lambda+2+2i\sigma r)\Delta]\mathscr{D}_0\\&+2iK(\lambda-6i\sigma r)-6i\sigma\Delta,\end{aligned}\tag{48}$$

and

$$\begin{aligned}\Delta^3\mathscr{D}_0\mathscr{D}_0\mathscr{D}_0\mathscr{D}_0 \equiv \{&-8iK[K^2+(r-M)^2]+[4iK(\lambda\!\!\!\bar{}+2)+8i\sigma^{+}r(r-M)]\Delta\\ &-8i\sigma^{+}\Delta^2\}\,\mathscr{D}_0\\ &+[(\lambda\!\!\!\bar{}+2+2i\sigma^{+}r)(\lambda\!\!\!\bar{}-6i\sigma^{+}r)-12i\sigma^{+}(iK-r+M)]\Delta\\ &+4iK(iK-r+M)(\lambda\!\!\!\bar{}-6i\sigma^{+}r)\\ \equiv\ &A_0\mathscr{D}_0+B_0 \quad \text{(say).}\end{aligned} \tag{49}$$

We must now apply the operator $\mathscr{D}^{\dagger}_{-1}$ four times, successively, to the identity (49); and we can do this with the aid of the recurrence relation

$$\begin{aligned}\mathscr{D}^{\dagger}_{-1}(A_n\mathscr{D}_0+B_n) &\equiv (A_{n,r}+B_n)\mathscr{D}_0+\frac{1}{\Delta}[(\lambda\!\!\!\bar{}-6i\sigma^{+}r)A_n-2(iK+r-M)B_n+\Delta B_{n,r}]\\ &\equiv A_{n+1}\mathscr{D}_0+B_{n+1} \quad \text{(say).}\end{aligned} \tag{50}$$

In this manner, we find

$$\left.\begin{aligned}A_1 &= [\lambda\!\!\!\bar{}(\lambda\!\!\!\bar{}+2)+4i\sigma^{+}r(\lambda\!\!\!\bar{}+3)+12\sigma^{+2}r^2-12i\sigma^{+}(iK+r-M)]\Delta\\ &\quad +4iK(iK+r-M)(\lambda\!\!\!\bar{}+6i\sigma^{+}r),\\ B_1 &= 2iK(\lambda\!\!\!\bar{}^2+36\sigma^{+2}r^2)-12i\sigma^{+}\lambda\!\!\!\bar{}\Delta;\end{aligned}\right\} \tag{51}$$

$$\left.\begin{aligned}A_2 &= -8i\sigma^{+}(\lambda\!\!\!\bar{}+6i\sigma^{+}r)\Delta-24i\sigma^{+}[K^2+(r-M)^2]\\ &\quad +2(iK+r-M)(\lambda\!\!\!\bar{}+6i\sigma^{+}r)(\lambda\!\!\!\bar{}+2+2i\sigma^{+}r),\\ B_2 &= (\lambda\!\!\!\bar{}^2+36\sigma^{+2}r^2)(\lambda\!\!\!\bar{}+2+2i\sigma^{+}r)+12i\sigma^{+}(\lambda\!\!\!\bar{}-6i\sigma^{+}r)(iK-r+M);\end{aligned}\right\} \tag{52}$$

$$\left.\begin{aligned}A_3 &= 48\sigma^{+2}\Delta+(\lambda\!\!\!\bar{}+6i\sigma^{+}r)[(\lambda\!\!\!\bar{}+2)^2+4\sigma^{+2}r^2],\\ &\quad -12i\sigma^{+}(r-M)(\lambda\!\!\!\bar{}+2-2i\sigma^{+}r)-4\sigma^{+}K(7\lambda\!\!\!\bar{}+6+18i\sigma^{+}r)\\ B_3 &= -6i\sigma^{+}(\lambda\!\!\!\bar{}+6i\sigma^{+}r)(\lambda\!\!\!\bar{}+2+2i\sigma^{+}r)+72\sigma^{+2}(iK-r+M);\end{aligned}\right\} \tag{53}$$

and

$$\left.\begin{aligned}A_4 &= 0,\\ B_4 &= \frac{1}{\Delta}\{\lambda\!\!\!\bar{}^2(\lambda\!\!\!\bar{}+2)^2-8\sigma^{+2}\lambda\!\!\!\bar{}[\alpha^2(5\lambda\!\!\!\bar{}+6)-12a^2]+144\sigma^{+2}(M^2+\sigma^{+2}\alpha^4)\}.\end{aligned}\right\} \tag{54}$$

The last line in the sequence of reductions establishes the value of the Starobinsky constant $|\mathscr{C}|^2$.

It is important to observe that while $|\mathscr{C}|^2$ has been determined, its argument remains to be ascertained.

THEOREM 3.

$$\mathscr{L}_{-1}\mathscr{L}_0\mathscr{L}_1\mathscr{L}_2 S_{+2} \quad \textit{is a constant multiple of } S_{-2}$$

and

$$\mathscr{L}^{\dagger}_{-1}\mathscr{L}^{\dagger}_0\mathscr{L}^{\dagger}_1\mathscr{L}^{\dagger}_2 S_{-2} \quad \textit{is a constant multiple of } S_{+2}.$$

Proof: The theorem is clearly equivalent to asserting the identity

$$\begin{aligned}&\mathscr{L}_{-1}\mathscr{L}_0\mathscr{L}_1\mathscr{L}_2(\mathscr{L}^\dagger_{-1}\mathscr{L}_2+6a\sigma\cos\theta)\\&=(\mathscr{L}_{-1}\mathscr{L}^\dagger_2-6a\sigma\cos\theta)\mathscr{L}_{-1}\mathscr{L}_0\mathscr{L}_1\mathscr{L}_2,\end{aligned}\tag{55}$$

and its adjoint obtained by replacing θ by $\pi-\theta$.

By considering the left-hand side of the identity (55), we obtain by successive reductions:

$$\begin{aligned}&\mathscr{L}_{-1}\mathscr{L}_0\mathscr{L}_1\mathscr{L}_2(\mathscr{L}^\dagger_{-1}\mathscr{L}_2+6a\sigma\cos\theta)\\
&=[\mathscr{L}_{-1}\mathscr{L}_0\mathscr{L}_1\mathscr{L}_2\mathscr{L}^\dagger_{-1}+(6a\sigma\cos\theta)\mathscr{L}_{-1}\mathscr{L}_0\mathscr{L}_1\\
&\quad-(24a\sigma\sin\theta)\mathscr{L}_0\mathscr{L}_1]\mathscr{L}_2\\
&=[\mathscr{L}_{-1}\mathscr{L}_0\mathscr{L}_1\mathscr{L}^\dagger_0+(2a\sigma\cos\theta)\mathscr{L}_{-1}\mathscr{L}_0-(12a\sigma\sin\theta)\mathscr{L}_0\\
&\quad+2\mathscr{L}_{-1}\mathscr{L}_0]\mathscr{L}_1\mathscr{L}_2\\
&=[\mathscr{L}_{-1}\mathscr{L}_0\mathscr{L}^\dagger_1-(2a\sigma\cos\theta)\mathscr{L}_{-1}+2\mathscr{L}_{-1}-4a\sigma\sin\theta]\mathscr{L}_0\mathscr{L}_1\mathscr{L}_2\\
&=(\mathscr{L}_{-1}\mathscr{L}^\dagger_2-6a\sigma\cos\theta)\mathscr{L}_{-1}\mathscr{L}_0\mathscr{L}_1\mathscr{L}_2,\end{aligned}\tag{56}$$

where, in the different steps, we have made use of the identities

$$\left.\begin{aligned}\mathscr{L}_2\mathscr{L}^\dagger_{-1}&=\mathscr{L}^\dagger_0\mathscr{L}_1-4a\sigma\cos\theta+2,\\
\mathscr{L}_1\mathscr{L}^\dagger_0&=\mathscr{L}^\dagger_1\mathscr{L}_0-4a\sigma\cos\theta,\\
\mathscr{L}_0\mathscr{L}^\dagger_1&=\mathscr{L}^\dagger_2\mathscr{L}_{-1}-4a\sigma\cos\theta-2,\\
\mathscr{L}_{-1}\mathscr{L}_0\mathscr{L}^\dagger_1&=\mathscr{L}_{-1}\mathscr{L}^\dagger_2\mathscr{L}_{-1}-(4a\sigma\cos\theta)\mathscr{L}_{-1}-2\mathscr{L}_{-1}+4a\sigma\sin\theta.\end{aligned}\right\}\tag{57}$$

The last step in the reductions (56) establishes the identity (55) and, therefore, also the theorem.

An alternative statement of Theorem 3 is the existence of relations of the form

$$\mathscr{L}_{-1}\mathscr{L}_0\mathscr{L}_1\mathscr{L}_2S_{+2}=D_1S_{-2}$$

and

$$\mathscr{L}^\dagger_{-1}\mathscr{L}^\dagger_0\mathscr{L}^\dagger_1\mathscr{L}^\dagger_2S_{-2}=D_2S_{+2},\tag{58}$$

where D_1 and D_2 are two real constants.

THEOREM 4. *If $S_{+2}(\theta)$ and $S_{-2}(\theta)$ are both normalized to unity, then $D_1=D_2$ in the relations* (58) *and we shall have*

$$\mathscr{L}_{-1}\mathscr{L}_0\mathscr{L}_1\mathscr{L}_2S_{+2}=DS_{-2}\quad\text{and}\quad\mathscr{L}^\dagger_{-1}\mathscr{L}^\dagger_0\mathscr{L}^\dagger_1\mathscr{L}^\dagger_2S_{-2}=DS_{+2}.\tag{59}$$

Proof: The proof follows as in Theorem 4 of Chapter 8 by four successive applications of Lemma 4 (§68) to the normalization integral:

$$D_1^2 = D_1^2 \int_0^\pi S_{-2}^2 \sin\theta \, d\theta$$

$$= \int_0^\pi (\mathscr{L}_{-1}\mathscr{L}_0\mathscr{L}_1\mathscr{L}_2 S_{+2})(\mathscr{L}_{-1}\mathscr{L}_0\mathscr{L}_1\mathscr{L}_2 S_{+2}) \sin\theta \, d\theta$$

$$= \int_0^\pi S_{+2}(\mathscr{L}_{-1}^\dagger\mathscr{L}_0^\dagger\mathscr{L}_1^\dagger\mathscr{L}_2^\dagger\mathscr{L}_{-1}\mathscr{L}_0\mathscr{L}_1\mathscr{L}_2 S_{+2}) \sin\theta \, d\theta$$

$$= D_1 D_2 \int_0^\pi S_{+2}^2 \sin\theta \, d\theta = D_1 D_2; \tag{60}$$

and the equality of D_1 and D_2 follows.

THEOREM 5. *The constant D of Theorem 3 has the value*

$$\left.\begin{aligned} D^2 &= \lambda^2(\lambda+2)^2 - 8\sigma^2\lambda[\alpha^2(5\lambda+6) - 12a^2] + 144\sigma^4\alpha^4 \\ &= |\mathscr{C}|^2 - 144\sigma^2 M^2. \end{aligned}\right\} \tag{61}$$

Proof: The value of D^2 must follow from a direct evaluation of the identity

$$\mathscr{L}_{-1}^\dagger\mathscr{L}_0^\dagger\mathscr{L}_1^\dagger\mathscr{L}_2^\dagger\mathscr{L}_{-1}\mathscr{L}_0\mathscr{L}_1\mathscr{L}_2 = D^2 \bmod \mathscr{L}_{-1}^\dagger\mathscr{L}_2 + 6a\sigma\cos\theta = -\lambda, \tag{62}$$

which follows from equations (58) (and which was in fact used in the penultimate line of the reduction (60)). And, again, the direct evaluation of the identity (62) is not an easy task. The principal steps in the evaluation are the following:

First, with the aid of the recurrence relation,

$$\mathscr{L}_n\mathscr{L}_2 \equiv -(\lambda + 6a\sigma\cos\theta) + [2Q + (n+1)\cot\theta]\mathscr{L}_2 \qquad \bmod \mathscr{L}_{-1}^\dagger\mathscr{L}_2 + 6a\sigma\cos\theta = -\lambda, \tag{63}$$

we find

$$\mathscr{L}_1\mathscr{L}_2 \equiv -(\lambda + 6a\sigma\cos\theta) + 2(Q + \cot\theta)\mathscr{L}_2, \tag{64}$$

$$\mathscr{L}_0\mathscr{L}_1\mathscr{L}_2 \equiv [6a\sigma\sin\theta - 2Q(\lambda + 6a\sigma\cos\theta)] + [-(\lambda+2) - 2a\sigma\cos\theta + 4Q(Q+\cot\theta)]\mathscr{L}_2, \tag{65}$$

and

$$\begin{aligned}\mathscr{L}_{-1}\mathscr{L}_0\mathscr{L}_1\mathscr{L}_2 &\equiv (\lambda + 6a\sigma\cos\theta)(\lambda + 2 - 4Q^2 + 4Q\cot\theta - 2a\sigma\cos\theta) \\ &+ 12a\sigma(Q\sin\theta - \cos\theta) + [8a\sigma\operatorname{cosec}\theta - 4Q(\lambda + 2 - 2Q^2 \\ &+ 2\cot^2\theta)]\mathscr{L}_2. \end{aligned} \tag{66}$$

We must now apply to the identity (66), successively, the operators, $\mathscr{L}_2^\dagger$, $\mathscr{L}_1^\dagger$, $\mathscr{L}_0^\dagger$, and $\mathscr{L}_{-1}^\dagger$. With the aid of the recurrence relation

$$\mathscr{L}_n^\dagger(A + B\mathscr{L}_2) \equiv [(n-2)A\cot\theta - 2QA + A_{,\theta} - B(\lambda + 6a\sigma\cos\theta)] + [(n+1)B\cot\theta + B_{,\theta} + A]\mathscr{L}_2, \tag{67}$$

we find, successively,

$$\begin{aligned}\mathscr{L}_2^\dagger\mathscr{L}_{-1}\mathscr{L}_0\mathscr{L}_1\mathscr{L}_2 = {} & 2Q(\lambda^2-36a^2\sigma^2\cos^2\theta)-12a\sigma\lambda\sin\theta \\ & +\{4Q[3a\sigma\sin\theta-(Q+\cot\theta)(\lambda-6a\sigma\cos\theta)] \\ & +\lambda(\lambda+2)-4\lambda a\sigma\cos\theta-12a^2\sigma^2\cos^2\theta\}\mathscr{L}_2,\end{aligned} \tag{68}$$

$$\begin{aligned}\mathscr{L}_1^\dagger\mathscr{L}_2^\dagger\mathscr{L}_{-1}\mathscr{L}_0\mathscr{L}_1\mathscr{L}_2 = {} & 4a\sigma(\lambda^2-36a^2\sigma\cos^2\theta)\cos\theta \\ & +(\lambda+6a\sigma\cos\theta)\{12a\sigma Q\sin\theta-[\lambda(\lambda+2)-4\lambda a\sigma\cos\theta-12a^2\sigma^2\cos^2\theta]\} \\ & +\{4(\lambda-6a\sigma\cos\theta)(Q-4a\sigma Q\cos\theta-2a\sigma\operatorname{cosec}\theta) \\ & \quad +2Q(\lambda^2-36a^2\sigma^2\cos^2\theta)-24a\sigma Q^2\sin\theta \\ & \quad +2[\lambda(\lambda+2)-4\lambda a\sigma\cos\theta-12a^2\sigma^2\cos^2\theta]\cot\theta\}\mathscr{L}_2,\end{aligned} \tag{69}$$

$$\begin{aligned}& \mathscr{L}_0^\dagger\mathscr{L}_1^\dagger\mathscr{L}_2^\dagger\mathscr{L}_{-1}\mathscr{L}_0\mathscr{L}_1\mathscr{L}_2 \\ & \equiv 6a\sigma(\lambda^2+2\lambda-8a\sigma\lambda\cos\theta+12a^2\sigma^2\cos^2\theta-12a\sigma Q\sin\theta)\sin\theta \\ & \quad +\{-48a^2\sigma^2-4\lambda-(\lambda-6a\sigma\cos\theta)(\lambda^2+2\lambda-4a\sigma\cos\theta-4a^2\sigma^2\cos^2\theta) \\ & \qquad -2(\lambda+6a\sigma\cos\theta)(\lambda-4a\sigma\cos\theta) \\ & \qquad +4a\sigma Q(6+7\lambda-18a\sigma\cos\theta)\sin\theta\}\mathscr{L}_2,\end{aligned} \tag{70}$$

and, finally,

$$\begin{aligned}& \mathscr{L}_{-1}^\dagger\mathscr{L}_0^\dagger\mathscr{L}_1^\dagger\mathscr{L}_2^\dagger\mathscr{L}_{-1}\mathscr{L}_0\mathscr{L}_1\mathscr{L}_2 \\ & \equiv \lambda^2(\lambda+2)^2-8\sigma^2\lambda[\alpha^2(5\lambda+6)-12a^2]+144\sigma^4\alpha^4 = D^2.\end{aligned} \tag{71}$$

(a) A collection of useful formulae

We shall write

$$\Delta^2 R_{+2} = P_{+2} \qquad \text{and} \qquad R_{-2} = P_{-2} \tag{72}$$

when the radial functions R_{+2} and R_{-2} are relatively normalized as required by equations (41) and (42).

It is now clear that, with the aid of the identities we have established, we can express (cf. Ch. 8, §70) as in the derivatives of the Teukolsky functions (both radial and angular) in terms of the functions themselves.

Since equation (49) is valid modulo the equation satisfied by P_{-2}, it is the reduced form of the equation

$$\Delta^3\mathscr{D}_0\mathscr{D}_0\mathscr{D}_0\mathscr{D}_0 P_{-2} = \Delta\mathscr{C}P_{+2}, \tag{73}$$

in which all the higher derivatives of P_{-2} have been eliminated with the aid of the second-order equation satisfied by P_{-2}. Therefore, equation (49) and its conjugate complex will enable us to express the derivatives of P_{+2} and P_{-2} as linear combinations of P_{+2} and P_{-2}. We find

$$iB_1\frac{dP_{-2}}{dr} = \Delta(\mathscr{C}_1 + i\mathscr{C}_2)P_{+2} - \left[\left(A_1 - \frac{B_1 K}{\Delta}\right) + iA_2\right]P_{-2}$$

and

$$-iB_1\frac{dP_{+2}}{dr} = \Delta(\mathscr{C}_1 - i\mathscr{C}_2)P_{-2} - \left[\left(A_1 - \frac{B_1 K}{\Delta}\right) - iA_2\right]P_{+2}, \tag{74}$$

where

$$\mathscr{C} = \mathscr{C}_1 + i\mathscr{C}_2,\ \Gamma_1 = \lambda(\lambda+2) - 12\sigma^2\alpha^2,\ \Gamma_2 = 12\sigma M, \tag{75}$$

$$\begin{aligned} A_1 &= +\Delta\Gamma_1 - 4\lambda K^2 + 24\sigma K(a^2 - Mr), \\ A_2 &= -\Delta\Gamma_2 - 4\lambda[K(r-M) + r\sigma\Delta] + 24\sigma r K^2 = -\tfrac{1}{2}B_{1,r}, \\ B_1 &= -8K[K^2 + (r-M)^2] + 4\Delta[K(\lambda+2) + 2\sigma r(r-M)] - 8\sigma\Delta^2 \\ &= -8K^3 + 8K(a^2 - M^2) - 8\sigma\Delta(a^2 - Mr) + 4\lambda K\Delta. \end{aligned} \tag{76}$$

Similarly, from equations (58) and (66) we find

$$\beta_1\mathscr{L}_2 S_{+2} = DS_{-2} - (\alpha_1 + \alpha_2)S_{+2}, \tag{77}$$

where

$$\begin{aligned} \beta_1 &= 8Q^3 - 8Q\cot^2\theta - 4(\lambda+2)Q + 8a\sigma\operatorname{cosec}\theta, \\ \alpha_1 &= \lambda(\lambda+2) - 12\alpha^2\sigma^2 + 24a\sigma Q\operatorname{cosec}\theta - 4\lambda Q^2, \\ \alpha_2 &= -24a\sigma Q^2\cos\theta + 4\lambda(Q\cot\theta + a\sigma\cos\theta). \end{aligned} \tag{78}$$

Combining equation (77) with its 'adjoint,' we obtain the pair of equations

$$\begin{aligned} \beta_1\frac{d(S_{+2} + S_{-2})}{d\theta} &= -(\alpha_2 + 2\beta_1\cot\theta)(S_{+2} + S_{-2}) \\ &\quad - (\alpha_1 + \beta_1 Q + D)(S_{+2} - S_{-2}), \end{aligned}$$

and

$$\begin{aligned} \beta_1\frac{d(S_{+2} - S_{-2})}{d\theta} &= -(\alpha_1 + \beta_1 Q - D)(S_{+2} + S_{-2}) \\ &\quad - (\alpha_2 + 2\beta_1\cot\theta)(S_{+2} - S_{-2}). \end{aligned} \tag{79}$$

Equations (74) and (79) play essential roles in the subsequent developments.

(*b*) *The bracket notation*

We shall find that the following notation is helpful in abbreviating the equations of the theory. We let

$$[P]^{\pm} = P_{+2} \pm P_{-2};\ [S]^{\pm} = S_{+2} \pm S_{-2}, \tag{80}$$

$$[\mathscr{D}P]^{\pm} = \mathscr{D}_0^{\dagger}P_{+2} \pm \mathscr{D}_0 P_{-2},\ [\mathscr{D}\mathscr{D}P]^{\pm} = \mathscr{D}_0^{\dagger}\mathscr{D}_0^{\dagger}P_{+2} \pm \mathscr{D}_0\mathscr{D}_0 P_{-2}; \tag{81}$$

$$[\mathscr{L}S]^{\pm} = \mathscr{L}_2 S_{+2} \pm \mathscr{L}_2^{\dagger}S_{-2};\ [\mathscr{L}\mathscr{L}S]^{\pm} = \mathscr{L}_1\mathscr{L}_2 S_{+2} \pm \mathscr{L}_1^{\dagger}\mathscr{L}_2^{\dagger}S_{-2}. \tag{82}$$

Elementary consequences of the foregoing definitions are

$$\frac{d}{dr}[P]^{\pm} = [\mathscr{D}P]^{\pm} + \frac{iK}{\Delta}[P]^{\mp}, \tag{83}$$

$$\frac{d}{dr}[\mathscr{D}P]^{\pm} = [\mathscr{D}\mathscr{D}P]^{\pm} + \frac{iK}{\Delta}[\mathscr{D}P]^{\mp}, \tag{84}$$

$$\frac{d}{d\theta}[S]^{\pm} = [\mathscr{L}S]^{\pm} - Q[S]^{\mp} - 2[S]^{\pm}\cot\theta, \tag{85}$$

and

$$\frac{d}{d\theta}[\mathscr{L}S]^{\pm} = [\mathscr{L}\mathscr{L}S]^{\pm} - Q[\mathscr{L}S]^{\mp} - [\mathscr{L}S]^{\pm}\cot\theta. \tag{86}$$

In this notation, equation (47) and its complex conjugate and equation (64) and its adjoint take the forms

$$\Delta[\mathscr{D}\mathscr{D}P]^{\pm} = -2iK[\mathscr{D}P]^{\mp} + 2(r-M)[\mathscr{D}P]^{\pm} + \lambda[P]^{\pm} + 6i\sigma r[P]^{\mp} \tag{87}$$

and

$$[\mathscr{L}\mathscr{L}S]^{\pm} = 2Q[\mathscr{L}S]^{\mp} + 2[\mathscr{L}S]^{\pm}\cot\theta - \lambda[S]^{\pm} - 6a\sigma[S]^{\mp}\cos\theta. \tag{88}$$

These are in fact Teukolsky's equations for the normalized functions.

82. Metric perturbations; a statement of the problem

In considerations relative to the perturbations of a space-time, one is principally interested in the changes induced in the metric coefficients. In the Newman–Penrose formalism, these changes are directly related to (and expressible in terms of) the changes, $\boldsymbol{l}^{(1)}$, $\boldsymbol{n}^{(1)}$, $\boldsymbol{m}^{(1)}$, and $\bar{\boldsymbol{m}}^{(1)}$, in the null vectors of the basis. Besides, one is also interested in the changes in the spin-coefficients and in the Weyl scalars. We must proceed, then, to a consideration of the quantities listed under (2). But before we can state precisely what it is we are seeking, it is important that we are clear as to what the analysis, hitherto, has accomplished and what it has not.

In § 79, we have obtained explicit solutions for the quantities listed under (1). In particular, the solutions for the spin-coefficients, κ, σ, λ, and ν in the gauge,

$$\Psi_1 = \Psi_3 = 0, \tag{89}$$

are given in equations (25)–(28). And the solutions for the Weyl scalars, Ψ_0 and Ψ_4 (which are independent of the choice of *any* gauge), are expressed in terms of the Teukolsky functions, $R_{\pm 2}$ and $S_{\pm 2}$; thus,

$$\Psi_0 = R_{+2}S_{+2} \quad \text{and} \quad \Psi_4 = R_{-2}S_{-2}/(\bar{\rho}^*)^4, \tag{90}$$

where it should be remembered that, if the angular functions S_{+2} and S_{-2} are both normalized to unity and the radial functions, $\Delta^2 R_{+2}$ and R_{-2}, are chosen to be P_{+2} and P_{-2} (consistently with equations (41) and (42)), there is yet an undetermined constant of proportionality in the solution for either Ψ_0 or Ψ_4. It may be recalled that, in the analogous context of the solution of Maxwell's equation in Chapter 8, §71, a similar indeterminateness was susceptible of easy resolution since the equations governing the Maxwell scalars ϕ_0 and ϕ_2 were not entirely decoupled as equations (7)–(12) are. Besides the relative normalization of the solutions for Ψ_0 and Ψ_4, the argument of the complex Starobinsky constant also remains indeterminate.

(a) *A matrix representation of the perturbations in the basis vectors*

It is convenient to introduce an index notation for the basis vectors, $(\boldsymbol{l}, \boldsymbol{n}, \boldsymbol{m}, \bar{\boldsymbol{m}})$. Thus, by letting

$$\boldsymbol{l}^1 = \boldsymbol{l},\ \boldsymbol{l}^2 = \boldsymbol{n},\ \boldsymbol{l}^3 = \boldsymbol{m}, \quad \text{and} \quad \boldsymbol{l}^4 = \bar{\boldsymbol{m}}, \tag{91}$$

we can express the perturbation $\boldsymbol{l}^{i(1)}$, in the vector $\boldsymbol{l}^i$, as a linear combination of the unperturbed basis-vectors, $\boldsymbol{l}^i$, in the manner

$$\boldsymbol{l}^{i(1)} = A^i_j \boldsymbol{l}^j. \tag{92}$$

The perturbations in the basis vectors are then fully described by the matrix $\boldsymbol{A}$.

Since $\boldsymbol{l}^1$ and $\boldsymbol{l}^2$ are real and $\boldsymbol{l}^3$ and $\boldsymbol{l}^4$ are complex conjugates, it follows that the matrix elements, A^1_1, A^2_2, A^1_2, and A^2_1, are real, while all the remaining elements are complex; also, that the elements in which the indices 3 and 4 are replaced, one by the other, are complex conjugates. On these accounts, the specification of $\boldsymbol{A}$ will require sixteen real functions.

We shall find that the following multiples of $A^i_j (i \neq j)$ occur very naturally in the subsequent analysis:

$$\left.\begin{aligned}
&F^1_2 = \frac{\Delta}{2\rho^2} A^1_2; \quad F^2_1 = \frac{2\rho^2}{\Delta} A^2_1; \quad F^3_1 = \frac{1}{\bar{\rho}^*} A^3_1; \quad F^4_1 = \frac{1}{\bar{\rho}} A^4_1,\\
&F^1_3 = \frac{\Delta}{2\rho^2 \bar{\rho}} A^1_3; \quad F^2_3 = \frac{1}{\bar{\rho}} A^2_3; \quad F^3_2 = \frac{\Delta}{2\rho^2 \bar{\rho}^*} A^3_2; \quad F^4_2 = \frac{\Delta}{2\rho^2 \bar{\rho}} A^4_2,\\
&F^1_4 = \frac{\Delta}{2\rho^2 \bar{\rho}^*} A^1_4; \quad F^2_4 = \frac{1}{\bar{\rho}^*} A^2_4; \quad F^3_4 = \frac{1}{(\bar{\rho}^*)^2} A^3_4; \quad F^4_3 = \frac{1}{(\bar{\rho})^2} A^4_3.
\end{aligned}\right\} \tag{93}$$

We shall also find that the following combinations of F^i_j and $A^i_j (i \neq j)$ play important roles in the theory:

$$\left.\begin{aligned}
F &= F^1_3 + F^1_4; \quad B_1 = (F^3_1 + F^3_2) + (F^4_1 + F^4_2),\\
G &= F^2_3 + F^2_4; \quad B_2 = (F^3_1 + F^3_2) - (F^4_1 + F^4_2),\\
H &= F^3_1 - F^3_2; \quad C_1 = (F^1_3 + F^2_3) - (F^1_4 + F^2_4),\\
J &= F^4_1 - F^4_2; \quad C_2 = (F^1_3 - F^2_3) - (F^1_4 - F^2_4),\\
U &= A^1_1 + A^2_2, \quad \text{and} \quad V = A^3_3 + A^4_4.
\end{aligned}\right\} \tag{94}$$

It will be observed that, as defined,

$$\left.\begin{array}{c} F^1_2,\ F^2_1,\ U,\ V,\ F,\ G, \text{ and } B_1 \text{ are real,} \\ B_2,\ C_1, \text{ and } C_2 \text{ are imaginary,} \\ \text{and } F^1_4 \text{ and } F^1_3,\ F^4_i \text{ and } F^3_i\ (i=1,2),\ F^4_3 \text{ and } F^3_4 \\ \text{and } H \text{ and } J \text{ are complex conjugates.} \end{array}\right\} \qquad (95)$$

(*b*) *The perturbation in the metric coefficients*

As we have stated, the central problem, in a theory of perturbations of a space-time geometry, is the specification of the normal modes of the perturbation in the metric coefficients. These metric pertubations are related to the perturbations in the null basis-vectors by (cf. Ch. 1, equation (246))

$$\begin{aligned} g^{\mu\nu(1)} &= l^\mu n^{\nu(1)} + l^{\mu(1)} n^\nu - m^\mu \bar{m}^{\nu(1)} - \bar{m}^{\mu(1)} m^\nu \\ &\quad + l^{\nu(1)} n^\mu + l^\nu n^{\mu(1)} - m^{\nu(1)} \bar{m}^\mu - \bar{m}^\nu m^{\mu(1)}. \end{aligned} \qquad (96)$$

Evaluating the various components of $g^{\mu\nu(1)}$ in accordance with the foregoing formula, equation (92), and Chapter 6, equation (170) we find

$$\begin{aligned} g^{tt(1)} &= \frac{(r^2+a^2)^2}{\rho^2\Delta}(A^1_1 + A^2_2) - \frac{a^2\sin^2\theta}{\rho^2}(A^3_3 + A^4_4) \\ &\quad + \frac{(r^2+a^2)^2}{\rho^2\Delta}(F^1_2 + F^2_1) + a^2(F^3_4 + F^4_3)\sin^2\theta \\ &\qquad + \frac{\sqrt{2}}{\Delta} ia(r^2+a^2)(C_1 + B_2)\sin\theta, \\ g^{rr(1)} &= -\frac{\Delta}{\rho^2}(A^1_1 + A^2_2) + \frac{\Delta}{\rho^2}(F^1_2 + F^2_1), \\ g^{\theta\theta(1)} &= -\frac{1}{\rho^2}(A^3_3 + A^4_4) - (F^3_4 + F^4_3), \\ g^{\varphi\varphi(1)} &= \frac{a^2}{\rho^2\Delta}(A^1_1 + A^2_2) - \frac{1}{\rho^2}(A^3_3 + A^4_4)\operatorname{cosec}^2\theta \\ &\quad + \frac{a^2}{\rho^2\Delta}(F^1_2 + F^2_1) + (F^3_4 + F^4_3)\operatorname{cosec}^2\theta + \frac{\sqrt{2}}{\Delta} ia(C_1 + B_2)\operatorname{cosec}\theta, \\ g^{tr(1)} &= -\frac{r^2+a^2}{\rho^2}(F^1_2 - F^2_1) + \frac{ia\sin\theta}{\sqrt{2}}(H - J - C_2), \\ g^{t\theta(1)} &= +ia(F^3_4 - F^4_3)\sin\theta + \frac{r^2+a^2}{\Delta\sqrt{2}}(F + G - B_1), \end{aligned}$$

$$g^{t\varphi(1)} = a\frac{r^2+a^2}{\Delta\rho^2}(A^1_1+A^2_2) - \frac{a}{\rho^2}(A^3_3+A^4_4)$$

$$+a\frac{r^2+a^2}{\rho^2\Delta}(F^1_2+F^2_1) + a(F^3_4+F^4_3)$$

$$+\frac{i}{\Delta\sqrt{2}}[(r^2+a^2)+a^2\sin^2\theta](C_1+B_2)\operatorname{cosec}\theta,$$

$$g^{r\theta(1)} = -\frac{1}{\sqrt{2}}(F-G+J+H),$$

$$g^{r\varphi(1)} = -\frac{a}{\rho^2}(F^1_2-F^2_1)+\frac{i}{\sqrt{2}}(H-J-C_2)\operatorname{cosec}\theta,$$

$$g^{\theta\varphi(1)} = i(F^3_4-F^4_3)\operatorname{cosec}\theta + \frac{a}{\Delta\sqrt{2}}(F+G-B_1). \tag{97}$$

We observe that the perturbations in the metric coefficients depend only on the following ten combinations of the elements of the matrix A:

$$\left.\begin{array}{l} A^1_1+A^2_2,\ A^3_3+A^4_4,\ F^1_2 \pm F^2_1,\ F^3_4+F^4_3,\ C_1+B_2, \\ H-J-C_2,\ F+G-B_1,\quad \text{and}\quad F-G+J+H. \end{array}\right\} \tag{98}$$

(c) *The enumeration of the quantities that have to be determined, the equations that are available, and the gauge freedom that we have*

The quantities that have to be determined (listed in (1) and (2)) require ten real functions to specify the five complex Weyl-scalars, twenty-four real functions to specify the twelve complex spin-coefficients, and sixteen real functions to specify the matrix A: all together, fifty real functions. These fifty functions are subject to ten degrees of gauge freedom. These ten degrees of freedom arise from the six degrees of freedom in setting up the local tetrad-frame and the four degrees of freedom from the general covariance of the theory.

As we have seen in detail in Chapter 1, § 8, the Newman–Penrose formalism provides three sets of equations: the Bianchi identities, the commutation relations, and the Ricci identities. By counting each complex equation as equivalent to two real equations, we have sixteen equations representing the Bianchi identities (for vacuum fields we are presently considering), twenty-four equations representing the commutation relations, and thirty-six equations representing the Ricci identities (or, only twenty if we allow for the sixteen eliminant relations in Chapter 1, equations (311)). These seventy-six equations are available for determining fifty real functions subject to ten degrees of gauge freedom.

We have already utilized four of the six degrees of tetrad freedom in assuming that

$$\Psi_1 = \Psi_3 = 0, \tag{99}$$

as we have in writing the solutions (25)–(28) for the spin-coefficients, κ, σ, λ, and ν. We shall further utilize two of the four coordinate degrees of freedom by assuming that the perturbation in the Weyl scalar also vanishes, i.e.,

$$\Psi_2^{(1)} = 0. \tag{100}$$

After these choices, we still have four degrees of gauge freedom remaining.

83. The linearization of the remaining Bianchi identities

Four of the eight Bianchi identities (namely those included in equations (3) and (4)) have already been used. The remaining four identities (Ch. 1, equations (321, b, c, f, and g)) are

$$D\Psi_2 = 3\tilde{\rho}\Psi_2, \quad \triangle\Psi_2 = -3\mu\Psi_2, \quad \delta\Psi_2 = 3\tau\Psi_2, \quad \text{and} \quad \delta^*\Psi_2 = -3\pi\Psi_2, \tag{101}$$

when quantities of the second order of smallness, such as $\lambda\Psi_0$, $\kappa\Psi_4$, $\nu\Psi_0$, and $\sigma\Psi_4$, are neglected. The remarkable feature of equations (101) is that, in the gauge $\Psi_1 = \Psi_3 = 0$, they are formally the same as in the stationary state: they are valid inclusive of quantities of the first order of smallness.

Since $\Psi_2 = -M(\bar{\rho}^*)^{-3}$ and we have chosen a coordinate gauge in which $\Psi_2^{(1)} = 0$, the linearized versions of equations (101) are

$$\left.\begin{aligned} &\tilde{\rho}^{(1)} = -\boldsymbol{l}^{1(1)}\lg\bar{\rho}^*; \quad \mu^{(1)} = \boldsymbol{l}^{2(1)}\lg\bar{\rho}^*; \quad \tau^{(1)} = -\boldsymbol{l}^{3(1)}\lg\bar{\rho}^*; \\ &\qquad \text{and } \pi^{(1)} = +\boldsymbol{l}^{4(1)}\lg\bar{\rho}^*. \end{aligned}\right\} \tag{102}$$

Making use of the relations

$$\left.\begin{aligned} &\boldsymbol{l}^1\lg\bar{\rho}^* = D\lg\bar{\rho}^* = \frac{1}{\bar{\rho}^*}; \quad \boldsymbol{l}^2\lg\bar{\rho}^* = \triangle\lg\bar{\rho}^* = \mu; \\ &\boldsymbol{l}^3\lg\bar{\rho}^* = \delta\lg\bar{\rho}^* = -\tau; \quad \boldsymbol{l}^4\lg\bar{\rho}^* = \delta^*\lg\bar{\rho}^* = +\pi, \end{aligned}\right\} \tag{103}$$

and remembering that

$$\boldsymbol{l}^{i(1)} = A^i_j\boldsymbol{l}^j, \tag{104}$$

we find that the expanded versions of equations (102) are

$$\left.\begin{aligned} \tilde{\rho}^{(1)} &= -\left(\frac{A^1_1}{\bar{\rho}^*} + A^1_2\mu - A^1_3\tau + A^1_4\pi\right) = -\frac{M^{(1)}}{\bar{\rho}^*}, \\ \mu^{(1)} &= +\left(\frac{A^2_1}{\bar{\rho}^*} + A^2_2\mu - A^2_3\tau + A^2_4\pi\right) = \frac{\Delta}{2\rho^2\bar{\rho}^*}M^{(2)}, \\ \tau^{(1)} &= -\left(\frac{A^3_1}{\bar{\rho}^*} + A^3_2\mu - A^3_3\tau + A^3_4\pi\right) = -M^{(3)}, \\ \pi^{(1)} &= +\left(\frac{A^4_1}{\bar{\rho}^*} + A^4_2\mu - A^4_3\tau + A^4_4\pi\right) = \frac{\bar{\rho}}{\bar{\rho}^*}M^{(4)}, \end{aligned}\right\} \tag{105}$$

where

$$\left.\begin{aligned} M^{(1)} &= +A^1_1 - F^1_2 + ia\sqrt{2}\frac{\rho^2}{\Delta}F\sin\theta,\\ M^{(2)} &= -A^2_2 + F^2_1 + ia\sqrt{2}\frac{\rho^2}{\Delta}G\sin\theta,\\ M^{(3)} &= H + \frac{ia\sin\theta}{\sqrt{2}}\left(F^3_4 + \frac{A^3_3}{\rho^2}\right),\\ M^{(4)} &= J + \frac{ia\sin\theta}{\sqrt{2}}\left(F^4_3 + \frac{A^4_4}{\rho^2}\right). \end{aligned}\right\} \tag{106}$$

We may note here for future reference the following equations which are direct consequences of equations (105):

$$\left.\begin{aligned} -(\tau^* + \pi)^{(1)} &= -\frac{2ia\cos\theta}{\rho^2}A^4_1 + \frac{ia\Delta\cos\theta}{\rho^4}A^4_2 - \sqrt{2}\frac{iar\sin\theta}{\rho^2}\left(\frac{A^4_4}{\bar{\rho}^*} + \frac{A^4_3}{\bar{\rho}}\right),\\ +(\mu^* - \mu)^{(1)} &= -\frac{2ia\cos\theta}{\rho^2}A^2_1 + \frac{ia\Delta\cos\theta}{\rho^4}A^2_2 - \sqrt{2}\frac{iar\sin\theta}{\rho^2}\left(\frac{A^2_4}{\bar{\rho}^*} + \frac{A^2_3}{\bar{\rho}}\right),\\ +(\tilde{\rho}^* - \tilde{\rho})^{(1)} &= +\frac{2ia\cos\theta}{\rho^2}A^1_1 - \frac{ia\Delta\cos\theta}{\rho^4}A^1_2 + \sqrt{2}\frac{iar\sin\theta}{\rho^2}\left(\frac{A^1_4}{\bar{\rho}^*} + \frac{A^1_3}{\bar{\rho}}\right),\\ +(\tau - \pi^*)^{(1)} &= -\frac{2r}{\rho^2}A^3_1 + \frac{r\Delta}{\rho^4}A^3_2 + \sqrt{2}\frac{a^2\sin\theta\cos\theta}{\rho^2}\left(\frac{A^3_3}{\bar{\rho}} + \frac{A^3_4}{\bar{\rho}^*}\right). \end{aligned}\right\} \tag{107}$$

We observe that the Bianchi identities (101) have enabled us to express the perturbations in the spin-coefficients, $\tilde{\rho}$, τ, μ, and π, directly in terms of the perturbations in the basic tetrad in the gauge $\Psi_2^{(1)} = 0$.

84. The linearization of the commutation relations. The three systems of equations

In our present notation, the commutation relations are the expanded versions of the equations included in

$$[l^i, l^j] = C^{ij}_k l^k, \tag{108}$$

where the l's (as tangent vectors) are interpreted as directional derivatives and the structure constants C^{ij}_k are expressed in terms of the spin-coefficients (as they are in Ch. 1, equation (307)). With the spin-coefficients given in Chapter 6, equations (175), the structure constants have, in the stationary background, the values listed below:

$$C_1^{21} = \gamma+\gamma^* = -\frac{r\Delta}{\rho^4}+\frac{r-M}{\rho^2}; \qquad C_1^{31} = \alpha^*+\beta-\pi^* = 0;$$

$$C_2^{21} = \varepsilon+\varepsilon^* = 0; \qquad C_2^{31} = \kappa = 0;$$

$$C_3^{21} = -(\tau^*+\pi) = -\sqrt{2}\frac{iar\sin\theta}{\rho^2\bar{\rho}^*}; \quad C_3^{31} = -(\tilde{\rho}^*+\varepsilon-\varepsilon^*) = \frac{1}{\bar{\rho}};$$

$$C_4^{21} = -(\tau+\pi^*) = +\sqrt{2}\frac{iar\sin\theta}{\rho^2\bar{\rho}}; \quad C_4^{31} = -\sigma = 0;$$

$$C_1^{32} = -\nu^* = 0; \qquad C_1^{43} = \mu^*-\mu = \frac{ia\Delta\cos\theta}{\rho^4};$$

$$C_2^{32} = \tau-\alpha^*-\beta = \sqrt{2}\frac{a^2\sin\theta\cos\theta}{\rho^2\bar{\rho}}; \; C_2^{43} = \tilde{\rho}^*-\tilde{\rho} = \frac{2ia\cos\theta}{\rho^2};$$

$$C_3^{32} = \mu-\gamma+\gamma^* = -\frac{\Delta}{2\rho^2\bar{\rho}}; \qquad C_3^{43} = \alpha-\beta^* = -\frac{\bar{\rho}^*\cot\theta-ia\sin\theta}{(\bar{\rho}^*)^2\sqrt{2}};$$

$$C_4^{32} = \lambda^* = 0; \qquad C_4^{43} = \beta-\alpha^* = +\frac{\bar{\rho}\cot\theta+ia\sin\theta}{(\bar{\rho}^2)\sqrt{2}}. \tag{109}$$

Using equation (92), we can write the linearized version of equation (108) in the form

$$[A^i_k l^k, l^j]+[l^i, A^j_k l^k] = C^{ij}_k A^k_m l^m + c^{ij}_m l^m, \tag{110}$$

where c^{ij}_m denotes the perturbation in C^{ij}_m. By expanding equation (110), we obtain

$$A^i_k C^{kj}_m l^m + A^j_k C^{ik}_m l^m + (l^i A^j_m) l^m - (l^j A^i_m) l^m = C^{ij}_k A^k_m l^m + c^{ij}_m l^m, \tag{111}$$

or, since the l's are linearly independent, we have

$$l^i A^j_m - l^j A^i_m = A^i_k C^{jk}_m - A^j_k C^{ik}_m + C^{ij}_k A^k_m + c^{ij}_m. \tag{112}$$

In many ways, equation (112) is the basic equation of the theory: it provides a basic set of inhomogeneous equations for the elements of A; and, moreover, the inhomogeneous terms are directly related to the perturbation in the spin-coefficients.

The twenty-four equations, which equation (112) represents, can be grouped into three systems of eight equations each.

It follows from the tabulation (109) that

$$\left.\begin{aligned} c_3^{21} &= -(\tau^*+\pi)^{(1)}; & c_1^{43} &= +(\mu^*-\mu)^{(1)},\\ c_4^{21} &= -(\tau+\pi^*)^{(1)}; & c_2^{43} &= +(\tilde{\rho}^*-\tilde{\rho})^{(1)},\\ c_1^{31}+c_2^{32} &= +(\tau-\pi^*)^{(1)}; & c_3^{31}+c_4^{41} &= -(\tilde{\rho}+\tilde{\rho}^*)^{(1)},\\ c_1^{41}+c_2^{42} &= +(\tau^*-\pi)^{(1)}; & c_3^{32}+c_4^{42} &= +(\mu+\mu^*)^{(1)}. \end{aligned}\right\} \tag{113}$$

Accordingly, with the aid of equations (105), we can write down a system of eight equations in which the inhomogeneous terms derived from c_m^{ij} in equation (112) are directly expressible in terms of the elements of $\boldsymbol{A}$. Equation (112), therefore, provides a system of eight *homogeneous* equations (which we shall call system I) for the elements of $\boldsymbol{A}$.

Next, we observe that

$$\left.\begin{aligned} c_2^{31} &= \kappa; \quad & c_4^{31} &= -\sigma; \quad & c_1^{32} &= -\nu^*; \quad & c_4^{32} &= \lambda^*, \\ c_2^{41} &= \kappa^*; \quad & c_3^{41} &= -\sigma^*; \quad & c_1^{42} &= -\nu; \quad & c_3^{42} &= \lambda. \end{aligned}\right\} \tag{114}$$

But we already have explicit solutions (given in equations (25)–(28)) for κ, σ, λ, and ν in terms of Teukolsky's functions (leaving aside the unspecified relative normalization of R_{+2} and R_{-2}). Therefore, equation (112) provides a further system of eight equations (which we shall call system II) for the elements of $\boldsymbol{A}$ in which we may consider the inhomogeneous terms as 'known.'

The two systems of equations—system I and system II—provide a total of sixteen equations for the sixteen functions required to specify $\boldsymbol{A}$. However, we shall find that these equations do not suffice to determine $\boldsymbol{A}$: we shall have to supplement them by the linearized versions of further Ricci identities.

Again, from the tabulation (109), it follows that

$$\left.\begin{aligned} c_1^{31} - c_2^{32} &= 2(\alpha^* + \beta)^{(1)} - (\tau + \pi^*)^{(1)}, \\ c_1^{41} - c_2^{42} &= 2(\alpha + \beta^*)^{(1)} - (\tau^* + \pi)^{(1)}, \\ c_4^{41} - c_3^{31} &= 2(\varepsilon - \varepsilon^*)^{(1)} - (\tilde{\rho} - \tilde{\rho}^*)^{(1)}, \\ c_3^{42} - c_3^{32} &= 2(\gamma - \gamma^*)^{(1)} - (\mu - \mu^*)^{(1)}, \\ c_1^{21} &= (\gamma + \gamma^*)^{(1)}; \quad c_3^{43} = (\alpha - \beta^*)^{(1)}, \\ c_2^{21} &= (\varepsilon + \varepsilon^*)^{(1)}; \quad c_4^{43} = (\beta - \alpha^*)^{(1)}. \end{aligned}\right\} \tag{115}$$

Accordingly, we can write down a further system of eight equations (which we shall call system III) in which the inhomogeneous terms derived from c_m^{ij} in equation (112) are directly related to the perturbations in the remaining spin-coefficients, α, β, γ, and ε. Therefore, if $\boldsymbol{A}$ has already been determined, this last system will serve to complete the solution.

We shall now write down the explicit forms of the three systems of equations. It should, however, be stated that the reduction of the equations to the forms given below is not a light task.

SYSTEM I.

$$\mathscr{D}_0^\dagger F_3^1 + \mathscr{D}_0 F_3^2 = -irT\sin\theta + \frac{2ia\cos\theta}{\rho^2}J \qquad (21, 3),$$

$$\mathscr{D}_0^\dagger F_4^1 + \mathscr{D}_0 F_4^2 = +irT\sin\theta - \frac{2ia\cos\theta}{\rho^2}H \qquad (21, 4),$$

$$\mathscr{L}_1 F^3_1 - \mathscr{L}^\dagger_1 F^4_1 = i\Delta T \cos\theta - \frac{2iar\sin\theta}{\rho^2} G \qquad (43, 1),$$

$$\mathscr{L}_1 F^3_2 - \mathscr{L}^\dagger_1 F^4_2 = i\Delta T \cos\theta + \frac{2iar\sin\theta}{\rho^2} F \qquad (43, 2),$$

$$+\frac{1}{\sqrt{2}}\mathscr{L}^\dagger_0 U - \rho^2 \mathscr{D}_0 F^3_1 + \rho^2 \mathscr{D}^\dagger_0 F^3_2 = +2ia(F^1_4 + F^2_4)\cos\theta$$

$$(31, 1) + (32, 2),$$

$$+\frac{1}{\sqrt{2}}\mathscr{L}_0 U - \rho^2 \mathscr{D}_0 F^4_1 + \rho^2 \mathscr{D}^\dagger_0 F^4_2 = -2ia(F^1_3 + F^2_3)\cos\theta$$

$$(41, 1) + (42, 2),$$

$$-\frac{1}{\sqrt{2}}\Delta \mathscr{D}_0 V + \rho^2 \mathscr{L}_1 F^1_4 + \rho^2 \mathscr{L}^\dagger_1 F^1_3 = +2iar(F^3_2 - F^4_2)\sin\theta$$

$$(31, 3) + (41, 4),$$

$$+\frac{1}{\sqrt{2}}\Delta \mathscr{D}^\dagger_0 V + \rho^2 \mathscr{L}_1 F^2_4 + \rho^2 \mathscr{L}^\dagger_1 F^2_3 = +2iar(F^4_1 - F^3_1)\sin\theta$$

$$(32, 3) + (42, 4), \qquad (116)$$

where

$$T = \frac{\sqrt{2}}{\rho^4} a(U - V), \quad U = A^1_1 + A^2_2, \quad V = A^3_3 + A^4_4. \qquad (117)$$

SYSTEM II.

$$\frac{1}{\sqrt{2}}\rho^2 \mathscr{L}^\dagger_0 F^1_2 - \mathscr{D}_{-1}\rho^4 F^3_2 - 2ia\rho^2 F^1_4 \cos\theta = \tfrac{1}{2}\Delta\bar{\rho}\kappa \qquad (31, 2),$$

$$\frac{1}{\sqrt{2}}\rho^2 \mathscr{L}_0 F^1_2 - \mathscr{D}_{-1}\rho^4 F^4_2 + 2ia\rho^2 F^1_3 \cos\theta = \tfrac{1}{2}\Delta\bar{\rho}^*\kappa^* \qquad (41, 2),$$

$$\frac{1}{\sqrt{2}}\mathscr{L}^\dagger_{-1}\rho^4 F^1_4 - \tfrac{1}{2}\rho^2\Delta\mathscr{D}_0\rho^2 F^3_4 + \sqrt{2}\cdot iar\rho^2 F^3_2 \sin\theta = -\tfrac{1}{2}\Delta(\bar{\rho})^2\sigma \qquad (31, 4),$$

$$\frac{1}{\sqrt{2}}\mathscr{L}_{-1}\rho^4 F^1_3 - \tfrac{1}{2}\rho^2\Delta\mathscr{D}_0\rho^2 F^4_3 - \sqrt{2}\cdot iar\rho^2 F^4_2 \sin\theta = -\tfrac{1}{2}\Delta(\bar{\rho}^*)^2\sigma^* \qquad (41, 3),$$

$$\frac{1}{\sqrt{2}}\rho^2 \mathscr{L}^\dagger_0 F^2_1 + \mathscr{D}^\dagger_{-1}\rho^4 F^3_1 - 2ia\rho^2 F^2_4 \cos\theta = -\frac{2\rho^4\bar{\rho}}{\Delta}\nu^* \qquad (32, 1),$$

$$\frac{1}{\sqrt{2}}\rho^2 \mathscr{L}_0 F^2_1 + \mathscr{D}^\dagger_{-1}\rho^4 F^4_1 + 2ia\rho^2 F^2_3 \cos\theta = -\frac{2\rho^4\bar{\rho}^*}{\Delta}\nu \qquad (42, 1),$$

$$\frac{1}{\sqrt{2}}\mathscr{L}^{\dagger}_{-1}\rho^4 F^{2}_{4}+\tfrac{1}{2}\rho^2\Delta\mathscr{D}^{\dagger}_{0}\rho^2 F^{3}_{4}-\sqrt{2}\cdot iar\rho^2 F^{3}_{1}\sin\theta=\rho^2(\bar{\rho})^2\lambda^* \qquad (32,4),$$

$$\frac{1}{\sqrt{2}}\mathscr{L}_{-1}\rho^4 F^{2}_{3}+\tfrac{1}{2}\rho^2\Delta\mathscr{D}^{\dagger}_{0}\rho^2 F^{4}_{3}+\sqrt{2}\cdot iar\rho^2 F^{4}_{1}\sin\theta=\rho^2(\bar{\rho}^*)^2\lambda \qquad (42,3).$$

(118)

SYSTEM III.

$$2(\alpha+\beta^*)^{(1)}=\frac{1}{\bar{\rho}^*\sqrt{2}}\mathscr{L}_0(A^1_1-A^2_2)-\frac{1}{\bar{\rho}}\mathscr{D}_0(\bar{\rho}^2F^4_1)-\frac{1}{\rho^2(\bar{\rho}^*)^3}\mathscr{D}^{\dagger}_{-2}[\rho^4(\bar{\rho}^*)^2F^4_2]$$
$$+\frac{2ia\cos\theta}{\bar{\rho}^*}(F^1_3-F^2_3)+\sqrt{2}\frac{ia\sin\theta}{\bar{\rho}^*}\bar{\rho}F^4_3+\sqrt{2}\frac{ia\sin\theta}{(\bar{\rho}^*)^2}A^4_4 \quad (41,1)-(42,2),$$

$$2(\varepsilon-\varepsilon^*)^{(1)}=\frac{\sqrt{2}}{\Delta\rho^2}\mathscr{L}_{-1}(\rho^4F^1_4)-\frac{\sqrt{2}}{\Delta\rho^2}\mathscr{L}^{\dagger}_{-1}(\rho^4F^1_3)-\mathscr{D}_0(A^4_4-A^3_3)$$
$$+2\sqrt{2}\frac{iar\sin\theta}{\Delta}(F^3_2+F^4_2) \qquad (41,4)-(31,3),$$

$$2(\gamma-\gamma^*)^{(1)}=\frac{1}{(\bar{\rho}^*)^4\sqrt{2}}\mathscr{L}_{-1}[(\bar{\rho}^*)^4F^2_4]-\frac{1}{(\bar{\rho})^4\sqrt{2}}\mathscr{L}^{\dagger}_{-1}[(\bar{\rho})^4F^2_3]$$
$$-\frac{\Delta}{2\rho^2}\mathscr{D}^{\dagger}_0(A^3_3-A^4_4)+\frac{2ia\Delta\cos\theta}{\rho^4}(F^2_1-A^2_2)-\sqrt{2}\frac{iar\sin\theta}{\rho^2}(F^3_1+F^4_1)$$
$$(42,3)-(32,3),$$

$$(\gamma+\gamma^*)^{(1)}=A^2_2\frac{\mathrm{d}}{\mathrm{d}r}\left(\frac{\Delta}{2\rho^2}\right)+\sqrt{2}\frac{iar\sin\theta}{\rho^2}(F^3_1-F^4_1)$$
$$-\frac{\Delta}{2\rho^2}\mathscr{D}^{\dagger}_0A^1_1-\mathscr{D}_0A^2_1 \qquad (21,1),$$

$$(\varepsilon+\varepsilon^*)^{(1)}=+2\sqrt{2}\frac{iar\sin\theta}{\Delta}(F^3_2-F^4_2)-2\sqrt{2}\frac{a^2\sin\theta\cos\theta}{\Delta}F$$
$$-\tfrac{1}{2}\Delta\mathscr{D}^{\dagger}_1\frac{A^1_2}{\rho^2}-\mathscr{D}_0A^2_2 \qquad (21,2)$$

$$(\alpha-\beta^*)^{(1)}=-\frac{2ia\cos\theta}{\bar{\rho}^*}(F^1_3+F^2_3)-J+\frac{1}{\bar{\rho}^*\sqrt{2}}\mathscr{L}_0A^3_3-\frac{1}{\sqrt{2}}\mathscr{L}^{\dagger}_1\frac{A^4_3}{\bar{\rho}}$$
$$-\frac{A^4_4}{(\bar{\rho}^*)^2\sqrt{2}}(\bar{\rho}^*\cot\theta-ia\sin\theta) \qquad (43,3).$$

(119)

(In the foregoing, we have not included the equations for $(\alpha^* + \beta)^{(1)}$ and $(\alpha^* - \beta)^{(1)}$, since they can be written down directly from the equations for $(\alpha + \beta^*)^{(1)}$ and $(\alpha - \beta^*)^{(1)}$.)

85. The reduction of system I

By eliminating T from the first four equations of system I (equations (116)) we obtain the pair of equations

$$\mathscr{D}_0^\dagger F + \mathscr{D}_0 G = \frac{2ia\cos\theta}{\rho^2}(J - H) \tag{120}$$

and

$$\mathscr{L}_1 H - \mathscr{L}_1^\dagger J = -\frac{2iar\sin\theta}{\rho^2}(F + G), \tag{121}$$

where we may recall that F, G, J, and H are defined in equations (94).

Next, by considering the difference of equations (21, 3) and (21, 4) and the sum of equations (43, 1) and (43, 2), we obtain the pair of equations

$$\mathscr{D}_0^\dagger(F^1_3 - F^1_4) + \mathscr{D}_0(F^2_3 - F^2_4) = -2irT\sin\theta + \frac{2ia\cos\theta}{\rho^2}(J + H), \tag{122}$$

and

$$\mathscr{L}_1(F^1_3 + F^3_2) - \mathscr{L}_1^\dagger(F^4_1 + F^4_2) = +2i\Delta T\cos\theta + \frac{2iar\sin\theta}{\rho^2}(F - G). \tag{123}$$

By similar additions and subtractions we can replace the second four equations of system I by the equivalent set

$$\left.\begin{aligned}
\mathscr{D}_0(F^3_1 + F^4_1) - \mathscr{D}_0^\dagger(F^3_2 + F^4_2) &= \frac{2ia\cos\theta}{\rho^2}C_1 + \frac{\sqrt{2}}{\rho^2}\frac{\partial U}{\partial\theta},\\
\mathscr{D}_0(F^3_1 - F^4_1) - \mathscr{D}_0^\dagger(F^3_2 - F^4_2) &= -\frac{2ia\cos\theta}{\rho^2}(F + G) - \frac{\sqrt{2}}{\rho^2}QU,\\
\mathscr{L}_1(F^1_4 + F^2_4) + \mathscr{L}_1^\dagger(F^1_3 + F^2_3) &= \frac{2iar\sin\theta}{\rho^2}(J - H) + \frac{\sqrt{2}}{\rho^2}iKV,\\
\mathscr{L}_1(F^1_4 - F^2_4) + \mathscr{L}_1^\dagger(F^1_3 - F^2_3) &= \frac{2iar\sin\theta}{\rho^2}B_2 + \frac{\sqrt{2}}{\rho^2}\Delta\frac{\partial V}{\partial r}.
\end{aligned}\right\} \tag{124}$$

It is useful to rewrite equations (122), (123), and (124) in the alternative forms

$$\frac{iK}{\Delta}C_2 = \frac{\partial C_1}{\partial r} + 2irT\sin\theta - \frac{2ia\cos\theta}{\rho^2}(J + H), \tag{125}$$

$$QB_1 = -\left(\frac{\partial}{\partial\theta} + \cot\theta\right)B_2 + 2i\Delta T\cos\theta + \frac{2iar\sin\theta}{\rho^2}(F - G), \tag{126}$$

$$\frac{iK}{\Delta}B_1 = \frac{2ia\cos\theta}{\rho^2}C_1 - \frac{\partial}{\partial r}(J+H) + \frac{\sqrt{2}}{\rho^2}\frac{\partial U}{\partial \theta}, \tag{127}$$

$$\frac{iK}{\Delta}B_2 = -\frac{2ia\cos\theta}{\rho^2}(F+G) + \frac{\partial}{\partial r}(J-H) - \frac{\sqrt{2}}{\rho^2}QU, \tag{128}$$

$$QC_1 = -\frac{2iar\sin\theta}{\rho^2}(J-H) + \left(\frac{\partial}{\partial\theta} + \cot\theta\right)(F+G) - \frac{\sqrt{2}}{\rho^2}iKV, \tag{129}$$

$$QC_2 = -\frac{2iar\sin\theta}{\rho^2}B_2 + \left(\frac{\partial}{\partial\theta} + \cot\theta\right)(F-G) - \frac{\sqrt{2}}{\rho^2}\Delta\frac{\partial V}{\partial r}, \tag{130}$$

where B_1, B_2, C_1, and C_2 are defined in equations (94).

It can be shown that not all of the eight equations, (120), (121), and (125)–(130), are independent: equations (127) and (130) can be derived, for example, from the other six equations. We conclude that *only six of the eight equations of system I are independent*. We shall take equations (127)–(130) together with equations (120) and (121) as our basic set of equations. The latter two equations can be written alternatively in the forms

$$\frac{iK}{\Delta}(F-G) = \frac{\partial}{\partial r}(F+G) - \frac{2ia\cos\theta}{\rho^2}(J-H), \tag{131}$$

$$Q(J+H) = \left(\frac{\partial}{\partial\theta} + \cot\theta\right)(J-H) - \frac{2iar\sin\theta}{\rho^2}(F+G). \tag{132}$$

From equations (127)–(132), it is evident that the quantities, B_1, B_2, C_1, C_2, $F-G$ and $J+H$ can all be expressed in terms of $F+G$, $J-H$, U, and V. It will eventually appear that the functions U and V are left indeterminate; they can, in fact, be set equal to zero by utilizing two of the remaining four degrees of gauge freedom.

It will appear that equations (131) and (132) play a very central role in the further developments.

86. The reduction of system II; an integrability condition

By addition and subtraction, we can replace the equations of system II (equations (118)) by the following four pairs of equations:

$$\sqrt{2}.Q\rho^2 F^1_2 + \mathscr{D}_{-1}\rho^4(F^3_2 - F^4_2) + 2ia\rho^2 F\cos\theta = \tfrac{1}{2}\Delta(\bar{\rho}^*\kappa^* - \bar{\rho}\kappa),$$

$$\sqrt{2}.Q\rho^2 F^2_1 - \mathscr{D}^{\dagger}_{-1}\rho^4(F^3_1 - F^4_1) + 2ia\rho^2 G\cos\theta = -\frac{2\rho^4}{\Delta}(\bar{\rho}^*\nu - \bar{\rho}\nu^*), \tag{a}$$

$$\sqrt{2}.\rho^2\frac{\partial F^1_2}{\partial\theta} - \mathscr{D}_{-1}\rho^4(F^3_2 + F^4_2) + 2ia\rho^2(F^1_3 - F^1_4)\cos\theta = \tfrac{1}{2}\Delta(\bar{\rho}^*\kappa^* + \bar{\rho}\kappa),$$

$$\sqrt{2}\,.\,\rho^2\frac{\partial F_1^2}{\partial\theta}+\mathscr{D}^\dagger_{-1}\rho^4(F_1^3+F_1^4)+2ia\rho^2(F_3^2-F_4^2)\cos\theta$$

$$=-\frac{2\rho^4}{\Delta}(\bar\rho^*\nu+\bar\rho\nu^*), \qquad (b)$$

$$-iK\rho^4F_4^3+\frac{1}{\sqrt{2}}\mathscr{L}^\dagger_{-1}\rho^4(F_4^1+F_4^2)-\sqrt{2}\,.\,iar\rho^2H\sin\theta$$

$$=(\bar\rho)^2(\rho^2\lambda^*-\tfrac{1}{2}\Delta\sigma),$$

$$-iK\rho^4F_3^4+\frac{1}{\sqrt{2}}\mathscr{L}_{-1}\rho^4(F_3^1+F_3^2)+\sqrt{2}\,.\,iar\rho^2J\sin\theta$$

$$=(\bar\rho^*)^2(\rho^2\lambda-\tfrac{1}{2}\Delta\sigma^*), \qquad (c)$$

$$-\rho^2\Delta\frac{\partial}{\partial r}\rho^2F_4^3+\frac{1}{\sqrt{2}}\mathscr{L}^\dagger_{-1}\rho^4(F_4^1-F_4^2)+\sqrt{2}\,.\,iar\rho^2(F_1^3+F_2^3)\sin\theta$$

$$=-(\bar\rho)^2(\rho^2\lambda^*+\tfrac{1}{2}\Delta\sigma),$$

$$-\rho^2\Delta\frac{\partial}{\partial r}\rho^2F_3^4+\frac{1}{\sqrt{2}}\mathscr{L}_{-1}\rho^4(F_3^1-F_3^2)-\sqrt{2}\,.\,iar\rho^2(F_1^4+F_2^4)\sin\theta$$

$$=-(\bar\rho^*)^2(\rho^2\lambda+\tfrac{1}{2}\Delta\sigma^*). \qquad (d)\quad (133)$$

Next, by adding the equations in each of the four pairs of equations included in the foregoing set and making use of equations (127)–(130) for further reductions, we obtain the four equations

$$\frac{Q}{\sqrt{2}}(F_2^1+F_1^2-U)+\sqrt{\Delta}\frac{\partial}{\partial r}\rho^2\frac{J-H}{\sqrt{\Delta}}$$

$$=\frac{1}{2\rho^2}\left[\tfrac{1}{2}\Delta(\bar\rho^*\kappa^*-\bar\rho\kappa)-\frac{2\rho^4}{\Delta}(\bar\rho^*\nu-\bar\rho\nu^*)\right], \qquad (134)$$

$$\frac{1}{\sqrt{2}}\frac{\partial}{\partial\theta}(F_2^1+F_1^2-U)+\sqrt{\Delta}\frac{\partial}{\partial r}\rho^2\frac{J+H}{\sqrt{\Delta}}$$

$$=\frac{1}{2\rho^2}\left[\tfrac{1}{2}\Delta(\bar\rho^*\kappa^*+\bar\rho\kappa)-\frac{2\rho^4}{\Delta}(\bar\rho^*\nu+\bar\rho\nu^*)\right], \qquad (135)$$

$$-\frac{iK}{\sqrt{2}}[\rho^2(F_4^3+F_3^4)+V]+\frac{\partial}{\partial\theta}\rho^2(F+G)$$

$$=\frac{+1}{\rho^2\sqrt{2}}[(\bar\rho)^2(\rho^2\lambda^*-\tfrac{1}{2}\Delta\sigma)+(\bar\rho^*)^2(\rho^2\lambda-\tfrac{1}{2}\Delta\sigma^*)], \qquad (136)$$

$$\frac{\Delta}{\sqrt{2}}\frac{\partial}{\partial r}[\rho^2(F_4^3+F_3^4)+V]-\frac{\partial}{\partial\theta}\rho^2(F-G)$$

$$=\frac{+1}{\rho^2\sqrt{2}}[(\bar\rho)^2(\rho^2\lambda^*+\tfrac{1}{2}\Delta\sigma)+(\bar\rho^*)^2(\rho^2\lambda+\tfrac{1}{2}\Delta\sigma^*)]. \qquad (137)$$

Eliminating $(F^1_2+F^2_1-U)$ from equations (134) and (135) and $[\rho^2(F^3_4+F^4_3)+V]$ from equations (136) and (137), we find (after some reductions in which use is made of equations (131) and (132))

$$\sqrt{\Delta}\frac{\partial}{\partial r}\rho^2\frac{J+H}{\sqrt{\Delta}}-\frac{\partial}{\partial\theta}\frac{\sqrt{\Delta}}{Q}\frac{\partial}{\partial r}\rho^2\frac{J-H}{\sqrt{\Delta}}=\frac{1}{2\rho^2}\Big[\tfrac{1}{2}\Delta(\bar\rho^*\kappa^*+\bar\rho\kappa)-\frac{2\rho^4}{\Delta}(\bar\rho^*\nu+\bar\rho\nu^*)\Big]$$
$$-\frac{\partial}{\partial\theta}\left\{\frac{1}{2\rho^2 Q}\left[\tfrac{1}{2}\Delta(\bar\rho^*\kappa^*-\bar\rho\kappa)-\frac{2\rho^4}{\Delta}(\bar\rho^*\nu-\bar\rho\nu^*)\right]\right\} \tag{138}$$

and

$$\frac{\partial}{\partial\theta}\rho^2(F-G)+i\Delta\frac{\partial}{\partial r}\frac{1}{K}\frac{\partial}{\partial\theta}\rho^2(F+G)=\frac{-1}{\rho^2\sqrt{2}}\big[(\bar\rho)^2(\rho^2\lambda^*+\tfrac{1}{2}\Delta\sigma)+(\bar\rho^*)^2(\rho^2\lambda+\tfrac{1}{2}\Delta\sigma^*)\big]$$
$$+\Delta\frac{\partial}{\partial r}\left\{\frac{i}{\rho^2K\sqrt{2}}\big[(\bar\rho)^2(\rho^2\lambda^*-\tfrac{1}{2}\Delta\sigma)+(\bar\rho^*)^2(\rho^2\lambda-\tfrac{1}{2}\Delta\sigma^*)\big]\right\}. \tag{139}$$

Equations (138) and (139) can be further reduced with the aid of equations (131) and (132). Thus, rewriting the left-hand side of equation (138) in the form

$$\sqrt{\Delta}\frac{\partial}{\partial r}Q\rho^2\frac{J+H}{\sqrt{\Delta}}-\sqrt{\Delta}\frac{\partial^2}{\partial r\,\partial\theta}\rho^2\frac{J-H}{\sqrt{\Delta}}+\sqrt{\Delta}\frac{Q_{,\theta}}{Q}\frac{\partial}{\partial r}\rho^2\frac{J-H}{\sqrt{\Delta}}$$
$$=\sqrt{\Delta}\frac{\partial}{\partial r}\frac{1}{\sqrt{\Delta}}\left[Q\rho^2(J+H)-\frac{\partial}{\partial\theta}\rho^2(J-H)+\frac{Q_{,\theta}}{Q}\rho^2(J-H)\right], \tag{140}$$

we consider the quantity in square brackets on the right-hand side; and making use of equation (132), we can reduce it successively in the manner

$$\rho^2\left(\frac{\partial}{\partial\theta}+\cot\theta\right)(J-H)-\rho^2\frac{\partial}{\partial\theta}(J-H)+2a^2(J-H)\sin\theta\cos\theta$$
$$+\frac{Q_{,\theta}}{Q}\rho^2(J-H)-2iar(F+G)\sin\theta$$
$$=\frac{1}{Q}(Q_{,\theta}+Q\cot\theta)\rho^2(J-H)+2a^2(J-H)\sin\theta\cos\theta-2iar(F+G)\sin\theta$$
$$=\frac{2a\sigma\cos\theta}{Q}\rho^2(J-H)+2a^2(J-H)\sin\theta\cos\theta-2iar(F+G)\sin\theta$$
$$=\frac{2a}{Q}(\rho^2\sigma+Qa\sin\theta)(J-H)\cos\theta-2iar(F+G)\sin\theta$$
$$=\frac{2a}{Q}[K(J-H)\cos\theta-irQ(F+G)\sin\theta]. \tag{141}$$

Therefore, letting

$$\Psi = K(J-H)\cos\theta - irQ(F+G)\sin\theta, \tag{142}$$

we find that equation (138) can be reduced to the form

$$\frac{2a\sqrt{\Delta}}{Q}\frac{\partial}{\partial r}\frac{\Psi}{\sqrt{\Delta}} = \frac{Q}{2\rho^2}\left[\tfrac{1}{2}\Delta(\bar{\rho}^*\kappa^*+\bar{\rho}\kappa)-\frac{2\rho^4}{\Delta}(\bar{\rho}^*\nu+\bar{\rho}\nu^*)\right]$$
$$-Q\frac{\partial}{\partial\theta}\left\{\frac{1}{2\rho^2 Q}\left[\tfrac{1}{2}\Delta(\bar{\rho}^*\kappa^*-\bar{\rho}\kappa)-\frac{2\rho^4}{\Delta}(\bar{\rho}^*\nu-\bar{\rho}\nu^*)\right]\right\}. \tag{143}$$

In a similar fashion, we find that equation (139) can be reduced to the form

$$-\frac{2ia}{K}\frac{\partial\Psi}{\partial\theta} = -\frac{iK}{\rho^2\Delta\sqrt{2}}[(\bar{\rho})^2(\rho^2\lambda^*+\tfrac{1}{2}\Delta\sigma)+(\bar{\rho}^*)^2(\rho^2\lambda+\tfrac{1}{2}\Delta\sigma^*)]$$
$$+iK\frac{\partial}{\partial r}\left\{\frac{i}{\rho^2 K\sqrt{2}}[(\bar{\rho})^2(\rho^2\lambda^*-\tfrac{1}{2}\Delta\sigma)+(\bar{\rho}^*)^2(\rho^2\lambda-\tfrac{1}{2}\Delta\sigma^*)]\right\}. \tag{144}$$

By replacing the derivative operators on the right-hand sides of equations (143) and (144) by the operators $\mathscr{L}$ and $\mathscr{D}$ defined in Chapter 8, §68, we obtain the pair of equations

$$2a\sqrt{\Delta}\frac{\partial}{\partial r}\frac{\Psi}{\sqrt{\Delta}} = -\frac{1}{2}Q^2\left\{\mathscr{L}_0^\dagger\frac{1}{Q}\left(\frac{1}{2}\Delta\frac{\kappa^*}{\bar{\rho}}-\frac{2\rho^2\bar{\rho}^*\nu}{\Delta}\right)-\mathscr{L}_0\frac{1}{Q}\left(\frac{1}{2}\Delta\frac{\kappa}{\bar{\rho}^*}-\frac{2\rho^2\bar{\rho}\nu^*}{\Delta}\right)\right\} \tag{145}$$

and

$$2a\frac{\partial\Psi}{\partial\theta} = \frac{iK^2}{\sqrt{2}}\left\{\mathscr{D}_0^\dagger\frac{1}{K}\left[\frac{1}{2}\Delta\left(\frac{\bar{\rho}}{\bar{\rho}^*}\sigma+\frac{\bar{\rho}^*}{\bar{\rho}}\sigma^*\right)\right]-\mathscr{D}_0\frac{1}{K}[(\bar{\rho})^2\lambda^*+(\bar{\rho}^*)^2\lambda]\right\}. \tag{146}$$

Equations (145) and (146) manifestly require that an integrability condition be satisfied: *the application of the operators,*

$$\frac{\partial}{\partial\theta}\quad \textit{and}\quad \sqrt{\Delta}\frac{\partial}{\partial r}\frac{1}{\sqrt{\Delta}}, \tag{147}$$

respectively, to the right-hand sides of equations (145) *and* (146) *must yield an equality*. This condition is crucial to the theory: it determines, as we shall see in §87 below, the relative normalization of the radial functions R_{+2} and R_{-2} and the argument of the Starobinsky constant, $\mathscr{C}$.

87. The solution of the integrability condition

It will be observed that on the right-hand side of equation (145) both κ and ν occur. But according to the solutions (25) and (28) for these spin coefficients, κ is expressed in terms of R_{+2} and S_{+2} while ν is expressed in terms of R_{-2} and S_{-2}; and, as we have repeatedly emphasized, the relative normalization of the functions R_{+2} and R_{-2} (when the angular functions S_{+2} and S_{-2} are both normalized to unity) has yet to be specified. In consequence of this lacuna in our information, the expression on the right-hand side of equation (145) is not fully defined. The same remark applies, equally, to the expression on the right-hand side of equation (146).

From our discussion of the Teukolsky–Starobinsky identities in § 81, it is apparent that the problem of specifying the relative normalization of the functions R_{+2} and R_{-2} is closely related to the problem of specifying, separately, the real and the imaginary parts of the Starobinsky constant, $\mathscr{C} = \mathscr{C}_1 + i\mathscr{C}_2$. Thus, if we should normalize $\Delta^2 R_{+2} (= P_{+2})$ and R_{-2} $(= P_{-2})$ so as to be consistent with equations (41) and (42), then only a numerical factor in the solution for Ψ_4 (or Ψ_0) needs to be specified. But the problem of determining $\mathscr{C}_1$ and $\mathscr{C}_2$, separately, will still remain.

The nature of the problem we are presented here becomes clearer when we examine the treatment of Maxwell's equations in Chapter 8, §§(70)–(71). There it was found that, with the choice of ΔR_{+1} and R_{-1} so as to be consistent with Chapter 8, equations (33) and (34), an additional factor $\frac{1}{2}$ had to be allowed for in the solution for ϕ_2 (cf. Ch. 8, equation (70)); and the determination of the factor $\frac{1}{2}$ was simultaneous with the establishment of the reality of the corresponding Starobinsky constant.

Turning to the integrability condition with which we are now confronted, we find very early in the analysis that, with the choice of P_{+2} and P_{-2} as the basic solutions for the radial functions, the solutions for Ψ_0 and Ψ_4 must, in fact, be

$$\Delta^2 \Psi_0 = P_{+2} S_{+2} \quad \text{and} \quad \Psi_4 = \frac{1}{4(\bar{\rho}^*)^4} P_{-2} S_{-2}, \tag{148}$$

—i.e., a factor $\frac{1}{4}$ as against a factor $\frac{1}{2}$ in the spin-1 case. It is convenient to assume the validity of the solutions (148) from the outset though it is strictly not necessary: we can include an additional factor—q, say—in the expression for Ψ_4 (in which case, the subsequent analysis will show that $q = 1$). Nothing essential is lost in suppressing the factor q: its restoration is manifest at all stages; but it will destroy the symmetry of the formulae we wish to display.

We shall assume then that the solutions for Ψ_0 and Ψ_4 are given by equations (148) with the understanding that $P_{+2} = \Delta^2 R_{+2}$ and $P_{-2} = R_{-2}$ are consistent with equations (41) and (42) and satisfy equations (74). Letting (cf. equations (80))

$$X = P_{+2} + P_{-2} = [P]^+ \quad \text{and} \quad iY = P_{+2} - P_{-2} = [P]^-, \tag{149}$$

we shall find it convenient to rewrite equations (74) in the forms

$$B_1 \frac{dX}{dr} = (\Delta\mathscr{C}_2 - A_2)X + \left(\Delta\mathscr{C}_1 + A_1 - \frac{B_1 K}{\Delta}\right)Y \tag{150}$$

and

$$B_1 \frac{dY}{dr} = \left(\Delta\mathscr{C}_1 - A_1 + \frac{B_1 K}{\Delta}\right)X - (\Delta\mathscr{C}_2 + A_2)Y, \tag{151}$$

where A_1, A_2, and B_1 have the same meanings as in equations (76).

With the additional factor $\frac{1}{4}$ in the solution for Ψ_4, the solutions (25)–(28) for κ, σ, λ, and ν become

$$\left.\begin{aligned}
\kappa &= -\frac{1}{6M}\frac{\sqrt{2}}{\Delta^2}(\bar{\rho}^*)^2 P_{+2}\left(\mathscr{L}_2 - \frac{3ia\sin\theta}{\bar{\rho}^*}\right)S_{+2},\\
\sigma &= +\frac{1}{6M}\frac{1}{\Delta}\frac{(\bar{\rho}^*)^2}{\bar{\rho}} S_{+2}\left(\mathscr{D}_0^\dagger - \frac{3}{\bar{\rho}^*}\right)P_{+2},\\
\lambda &= +\frac{1}{6M}\frac{1}{2\bar{\rho}^*} S_{-2}\left(\mathscr{D}_0 - \frac{3}{\bar{\rho}^*}\right)P_{-2},\\
\nu &= +\frac{1}{6M}\frac{\sqrt{2}}{4\rho^2} P_{-2}\left(\mathscr{L}_2^\dagger - \frac{3ia\sin\theta}{\bar{\rho}^*}\right)S_{-2}.
\end{aligned}\right\} \tag{152}$$

Returning to equations (145) and (146) and inserting the foregoing solutions for κ, σ, λ, and ν in the expression on the right-hand sides of the equations, we find, after some considerable reductions in which we make use of equations (87) and (88)),

$$\begin{aligned}
\frac{1}{12M\sqrt{2}}\frac{Q}{\Delta}\Bigg\{ &-\bar{\rho}^*\Bigg[-\left(\lambda\!\!\!^{-} + 6a\sigma^{+}\cos\theta - \frac{6ia^2\sigma^{+}\sin\theta\cos\theta}{Q\bar{\rho}^*}\right)S_{+2}\\
&\qquad +2\left(\cot\theta + Q - \frac{ia\sin\theta}{\bar{\rho}^*} - \frac{a\sigma^{+}\cos\theta}{Q}\right)\mathscr{L}_2 S_{+2}\Bigg]P_{+2}\\
&+\bar{\rho}\Bigg[-\left(\lambda\!\!\!^{-} - 6a\sigma^{+}\cos\theta + \frac{6ia^2\sigma^{+}\sin\theta\cos\theta}{Q\bar{\rho}}\right)S_{-2}\\
&\qquad +2\left(\cot\theta - Q + \frac{ia\sin\theta}{\bar{\rho}} - \frac{a\sigma^{+}\cos\theta}{Q}\right)\mathscr{L}_2^\dagger S_{-2}\Bigg]P_{+2}\\
&+\bar{\rho}^*\Bigg[-\left(\lambda\!\!\!^{-} - 6a\sigma^{+}\cos\theta - \frac{6ia^2\sigma^{+}\sin\theta\cos\theta}{Q\bar{\rho}^*}\right)S_{-2}\\
&\qquad +2\left(\cot\theta - Q - \frac{ia\sin\theta}{\bar{\rho}^*} - \frac{a\sigma^{+}\cos\theta}{Q}\right)\mathscr{L}_2^\dagger S_{-2}\Bigg]P_{-2}
\end{aligned}$$

$$-\bar{\rho}\left[-\left(\lambda+6a\sigma^{+}\cos\theta+\frac{6ia^{2}\sigma^{+}\sin\theta\cos\theta}{Q\bar{\rho}}\right)S_{+2}\right.$$
$$\left.+2\left(\cot\theta+Q+\frac{ia\sin\theta}{\bar{\rho}}-\frac{a\sigma^{+}\cos\theta}{Q}\right)\mathscr{L}_{2}S_{+2}\right]P_{-2}\Bigg\}$$
$$=2a\sqrt{\Delta}\frac{\partial}{\partial r}\frac{\Psi}{\sqrt{\Delta}},\tag{153}$$

and

$$\frac{i}{12M\sqrt{2}}\frac{K}{\Delta}\Bigg\{\bar{\rho}^{*}\left[\left(\lambda+6i\sigma^{+}r+\frac{6r\sigma^{+}\Delta}{K\bar{\rho}^{*}}\right)P_{+2}\right.$$
$$\left.+2\left(r-M-iK-\frac{r\sigma^{+}\Delta}{K}-\frac{\Delta}{\bar{\rho}^{*}}\right)\mathscr{D}_{0}^{\dagger}P_{+2}\right]S_{+2}$$
$$+\bar{\rho}\left[\left(\lambda+6i\sigma^{+}r+\frac{6r\sigma^{+}\Delta}{K\bar{\rho}}\right)P_{+2}+2\left(r-M-iK-\frac{r\sigma^{+}\Delta}{K}-\frac{\Delta}{\bar{\rho}}\right)\mathscr{D}_{0}^{\dagger}P_{+2}\right]S_{-2}$$
$$-\bar{\rho}^{*}\left[\left(\lambda-6i\sigma^{+}r+\frac{6r\sigma^{+}\Delta}{K\bar{\rho}^{*}}\right)P_{-2}+2\left(r-M+iK-\frac{r\sigma^{+}\Delta}{K}-\frac{\Delta}{\bar{\rho}^{*}}\right)\mathscr{D}_{0}P_{-2}\right]S_{-2}$$
$$-\bar{\rho}\left[\left(\lambda-6i\sigma^{+}r+\frac{6r\sigma^{+}\Delta}{K\bar{\rho}}\right)P_{-2}+2\left(r-M+iK-\frac{r\sigma^{+}\Delta}{K}-\frac{\Delta}{\bar{\rho}}\right)\mathscr{D}_{0}P_{-2}\right]S_{+2}\Bigg\}$$
$$=2a\frac{\partial\Psi}{\partial\theta}.\tag{154}$$

After some further reductions, the foregoing equations can be brought to the forms

$$\frac{1}{12M\Delta\sqrt{2}}\Bigg[\!\!\Bigg[ia\{[Q\mathscr{L}_{1}\mathscr{L}_{2}S_{+2}-(2a\sigma^{+}\cos\theta)\mathscr{L}_{2}S_{+2}]\cos\theta$$
$$-2[(3a\sigma^{+}\cos\theta)S_{+2}-Q\mathscr{L}_{2}S_{+2}]\sin\theta$$
$$+[Q\mathscr{L}_{1}^{\dagger}\mathscr{L}_{2}^{\dagger}S_{-2}-(2a\sigma^{+}\cos\theta)\mathscr{L}_{2}^{\dagger}S_{-2}]\cos\theta$$
$$-2[(3a\sigma^{+}\cos\theta)S_{-2}-Q\mathscr{L}_{2}^{\dagger}S_{-2}]\sin\theta\}(P_{+2}-P_{-2})$$
$$-\{[Q\mathscr{L}_{1}\mathscr{L}_{2}S_{+2}-(2a\sigma^{+}\cos\theta)\mathscr{L}_{2}S_{+2}]$$
$$-[Q\mathscr{L}_{1}^{\dagger}\mathscr{L}_{2}^{\dagger}S_{-2}-(2a\sigma^{+}\cos\theta)\mathscr{L}_{2}^{\dagger}S_{-2}\}r(P_{+2}+P_{-2})]\Bigg]\!\!\Bigg]$$
$$=2a\sqrt{\Delta}\frac{\partial}{\partial r}\frac{\Psi}{\sqrt{\Delta}},\tag{155}$$

and

$$\frac{i}{12M\sqrt{2}}\Big\{[r(K\mathscr{D}_0^\dagger\mathscr{D}_0^\dagger P_{+2}-2r\sigma^+\mathscr{D}_0^\dagger P_{+2})+2(3r\sigma^+P_{+2}-K\mathscr{D}_0^\dagger P_{+2})$$

$$-r(K\mathscr{D}_0\mathscr{D}_0P_{-2}-2r\sigma^+\mathscr{D}_0P_{-2})-2(3r\sigma^+P_{-2}-K\mathscr{D}_0P_{-2})](S_{+2}+S_{-2})$$

$$-ia[(K\mathscr{D}_0^\dagger\mathscr{D}_0^\dagger P_{+2}-2r\sigma^+\mathscr{D}_0^\dagger P_{+2})+(K\mathscr{D}_0\mathscr{D}_0P_{-2}-2r\sigma^+\mathscr{D}_0P_{-2})]$$

$$\times(S_{+2}-S_{-2})\cos\theta\Big\}$$

$$=2a\frac{\partial\Psi}{\partial\theta};\tag{156}$$

or, in the bracket notation of §81(*b*), we have

$$2a\sqrt{\Delta}\frac{\partial}{\partial r}\frac{\Psi}{\sqrt{\Delta}}=-\frac{1}{12M\Delta\sqrt{2}}\Big[\!\Big[\{Q[\mathscr{L}\mathscr{L}S]^- -(2a\sigma^+\cos\theta)[\mathscr{L}S]^-\}rX$$

$$+a\{Q[\mathscr{L}\mathscr{L}S]^+\cos\theta+2(Q\sin\theta-a\sigma^+\cos^2\theta)[\mathscr{L}S]^+$$

$$-6a\sigma^+(\sin\theta\cos\theta)[S]^+\}Y\Big]\!\Big],\tag{157}$$

and

$$2a\frac{\partial\Psi}{\partial\theta}=\frac{i}{12M\sqrt{2}}\Big[\!\Big[\{rK[\mathscr{D}\mathscr{D}P]^- -2(r^2\sigma^++K)[\mathscr{D}P]^- +6r\sigma^+[P]^-\}[S]^+$$

$$-ia\{K[\mathscr{D}\mathscr{D}P]^+ -2r\sigma^+[\mathscr{D}P]^+\}[S]^-\cos\theta\Big]\!\Big].\tag{158}$$

We must now require that the result of applying the operator ∂_θ to the expression on the left-hand side of equation (155) is the same as applying the operator $\Delta^{1/2}\partial_r\Delta^{-1/2}$ to the expression on the left-hand side of equation (156). In the first instance one is at a complete loss to know how one is to proceed: the expressions to be equalled involve up to the third derivatives of the radial and the angular functions defined only by the Teukolsky equations which they satisfy. However, one soon realizes that the only way the required condition can be found is to replace at each stage the derivatives of $P_{\pm 2}(r)$ and $S_{\pm 2}(\theta)$ by the appropriate combinations of the functions with the aid of equations (79), (150), and (151). By such replacements equations (155) and (156) become

$$-\frac{1}{12M\Delta\beta_1\sqrt{2}}\Big[\!\Big[\{2(Q\cot\theta-a\sigma^+\cos\theta)(D-\Gamma_1)$$

$$+8a\sigma^+\lambda Q^2\cos\theta\}(S_{+2}+S_{-2})$$

$$+\{-2Q^2(D-\Gamma_1)-8a\sigma^+\lambda(Q\operatorname{cosec}\theta-a\sigma^+\cos^2\theta)\}(S_{+2}-S_{-2})\Big]\!\Big]rX$$

$$-\frac{a}{12M\Delta\beta_1\sqrt{2}}\Bigg[\!\!\Bigg[\{2Q^2(D+\Gamma_1)$$

$$-8a\sigma^+\lambda(Q\operatorname{cosec}\theta-a\sigma^+\cos^2\theta)\}(S_{+2}+S_{-2})\cos\theta$$

$$+\{-2(Q\cot\theta-a\sigma^+\cos\theta)(D+\Gamma_1)+8a\sigma^+\lambda Q^2\cos\theta\}(S_{+2}-S_{-2})\cos\theta$$

$$-2\{[(3a\sigma^+\cos\theta)\beta_1+\alpha_2 Q](S_{+2}+S_{-2})$$

$$+Q(\alpha_1+D)(S_{+2}-S_{-2})\}\sin\theta\Bigg]\!\!\Bigg]Y$$

$$=2a\sqrt{\Delta}\frac{\partial}{\partial r}\frac{\Psi}{\sqrt{\Delta}} \tag{159}$$

and

$$\frac{1}{12MB_1\sqrt{2}}\Bigg[\!\!\Bigg[\{2K^2(\mathscr{C}_2-\Gamma_2-4\lambda\sigma^+r)-2[K(r-M)-\sigma^+r\Delta](\mathscr{C}_1-\Gamma_1)\}rX$$

$$+\{2K^2(\mathscr{C}_1-\Gamma_1)+2[K(r-M)-\sigma^+r\Delta](\mathscr{C}_2-\Gamma_2-4\lambda\sigma^+r)+8\lambda\sigma^+K\Delta\}rY$$

$$+2K(\Delta\mathscr{C}_1-A_1)X-2[3r\sigma^+B_1+K(\Delta\mathscr{C}_2+A_2)]Y\Bigg]\!\!\Bigg](S_{+2}+S_{-2})$$

$$+\frac{a}{12MB_1\sqrt{2}}\Bigg[\!\!\Bigg[\{2K^2(\mathscr{C}_1+\Gamma_1)+2[K(r-M)-\sigma^+r\Delta](\mathscr{C}_2+\Gamma_2+4\lambda\sigma^+r)$$

$$-8\lambda\sigma^+K\Delta\}X$$

$$+\{-2K^2(\mathscr{C}_2+\Gamma_2+4\lambda\sigma^+r)$$

$$+2[K(r-M)-\sigma^+r\Delta](\mathscr{C}_1+\Gamma_1)\}Y\Bigg]\!\!\Bigg](S_{+2}-S_{-2})\cos\theta$$

$$=2a\frac{\partial\Psi}{\partial\theta}, \tag{160}$$

where it may be recalled that (cf. equations (43), (71) and (75))

$$\left.\begin{aligned}\mathscr{C}=\mathscr{C}_1+i\mathscr{C}_2,\ \Gamma_1=\lambda(\lambda+2)-12\sigma^{+2}\alpha^2,\ \Gamma_2=12\sigma^+M,\\ D^2=\lambda^2(\lambda+2)^2-8\sigma^{+2}\lambda[\alpha^2(5\lambda+6)-12a^2]+144\sigma^{+4}\alpha^4,\\ |\mathscr{C}|^2=D^2+144\sigma^{+2}M^2,\quad\text{and}\quad\alpha^2=a^2+am/\sigma^+,\end{aligned}\right\} \tag{161}$$

and the remaining symbols, $A_1, A_2, B_1, \alpha_1, \alpha_2$, and β_1 have the same meanings as in equations (76) and (78). Also, we may note that in the reductions leading to equations (159) and (160), use has been made of equations (47), (48), (64), and (65) (and their complex conjugates and their adjoints).

Now applying, respectively, the operators ∂_θ and $\Delta^{1/2}\partial_r\Delta^{-1/2}$ to the expressions on the left-hand sides of equations (159) and (160) and making the same replacements of the derivatives in terms of the functions, we find after some very considerable reductions and equally remarkable cancellations, that we are finally left with

$$2a\frac{\partial}{\partial\theta}\sqrt{\Delta}\frac{\partial}{\partial r}\frac{\Psi}{\sqrt{\Delta}}=\frac{1}{48M\Delta\sqrt{2}}\Bigg[\!\!\Bigg[a\{(D+\Gamma_1)(S_{+2}-S_{-2})\cos\theta$$

$$+\frac{4\sigma^+}{a}(\lambda\alpha^2-6a^2)(S_{+2}+S_{-2})\}Y$$

$$+\{-(D-\Gamma_1)(S_{+2}+S_{-2})+4a\sigma^+\lambda(S_{+2}-S_{-2})\cos\theta\}rX\Bigg]\!\!\Bigg] \tag{162}$$

and

$$\sqrt{\Delta}\frac{\partial}{\partial r}\frac{2a\partial\Psi}{\Delta\,\partial\theta}=\frac{1}{48M\Delta\sqrt{2}}\Bigg[\!\!\Bigg[a\{[(\mathscr{C}_2+\Gamma_2)+4\lambda\sigma^+r]X$$

$$+(\mathscr{C}_1+\Gamma_1)Y\}(S_{+2}-S_{-2})\cos\theta$$

$$+\{-(\mathscr{C}_1-\Gamma_1)rX+[r(\mathscr{C}_2+\Gamma_2)+4\sigma^+(\lambda\alpha^2-6a^2)]Y\}(S_2+S_{-2})\Bigg]\!\!\Bigg]. \tag{163}$$

A comparison of the expressions on the right-hand sides of equations (162) and (163) shows that their equality requires only (!) that

$$\mathscr{C}_1=D \text{ and } \mathscr{C}_2=-\Gamma_2=-12\sigma^+M \tag{164}$$

—remarkably simple results to arrive at, after such a long road.

Now defining the functions

$$\left.\begin{aligned}\mathscr{S}_+&=\int(S_{+2}+S_{-2})\mathrm{d}\theta,\ \mathscr{S}_-=\int(S_{+2}-S_{-2})\cos\theta\,\mathrm{d}\theta,\\ \mathscr{R}_+&=\Delta^{1/2}\int\frac{rX}{\Delta^{3/2}}\mathrm{d}r,\text{ and }\mathscr{R}_-=\Delta^{1/2}\int\frac{Y}{\Delta^{3/2}}\mathrm{d}r,\end{aligned}\right\} \tag{165}$$

we can write the integral of equations (162) and (163) in the form

$$\Psi=\frac{1}{96M\sqrt{2}}\Bigg[(\mathscr{C}_1+\Gamma_1)\mathscr{R}_-\mathscr{S}_-+\frac{4\sigma^+}{a}(\lambda\alpha^2-6a^2)\mathscr{R}_-\mathscr{S}_+$$

$$+4\lambda\sigma^+\mathscr{R}_+\mathscr{S}_--\frac{1}{a}(\mathscr{C}_1-\Gamma_1)\mathscr{R}_+\mathscr{S}_+\Bigg], \tag{166}$$

where it may be recalled that

$$\Psi=K(J-H)\cos\theta-irQ(F+G)\sin\theta. \tag{167}$$

Equation (166) represents the solution of the integrability condition presented by equations (145) and (146). And in deriving this solution, we have resolved, at the same time, the problems associated with the real and the imaginary parts of the Starobinsky constant and with the relative normalization of the radial Teukolsky functions.

We conclude this section by noting that with the definition (167), equations (131) and (132) have the alternative forms

$$\left.\begin{aligned} \rho^2(J+H) &= \frac{\partial}{\partial\theta}\rho^2\frac{J-H}{Q}+\frac{2a\Psi}{Q^2} \\ \text{and} \qquad i\rho^2\frac{F-G}{\Delta} &= \frac{\partial}{\partial r}\rho^2\frac{F+G}{K}-\frac{2ia\Psi}{K^2}. \end{aligned}\right\} \tag{168}$$

88. The separability of Ψ and the functions $\mathscr{R}$ and $\mathscr{S}$

It is a remarkable fact that the solution for Ψ found in §87 is separable in the variables and we have an explicit evaluation of the integrals in terms of which the solution is expressed.

We first observe that the finiteness of Ψ, for $a\to 0$, requires that $\mathscr{C}_1$ be defined as the positive square root of D^2 (cf. equations (61) and (164)); for only then will $\mathscr{C}_1-\Gamma_1\to 0$, as $a\to 0$. We have in fact the identity

$$\mathscr{C}_1^2-\Gamma_1^2 = -16\sigma^{+2}\lambda(\lambda\alpha^2-6a^2). \tag{169}$$

Accordingly, we may write

$$\frac{1}{a}(\mathscr{C}_1-\Gamma_1)=\frac{\mathscr{C}_1^2-\Gamma_1^2}{a(\mathscr{C}_1+\Gamma_1)}=-\frac{16\sigma^{+2}\lambda(\lambda\alpha^2-6a^2)}{a(\mathscr{C}_1+\Gamma_1)}. \tag{170}$$

Making use of this last relation, we can rewrite the solution (166) for Ψ in the form

$$\Psi=\frac{1}{96M\sqrt{2}}\left\{[(\mathscr{C}_1+\Gamma_1)\mathscr{R}_-+4\lambda\sigma^+\mathscr{R}_+]\mathscr{S}_- +\frac{4\sigma^+}{a}(\lambda\alpha^2-6a^2)\left[\mathscr{R}_-+\frac{4\lambda\sigma^+}{\mathscr{C}_1+\Gamma_1}\mathscr{R}_+\right]\mathscr{S}_+\right\}, \tag{171}$$

or

$$\Psi=\frac{\mathscr{C}_1+\Gamma_1}{96M\sqrt{2}}\left[\mathscr{R}_-+\frac{4\lambda\sigma^+}{\mathscr{C}_1+\Gamma_1}\mathscr{R}_+\right]\left[\mathscr{S}_-+\frac{4\sigma^+(\lambda\alpha^2-6a^2)}{a(\mathscr{C}_1+\Gamma_1)}\mathscr{S}_+\right]. \tag{172}$$

The separability of Ψ is manifest:

$$\Psi=\frac{\mathscr{C}_1+\Gamma_1}{96M\sqrt{2}}\mathscr{R}(r)\mathscr{S}(\theta), \tag{173}$$

where

$$\left.\begin{aligned} \mathscr{R}(r) &= \mathscr{R}_-+\frac{4\lambda\sigma^+}{\mathscr{C}_1+\Gamma_1}\mathscr{R}_+ \\ \text{and} \qquad \mathscr{S}(\theta) &= \mathscr{S}_-+\frac{4\sigma^+(\lambda\alpha^2-6a^2)}{\mathscr{C}_1+\Gamma_1}\mathscr{S}_+. \end{aligned}\right. \tag{174}$$

From the definitions of $\mathscr{R}$ and $\mathscr{S}$ it follows that

$$\sqrt{\Delta}\frac{\mathrm{d}}{\mathrm{d}r}\frac{\mathscr{R}}{\sqrt{\Delta}}=\frac{1}{\Delta}\left(Y+\frac{4\lambda\sigma^{+}}{\mathscr{C}_1+\Gamma_1}rX\right) \tag{175}$$

and

$$\frac{\mathrm{d}\mathscr{S}}{\mathrm{d}\theta}=[S]^{-}\cos\theta+\frac{4\sigma^{+}(\lambda\alpha^2-6a^2)}{a(\mathscr{C}_1+\Gamma_1)}[S]^{+}. \tag{176}$$

(a) The expression of $\mathscr{R}$ and $\mathscr{S}$ in terms of the Teukolsky functions

In accordance with equations (173) and (176), equation (158) gives

$$\begin{aligned}2a\frac{\partial\Psi}{\partial\theta}&=\frac{a(\mathscr{C}_1+\Gamma_1)}{48M\sqrt{2}}\mathscr{R}\left\{[S]^{-}\cos\theta+\frac{4\sigma^{+}(\lambda\alpha^2-6a^2)}{a(\mathscr{C}_1+\Gamma_1)}[S]^{+}\right\}\\&=\frac{i}{12M\sqrt{2}}\Big[\!\Big[\{rK[\mathscr{D}\mathscr{D}P]^{-}-2(r^2\sigma^{+}+K)[\mathscr{D}P]^{-}+6r\sigma^{+}[P]^{-}\}[S]^{+}\\&\qquad -ia\{K[\mathscr{D}\mathscr{D}P]^{+}-2r\sigma^{+}[\mathscr{D}P]^{+}\}[S]^{-}\cos\theta\Big]\!\Big].\end{aligned} \tag{177}$$

Equating now the coefficients of $[S]^{-}\cos\theta$ and $[S]^{+}$ in the two expressions on the right-hand side, we obtain the following two alternative expressions for $\mathscr{R}$:

$$\tfrac{1}{4}(\mathscr{C}_1+\Gamma_1)\mathscr{R}=K[\mathscr{D}\mathscr{D}P]^{+}-2r\sigma^{+}[\mathscr{D}P]^{+}, \tag{178}$$

and

$$\sigma^{+}(\lambda\alpha^2-6a^2)\mathscr{R}=i\{rK[\mathscr{D}\mathscr{D}P]^{-}-2(r^2\sigma^{+}+K)[\mathscr{D}P]^{-}+6r\sigma^{+}[P]^{-}\}. \tag{179}$$

From a comparison of equations (158) and (160), we can now read off the explicit evaluations of the expressions on the right-hand sides of equations (178) and (179), in terms of the functions X and Y. We thus obtain

$$\begin{aligned}\tfrac{1}{4}(\mathscr{C}_1+\Gamma_1)\mathscr{R}=\frac{1}{B_1}\Big[\!\Big[&\{2K^2(\mathscr{C}_1+\Gamma_1)+8\lambda\sigma^{+}r[K(r-M)-\sigma^{+}r\Delta]\\&-8\lambda\sigma^{+}K\Delta\}X\\&+\{-8K^2\lambda\sigma^{+}r+2(\mathscr{C}_1+\Gamma_1)[K(r-M)-\sigma^{+}r\Delta]\}Y\Big]\!\Big]\end{aligned} \tag{180}$$

and

$$\begin{aligned}\sigma^{+}(\lambda\alpha^2-6a^2)\mathscr{R}=\frac{1}{B_1}\Big[\!\Big[&\{-8K^2\sigma^{+}(6M+\lambda r)\\&-2[K(r-M)-\sigma^{+}r\Delta](\mathscr{C}_1-\Gamma_1)\}rX\\&+\{2K^2(\mathscr{C}_1-\Gamma_1)-8\sigma^{+}(6M+\lambda r)[K(r-M)-\sigma^{+}r\Delta]+8\lambda\sigma^{+}K\Delta\}rY\\&+2K(\Delta\mathscr{C}_1-A_1)X-2[3r\sigma^{+}B_1+K(\Delta\mathscr{C}_2+A_2)]Y\Big]\!\Big].\end{aligned} \tag{181}$$

Again, in accordance with equations (173) and (175), equation (157) gives

$$2a\sqrt{\Delta}\frac{\partial}{\partial r}\frac{\Psi}{\sqrt{\Delta}} = \frac{a(\mathscr{C}_1+\Gamma_1)}{48M\Delta\sqrt{2}}\mathscr{S}\left(Y+\frac{4\lambda\sigma^+}{\mathscr{C}_1+\Gamma_1}rX\right)$$

$$= -\frac{1}{12M\Delta\sqrt{2}}\Big[\!\Big[\{Q[\mathscr{L}\mathscr{L}S]^- - (2a\sigma^+\cos\theta)[\mathscr{L}S]^-\}rX$$

$$+a\{Q[\mathscr{L}\mathscr{L}S]^+\cos\theta + 2(Q\sin\theta - a\sigma^+\cos^2\theta)[\mathscr{L}S]^+$$

$$-6a\sigma^+(\sin\theta\cos\theta)[S]^+\}Y\Big]\!\Big]. \quad (182)$$

Equating now the coefficients of rX and Y in the two expressions on the right-hand side, we obtain the following two alternative expressions for $\mathscr{S}$:

$$-\lambda a\sigma^+\mathscr{S} = Q[\mathscr{L}\mathscr{L}S]^- - 2a\sigma^+[\mathscr{L}S]^-\cos\theta \quad (183)$$

and

$$-\tfrac{1}{4}(\mathscr{C}_1+\Gamma_1)\mathscr{S} = Q[\mathscr{L}\mathscr{L}S]^+\cos\theta + 2(Q\sin\theta - a\sigma^+\cos^2\theta)[\mathscr{L}S]^+$$
$$-6a\sigma^+[S]^+\sin\theta\cos\theta. \quad (184)$$

From a comparison of equations (157) and (159), we can read off the explicit evaluations of the expressions on the right-hand sides of equations (183) and (184) in terms of $[S]^+$ and $[S]^-$. We thus obtain

$$-\lambda a\sigma^+\mathscr{S} = \frac{1}{\beta_1}\Big[\!\Big[\{2(Q\cot\theta - a\sigma^+\cos\theta)(\mathscr{C}_1-\Gamma_1) + 8a\sigma^+\lambda Q^2\cos\theta\}[S]^+$$

$$+\{-2Q^2(\mathscr{C}_1-\Gamma_1) - 8a\sigma^+\lambda(Q\operatorname{cosec}\theta - a\sigma^+\cos^2\theta)\}[S]^-\Big]\!\Big] \quad (185)$$

and

$$-\tfrac{1}{4}(\mathscr{C}_1+\Gamma_1)\mathscr{S} = \frac{1}{\beta_1}\Big[\!\Big[\{2Q^2(\mathscr{C}_1+\Gamma_1) - 8a\sigma^+\lambda(Q\operatorname{cosec}\theta - a\sigma^+\cos^2\theta)\}[S]^+\cos\theta$$

$$+\{-2(Q\cot\theta - a\sigma^+\cos\theta)(\mathscr{C}_1+\Gamma_1) + 8a\sigma^+\lambda Q^2\cos\theta\}[S]^-\cos\theta$$

$$-2\{[(3a\sigma^+\cos\theta)\beta_1 + \alpha_2 Q][S]^+ + Q(\alpha_1+\mathscr{C}_1)[S]^-\}\sin\theta\Big]\!\Big]. \quad (186)$$

The reader does not need to be told that the relations (178), (179), (183), and (184) are astonishing—not only in themselves but also in providing explicit evaluations for the indefinite integrals in terms of which Ψ was originally defined. We shall return to these aspects of the solution of the integrability condition in §94.

89. The completion of the reduction of system II and the differential equations satisfied by $\mathscr{R}$ and $\mathscr{S}$

In §§86–88, our considerations were restricted to the four equations (134)–(137) derived from the eight equations (133) (equivalent to those of system II). After satisfying the integrability condition, which follows from these equations, we are left with two equations, namely (134) and (136), besides the solution for Ψ we have found. It remains to consider a further set of four equations, independent of the ones already considered.

By subtracting the second equation from the first (instead of adding them) in each of the four pairs of equations included in (133), we obtain

$$\sqrt{2}.Q\rho^2(F\tfrac{1}{2}-F\tfrac{2}{1})+\Delta\frac{\partial}{\partial r}\frac{\rho^4}{\Delta}B_2+\frac{iK}{\Delta}\rho^4(J-H)$$
$$+2ia\rho^2(F-G)\cos\theta=\tfrac{1}{2}\Delta(\bar{\rho}^*\kappa^*-\bar{\rho}\kappa)+\frac{2\rho^4}{\Delta}(\bar{\rho}^*\nu-\bar{\rho}\nu^*), \tag{187}$$

$$\sqrt{2}.\rho^2\frac{\partial}{\partial\theta}(F\tfrac{1}{2}-F\tfrac{2}{1})-\Delta\frac{\partial}{\partial r}\frac{\rho^4}{\Delta}B_1+\frac{iK}{\Delta}\rho^4(J+H)$$
$$+2ia\rho^2C_2\cos\theta=\tfrac{1}{2}\Delta(\bar{\rho}^*\kappa^*+\bar{\rho}\kappa)+\frac{2\rho^4}{\Delta}(\bar{\rho}^*\nu+\bar{\rho}\nu^*), \tag{188}$$

$$-iK\rho^4(F\tfrac{3}{4}-F\tfrac{4}{3})-\frac{1}{\sqrt{2}}\left(\frac{\partial}{\partial\theta}-\cot\theta\right)\rho^4C_1-\frac{Q}{\sqrt{2}}\rho^4(F+G)$$
$$-\sqrt{2}.iar\rho^2(J+H)\sin\theta=(\bar{\rho})^2(\rho^2\lambda^*-\tfrac{1}{2}\Delta\sigma)-(\bar{\rho}^*)^2(\rho^2\lambda-\tfrac{1}{2}\Delta\sigma^*), \tag{189}$$

$$-\rho^2\Delta\frac{\partial}{\partial r}\rho^2(F\tfrac{3}{4}-F\tfrac{4}{3})-\frac{1}{\sqrt{2}}\left(\frac{\partial}{\partial\theta}-\cot\theta\right)\rho^4C_2-\frac{Q}{\sqrt{2}}\rho^4(F-G)$$
$$+\sqrt{2}.iar\rho^2B_1\sin\theta=-(\bar{\rho})^2(\rho^2\lambda^*+\tfrac{1}{2}\Delta\sigma)+(\bar{\rho}^*)^2(\rho^2\lambda+\tfrac{1}{2}\Delta\sigma^*). \tag{190}$$

By eliminating $(F\tfrac{1}{2}-F\tfrac{2}{1})$ from equations (187) and (188), we obtain

$$\frac{\partial}{\partial\theta}\frac{1}{Q\rho^2}\left[\tfrac{1}{2}\Delta(\bar{\rho}^*\kappa^*-\bar{\rho}\kappa)+\frac{2\rho^4}{\Delta}(\bar{\rho}^*\nu-\bar{\rho}\nu^*)\right]$$
$$-\frac{\Delta}{2\rho^2}(\bar{\rho}^*\kappa^*+\bar{\rho}\kappa)-\frac{2\rho^2}{\Delta}(\bar{\rho}^*\nu+\bar{\rho}\nu^*)$$
$$=\frac{2ia\cos\theta}{Q^2\rho^2}\Delta\left[\frac{\partial}{\partial r}\rho^2\frac{\partial}{\partial r}(J-H)+\frac{4a^2\cos^2\theta}{\rho^2}(J-H)\right]$$
$$+\frac{2ia\sin\theta}{rQ}\Delta\frac{\partial}{\partial r}r^2\frac{F-G}{\Delta}-\frac{2iaK}{Q^2\Delta}\Psi$$
$$=\frac{2ia\Delta}{Q^2K}\left[\frac{\partial}{\partial r}K\frac{\partial}{\partial r}\frac{\Psi}{K}+\left(\frac{4\alpha^2\sigma^2}{K^2}-\frac{K^2}{\Delta^2}\right)\Psi\right]. \tag{191}$$

Similarly, eliminating $\rho^2(F_4^3 - F_3^4)$ from equations (189) and (190), we obtain

$$\Delta\frac{\partial}{\partial r}\frac{i\sqrt{2}}{K\rho^2}[(\bar\rho)^2(\rho^2\lambda^* - \tfrac{1}{2}\Delta\sigma) - (\bar\rho^*)^2(\rho^2\lambda - \tfrac{1}{2}\Delta\sigma^*)]$$

$$+\frac{\sqrt{2}}{\rho^2}[-(\bar\rho)^2(\rho^2\lambda^* + \tfrac{1}{2}\Delta\sigma) + (\bar\rho^*)^2(\rho^2\lambda + \tfrac{1}{2}\Delta\sigma^*)]$$

$$= -\frac{2iar\Delta\sin\theta}{K^2\rho^2}\left[\frac{\partial}{\partial\theta}\rho^2\left(\frac{\partial}{\partial\theta}+\cot\theta\right)(F+G) + \frac{4a^2r^2\sin^2\theta}{\rho^2}(F+G)\right]$$

$$+\frac{2a\Delta\sin\theta}{K\cos\theta}\frac{\partial}{\partial\theta}\frac{\cos^2\theta}{\sin\theta}(J+H) + \frac{2a\Delta Q}{K^2}\Psi$$

$$= \frac{2a\Delta}{K^2Q}\left[\frac{\partial}{\partial\theta}Q\frac{\partial}{\partial\theta}\frac{\Psi}{Q} - \left(\frac{4\alpha^2\sigma^{+2}}{Q^2} - Q^2\right)\Psi\right]. \tag{192}$$

It may be noted that equations (168) are crucial for the reductions leading to the final results in equations (191) and (192).

Equations (191) and (192) may be written in the alternative forms

$$\frac{ia\Delta}{K}\left\{\frac{\partial}{\partial r}\left(K\frac{\partial}{\partial r}\frac{\Psi}{K}\right) + \left(\frac{4\alpha^2\sigma^{+2}}{K^2} - \frac{K^2}{\Delta^2}\right)\Psi\right\}$$

$$= \frac{1}{2}Q^2\left\{\mathscr{L}_0^\dagger\frac{1}{Q}\left(\frac{1}{2}\Delta\frac{\kappa^*}{\bar\rho} + \frac{2\rho^2\bar\rho^*}{\Delta}\nu\right) - \mathscr{L}_0\frac{1}{Q}\left(\frac{1}{2}\Delta\frac{\kappa}{\bar\rho^*} + \frac{2\rho^2\bar\rho}{\Delta}\nu^*\right)\right\} \tag{193}$$

and

$$\frac{a}{Q}\left\{\frac{\partial}{\partial\theta}\left(Q\frac{\partial}{\partial\theta}\frac{\Psi}{Q}\right) - \left(\frac{4\alpha^2\sigma^{+2}}{Q^2} - Q^2\right)\Psi\right\}$$

$$= \frac{i}{\sqrt{2}}K^2\left\{\mathscr{D}_0^\dagger\frac{1}{K}\left[-\frac{1}{2}\Delta\frac{\bar\rho}{\bar\rho^*}\sigma + \frac{1}{2}\Delta\frac{\bar\rho^*}{\bar\rho}\sigma^*\right] - \mathscr{D}_0\frac{1}{K}\left[-(\bar\rho)^2\lambda^* + (\bar\rho^*)^2\lambda\right]\right\}. \tag{194}$$

By inserting the known solutions for κ and ν on the right-hand side of equation (193), we find, by reductions analogous to those included in equation (153), that

$$\frac{i}{12M\Delta\sqrt{2}}\Big[\!\Big[\{Q[\mathscr{L}\mathscr{L}S]^- - 2a\sigma^+[\mathscr{L}S]^-\cos\theta\}rY$$

$$-a\{Q[\mathscr{L}\mathscr{L}S]^+\cos\theta + 2(Q\sin\theta - a\sigma^+\cos^2\theta)[\mathscr{L}S]^+$$

$$-6a\sigma^+[S]^+\cos\theta\sin\theta\}X\Big]\!\Big]$$

$$= ia\frac{\mathscr{C}_1+\Gamma_1}{48M\Delta\sqrt{2}}\left(X - \frac{4\lambda\sigma^+}{\mathscr{C}_1+\Gamma_1}rY\right)\mathscr{S}, \tag{195}$$

where, in the last step, we have made use of both the relations (183) and (184). On the other hand, the fact that Ψ is separable (as expressed in equation (173)) enables us to write the right-hand side of equation (193) in the form

$$\frac{ia\Delta}{K}\frac{\mathscr{C}_1+\Gamma_1}{96M\sqrt{2}}\mathscr{S}\left[\frac{\mathrm{d}}{\mathrm{d}r}\left(K\frac{\mathrm{d}}{\mathrm{d}r}\frac{\mathscr{R}}{K}\right)-\left(\frac{4\alpha^2\sigma^{+2}}{K^2}-\frac{K^2}{\Delta^2}\right)\mathscr{R}\right]. \tag{196}$$

We are thus led to the following differential equation for the radial function, $\mathscr{R}$:

$$\frac{\Delta^2}{K}\left[\frac{\mathrm{d}}{\mathrm{d}r}\left(K\frac{\mathrm{d}}{\mathrm{d}r}\frac{\mathscr{R}}{K}\right)+\left(\frac{4\alpha^2\sigma^{+2}}{K^2}-\frac{K^2}{\Delta^2}\right)\mathscr{R}\right]=2\left(X-\frac{4\lambda\sigma^+}{\mathscr{C}_1+\Gamma_1}rY\right). \tag{197}$$

Considering the next equation (194), we similarly find that the right-hand side of the equation can be reduced to give

$$\begin{aligned}&\frac{i}{12M\sqrt{2}}\Big[\!\Big[-\{rK[\mathscr{D}\mathscr{D}P]^- -2(r^2\sigma^+ +K)[\mathscr{D}P]^- +6r\sigma^+[P]^-\}[S]^-\\&\qquad +ia\{K[\mathscr{D}\mathscr{D}P]^+ -2r\sigma^+[\mathscr{D}P]^+\}[S]^+\cos\theta\Big]\!\Big]\\&=\frac{i}{12M\sqrt{2}}\{i\sigma^+(\lambda\alpha^2-6a^2)\mathscr{R}[S]^- +\tfrac{1}{4}ia(\mathscr{C}_1+\Gamma_1)\mathscr{R}[S]^+\cos\theta\}\\&=-\frac{a(\mathscr{C}_1+\Gamma_1)}{48M\sqrt{2}}\mathscr{R}\left\{[S]^+\cos\theta+\frac{4\sigma^+(\lambda\alpha^2-6a^2)}{a(\mathscr{C}_1+\Gamma_1)}[S]^-\right\},\end{aligned} \tag{198}$$

where, in the last step, we have made use of both the relations (178) and (179); while the left-hand side gives

$$\frac{a}{Q}\frac{\mathscr{C}_1+\Gamma_1}{96M\sqrt{2}}\mathscr{R}\left[\frac{\mathrm{d}}{\mathrm{d}\theta}\left(Q\frac{\mathrm{d}}{\mathrm{d}\theta}\frac{\mathscr{S}}{Q}\right)-\left(\frac{4\alpha^2\sigma^{+2}}{Q^2}-Q^2\right)\mathscr{S}\right]. \tag{199}$$

We are thus led to the following differential equation for the angular function, $\mathscr{S}$:

$$\begin{aligned}&\frac{1}{Q}\left[\frac{\mathrm{d}}{\mathrm{d}\theta}\left(Q\frac{\mathrm{d}}{\mathrm{d}\theta}\frac{\mathscr{S}}{Q}\right)-\left(\frac{4\alpha^2\sigma^{+2}}{Q^2}-Q^2\right)\mathscr{S}\right]\\&\qquad=-2\left\{[S]^+\cos\theta+\frac{4\sigma^+(\lambda\alpha^2-6a^2)}{a(\mathscr{C}_1+\Gamma_1)}[S]^-\right\}.\end{aligned} \tag{200}$$

Alternative forms of equations (197) and (200) are

$$\frac{\Delta^2}{K}\left(\frac{\mathrm{d}^2\mathscr{R}}{\mathrm{d}r^2}-\frac{2r\sigma^+}{K}\frac{\mathrm{d}\mathscr{R}}{\mathrm{d}r}+\frac{2\sigma^+\Delta^2-K^3}{K\Delta^2}\mathscr{R}\right)=2\left(X-\frac{4\lambda\sigma^+}{\mathscr{C}_1+\Gamma_1}rY\right) \tag{201}$$

and

$$\frac{1}{Q}\left[\frac{d^2\mathscr{S}}{d\theta^2}-\left(\frac{2a\sigma^+}{Q}\cos\theta-\cot\theta\right)\frac{d\mathscr{S}}{d\theta}+\left(Q^2-\operatorname{cosec}^2\theta-\frac{2a\sigma^+}{Q}\operatorname{cosec}\theta\right)\mathscr{S}\right]$$
$$=-2\left\{[S]^+\cos\theta+\frac{4\sigma^+(\lambda\!\!\!^{-}a^2-6a^2)}{a(\mathscr{C}_1+\Gamma_1)}[S]^-\right\}. \quad (202)$$

Thus, the integrability conditions of equations (187)–(190) lead to the remarkable identities expressed by equations (201) and (202). Apart from these identities, we are left with equations (187) and (189) which may be considered as equations for $(F^1_2-F^2_1)$ and $\rho^2(F^3_4-F^4_3)$ in terms of the other quantities.

90. Four linearized Ricci-identities

Before we proceed further, we shall pause to ascertain what still remains to be resolved.

We have seen that the equations of system I enable us to express the functions, $B_1, B_2, C_1, C_2, F-G$, and $J+H$ in terms of $F+G, J-H, U$, and V.

The equations of system II, besides determining the relative normalization of the radial functions, R_{+2} and R_{-2}, and the real and the imaginary parts, separately, of the Starobinsky constant $\mathscr{C}$, provide a separable solution for

$$\Psi=K(J-H)\cos\theta-irQ(F+G)\sin\theta, \quad (203)$$

and four equations (namely, (134), (136), (187), and (189)) which express $(F^1_2\pm F^2_1)$ and $\rho^2(F^3_4\pm F^4_3)$ in terms of the other functions we have enumerated.

The equations of system III serve only to determine the perturbations in the spin-coefficients, α, β, γ, and ε, once the solution for A has been completed. And we shall find that the functions U and V are left indeterminate till the end so that we may utilize two of the remaining four degrees of gauge freedom to set them equal to zero.

Therefore, what remains to complete the solution is to determine $F+G$ and $J-H$, separately.

Since the sixteen Bianchi identities and the twenty-four commutation relations have all been used, we must, of necessity, turn to the Ricci identities. An examination of these identities listed in Chapter 1 (equations (310)) suggests that we consider Chapter 1, equations (310, a, n, g, and p). The linearized versions of these identities are

$$D\tilde{\rho}^{(1)}+2\tilde{\rho}^{(1)}/\bar{\rho}^*-D^{(1)}(\bar{\rho}^*)^{-1}+(\varepsilon+\varepsilon^*)^{(1)}/\bar{\rho}^*=\delta^*\kappa-\kappa(3\alpha+\beta^*-\pi)-\kappa^*\tau,$$
$$\Delta\mu^{(1)}+(2\mu+\gamma+\gamma^*)\mu^{(1)}+\Delta^{(1)}\mu+(\gamma+\gamma^*)^{(1)}\mu=\delta\nu-\nu(\tau-3\beta-\alpha^*)+\nu^*\pi,$$
$$\delta^*\pi^{(1)}+(2\pi+\alpha-\beta^*)\pi^{(1)}+\delta^{*(1)}\pi+(\alpha-\beta^*)^{(1)}\pi=D\lambda+\lambda/\bar{\rho}^*-\sigma^*\mu,$$
$$\delta\tau^{(1)}-(2\tau+\beta-\alpha^*)\tau^{(1)}+\delta^{(1)}\tau+(\alpha^*-\beta)^{(1)}\tau=\Delta\sigma+\sigma\mu-\lambda^*/\bar{\rho}^*-\sigma(3\gamma-\gamma^*).$$
$$(204)$$

Inserting in the foregoing equations, the solutions for κ, σ, λ, and ν given in equations (152), for $\tilde{\rho}^{(1)}$, $\tau^{(1)}$, $\mu^{(1)}$, and $\pi^{(1)}$ given in equations (105), and for $\alpha^{(1)}$, $\beta^{(1)}$, $\gamma^{(1)}$, and $\varepsilon^{(1)}$ given in the equations of system III (equations (119)), we find, after considerable reductions, that they yield

$$-iKF^1_2+\tfrac{1}{2}\Delta\mathscr{D}_0U+\frac{1}{\sqrt{2}}(ia\sin\theta)\mathscr{D}_{-1}\rho^2F+\sqrt{2}.\,(a^2\sin\theta\cos\theta)F$$

$$-\sqrt{2}.(iar\sin\theta)(F^3_2-F^4_2)=-\frac{\Delta}{2\sqrt{2}}\left[(\bar{\rho}^*)^2\mathscr{L}_1\frac{\kappa}{(\bar{\rho}^*)^2}+\frac{ia\sin\theta}{\bar{\rho}}\kappa^*\right], \tag{205}$$

$$+iKF^2_1+\tfrac{1}{2}\Delta\mathscr{D}^\dagger_0U-\frac{1}{\sqrt{2}}(ia\sin\theta)\mathscr{D}^\dagger_{-1}\rho^2G-\sqrt{2}.(a^2\sin\theta\cos\theta)G$$

$$-\sqrt{2}.(iar\sin\theta)(F^3_1-F^4_1)=\sqrt{2}\frac{\rho^4\bar{\rho}^*}{\Delta}\left[\frac{1}{\rho^2\bar{\rho}}\mathscr{L}^\dagger_1(\rho^2\nu)+\frac{ia\sin\theta}{(\bar{\rho}^*)^2}\nu^*\right], \tag{206}$$

$$(iaQ\sin\theta)F^4_3+\frac{ia\sin\theta}{2\rho^2}\mathscr{L}_0V+\frac{1}{\sqrt{2}}\left(\mathscr{L}_{-1}J-\frac{2ia\sin\theta}{\bar{\rho}}J\right)$$

$$+\sqrt{2}\frac{a^2\sin\theta\cos\theta}{\rho^2}(F^1_3+F^2_3)=\frac{\bar{\rho}^*}{\bar{\rho}}\left[\mathscr{D}_0(\bar{\rho}^*\lambda)+\frac{\Delta}{2\rho^2}\sigma^*\right], \tag{207}$$

$$(iaQ\sin\theta)F^3_4-\frac{ia\sin\theta}{2\rho^2}\mathscr{L}^\dagger_0V-\frac{1}{\sqrt{2}}\left(\mathscr{L}^\dagger_{-1}H-\frac{2ia\sin\theta}{\bar{\rho}}H\right)$$

$$+\sqrt{2}\frac{a^2\sin\theta\cos\theta}{\rho^2}(F^1_4+F^2_4)=-\left[\frac{\Delta\bar{\rho}^*}{2\bar{\rho}}\mathscr{D}^\dagger_1\frac{\bar{\rho}\sigma}{(\bar{\rho}^*)^2}+\frac{\bar{\rho}}{\bar{\rho}^*}\lambda^*\right]. \tag{208}$$

By considering the difference of equations (205) and (206) and the sum of equations (207) and (208) we obtain the pair of equations

$$iK(F^1_2+F^2_1-U)-\sqrt{2}.(a^2\sin\theta\cos\theta)(F+G)+\sqrt{2}.(iar\sin\theta)(J-H)$$

$$-\frac{1}{\sqrt{2}}(ia\sin\theta)(\mathscr{D}_{-1}\rho^2F+\mathscr{D}^\dagger_{-1}\rho^2G)=\sqrt{2}\frac{\rho^4\bar{\rho}^*}{\Delta}$$

$$\times\left[\frac{1}{\rho^2\bar{\rho}}\mathscr{L}^\dagger_1(\rho^2\nu)+\frac{ia\sin\theta}{(\bar{\rho}^*)^2}\nu^*\right]+\frac{\Delta}{2\sqrt{2}}\left[(\bar{\rho}^*)^2\mathscr{L}_1\frac{\kappa}{(\bar{\rho}^*)^2}+\frac{ia\sin\theta}{\bar{\rho}}\kappa^*\right] \tag{209}$$

and

$$\frac{iaQ\sin\theta}{\rho^2}[\rho^2(F^3_4+F^4_3)+V]+\sqrt{2}\frac{a^2\sin\theta\cos\theta}{\rho^2}(F+G)$$

$$+\frac{1}{\sqrt{2}}\left[\mathscr{L}_{-1}J-\mathscr{L}^\dagger_{-1}H-\frac{2ia\sin\theta}{\bar{\rho}}(J-H)\right]=\frac{\bar{\rho}^*}{\bar{\rho}}\left[\mathscr{D}_0(\bar{\rho}^*\lambda)+\frac{\Delta}{2\rho^2}\sigma^*\right]$$

$$-\left[\frac{\Delta\bar{\rho}^*}{2\bar{\rho}}\mathscr{D}^\dagger_1\frac{\bar{\rho}\sigma}{(\bar{\rho}^*)^2}+\frac{\bar{\rho}}{\bar{\rho}^*}\lambda^*\right]. \tag{210}$$

And similarly, by considering the sum of equations (205) and (206) and the difference of equations (207) and (208), we obtain the complementary pair of equations

$$-iK(F^{1}_{2}-F^{2}_{1})+\tfrac{1}{2}\Delta(\mathscr{D}_0+\mathscr{D}_0^{\dagger})U+\frac{1}{\sqrt{2}}(ia\sin\theta)(\mathscr{D}_{-1}\rho^2F-\mathscr{D}^{\dagger}_{-1}\rho^2G)$$
$$+\sqrt{2}.(a^2\sin\theta\cos\theta)(F-G)-\sqrt{2}.(iar\sin\theta)B_2$$
$$=-\frac{\Delta}{2\sqrt{2}}\left[(\bar\rho^*)^2\mathscr{L}_1\frac{\kappa}{(\bar\rho^*)^2}+\frac{ia\sin\theta}{\bar\rho}\kappa^*\right]$$
$$+\sqrt{2}\frac{\rho^4\bar\rho^*}{\Delta}\left[\frac{1}{\rho^2\bar\rho}\mathscr{L}^{\dagger}_1(\rho^2\nu)+\frac{ia\sin\theta}{(\bar\rho^*)^2}\nu^*\right], \tag{211}$$

and

$$-(iaQ\sin\theta)(F^{3}_{4}-F^{4}_{3})+\frac{ia\sin\theta}{2\rho^2}(\mathscr{L}_0+\mathscr{L}_0^{\dagger})V+\sqrt{2}\frac{a^2\sin\theta\cos\theta}{\rho^2}C_1$$
$$+\frac{1}{\sqrt{2}}\left(\mathscr{L}_{-1}J+\mathscr{L}^{\dagger}_{-1}H-\frac{2ia\sin\theta}{\bar\rho}J-\frac{2ia\sin\theta}{\bar\rho}H\right)$$
$$=\frac{\bar\rho^*}{\bar\rho}\left[\mathscr{D}_0(\bar\rho^*\lambda)+\frac{\Delta}{2\rho^2}\sigma^*\right]+\left[\frac{\Delta\bar\rho^*}{2\bar\rho}\mathscr{D}^{\dagger}_1\frac{\bar\rho\sigma}{(\bar\rho^*)^2}+\frac{\bar\rho}{\bar\rho^*}\lambda^*\right]. \tag{212}$$

91. The solution of equations (209) and (210)

We shall now show how, by eliminating $(F^{1}_{2}+F^{2}_{1}-U)$ and $[\rho^2(F^{3}_{4}+F^{4}_{3})+V]$ from equations (209) and (210) (with the aid of equations (134) and (136)), we can obtain a remarkably simple set of equations—equations (242)–(245) below—governing $K(J-H)\cos\theta\ (=Z_1)$ and $-irQ(F+G)\sin\theta\ (=Z_2)$.

Considering first the imaginary parts of equations (209) and (210), we have

$$(ia\sin\theta)\left\{\sqrt{\Delta}\frac{\partial}{\partial r}\left(\rho^2\frac{F+G}{\sqrt{\Delta}}\right)-[r(F+G)+(ia\cos\theta)(J-H)]\right\}$$
$$=[\kappa,\nu]-[\kappa,\nu]^* \tag{213}$$

and

$$\frac{\partial}{\partial\theta}\rho^2(J-H)-(ia\sin\theta)[r(F+G)+(ia\cos\theta)(J-H)]=[\lambda,\sigma]-[\lambda,\sigma]^*, \tag{214}$$

where

$$[\kappa,\nu]=\frac{\rho^4\bar\rho}{2\Delta}\left[\frac{1}{\rho^2\bar\rho^*}\mathscr{L}_1(\rho^2\nu^*)-\frac{ia\sin\theta}{\bar\rho^2}\nu\right]+\frac{1}{8}\Delta\left[\bar\rho^2\mathscr{L}^{\dagger}_1\frac{\kappa^*}{\bar\rho^2}-\frac{ia\sin\theta}{\bar\rho^*}\kappa\right] \tag{215}$$

and

$$[\lambda, \sigma] = \frac{1}{2\sqrt{2}}(\bar{\rho}^*)^2\left[\mathscr{D}_0(\bar{\rho}^*\lambda) + \frac{\Delta}{2\rho^2}\sigma^*\right] - \frac{1}{2\sqrt{2}}\rho^2\left[\frac{\Delta\bar{\rho}^*}{2\bar{\rho}}\mathscr{D}_1^\dagger\frac{\bar{\rho}\sigma}{(\bar{\rho}^*)^2} + \frac{\bar{\rho}}{\bar{\rho}^*}\lambda^*\right]. \tag{216}$$

We observe that, in both equations (213) and (214), the combination $[r(F+G)+ia(J-H)\cos\theta]$ occurs. In equation (213) we replace it by a combination of $F+G$ and Ψ and in equation (214) by a combination of $J-H$ and Ψ by making use of one or the other of the two relations included in

$$\begin{aligned} r(F+G)+(ia\cos\theta)(J-H) &= \frac{r\sigma^+}{K}\rho^2(F+G)+\frac{ia\Psi}{K} \\ &= -\frac{i\sigma^+}{Q\sin\theta}\rho^2(J-H)\cos\theta+\frac{i\Psi}{Q\sin\theta}. \end{aligned} \tag{217}$$

With these replacements, equations (213) and (214) can be brought to the forms

$$(ia\sin\theta)\left[\sqrt{(\Delta K)}\frac{\partial}{\partial r}\left(\rho^2\frac{F+G}{\sqrt{(\Delta K)}}\right)-\frac{ia\Psi}{K}\right] = [\kappa, \nu]-[\kappa, \nu]^* \tag{218}$$

and

$$\sqrt{(Q\sin\theta)}\frac{\partial}{\partial\theta}\left(\rho^2\frac{J-H}{\sqrt{(Q\sin\theta)}}\right)+\frac{a\Psi}{Q} = [\lambda, \sigma]-[\lambda, \sigma]^*. \tag{219}$$

Returning to equations (209) and (210), we next consider their real parts:

$$\frac{1}{\sqrt{2}}iK(F_2^1+F_1^2-U)+(a\sin\theta)[ir(J-H)-a(F+G)\cos\theta] = [\kappa, \nu]+[\kappa, \nu]^* \tag{220}$$

and

$$\frac{1}{\sqrt{2}}iQ(a\sin\theta)[\rho^2(F_4^3+F_3^4)+V]-(a\sin\theta)[ir(J-H)-a(F+G)\cos\theta] = [\lambda, \sigma]+[\lambda, \sigma]^*. \tag{221}$$

By making use of one or the other of the two relations included in

$$\begin{aligned} ir(J-H)-a(F+G)\cos\theta &= \frac{i\alpha^2\sigma^+}{arQ\sin\theta}\rho^2(J-H)-\frac{ia\cos\theta}{rQ\sin\theta}\Psi \\ &= -\frac{\alpha^2\sigma^+}{aK\cos\theta}\rho^2(F+G)+\frac{ir}{K\cos\theta}\Psi, \end{aligned} \tag{222}$$

we can bring equations (220) and (221) to the forms

$$\frac{1}{\sqrt{2}}iK(F_2^1+F_1^2-U)+\frac{i\alpha^2\sigma^+}{rQ}\rho^2(J-H)-\frac{ia^2\cos\theta}{rQ}\Psi = [\kappa, \nu]+[\kappa, \nu]^* \tag{223}$$

and

$$\frac{1}{\sqrt{2}} iQ(a\sin\theta)[\rho^2(F_4^3+F_3^4)+V]+\frac{\alpha^2\sigma^+\sin\theta}{K\cos\theta}\rho^2(F+G)-\frac{iar\sin\theta}{K\cos\theta}\Psi$$
$$= [\lambda,\sigma]+[\lambda,\sigma]^*. \quad (224)$$

We can now eliminate $(F_2^1+F_1^2-U)$ and $[\rho^2(F_4^3+F_3^4)+V]$ from equations (223) and (224) with the aid of equations (134) and (136). We write these latter equations in the forms

$$\frac{1}{\sqrt{2}}Q(F_2^1+F_1^2-U)+\sqrt{\Delta}\frac{\partial}{\partial r}\left(\rho^2\frac{J-H}{\sqrt{\Delta}}\right)=(\kappa,\nu) \quad (225)$$

and

$$-\frac{1}{\sqrt{2}}iK[\rho^2(F_4^3+F_3^4)+V]+\frac{\partial}{\partial\theta}\rho^2(F+G)=(\lambda,\sigma), \quad (226)$$

where

$$(\kappa,\nu)=\frac{1}{2\rho^2}\left[\tfrac{1}{2}\Delta(\bar{\rho}^*\kappa^*-\bar{\rho}\kappa)-\frac{2\rho^4}{\Delta}(\bar{\rho}^*\nu-\bar{\rho}\nu^*)\right] \quad (227)$$

and

$$(\lambda,\sigma)=\frac{1}{\rho^2\sqrt{2}}[(\bar{\rho})^2(\rho^2\lambda^*-\tfrac{1}{2}\Delta\sigma)+(\bar{\rho}^*)^2(\rho^2\lambda-\tfrac{1}{2}\Delta\sigma^*)]. \quad (228)$$

Carrying out the stated eliminations, we obtain the pair of equations

$$K\sqrt{\Delta}\frac{\partial}{\partial r}\left(\rho^2\frac{J-H}{\sqrt{\Delta}}\right)-\frac{\alpha^2\sigma^+}{r}\rho^2(J-H)+\frac{a^2\cos\theta}{r}\Psi$$
$$=iQ\{[\kappa,\nu]+[\kappa,\nu]^*\}+K(\kappa,\nu), \quad (229)$$

and

$$(Qa\sin\theta)\frac{\partial}{\partial\theta}\rho^2(F+G)+\frac{\alpha^2\sigma^+\sin\theta}{\cos\theta}\rho^2(F+G)-\frac{iar\sin\theta}{\cos\theta}\Psi$$
$$=K\{[\lambda,\sigma]+[\lambda,\sigma]^*\}+(Qa\sin\theta)(\lambda,\sigma), \quad (230)$$

where it may be noted that the left-hand sides of these equations have the alternative forms

$$r\sqrt{(\Delta K)}\left[\frac{\partial}{\partial r}\left(\frac{\sqrt{K}}{r\sqrt{\Delta}}\rho^2(J-H)\right)+\frac{a^2\cos\theta}{r^2\sqrt{(\Delta K)}}\Psi\right] \quad (231)$$

and

$$(a\cos\theta)\sqrt{(Q\sin\theta)}\left[\frac{\partial}{\partial\theta}\left(\frac{\sqrt{(Q\sin\theta)}}{\cos\theta}\rho^2(F+G)\right)-\frac{ir\sin\theta}{(\cos^2\theta)\sqrt{(Q\sin\theta)}}\Psi\right]. \quad (232)$$

The four equations governing $F+G$ and $J-H$, which we have derived are

$$\frac{\partial}{\partial r}\left[\frac{1}{\sqrt{(\Delta K)}}\rho^2(F+G)\right]-\frac{ia}{\sqrt{(K^3\Delta)}}\Psi$$
$$=\frac{1}{(ia\sin\theta)\sqrt{(\Delta K)}}\{[\kappa,\nu]-[\kappa,\nu]^*\},\tag{233}$$

$$\frac{\partial}{\partial\theta}\left[\frac{\sqrt{(Q\sin\theta)}}{\cos\theta}\rho^2(F+G)\right]-\frac{ir\sin\theta}{(\cos^2\theta)\sqrt{(Q\sin\theta)}}\Psi$$
$$=\frac{1}{(a\cos\theta)\sqrt{(Q\sin\theta)}}\left[\!\left[K\{[\lambda,\sigma]+[\lambda,\sigma]^*\}+(Qa\sin\theta)(\lambda,\sigma)\right]\!\right],\tag{234}$$

$$\frac{\partial}{\partial r}\left[\frac{\sqrt{K}}{r\sqrt{\Delta}}\rho^2(J-H)\right]+\frac{a^2\cos\theta}{r^2\sqrt{(\Delta K)}}\Psi$$
$$=\frac{1}{r\sqrt{(\Delta K)}}\left[\!\left[iQ\{[\kappa,\nu]+[\kappa,\nu]^*\}+K(\kappa,\nu)\right]\!\right],\tag{235}$$

$$\frac{\partial}{\partial\theta}\left[\frac{1}{\sqrt{(Q\sin\theta)}}\rho^2(J-H)\right]+\frac{a}{\sqrt{(Q^3\sin\theta)}}\Psi$$
$$=\frac{1}{\sqrt{(Q\sin\theta)}}\{[\lambda,\sigma]-[\lambda,\sigma]^*\}.\tag{236}$$

(a) The reduction of equations (233)–(236)

Letting

$$Z_1=K(J-H)\cos\theta\quad\text{and}\quad Z_2=-irQ(F+G)\sin\theta,\tag{237}$$

so that

$$\Psi=Z_1+Z_2,\tag{238}$$

we shall seek a transformation of equations (233)–(236) such that $J-H$ and $F+G$ occur in the combinations Z_1 and Z_2. The desired transformation can be effected by multiplying the equations (233), (234), (235), and (236) by $-i\sqrt{(Q\sin\theta)}/\cos\theta$, $-i/\sqrt{(\Delta K)}$, $1/\sqrt{(Q\sin\theta)}$, and $\sqrt{K}/r\sqrt{\Delta}$, respectively. Thus, with the definitions

$$E=\frac{\rho^2}{(r\cos\theta)\sqrt{(\Delta KQ\sin\theta)}},\tag{239}$$

$$A_1=\frac{r^2\sin\theta}{\rho^2\cos\theta},\quad B_1=\frac{aK\cos\theta}{\rho^2 Q},$$
$$A_2=\frac{arQ\sin\theta}{K\rho^2},\quad B_2=\frac{a^2\cos^2\theta}{r\rho^2},\tag{240}$$

and

$$\left.\begin{aligned} X_1 &= \frac{rQ}{a\rho^2}\{[\kappa, \nu]-[\kappa, \nu]^*\}, \\ Y_1 &= \frac{K\cos\theta}{\rho^2}\{[\lambda, \sigma]-[\lambda, \sigma]^*\}, \\ X_2 &= \frac{\cos\theta}{\rho^2}\left[\!\left[iQ\{[\kappa, \nu]+[\kappa, \nu]^*\}+K(\kappa, \nu)\right]\!\right], \\ Y_2 &= \frac{ir}{a\rho^2}\left[\!\left[K\{[\lambda, \sigma]+[\lambda, \sigma]^*\}+(Qa\sin\theta)(\lambda, \sigma)\right]\!\right], \end{aligned}\right\} \tag{241}$$

we find, after the stated multiplications, that the equations take the remarkably simple forms

$$\frac{\partial}{\partial r}EZ_2 - A_2 E\Psi = -EX_1, \tag{242}$$

$$\frac{\partial}{\partial \theta}EZ_2 - A_1 E\Psi = -EY_2, \tag{243}$$

$$\frac{\partial}{\partial r}EZ_1 + B_2 E\Psi = +EX_2, \tag{244}$$

and

$$\frac{\partial}{\partial \theta}EZ_1 + B_1 E\Psi = +EY_1. \tag{245}$$

We may note here, for future reference, that the various bracket expressions involving the spin-coefficients, which occur in the definitions of X_1, X_2, Y_1, and Y_2, are given by

$$(\kappa, \nu) = \frac{1}{12M\Delta\sqrt{2}}\left[\!\left[rX[\mathscr{L}S]^- + aY\{[\mathscr{L}S]^+\cos\theta + 3[S]^+\sin\theta\}\right]\!\right],$$

$$\begin{aligned}[\kappa, \nu]-[\kappa, \nu]^* = \frac{1}{24M\Delta\sqrt{2}}\Big[\!\Big[& X\{(r^2-a^2\cos^2\theta)[\mathscr{L}\mathscr{L}S]^- \\ & -2a^2[\mathscr{L}S]^-\sin\theta\cos\theta\} \\ & +2arY\{[\mathscr{L}\mathscr{L}S]^+\cos\theta + [\mathscr{L}S]^+\sin\theta\}\Big]\!\Big],\end{aligned}$$

$$\begin{aligned}[\kappa, \nu]+[\kappa, \nu]^* = \frac{i}{24M\Delta\sqrt{2}}\Big[\!\Big[& 2arX\{[\mathscr{L}\mathscr{L}S]^-\cos\theta + 2[\mathscr{L}S]^-\sin\theta\} \\ & +Y\{-(r^2-a^2\cos^2\theta)[\mathscr{L}\mathscr{L}S]^+ \\ & +4a^2[\mathscr{L}S]^+\sin\theta\cos\theta \\ & +6a^2[S]^+\sin^2\theta\}\Big]\!\Big]; \end{aligned} \tag{246}$$

$$(\lambda, \sigma) = \frac{1}{12M\sqrt{2}} \left[\!\left[\{-r[\mathscr{D}P]^- + 3[P]^-\}[S]^+ + ia[\mathscr{D}P]^+[S]^- \cos\theta \right]\!\right],$$

$$[\lambda, \sigma] - [\lambda, \sigma]^* = \frac{1}{24M\sqrt{2}} \left[\!\left[\{-(r^2 - a^2\cos^2\theta)[\mathscr{D}\mathscr{D}P]^+ + 2r[\mathscr{D}P]^+\}[S]^- + 2ia\{r[\mathscr{D}\mathscr{D}P]^- - [\mathscr{D}P]^-\}[S]^+ \cos\theta \right]\!\right],$$

$$[\lambda, \sigma] + [\lambda, \sigma]^* = \frac{1}{24M\sqrt{2}} \left[\!\left[2ia\{r[\mathscr{D}\mathscr{D}P]^+ - 2[\mathscr{D}P]^+\}[S]^- \cos\theta + \{-(r^2 - a^2\cos^2\theta)[\mathscr{D}\mathscr{D}P]^- + 4r[\mathscr{D}P]^- - 6[P]^-\}[S]^+ \right]\!\right]. \quad (247)$$

There are some unexpected relations among the coefficients, E, A_1, A_2, B_1, and B_2, that play crucial roles for the solvability of equations (242)–(245). We enumerate these relations in the following lemma.

LEMMA:

(i) $A_1 B_2 = A_2 B_1$; $(A_1 + B_1)(A_2 - B_2) = (A_1 - B_1)(A_2 + B_2)$;

(ii) $A_1 + B_1 = \alpha^2 \sigma^+/(aQ\cos\theta)$ (a function of θ only);

(iii) $A_2 + B_2 = \alpha^2 \sigma^+/rK$ (a function of r only);

(iv) $\partial A_1/\partial r = \partial A_2/\partial\theta$; $\partial B_1/\partial r = \partial B_2/\partial\theta$;

(v) $\partial \lg E/\partial r = A_2 - B_2 - (r-M)/\Delta$; $\partial \lg E/\partial\theta = A_1 - B_1$;

(vi) $A_1\, \partial \lg E/\partial r - A_2\, \partial \lg E/\partial\theta = -A_1(r-M)/\Delta$;

(vii) $B_1\, \partial \lg E/\partial r - B_2\, \partial \lg E/\partial\theta = -B_1(r-M)/\Delta$. (248)

Since $\Psi = Z_1 + Z_2$, the essential content of equations (242)–(245) is that the derivatives of Z_1 and Z_2, with respect to r and θ, are expressible as linear combinations of Z_1 and Z_2. Thus, with the aid of the relations included in the lemma, we find from equations (242)–(245) that

$$\frac{\partial Z_1}{\partial r} = -\left(A_2 - \frac{r-M}{\Delta}\right) Z_1 - B_2 Z_2 + X_2, \quad (249)$$

$$\frac{\partial Z_1}{\partial\theta} = -A_1 Z_1 - B_1 Z_2 + Y_1, \quad (250)$$

$$\frac{\partial Z_2}{\partial r} = +A_2 Z_1 + \left(B_2 + \frac{r-M}{\Delta}\right) Z_2 - X_1, \quad (251)$$

$$\frac{\partial Z_2}{\partial\theta} = +A_1 Z_1 + B_1 Z_2 - Y_2. \quad (252)$$

(b) The integrability conditions

Equations (242)–(245) (or, equivalently, equations (249)–(252)) lead to integrability conditions which are of some interest. Thus, remembering that $\Psi = Z_1 + Z_2$, we obtain by adding equations (249) and (251), and similarly equations (250) and (252), the pair of equations

$$\sqrt{\Delta}\frac{\partial}{\partial r}\frac{\Psi}{\sqrt{\Delta}} = -(X_1 - X_2), \quad \frac{\partial \Psi}{\partial \theta} = +(Y_1 - Y_2). \tag{253}$$

By virtue of the known properties of Ψ (cf. equations (173), (175), and (176)), these equations imply that

$$X_1 - X_2 = -\frac{\mathscr{C}_1 + \Gamma_1}{96 M \Delta \sqrt{2}}\left(Y + \frac{4\lambda\sigma}{\mathscr{C}_1 + \Gamma_1} r X\right)\mathscr{S} \tag{254}$$

and

$$Y_1 - Y_2 = +\frac{\mathscr{C}_1 + \Gamma_1}{96 M \sqrt{2}}\left\{[S]^- \cos\theta + \frac{4\sigma(\lambda\alpha^2 - 6a^2)}{a(\mathscr{C}_1 + \Gamma_1)}[S]^+\right\}\mathscr{R}. \tag{255}$$

In view of the complexity of the equations defining $X_1, \ldots, Y_2$, the foregoing relations are remarkable identities: they could not have been foreseen. However, in this instance, they can be verified directly (with some effort) if proper use is made of the relations (178), (179), (183), and (184).

Again, equations (242) and (243) and equations (244) and (245) lead to the pair of integrability conditions

$$\frac{\partial}{\partial r}(A_1 E \Psi) - \frac{\partial}{\partial \theta}(A_2 E \Psi) = \frac{\partial}{\partial r}(E Y_2) - \frac{\partial}{\partial \theta}(E X_1) \tag{256}$$

and

$$\frac{\partial}{\partial r}(B_1 E \Psi) - \frac{\partial}{\partial \theta}(B_2 E \Psi) = \frac{\partial}{\partial r}(E Y_1) - \frac{\partial}{\partial \theta}(E X_2). \tag{257}$$

By making use of the relations included in the lemma we find that the foregoing equations give

$$A_1 \sqrt{\Delta}\frac{\partial}{\partial r}\left(\frac{\Psi}{\sqrt{\Delta}}\right) - A_2 \frac{\partial \Psi}{\partial \theta} = \frac{1}{E}\left(\frac{\partial}{\partial r} E Y_2 - \frac{\partial}{\partial \theta} E X_1\right) \tag{258}$$

and

$$B_1 \sqrt{\Delta}\frac{\partial}{\partial r}\left(\frac{\Psi}{\sqrt{\Delta}}\right) - B_2 \frac{\partial \Psi}{\partial \theta} = \frac{1}{E}\left(\frac{\partial}{\partial r} E Y_1 - \frac{\partial}{\partial \theta} E X_2\right). \tag{259}$$

Now, substituting from equations (253) for the quantities on the right-hand sides of equations (258) and (259), we obtain the pair of equations

$$\sqrt{\Delta}\frac{\partial}{\partial r}\left(\frac{Y_1}{\sqrt{\Delta}}\right) - \frac{\partial X_2}{\partial \theta} = -B_1 X_1 + A_1 X_2 - A_2 Y_1 + B_2 Y_2 \tag{260}$$

and

$$\sqrt{\Delta}\frac{\partial}{\partial r}\left(\frac{Y_2}{\sqrt{\Delta}}\right)-\frac{\partial X_1}{\partial\theta} = -B_1X_1+A_1X_2-A_2Y_1+B_2X_2. \quad (261)$$

The difference of these equations gives

$$\sqrt{\Delta}\frac{\partial}{\partial r}\left(\frac{Y_1-Y_2}{\sqrt{\Delta}}\right)+\frac{\partial}{\partial\theta}(X_1-X_2)=0; \quad (262)$$

and this relation is no more than what is required by equation (253). But the sum of equations (260) and (261) (after some reductions in which use is made of the relations included in the lemma) provides a new relation:

$$\frac{\partial}{\partial r}E[(Y_1+Y_2)-(A_1+B_1)\Psi]=\frac{\partial}{\partial\theta}E[(X_1+X_2)-(A_2+B_2)\Psi]. \quad (263)$$

This relation must be an identity since all the quantities which occur in this equation are known functions. We shall consider the implications of this identity in §94.

Returning to equations (242)–(245), we now find that the difference of equations (242) and (244) and of (243) and (245) yield the pair of equations

$$\left.\begin{aligned}\frac{\partial}{\partial r}E(Z_1-Z_2)&=E[(X_1+X_2)-(A_2+B_2)\Psi]\\ \text{and}\qquad \frac{\partial}{\partial\theta}E(Z_1-Z_2)&=E[(Y_1+Y_2)-(A_1+B_1)\Psi].\end{aligned}\right\} \quad (264)$$

Equation (263) guarantees the integrability of these equations so that the solution for $E(Z_1-Z_2)$ is correctly given by

$$E(Z_1-Z_2)=\int^r E[(X_1+X_2)-(A_2+B_2)\Psi]\,dr, \quad (265)$$

or by

$$E(Z_1-Z_2)=\int^\theta E[(Y_1+Y_2)-(A_1+B_1)\Psi]\,d\theta. \quad (266)$$

This is only a formal solution since it involves integrals over many functions. We shall show in §92 below how explicit solutions for Z_1 and Z_2 can be obtained from the remaining pair of equations (211) and (212) with the aid of the equations (249)–(252).

92. Explicit solutions for Z_1 and Z_2

The principal results of our consideration of the pair of equations (209) and (210) (derived from the linearized version of the four Ricci identities (204)) are equations (249)–(252) which express the derivatives of Z_1 and Z_2 as linear combinations of the functions themselves. We shall now show how with the remaining pair of equations, (211) and (212), we can complete the solution for Z_1 and Z_2.

But first we observe that the fact that the derivatives of Z_1 and Z_2 are expressible as linear combinations of the functions themselves means that, in any equation in which the derivatives of Z_1 and Z_2 occur (regardless of how high are the orders), we can replace them by the functions. As examples, consider equations (131) and (132). By rewriting the right-hand sides of these equations in terms of Z_1 and Z_2, we have

$$\frac{rKQ\sin\theta}{\Delta}(F-G)=\frac{\partial Z_2}{\partial r}-\frac{Z_2}{r}-\frac{2arQ\sin\theta}{K\rho^2}Z_1 \tag{267}$$

and

$$KQ(J+H)\cos\theta=\frac{\partial Z_1}{\partial\theta}+\frac{Z_1}{\cos\theta\sin\theta}+\frac{2aK\cos\theta}{Q\rho^2}Z_2. \tag{268}$$

Now substituting for the derivatives of Z_1 and Z_2, which occur on the right-hand sides of these equations, from equations (250) and (251), we obtain

$$\frac{rKQ\sin\theta}{\Delta}(F-G)=-\frac{arQ\sin\theta}{K\rho^2}Z_1+\left(\frac{r-M}{\Delta}-\frac{r}{\rho^2}\right)Z_2-X_1, \tag{269}$$

and

$$KQ(J+H)\cos\theta=\frac{(r^2+a^2)\cot\theta}{\rho^2}Z_1+\frac{aK\cos\theta}{Q\rho^2}Z_2+Y_1. \tag{270}$$

The basic equations, for our further consideration, are the imaginary part of equation (211) and the real part of equation (212). These equations can be written in the forms

$$\mathscr{D}_{-1}\rho^2F-\mathscr{D}^\dagger_{-1}\rho^2G=\frac{2}{ia\sin\theta}[\{\kappa,\nu\}-\{\kappa,\nu\}^*] \tag{271}$$

and

$$\rho^2\left[\mathscr{L}_{-1}J+\mathscr{L}^\dagger_{-1}H-\frac{2a^2\sin\theta\cos\theta}{\rho^2}(J+H)\right]=2[\{\lambda,\sigma\}+\{\lambda,\sigma\}^*], \tag{272}$$

where

$$\{\kappa,\nu\}=\tfrac{1}{8}\Delta\left[\bar\rho^2\mathscr{L}^\dagger_1\frac{\kappa^*}{\bar\rho^2}-\frac{ia\sin\theta}{\bar\rho^*}\kappa\right]-\frac{\rho^4\bar\rho}{2\Delta}\left[\frac{1}{\rho^2\bar\rho^*}\mathscr{L}_1(\rho^2\nu^*)-\frac{ia\sin\theta}{\bar\rho^2}\nu\right] \tag{273}$$

and

$$\{\lambda, \sigma\} = \frac{1}{2\sqrt{2}}(\bar{\rho}^*)^2\left[\mathscr{D}_0(\bar{\rho}^*\lambda) + \frac{\Delta}{2\rho^2}\sigma^*\right]$$

$$+\frac{1}{2\sqrt{2}}\rho^2\left[\frac{\Delta\bar{\rho}^*}{2\bar{\rho}}\mathscr{D}_1^\dagger\frac{\bar{\rho}\sigma}{(\bar{\rho}^*)^2} + \frac{\bar{\rho}}{\bar{\rho}^*}\lambda^*\right]. \quad (274)$$

Rewriting equation (271) in the form

$$\Delta\frac{\partial}{\partial r}\rho^2\frac{F-G}{\Delta} + \frac{iK}{\Delta}\rho^2(F+G)$$

$$= \Delta\frac{\partial}{\partial r}\frac{\rho^2}{rKQ\sin\theta}\left[\frac{rKQ\sin\theta}{\Delta}(F-G)\right] - \frac{K\rho^2}{r\Delta Q\sin\theta}Z_2$$

$$= -\frac{2i}{a\sin\theta}[\{\kappa, \nu\} - \{\kappa, \nu\}^*], \quad (275)$$

and making use of equation (269), we obtain

$$rK\Delta\frac{\partial}{\partial r}\frac{\rho^2}{rK}\left[-\frac{arQ\sin\theta}{K\rho^2}Z_1 + \left(\frac{r-M}{\Delta} - \frac{r}{\rho^2}\right)Z_2 - X_1\right] - \frac{K^2\rho^2}{\Delta}Z_2$$

$$= -2i\frac{rKQ}{a}[\{\kappa, \nu\} - \{\kappa, \nu\}^*]. \quad (276)$$

Similarly, rewriting equation (272) in the form

$$\sin\theta\frac{\partial}{\partial\theta}\rho^2\frac{J+H}{\sin\theta} + Q\rho^2(J-H) = 2[\{\lambda, \sigma\} + \{\lambda, \sigma\}^*] \quad (277)$$

and making use of equation (270), we obtain

$$\sin\theta\frac{\partial}{\partial\theta}\frac{\rho^2}{KQ\cos\theta\sin\theta}\left[\frac{(r^2+a^2)\cot\theta}{\rho^2}Z_1 + \frac{aK\cos\theta}{Q\rho^2}Z_2 + Y_1\right]$$

$$+\frac{Q\rho^2}{K\cos\theta}Z_1 = 2[\{\lambda, \sigma\} + \{\lambda, \sigma^*\}]. \quad (278)$$

Now expanding the left-hand sides of equations (276) and (278) and substituting, once again, for the derivatives of Z_1 and Z_2 which occur, from equations (249)–(252), we obtain, after some considerable reductions, the pair of equations

$$\left[\frac{\rho^2}{\Delta}(a^2 - M^2 - K^2) - \frac{\sigma^+}{K}\{\Delta a^2\cos^2\theta + 2r\rho^2(r-M) - 2r^2\Delta\}\right]Z_2$$

$$+\frac{3a\sigma^+}{K^2}(r^2\Delta Q\sin\theta)Z_1 = \frac{arQ\sin\theta}{K}\Delta X_2 + [\rho^2(r-M) - r\Delta]X_1$$

$$+\frac{rKQ}{a}\left\{\Delta\frac{\partial}{\partial r}\frac{[\kappa, \nu] - [\kappa, \nu]^*}{K} - 2i[\{\kappa, \nu\} - \{\kappa, \nu\}^*]\right\} \quad (279)$$

and

$$\left[\rho^2(Q^2-\operatorname{cosec}^2\theta)+\frac{a\sigma^{+}}{Q\sin\theta}(3r^2\sin^2\theta-2\rho^2)\right]Z_1-\frac{3a^2\sigma^{+}}{Q^2}(K\cos^2\theta)Z_2$$

$$=-[(r^2+a^2)\cot\theta]Y_1+\frac{aK\cos\theta}{Q}Y_2$$

$$+(QK\cos\theta)\left\{-\sin\theta\frac{\partial}{\partial\theta}\frac{[\lambda,\sigma]-[\lambda,\sigma]^*}{Q\sin\theta}+2[\{\lambda,\sigma\}+\{\lambda,\sigma\}^*]\right\}, \qquad (280)$$

where it may be noted that

$$\{\kappa,\nu\}-\{\kappa,\nu\}^*=\frac{i}{24M\Delta\sqrt{2}}\left[\!\!\left[Y\{(r^2-a^2\cos^2\theta)[\mathscr{L}\mathscr{L}S]^- -2a^2[\mathscr{L}S]^-\sin\theta\cos\theta\}\right.\right.$$

$$\left.\left.-2arX\{[\mathscr{L}\mathscr{L}S]^+\cos\theta+[\mathscr{L}S]^+\sin\theta\}\right]\!\!\right], \qquad (281)$$

$$\{\lambda,\sigma\}+\{\lambda,\sigma\}^*=\frac{1}{24M\sqrt{2}}\left[\!\!\left[\{(r^2-a^2\cos^2\theta)[\mathscr{D}\mathscr{D}P]^+-2r[\mathscr{D}P]^+\}[S]^+\right.\right.$$

$$\left.\left.-2ia\{r[\mathscr{D}\mathscr{D}P]^- -[\mathscr{D}P]^-\}[S]^-\cos\theta\right]\!\!\right]. \qquad (282)$$

It is now evident that either of the two equations, (279) and (280), will suffice to determine Z_1 and Z_2 since their sum is known to be Ψ. We shall, however, find it more convenient to treat the two equations symmetrically as in §(*a*) below.

(*a*) *The reduction of the solutions for Z_1 and Z_2*

In equation (279), we replace Z_1 on the left-hand side by $\Psi-Z_2$, while in equation (280) we replace Z_2 by $\Psi-Z_1$. Also, we replace X_2 and Y_2 (on the right-hand sides of the equations) by X_1 and Y_1, respectively, by making use of the relations (254) and (255). In this manner we reduce equations (279) and (280) to the forms

$$\rho^2\left[\frac{1}{\Delta}(a^2-M^2-K^2)-\frac{\sigma^{+}\Delta}{K}\left\{1+\frac{2r(r-M)}{\Delta}-\frac{3r^2\sigma^{+}}{K}\right\}\right]Z_2$$

$$=\frac{arQ\sin\theta}{K}\Delta\left(\sqrt{\Delta}\frac{\partial}{\partial r}\frac{\Psi}{\sqrt{\Delta}}-\frac{3r\sigma^{+}}{K}\Psi\right)$$

$$+\frac{rKQ}{a}\left\{\frac{\Delta}{K}\left(\frac{\partial}{\partial r}+\frac{r-M}{\Delta}-\frac{3r\sigma^{+}}{K}\right)\{[\kappa,\nu]-[\kappa,\nu]^*\}\right.$$

$$\left.-2i[\{\kappa,\nu\}-\{\kappa,\nu\}^*]\right\} \qquad (283)$$

and

$$\rho^2\left[(Q^2-\operatorname{cosec}^2\theta)+\frac{a\sigma^+}{Q\sin\theta}\left(3\sin^2\theta-2+\frac{3a\sigma^+}{Q}\cos^2\theta\sin\theta\right)\right]Z_1$$

$$=-\frac{aK\cos\theta}{Q}\left(\frac{\partial\Psi}{\partial\theta}-\frac{3a\sigma^+\cos\theta}{Q}\Psi\right)$$

$$+(QK\cos\theta)\left\{\frac{1}{Q}\left[-\left(\frac{\partial}{\partial\theta}+\cot\theta\right)+\frac{3a\sigma^+}{Q}\cos\theta\right]\{[\lambda,\sigma]-[\lambda,\sigma]^*\}\right.$$

$$\left.+2[\{\lambda,\sigma\}+\{\lambda,\sigma\}^*]\right\}. \tag{284}$$

Now inserting for the terms in the spin-coefficients in the foregoing equations from equations (246), (247), (281), and (282), we obtain the following solutions for Z_1 and Z_2 where we have suppressed a common factor $1/(24M\sqrt{2})$:

$$\rho^2\left[(Q^2-\operatorname{cosec}^2\theta)+\frac{a\sigma^+}{Q\sin\theta}\left(3\sin^2\theta-2+\frac{3a\sigma^+}{Q}\cos^2\theta\sin\theta\right)\right]Z_1$$

$$=-\tfrac{1}{4}(\mathscr{C}_1+\Gamma_1)\frac{aK\cos\theta}{Q}\left\{\mathscr{R}\left[[S]^-\cos\theta+\frac{4\sigma^+(\lambda\alpha^2-6a^2)}{a(\mathscr{C}_1+\Gamma_1)}[S]^+\right]\right.$$

$$\left.-\frac{3a\sigma^+\cos\theta}{Q}\mathscr{R}\mathscr{S}\right\}$$

$$+K\cos\theta\left[\!\!\left[\left\{[\mathscr{L}S]^--\left(\cot\theta+\frac{3a\sigma^+}{Q}\cos\theta\right)[S]^-+Q[S]^+\right\}\right.\right.$$

$$\times\{(r^2-a^2\cos^2\theta)[\mathscr{D}\mathscr{D}P]^+-2r[\mathscr{D}P]^+\}$$

$$-2ia\left\{[\mathscr{L}S]^+-\left(\frac{1}{\sin\theta\cos\theta}+\frac{3a\sigma^+\cos\theta}{Q}\right)[S]^+\right.$$

$$\left.+Q[S]^-\right\}\{r[\mathscr{D}\mathscr{D}P]^--[\mathscr{D}P]^-\}\cos\theta$$

$$\left.\left.+2a^2[\mathscr{D}\mathscr{D}P]^+[S]^-\sin\theta\cos\theta\right]\!\!\right] \tag{285}$$

and

$$\rho^2\left[\frac{1}{\Delta}(a^2-M^2-K^2)-\frac{\sigma^+\Delta}{K}\left\{1+\frac{2r(r-M)}{\Delta}-\frac{3r^2\sigma^+}{K}\right\}\right]Z_2$$

$$=\tfrac{1}{4}(\mathscr{C}_1+\Gamma_1)\frac{arQ\sin\theta}{K}\Delta\left\{\frac{\mathscr{S}}{\Delta}\left(Y+\frac{4\lambda\sigma^+}{\mathscr{C}_1+\Gamma_1}rX\right)-\frac{3r\sigma^+}{K}\mathscr{R}\mathscr{S}\right\}$$

$$+\frac{rQ}{a}\left[\!\left[\left\{[\mathscr{D}P]^{+}-\left(\frac{r-M}{\Delta}+\frac{3r\sigma^{+}}{K}\right)X+K\frac{Y}{\Delta}\right\}\right.\right.$$

$$\times\{(r^2-a^2\cos^2\theta)[\mathscr{L}\mathscr{L}S]^{-}-2a^2[\mathscr{L}S]^{-}\sin\theta\cos\theta\}$$

$$-2ar\left\{i[\mathscr{D}P]^{-}+\left(\frac{Mr-a^2}{r\Delta}+\frac{3r\sigma^{+}}{K}\right)Y+K\frac{X}{\Delta}\right\}$$

$$\left.\left.\times\{[\mathscr{L}\mathscr{L}S]^{+}\cos\theta+[\mathscr{L}S]^{+}\sin\theta\}+2r[\mathscr{L}\mathscr{L}S]^{-}X\right]\!\right]. \quad (286)$$

It is, of course, a necessary identity that Z_1 and Z_2, determined in accordance with equations (285) and (286), are consistent with the requirement

$$Z_1+Z_2=\tfrac{1}{4}(\mathscr{C}_1+\Gamma_1)\mathscr{R}\mathscr{S}. \quad (287)$$

(b) Further implications of equations (211) and (212)

We have considered the imaginary part of equation (211) and the real part of equation (212). It remains to consider the real part of equation (211) and the imaginary part of equation (212); and they are:

$$-iK(F^1_2-F^2_1)+\Delta\frac{\partial U}{\partial r}-\sqrt{2}\cdot(iar\sin\theta)B_2$$
$$+\sqrt{2}\cdot(a^2\sin\theta\cos\theta)(F-G)=-\sqrt{2}\cdot[\{\kappa,\nu\}+\{\kappa,\nu\}^*] \quad (288)$$

and

$$-(iaQ\sin\theta)\rho^2(F^3_4-F^4_3)+ia\sin\theta\frac{\partial V}{\partial\theta}+\sqrt{2}\cdot(a^2\sin\theta\cos\theta)C_1$$
$$-\sqrt{2}\cdot(iar\sin\theta)(J+H)=\sqrt{2}\cdot[\{\lambda,\sigma\}-\{\lambda,\sigma\}^*]. \quad (289)$$

It would appear that equations (288) and (289) determine U and V which have so far remained undetermined: in the analysis beginning with §86, U and V have always appeared in the combination $(F^1_2+F^2_1-U)$ and $[\rho^2(F^3_4+F^4_3)+V]$ even though they appear explicitly in the expressions (127)–(130) for B_1, B_2, C_1, and C_2. But we shall verify that the terms in U and V in equations (288) and (289) cancel identically. Thus, in equation (288), besides $\Delta U_{,r}$, terms in U arise from the solutions for B_2 and $(F^1_2-F^2_1)$ given in equations (128) and (187). Writing out these terms, we find

$$\frac{iK}{Q\rho^2\sqrt{2}}\Delta\frac{\partial}{\partial r}\left[\frac{\rho^4}{\Delta}\frac{\Delta}{iK}\frac{-Q\sqrt{2}}{\rho^2}U\right]-\sqrt{2}\cdot(iar\sin\theta)\frac{\Delta}{iK}\frac{-Q\sqrt{2}}{\rho^2}U$$
$$+\Delta\frac{\partial U}{\partial r}=-\frac{K\Delta}{\rho^2}\frac{\partial}{\partial r}\left(\frac{\rho^2}{K}U\right)+\frac{2\Delta r}{\rho^2K}(aQ\sin\theta)U+\Delta\frac{\partial U}{\partial r}=0. \quad (290)$$

Similarly, in equation (289), besides $(ia\sin\theta)V_{,\theta}$, terms in V arise from the solutions for C_1 and $(F_4^3 - F_3^4)$ given in equations (129) and (189). Writing out these terms, we again find

$$+(iaQ\sin\theta)\left[\frac{1}{iK\rho^2}\frac{1}{\sqrt{2}}\left(\frac{\partial}{\partial\theta}-\cot\theta\right)\left(\rho^4\frac{-\sqrt{2}}{Q\rho^2}iKV\right)\right]$$

$$+\sqrt{2}\cdot(a^2\sin\theta\cos\theta)\frac{-\sqrt{2}}{Q\rho^2}iKV+ia\sin\theta\frac{\partial V}{\partial\theta}$$

$$=-\frac{iaQ\sin\theta}{\rho^2}V\left(\frac{\partial}{\partial\theta}-\cot\theta\right)\frac{\rho^2}{Q}-\frac{2ia^2\sin\theta\cos\theta}{Q\rho^2}KV=0. \tag{291}$$

We conclude that *in the chosen gauge,*

$$\Psi_1=\Psi_3=\Psi_2^{(1)}=0, \tag{292}$$

the functions,

$$U=A_1^1+A_2^2 \quad \textit{and} \quad V=A_3^3+A_4^4, \tag{293}$$

are left indeterminate. This is consistent with the fact, that after the choice (292), we have four degrees of gauge freedom still remaining. We may utilize this freedom to set

$$U=V=0; \tag{294}$$

indeed, *we may set all four diagonal elements of the matrix **A** to be equal to zero.*

93. The completion of the solution

Having found explicit solutions for

$$Z_1=K(J-H)\cos\theta \quad \text{and} \quad Z_2=-irQ(F+G)\sin\theta, \tag{295}$$

we can now write, in terms of Z_1 and Z_2, the solutions for all the other functions which determine the metric perturbations. Thus, equations (269) and (270), namely,

$$\frac{rKQ\sin\theta}{\Delta}(F-G)=-\frac{arQ\sin\theta}{K\rho^2}Z_1+\left(\frac{r-M}{\Delta}-\frac{r}{\rho^2}\right)Z_2-X_1 \tag{296}$$

and

$$KQ(J+H)\cos\theta=\frac{(r^2+a^2)\cot\theta}{\rho^2}Z_1+\frac{aK\cos\theta}{\rho^2Q}Z_2+Y_1, \tag{297}$$

provide the solutions for $F-G$ and $J+H$.

Equations (128) and (129) written out in terms of Z_1 and Z_2 provide the solutions for B_2 and C_1. We find (in the gauge $U=V=0$)

$$\frac{iK}{\Delta}B_2=\frac{a(K+\rho^2\sigma^+)\cos\theta}{r\rho^2KQ\sin\theta}Z_2+\frac{1}{K\cos\theta}\left[X_2-\left(\frac{r\sigma^+}{K}+\frac{r}{\rho^2}-\frac{r-M}{\Delta}\right)Z_1\right] \tag{298}$$

and

$$QC_1 = \frac{ir(\rho^2\sigma^+ - aQ\sin\theta)}{\rho^2 KQ\cos\theta} Z_1$$

$$-\frac{i}{rQ\sin\theta}\left[Y_2 + \left(\frac{a\sigma^+\cos\theta}{Q} - \frac{(r^2+a^2)\cot\theta}{\rho^2}\right) Z_2 \right]. \quad (299)$$

Explicit expressions for $X_1 + X_2$ and $Y_1 + Y_2$ are given in §94 below (equations (319) and (320)); these equations together with equations (254) and (255) giving $X_1 - X_2$ and $Y_1 - Y_2$ determine $X_1, \ldots, Y_2$ explicitly.

With B_2 and C_1 given by equations (298) and (299), the solutions for B_1 and C_2 follow from equations (127) and (130), namely

$$\frac{iK}{\Delta}B_1 = +\frac{2ia\cos\theta}{\rho^2}C_1 - \frac{\partial}{\partial r}(J+H) \quad (300)$$

and

$$QC_2 = -\frac{2iar\sin\theta}{\rho^2}B_2 + \frac{1}{\sin\theta}\frac{\partial}{\partial\theta}(F-G)\sin\theta. \quad (301)$$

And finally, equations (134) and (136) (or (220) and (221)) and (187) and (189) complete the solution. In terms of Z_1 and Z_2 these equations are

$$\frac{iK}{\sqrt{2}}(F^1_2 + F^2_1) = -\frac{ia}{rKQ\cos\theta}[(r^2Q\sin\theta)Z_1 - (aK\cos^2\theta)Z_2]$$

$$+[\kappa, \nu] + [\kappa, \nu]^*, \quad (302)$$

$$\frac{iaQ\sin\theta}{\sqrt{2}}\rho^2(F^4_3 + F^3_4) = +\frac{ia}{rKQ\cos\theta}[(r^2Q\sin\theta)Z_1 - (aK\cos^2\theta)Z_2]$$

$$+[\lambda, \sigma] + [\lambda, \sigma]^*, \quad (303)$$

$$\frac{Q}{\sqrt{2}}(F^1_2 - F^2_1) + i\frac{K^2\rho^4 - 2a^2\Delta^2\cos^2\theta}{2\Delta K^2\rho^2\cos\theta}Z_1$$

$$+\frac{ia\Delta\cos\theta}{rKQ\sin\theta}\left[\left(\frac{r-M}{\Delta} - \frac{r}{\rho^2}\right)Z_2 - X_1\right] + \frac{\Delta}{2\rho^2}\frac{\partial}{\partial r}\frac{\rho^4}{\Delta}B_2 = \langle \kappa, \nu \rangle, \quad (304)$$

and

$$-\frac{iK}{\sqrt{2}}\rho^2(F^3_4 - F^4_3) - i\frac{Q^2\rho^4 + 2a^2r^2\sin^2\theta}{2r\rho^2Q^2\sin\theta}Z_2$$

$$-\frac{iar\sin\theta}{KQ\cos\theta}\left[\frac{(r^2+a^2)\cot\theta}{\rho^2}Z_1 + Y_1\right] - \frac{1}{2\rho^2}\left(\frac{\partial}{\partial\theta} - \cot\theta\right)\rho^4 C_1$$

$$= \langle \lambda, \sigma \rangle. \quad (305)$$

The functions of the spin-coefficients which occur on the right-hand sides of equations (302) and (303) have already been listed in equations (246) and (247);

the additional functions which appear in equations (304) and (305) are

$$\langle \kappa, \nu \rangle = \frac{1}{2\rho^2}\left[\tfrac{1}{2}\Delta(\bar{\rho}^*\kappa^* - \bar{\rho}\kappa) + \frac{2\rho^4}{\Delta}(\bar{\rho}^*\nu - \bar{\rho}\nu^*)\right]$$

$$= \frac{i}{12M\Delta\sqrt{2}}\left[\!\left[rY[\mathscr{L}S]^- - aX\{[\mathscr{L}S]^+\cos\theta + 3[S]^+\sin\theta\}\right]\!\right] \quad (306)$$

and

$$\langle \lambda, \sigma \rangle = \frac{1}{\rho^2\sqrt{2}}[\bar{\rho}^2(\rho^2\lambda^* - \tfrac{1}{2}\Delta\sigma) - (\bar{\rho}^*)^2(\rho^2\lambda - \tfrac{1}{2}\Delta\sigma^*)]$$

$$= \frac{1}{12M\sqrt{2}}\left[\!\left[\{-r[\mathscr{D}P]^- + 3[P]^-\}[S]^- + ia[\mathscr{D}P]^+[S]^+\cos\theta\right]\!\right]. \quad (307)$$

Among the foregoing solutions for the metric functions in the gauge $U = V = 0$, we have the identities

$$\frac{iK}{\sqrt{2}}(F^1_2 - F^2_1) + (iar\sin\theta)B_2 - (a^2\sin\theta\cos\theta)(F - G)$$

$$= [\{\kappa, \nu\} + \{\kappa, \nu\}^*] \quad (308)$$

and

$$\frac{Q\rho^2}{\sqrt{2}}(F^3_4 - F^4_3) + (ia\cos\theta)C_1 + r(J + H) = \frac{i}{a\sin\theta}[\{\lambda, \sigma\} - \{\lambda, \sigma\}^*]. \quad (309)$$

These identities follow from equations (288) and (289); for, as we have seen, the terms in U and V in these equations vanish identically.

94. Integral identities

In our pursuit of the solution of the Newman–Penrose equations, we have hurried past several facets of interest. One of these is that the functions $\mathscr{R}$ and $\mathscr{S}$ provide explicit evaluations of the indefinite integrals

$$\left.\begin{aligned} \mathscr{R} &= \sqrt{\Delta}\int^r \left(Y + \frac{4\lambda\sigma}{\mathscr{C}_1 + \Gamma_1} rX\right)\frac{dr}{\sqrt{\Delta^3}}, \\ \text{and} \quad \mathscr{S} &= \int^\theta \{[S]^-\cos\theta + \frac{4\sigma(\lambda\alpha^2 - 6a^2)}{a(\mathscr{C}_1 + \Gamma_1)}[S]^+\}d\theta. \end{aligned}\right\} \quad (310)$$

If one had been presented with these integrals, would one have known that they could be explicitly evaluated?

A second facet concerns the differential equations (201) and (202) relating the functions $\mathscr{R}$ and $\mathscr{S}$ with the Teukolsky functions. The real 'depth' of these equations becomes apparent only when we ask how one would verify them.

The terms on the left-hand sides of these equations can be expressed in terms of $X(=P_{+2}+P_{-2})$ and $Y(=-iP_2+iP_{-2})$ or $[S]^+ (=S_{+2}+S_{-2})$ and $[S]^- (=S_{+2}-S_{-2})$ with the aid of the equations

$$\frac{d\mathscr{R}}{dr}=\frac{1}{\Delta}\left(Y+\frac{4\lambda\sigma^+}{\mathscr{C}_1+\Gamma_1}rX\right)+\frac{r-M}{\Delta}\mathscr{R}, \tag{311}$$

$$\frac{d^2\mathscr{R}}{dr^2}=\frac{1}{\Delta}\left(\frac{dY}{dr}+\frac{4\lambda\sigma^+}{\mathscr{C}_1+\Gamma_1}r\frac{dX}{dr}+\frac{4\lambda\sigma^+}{\mathscr{C}_1+\Gamma_1}X\right)-\frac{r-M}{\Delta^2}\left(Y+\frac{4\lambda\sigma^+}{\mathscr{C}_1+\Gamma_1}rX\right)+\frac{a^2-M^2}{\Delta^2}\mathscr{R}, \tag{312}$$

$$\frac{d\mathscr{S}}{d\theta}=[S]^-\cos\theta+\frac{4\sigma^+(\lambda\alpha^2-6a^2)}{a(\mathscr{C}_1+\Gamma_1)}[S]^+, \tag{313}$$

and

$$\frac{d^2\mathscr{S}}{d\theta^2}=-[S]^-\sin\theta+\frac{d[S]^-}{d\theta}\cos\theta+\frac{4\sigma^+(\lambda\alpha^2-6a^2)}{a(\mathscr{C}_1+\Gamma_1)}\frac{d[S]^+}{d\theta}, \tag{314}$$

since $\mathscr{R}$ and $\mathscr{S}$ are known in terms of X and Y and $[S]^+$ and $[S]^-$ (by equations (180) (or (181)) and (185) (or (186)), and the derivatives of X and Y and of $[S]^+$ and $[S]^-$ are expressible in terms of the functions themselves (by equations (74) and (79)). The complexity of these various relations which must be inserted in equations (201) and (202) to reduce them to identities is a measure of their 'depth.'

Conversely, we can write differential equations for $\mathscr{R}$ and $\mathscr{S}$ which contain no reference to the Teukolsky functions. Thus, by expressing the derivatives of X and Y which occur in equation (312) in terms of the functions themselves (with the aid of equations (150) and (151)), we obtain a second equation relating X and Y to the derivatives of $\mathscr{R}$; this second equation together with equation (311) will enable us to express X and Y, separately, in terms of $\mathscr{R}$ and its first and second derivatives. The substitution of these latter relations on the right-hand side of equation (197) will provide a second-order differential equation for $\mathscr{R}$ with no reference to X and Y. The equation which we obtain in this manner is

$$\begin{aligned}\frac{\Delta^2}{K}\left[\frac{d}{dr}\left(K\frac{d}{dr}\frac{\mathscr{R}}{K}\right)+\left(\frac{4\alpha^2\sigma^{+2}}{K^2}-\frac{K^2}{\Delta^2}\right)\mathscr{R}\right] &= -\frac{2}{[\det]_r}\left\{(\not{p}^2r^2+1)\left[\Delta^2\frac{d^2\mathscr{R}}{dr^2}-(a^2-M^2)\mathscr{R}\right]\right.\\ &\quad-\left[2\not{p}r\frac{\Delta^2\mathscr{C}_1}{B_1}+(\not{p}^2r^2-1)\left(\frac{\Delta^2\mathscr{C}_2}{B_1}-M\right)-(\not{p}^2r^2+1)\frac{\Delta A_2}{B_1}+r(\not{p}^2a^2-1)\right]\\ &\quad\left.\times\left[\Delta\frac{d\mathscr{R}}{dr}-(r-M)\mathscr{R}\right]\right\}, \end{aligned}\tag{315}$$

where

$$[\det]_r = \Delta\left[(\not{p}^2 r^2 - 1)\frac{\Delta\mathscr{C}_1}{B_1} + (\not{p}^2 r^2 + 1)\left(\frac{A_1}{B_1} - \frac{K}{\Delta}\right) - 2\not{p}r\frac{\Delta\mathscr{C}_2}{B_1} - \not{p}\right], \quad (316)$$

and $\not{p} = 4\lambda\sigma^{+}/(\mathscr{C}_1 + \Gamma_1)$.

Similarly, we obtain for $\mathscr{S}$ the equation

$$\frac{1}{Q}\left[\frac{d}{d\theta}\left(Q\frac{d}{d\theta}\frac{\mathscr{S}}{Q}\right) - \left(\frac{4\alpha^2\sigma^{+2}}{Q^2} - Q^2\right)\mathscr{S}\right]$$

$$= \frac{2}{[\det]_\theta}\left\{(\not{q}^2 - \cos^2\theta)\frac{d^2\mathscr{S}}{d\theta^2} + \frac{1}{\beta_1}[(\alpha_2 + 2\beta_1\cot\theta)(\not{q}^2 - \cos^2\theta)\right.$$

$$\left. - 2\mathscr{C}_1\not{q}\cos\theta - \beta_1\cos\theta\sin\theta]\frac{d\mathscr{S}}{d\theta}\right\}, \quad (317)$$

where

$$[\det]_\theta = \frac{1}{\beta_1}[\beta_1\not{q}\sin\theta + (\alpha_1 + \beta_1 Q)(\not{q}^2 - \cos^2\theta) + \mathscr{C}_1(\not{q}^2 + \cos^2\theta)], \quad (318)$$

and $\not{q} = 4\sigma^{+}(\lambda\alpha^2 - 6a^2)/a(\mathscr{C}_1 + \Gamma_1)$.

(*a*) *Further identities derived from the integrability condition (263)*

It will be recalled that the explicit expression of $\mathscr{R}$ and $\mathscr{S}$, in terms of the Teukolsky functions, was accomplished via the identities implied by the integrability condition considered in §87. It will, therefore, appear that the integrability condition (263) of equations (265) and (266) must lead to similar identities. To find them, we need explicit expressions for $X_1 + X_2$ and $Y_1 + Y_2$. Evaluating them in accordance with their definitions (241) and the expressions for the various 'brackets' given in equations (246) and (247), we find, after some considerable reductions, that they can be written in the forms

$$X_1 + X_2 = \frac{rX}{a\Delta}\left[G_1(\theta) - \frac{4a^2 Q\cos\theta}{\rho^2}G_2(\theta)\right] + \frac{Y}{\Delta}\left[G_3(\theta) - \frac{4a^2 Q\cos\theta}{\rho^2}G_4(\theta)\right] \quad (319)$$

and

$$Y_1 + Y_2 = [S]^-(\cos\theta)\left[F_1(r) - \frac{4rK}{\rho^2}F_2(r)\right] + \frac{i}{a}[S]^+\left[F_3(r) - \frac{4rK}{\rho^2}F_4(r)\right], \quad (320)$$

where a common factor $1/(24M\sqrt{2})$ has been suppressed and

$F_1(r) = K[\mathscr{D}\mathscr{D}P]^+ + 2r\sigma^{+}[\mathscr{D}P]^+$; $F_2(r) = r[\mathscr{D}\mathscr{D}P]^+ - [\mathscr{D}P]^+$;

$F_3(r) = 3Kr[\mathscr{D}\mathscr{D}P]^- - 2\alpha^2\sigma^{+}[\mathscr{D}P]^- - 6r\sigma^{+}[P]^-$;

$F_4(r) = r^2[\mathscr{D}\mathscr{D}P]^- - r[\mathscr{D}P]^-$;

$$G_1(\theta) = Q[\mathscr{L}\mathscr{L}S]^- + 2a\sigma^+[\mathscr{L}S]^- \cos\theta;$$
$$G_2(\theta) = [\mathscr{L}\mathscr{L}S]^- \cos\theta + [\mathscr{L}S]^- \sin\theta;$$
$$G_3(\theta) = 3Q[\mathscr{L}\mathscr{L}S]^+ \cos\theta + (2\alpha^2\sigma^+/a)[\mathscr{L}S]^+ + 6a\sigma^+[S]^+ \cos\theta \sin\theta;$$
$$G_4(\theta) = [\mathscr{L}\mathscr{L}S]^+ \cos^2\theta + [\mathscr{L}S]^+ \cos\theta \sin\theta. \tag{321}$$

It can be verified that among the functions $F_1, \ldots, G_4$ the following relations exist:

$$\left.\begin{aligned} &F_1 - 2K[\mathscr{D}\mathscr{D}P]^+ = -\tfrac{1}{4}(\mathscr{C}_1 + \Gamma_1)\,\mathscr{R};\; G_1 - 2Q[\mathscr{L}\mathscr{L}S]^- = \lambda a\sigma^+\mathscr{S};\\ &F_3 - 4KF_4/r = i\sigma^+(\lambda\alpha^2 - 6a^2)\,\mathscr{R};\; G_3 - 4QG_4/\cos\theta = \tfrac{1}{4}(\mathscr{C}_1 + \Gamma_1)\mathscr{S}. \end{aligned}\right\} \tag{322}$$

The appearance of the functions $\mathscr{R}$ and $\mathscr{S}$ in these relations is noteworthy.

Turning to the integrability condition (263), we observe that it requires the equality of the two expressions

$$\frac{1}{E}\frac{\partial}{\partial r}\left[\frac{\partial}{\partial\theta}E(Z_1 - Z_2)\right]$$
$$= \sqrt{\Delta}\frac{\partial}{\partial r}\left(\frac{Y_1 + Y_2}{\sqrt{\Delta}}\right) + (A_2 - B_2)(Y_1 + Y_2) + (A_2 + B_2)\frac{\partial\Psi}{\partial\theta}$$
$$-\left\{(A_1 + B_1)\sqrt{\Delta}\frac{\partial}{\partial r}\left(\frac{\Psi}{\sqrt{\Delta}}\right) + (A_2 + B_2)\frac{\partial\Psi}{\partial\theta} + (A_1 + B_1)(A_2 - B_2)\Psi\right\} \tag{323}$$

and

$$\frac{1}{E}\frac{\partial}{\partial\theta}\left[\frac{\partial}{\partial r}E(Z_1 - Z_2)\right]$$
$$= \frac{\partial}{\partial\theta}(X_1 + X_2) + (A_1 - B_1)(X_1 + X_2) + (A_1 + B_1)\sqrt{\Delta}\frac{\partial}{\partial r}\left(\frac{\Psi}{\sqrt{\Delta}}\right)$$
$$-\left\{(A_1 + B_1)\sqrt{\Delta}\frac{\partial}{\partial r}\left(\frac{\Psi}{\sqrt{\Delta}}\right) + (A_2 + B_2)\frac{\partial\Psi}{\partial\theta} + (A_1 - B_1)(A_2 + B_2)\Psi\right\}. \tag{324}$$

We notice that the quantities in the third lines of both equations are the same. Consequently, it will suffice to consider the required equality of the quantities in the second line only.

On substituting for $X_1 + X_2$ and $Y_1 + Y_2$ from equations (319) and (320), we find that the terms in the first lines on the right-hand sides of equations (323) and (324) can be brought to the following forms if appropriate use is made of the various relations included in equations (248) and (322), as well as the identities involving the functions $\mathscr{R}$ and $\mathscr{S}$:

$$\left[\left\{\sqrt{\Delta}\frac{d}{dr}\left(\frac{F_1}{\sqrt{\Delta}}\right) - \frac{2}{r}F_1 + \frac{2\alpha^2\sigma^+}{r}[\mathscr{D}\mathscr{D}P]^+\right\}\right.$$

$$-\frac{4r}{\rho^2}\left\{\sqrt{\Delta}\frac{d}{dr}\left(\frac{KF_2}{\sqrt{\Delta}}\right)-r\sigma^+F_2-\tfrac{1}{2}F_1\right\}\Bigg][S]^-\cos\theta$$

$$+\Bigg[\left\{\sqrt{\Delta}\frac{d}{dr}\left(\frac{F_3}{\sqrt{\Delta}}\right)-\frac{2}{r}F_3+\frac{4\alpha^2\sigma^+}{r^2}F_4\right\}$$

$$-\frac{4r}{\rho^2}\left\{\sqrt{\Delta}\frac{d}{dr}\left(\frac{KF_4}{\sqrt{\Delta}}\right)-r\sigma^+F_4-\tfrac{1}{2}F_3\right\}\Bigg]\frac{i[S]^+}{a} \tag{325}$$

and

$$\Bigg[\left\{\frac{dG_1}{d\theta}+2\frac{\sin\theta}{\cos\theta}G_1-\frac{2\alpha^2\sigma^+}{a\cos\theta}[\mathscr{L}\mathscr{L}S]^-\right\}$$

$$-\frac{4a^2\cos\theta}{\rho^2}\left[\frac{d}{d\theta}(QG_2)-a\sigma^+G_2\cos\theta+\tfrac{1}{2}G_1\sin\theta\right]\Bigg\}\frac{rX}{a\Delta}$$

$$+\Bigg[\left\{\frac{dG_3}{d\theta}+2\frac{\sin\theta}{\cos\theta}G_3-\frac{4\alpha^2\sigma^+}{a\cos^2\theta}G_4\right\}$$

$$-\frac{4a^2\cos\theta}{\rho^2}\left[\frac{d}{d\theta}(QG_4)-a\sigma^+G_4\cos\theta+\tfrac{1}{2}G_3\sin\theta\right]\Bigg\}\frac{Y}{\Delta}. \tag{326}$$

Our aim is to reduce the expressions (325) and (326) to forms in which, apart from simple factors such as r^2 or $\cos^2\theta$, only the functions $X, Y, [S]^+$, and $[S]^-$ occur and none of their derivatives. The necessary reductions, while not as massive as those required in the context of the integrability condition considered in §87, are by no means light. We find

$$\frac{[S]^-\cos\theta}{\rho^2\Delta}\Bigg[\!\!\Bigg[\tfrac{1}{4}(\mathscr{C}_1+\Gamma_1)(-3r^2+a^2\cos^2\theta)\left(Y+\frac{4\lambda\sigma^+}{\mathscr{C}_1+\Gamma_1}rX\right)$$

$$+2\sigma^+r(\alpha^2+a^2\cos^2\theta)(\lambda X-6\sigma^+rY)\Bigg]\!\!\Bigg]$$

$$+\frac{[S]^+}{a\rho^2\Delta}\Bigg[\!\!\Bigg[\sigma^+(\lambda\alpha^2-6a^2)(-r^2+3a^2\cos^2\theta)\left(Y+\frac{4\lambda\sigma^+}{\mathscr{C}_1+\Gamma_1}rX\right)$$

$$-2a^2\sigma^+\{6\sigma^+r(r^2-\alpha^2)X+[(\lambda\alpha^2-6a^2)+3(\lambda+2)r^2]Y\}\cos^2\theta\Bigg]\!\!\Bigg] \tag{327}$$

and

$$\frac{rX}{a\rho^2\Delta}\Bigg[\!\!\Bigg[\lambda a\sigma^+(-r^2+3a^2\cos^2\theta)\left\{[S]^-\cos\theta+\frac{4\sigma^+(\lambda\alpha^2-6a^2)}{a(\mathscr{C}_1+\Gamma_1)}[S]^+\right\}$$

$$-2a\sigma^+(r^2-\alpha^2)\{\lambda[S]^-+6a\sigma^+[S]^+\cos\theta\}\cos\theta\Bigg]\!\!\Bigg]$$

$$+\frac{Y}{\rho^2\Delta}\left[\!\!\left[\tfrac{1}{4}(\mathscr{C}_1+\Gamma_1)(-3r^2+a^2\cos^2\theta)\left\{[S]^-\cos\theta+\frac{4\sigma^+(\lambda\alpha^2-6a^2)}{a(\mathscr{C}_1+\Gamma_1)}[S]^+\right\}\right.\right.$$

$$-\frac{2\sigma^+}{a}r^2\{6a\sigma^+(\alpha^2+a^2\cos^2\theta)[S]^-\cos\theta$$

$$\left.\left.-[(\lambda\alpha^2-6a^2)-3a^2(\lambda+2)\cos^2\theta][S]^+\}\right]\!\!\right]. \tag{328}$$

A comparison of these expressions shows that they are indeed the same.

After some rearrangements of the terms in (327) and (328), we can now write in place of (323) and (324), the single equation

$$\frac{1}{E}\frac{\partial^2}{\partial r\partial\theta}E(Z_1-Z_2)$$

$$=\lambda\sigma^+[-3(r^2-a^2\cos^2\theta)+2\alpha^2]\frac{rX[S]^-\cos\theta}{\rho^2\Delta}$$

$$+\{\tfrac{1}{4}(\mathscr{C}_1+\Gamma_1)(-3r^2+a^2\cos^2\theta)-12\sigma^{+2}r^2(\alpha^2+a^2\cos^2\theta)\}\frac{Y[S]^-\cos\theta}{\rho^2\Delta}$$

$$+\left\{\frac{4\lambda\sigma^+(\lambda\alpha^2-6a^2)}{\mathscr{C}_1+\Gamma_1}(-r^2+3a^2\cos^2\theta)\right.$$

$$\left.-12a^2\sigma^{+2}(r^2-\alpha^2)\cos^2\theta\right\}\frac{rX[S]^+}{a\rho^2\Delta}$$

$$+\{\sigma^+(\lambda\alpha^2-6a^2)(-r^2+a^2\cos^2\theta)-6a^2\sigma^+(\lambda+2)r^2\cos^2\theta\}\frac{Y[S]^+}{a\rho^2\Delta}$$

$$-\left\{(A_1+B_1)\sqrt{\Delta}\frac{\partial}{\partial r}\frac{\Psi}{\sqrt{\Delta}}+(A_2+B_2)\frac{\partial\Psi}{\partial\theta}+(A_1+B_1)(A_2-B_2)\Psi\right\}. \tag{329}$$

Letting

$$\not p=\frac{4\lambda\sigma^+}{\mathscr{C}_1+\Gamma_1}\quad\text{and}\quad \not q=\frac{4\sigma^+(\lambda\alpha^2-6a^2)}{a(\mathscr{C}_1+\Gamma_1)}, \tag{330}$$

we can rewrite equation (329) in the more convenient form

$$\frac{\rho^2\Delta}{E}\frac{\partial^2}{\partial r\partial\theta}E(Z_1-Z_2)$$

$$=-\tfrac{3}{4}(\mathscr{C}_1+\Gamma_1)r^2(Y+\not p rX)\{[S]^-\cos\theta+\tfrac{1}{3}\not q[S]^+\}$$
$$+\tfrac{1}{4}a^2(\mathscr{C}_1+\Gamma_1)(Y+3\not p rX)\{[S]^-\cos\theta+\not q[S]^+\}\cos^2\theta$$
$$-12\sigma^{+2}r^2Y(\alpha^2+a^2\cos^2\theta)[S]^-\cos\theta-12a\sigma^{+2}(r^2-\alpha^2)rX[S]^+\cos^2\theta$$

$$+2\alpha^2 \lambda \sigma^+ rX[S]^- \cos\theta - 6a\sigma^+(\lambda+2)r^2 Y[S]^+ \cos^2\theta$$

$$-\tfrac{1}{4}(\mathscr{C}_1+\Gamma_1)\rho^2\Delta \left[\!\!\left[\left\{ \frac{\alpha^2\sigma^+}{aQ\cos\theta}\frac{\mathscr{S}}{E}\frac{\partial}{\partial r}E\mathscr{R} + \frac{\alpha^2\sigma^+}{rK}\mathscr{R}\{[S]^-\cos\theta + \not{q}[S]^+\}\right\}\right.\right.$$

$$\text{or} \quad \left.\left. \left\{ \frac{\alpha^2\sigma^+}{rK}\frac{\mathscr{R}}{E}\frac{\partial}{\partial\theta}E\mathscr{S} + \frac{\alpha^2\sigma^+}{aQ\cos\theta}\frac{\mathscr{S}}{\Delta}(Y+\not{p}rX)\right\}\right]\!\!\right], \tag{331}$$

where the alternative forms for the terms in the last two lines of the foregoing equation follow from the relations included in equation (248).

The solution for $(Z_1 - Z_2)$ can now be directly written down by integrating over r and θ the expression on the right-hand side of equation (331) after multiplication by

$$\frac{E}{\rho^2\Delta} = (r^2\Delta^3 K)^{-1/2}(Q\cos^2\theta\sin\theta)^{-1/2}. \tag{332}$$

Since $E/\rho^2\Delta$ is a product of a function of r and a function of θ, it is clear that the solution for $Z_1 - Z_2$, obtained by such integration over r and θ, is a sum of products of an integral over r and an integral over θ. The radial integrals which occur are of six kinds:

$$\text{over } rX,\ r^3X,\ Y, \text{ and } r^2Y \text{ with weight-function } (r^2K\Delta^3)^{-1/2}, \tag{333}$$

and

$$\text{over } \mathscr{R} \text{ and } r^2\mathscr{R} \text{ with weight-function } (r^4K^3\Delta)^{-1/2}. \tag{334}$$

Similarly, the angular integrals which occur are also of six kinds:

$$\text{over } [S]^+,\ [S]^+\cos^2\theta,\ [S]^-\cos\theta, \text{ and } [S]^-\cos^3\theta$$
$$\text{with weight-function } (Q\cos^2\theta\sin\theta)^{-1/2}, \tag{335}$$

and

$$\text{over } \mathscr{S} \text{ and } \mathscr{S}\cos^2\theta \text{ with weight-function } (a^2Q^3\cos^4\theta\sin\theta)^{-1/2}. \tag{336}$$

We shall denote the various integrals (with the specified weight-function) with angular brackets enclosing the quantity which is integrated. Thus, in this notation,

$$\langle r^3X\rangle = \int^r \frac{r^3X}{\sqrt{(r^2K\Delta^3)}}\mathrm{d}r \quad \text{and} \quad \langle \mathscr{S}\cos^2\theta\rangle = \int^\theta \frac{(\mathscr{S}\cos^2\theta)\mathrm{d}\theta}{\sqrt{(a^2Q^3\cos^4\theta\sin\theta)}}. \tag{337}$$

The solution for $(Z_1 - Z_2)$ in the notation adopted is

$$E(Z_1 - Z_2)$$
$$= -\tfrac{3}{4}(\mathscr{C}_1+\Gamma_1)\{\langle r^2Y\rangle + \not{p}\langle r^3X\rangle\}\{\langle [S]^-\cos\theta\rangle + \tfrac{1}{3}\not{q}\langle [S]^+\rangle\}$$
$$+\tfrac{1}{4}a^2(\mathscr{C}_1+\Gamma_1)\{\langle Y\rangle + 3\not{p}\langle rX\rangle\}\{\langle [S]^-\cos^3\theta\rangle + \not{q}\langle [S]^+\cos^2\theta\rangle\}$$
$$-12\sigma^{+2}\langle r^2Y\rangle\{\alpha^2\langle [S]^-\cos\theta\rangle + a^2\langle [S]^-\cos^3\theta\rangle\}$$

$$
\begin{aligned}
&-12a\sigma^{+2}\langle [S]^+\cos^2\theta\rangle\{\langle r^3X\rangle-\alpha^2\langle rX\rangle\}\\
&+2\alpha^2\lambda\sigma^+\langle rX\rangle\langle [S]^-\cos\theta\rangle-6a\sigma^+(\lambda+2)\langle r^2Y\rangle\langle [S]^+\cos^2\theta\rangle\\
&-\tfrac{1}{4}\alpha^2\sigma^+(\mathscr{C}_1+\Gamma_1)\left\{\left[\!\!\left[\frac{\mathscr{R}}{r\sqrt{(\Delta K)}}\{r^2\langle\mathscr{S}\rangle+a^2\langle\mathscr{S}\cos^2\theta\rangle\}\right.\right.\right.\\
&+\langle r^2\mathscr{R}\rangle\{\langle [S]^-\cos\theta+\not{q}\langle [S]^+\rangle\}+a^2\langle\mathscr{R}\rangle\{\langle [S]^-\cos^3\theta\rangle\\
&\left.\left.+\not{q}\langle [S]^+\cos^2\theta\rangle\}\right]\!\!\right]\\
&\qquad\text{or}\qquad\left[\!\!\left[\frac{\mathscr{S}}{\sqrt{(Q\cos^2\theta\sin\theta)}}\{\langle r^2\mathscr{R}\rangle+a^2\langle\mathscr{R}\rangle\cos^2\theta\}\right.\right.\\
&\left.\left.\left.+\langle\mathscr{S}\rangle\{\langle r^2Y\rangle+\not{p}\langle r^3X\rangle\}+a^2\langle\mathscr{S}\cos^2\theta\rangle\{\langle Y\rangle+\not{p}\langle rX\rangle\}\right]\!\!\right]\right\}.
\end{aligned}
\tag{338}
$$

Integral identities follow from the solution (338) for (Z_1-Z_2) even as integral identities followed from the solution (166) for Ψ. To obtain the identities, we differentiate the solution (338) with respect to r, and with respect to θ, and equate the resulting expressions with what we know them to be, namely, those given by equations (264). In this manner we obtain the equations

$$
\begin{aligned}
&\frac{1}{E}\frac{\partial}{\partial r}E(Z_1-Z_2)=X_1+X_2-\frac{\alpha^2\sigma^+}{rK}\Psi\\
&=(\rho^2G_1-4a^2QG_2\cos\theta)\frac{rX}{a\Delta\rho^2}+(\rho^2G_3-4a^2QG_4\cos\theta)\frac{Y}{\Delta\rho^2}\\
&\quad-\frac{\alpha^2\sigma^+(\mathscr{C}_1+\Gamma_1)}{4rK}\mathscr{R}\mathscr{S}\\
&=\left[\!\!\left[-\tfrac{3}{4}(\mathscr{C}_1+\Gamma_1)r^2(Y+\not{p}rX)\{\langle [S]^-\cos\theta\rangle+\tfrac{1}{3}\not{q}\langle [S]^+\rangle\}\right.\right.\\
&\quad+\tfrac{1}{4}a^2(\mathscr{C}_1+\Gamma_1)(Y+3\not{p}rX)\{\langle [S]^-\cos^3\theta\rangle+\not{q}\langle [S]^+\cos^2\theta\rangle\}\\
&\quad-12\sigma^{+2}r^2Y\{\alpha^2\langle [S]^-\cos\theta\rangle+a^2\langle [S]^-\cos^3\theta\rangle\}\\
&\quad-12a\sigma^{+2}(r^2-\alpha^2)rX\langle [S]^+\cos^2\theta\rangle\\
&\quad+2\lambda\alpha^2\sigma^+rX\langle [S]^-\cos\theta\rangle-6a\sigma^+(\lambda+2)r^2Y\langle [S]^+\cos^2\theta\rangle\\
&\quad-\tfrac{1}{4}\alpha^2\sigma^+(\mathscr{C}_1+\Gamma_1)(Y+\not{p}rX)\\
&\quad\left.\left.\times\{r^2\langle\mathscr{S}\rangle+a^2\langle\mathscr{S}\cos^2\theta\rangle\}\right]\!\!\right]\frac{(Q\sin\theta)^{1/2}\cos\theta}{\Delta\rho^2}\\
&\quad-\frac{\alpha^2\sigma^+(\mathscr{C}_1+\Gamma_1)}{4rK}\mathscr{R}\mathscr{S},
\end{aligned}
\tag{339}
$$

and

$$\frac{1}{E}\frac{\partial}{\partial\theta}E(Z_1-Z_2)=Y_1+Y_2-\frac{\alpha^2\sigma^+}{aQ\cos\theta}\Psi$$

$$=(\rho^2F_1-4rKF_2)\frac{[S]^-\cos\theta}{\rho^2}+(\rho^2F_3-4rKF_4)\frac{i[S]^+}{a\rho^2}$$
$$-\frac{\alpha^2\sigma^+(\mathscr{C}_1+\Gamma_1)}{4aQ\cos\theta}\mathscr{R}\mathscr{S}$$

$$=\Big[-\tfrac{3}{4}(\mathscr{C}_1+\Gamma_1)\{\langle r^2Y\rangle+\not{p}\langle r^3X\rangle\}\{[S]^-\cos\theta+\tfrac{1}{3}\not{q}[S]^+\}$$
$$+\tfrac{1}{4}a^2(\mathscr{C}_1+\Gamma_1)\{\langle Y\rangle+3\not{p}\langle rX\rangle\}\{[S]^-\cos^3\theta+\not{q}[S]^+\cos^2\theta\}$$
$$-12\sigma^{+2}\langle r^2Y\rangle(\alpha^2+a^2\cos^2\theta)[S]^-\cos\theta$$
$$-12a\sigma^{+2}\{\langle r^3X\rangle-\alpha^2\langle rX\rangle\}[S]^+\cos^2\theta$$
$$+2\lambda\alpha^2\sigma^+\langle rX\rangle[S]^-\cos\theta-6a\sigma^+(\lambda+2)\langle r^2Y\rangle[S]^+\cos^2\theta$$
$$-\tfrac{1}{4}\alpha^2\sigma^+(\mathscr{C}_1+\Gamma_1)\{\langle r^2\mathscr{R}\rangle+a^2\langle\mathscr{R}\rangle\cos^2\theta\}$$
$$\times\{[S]^-\cos\theta+\not{q}[S]^+\}\Big]\frac{r\sqrt{(\Delta K)}}{\rho^2}-\frac{\alpha^2\sigma^+(\mathscr{C}_1+\Gamma_1)}{4aQ\cos\theta}\mathscr{R}\mathscr{S}. \tag{340}$$

We observe that the terms in $\mathscr{R}\mathscr{S}$ cancel in both equations (339) and (340). The remaining terms in (339) consist of functions of θ with the factors r^3X, rX, Y, and r^2Y. We may, therefore, equate the functions of θ which occur with these radial factors. Similarly, we may equate in equation (340), the functions of r which occur with the angular factors $[S]^+$, $[S]^+\cos^2\theta$, $[S]^-\cos\theta$, and $[S]^-\cos^3\theta$. In this manner we obtain the following eight identities:

$$E_\theta G_1=-3\lambda a\sigma^+\{\langle[S]^-\cos\theta\rangle+\tfrac{1}{3}\not{q}\langle[S]^+\rangle+\tfrac{1}{3}\alpha^2\sigma^+\langle\mathscr{S}\rangle\}$$
$$-12a^2\sigma^{+2}\langle[S]^+\cos^2\theta\rangle, \tag{341}$$

$$E_\theta(G_1\cos^2\theta-4QG_2\cos\theta)=3\lambda a\sigma^+\{\langle[S]^-\cos^3\theta\rangle$$
$$+\not{q}\langle[S]^+\cos^2\theta\rangle-\tfrac{1}{3}\alpha^2\sigma^+\langle\mathscr{S}\cos^2\theta\rangle\}+12\alpha^2\sigma^{+2}\langle[S]^+\cos^2\theta\rangle$$
$$+2\lambda(\alpha^2\sigma^+/a)\langle[S]^-\cos\theta\rangle, \tag{342}$$

$$E_\theta G_3=-\tfrac{3}{4}(\mathscr{C}_1+\Gamma_1)\{\langle[S]^-\cos\theta\rangle+\tfrac{1}{3}\not{q}\langle[S]^+\rangle$$
$$+\tfrac{1}{3}\alpha^2\sigma^+\langle\mathscr{S}\rangle\}-6a\sigma^+(\lambda+2)\langle[S]^+\cos^2\theta\rangle$$
$$-12\sigma^{+2}\{\alpha^2\langle[S]^-\cos\theta\rangle+a^2\langle[S]^-\cos^3\theta\rangle\}, \tag{343}$$

$$E_\theta(G_3\cos^2\theta-4QG_4\cos\theta)=\tfrac{1}{4}E_\theta(\mathscr{C}_1+\Gamma_1)\mathscr{S}\cos^2\theta$$
$$=\tfrac{1}{4}(\mathscr{C}_1+\Gamma_1)\{\langle[S]^-\cos^3\theta\rangle+\not{q}\langle[S]^+\cos^2\theta\rangle$$
$$-\alpha^2\sigma^+\langle\mathscr{S}\cos^2\theta\rangle\}, \tag{344}$$

$$E_rF_1=\tfrac{1}{4}(\mathscr{C}_1+\Gamma_1)\{\langle Y\rangle+3\not{p}\langle rX\rangle-\alpha^2\sigma^+\langle\mathscr{R}\rangle\}-12\sigma^{+2}\langle r^2Y\rangle, \tag{345}$$

$$E_r(r^2F_1 - 4rKF_2) = -\tfrac{3}{4}(\mathscr{C}_1 + \Gamma_1)\{\langle r^2 Y\rangle + \not{p}\langle r^3 X\rangle + \tfrac{1}{3}\alpha^2\sigma\!\!\!^{+}\langle r^2\mathscr{R}\rangle\} - 12\alpha^2\sigma\!\!\!^{+2}\langle r^2 Y\rangle + 2\lambda\!\!\!^{-}\alpha^2\sigma\!\!\!^{+}\langle rX\rangle, \tag{346}$$

$$iE_rF_3 = \tfrac{1}{4}a(\mathscr{C}_1 + \Gamma_1)\not{q}\{\langle Y\rangle + 3\not{p}\langle rX\rangle - \alpha^2\sigma\!\!\!^{+}\langle\mathscr{R}\rangle\} - 12\sigma\!\!\!^{+2}\{\langle r^3 X\rangle - \alpha^2\langle rX\rangle\} - 6\sigma\!\!\!^{+}(\lambda\!\!\!^{-} + 2)\langle r^2 Y\rangle, \tag{347}$$

and

$$iE_r(r^2F_3 - 4rKF_4) = -\tfrac{1}{4}E_r a(\mathscr{C}_1 + \Gamma_1)\not{q}r^2\mathscr{R} = -\tfrac{1}{4}a(\mathscr{C}_1 + \Gamma_1)\not{q}\{\langle r^2 Y\rangle + \not{p}\langle r^3 X\rangle + \alpha^2\sigma\!\!\!^{+}\langle r^2\mathscr{R}\rangle\}, \tag{348}$$

where

$$E_r = (r^2\Delta K)^{-1/2},\ E_\theta = (Q\cos^2\theta\sin\theta)^{-1/2},\ \text{and}\ E = E_r E_\theta \rho^2. \tag{349}$$

It may be noted that in equations (344) and (348) we have incorporated two of the relations included in equations (322).

Additional identities follow from integrating the equations

$$\frac{\partial}{\partial\theta}E\mathscr{S} = E\left[\frac{\mathrm{d}\mathscr{S}}{\mathrm{d}\theta} + (A_1 - B_1)\mathscr{S}\right] = E\left[\{[S]^-\cos\theta + \not{q}[S]^+\} + \frac{\sigma\!\!\!^{+}}{\rho^2 aQ\cos\theta}(r^2\alpha^2 - a^2\alpha^2\cos^2\theta - 2a^2r^2\cos^2\theta)\mathscr{S}\right] \tag{350}$$

and

$$\frac{\partial}{\partial r}E\mathscr{R} = E\left[\sqrt{\Delta}\frac{\mathrm{d}}{\mathrm{d}r}\frac{\mathscr{R}}{\sqrt{\Delta}} + (A_2 - B_2)\mathscr{R}\right] = E\left[\frac{1}{\Delta}(Y + \not{p}rX) + \frac{\sigma\!\!\!^{+}}{\rho^2 rK}(r^2\alpha^2 - a^2\alpha^2\cos^2\theta - 2a^2r^2\cos^2\theta)\mathscr{R}\right]. \tag{351}$$

We find

$$E_\theta\mathscr{S} = \langle[S]^-\cos\theta\rangle + \not{q}\langle[S]^+\rangle + \sigma\!\!\!^{+}[\alpha^2\langle\mathscr{S}\rangle - 2a^2\langle\mathscr{S}\cos^2\theta\rangle], \tag{352}$$

$$E_\theta\mathscr{S}\cos^2\theta = \langle[S]^-\cos^3\theta\rangle + \not{q}\langle[S]^+\cos^2\theta\rangle - \alpha^2\sigma\!\!\!^{+}\langle\mathscr{S}\cos^2\theta\rangle, \tag{353}$$

$$E_r\mathscr{R} = \langle Y\rangle + \not{p}\langle rX\rangle - \sigma\!\!\!^{+}[\alpha^2\langle\mathscr{R}\rangle + 2\langle r^2\mathscr{R}\rangle], \tag{354}$$

and

$$r^2E_r\mathscr{R} = \langle r^2 Y\rangle + \not{p}\langle r^3 X\rangle + \alpha^2\sigma\!\!\!^{+}\langle r^2\mathscr{R}\rangle. \tag{355}$$

But two of these, namely, (353) and (355), are the same as (344) and (348) included in the earlier set of eight identities.

Finally, we may note the following two alternative ways in which the solution (338) for $(Z_1 - Z_2)$ can be written by making use of the eight identities (and restoring also the factor $1/(24M\sqrt{2})$ which had been suppressed in equations (319) and (320)):

$$Z_1 - Z_2 = \frac{\sqrt{(r^2\Delta K)}}{(24M\sqrt{2})\rho^2}\left\{\frac{G_1}{a}\langle r^3 X\rangle + a(G_1\cos^2\theta - 4QG_2\cos\theta)\langle rX\rangle + G_3\langle r^2 Y\rangle + \tfrac{1}{4}a^2(\mathscr{C}_1+\Gamma_1)\langle Y\rangle\mathscr{S}\cos^2\theta - \tfrac{1}{4}\alpha^2\sigma^+(\mathscr{C}_1+\Gamma_1)\mathscr{S}[\langle r^2\mathscr{R}\rangle + a^2\langle\mathscr{R}\rangle\cos^2\theta]\right\}$$

$$= \frac{\sqrt{(Q\cos^2\theta\sin\theta)}}{(24M\sqrt{2})\rho^2}\{a^2F_1\langle[S]^-\cos^3\theta\rangle + (r^2F_1 - 4rKF_2)\langle[S]^-\cos\theta\rangle + iaF_3\langle[S]^+\cos^2\theta\rangle - \tfrac{1}{4}(\mathscr{C}_1+\Gamma_1)\varphi\langle[S]^+\rangle r^2\mathscr{R} - \tfrac{1}{4}\alpha^2\sigma^+(\mathscr{C}_1+\Gamma_1)\mathscr{R}[r^2\langle\mathscr{S}\rangle + a^2\langle\mathscr{S}\cos^2\theta\rangle\}. \quad (356)$$

For the six integrals which appear in each of the two foregoing alternative ways of expressing the solution for $(Z_1 - Z_2)$, we have only five equations relating them to known functions. The situation in this context is, therefore, different from what obtained in the context of the solution for Ψ: there the solution was uniquely determined. Fortunately, explicit solutions for Z_1 and Z_2 have been found by independent considerations. Nevertheless, the emergence of the various integral identities express relations whose origins are shrouded in mystery.

95. A retrospect

In view of the bewildering complexity of the analysis just completed, it may be useful to unravel the principal strands.

The analysis is addressed to the problem of the gravitational perturbations of the Kerr space-time via the equations of the Newman–Penrose formalism. The central problem here is the determination of the changes in the metric coefficients induced by the perturbation. In the Newman–Penrose formalism, these metric perturbations are directly related to the changes, $\boldsymbol{l}^{i(1)}$, in the basis vectors $\boldsymbol{l}^i\,(=\boldsymbol{l}, \boldsymbol{n}, \boldsymbol{m}, \bar{\boldsymbol{m}})$. These changes can be represented by the matrix A of the transformation

$$\boldsymbol{l}^{i(1)} = A^i_j \boldsymbol{l}^j. \quad (357)$$

Besides the elements of A, we have to solve for the perturbations in the five Weyl scalars and in the twelve spin-coefficients—altogether fifty real quantities. For the solution of these quantities, we have eight Bianchi identities, twelve commutation relations, and eighteen Ricci identities—altogether seventy-six real equations. The solutions we seek must be consistent with ten degrees of gauge freedom—six degrees arising from the admissibility of infinitesimal Lorentz transformations of the tetrad basis and four degrees of freedom arising from the admissibility of infinitesimal coordinate transformations.

The analysis shows that a gauge which appears most natural for the problem is the one in which

$$\Psi_1 = \Psi_3 = \Psi_2^{(1)} = 0 \quad \text{and} \quad A_1^1 = A_2^2 = A_3^3 = A_4^4 = 0. \qquad (358)$$

This choice exhausts all the ten degrees of freedom allowed and fixes the gauge completely. The following description of the analysis is in the context of this choice of gauge.

Four of the Bianchi identities and two of the Ricci identities enable an explicit solution of the Weyl scalars Ψ_0 and Ψ_4 and the spin-coefficients, κ, σ, λ, and ν, i.e., of the quantities which vanish in the background Kerr-geometry by virtue of its type-D character. The separability of the equations governing Ψ_0 and Ψ_4 and the expression of their solution in terms of Teukolsky's radial and angular functions, $R_{\pm 2}$ and $S_{\pm 2}$, are what makes possible, even, a contemplation of solving for the remaining quantities. Almost as important for the subsequent analysis, are the remarkable identities of Teukolsky and Starobinsky. But the analysis leaves the relative normalization of the solutions for R_{+2} and R_{-2} and the argument of the complex Starobinsky-constant undetermined—serious lacunae in the needed information.

The four remaining Bianchi identities enable the expression of the perturbations in the spin-coefficients, $\tilde{\rho}$, τ, μ, and π, directly in terms of the elements of the matrix A.

The principal equations for the determination of the elements of A derive from the linearization of the twelve commutation relations. The equations, one derives, are linear and inhomogeneous; the inhomogeneous terms are linear combinations of the perturbed spin-coefficients. This last fact enables the grouping of the twenty-four equations one obtains into three systems (system I, II, and III) of eight equations each. Thus, with the solutions for the perturbations in the spin-coefficients, $\tilde{\rho}$, τ, μ, and π, expressed in terms of the elements of A, and the solutions for the spin-coefficients, κ, σ, λ, and ν, expressed in terms of the Teukolsky functions, two of the systems of equations—a homogeneous system-I and an inhomogeneous system-II—become equations for twelve linear combinations of the elements of A designated by

$$F+G,\, J-H,\, F-G,\, J+H,\, B_1,\, B_2,\, C_1,\, C_2,\, F_2^1 \pm F_1^2, \text{ and } F_4^3 \pm F_3^4. \qquad (359)$$

The last remaining eight equations of system III serve to determine the perturbations in the spin-coefficients, α, β, γ, and ε, once the solution for A has been completed.

Only six of the eight equations of system I are independent. But they enable one to express six of the linear combinations (359), namely,

$$F-G,\ J+H,\ B_1,\ B_2,\ C_1,\ \text{and}\ C_2, \tag{360}$$

in terms of $F+G$ and $J-H$.

The eight equations of system II (in their alternative forms (133)) provide two sets of equations of four equations each (equations (134)–(137) and (187)–(190)). The first set leads to a crucial integrability condition for the solution of

$$\Psi = K(J-H)\cos\theta - irQ(F+G)\sin\theta. \tag{361}$$

The solution of the integrability condition presents the single most massive piece of reduction of the entire analysis. It is also the central fulcrum of the theory: it resolves the two principal ambiguities—the relative normalization of the solutions for Ψ_0 and Ψ_4 and the argument of the Starobinsky constant. And even more than resolving the ambiguities, it lets us into a realm of the theory well beyond imagination: the separability of Ψ and its expressibility as the product of a radial function $\mathscr{R}$ and an angular function $\mathscr{S}$; the existence of two sets of formulae relating $\mathscr{R}$ and $\mathscr{S}$ to the Teukolsky functions; and, most surprising of all, the explicit evaluations that $\mathscr{R}$ and $\mathscr{S}$ provide for certain indefinite integrals over the Teukolsky functions.

The second set of four equations of system II provides, as integrability conditions, differential equations for $\mathscr{R}$ and $\mathscr{S}$ relating them very simply to the Teukolsky functions. By making use of these equations, one can obtain differential equations for $\mathscr{R}$ and $\mathscr{S}$ which contain no reference to the Teukolsky functions—differential equations which, in principle, make it possible to eliminate any direct reference to the Teukolsky functions altogether! The remaining equations of the set, together with a similar pair of equations from the earlier set, enable one to express the remaining four functions, namely,

$$F^{1}_{2} \pm F^{2}_{1}\ \text{and}\ F^{3}_{4} \pm F^{4}_{3}, \tag{362}$$

also in terms of $F+G$ and $J-H$.

With the completion of the reduction of system II, there remains only one lacuna in the information needed to specify all the functions listed in (359): to determine $F+G$ and $J-H$ separately. More precisely, letting

$$Z_1 = K(J-H)\cos\theta \quad \text{and} \quad Z_2 = -irQ(F+G)\sin\theta, \tag{363}$$

we know $Z_1+Z_2\,(=\Psi)$. We need to solve for another linear combination of Z_1 and Z_2.

At this stage, having made full use of the Bianchi identities and the commutation relations, one must, of necessity, turn to the Ricci identities.

The four Ricci identities selected for consideration separate into two pairs of equations. The first pair enables the expression of the derivatives of Z_1 and Z_2, with respect to r and θ, as linear combinations of Z_1 and Z_2; it provides, in addition, a further integrability condition. The second pair of equations, together with the relations obtained with the first pair, enables the completion of the solution by providing explicit expressions for Z_1 and Z_2, separately. At the same time, the integrability condition yields eight integral identities of great complexity.

In summary, a total of twenty-six complex equations (including the eight Bianchi identities, the twelve commutation relations, and six Ricci identities) was needed to specify the solution of forty different quantities and ten integral identities. Two of these integral identities relating to $\mathcal{R}$ and $\mathcal{S}$ are essential for the analysis: without them the solution could not have been completed.

A major question which emerges from the analysis (particularly of §94) is this: can the superfluity of the Newman–Penrose equations—sixty equations for fifty quantities subject to ten degrees of gauge freedom—enable one to discover new classes of integral identities among the functions of mathematical physics when they occur in the solution of Einstein's equations?

Details aside, the simple fact is that it has been possible to take a system of sixty equations governing some basic aspects of a space-time and solve them in their entirety.

96. The form of the solution in the Schwarzschild limit, $a \to 0$

The form of the solution in the Schwarzschild limit, $a \to 0$, is of some general interest: among other things, it will enable us to present the basic identities in their simplest and, in some instances, in their directly verifiable forms. Moreover, as is manifest from some of the equations of the theory (e.g., equations (179) and (185)), the passage to the limit, $a \to 0$ will require some caution.

First, we observe that in the limit $a \to 0$,

$$\left.\begin{aligned} Q = m\operatorname{cosec}\theta,\ K = r^2\sigma^+,\ \bar{\rho} = \bar{\rho}^* = r,\ \rho^2 = r^2, \\ \Delta = r^2 - 2Mr,\ \text{and}\ \mathscr{C}_1 = \Gamma_1 = \lambda\!\!\!^-(\lambda\!\!\!^- + 2). \end{aligned}\right\} \tag{364}$$

And from equation (170) we conclude that

$$\mathscr{C}_1 - \Gamma_1 = -\frac{8\lambda\!\!\!^-\sigma^+ m}{\lambda\!\!\!^- + 2}a + O(a^2). \tag{365}$$

Also, the limiting forms of the coefficients listed in equations (76) and (78) are

$$\left.\begin{aligned} A_1 &= \lambda\!\!\!^-(\lambda\!\!\!^-+2)\Delta - 4\lambda\!\!\!^-\sigma^{+2}r^4 - 24M\sigma^{+2}r^3, \\ A_2 &= -12M\sigma^+\Delta - 4\lambda\!\!\!^-\sigma^+ r^2(2r-3M) + 24\sigma^{+3}r^5, \\ B_1 &= 4\sigma^+ r^2[\lambda\!\!\!^-\Delta + 2M(r-3M) - 2\sigma^{+2}r^4]; \end{aligned}\right\} \tag{366}$$

and

$$\left.\begin{aligned} \alpha_1 &= \lambda(\lambda+2) - 4\lambda m^2 \operatorname{cosec}^2\theta, \\ \alpha_2 &= 4\lambda m \cos\theta \operatorname{cosec}^2\theta, \\ \beta_1 &= 4m[2(m^2-1) - \lambda\sin^2\theta]\operatorname{cosec}^3\theta. \end{aligned}\right\} \tag{367}$$

Greatest interest is clearly attached to the limiting form of the expressions for $\mathscr{R}, \mathscr{S}, Z_1$, and Z_2. By making use of equations (366) and (367), we find that equations (180) and (181), and, similarly, equations (185) and (186) provide, consistently with each other, the following expressions for $\mathscr{R}$ and $\mathscr{S}$:

$$\tfrac{1}{2}[\lambda\Delta + 2M(r-3M) - 2\sigma^2 r^4]\,\mathscr{R} = \left[r - \frac{2}{\lambda+2}(r-3M)\right]\sigma r X + \left[M - \frac{2\sigma^2}{\lambda+2}r^3\right]Y \tag{368}$$

and

$$\tfrac{1}{2}[\lambda\sin^2\theta - 2(m^2-1)]\,\mathscr{S} = \frac{m\lambda}{\lambda+2}[S]^+\cos\theta\sin\theta + \left(\frac{2m^2}{\lambda+2} - 1\right)[S]^-\sin\theta. \tag{369}$$

These expressions for $\mathscr{R}$ and $\mathscr{S}$ evaluate the integrals,

$$\mathscr{R} = \sqrt{\Delta}\int\left(Y + \frac{2\sigma r}{\lambda+2}X\right)\frac{dr}{\Delta^{3/2}} \tag{370}$$

and

$$\mathscr{S} = \int\left\{[S]^-\cos\theta + \frac{2m}{\lambda+2}[S]^+\right\}d\theta. \tag{371}$$

It is, as we have stated in §95, a striking aspect of the Newman–Penrose formalism, when applied to the theory of perturbations of a space-time, that it yields novel integral relations, such as (370) and (371), among the classical functions of analysis.

Turning to the solutions for Z_1 and Z_2 given by equations (285) and (286), we find after some careful analysis that, in the limit of $a \to 0$, they give (omitting again the factor $1/(24M\sqrt{2})$)

$$Z_1 = -\tfrac{1}{2}\lambda(\lambda+2)\,\mathscr{R}\,\frac{2\sin\theta\cos\theta}{\lambda\sin^2\theta - 2(m^2-1)}\{m[S]^+ - [S]^-\cos\theta\} \tag{372}$$

and

$$Z_2 = \mathscr{R}\,\frac{2\lambda\sin\theta}{\lambda\sin^2\theta - 2(m^2-1)}\{m(\lambda+1)[S]^+\cos\theta + [m^2 - \tfrac{1}{2}(\lambda+2)(1+\cos^2\theta)][S]^-\}. \tag{373}$$

In obtaining the foregoing formulae, the following relations, valid in the limit

$a \to 0$, were used:

$$m[\mathscr{L}\mathscr{L}S]^- \operatorname{cosec}\theta = -\sigma^+(\lambda\mathscr{S} - 2[\mathscr{L}S]^- \cos\theta)a + O(a^2), \quad (374)$$

$$m\{[\mathscr{L}\mathscr{L}S]^+ \cos\theta + 2[\mathscr{L}S]^+ \sin\theta\} \operatorname{cosec}\theta = -\tfrac{1}{2}\lambda(\lambda+2)\mathscr{S}, \quad (375)$$

$$r\sigma^+\{r[\mathscr{D}\mathscr{D}P]^+ - 2[\mathscr{D}P]^+\} = \tfrac{1}{2}\lambda(\lambda+2)\mathscr{R}, \quad (376)$$

and

$$\begin{aligned}\tfrac{1}{2}\lambda(\lambda+2)\mathscr{S} + m[\mathscr{L}S]^+ &= -\tfrac{1}{4}(\lambda+2)\{-\lambda\mathscr{S} + 2[\mathscr{L}S]^- \cos\theta\} \\ &= \frac{\lambda\sin\theta}{\lambda\sin^2\theta - 2(m^2-1)}\{m(\lambda+1)[S]^+\cos\theta \\ &\qquad + [m^2 - \tfrac{1}{2}\lambda(\lambda+2)(1+\cos^2\theta)[S]^-\}. \quad (377)\end{aligned}$$

With Z_1 and Z_2, given by equations (372) and (373), it can be directly verified that, as required,

$$Z_1 + Z_2 = \tfrac{1}{2}\lambda(\lambda+2)\,\mathscr{R}\mathscr{S}. \quad (378)$$

The solutions for the remaining metric functions can be obtained by appropriately passing to the limit, $a \to 0$, of the various expressions listed in §93.

97. The transformation theory and potential barriers for incident gravitational waves

We now turn to the problem of how incident gravitational waves will be reflected and absorbed by the Kerr black-hole. As in the case of incident electromagnetic waves considered in Chapter 8, we shall first reduce the problem to one of reflexion and transmission by one-dimensional potential-barriers by the procedure described in §§72 and 73.

We recall that, by the change of variables (cf. Ch. 8, equations (100) and (105)),

$$\frac{d}{dr_*} = \frac{\Delta}{\varpi^2}\frac{d}{dr} \quad \text{and} \quad Y = |\varpi^2|^{-3/2} P_{+2}$$

where

$$\varpi^2 = r^2 + \alpha^2 = r^2 + a^2 + (am/\sigma^+), \quad (379)$$

Teukolsky's equation (24) governing P_{+2}, becomes (cf. Ch. 8, equations (108)–(110))

$$\Lambda^2 Y + P\Lambda_- Y - QY = 0, \quad (380)$$

where

$$P = \frac{d}{dr_*}\lg\frac{\varpi^8}{\Delta^2}, \quad (381)$$

and

$$Q = \frac{\Delta}{\varpi^4}\left\{\lambda - 3\left[\frac{\Delta - 2r(r-M)}{\varpi^2} + \frac{r^2\Delta}{\varpi^4}\right]\right\}$$

$$= \frac{\Delta}{\varpi^8}[\lambda\varpi^4 + 3\varpi^2(r^2-a^2) - 3r^2\Delta]. \tag{382}$$

And we seek to transform equation (380) to a one-dimensional wave-equation of the form

$$\Lambda^2 Z = VZ. \tag{383}$$

The transformation theory developed in §73 is applicable to the present problem: we need only put $s = 2$ in the relevant equations. Thus, making the substitutions Chapter 8, equations (120) and (122) (with $s = 2$), we are led to consider the following equations (cf. Ch. 8, equations (130)–(133)):

$$f V = R - \frac{dT}{dr_*}, \qquad T = W + 2i\sigma f, \tag{384}$$

$$\frac{d}{dr_*}\left(\frac{\varpi^8}{\Delta^2}R\right) = \frac{\varpi^8}{\Delta^2}(QT - 2i\sigma R) + \beta, \tag{385}$$

$$R\left(R - \frac{dT}{dr_*}\right) + \frac{\Delta^2}{\varpi^8}\beta T = \frac{\Delta^2}{\varpi^8}K, \tag{386}$$

and

$$RV - Q\left(R - \frac{dT}{dr_*}\right) = \frac{\Delta^2}{\varpi^8}\frac{d\beta}{dr_*}. \tag{387}$$

(a) An explicit solution

We shall seek solutions of equations (385)–(387) which are of the forms

$$T = T_1(r_*) + 2i\sigma, \quad \beta = \beta_1(r_*) + 2i\sigma\beta_2, \tag{388}$$

and

$$K = \kappa_1 + 2i\sigma\kappa_2, \tag{389}$$

where β_2, κ_1, and κ_2 are constants (to be specified) and R and V are explicitly independent of σ in that they do not contain any term linear in $i\sigma$ (as T and β do). We shall, further, assume that, when the solutions of the chosen form are substituted in equations (385)–(387), we can equate separately the terms which occur with $i\sigma$ as a factor and the terms which do not. In making these various assumptions, we are imposing more restrictions than we are permitted by the latitude that we have. Nevertheless, we shall find that solutions of the assumed forms do exist by virtue of Q (rather $F = \varpi^8 Q/\Delta^2$) satisfying a certain non-linear differential equation (equation (404) below) as in earlier contexts (see Ch. 4, equation (311) and Ch. 5, equation (267)).

Now substituting for T and β from equations (388) in equation (385) and equating (as we said we would) the terms with and without the factor $i\sigma^{+}$, we obtain the pair of equations

$$R = Q + \frac{\Delta^2}{\varpi^8}\beta_2, \tag{390}$$

and

$$\frac{d}{dr_*}\left(\frac{\varpi^8}{\Delta^2}R\right) = \frac{\varpi^8}{\Delta^2}QT_1 + \beta_1. \tag{391}$$

Similarly, from equation (386), we obtain the pair of equations

$$\beta_1 + \beta_2 T_1 = \kappa_2 \tag{392}$$

and

$$R^2 - R\frac{dT_1}{dr_*} + \frac{\Delta^2}{\varpi^8}\beta_1 T_1 = \frac{\Delta^2}{\varpi^8}\kappa, \tag{393}$$

where

$$\kappa = \kappa_1 + 4\sigma^{+2}\beta_2. \tag{394}$$

In view of this last relation, we may write (cf. equation (389))

$$K = (\kappa - 4\sigma^{+2}\beta_2) + 2i\sigma^{+}\kappa_2. \tag{395}$$

Substituting for R and β_1, in accordance with equations (390) and (392), in equation (391), we obtain

$$\frac{d}{dr_*}\left(\frac{\varpi^8}{\Delta^2}Q\right) = \frac{\varpi^8}{\Delta^2}QT_1 - \beta_2 T_1 + \kappa_2. \tag{396}$$

Let

$$F = \frac{\varpi^8}{\Delta^2}Q = \frac{1}{\Delta}[\lambda\varpi^4 + 3\varpi^2(r^2 - a^2) - 3r^2\Delta]. \tag{397}$$

Then,

$$R = \frac{\Delta^2}{\varpi^8}(F + \beta_2), \tag{398}$$

and equation (396) becomes

$$\frac{dF}{dr_*} = T_1(F - \beta_2) + \kappa_2, \tag{399}$$

or, alternatively,

$$T_1 = \frac{F' - \kappa_2}{F - \beta_2} \quad \text{where} \quad F' = F_{,r_*}. \tag{400}$$

Returning to equation (393) and writing it in terms of F, we have

$$\frac{\Delta^2}{\varpi^8}(F + \beta_2)^2 - (F + \beta_2)\frac{dT_1}{dr_*} + T_1(\kappa_2 - \beta_2 T_1) = \kappa. \tag{401}$$

We now eliminate T_1 from this equation with the aid of equation (400) and the further equation,

$$T_1' = \frac{1}{(F-\beta_2)^2}[(F-\beta_2)F'' - (F'-\kappa_2)F']; \tag{402}$$

and we find

$$\frac{\Delta^2}{\varpi^8}(F^2-\beta_2^2)^2 - (F^2-\beta_2^2)F'' + (F'^2-\kappa_2^2)F = \kappa(F-\beta_2)^2. \tag{403}$$

On expanding this last equation, we obtain

$$\left(\frac{\mathrm{d}F}{\mathrm{d}r_*}\right)^2 - F\frac{\mathrm{d}^2F}{\mathrm{d}r_*^2} + \frac{\Delta^2}{\varpi^8}F^3 = (\kappa_2^2 - 2\beta_2\kappa) + \left(\kappa + 2\frac{\Delta^2}{\varpi^8}\beta_2^2\right)F$$
$$-\frac{\beta_2^2}{F}\left(\frac{\mathrm{d}^2F}{\mathrm{d}r_*^2} + \beta_2^2\frac{\Delta^2}{\varpi^8} - \kappa\right), \tag{404}$$

where it should be remembered that F is a *known* function. Equation (404) is, therefore, a condition on F if solutions of the chosen form are to exist. Equation (404) is a generalization of Chapter 4, equation (311) to which it reduces when $\beta_2 = 0$.

It is a remarkable fact that F as defined in equation (397) does satisfy equation (404) with the choice

$$\kappa = \lambda(\lambda+2), \qquad \beta_2 = \pm 3\alpha^2, \tag{405}$$

and

$$\kappa_2 = \pm\{36M^2 - 2\lambda[\alpha^2(5\lambda+6) - 12a^2] + 2\beta_2\lambda(\lambda+2)\}^{1/2}, \tag{406}$$

where the signs of β_2 and κ_2 may be assigned independently; but the task of verification is a formidable one.

The solution for V can now be found from the last equation (387). Rewriting this equation in the form

$$R(V-Q) = -Q\frac{\mathrm{d}T_1}{\mathrm{d}r_*} + \frac{\Delta^2}{\varpi^8}\frac{\mathrm{d}\beta_1}{\mathrm{d}r_*}, \tag{407}$$

and making use of equations (390) and (392), we obtain

$$R(V-Q) = -\left(Q + \frac{\Delta^2}{\varpi^8}\beta_2\right)\frac{\mathrm{d}T_1}{\mathrm{d}r_*} = -R\frac{\mathrm{d}T_1}{\mathrm{d}r_*}. \tag{408}$$

Therefore,

$$V = Q - \frac{\mathrm{d}T_1}{\mathrm{d}r_*}, \tag{409}$$

where we may insert for $T_1{}'$ from equation (402). It is, however, convenient to eliminate F'' in equation (402) with the aid of the differential equation (403) satisfied by F; and we find, after the elimination,

$$T_1' = \frac{\Delta^2}{\varpi^8}(F+\beta_2) - \frac{\kappa}{F+\beta_2} + \frac{(F'-\kappa_2)(\kappa_2 F - \beta_2 F')}{(F-\beta_2)(F^2-\beta_2^2)}. \tag{410}$$

The expression for V now becomes

$$V = -\frac{\Delta^2}{\varpi^8}\beta_2 + \frac{\kappa}{F+\beta_2} - \frac{(F'-\kappa_2)(\kappa_2 F - \beta_2 F')}{(F-\beta_2)(F^2-\beta_2^2)}. \tag{411}$$

(b) The distinction between $Z^{(+\sigma)}$ *and* $Z^{(-\sigma)}$

As in §74(a), we shall distinguish Y and the function satisfying the complex-conjugate equation by $Y^{(+\sigma)}$ and $Y^{(-\sigma)}$ and the functions satisfying the associated one-dimensional wave-equations by $Z^{(+\sigma)}$ and $Z^{(-\sigma)}$. And we find (again, as in § 74(a)) that $Z^{(+\sigma)}$ *and* $Z^{(-\sigma)}$ *satisfy wave equations with the same potential* (411).

The various equations expressing $Y^{(\pm\sigma)}$ in terms of $Z^{(\pm\sigma)}$, and conversely, are

$$\left.\begin{aligned}
Y^{(\pm\sigma)} &= f\, VZ^{(\pm\sigma)} + (T_1 \pm 2i\sigma)\Lambda_{\pm} Z^{(\pm\sigma)},\\
\Lambda_{\mp} Y^{(\pm\sigma)} &= -\frac{\Delta^2}{\varpi^8}(\beta_1 \pm 2i\sigma\beta_2)Z^{(\pm\sigma)} + R\Lambda_{\pm} Z^{(\pm\sigma)},\\
K^{(\pm\sigma)}Z^{(\pm\sigma)} &= \frac{\varpi^8}{\Delta^2}RY^{(\pm\sigma)} - \frac{\varpi^8}{\Delta^2}(T_1 \pm 2i\sigma)\Lambda_{\mp} Y^{(\pm\sigma)},\\
K^{(\pm\sigma)}\Lambda_{\pm} Z^{(\pm\sigma)} &= (\beta_1 \pm 2i\sigma\beta_2)Y^{(\pm\sigma)} + \frac{\varpi^8}{\Delta^2} f\, V\Lambda_{\mp} Y^{(\pm\sigma)},
\end{aligned}\right\} \tag{412}$$

where it may be noted that (cf. equations (395) and (405))

$$K^{(\pm\sigma)} = [\lambda(\lambda+2) - 4\sigma^2\beta_2] \pm 2i\sigma\kappa_2. \tag{413}$$

The constants $K^{(+\sigma)}$ and $K^{(-\sigma)}$ are related in an important way with the Starobinsky constant $\mathscr{C}$; thus

$$\begin{aligned}
K^{(+\sigma)}K^{(-\sigma)} &= [\lambda(\lambda+2) - 4\sigma^2\beta_2]^2 + 4\sigma^2\kappa_2^2\\
&= \lambda^2(\lambda+2)^2 - 8\sigma^2\lambda[\alpha^2(5\lambda+6) - 12a^2] + 144\sigma^2(M^2+\alpha^4\sigma^2) = |\mathscr{C}|^2
\end{aligned} \tag{414}$$

—a relation free of the ambiguities in the signs of β_2 and κ_2. The existence of a relation such as the foregoing between the Starobinsky constants, $\mathscr{C}$ and $\mathscr{C}^*$ and $K^{(+\sigma)}$ and $K^{(-\sigma)}$, is perhaps to be expected: both pairs of constants in different ways relate the functions belonging to $s = +2$ and $+\sigma$ and $s = -2$ and $-\sigma$. That the relation (414) is, in fact, needed for the consistency of the entire theory will emerge more clearly in § 98 below.

(c) *The nature of the potentials*

Since β_2 and κ_2 can be chosen to be of either sign, independently of one another, equation (411) yields, in general, four possible potentials for use in equation (383). It is also clear that, depending on the sign of the quantity under the square root in the definition (406) of κ_2, we can have a pair of complex-conjugate potentials (even as we had for incident electromagnetic waves in Chapter 8).

The explicit form of the potential given by equation (411) is

$$V = \Delta\left\{ -\beta_2 \frac{\Delta}{\varpi^8} + \frac{\lambda(\lambda+2)}{q+\beta_2\Delta} + \frac{[\kappa_2\varpi^2\Delta - (q'\Delta - \Delta' q)][\kappa_2\varpi^2 q - \beta_2(q'\Delta - \Delta' q)]}{\varpi^4(q+\beta_2\Delta)(q-\beta_2\Delta)^2}\right\}, \quad (415)$$

where

$$q = \lambda\varpi^4 + 3\varpi^2(r^2-a^2) - 3r^2\Delta = \Delta F, \quad (416)$$

$$\left.\begin{aligned} q-\beta_2\Delta &= \varpi^2(\lambda\varpi^2+6Mr-6a^2) & (\beta_2 = +3\alpha^2),\\ q-\beta_2\Delta &= \lambda\varpi^4+6r^2(\alpha^2-a^2)+6Mr(r^2-\alpha^2) & (\beta_2 = -3\alpha^2),\end{aligned}\right\} \quad (417)$$

and

$$q'\Delta - \Delta' q = -2(r-M)\lambda\varpi^4 + 2\varpi^2(2\lambda r\Delta - 3Mr^2 - 3Ma^2 + 6ra^2) + 12r\Delta(Mr-a^2). \quad (418)$$

The runs of the potentials, given by equation (415), are exhibited in Figs. 43 and 44 for a number of typical cases.

First, we observe that in the Schwarzschild limit, ($a = 0$),

$$\left.\begin{aligned} &\beta_2 = 0,\ \kappa_2 = \pm 6M,\ \lambda = (l-1)(l+2) = \mu^2 = 2n,\\ \text{and}\quad &K^{(\pm\sigma)} = \mu^2(\mu^2+2) \pm 12i\sigma M,\end{aligned}\right\} \quad (419)$$

in agreement with Chapter 4, equation (327). The four potentials, therefore, degenerate into two in this limit; and the two potentials are the Regge–Wheeler and the Zerilli potentials appropriate for the axial and the polar perturbations of the Schwarzschild black-hole. The question occurs, then, whether the four potentials we have, in general, for the Kerr black-hole may not, in fact, point to the existence of an internal symmetry besides parity.

Next, we observe that all the potentials have the following common feature: they all have an inverse-square behaviour for $r \to \infty$; and they all tend to zero exponentially for $r_* \to -\infty$ (in case $\sigma > \sigma_s$) and $r_* \to +\infty$ (in case $\sigma < \sigma_s$) as we approach the horizon at $r_+ + 0$. We conclude that in all cases the solutions of the one-dimensional wave-equations have the asymptotic behaviour

$$Z \to e^{\pm i\sigma r_*} \qquad (r \to \infty \quad \text{and} \quad r \to r_+ + 0). \quad (420)$$

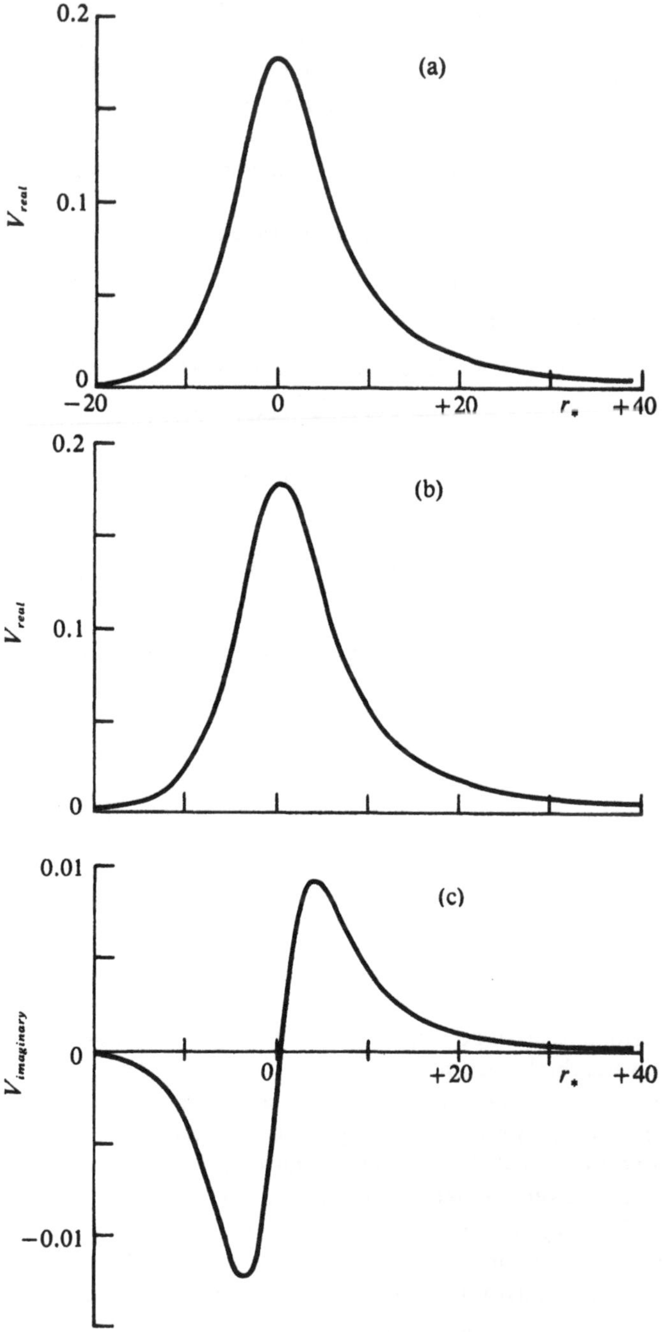
0.2
0.1
0
V_{real}
(a)
−20
0
+20
r_*
+40
0.2
0.1
0
V_{real}
(b)
0.01
0
−0.01
$V_{imaginary}$
(c)
0
+20
r_*
+40

For $\sigma^+ > \sigma^+_s$, when the $r_*(r)$-relation is single-valued, all four potentials are continuous, bounded, and of short range. Moreover, from equation (409) it follows that the integrals of all four potentials, over the entire range, $(-\infty, +\infty)$, of r_*, are the same:

$$\int_{-\infty}^{+\infty} V \mathrm{d}r_* = \int_{-\infty}^{+\infty} Q \mathrm{d}r_* + (T_1)_{\Delta=0} \qquad (\sigma^+ > \sigma^+_s), \tag{421}$$

where

$$(T_1)_{\Delta=0} = -2\frac{r_+ - M}{\varpi_+^2} = -\frac{\sqrt{(M^2-a^2)}}{Mr_+(1-\sigma^+_s/\sigma^+)}. \tag{422}$$

In particular, the integral is real even in cases where the potential is complex. The equality of the integrals of the potentials is reminiscent of a similar equality we found for the potentials surrounding the Schwarzschild and the Reissner–Nordström black-holes. Indeed, when $\beta_2 = 0$, the expression for V belongs to the special class, Chapter 4, equation (133), which guarantees the equality of the reflexion and the transmission coefficients for *all* f's, κ's, β's— not only for the particular f's, κ's, and β's which distinguish the problem on hand. We shall find in §98 below that all four potentials given by equation (415) also yield the same reflexion and transmission coefficients. The question occurs whether equation (411) isolates a similar (and larger) class of potentials which guarantees the equality of the reflexion and the transmission coefficients.

The case of axisymmetry, when $m = 0$ and $\alpha^2 = a^2$, is clearly the simplest. In this case all four potentials (bounded and of short range) can be real (for sufficiently small values of a/M, as we shall presently see) or a pair can be real and the other, a complex-conjugate pair, or they can be two complex-conjugate pairs. The condition for the occurrence of a complex-conjugate pair of potentials is (cf. equation (406))

$$2a^2\lambda(5\lambda-6) - 2\beta_2\lambda(\lambda+2) > 36M^2. \tag{423}$$

For the two allowed values of β_2, namely $-3\alpha^2$ and $+3\alpha^2$, the condition (423) gives

$$\lambda > 3 + 3(1+M^2/a^2)^{1/2} \quad \text{and} \quad \lambda > 3M/2a, \tag{424}$$

respectively. Since the least value of λ is $(l-1)(l+2)$, it follows that all four potentials will be real for

$$a/M < 3/[2(l-1)(l+2)], \tag{425}$$

FIG. 43. The potential barriers surrounding a Kerr black-hole for the incidence of gravitational waves. Of the four potentials belonging to $l = 2$, $m = 0$, $a = 0.9$, and $\sigma^{+2} = 0.13$, two are real (associated with the values $\kappa_2 = \pm 7.848$ and $\beta_2 = +2.43$); and two are complex conjugates (associated with the values $\kappa_2 = \pm 13.288i$ and $\beta_2 = -2.43$). The real potential belonging to $\kappa_2 = +7.848$ and $\beta_2 = +2.43$ is illustrated in (*a*): that belonging to $\kappa_2 = -7.848$ and $\beta_2 = +2.43$ cannot be distinguished from the one illustrated in the scale of the graph. In (*b*) and (*c*) the real and the imaginary parts of the potential belonging to $\kappa_2 = +13.288i$ and $\beta_2 = -2.43$ are illustrated.

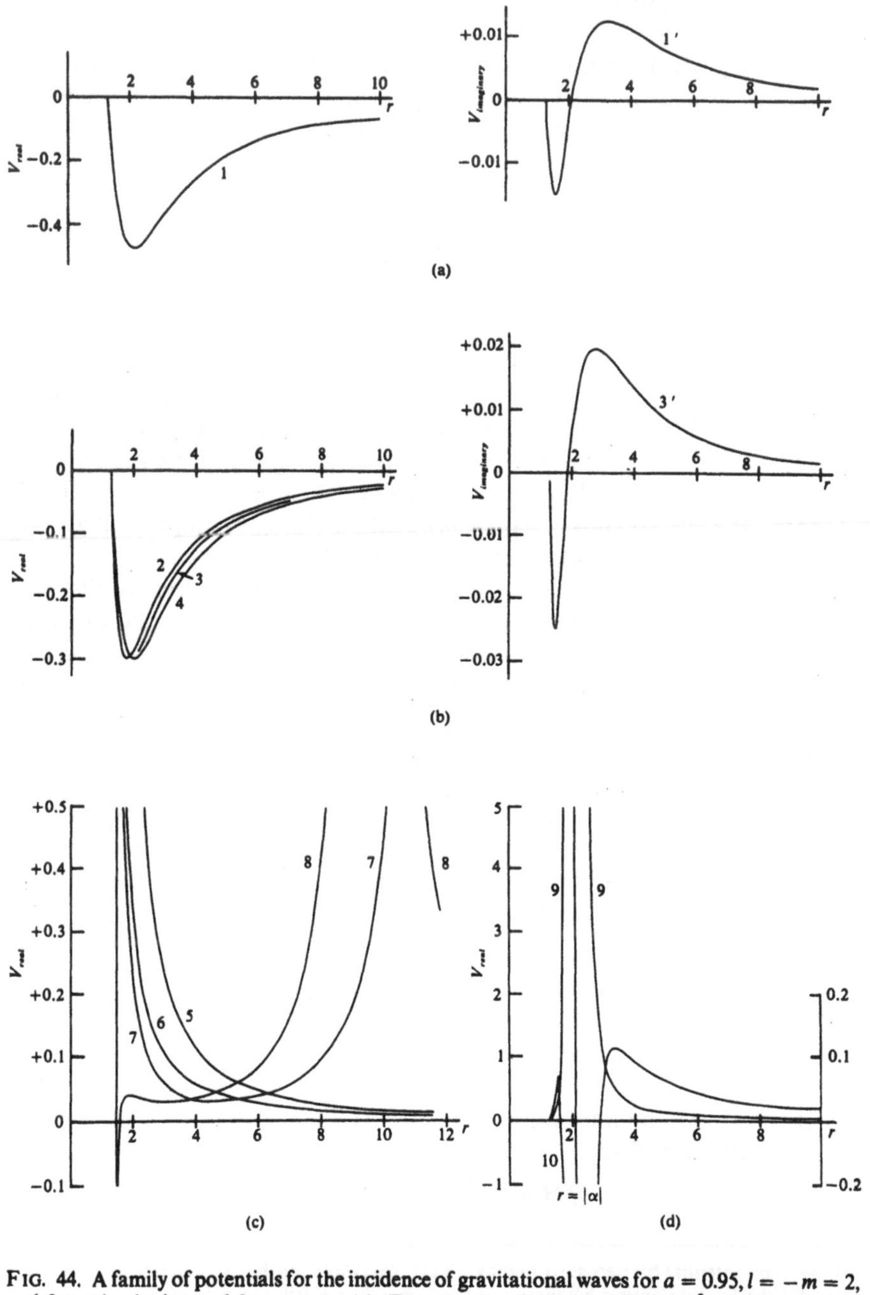

FIG. 44. A family of potentials for the incidence of gravitational waves for $a = 0.95, l = -m = 2$, and for σ^+ in the interval $0 < \sigma^+ \leqslant \sigma_s^+$. (*a*). For $\sigma^+ = \sigma_c^+ = 2.11$, $\kappa_2 = 12.31\,i$, $\alpha^2 = 0$, $\beta_2 = 0$, and $\lambda = -8.66$. The potential (curves labelled 1 and 1′) is complex and it has a real and an imaginary part. (*b*). For a value of σ between co-rotation and σ_s^+; $\sigma^+ = 1.41$, $\alpha^2 = -0.4406$, and $\lambda = -4.608$. Two of the potentials (curves labelled 2 and 4) are real and belong to $\beta_2 = 1.322$ and $\kappa_2 = \pm 6.096$, respectively; and two of the remaining potentials are complex conjugates and belong to $\kappa_2 = \pm 5.137\,i$ and $\beta_2 = -1.322$; the curve labelled 3 belongs to one of these. (*c*). For $\sigma^+ = \sigma_s^+$;

i.e., only for a/M sufficiently small. From Figs. 43 and 44 it appears that, in general, the real parts of the potentials dominate and are nearly equal, their small differences being 'compensated' by their differing (small) imaginary parts. In any event, as we have shown, the integrals over the imaginary parts must vanish while the integrals over the real parts must be equal.

The potentials for the non-axisymmetric case, for $\sigma^+ > \sigma^+_c = -a/m$, have very much the same behaviours as the potentials for the axisymmetric case.

At co-rotation, when $\sigma^+ = \sigma^+_c$ and $\alpha^2 = 0$ (and, therefore, $\beta^2 = \pm 3\alpha^2 = 0$) we have only two potentials; they are given by

$$V = \frac{\Delta}{r^2(\lambda r^2 + 6Mr - 6a^2)}\left\{\lambda(\lambda+2) + \kappa_2 \frac{\kappa_2 r^2 \Delta - (q'\Delta - \Delta' q)}{r^4(\lambda r^2 + 6Mr - 6a^2)}\right\}, \quad (426)$$

where

$$\kappa_2 = \pm(36M^2 + 24a^2\lambda)^{1/2} \quad \text{and} \quad q = r^2(\lambda r^2 + 6Mr - 6a^2); \quad (427)$$

and depending on whether κ_2 is real or imaginary, they will be real or complex. For $\sigma^+_s < \sigma^+ < \sigma^+_c$, the potentials continue to have the same behaviours. However, when $\sigma^+ \to \sigma^+_s + 0$, they become singular exactly on the horizon; and they lead to unit reflexion coefficients, predicting the onset of super-radiance (see §98 below).

For $\sigma^+ < \sigma^+_s$, the potentials have generic singularities at $r = |\alpha| > r_+$. It can be readily verified that the behaviours of the potentials at $r = |\alpha|$, in these cases, is given by

$$\left.\begin{aligned} V &\simeq -3\frac{\Delta^2_{|\alpha|}}{16|\alpha|^2(r-|\alpha|)^4} \qquad (\beta_2 = +3|\alpha|^2), \\ &\text{and} \\ V &\simeq +5\frac{\Delta_{|\alpha|^2}}{16|\alpha|^2(r-|\alpha|)^4} \qquad (\beta_2 = -3|\alpha|^2), \end{aligned}\right\} \quad (428)$$

i.e., the same behaviours as the potentials for the electromagnetic perturbations (cf. Ch. 8, equation (193)). Besides the singularity at $r = |\alpha|$, it will be observed from Fig. 44(c) that in some cases, one or the other of the potentials exhibits a further singularity; these additional singularities derive from a zero of $q - \beta_2\Delta$, in the denominator of the expression (415) for V.

$\sigma^+ = 0.72$, $\alpha^2 = -1.722$, and $\lambda = -0.478$. Two of the potentials (curve labelled 5 belonging to $\kappa_2 = -3.4941$ and $\beta_2 = 5.166$ and curve labelled 6 belonging to $\kappa_2 = -5.217$ and $\beta_2 = -5.166$) are well-behaved except that they become infinite at the horizon. The remaining two (belonging to $\kappa_2 = 5.217$, $\beta_2 = -5.166$ and $\kappa_2 = 3.491$, $\beta_2 = 5.166$) have singularities arising from a zero of $q - \beta_2\Delta$. (d). For a value of σ inside the super-radiant interval; $\sigma = 0.36$, $\alpha^2 = -4.3465$, and $\lambda = 1.736$. The curve labelled 9 belongs to $\kappa_2 = 11.226$ and $\beta_2 = -13.039$ and the curve labelled 10 belongs to $\kappa_2 = 21.551$ and $\beta_2 = 13.039$; both these potentials have a singularity at $r = |\alpha| = 2.085$. (Note that for the curve labelled 9 the scale of the ordinates is on the right.) The remaining two potentials have singularities arising from zeros of $q - \beta_2\Delta$.

(d) The relation between the solutions belonging to the different potentials

Denoting by $V^j (j = 1, \ldots, 4)$ the different potentials, and by Z^j the solutions belonging to them, we can show (as in earlier similar contexts) with the aid of the transformation equations (412) how, given a solution Z^j belonging to V^j, we can derive a solution Z^i belonging to V^i. Thus, by substituting for Y and $\Lambda_- Y$, in the expression for $K^i Z^i$, from the equations relating them to Z^j and $\Lambda_+ Z^j$, we obtain

$$\begin{aligned} K^i Z^i &= \frac{\varpi^8}{\Delta^2} R^i Y - \frac{\varpi^8}{\Delta^2}(T_1^i + 2i\sigma^+)\Lambda_- Y \\ &= \left[\frac{\varpi^8}{\Delta^2} R^i f^j V^j + (T_1^i + 2i\sigma^+)(\beta_1^j + 2i\sigma^+\beta_2^j)\right] Z^j \\ &\quad + \frac{\varpi^8}{\Delta^2}[R^i(T_1^j + 2i\sigma^+) - R^j(T_1^i + 2i\sigma^+)]\Lambda_+ Z^j. \end{aligned} \tag{429}$$

The expression on the right-hand side of this equation can be simplified by making use of equations (398), (400), and (384) and (387). Thus,

$$\begin{aligned} &\frac{\varpi^8}{\Delta^2}[R^i(T_1^j + 2i\sigma^+) - R^j(T_1^i + 2i\sigma^+)] \\ &\qquad = 2i\sigma^+(\beta_2^i - \beta_2^j) + (\kappa_2^i - \kappa_2^j)\frac{F^2 - \beta_2^2}{(F - \beta_2^i)(F - \beta_2^j)} = D^{ij} \text{ (say)}, \end{aligned} \tag{430}$$

and

$$\begin{aligned} &\frac{\varpi^8}{\Delta^2}\frac{R^i}{R^j} R^j f^j V^j + (T_1^i + 2i\sigma^+)(\beta_1^j + 2i\sigma^+\beta_2^j) \\ &\quad = \frac{R^i}{R^j}[K^j - (\beta_1^j + 2i\sigma^+\beta_2^j)(T_1^j + 2i\sigma^+)] + (T_1^i + 2i\sigma^+)(\beta_1^j + 2i\sigma^+\beta_2^j) \\ &\quad = \frac{R^i}{R^j}K^j - \frac{\beta_1^j + 2i\sigma^+\beta_2^j}{\varpi^8 R^j/\Delta^2}\frac{\varpi^8}{\Delta^2}[R^i(T_1^j + 2i\sigma^+) - R^j(T_1^i + 2i\sigma^+)] \\ &\quad = \frac{R^i}{R^j}K^j - \frac{\beta_1^j + 2i\sigma^+\beta_2^j}{F + \beta_2^j}D^{ij}; \end{aligned} \tag{431}$$

and equation (429) becomes

$$K^i Z^i = \left\{\frac{F + \beta_2^i}{F + \beta_2^j}K^j - \frac{\beta_1^j + 2i\sigma^+\beta_2^j}{F + \beta_2^j}D^{ij}\right\}Z^j + D^{ij}\Lambda_+ Z^j. \tag{432}$$

Again, from equations (392) and (400), we find

$$\frac{\beta_1^j + 2i\sigma^+\beta_2^j}{F + \beta_2^j} = \frac{1}{F^2 - \beta_2^2}[(\kappa_2^j + 2i\sigma^+\beta_2^j)F - \beta_2^j F' - 2i\sigma^+\beta_2^2]. \tag{433}$$

Inserting this expression in equation (432), we finally obtain the relation

$$K^i Z^i = \left\{ \frac{F+\beta_2^i}{F+\beta_2^j} K^j + \left[i\sigma^+ \frac{F-\beta_2^j}{F+\beta_2^j} - \frac{\kappa_2^j F - \beta_2^j F'}{F^2 - \beta_2^2} \right] D^{ij} \right\} Z^j + D^{ij} \frac{dZ^j}{dr_*}. \quad (434)$$

This relation clearly enables us to derive a solution Z^i, belonging to V^i, from a solution Z^j, belonging to V^j.

Since F tends to infinity both for $r \to \infty$ and for $r \to r_+ + 0$, it follows from equation (434) that

$$K^i Z^i \to (K^j + i\sigma^+ D^{ij}) Z^j + D^{ij} Z^j_{,r_*} \quad (r \to \infty; r \to r_+ + 0). \quad (435)$$

This relation can be used to relate the asymptotic behaviours of the solutions belonging to the different potentials. In deriving these relations the following alternative formula for D^{ij} is useful. Since

$$\begin{aligned} K^i - K^j &= -4\sigma^{+2}(\beta_2^i - \beta_2^j) + 2i\sigma^+(\kappa_2^i - \kappa_2^j) \\ &= 2i\sigma^+[2i\sigma^+(\beta_2^i - \beta_2^j) + (\kappa_2^i - \kappa_2^j)], \end{aligned} \quad (436)$$

and $(\beta_2^i)^2 = \beta_2^2 (= 9\alpha^4)$, we can write (cf. equation (430))

$$D^{ij} = \frac{1}{2i\sigma^+}(K^i - K^j) + (\kappa_2^i - \kappa_2^j)(\beta_2^i + \beta_2^j) \frac{1}{(F - \beta_2^j)}. \quad (437)$$

From this equation, it follows that

$$D^{ij} \to (K^i - K^j)/2i\sigma^+ \quad (r \to \infty; r \to r_+ + 0). \quad (438)$$

We have already seen that the solutions of the one-dimensional wave-equations have the asymptotic behaviours, $\exp(\pm i\sigma^+ r_*)$, both for $r \to \infty$ and for $r \to r_+ + 0$. Therefore, if Z^j has the asymptotic behaviours

$$Z^j \to e^{-i\sigma^+ r_*} \quad \text{and} \quad Z^j \to e^{+i\sigma^+ r_*} \quad (r \to \infty; r \to r_+ + 0), \quad (439)$$

then it follows from equations (435) and (438) that the corresponding solutions for Z^i, derived from Z^j, have, respectively, the asymptotic behaviours

$$Z^i \to \frac{K^j}{K^i} e^{-i\sigma r_*} \quad \text{and} \quad Z^i \to e^{+i\sigma^+ r_*} \quad (r \to \infty; r \to r_+ + 0). \quad (440)$$

(*e*) *The asymptotic behaviours of the solutions*

Returning to equations (412) relating the solutions for $Y^{(\pm\sigma^+)}$ and $Z^{(\pm\sigma^+)}$, we can now derive (as in earlier contexts) the asymptotic behaviours of $Y^{(\pm\sigma^+)}$ from the known asymptotic behaviours of $Z^{(\pm\sigma^+)}$. We find

$$\left.\begin{aligned} &Z^{(+\sigma^+)} \to e^{+i\sigma^+ r_*} && (r \to \infty; r \to r_+ + 0): \\ &Y^{(+\sigma^+)}_{+2} \to -4\sigma^{+2} e^{+i\sigma^+ r_*} && (r \to \infty), \\ &Y^{(+\sigma^+)}_{+2} \to -\frac{4\sigma^{+2}}{\sigma^+ - \sigma^+_s}[(\sigma^+ - \sigma^+_s) + 2i\varepsilon_0] e^{+i\sigma^+ r_*} && (r \to r_+ + 0); \end{aligned}\right\} \quad (441)$$

$$\left.\begin{aligned}
&Z^{(+\sigma)} \to e^{-i\sigma r_*} && (r \to \infty;\ r \to r_+ + 0):\\
&Y^{(+\sigma)}_{+2} \to -\frac{K^{(+\sigma)}}{4\sigma^2}\frac{e^{-i\sigma r_*}}{r^4} && (r \to \infty),\\
&Y^{(+\sigma)}_{+2} \to -\frac{K^{(+\sigma)}(\sigma - \sigma_s)^2 \Delta^2 e^{-i\sigma r_*}}{4(\varpi_+^2)^4 \sigma^2[(\sigma - \sigma_s) + 2i\varepsilon_0][(\sigma - \sigma_s) + 4i\varepsilon_0]} && (r \to r_+ + 0);
\end{aligned}\right\} \qquad (442)$$

$$\left.\begin{aligned}
&Z^{(-\sigma^*)} \to e^{+i\sigma r_*} && (r \to \infty;\ r \to r_+ + 0):\\
&Y^{(-\sigma)}_{-2} \to -\frac{K^{(-\sigma)}}{4\sigma^2}\frac{e^{+i\sigma r_*}}{r^4} && (r \to \infty),\\
&Y^{(-\sigma)}_{-2} \to -\frac{K^{(-\sigma)}(\sigma - \sigma_s)^2 \Delta^2 e^{+i\sigma r_*}}{4(\varpi_+^2)^4 \sigma^2[(\sigma - \sigma_s) - 2i\varepsilon_0][(\sigma - \sigma_s) - 4i\varepsilon_0]} && (r \to r_+ + 0);
\end{aligned}\right\} \qquad (443)$$

and

$$\left.\begin{aligned}
&Z^{(-\sigma)} \to e^{-i\sigma r_*} && (r \to \infty;\ r \to r_+ + 0):\\
&Y^{(-\sigma)}_{-2} \to -4\sigma^2 e^{-i\sigma r_*} && (r \to \infty),\\
&Y^{(-\sigma)}_{-2} \to -\frac{4\sigma^2}{\sigma - \sigma_s}[(\sigma - \sigma_s) - 2i\varepsilon_0]e^{-i\sigma r_*} && (r \to r_+ + 0);
\end{aligned}\right\} \qquad (444)$$

where

$$\varepsilon_0 = \frac{r_+ - M}{2(r_+^2 + a^2)} = \frac{\sqrt{(M^2 - a^2)}}{4Mr_+}. \qquad (445)$$

98. The problem of reflexion and transmission

With the reduction of Teukolsky's radial equations to the form of one-dimensional wave-equations, we can now complete our consideration of the problem of reflexion and transmission of incident gravitational waves along the same lines as in §§75 and 76 for incident electromagnetic waves. The principal difference in the two cases is that we now have four potentials instead of two—a difference that is inconsequential in the present context since, as we shall see presently, all four potentials yield the same reflexion and transmission coefficients.

Since the potentials can be complex, it is convenient to have a formulation of the problem of reflexion and transmission which will apply, equally, to real and to complex potentials.

Let $Z^{(\pm\sigma)}$ denote solutions of a one-dimensional wave-equation, with a bounded short-range potential V (real or complex), which have the asymptotic behaviours

$$\begin{aligned}
Z^{(\pm\sigma)} &\to e^{\pm i\sigma r_*} + A^{(\pm\sigma)}e^{\mp i\sigma r_*} && (r_* \to +\infty)\\
&\to B^{(\pm\sigma)}e^{\pm i\sigma r_*} && (r_* \to -\infty).
\end{aligned} \qquad (446)$$

If the potential V is real, $A^{(+\sigma)}$ and $B^{(+\sigma)}$ will be the complex conjugates of $A^{(-\sigma)}$ and $B^{(-\sigma)}$; but this will not be the case if V is complex. But in all cases, with the *definitions*

$$\mathbb{R} = A^{(+\sigma)}A^{(-\sigma)} \qquad \text{and} \qquad \mathbb{T} = B^{(+\sigma)}B^{(-\sigma)}, \tag{447}$$

we will have the conservation law

$$\mathbb{R}+\mathbb{T} = 1; \tag{448}$$

but the reality of $\mathbb{R}$ and $\mathbb{T}$ is not in general guaranteed.

We shall now prove two theorems with respect to the particular potentials which characterize the problem on hand.

For the sake of convenience, we shall restrict ourselves to the case $\sigma > \sigma_s$ when the $r_*(r)$-relation is monotonic and single-valued and the potentials are bounded and of short range. The generalization to include $\sigma < \sigma_s$ requires only some additional remarks concerning how the singularities in V, which then occur, are to be crossed (cf. Ch. 8, §75(c)).

THEOREM 1. *All potentials yield the same reflexion and transmission coefficients.*

Proof. Consider the solutions $Z^{(j,\pm\sigma)}$, belonging to V^j, whose asymptotic behaviours are

$$\left.\begin{aligned} Z^{(j,\pm\sigma)} &\to C^{(j,\pm\sigma)}e^{\pm i\sigma r_*} + A^{(j,\pm\sigma)}e^{\mp i\sigma r_*} \qquad && (r_* \to +\infty) \\ &\to \qquad\qquad\qquad B^{(j,\pm\sigma)}e^{\pm i\sigma r_*} && (r_* \to -\infty). \end{aligned}\right\} \tag{449}$$

According to the relations (439) and (440) established in §97(d), the solutions for $Z^{(i,\pm\sigma)}$, belonging to V^i, derived from solutions belonging to V^j with the foregoing asymptotic behaviours, will have similar behaviours with coefficients given by

$$\left.\begin{aligned} &C^{(i,+\sigma)} = C^{(j,+\sigma)}, \qquad B^{(i,+\sigma)} = B^{(j,+\sigma)}, \qquad A^{(i,+\sigma)} = \frac{K^j}{K^i}A^{(j,+\sigma)}; \\ &C^{(i,-\sigma)} = \frac{K^j}{K^i}C^{(j,-\sigma)}, \quad B^{(i,-\sigma)} = \frac{K^j}{K^i}B^{(j,-\sigma)}, \quad A^{(i,-\sigma)} = A^{(j,-\sigma)}. \end{aligned}\right\} \tag{450}$$

Therefore,

$$\mathbb{R} = \frac{A^{(i,+\sigma)}A^{(i,-\sigma)}}{C^{(i,+\sigma)}C^{(i,-\sigma)}} = \frac{A^{(j,+\sigma)}A^{(j,-\sigma)}}{C^{(j,+\sigma)}C^{(j,-\sigma)}}, \tag{451}$$

and

$$\mathbb{T} = \frac{B^{(i,+\sigma)}B^{(i,-\sigma)}}{C^{(i,+\sigma)}C^{(i,-\sigma)}} = \frac{B^{(j,+\sigma)}B^{(j,-\sigma)}}{C^{(j,+\sigma)}C^{(j,-\sigma)}}; \tag{452}$$

and the theorem stated follows.

THEOREM 2. *If $Z^{(j,\pm\sigma)}$ are solutions, belonging to a complex potential V^j, with the asymptotic behaviours*

$$\left.\begin{aligned} Z^{(j,\pm\sigma)} &\to e^{\pm i\sigma r_*} + A^{(j,\pm\sigma)} e^{\mp i\sigma r_*} && (r_* \to +\infty), \\ &\to B^{(j,\pm\sigma)} e^{\pm i\sigma r_*} && (r_* \to -\infty), \end{aligned}\right\} \tag{453}$$

then the reflexion and the transmission coefficients are given by

$$\mathbb{R} = |A^{(j,+\sigma)}|^2 \frac{[K^{(j,+\sigma)}]^2}{|\mathscr{C}|^2} = |A^{(j,-\sigma)}|^2 \frac{[K^{(j,-\sigma)}]^2}{|\mathscr{C}|^2},$$

and

$$\mathbb{T} = |B^{(j,+\sigma)}|^2 = |B^{(j,-\sigma)}|^2, \tag{454}$$

where $|\mathscr{C}|^2 = K^{(j,+\sigma)} K^{(j,-\sigma)}$ is the absolute square of the Starobinsky constant.

Proof. First, we observe that the potential V^j can be complex if, and only if, κ_2 given by equation (406) is purely imaginary; and in this case, we can write

$$\kappa_2 = \pm ik_2 \quad \text{where } k_2 \text{ is real.} \tag{455}$$

Let V^j and V^i be the potentials belonging to $-ik_2$ and $+ik_2$, respectively, and the same value of β_2. The potentials V^j and V^i are then complex-conjugate functions and the associated constants K^j and K^i are

$$\left.\begin{aligned} K^{(j,\sigma)} &= [\lambda(\lambda+2) - 4\sigma^2\beta_2] + 2\sigma k_2 \\ \text{and} \quad K^{(i,\sigma)} &= [\lambda(\lambda+2) - 4\sigma^2\beta_2] - 2\sigma k_2 = K^{(j,-\sigma)}, \end{aligned}\right\} \tag{456}$$

and are real.

Since V^i and V^j are complex-conjugate functions, the equations satisfied by Z^i and the complex-conjugate of Z^j are the same. Accordingly, the solution, belonging to V^i, with the asymptotic behaviours

$$\left.\begin{aligned} Z^{(i,-\sigma)} &\to e^{-i\sigma r_*} + A^{(j,-\sigma)} \frac{K^i}{K^j} e^{+i\sigma r_*} && (r_* \to +\infty), \\ &\to B^{(j,-\sigma)} e^{-i\sigma r_*} && (r_* \to -\infty) \end{aligned}\right\} \tag{457}$$

will be the complex-conjugate of the solution $Z^{(j,+\sigma)}$ with the asymptotic behaviour specified in equation (453). (One verifies this fact by considering the solution with the asymptotic behaviour of $Z^{(j,-\sigma)}$ and deriving from it a solution $Z^{(i,-\sigma)}$ in conformity with the relations (450), and multiplying by K^i/K^j to make the coefficient of $e^{-i\sigma r_*}$, in its asymptotic behaviour for $r_* \to \infty$, unity.) Therefore (cf. equation (456)),

$$[A^{(j,+\sigma)}]^* = A^{(j,-\sigma)} \frac{K^i}{K^j} = A^{(j,-\sigma)} \frac{K^{(j,-\sigma)}}{K^{(j,+\sigma)}}, \tag{458}$$

and

$$[B^{(j,\,+\sigma)}]^* = B^{(j,\,-\sigma)}. \tag{459}$$

By virtue of these relations,

$$\mathbb{R} = A^{(j,\,+\sigma)}A^{(j,\,-\sigma)} = |A^{(j,\,+\sigma)}|^2\frac{K^{(j,\,+\sigma)}}{K^{(j,\,-\sigma)}}$$

$$= |A^{(j,\,+\sigma)}|^2\frac{[K^{(j,\,+\sigma)}]^2}{|\mathscr{C}|^2} = |A^{(j,\,-\sigma)}|^2\frac{[K^{(j,\,-\sigma)}]^2}{|\mathscr{C}|^2}, \tag{460}$$

and

$$\mathbb{T} = |B^{(j,\,+\sigma)}|^2 = |B^{(j,\,-\sigma)}|^2. \tag{461}$$

This completes the proof of the theorem.

COROLLARY. *Quite generally, we may write*

$$\mathbb{R} = |A^{(j,\,+\sigma)}|^2\frac{[K^{(j,\,+\sigma)}]\,[K^{(j,\,+\sigma)}]^*}{|\mathscr{C}|^2} \quad \textit{and} \quad \mathbb{T} = |B^{(j,\,+\sigma)}|^2. \tag{462}$$

The relations are manifest, since when V^j is real K^j is complex and when V^j is complex,

$$[K^{(j,\,+\sigma)}]^* = K^{(j,\,+\sigma)} \qquad (V^j \text{ complex}). \tag{463}$$

It is now clear that the different cases, $\sigma \geqslant \sigma_c$, $\sigma_c > \sigma > \sigma_s$, and $\sigma < \sigma_s$, can be treated exactly as in Chapter 8, §75; in particular, the phenomenon of super-radiance emerges as naturally.

In Table X, we list the reflexion and transmission coefficients for gravitational waves of various frequencies incident on a Kerr black-hole with $a = 0.95$.

(*a*) *The expression of* $\mathbb{R}$ *and* $\mathbb{T}$ *in terms of solutions of Teukolsky's equations with appropriate boundary conditions*

First, we recall that (cf. equation (6))

$$\left.\begin{aligned} \Psi_0 &= \Phi_0 = R_{+2}S_{+2}e^{i(\sigma t + m\varphi)} \\ \text{and} \quad (\bar\rho^*)^4\Psi_4 &= \Phi_4 = R_{-2}S_{-2}e^{i(\sigma t + m\varphi)}, \end{aligned}\right\} \tag{464}$$

where we have restored the t- and the φ-dependent factors. The Teukolsky functions, R_{+2} and R_{-2}, are in turn related to $Y^{(+\sigma)}_{+2}$ and $Y^{(-\sigma)}_{-2}$ by (cf. equation (379))

$$R_{+2} = \frac{(\varpi^2)^{3/2}}{\Delta^2}Y^{(+\sigma)}_{+2} \quad \text{and} \quad R_{-2} = (\varpi^2)^{3/2}Y^{(-\sigma)}_{-2}. \tag{465}$$

Since all four potentials lead to the same reflexion and transmission coefficients, we shall not distinguish the solutions by any superscripts. Also, to avoid ambiguities, we shall explicitly restrict ourselves to the case when the potential is complex and $\sigma^{+} > \sigma^{+}_{s}$. The restrictions do not imply any essential loss of generality: the modifications necessary to allow for $\alpha^2 < 0$ and real potentials are minor and mostly *pro forma*. And one can readily verify that the final results are not affected.

With the aid of the relations (441)–(444), listed in §97(*e*), we can write down the asymptotic behaviours of the solutions for R_{+2} and R_{-2} which follow from equations (412) of the transformation theory when the solutions for $Z^{(\pm\sigma^{+})}$ have the asymptotic behaviours specified in equation (446). We find

$$\left.\begin{aligned} R_{+2} &\to R^{(\text{inc})}_{+2}\frac{e^{+i\sigma^{+}r_*}}{r} + R^{(\text{ref})}_{+2}\frac{e^{-i\sigma^{+}r_*}}{r^5} && (r\to\infty),\\ &\to R^{(\text{trans})}_{+2}\frac{e^{+i\sigma^{+}r_*}}{\Delta^2} && (r\to r_+ + 0);\end{aligned}\right\}\qquad(466)$$

and

$$\left.\begin{aligned} R_{-2} &\to R^{(\text{inc})}_{-2}\frac{e^{+i\sigma^{+}r_*}}{r} + R^{(\text{ref})}_{-2}r^3 e^{-i\sigma^{+}r_*} && (r\to\infty),\\ &\to R^{(\text{trans})}_{-2}\Delta^2 e^{+i\sigma^{+}r_*} && (r\to r_+ + 0);\end{aligned}\right\}\qquad(467)$$

where

$$\left.\begin{aligned} R^{(\text{inc})}_{+2} &= -4\sigma^{+2}; \qquad R^{(\text{inc})}_{-2} = -\frac{K^{(-\sigma^{+})}}{4\sigma^{+2}},\\ R^{(\text{ref})}_{+2} &= -\frac{K^{(+\sigma^{+})}}{4\sigma^{+2}}A^{(+\sigma^{+})}; \quad R^{(\text{ref})}_{-2} = -4\sigma^{+2}A^{(-\sigma^{+})},\\ R^{(\text{trans})}_{+2} &= -4\frac{(\varpi_+^2)^{3/2}}{\sigma^{+}-\sigma^{+}_{s}}\sigma^{+2}\left[(\sigma^{+}-\sigma^{+}_{s})+2i\varepsilon_0\right]B^{(+\sigma^{+})};\\ R^{(\text{trans})}_{-2} &= -\frac{K^{(-\sigma^{+})}(\sigma^{+}-\sigma^{+}_{s})^2 B^{(-\sigma^{+})}}{4(\varpi_+^2)^{5/2}\sigma^{+2}\left[(\sigma^{+}-\sigma^{+}_{s})-2i\varepsilon_0\right]\left[(\sigma^{+}-\sigma^{+}_{s})-4i\varepsilon_0\right]}.\end{aligned}\right\}\qquad(468)$$

It now follows from the foregoing relations that

$$\left|\frac{R^{(\text{ref})}_{+2}}{R^{(\text{inc})}_{+2}}\right|^2 = \frac{[K^{(+\sigma^{+})}]^2|A^{(+\sigma^{+})}|^2}{256\sigma^{+8}} \quad\text{and}\quad \left|\frac{R^{(\text{ref})}_{-2}}{R^{(\text{inc})}_{-2}}\right|^2 = \frac{256\sigma^{+8}}{[K^{(-\sigma^{+})}]^2}|A^{(-\sigma^{+})}|^2, \qquad(469)$$

or, by virtue of the relation (458) between $A^{(+\sigma^{+})}$ and $A^{(-\sigma^{+})}$,

$$\left|\frac{R^{(\text{ref})}_{+2}}{R^{(\text{inc})}_{+2}}\right|^2 = \frac{|\mathscr{C}|^2}{256\sigma^{+8}}A^{(+\sigma^{+})}A^{(-\sigma^{+})} \quad\text{and}\quad \left|\frac{R^{(\text{ref})}_{-2}}{R^{(\text{inc})}_{-2}}\right|^2 = \frac{256\sigma^{+8}}{|\mathscr{C}|^2}A^{(+\sigma^{+})}A^{(-\sigma^{+})}. \qquad(470)$$

TABLE X

Reflexion coefficients for gravitational waves incident on a Kerr black-hole with $a = 0.95$

$(l = 2, m = -2)$

σ^+	σ^+/σ^+_s	R	σ^+	σ^+/σ^+_s	R
0.50	0.6907	1.04422	0.77	1.0636	0.27057
0.55	0.7597	1.07031	0.78	1.0774	0.16754
0.60	0.8288	1.10693	0.79	1.0912	0.10007
0.65	0.8979	1.15358	0.80	1.1051	0.05844
0.70	0.9669	1.15101	0.81	1.1189	0.03372
0.73	1.0084	0.92530	0.82	1.1327	0.01932
0.74	1.0222	0.76882	0.83	1.1465	0.01103
0.75	1.0360	0.58828	0.84	1.1603	0.00628
0.76	1.0498	0.41376	0.85	1.1741	0.00357

By equation (447), the reflexion coefficient is given by either of the two formulae:

$$\mathbb{R} = \frac{256\sigma^{+8}}{|\mathscr{C}|^2}\left|\frac{R^{(\text{ref})}_{+2}}{R^{(\text{inc})}_{+2}}\right|^2 \quad \text{and} \quad \mathbb{R} = \frac{|\mathscr{C}|^2}{256\sigma^{+8}}\left|\frac{R^{(\text{ref})}_{-2}}{R^{(\text{inc})}_{-2}}\right|^2. \tag{471}$$

Similarly, for the transmission coefficient, we find the two formulae

$$\mathbb{T} = |B^{(+\sigma^+)}|^2 = \frac{\sigma^{+3}}{(2Mr_+)^3\,(\sigma^+ - \sigma^+_s)\,[(\sigma^+ - \sigma^+_s)^2 + 4\varepsilon_0^2]}\left|\frac{R^{(\text{trans})}_{+2}}{R^{(\text{inc})}_{+2}}\right|^2 \tag{472}$$

and

$$\mathbb{T} = |B^{(-\sigma^+)}|^2$$
$$= \frac{(2Mr_+)^5}{\sigma^{+5}}(\sigma^+ - \sigma^+_s)\,[(\sigma^+ - \sigma^+_s)^2 + 4\varepsilon_0^2]\,[(\sigma^+ - \sigma^+_s)^2 + 16\varepsilon_0^2]\left|\frac{R^{(\text{trans})}_{-2}}{R^{(\text{inc})}_{-2}}\right|^2. \tag{473}$$

The conservation law (448) now requires that

$$|R^{(\text{inc})}_{+2}|^2 = \frac{256\sigma^{+8}}{|\mathscr{C}|^2}|R^{(\text{ref})}_{+2}|^2 + \frac{\sigma^{+3}}{(2Mr_+)^3\,(\sigma^+ - \sigma^+_s)\,[(\sigma^+ - \sigma^+_s)^2 + 4\varepsilon_0^2]}|R^{(\text{trans})}_{+2}|^2, \tag{474}$$

and

$$|R^{(\text{inc})}_{-2}|^2 = \frac{|\mathscr{C}|^2}{256\sigma^{+8}}|R^{(\text{ref})}_{-2}|^2 + \frac{(2Mr_+)^5}{\sigma^{+5}}(\sigma^+ - \sigma^+_s)[(\sigma^+ - \sigma^+_s)^2 + 4\varepsilon_0^2]\,[(\sigma^+ - \sigma^+_s)^2 + 16\varepsilon_0^2]\,|R^{(\text{trans})}_{-2}|^2. \tag{475}$$

While we have explicitly restricted ourselves to $\alpha^2 > 0$ and complex potentials, it can be readily verified that equations (471)–(473) are in fact valid without the restrictions. And we observe that in agreement with what we find from the one-dimensional wave-equations that

$$T < 0 \qquad \text{for} \qquad \sigma^{+} < \sigma^{+}_{s}, \tag{476}$$

i.e., we have super-radiance as predicted.

As in §76(*a*), we can obtain alternative expressions for the reflexion and the transmission coefficients by exploiting unitarity which obtains equally in the present problem. Thus, seeking solutions of Teukolsky's equations with the asymptotic behaviours

$$\left.\begin{aligned} R_{+2} &\to R^{(\text{inc})}_{+2} e^{-i\sigma^{+} r_*} + R^{(\text{ref})}_{+2} \frac{e^{+i\sigma^{+} r_*}}{\Delta^2} && (r \to r_+ + 0), \\ &\to \qquad R^{(\text{trans})}_{+2} \frac{e^{-i\sigma^{+} r_*}}{r^3} && (r \to \infty); \end{aligned}\right\} \tag{477}$$

and

$$\left.\begin{aligned} R_{-2} &\to R^{(\text{inc})}_{-2} e^{-i\sigma^{+} r_*} + R^{(\text{ref})}_{-2} \Delta^2 e^{+i\sigma^{+} r_*} && (r \to r_+ + 0), \\ &\to \qquad R^{(\text{trans})}_{-2} r^3 e^{-i\sigma^{+} r_*} && (r \to \infty), \end{aligned}\right\} \tag{478}$$

we find

$$\begin{aligned} \mathbb{R} &= \frac{|\mathscr{C}|^2}{256\,(2Mr_+)^8\,(\sigma^{+}-\sigma^{+}_s)^2\,[(\sigma^{+}-\sigma^{+}_s)^2+4\varepsilon_0^2]^2\,[(\sigma^{+}-\sigma^{+}_s)^2+16\varepsilon_0^2]} \left|\frac{R^{(\text{ref})}_{+2}}{R^{(\text{inc})}_{+2}}\right|^2, \\ &= \frac{256}{|\mathscr{C}|^2}(2Mr_+)^8\,(\sigma^{+}-\sigma^{+}_s)^2\,[(\sigma^{+}-\sigma^{+}_s)^2+4\varepsilon_0^2]^2\,[(\sigma^{+}-\sigma^{+}_s)^2+16\varepsilon_0^2] \left|\frac{R^{(\text{ref})}_{-2}}{R^{(\text{inc})}_{-2}}\right|^2 \end{aligned} \tag{479}$$

and

$$\begin{aligned} \mathbb{T} &= \frac{\sigma^{+5}}{(2Mr_+)^5\,(\sigma^{+}-\sigma^{+}_s)\,[(\sigma^{+}-\sigma^{+}_s)^2+4\varepsilon_0^2]\,[(\sigma^{+}-\sigma^{+}_s)^2+16\varepsilon_0^2]} \left|\frac{R^{(\text{trans})}_{+2}}{R^{(\text{inc})}_{+2}}\right|^2 \\ &= (2Mr_+)^3\,[(\sigma^{+}-\sigma^{+}_s)^2+4\varepsilon_0^2]\,\frac{\sigma^{+}-\sigma^{+}_s}{\sigma^{+3}} \left|\frac{R^{(\text{trans})}_{-2}}{R^{(\text{inc})}_{-2}}\right|^2. \end{aligned} \tag{480}$$

In the Schwarzschild limit ($a = 0$), the expressions for $\mathbb{R}$ and $\mathbb{T}$ given in equations (471)–(473), (479), and (480) reduce to those given in Chapter 4 (equations (371), (372), (377), and (378)).

(*b*) *A direct evaluation of the flux of radiation at infinity*

It remains to justify that the reflexion and the transmission coefficients we have derived in §(*a*) above have the physical meanings that were natural to attribute to them. The problem that we are now presented with is somewhat more subtle than the one we considered in Chapter 8, §76(*b*): we do not now have a unique and a generally applicable definition of an energy-momentum tensor for the gravitational field as we have for the electromagnetic field. However, if the background space-time is stationary and asymptotically flat, then, for the perturbed space-time, there are a number of ways in which we can isolate quantities which have the requisite physical meanings to be interpreted as the incident and the reflected fluxes of gravitational energy. It will take us too far afield to go into these matters with the detail necessary for a rigorous treatment. We shall, therefore, be content to outline an argument which can be sustained by a more careful discussion.

It can be shown that with a suitable choice of gauge—the de Donder or the harmonic gauge—Einstein's equations for the vacuum, linearized about the flat Minkowskian space-time, allow plane-wave solutions with the line-element

$$ds^2 = (dx^0)^2 - (dx^2)^2 - (1-h_{11})(dx^1)^2 - (1+h_{11})(dx^3)^2 + 2h_{31}\,dx^3\,dx^1, \tag{481}$$

where, by assumption, h_{11} and h_{31} are small quantities of the first order which are functions of the argument, $x^0 \pm x^2$:

$$h_{11} = h_{11}(x^0 \pm x^2) \qquad \text{and} \qquad h_{31} = h_{31}(x^0 \pm x^2); \tag{482}$$

in other words, solutions representing plane waves progressing in the inward (+sign) or the outward (−sign) directions. Indeed, with the chosen form for the metric, the only non-vanishing components of the Reimann tensor, evaluated with the aid of the formula

$$\delta R_{ijkl} = \frac{1}{2}\left(\frac{\partial^2 h_{il}}{\partial x^j \partial x^k} + \frac{\partial^2 h_{jk}}{\partial x^i \partial x^l} - \frac{\partial^2 h_{ik}}{\partial x^j \partial x^l} - \frac{\partial^2 h_{jl}}{\partial x^i \partial x^k}\right) + O(h^2) \tag{483}$$

are:

$$\left.\begin{aligned}
\delta R_{0303} &= \delta R_{2323} = -\delta R_{0101} = -\delta R_{2121} = \tfrac{1}{2}\ddot{h}_{11},\\
\delta R_{0301} &= \delta R_{2321} = -\tfrac{1}{2}\ddot{h}_{31},\\
\delta R_{0323} &= -\delta R_{0121} = \pm\tfrac{1}{2}\ddot{h}_{11},\\
\delta R_{2301} &= \delta R_{0321} = \mp\tfrac{1}{2}\ddot{h}_{31},
\end{aligned}\right\} \tag{484}$$

where the dots denote differentiations with respect to the argument $x^0 \pm x^2$; and by contracting the Riemann tensor appropriately to obtain the components of the Ricci tensor, we find that they all vanish, confirming that the metric, defined by equations (481) and (482), does satisfy Einstein's vacuum equations.

We can now evaluate the Weyl scalars, Ψ_0 and Ψ_4, belonging to the solution (481), with respect to the null basis (cf. Ch. 4, equation (333)),

$$\boldsymbol{l}^i = (1, 0, 1, 0), \quad \boldsymbol{n}^i = \tfrac{1}{2}(1, 0, -1, 0) \quad \text{and}$$

$$\boldsymbol{m}^i = \frac{1}{\sqrt{2}}(0, i, 0, 1). \tag{485}$$

For this basis (cf. Ch. 4, equation (334))

$$\begin{aligned}\Psi_0 = & -i(\delta R_{0301}+\delta R_{2321}+\delta R_{2301}+\delta R_{0321}) \\ & -\tfrac{1}{2}(\delta R_{0303}+2\delta R_{0323}+\delta R_{2323} \\ & \qquad -\delta R_{0101}-2\delta R_{0121}-\delta R_{2121});\end{aligned} \tag{486}$$

and inserting for the components of the Riemann tensor from equations (484), we find

$$\left.\begin{aligned} &\Psi_0 = 0 \quad \text{for outgoing waves} \\ \text{and} \\ &\Psi_0 = -2(\ddot{h}_{11}-i\ddot{h}_{31}) \quad \text{for ingoing waves.}\end{aligned}\right\} \tag{487}$$

Similarly, we find

$$\left.\begin{aligned} &\Psi_4 = -\tfrac{1}{2}(\ddot{h}_{11}+i\ddot{h}_{31}) \quad \text{for outgoing waves} \\ \text{and} \\ &\Psi_4 = 0 \quad \text{for ingoing waves.}\end{aligned}\right\} \tag{488}$$

We observe that while Ψ_0 is non-vanishing only for ingoing waves, Ψ_4 is non-vanishing only for outgoing waves,

The flux of energy per unit area, associated with the plane waves

$$e^{i\sigma^+(x^0 \pm x^2)}, \tag{489}$$

is given by

$$\mathscr{F} = \tfrac{1}{4}\sigma^{+2}(|h_{11}|^2+|h_{31}|^2). \tag{490}$$

The corresponding expressions, in terms of the Weyl scalars, for the fluxes of energy in the ingoing and in the outgoing waves, are

$$\mathscr{F}^{(\text{in})} = \frac{1}{16\sigma^{+2}}|\Psi_0^{(\text{in})}|^2 \qquad \text{and} \qquad \mathscr{F}^{(\text{out})} = \frac{1}{\sigma^{+2}}|\Psi_4^{(\text{out})}|^2. \tag{491}$$

The foregoing analysis applies to plane waves. It is, however, clear that the analysis will apply to spherical waves (locally) progressing radially inward (in the direction of incidence) or outward (in the direction of reflexion) in asymptotically flat regions. We should then write, in place of equation (490),

$$\frac{\mathrm{d}^2 E}{\mathrm{d}t\,\mathrm{d}\Omega} = \frac{\sigma^{+2}}{16\pi}\lim_{r\to\infty} r^2\,(|h_{11}|^2+|h_{31}|^2). \tag{492}$$

The corresponding expressions, in terms of the Weyl scalars, for the energies in the incident and in the reflected gravitational waves, are

$$\frac{\mathrm{d}^2 E^{(\text{inc})}}{\mathrm{d}t\,\mathrm{d}\Omega} = \frac{1}{64\pi\sigma^{+2}}\lim_{r\to\infty} r^2\,|\Psi_0^{(\text{inc})}|^2 \tag{493}$$

and

$$\frac{\mathrm{d}^2 E^{(\text{ref})}}{\mathrm{d}t\,\mathrm{d}\Omega} = \frac{1}{4\pi\sigma^{+2}}\lim_{r\to\infty} r^2\,|\Psi_4^{(\text{ref})}|^2. \tag{494}$$

For Ψ_0 and Ψ_4 we now have the solutions (cf. equation (148))

$$\left.\begin{aligned} \Psi_0 &= R_{+2}S_{+2}e^{i(\sigma t+m\varphi)} \\ \text{and}\qquad 4(\bar{\rho}^*)^4\Psi_4 &= R_{-2}S_{-2}e^{i(\sigma t+m\varphi)}, \end{aligned}\right\} \tag{495}$$

where the angular functions, S_{+2} and S_{-2}, are normalized to unity and R_{+2} and R_{-2} are relatively normalized to be in accord with equations (41) and (42)—a full knowledge which is a prerequisite to the use of the solutions (495).

With the solutions for Ψ_0 and Ψ_4 given in equations (495), equations (493) and (494) give

$$\frac{d^2E^{(\text{inc})}}{dt\,d\Omega} = \frac{S_{+2}^2}{64\pi\sigma^2}\lim_{r\to\infty} r^2[R_{+2}^{(+\sigma)}R_{+2}^{(-\sigma)}] \tag{496}$$

and

$$\frac{d^2E^{(\text{ref})}}{dt\,d\Omega} = \frac{S_{-2}^2}{64\pi\sigma^2}\lim_{r\to\infty} \frac{1}{r^6}[R_{-2}^{(+\sigma)}R_{-2}^{(-\sigma)}]. \tag{497}$$

Equations (466) and (467) show that Teukolsky's equations do allow solutions for which the limits (496) and (497) exist. Indeed, with the definitions of the coefficients $R_{+2}^{(\text{inc})}$ and $R_{-2}^{(\text{ref})}$ in (466) and (467), we can now write

$$\frac{d^2E^{(\text{inc})}}{dt\,d\Omega} = \frac{S_{+2}^2}{64\pi\sigma^2}[R_{+2}^{(\text{inc}+\sigma)}R_{+2}^{(\text{inc}-\sigma)}] \tag{498}$$

and

$$\frac{d^2E^{(\text{ref})}}{dt\,d\Omega} = \frac{S_{-2}^2}{64\pi\sigma^2}[R_{-2}^{(\text{ref}+\sigma)}R_{-2}^{(\text{ref}-\sigma)}]. \tag{499}$$

Inserting for the coefficients, $R_{+2}^{(\text{inc}+\sigma)}$, etc., listed in equations (468), we find from the foregoing equations, after integration over the angles, that the reflexion coefficient is given by

$$\mathbb{R} = A^{(+\sigma)}A^{(-\sigma)}, \tag{500}$$

in agreement with the expression (447) we derived earlier from considerations pertaining to the one-dimensional wave-equations satisfied by $Z^{(\pm\sigma)}$.

(c) *The flow of energy across the event horizon*

It can be shown quite generally that one can define, without any essential ambiguity, a conserved energy-momentum tensor for the gravitational field pervading a space-time perturbed about a stationary background. Consequently, from the fact that the expression for the reflexion coefficient, $\mathbb{R}$, does represent the fraction of an incident flux of gravitational energy that is reflected, we can conclude that the transmission coefficient, $\mathbb{T}$, which is consistent with the requirement (448) for the conservation of energy, necessarily represents the fraction of the incident energy that crosses the event horizon and is absorbed by the black hole. Nevertheless, it is of interest to verify that the expression we have obtained for the rate of flow of energy across the event horizon is consistent with Hawking's area-theorem (cf. Ch. 7, §65 (*d*)) that a flow of energy across the event horizon is directly related to the rate of increase of the surface area of the horizon.

Now, the surface area, Σ, of the horizon of the Kerr black-hole, expressed in terms of its mass, M, and the angular momentum L_z ($= aM$) about its axis of rotation, is

$$\Sigma = 4\pi(r_+^2 + a^2) = 8\pi[M^2 + \surd(M^4 - L_z^2)]. \tag{501}$$

Therefore, the change in area, $\mathrm{d}\Sigma$, consequent to changes $\mathrm{d}M$, in its mass, and $\mathrm{d}L_z$, in its angular momentum, is

$$\mathrm{d}\Sigma = \frac{8\pi}{\surd(M^4 - L_z^2)}\left\{2[M^2 + \surd(M^4 - L_z^2)]\, M\mathrm{d}M - L_z\mathrm{d}L_z\right\} \tag{502}$$

or

$$\mathrm{d}\Sigma = \frac{8\pi}{M\surd(M^2 - a^2)}(2M^2 r_+\mathrm{d}M - L_z\mathrm{d}L_z). \tag{503}$$

Since, in our present context (cf. Ch. 8, equation 253)),

$$\mathrm{d}L_z = -m\mathrm{d}M/\sigma \tag{504}$$

(minus, because $\mathrm{d}L_z$ now refers to the change in the angular momentum of the black hole, not that of the field), it follows from equation (503) that

$$\mathrm{d}\Sigma = \frac{4\pi}{\varepsilon_0}(1 - \sigma_s/\sigma)\,\mathrm{d}M, \tag{505}$$

where (cf. equation (445))

$$\varepsilon_0 = \frac{\surd(M^2 - a^2)}{4Mr_+}. \tag{506}$$

We establish in § (*d*), below, the formula

$$\mathrm{d}\Sigma = \frac{2Mr_+}{\varepsilon_0}|\sigma_{r_+}^{HH}|^2, \tag{507}$$

where $\sigma_{r_+}^{HH}$ denotes the first-order perturbation in the spin coefficient, σ, at the horizon, in the Hawking–Hartle basis (Ch. 8, equation (243)). Combining equations (505) and (507), we obtain, in accordance with the area-theorem,

$$\frac{\mathrm{d}^2 E^{(\mathrm{trans})}}{\mathrm{d}t\,\mathrm{d}\Omega} = \frac{Mr_+\sigma}{2\pi(\sigma - \sigma_s)}|\sigma_{r_+}^{HH}|^2. \tag{508}$$

We now evaluate $\sigma_{r_+}^{HH}$ by considering the Ricci identity (Ch. 1, equation (310, b); see also equation (3)),

$$D\sigma - \delta\kappa = \sigma(\tilde{\rho} + \tilde{\rho}^* + 3\varepsilon - \varepsilon^*) - \kappa(\tau - \pi^* + \alpha^* + 3\beta) + \Psi_0, \tag{509}$$

in the Hawking–Hartle basis, as it applies at the horizon. Since the Hawking–Hartle basis is obtained by subjecting our standard basis (Ch. 6, (170)) to a rotation of class III (Ch. 1, §8(g)) with a parameter $A = 2(r^2+a^2)/\Delta$, it follows from Chapter 1, equation (347) that

$$\varepsilon^{HH} = -\tfrac{1}{2}A^{-2}DA = \tfrac{1}{2}DA^{-1} = \frac{1}{4}\frac{\mathrm{d}}{\mathrm{d}r}\frac{\Delta}{r^2+a^2}, \tag{510}$$

(since $\varepsilon = 0$ in the original basis). Therefore, at the horizon,

$$\varepsilon^{HH}_{r_+} = \frac{r_+ - M}{2(r_+^2+a^2)} = \frac{\sqrt{(M^2-a^2)}}{4Mr_+} = \varepsilon_0. \tag{511}$$

Also $\tilde{\rho}^{HH}$ vanishes at the horizon; and $\kappa = 0$ by definition. We thus obtain the equation

$$D^{HH}_{r_+}\sigma^{HH}_{r_+} = 2\varepsilon_0\sigma^{HH}_{r_+} + (\Psi^{HH}_0)_{r_+}. \tag{512}$$

On the other hand, at the horizon (cf. Ch. 8, equations (250) and (252)),

$$D^{HH}_{r_+} = i\sigma + i\frac{am}{2Mr_+} = i(\sigma - \sigma_s). \tag{513}$$

Therefore,

$$\sigma^{HH}_{r_+} = \frac{(\Psi^{HH}_0)_{r_+}}{i[(\sigma-\sigma_s)+2i\varepsilon_0]}. \tag{514}$$

We also have

$$(\Psi^{HH}_0)_{r_+} = \left[\frac{\Delta^2}{4(r^2+a^2)^2}\Psi_0\right]_{r_+} = \frac{(\Delta^2\Psi_0)_{r_+}}{4(2Mr_+)^2}. \tag{515}$$

Inserting the expressions (514) and (515) in equation (508), we obtain

$$\frac{\mathrm{d}^2E^{(\mathrm{trans})}}{\mathrm{d}t\,\mathrm{d}\Omega} = \frac{1}{64\pi}\frac{1}{(2Mr_+)^3[(\sigma-\sigma_s)^2+4\varepsilon_0^2]}\frac{\sigma}{(\sigma-\sigma_s)}|\Delta^2\Psi_0|^2_{r_+}. \tag{516}$$

Substituting for Ψ_0 its solution, we have

$$\frac{\mathrm{d}^2E^{(\mathrm{trans})}}{\mathrm{d}t\,\mathrm{d}\Omega} = \frac{S^2_{+2}}{64\pi}\frac{1}{(2Mr_+)^3[(\sigma-\sigma_s)^2+4\varepsilon_0^2]}\frac{\sigma}{(\sigma-\sigma_s)}|\Delta^2R_{+2}|^2_{r_+}, \tag{517}$$

or, expressing the asymptotic behaviour of the solution R_{+2} at the horizon in the manner we have in (466), we finally obtain

$$\frac{\mathrm{d}^2E^{(\mathrm{trans})}}{\mathrm{d}t\,\mathrm{d}\Omega} = \frac{S^2_{+2}}{64\pi}\frac{1}{(2Mr_+)^3[(\sigma-\sigma_s)^2+4\varepsilon_0^2]}\frac{\sigma}{(\sigma-\sigma_s)}|R^{(\mathrm{trans})}_{+2}|^2. \tag{518}$$

With the incident flux of energy given by equation (498), equation (518) provides an expression for the transmission coefficient which is in agreement with the expression (472) derived earlier.

The justification is now complete that the reflexion and the transmission coefficients derived in § (*a*) have the physical meanings that were attributed to them.

(*d*) *The Hawking–Hartle formula*

The Hawking–Hartle formula (507) follows from the equations for the optical scalars considered in Chapter 1, § 9(*a*).

It will be recalled that the spin coefficients, $\tilde{\rho}$ and σ, measure the convergence and the shear of a null congruence, *l*. If the congruence is geodesic, as well, $\kappa = 0$ and the Ricci identity, Chapter 1, equation (310, a), gives

$$D\tilde{\rho} = (\tilde{\rho}^2 + |\sigma|^2) + (\varepsilon + \varepsilon^*)\tilde{\rho}. \tag{519}$$

We consider this equation on the horizon in the Hawking–Hartle basis. Since we shall be restricting ourselves to this basis exclusively in this section, we shall not distinguish the various quantities by the superscripts "*HH*" as we have in the preceding section.

At the horizon, $\varepsilon = \varepsilon_0$ = a constant (given by equation (511) for the Kerr black-hole) and equation (519) gives

$$D\tilde{\rho} = (\tilde{\rho}^2 + |\sigma|^2) + 2\varepsilon_0\tilde{\rho}, \tag{520}$$

where $D = \mathrm{d}/\mathrm{d}v$ (as in Ch. 8, equation (242)).

Since $\tilde{\rho}$ and σ both vanish on the horizon in the stationary state, equation (520), linearized to the first order in the perturbation, gives

$$D\tilde{\rho}^{(1)} = 2\varepsilon_0\tilde{\rho}^{(1)}. \tag{521}$$

The only admissible solution of this equation is

$$\tilde{\rho}^{(1)} = 0, \tag{522}$$

since $\tilde{\rho}^{(1)}$ (and $\sigma^{(1)}$) must be periodic in v for a rotating black-hole; and equation (521) allows no such solution. We must accordingly consider equation (520) expanded to the second order, when we obtain

$$D\tilde{\rho}^{(2)} = |\sigma^{(1)}|^2 + 2\varepsilon_0\tilde{\rho}^{(2)}. \tag{523}$$

We write the solution of this equation as

$$\tilde{\rho}^{(2)} = -\int_v^\infty e^{2\varepsilon_0(v-v')}|\sigma^{(1)}(v')|^2\,\mathrm{d}v', \tag{524}$$

compatible with the requirement that $\tilde{\rho}^{(2)}$ is zero in the final state. On the other hand, if $\mathrm{d}\Sigma$ is an element of surface of the horizon, its rate of change is determined by the convergence of the null congruence emanating from it; therefore,

$$\frac{\mathrm{d}}{\mathrm{d}v}\mathrm{d}\Sigma = -2\tilde{\rho}^{(2)}\mathrm{d}\Sigma, \tag{525}$$

or

$$\lg\left[\frac{d\Sigma(v_1)}{d\Sigma(0)}\right] = -2\int_0^{v_1} \tilde{\rho}^{(2)}(v)\,dv. \tag{526}$$

Inserting for $\tilde{\rho}^{(2)}$ its solution (524), we obtain

$$\lg\left[\frac{d\Sigma(v_1)}{d\Sigma(0)}\right] = 2\int_0^{v_1} dv \int_v^{\infty} dv'|\sigma^{(1)}(v')|^2\, e^{2\varepsilon_0(v-v')}, \tag{527}$$

or, inverting the order of the integrations, we have

$$\lg\left[\frac{d\Sigma(v_1)}{d\Sigma(0)}\right] = 2\int_0^{v_1} dv'|\sigma^{(1)}(v')|^2\, e^{-2\varepsilon_0 v'} \int_0^{v'} dv e^{2\varepsilon_0 v}.$$

$$+2\int_{v_1}^{\infty} dv'\,|\sigma^{(1)}(v')|^2\, e^{-2\varepsilon_0 v'} \int_0^{v_1} dv e^{2\varepsilon_0 v}. \tag{528}$$

After the integrations over v, we obtain

$$\lg\left[\frac{d\Sigma(v_1)}{d\Sigma(0)}\right] = \frac{1}{\varepsilon_0}\int_0^{v_1} (1-e^{-2\varepsilon_0 v})|\sigma^{(1)}(v)|^2\, dv$$

$$+\frac{1}{\varepsilon_0}(e^{2\varepsilon_0 v_1}-1)\int_{v_1}^{\infty} |\sigma^{(1)}(v)|^2\, e^{-2\varepsilon_0 v}\, dv. \tag{529}$$

As Carter has pointed out, the second term on the right-hand side of this formula "has a teleological (as opposed to causal) character: it shows that the behaviour of the horizon during the time interval from 0 to v_1 depends on what happens subsequent to the time v_1." And as Carter explains, "this bizarre feature is a consequence of the teleological way in which a black hole is defined as a region from which light cannot escape to infinity: it results from the fact that local information can never guarantee the possibility of escape." We shall eliminate the teleological term by supposing that

$$|\sigma^{(1)}(v)|^2 = 0 \quad \text{for} \quad v > v_1 \quad \text{and} \quad v_1 \gg 1/2\varepsilon_0. \tag{530}$$

Equation (529) then simplifies to give

$$\lg\left[\frac{d\Sigma(v_1)}{d\Sigma(0)}\right] \simeq \frac{1}{\varepsilon_0}\int_0^{v_1} |\sigma^{(1)}(v)|^2\, dv. \tag{531}$$

If we now further suppose that $\Sigma(v)$ changes only a little during the interval $(0, v_1)$, we may write, in this limit,

$$\delta(d\Sigma) = \frac{d\Sigma}{\varepsilon_0}\int_0^{v_1} |\sigma^{(1)}(v)|^2\, dv, \tag{532}$$

or, equivalently, for the Kerr black-hole,

$$\frac{d^2\Sigma}{dv\,d\Omega} = \frac{2Mr_+}{\varepsilon_0}|\sigma^{(1)}(v)|^2, \tag{533}$$

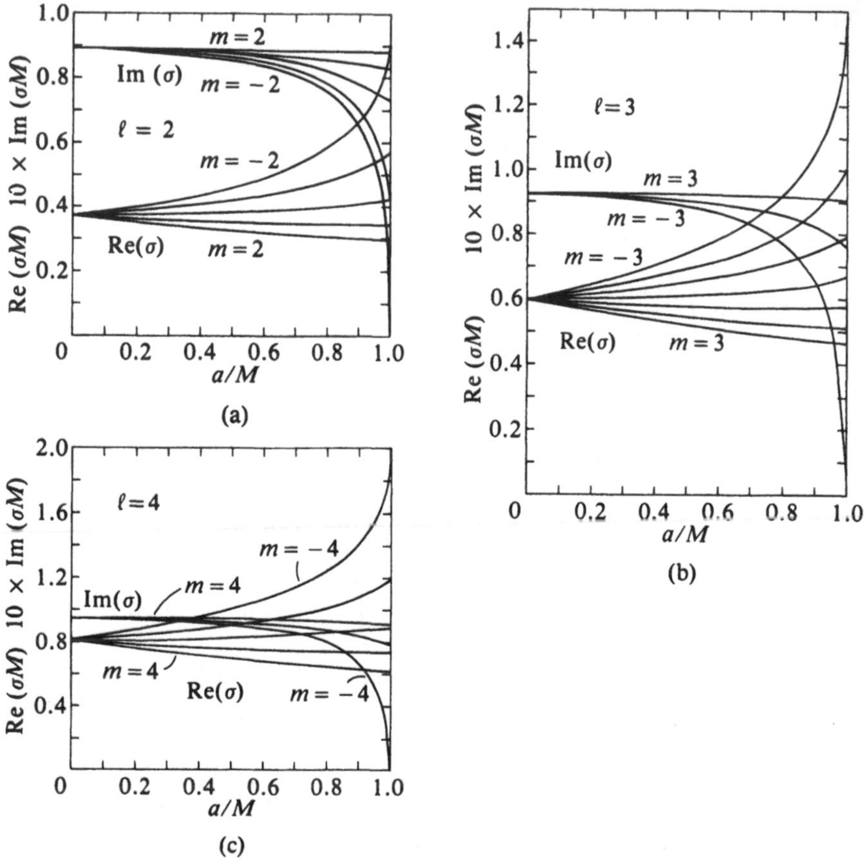

FIG. 45. The real and imaginary parts of the resonant frequency of a Kerr black-hole as a function of the parameter, a, for various values of l and m. (*a*). The case $l = 2$; for all values of m between -2 and 2. (*b*). The case $l = 3$; the imaginary part only for $m = -3$, 0, and 3 are illustrated. (*c*). The case $l = 4$; the real part only for even values of m are illustrated; and the imaginary part only for $m = -4$, 0, and 4 are illustrated.

which is the formula to be established. It is of interest to note that this formula (which requires so much sophisticated reasoning to establish) is implicitly contained in the analysis leading directly to the reflexion and the transmission coefficients in §(a).

99. The quasi-normal modes of the Kerr black-hole

In Chapters 4 (§35) and 5 (§48) we have considered the quasi-normal modes of the Schwarzschild and the Reissner-Nordström black-holes. As we have explained, these modes determine the pure dying tones of a perturbed black-hole. Clearly, such modes exist for the Kerr black-hole; and they have been determined by Detweiler. The real and the imaginary parts of the characteristic frequencies of the quasi-normal modes (as determined by Detweiler) are exhibited in Fig. 45.

100. A last observation

The treatment of the perturbations of the Kerr space-time in this chapter has been prolixious in its complexity. Perhaps, at a later time, the complexity will be unravelled by deeper insights. But mean time, the analysis has led us into a realm of the rococo: splendorous, joyful, and immensely ornate.

BIBLIOGRAPHICAL NOTES

It was Teukolsky's decoupling and separation of the equations of the Newman–Penrose formalism governing the Weyl scalars that brought the problem of the perturbations of the Kerr black-hole into the realm of the practicable:

1. S. A. Teukolsky, *Phys. Rev. Lett.*, 29, 1114–8 (1972).
2. ———, *Astrophys. J.*, 185, 635–49 (1973).
3. W. H. Press and S. A. Teukolsky, ibid., 649–73 (1973).
4. S. A. Teukolsky and W. H. Press, ibid., 193, 443–61 (1974).

What we have called the Teukolsky–Starobinsky identities are stated in paper 4 and in:

5. A. A. Starobinsky and S. M. Churilov, *Zh. Exp. i. Teoret. Fiz.*, 65, 3–8 (1973); translated in *Soviet Phys. JETP*, 38, 1–5 (1973).

§§79–96. The analysis in these sections, in its entirety, is derived from:

6. S. Chandrasekhar, *Proc. Roy. Soc.* (*London*) A, 358, 421–39 (1978).
7. ———, ibid., 441–65 (1978).
8. ———, ibid., 365, 425–51 (1979).
9. ———, ibid., 372, 475–84 (1980).

A more coherent presentation of the theory is given than was possible in the original papers since, at the time these papers were written, the end was not known at the beginning. It has not, however, been possible to reduce to any substantial extent the complexity of the analysis: it appears intrinsic to the problem.

§97. The transformation theory presented in this section follows in main:

10. S. Chandrasekhar and S. Detweiler, *Proc. Roy. Soc.* (*London*) A, 345, 145–67 (1975).
11. ———, ibid., 350, 166–74 (1976).

See also:

12. S. Chandrasekhar in *General Relativity—An Einstein Centenary Survey*, chapter 7, 371–91, edited by S. W. Hawking and W. Israel, Cambridge, England, 1979.

§98(*b*). The solution of Einstein's equations for the vacuum, linearized about the flat Minkowskian space-time, is considered in every textbook on relativity. The author found most attractive the treatment by H. P. Robertson in:

13. H. P. Robertson and T. W. Noonan, *Relativity and Cosmology*, 255–62, W. B. Saunders and Co., Philadelphia, 1968.

For a more careful discussion of the delicate issues involved see:

14. A. Trautman, *Lectures on General Relativity*, mimeographed notes, Kings College, London, 1958.
15. A. Trautman, *Bulletin de L'Academie Polonaise des Sciences*, 6, 407–12 (1958).

See also:

16. F. H. J. Cornish, *Proc. Roy. Soc.* (*London*) A, 282, 358–71 (1964).
17. ———, ibid., 372–9 (1964).

§98(*c*). The arguments in this section follow in main that given in paper 4.

§98(*d*). The basic references to the theorem of Hawking and Hartle are:

18. S. W. Hawking and J. B. Hartle, *Commun. Math. Phys.*, 27, 283–90 (1972).
19. J. B. Hartle, *Phys. Rev. D*, 8, 1010–24 (1973).
20. ———, ibid., 9, 2749–59 (1974).

The account in the text follows:

21. B. Carter in *General Relativity—An Einstein Centenary Survey*, 310–2, edited by S. W. Hawking and W. Israel, Cambridge, England, 1979.

§99. The complex charactistic-frequencies of the quasi-normal modes of the Kerr black-hole were determined by Detweiler:

22. S. Detweiler, *Astrophys. J.*, 239, 292–5 (1980).

See also:

23. S. Detweiler, *Proc. Roy. Soc.* (*London*) A, 352, 381–95 (1977).
24. ———, *Astrophys. J.*, 225, 687–93 (1978).
25. ——— and E. Szedenits, Jr., ibid., 231, 211–8 (1979).
26. ——— in *Sources of Gravitational Radiation*, 211–30, edited by L. Smarr, Cambridge, England, 1979.

Every effort has been taken to present the mathematical developments in this chapter in a comprehensible logical sequence. But the nature of the developments simply does not allow a presentation that can be followed in detail with modest effort: the reductions that are necessary to go from one step to another are often very elaborate and, on occasion, may require as many as ten, twenty, or even fifty pages. In the event that some reader may wish to undertake a careful scrutiny of the entire development, the author's derivations (in some 600 legal-size pages and in six additional notebooks) have been deposited in the Joseph Regenstein Library of the University of Chicago.

10

SPIN $-\frac{1}{2}$ PARTICLES IN KERR GEOMETRY

101. Introduction

Considerations pertaining to the propagation of waves in Kerr geometry have so far been restricted to electromagnetic waves and gravitational waves—both massless and of integral spins (one and two). We now turn to a consideration of fields of spin-$\frac{1}{2}$, both massive and massless, such as the electrons and the two component neutrinos. On the mathematical side, the interest in these fields derives principally from the separability of Dirac's equation in Kerr geometry. And on the physical side, the interest derives from the lack of super-radiance exhibited by the spin-$\frac{1}{2}$ waves.

Since the most satisfactory way of writing Dirac's equation is in the framework of the spinor formalism, we shall begin with a brief account of the spinor analysis and the spinorial basis of the Newman–Penrose formalism to the extent that we shall need them in our present context.

102. Spinor analysis and the spinorial basis of the Newman–Penrose formalism

The notion of spinors originates in the observation that a four-vector in Minkowski space can be represented equally by a Hermitian matrix and that a unimodular transformation in the complex two-dimensional space (see equations (10) and (11) below) induces a Lorentz transformation in the Minkowski space.

Consider a point x^i $(i = 0, 1, 2, 3)$ in Minkowski space; and let

$$(x^0)^2 - (x^1)^2 - (x^2)^2 - (x^3)^2 = 0. \tag{1}$$

We now represent the point (x^i) in terms of two complex numbers ξ^0 and ξ^1 (say), and their complex conjugates, $\bar{\xi}^{0'}$ and $\bar{\xi}^{1'}$, in the manner

$$\left.\begin{aligned}
x^0 &= +\frac{1}{\sqrt{2}}(\xi^0\bar{\xi}^{0'} + \xi^1\bar{\xi}^{1'}),\\
x^1 &= +\frac{1}{\sqrt{2}}(\xi^0\bar{\xi}^{1'} + \xi^1\bar{\xi}^{0'}),\\
x^2 &= -\frac{i}{\sqrt{2}}(\xi^0\bar{\xi}^{1'} - \xi^1\bar{\xi}^{0'}),\\
x^3 &= +\frac{1}{\sqrt{2}}(\xi^0\bar{\xi}^{0'} - \xi^1\bar{\xi}^{1'}),
\end{aligned}\right\} \tag{2}$$

or, inversely,

$$\left.\begin{aligned} \xi^0\bar{\xi}^{0'} &= \frac{1}{\sqrt{2}}(x^0+x^3); & \xi^0\bar{\xi}^{1'} &= \frac{1}{\sqrt{2}}(x^1+ix^2),\\ \xi^1\bar{\xi}^{0'} &= \frac{1}{\sqrt{2}}(x^1-ix^2); & \xi^1\bar{\xi}^{1'} &= \frac{1}{\sqrt{2}}(x^0-x^3). \end{aligned}\right\} \tag{3}$$

By these equations,

$$\begin{aligned}(x^0)^2-(x^1)^2-(x^2)^2-(x^3)^2 &= (x^0+x^3)(x^0-x^3)-(x^1+ix^2)(x^1-ix^2)\\ &= 2(\xi^0\bar{\xi}^{0'}\xi^1\bar{\xi}^{1'}-\xi^0\bar{\xi}^{1'}\xi^1\bar{\xi}^{0'})=0. \end{aligned}\tag{4}$$

Therefore, the representation (2) guarantees that it is a point on a null ray in Minkowski space, joining the origin with the point (x^i); and it also guarantees that the light ray is future directed since x^0, by this representation, is necessarily positive.

Now let

$$\xi_*^{\ A} = \alpha^A{}_B\xi^B \quad \text{and} \quad \bar{\xi}_*^{\ A'} = \bar{\alpha}^{A'}{}_{B'}\bar{\xi}^{B'} \quad (A, B, A', B' = 0, 1) \tag{5}$$

represent linear transformations of the complex two-dimensional spaces, (ξ^0, ξ^1) and $(\bar{\xi}^{0'}, \bar{\xi}^{1'})$, where $(\alpha^A{}_B)$ and $(\bar{\alpha}^{A'}{}_{B'})$ are two complex-conjugate matrices. With $x_*^{\ i}$ defined in terms of $\xi_*^{\ A}$ and $\bar{\xi}_*^{\ A'}$ in the identical manner (2), the linear transformations (5) will induce a linear transformation,

$$x_*^{\ i} = \beta^i{}_j x^j, \tag{6}$$

in the Minkowskian space with the coefficients $\beta^i{}_j$ given by certain bilinear combinations of the α's and the $\bar{\alpha}$'s. We ask for the conditions on the transformations (5) which will ensure that the induced transformation (6) in the Minkowskian space is Lorentzian.

By the transformations (5), we have in particular

$$\begin{aligned} x_*^{\ 0} &= \frac{1}{\sqrt{2}}(\alpha^0{}_0\xi^0+\alpha^0{}_1\xi^1)(\bar{\alpha}^{0'}{}_{0'}\bar{\xi}^{0'}+\bar{\alpha}^{0'}{}_{1'}\bar{\xi}^{1'})\\ &\quad+\frac{1}{\sqrt{2}}(\alpha^1{}_0\xi^0+\alpha^1{}_1\xi^1)(\bar{\alpha}^{1'}{}_{0'}\bar{\xi}^{0'}+\bar{\alpha}^{1'}{}_{1'}\bar{\xi}^{1'})\\ &= \tfrac{1}{2}(\alpha^0{}_0\bar{\alpha}^{0'}{}_{0'}+\alpha^1{}_0\bar{\alpha}^{1'}{}_{0'})(x^0+x^3)\\ &\quad+\tfrac{1}{2}(\alpha^0{}_1\bar{\alpha}^{0'}{}_{1'}+\alpha^1{}_1\bar{\alpha}^{1'}{}_{1'})(x^0-x^3)\\ &\quad+\tfrac{1}{2}(\alpha^0{}_0\bar{\alpha}^{0'}{}_{1'}+\alpha^1{}_0\bar{\alpha}^{1'}{}_{1'})(x^1+ix^2)\\ &\quad+\tfrac{1}{2}(\alpha^0{}_1\bar{\alpha}^{0'}{}_{0'}+\alpha^1{}_1\bar{\alpha}^{1'}{}_{0'})(x^1-ix^2). \end{aligned}\tag{7}$$

Therefore,

$$\left.\begin{aligned}\beta^0{}_0+\beta^0{}_3 &= \alpha^0{}_0\bar{\alpha}^{0'}{}_{0'}+\alpha^1{}_0\bar{\alpha}^{1'}{}_{0'},\\ \beta^0{}_0-\beta^0{}_3 &= \alpha^0{}_1\bar{\alpha}^{0'}{}_{1'}+\alpha^1{}_1\bar{\alpha}^{1'}{}_{1'},\\ \beta^0{}_1-i\beta^0{}_2 &= \alpha^0{}_0\bar{\alpha}^{0'}{}_{1'}+\alpha^1{}_0\bar{\alpha}^{1'}{}_{1'},\\ \beta^0{}_1+i\beta^0{}_2 &= \alpha^0{}_1\bar{\alpha}^{0'}{}_{0'}+\alpha^1{}_1\bar{\alpha}^{1'}{}_{0'}.\end{aligned}\right\} \tag{8}$$

A requirement that the transformation (6) is Lorentzian is (in particular) that

$$(\beta^0{}_0)^2-(\beta^0{}_1)^2-(\beta^0{}_2)^2-(\beta^0{}_3)^2 = 1. \tag{9}$$

By equations (8), this condition requires

$$\begin{aligned}&\left\|\begin{matrix}\alpha^0{}_0\bar{\alpha}^{0'}{}_{0'}+\alpha^1{}_0\bar{\alpha}^{1'}{}_{0'} & \alpha^0{}_0\bar{\alpha}^{0'}{}_{1'}+\alpha^1{}_0\bar{\alpha}^{1'}{}_{1'}\\ \alpha^0{}_1\bar{\alpha}^{0'}{}_{0'}+\alpha^1{}_1\bar{\alpha}^{1'}{}_{0'} & \alpha^0{}_1\bar{\alpha}^{0'}{}_{1'}+\alpha^1{}_1\bar{\alpha}^{1'}{}_{1'}\end{matrix}\right\|\\ &= \left\|\begin{matrix}\alpha^0{}_0 & \alpha^0{}_1\\ \alpha^1{}_0 & \alpha^1{}_1\end{matrix}\right\| \left\|\begin{matrix}\bar{\alpha}^{0'}{}_{0'} & \bar{\alpha}^{0'}{}_{1'}\\ \bar{\alpha}^{1'}{}_{0'} & \bar{\alpha}^{1'}{}_{1'}\end{matrix}\right\| = \Delta\bar{\Delta} = 1,\end{aligned} \tag{10}$$

where Δ and $\bar{\Delta}$ denote the determinants of the transformations (5). Therefore, a necessary condition that the transformations (5) represent a Lorentz transformation is that their determinants be of modulus 1, i.e:, *unimodular*; and it is clear that it is also sufficient. In our further considerations, we shall suppose that

$$\Delta = \bar{\Delta} = 1, \tag{11}$$

i.e., we shall restrict ourselves to transformations with unit determinants. It is clear that these transformations (which preserve the time-direction) include all Lorentz transformations exclusive of reflexions. But, by including transformations which are the negatives of the ones we have considered, we can recover all Lorentz transformations.

We now define *spinors* ξ^A *and* $\eta^{A'}$ *of rank* 1 *as complex vectors in a two-dimensional space* (A, $A' = 0, 1$) *subject to the transformations,*

$$\xi_*{}^A = \alpha^A{}_B\xi^B \quad\text{and}\quad \eta_*{}^{A'} = \bar{\alpha}^{A'}{}_{B'}\eta^{B'} \quad (A, A', B, B' = 0, 1), \tag{12}$$

where $(\alpha^A{}_B)$ *and* $(\bar{\alpha}^{A'}{}_{B'})$ *are complex-conjugate matrices with unit determinants:*

$$\|\alpha^A{}_B\| = \|\bar{\alpha}^{A'}{}_{B'}\| = 1. \tag{13}$$

It is important that we distinguish spinors of the two classes: those with the unprimed and those with the primed indices, which are subject to complex-conjugate transformations. Also, we shall restrict capital Latin alphabets for spinor indices.

If ξ^A and η^A are two spinors of the same class, then their determinant,

$$\left\|\begin{matrix}\xi^0 & \xi^1\\ \eta^0 & \eta^1\end{matrix}\right\| = \xi^0\eta^1-\xi^1\eta^0, \tag{14}$$

is invariant to unimodular transformations. Therefore, we may define a *skew metric*, ε_{AB}, for the space such that

$$\varepsilon_{AB}\,\xi^A\eta^B \text{ is invariant.} \tag{15}$$

By comparison with (14) it follows that

$$\varepsilon_{00} = \varepsilon_{11} = 0 \qquad \text{and} \qquad \varepsilon_{01} = -\varepsilon_{10} = 1; \tag{16}$$

i.e., ε_{AB} is the two-dimensional Levi–Civita symbol. We may, of course, similarly define a metric, $\varepsilon_{A'B'}$, for the primed spinors; it will again be the Levi–Civita symbol.

As in tensor analysis, we may use the metrics, ε_{AB} and $\varepsilon_{A'B'}$, to *lower* the spinor indices; thus,

$$\xi_A = \xi^C\varepsilon_{CA}, \tag{17}$$

or, explicitly,

$$\xi_0 = -\xi^1 \quad \text{and} \quad \xi_1 = \xi^0. \tag{18}$$

Accordingly, indices can be *raised* by the Levi-Civita symbol ε^{AB} in the manner

$$\xi^A = \varepsilon^{AC}\xi_C. \tag{19}$$

In view of the antisymmetry of ε^{AC} and ε_{CA}, it is important to preserve the order of the indices in equations (17) and (19) with respect to the index which is contracted. Since

$$\xi_A = \xi^C\varepsilon_{CA} = \varepsilon^{CB}\xi_B\varepsilon_{CA}, \tag{20}$$

it follows that

$$\delta_A^B = \varepsilon^{CB}\varepsilon_{CA} = \varepsilon_A{}^B = -\varepsilon^B{}_A. \tag{21}$$

It is, of course, clear that by considering spinors with the primed indices, we shall obtain the same formulae (17)–(21) with the indices primed.

As in tensor analysis, we can construct spinors of higher rank,

$$\xi^{AB},\ \xi_{AB'C},\ \xi_{ABC'}{}^{D'E}{}_{F'G},\ \text{etc.}, \tag{22}$$

with their appropriate transformation properties. Thus,

$$\xi_*^{AB'} = \alpha^A{}_C\bar{\alpha}^{B'}{}_{D'}\,\xi^{CD'}. \tag{23}$$

It is important to observe that while the order of the indices of *each kind* is relevant and must be preserved, the *relative* ordering of the primed and the unprimed indices is of no consequence.

Again, as in tensor analysis, contraction of spinors with respect to a pair of primed, or unprimed, indices can be effected with the metric ε_{AB}, or $\varepsilon_{A'B'}$; but contraction of a primed and unprimed index is, of course, forbidden. Thus,

$$\xi_{A'}{}^{A'} = \varepsilon^{A'B'}\xi_{A'B'}. \tag{24}$$

On the other hand,

$$\xi^{A'}{}_{A'} = \varepsilon^{A'B'}\xi_{B'A'}. \tag{25}$$

Therefore,

$$\xi_{A'}{}^{A'} = -\xi^{A'}{}_{A'}. \tag{26}$$

In particular,

$$\xi_{A'}\xi^{A'} = 0. \tag{27}$$

Equation (26) is an example of Penrose's '*see-saw*',

$$\xi^{\cdots A \cdots}{}_{\cdots A \cdots} = -\xi^{\cdots}{}_{A}{}^{\cdots}{}_{\cdots}{}^{A}{}_{\cdots}. \tag{28}$$

(a) The representation of vectors and tensors in terms of spinors

Equations (2) and (3) provide a representation of the position vector, x^i, in terms of a pair of complex-conjugate spinors, ξ^A and $\bar{\xi}^{A'}$, which we can express in the manner

$$x^i \leftrightarrow \begin{vmatrix} \xi^0\bar{\xi}^{0'} & \xi^0\bar{\xi}^{1'} \\ \xi^1\bar{\xi}^{0'} & \xi^1\bar{\xi}^{1'} \end{vmatrix} = \frac{1}{\sqrt{2}}\begin{vmatrix} x^0+x^3 & x^1+ix^2 \\ x^1-ix^2 & x^0-x^3 \end{vmatrix}. \tag{29}$$

Quite generally, we associate any four vector X^i with a spinor of the second rank $\xi^{AB'}$ in the manner

$$X^i \leftrightarrow \begin{vmatrix} \xi^{00'} & \xi^{01'} \\ \xi^{10'} & \xi^{11'} \end{vmatrix} = \frac{1}{\sqrt{2}}\begin{vmatrix} X^0+X^3 & X^1+iX^2 \\ X^1-iX^2 & X^0-X^3 \end{vmatrix} = X^{AB'} \quad \text{(say).} \tag{30}$$

Thus, *a four vector is associated with a Hermitian matrix.*

The invariant associated with the four vector is

$$\begin{aligned}(X^0)^2-(X^1)^2-(X^2)^2-(X^3)^2 &= (X^0+X^3)(X^0-X^3) \\ &\quad -(X^1+iX^2)(X^1-iX^2) \\ &= 2(\xi^{00'}\xi^{11'}-\xi^{01'}\xi^{10'}) \\ &= (\xi^{00'}\xi_{00'}+\xi_{11'}\xi^{11'}+\xi_{10'}\xi^{10'}+\xi^{01'}\xi_{01'}) \\ &= X_{AB'}X^{AB'};\end{aligned} \tag{31}$$

or, expressed in terms of the metrics, g_{ij}, ε_{AB}, and $\varepsilon_{A'B'}$ of the Minkowskian and the spinor spaces,

$$g_{ij}X^iX^j = \varepsilon_{AC}\varepsilon_{B'D'}X^{AB'}X^{CD'}. \tag{32}$$

The relationship,

$$X^i \leftrightarrow X^{AB'}, \tag{33}$$

is now expressed in the form

$$X^i = \sigma^i{}_{AB'}X^{AB'}, \tag{34}$$

or, in its inverse form,

$$X^{AB'} = \sigma^{AB'}{}_i X^i, \tag{35}$$

where $\sigma^i{}_{AB'}$ and $\sigma^{AB'}{}_i$, for each i, are constant Hermitian matrices. Relations which are immediate consequences of the foregoing definitions are

$$\left.\begin{aligned} \sigma^{AB'}{}_i \sigma^i{}_{CD'} &= \delta^A_C \delta^{B'}_{D'} \\ \text{and} \qquad \sigma^i{}_{AB'} \sigma_j{}^{AB'} &= \delta^i_j. \end{aligned}\right\} \tag{36}$$

And, finally, we deduce from equations (32) and (35) that

$$g_{ij} = \varepsilon_{AC} \varepsilon_{B'D'} \sigma^{AB'}{}_i \sigma^{CD'}{}_j \tag{37}$$

and

$$\varepsilon_{AC} \varepsilon_{B'D'} = g_{ij} \sigma^i{}_{AB'} \sigma^j{}_{CD'}. \tag{38}$$

It is of interest to note that the matrices, $\sigma^i{}_{AB'}$ and $\sigma^{AB'}{}_i$, defined by the representation (29), are

$$\sigma^{AB'}{}_0 = \frac{1}{\sqrt{2}}\begin{vmatrix} 1 & 0 \\ 0 & 1 \end{vmatrix}, \sigma^{AB'}{}_1 = \frac{1}{\sqrt{2}}\begin{vmatrix} 0 & 1 \\ 1 & 0 \end{vmatrix}, \sigma^{AB'}{}_2 = \frac{1}{\sqrt{2}}\begin{vmatrix} 0 & i \\ -i & 0 \end{vmatrix},$$

$$\sigma^{AB'}{}_3 = \frac{1}{\sqrt{2}}\begin{vmatrix} 1 & 0 \\ 0 & -1 \end{vmatrix} \tag{39}$$

and

$$\sigma^0{}_{AB'} = \frac{1}{\sqrt{2}}\begin{vmatrix} 1 & 0 \\ 0 & 1 \end{vmatrix}, \sigma^1{}_{AB'} = \frac{1}{\sqrt{2}}\begin{vmatrix} 0 & 1 \\ 1 & 0 \end{vmatrix}, \sigma^2{}_{AB'} = \frac{1}{\sqrt{2}}\begin{vmatrix} 0 & -i \\ i & 0 \end{vmatrix},$$

$$\sigma^3{}_{AB'} = \frac{1}{\sqrt{2}}\begin{vmatrix} 1 & 0 \\ 0 & -1 \end{vmatrix}. \tag{40}$$

It will be observed that, apart from the normalization factor $1/\sqrt{2}$, $\sigma^1{}_{AB'}$, $\sigma^2{}_{AB'}$, and $\sigma^3{}_{AB'}$ are the *Pauli spin-matrices*.

In terms of the σ-matrices, we can now relate tensors of arbitrary ranks with their spinor equivalents. Thus,

$$\left.\begin{aligned} Y^{ij}{}_k &= \sigma^i{}_{AB'} \sigma^j{}_{CD'} \sigma_k{}^{EF'} Y^{AB'CD'}{}_{EF'} \\ \text{and} \qquad Y^{AB'CD'}{}_{EF'} &= \sigma^{AB'}{}_i \sigma^{CD'}{}_j \sigma^k{}_{EF'} Y^{ij}{}_k. \end{aligned}\right\} \tag{41}$$

By virtue of these relations, we have the correspondence

$$Y^{AB'CD'}{}_{EF'} \leftrightarrow Y^{ij}{}_k. \tag{42}$$

In this sense, equation (38) expresses the correspondence

$$\varepsilon_{AC} \varepsilon_{B'D'} \leftrightarrow g_{ij}. \tag{43}$$

(*b*) *Penrose's pictorial representation of a spinor ξ^A as a 'flag'*

Consider a null vector, u_i, and its spinor representation, $U_{AB'}$, in terms of a spinor, ξ_A, and its complex conjugate, $\bar{\xi}_{B'}$:

$$u_i \leftrightarrow U_{AB'} = \xi_A \bar{\xi}_{B'}. \tag{44}$$

(The vector u_i *is* null, since by equation (27)

$$u_i u^i = \xi_A \bar{\xi}_{B'} \xi^A \bar{\xi}^{B'} = 0.) \tag{45}$$

Let η_A be an associated spinor with the property

$$\xi_A \eta^A = \varepsilon^{AB} \eta_B \xi_A = \xi_0 \eta_1 - \xi_1 \eta_0 = 1. \tag{46}$$

Further, let

$$w_i \leftrightarrow W_{AB'} = \xi_A \bar{\eta}_{B'} + \eta_A \bar{\xi}_{B'}; \tag{47}$$

and consider the antisymmetric tensor

$$p_{ij} = u_i w_j - u_j w_i. \tag{48}$$

The spinor representation of p_{ij} is

$$\begin{aligned} P_{AB'CD'} &= \xi_A \bar{\xi}_{B'} (\xi_C \bar{\eta}_{D'} + \eta_C \bar{\xi}_{D'}) - \xi_C \bar{\xi}_{D'} (\xi_A \bar{\eta}_{B'} + \eta_A \bar{\xi}_{B'}) \\ &= \xi_A \xi_C (\bar{\xi}_{B'} \bar{\eta}_{D'} - \bar{\xi}_{D'} \bar{\eta}_{B'}) + \bar{\xi}_{B'} \bar{\xi}_{D'} (\xi_A \eta_C - \xi_C \eta_A). \end{aligned} \tag{49}$$

On the other hand,

$$\xi_A \eta_C - \eta_A \xi_C = \varepsilon_{AC}, \tag{50}$$

since the left-hand side is antisymmetric in A and C and, by virtue of the assumption (46), it takes the value of $+1$ or -1 according as the pair (A, C) is (0, 1) or (1, 0), respectively. Accordingly, $P_{AB'CD'}$ has the alternative form

$$P_{AB'CD'} = \xi_A \xi_C \varepsilon_{B'D'} + \varepsilon_{AC} \bar{\xi}_{B'} \bar{\xi}_{D'}. \tag{51}$$

In this form, it is manifest that p_{ij} is also determined by the spinor ξ_A.

By definition, the vector w_i is real; and it is orthogonal to u_i since

$$u^i w_i = \xi^A \bar{\xi}^{B'} (\xi_A \bar{\eta}_{B'} + \eta_A \bar{\xi}_{B'}) = 0. \tag{52}$$

And, further, w_i is space-like since

$$\begin{aligned} w_i w^i &= (\xi_A \bar{\eta}_{B'} + \eta_A \bar{\xi}_{B'})(\xi^A \bar{\eta}^{B'} + \eta^A \bar{\xi}^{B'}) \\ &= \bar{\eta}_{B'} \bar{\xi}^{B'} - \bar{\xi}_{B'} \bar{\eta}^{B'} = -2, \end{aligned} \tag{53}$$

where repeated use has been made of the relation (46). Thus, the spinor ξ^A, besides determining the null vector u^i, determines, also, a real, space-like, orthogonal vector, w^i, apart from a multiple of u^i. Therefore, ξ^A defines a null vector u^i—the *'flag-pole'*—and an orthogonal, two-surface, w^i + a multiple of u^i—the *'flag'*; and, conversely, the flag specifies the spinor, apart from sign.

(c) *The dyad formalism*

Since the space-time of general relativity is locally Minkowskian, we can set up, at each point of the space-time, an orthonormal *dyad basis*, $\zeta_{(a)}{}^A$ and $\zeta_{(a')}{}^{A'}$ (a, $a' = 0, 1$ and A, $A' = 0, 1$), for spinors even as we set up an orthonormal *tetrad basis*, $e^i_{(a)}$ ($a = 0, 1, 2, 3$ and $i = 0, 1, 2, 3$) for tensors in a tetrad formalism (cf. Chapter 1, §7). And, as in the tetrad formalism, we shall enclose the dyad indices—the lowercase letters of the earlier part of the Latin alphabet—in parentheses. It is, however, convenient to have special symbols for the two basis spinors, $\zeta_{(0)}{}^A$ and $\zeta_{(1)}{}^A$. We shall write

$$\zeta_{(0)}{}^A = o^A \quad \text{and} \quad \zeta_{(1)}{}^A = \iota^A. \tag{54}$$

And the condition of orthonormality is

$$\varepsilon_{AB}\, o^A \iota^B = o^0 \iota^1 - o^1 \iota^0 = o_B \iota^B = -o^A \iota_A = 1. \tag{55}$$

Elementary consequences of these definitions are

$$\varepsilon_{AB}\zeta_{(a)}{}^A\zeta_{(b)}{}^B = \zeta_{(a)B}\zeta_{(b)}{}^B = \varepsilon_{(a)(b)} \tag{56}$$

and

$$\begin{aligned}\varepsilon^{(a)(b)}\zeta_{(a)}{}^A\zeta_{(b)}{}^B &= \zeta_{(0)}{}^A\zeta_{(1)}{}^B - \zeta_{(1)}{}^A\zeta_{(0)}{}^B \\ &= o^A\iota^B - \iota^A o^B = \varepsilon^{AB}.\end{aligned} \tag{57}$$

It is also clear that we can raise and lower the dyad indices by $\varepsilon^{(a)(b)}$ and $\varepsilon_{(a)(b)}$. Thus,

$$\zeta^{(c)A}\varepsilon_{(c)(a)} = \zeta_{(a)}{}^A \quad \text{and} \quad \varepsilon^{(a)(c)}\zeta_{(c)}{}^A = \zeta^{(a)A}. \tag{58}$$

Further consequences are

$$\left.\begin{aligned}\zeta_{(a)A}\zeta^{(b)A} &= -\zeta_{(a)}{}^A\zeta^{(b)}{}_A = \delta^{(b)}_{(a)} \\ \text{and} \qquad \zeta_{(a)A}\zeta_{(b)}{}^A &= -\zeta_{(a)}{}^A\zeta_{(b)A} = \varepsilon_{(a)(b)}.\end{aligned}\right\} \tag{59}$$

As in the tetrad formalism, we can project any spinor ξ_A on to the dyad basis:

$$\xi_{(a)} = \xi_A\zeta_{(a)}{}^A, \tag{60}$$

or, explicitly,

$$\xi_{(0)} = \xi_A o^A \quad \text{and} \quad \xi_{(1)} = \xi_A \iota^A. \tag{61}$$

We also have

$$\xi_A = \xi^{(a)}\zeta_{(a)A} = \xi^{(0)} o_A + \xi^{(1)}\iota_A. \tag{62}$$

The spinors o^A and ι^A and their complex conjugates determine the null vectors $\boldsymbol{l}$, $\boldsymbol{n}$, $\boldsymbol{m}$, and $\bar{\boldsymbol{m}}$ by the correspondence

$$\left.\begin{aligned}l^i &\leftrightarrow o^A\bar{o}^{B'}, & m^i &\leftrightarrow o^A\bar{\iota}^{B'}, \\ \bar{m}^i &\leftrightarrow \iota^A\bar{o}^{B'}, & n^i &\leftrightarrow \iota^A\bar{\iota}^{B'}.\end{aligned}\right\} \tag{63}$$

The null vectors satisfy the orthogonality conditions,

$$l^i n_i = o^A \bar{o}^{B'} \iota_A \bar{\iota}_{B'} = 1 \quad \text{and} \quad m^i \bar{m}_i = o^A \bar{\iota}^{B'} \iota_A o_{B'} = -1, \tag{64}$$

while all the remaining scalar products are zero. Thus, the dyad basis determines four null vectors which can be used as a basis for a Newman–Penrose formalism as described in Chapter 1, §6.

The representation (63) yields the Hermitian matrices,

$$\sigma^i{}_{AB'} \quad \text{and} \quad \sigma^{AB'}{}_i, \tag{65}$$

such that

$$\left.\begin{aligned} l^i &= \sigma^i{}_{AB'} \zeta_{(0)}{}^A \bar{\zeta}_{(0')}{}^{B'} = \sigma^i{}_{AB'} o^A \bar{o}^{B'}, \\ m^i &= \sigma^i{}_{AB'} \zeta_{(0)}{}^A \bar{\zeta}_{(1')}{}^{B'} = \sigma^i{}_{AB'} o^A \bar{\iota}^{B'}, \\ \bar{m}^i &= \sigma^i{}_{AB'} \zeta_{(1)}{}^A \bar{\zeta}_{(0')}{}^{B'} = \sigma^i{}_{AB'} \iota^A \bar{o}^{B'}, \\ \text{and} \qquad n^i &= \sigma^i{}_{AB'} \zeta_{(1)}{}^A \bar{\zeta}_{(1')}{}^{B'} = \sigma^i{}_{AB'} \iota^A \bar{\iota}^{B'}. \end{aligned}\right\} \tag{66}$$

Accordingly, we may write

$$\sigma^i{}_{AB'} = \frac{1}{\sqrt{2}} \begin{vmatrix} l^i & m^i \\ \bar{m}^i & n^i \end{vmatrix} \quad \text{and} \quad \sigma^{AB'}{}_i = \frac{1}{\sqrt{2}} \begin{vmatrix} n_i & -\bar{m}_i \\ -m_i & l_i \end{vmatrix}. \tag{67}$$

Comparison with equations (39) and (40) shows that the foregoing definitions provide the natural generalizations of the Pauli spin-matrices.

And finally, associated with the directional derivatives (Ch. 1, equations (285)),

$$D = l^i \partial_i, \ \Delta = n^i \partial_i, \ \delta = m^i \partial_i \quad \text{and} \quad \delta^* = \bar{m}^i \partial_i, \tag{68}$$

of the Newman–Penrose formalism, we have the spinor equivalents

$$\partial_{00'} = D, \quad \partial_{11'} = \Delta, \quad \partial_{01'} = \delta, \quad \text{and} \quad \partial_{10'} = \delta^*. \tag{69}$$

(d) *Covariant differentiation of spinor fields and spin coefficients*

We now wish to define covariant differentiation of spinor fields. Consistency requires that the definition must be based on the correspondences

$$\nabla_i \leftrightarrow \nabla_{AB'} \tag{70}$$

and

$$\nabla_i X_j = X_{j;i} \leftrightarrow \nabla_{AB'} X_{CD'} = X_{CD';AB'}. \tag{71}$$

In accordance with equation (41), this last equation requires

$$X_{CD';AB'} = \sigma^j{}_{CD'} \sigma^i{}_{AB'} X_{j;i}. \tag{72}$$

Besides, we shall require that the covariant differentiation of spinor fields satisfies the Leibnitz rule, namely,

$$\nabla_{AB'}(S^{\cdots}{}_{\cdots} \times T^{\cdots}{}_{\cdots}) = T^{\cdots}{}_{\cdots} \nabla_{AB'}(S^{\cdots}{}_{\cdots}) + S^{\cdots}{}_{\cdots} \nabla_{AB'}(T^{\cdots}{}_{\cdots}), \tag{73}$$

where $S^{\cdots}_{\;\cdots}$ and $T^{\cdots}_{\;\cdots}$ are any two spinor fields. And we shall also require that the operator $\nabla_{AB'}$ is real, i.e.,

$$\nabla_{AB'} = \bar{\nabla}_{A'B}. \tag{74}$$

We shall show presently how the foregoing postulates suffice to define uniquely the operation of covariant differentiation. But first, we note that, as in the tetrad formalism, we can define, in analogous fashion, the notion of *intrinsic differentiation*. Thus, we define the *intrinsic derivative* of the dyadic component, $\xi_{(a)}$, of a spinor along the 'direction' $(a)(b')$ by

$$\xi_{(c)|(a)(b')} = \zeta_{(c)}{}^{C}\xi_{C;AB'}\zeta_{(a)}{}^{A}\xi_{(b)}{}^{B'}, \tag{75}$$

or, equivalently,

$$\xi_{(c)|AB'} = \zeta_{(c)}{}^{C}\xi_{C;AB'}. \tag{76}$$

We shall now state two elementary consequences of the foregoing definitions and requirements.

LEMMA 1. $$\sigma^{j}{}_{CD';AB'} = 0 \quad \text{and} \quad \sigma^{CD'}{}_{j;AB'} = 0. \tag{77}$$

Proof. The lemma follows from equation (72). The left-hand side of this equation can be written as

$$X_{CD';AB'} = (\sigma^{j}{}_{CD'}X_{j})_{;AB'}, \tag{78}$$

while the right-hand side has the alternative form

$$\sigma^{j}{}_{CD'}(\sigma^{i}{}_{AB'}X_{j;i}) = \sigma^{j}{}_{CD'}X_{j;AB'}. \tag{79}$$

From the required equality of the right-hand sides of equations (78) and (79) the first of the two results stated in the lemma follows; and the second follows from the orthogonality relations (36) and the Leibnitz rule.

LEMMA 2. $$\varepsilon_{CD;AB'} = 0. \tag{80}$$

Proof. From the correspondence (71) and the Leibnitz rule valid for both tensor and spinor fields, it follows from equation (37) that

$$0 = g_{jk;i} = \sigma_{i}{}^{AB'}(\varepsilon_{CD}\varepsilon_{E'F'}\sigma_{j}{}^{CE'}\sigma_{k}{}^{DF'})_{;AB'}, \tag{81}$$

or, by Lemma 1, that

$$\sigma_{i}{}^{AB'}(\varepsilon_{CD}\varepsilon_{E'F'})_{;AB'} = 0; \tag{82}$$

and equation (80) follows.

DEFINITION: *The spin coefficients,* $\Gamma_{(a)(b)(c)(d')}$, *are defined in the dyad formalism by*

$$\Gamma_{(a)(b)(c)(d')} = [\zeta_{(a)F}]_{;CD'}\zeta_{(b)}{}^{F}\zeta_{(c)}{}^{C}\zeta_{(d')}{}^{D'}. \tag{83}$$

It is convenient, for the sake of brevity, to write instead the formula (in spite of its ugliness!)

$$\Gamma_{(a)(b)CD'} = [\zeta_{(a)F}]_{;CD'}\zeta_{(b)}{}^{F}. \tag{84}$$

The spin coefficients, as defined, are symmetric in the first pair of its indices. This symmetry follows from the relation (cf. equation (59))

$$\zeta_{(a)F}\zeta_{(b)}{}^{F} = \varepsilon_{(a)(b)}. \tag{85}$$

For, by this relation and Lemma 2, we have

$$\begin{aligned}\Gamma_{(a)(b)CD'} &= -[\zeta_{(b)}{}^{F}]_{;CD'}\zeta_{(a)F}\\ &= +[\zeta_{(b)F}]_{;CD'}\zeta_{(a)}{}^{F} = \Gamma_{(b)(a)CD'}.\end{aligned} \tag{86}$$

An alternative form of equation (84) follows by contraction with $\zeta^{(b)}{}_{E}$ and making use of the relations included in equations (59). Thus,

$$[\zeta_{(a)E}]_{;CD'} = -\zeta^{(b)}{}_{E}\Gamma_{(b)(a)CD'} = \zeta_{(b)E}\Gamma^{(b)}{}_{(a)CD'}. \tag{87}$$

We shall now show how, with the spin coefficients as defined, the intrinsic derivatives of the dyadic components of the spinors of the first rank, $\xi_{(a)}$ and $\xi^{(a)}$, can be expressed; and this, by the Leibnitz rule, will clearly suffice to obtain the covariant derivatives of spinors of arbitrary ranks. Thus,

$$\begin{aligned}\xi_{(a)|BC'} &= \zeta_{(a)}{}^{A}\xi_{A;BC'}\\ &= [\xi_{A}\zeta_{(a)}{}^{A}]_{;BC'} - \xi_{A}[\zeta_{(a)}{}^{A}]_{;BC'}.\end{aligned} \tag{88}$$

The quantity in the first square brackets on the right-hand side is a scalar, namely, $\xi_{(a)}$. Therefore, by equation (87)),

$$\xi_{(a)|BC'} = \xi_{(a),BC'} - \xi_{A}\Gamma^{(d)}{}_{(a)BC'}\zeta_{(d)}{}^{A}, \tag{89}$$

or, equivalently,

$$\xi_{(a)|BC'} = \xi_{(a),BC'} + \Gamma_{(d)(a)BC'}\xi^{(d)}. \tag{90}$$

Similarly, we find

$$\xi^{(a)}{}_{|BC'} = \xi^{(a)}{}_{,BC'} + \Gamma^{(a)}{}_{(d)BC'}\xi^{(d)}. \tag{91}$$

In view of the symmetry of the spin coefficients in the first pair of the dyadic indices, it is clear that twelve coefficients will have to be specified. In the Newman–Penrose formalism, these coefficients are assigned special symbols which are listed in the tabulation below.

$\Gamma_{(a)(b)(c)(d')}$: (92)

$(c)(d')$ \ $(a)(b)$	00	01 or 10	11
00′	κ	ε	π
10′	ρ	α	λ
01′	σ	β	μ
11′	τ	γ	ν

It remains to verify that these definitions of the spin coefficients are in agreement with those defined in terms of the Ricci rotation-coefficients, γ_{ijk}, in Chapter 1, equations (286).

First, we prove the following lemma due to J. Friedman.

LEMMA 3.

$$\Gamma_{(a)(b)CD'} = \tfrac{1}{2}\varepsilon^{(k')(f')}\zeta_{(a)}{}^{E}\bar{\zeta}_{(f')}{}^{F'}[\zeta_{(b)E}\bar{\zeta}_{(k')F'}]_{;CD'}. \tag{93}$$

(Notice that by equations (63), $\zeta_{(a)}{}^{E}\bar{\zeta}_{(f')}{}^{F'}$ and $\zeta_{(b)E}\bar{\zeta}_{(k')F'}$ both represent basis null-vectors.)

Proof. Expanding the right-hand side of equation (93) by the Leibnitz rule and reducing it with the aid of the relations included in (59), we successively obtain

$$\begin{aligned}
&\tfrac{1}{2}\varepsilon^{(k')(f')}\{\zeta_{(a)}{}^{E}\bar{\zeta}_{(f')}{}^{F'}\zeta_{(b)E}[\bar{\zeta}_{(k')F'}]_{;CD'} \\
&\qquad + \zeta_{(a)}{}^{E}\bar{\zeta}_{(f')}{}^{F'}\bar{\zeta}_{(k')F'}[\zeta_{(b)E}]_{;CD'}\} \\
&= \tfrac{1}{2}\varepsilon^{(k')(f')}\{-\varepsilon_{(a)(b)}\bar{\zeta}_{(f')}{}^{F'}\Gamma^{(d')}{}_{(k')CD'}\bar{\zeta}_{(d')F'} \\
&\qquad + \varepsilon_{(k')(f')}\zeta_{(a)}{}^{E}\Gamma^{(d)}{}_{(b)CD'}\zeta_{(d)E}\} \\
&= \tfrac{1}{2}\varepsilon^{(k')(f')}\{-\varepsilon_{(a)(b)}\delta^{(d')}_{(f')}\Gamma_{(d')(k')CD'} + \varepsilon_{(k')(f')}\delta^{(d)}_{(a)}\Gamma_{(d)(b)CD'}\} \\
&= \tfrac{1}{2}\varepsilon^{(k')(f')}\{-\varepsilon_{(a)(b)}\Gamma_{(k')(f')CD'} + \varepsilon_{(k')(f')}\Gamma_{(a)(b)CD'}\} \\
&= \tfrac{1}{2}\varepsilon^{(k')(f')}\varepsilon_{(k')(f')}\Gamma_{(a)(b)CD'} = \Gamma_{(a)(b)CD'};
\end{aligned} \tag{94}$$

and the lemma is established.

Using Friedman's lemma, we can express the various spin coefficients $\Gamma_{(a)(b)(c)(d')}$ in terms of the covariant (tensorial) derivatives of the basis null-vectors, $\boldsymbol{l}$, $\boldsymbol{n}$, $\boldsymbol{m}$, and $\bar{\boldsymbol{m}}$, and verify that they are in agreement with their definitions in terms of the rotation coefficients, γ_{ijk}. Thus, with the aid of the definitions (54), (63), and (69), we find, suppressing the parentheses around the dyadic indices when they are assigned the particular values 0 or 1,

$$\begin{aligned}
\Gamma_{0000'} &= \tfrac{1}{2}\varepsilon^{(k')(f')}\zeta_{0}{}^{E}\bar{\zeta}_{(f')}{}^{F'}[\zeta_{0E}\bar{\zeta}_{(k')F'}]_{;00'} \\
&= \tfrac{1}{2}\varepsilon^{(k')(f')}o^{E}\bar{\zeta}_{(f')}{}^{F'}[o_{E}\bar{\zeta}_{(k')F'}]_{;00'} \\
&= \tfrac{1}{2}[o^{E}\bar{\imath}^{F'}l^{i}(o_{E}\bar{o}_{F'})_{;i} - o^{E}\bar{o}^{F'}l^{i}(o_{E}\bar{\imath}_{F'})_{;i}] \\
&= \tfrac{1}{2}[m^{j}l^{i}l_{j;i} - l^{j}l^{i}m_{j;i}] \\
&= \tfrac{1}{2}[\gamma_{311} - \gamma_{131}] = \gamma_{311} = \kappa,
\end{aligned} \tag{95}$$

in agreement with the definition of κ in Chapter 1, equation (286). Similarly,

$$\begin{aligned}\Gamma_{1101'} &= \tfrac{1}{2}\varepsilon^{(k')(f')}\iota^{E}\bar{\zeta}_{(f')}{}^{F'}[\iota_E\bar{\zeta}_{(k')F'}]_{;01'} \\ &= \tfrac{1}{2}[\iota^E\bar{\iota}^{F'}(\iota_E\bar{o}_{F'})_{;01'} - \iota^E\bar{o}^{F'}(\iota_E\bar{\iota}_{F'})_{;01'}] \\ &= \tfrac{1}{2}[n^j m^i \bar{m}_{j;i} - \bar{m}^j m^i n_{j;i}] \\ &= \tfrac{1}{2}(\gamma_{243} - \gamma_{423}) = \gamma_{243} = \mu,\end{aligned} \tag{96}$$

again in agreement with the definition of μ in Chapter 1, equation (286). The remaining coefficients can be evaluated in the same fashion and shown to be in agreement with the earlier definitions.

This completes our account of spinor analysis and the spinorial basis of the Newman–Penrose formalism.

103. Dirac's equation in the Newman–Penrose formalism

As is well known, in the relativistic theory of spin-$\frac{1}{2}$ particles, the wave function is represented by a pair of spinors, P^A and $\bar{Q}^{A'}$; and in Minkowski space, Dirac's equations governing them are

$$\sigma^i{}_{AB'}\partial_i P^A + i\mu_*\bar{Q}_{B'} = 0 \tag{97}$$

and

$$\sigma^i{}_{AB'}\partial_i Q^A + i\mu_*\bar{P}_{B'} = 0, \tag{98}$$

where $\sigma^i{}_{AB'}$ are the Pauli-matrices and $\mu_*\sqrt{2}$ is the mass of the particle (expressed as the inverse of its Compton wave-length). The factor $\sqrt{2}$ in the definition of the mass arises from the fact that the Pauli matrices as defined in equation (40) differ from their customary definitions by the factor $1/\sqrt{2}$.

In the Newman–Penrose formalism in a curved space-time, we take over equations (97) and (98) with the covariant derivatives replacing the ordinary derivatives and the σ-matrices, defined as in equations (67), replacing the Pauli matrices. The required equations are, therefore,

$$\sigma^i{}_{AB'}P^A{}_{;i} + i\mu_*\bar{Q}^{C'}\varepsilon_{C'B'} = 0 \tag{99}$$

and

$$\sigma^i{}_{AB'}Q^A{}_{;i} + i\mu_*\bar{P}^{C'}\varepsilon_{C'B'} = 0, \tag{100}$$

where

$$\sigma^i{}_{AB'} = \frac{1}{\sqrt{2}}\begin{vmatrix} l^i & m^i \\ \bar{m}^i & n^i \end{vmatrix}. \tag{101}$$

We shall now write out the explicit forms of these equations in terms of the spin coefficients we have defined.

Consider equation (99) for $B' = 0$. We have

$$\sigma^i{}_{00'}P^0{}_{;i} + \sigma^i{}_{10'}P^1{}_{;i} - i\mu_*\bar{Q}^{1'} = 0, \tag{102}$$

or, by virtue of equations (68), (69), (91), and (101),

$$(\partial_{00'}P^0+\Gamma^0{}_{b00'}P^b)+(\partial_{10'}P^1+\Gamma^1{}_{b10'}P^b)-i\mu_*\bar{Q}^{1'}=0, \tag{103}$$

or, more explicitly,

$$(D+\Gamma_{1000'}-\Gamma_{0010'})P^0+(\delta^*+\Gamma_{1100'}-\Gamma_{0110'})P^1-i\mu_*\bar{Q}^{1'}=0. \tag{104}$$

Now replacing the various spin coefficients in equation (104) by their named symbols listed in (92), we obtain

$$(D+\varepsilon-\rho)P^0+(\delta^*+\pi-\alpha)P^1-i\mu_*\bar{Q}^{1'}=0. \tag{105}$$

Similarly, equation (99) for $B'=1$, yields

$$(\triangle+\mu-\gamma)P^1+(\delta+\beta-\tau)P^0+i\mu_*\bar{Q}^{0'}=0. \tag{106}$$

Equation (100) provides the same equations (105) and (106) with P and Q interchanged. It is, however, convenient to consider the complex conjugate of equation (100) and further write

$$F_1=P^0,\ F_2=P^1,\ G_1=\bar{Q}^{1'},\quad\text{and}\quad G_2=-\bar{Q}^{0'}. \tag{107}$$

The resulting equations are

$$\left.\begin{aligned}(D+\varepsilon-\rho)F_1+(\delta^*+\pi-\alpha)F_2&=i\mu_*G_1;\\(\triangle+\mu-\gamma)F_2+(\delta+\beta-\tau)F_1&=i\mu_*G_2;\\(D+\varepsilon^*-\rho^*)G_2-(\delta+\pi^*-\alpha^*)G_1&=i\mu_*F_2;\\(\triangle+\mu^*-\gamma^*)G_1-(\delta^*+\beta^*-\tau^*)G_2&=i\mu_*F_1.\end{aligned}\right\} \tag{108}$$

These are the Dirac equations in the Newman–Penrose formalism.

104. Dirac's equations in Kerr geometry and their separation

Assuming that the four components of the wave function—F_1, F_2, G_1, and G_2 in our present context—have the customary dependence, $e^{i(\sigma t+m\varphi)}$, on t and φ and inserting for the spin coefficients and the directional derivatives their values given in Chapter 6, equations (170)–(173) and (175) and Chapter 8, equations (2)–(5), we find that equations (108) reduce to the forms

$$\left.\begin{aligned}\left(\mathscr{D}_0+\frac{1}{\bar\rho^*}\right)F_1+\frac{1}{\bar\rho^*\sqrt{2}}\mathscr{L}_{\frac{1}{2}}F_2&=+i\mu_*G_1,\\\frac{\Delta}{2\rho^2}\mathscr{D}^\dagger_{\frac{1}{2}}F_2-\frac{1}{\bar\rho\sqrt{2}}\left(\mathscr{L}^\dagger_{\frac{1}{2}}+\frac{ia\sin\theta}{\bar\rho^*}\right)F_1&=-i\mu_*G_2,\\\left(\mathscr{D}_0+\frac{1}{\bar\rho}\right)G_2-\frac{1}{\bar\rho\sqrt{2}}\mathscr{L}^\dagger_{\frac{1}{2}}G_1&=+i\mu_*F_2,\\\frac{\Delta}{2\rho^2}\mathscr{D}^\dagger_{\frac{1}{2}}G_1+\frac{1}{\bar\rho^*\sqrt{2}}\left(\mathscr{L}_{\frac{1}{2}}-\frac{ia\sin\theta}{\bar\rho}\right)G_2&=-i\mu_*F_1.\end{aligned}\right\} \tag{109}$$

The forms of equations (109) suggest that in place of F_1 and G_2 we define

$$f_1 = \bar{\rho}^* F_1 = (r - ia\cos\theta)F_1 \quad (= \bar{\rho}^* P^0)$$

and (110)

$$g_2 = \bar{\rho} G_2 = (r + ia\cos\theta)G_2 \quad (= -\bar{\rho}\bar{Q}^{0'}).$$

Also writing f_2 and g_1 in place of $F_2\,(= P^1)$ and $G_1\,(= \bar{Q}^{1'})$ (for symmetry in form of the resulting equations) we find that equations (109) become

$$\left.\begin{aligned}
\mathscr{D}_0 f_1 + 2^{-\frac{1}{2}}\mathscr{L}_{\frac{1}{2}} f_2 &= +(i\mu_* r + a\mu_*\cos\theta)g_1,\\
\Delta\mathscr{D}^\dagger_{\frac{1}{2}} f_2 - 2^{+\frac{1}{2}}\mathscr{L}^\dagger_{\frac{1}{2}} f_1 &= -2(i\mu_* r + a\mu_*\cos\theta)g_2,\\
\mathscr{D}_0 g_2 - 2^{-\frac{1}{2}}\mathscr{L}^\dagger_{\frac{1}{2}} g_1 &= +(i\mu_* r - a\mu_*\cos\theta)f_2,\\
\Delta\mathscr{D}^\dagger_{\frac{1}{2}} g_1 + 2^{+\frac{1}{2}}\mathscr{L}_{\frac{1}{2}} g_2 &= -2(i\mu_* r - a\mu_*\cos\theta)f_1.
\end{aligned}\right\} \tag{111}$$

It is now apparent that the variables can be separated by the substitutions

$$\left.\begin{aligned}
f_1(r,\theta) = R_{-\frac{1}{2}}(r)S_{-\frac{1}{2}}(\theta), \quad & f_2(r,\theta) = R_{+\frac{1}{2}}(r)S_{+\frac{1}{2}}(\theta),\\
g_1(r,\theta) = R_{+\frac{1}{2}}(r)S_{-\frac{1}{2}}(\theta), \quad \text{and} \quad & g_2(r,\theta) = R_{-\frac{1}{2}}(r)S_{+\frac{1}{2}}(\theta),
\end{aligned}\right\} \tag{112}$$

where $R_{\pm\frac{1}{2}}(r)$ and $S_{\pm\frac{1}{2}}(\theta)$ are functions, respectively, of r and θ only. With these substitutions, equations (111) become

$$\begin{aligned}
(\mathscr{D}_0 R_{-\frac{1}{2}} - i\mu_* r R_{+\frac{1}{2}})S_{-\frac{1}{2}} + [2^{-\frac{1}{2}}\mathscr{L}_{\frac{1}{2}} S_{+\frac{1}{2}} - (a\mu_*\cos\theta)S_{-\frac{1}{2}}]R_{+\frac{1}{2}} &= 0,\\
(\Delta\mathscr{D}^\dagger_{\frac{1}{2}} R_{+\frac{1}{2}} + 2i\mu_* r R_{-\frac{1}{2}})S_{+\frac{1}{2}} - [2^{+\frac{1}{2}}\mathscr{L}^\dagger_{\frac{1}{2}} S_{-\frac{1}{2}} - 2(a\mu_*\cos\theta)S_{+\frac{1}{2}}]R_{-\frac{1}{2}} &= 0,\\
(\mathscr{D}_0 R_{-\frac{1}{2}} - i\mu_* r R_{+\frac{1}{2}})S_{+\frac{1}{2}} - [2^{-\frac{1}{2}}\mathscr{L}^\dagger_{\frac{1}{2}} S_{-\frac{1}{2}} - (a\mu_*\cos\theta)S_{+\frac{1}{2}}]R_{+\frac{1}{2}} &= 0,\\
(\Delta\mathscr{D}^\dagger_{\frac{1}{2}} R_{+\frac{1}{2}} + 2i\mu_* r R_{-\frac{1}{2}})S_{-\frac{1}{2}} + [2^{+\frac{1}{2}}\mathscr{L}_{\frac{1}{2}} S_{+\frac{1}{2}} - 2(a\mu_*\cos\theta)S_{-\frac{1}{2}}]R_{-\frac{1}{2}} &= 0.
\end{aligned} \tag{113}$$

These equations imply that

$$\begin{aligned}
\mathscr{D}_0 R_{-\frac{1}{2}} - i\mu_* r R_{+\frac{1}{2}} &= \lambda_1 R_{+\frac{1}{2}}; & 2^{-\frac{1}{2}}\mathscr{L}_{\frac{1}{2}} S_{+\frac{1}{2}} - (a\mu_*\cos\theta)S_{-\frac{1}{2}} &= -\lambda_1 S_{-\frac{1}{2}},\\
\Delta\mathscr{D}^\dagger_{\frac{1}{2}} R_{+\frac{1}{2}} + 2i\mu_* r R_{-\frac{1}{2}} &= \lambda_2 R_{-\frac{1}{2}}; & 2^{+\frac{1}{2}}\mathscr{L}^\dagger_{\frac{1}{2}} S_{-\frac{1}{2}} - 2(a\mu_*\cos\theta)S_{+\frac{1}{2}} &= +\lambda_2 S_{+\frac{1}{2}},\\
\mathscr{D}_0 R_{-\frac{1}{2}} - i\mu_* r R_{+\frac{1}{2}} &= \lambda_3 R_{+\frac{1}{2}}; & 2^{-\frac{1}{2}}\mathscr{L}^\dagger_{\frac{1}{2}} S_{-\frac{1}{2}} - (a\mu_*\cos\theta)S_{+\frac{1}{2}} &= +\lambda_3 S_{+\frac{1}{2}},\\
\Delta\mathscr{D}^\dagger_{\frac{1}{2}} R_{+\frac{1}{2}} + 2i\mu_* r R_{-\frac{1}{2}} &= \lambda_4 R_{-\frac{1}{2}}; & 2^{+\frac{1}{2}}\mathscr{L}_{\frac{1}{2}} S_{+\frac{1}{2}} - 2(a\mu_*\cos\theta)S_{-\frac{1}{2}} &= -\lambda_4 S_{-\frac{1}{2}},
\end{aligned} \tag{114}$$

where $\lambda_1, \ldots, \lambda_4$ are four constants of separation. However, it is manifest that the consistency of the foregoing equations requires that

$$\lambda_1 = \lambda_3 = \tfrac{1}{2}\lambda_2 = \tfrac{1}{2}\lambda_4 = \lambda \quad \text{(say).} \tag{115}$$

We are thus left with the two pairs of equations,

$$\left.\begin{aligned}\mathscr{D}_0 R_{-\frac{1}{2}} &= (\lambda + i\mu_* r) R_{+\frac{1}{2}},\\ \Delta \mathscr{D}_{\frac{1}{2}}^{\dagger} R_{+\frac{1}{2}} &= 2(\lambda - i\mu_* r) R_{-\frac{1}{2}},\end{aligned}\right\} \tag{116}$$

and

$$\left.\begin{aligned}\mathscr{L}_{\frac{1}{2}} S_{+\frac{1}{2}} &= -2^{\frac{1}{2}}(\lambda - a\mu_* \cos\theta) S_{-\frac{1}{2}},\\ \mathscr{L}_{\frac{1}{2}}^{\dagger} S_{-\frac{1}{2}} &= +2^{\frac{1}{2}}(\lambda + a\mu_* \cos\theta) S_{+\frac{1}{2}}.\end{aligned}\right\} \tag{117}$$

It is convenient at this stage to:

$$\text{replace } 2^{\frac{1}{2}}\lambda \text{ by } \lambda,\ 2^{\frac{1}{2}}\mu_* \text{ by } m_e \quad\text{and}\quad 2^{\frac{1}{2}}R_{-\frac{1}{2}} \text{ by } R_{-\frac{1}{2}}. \tag{118}$$

With these replacements, equations (116) and (117) become

$$\Delta^{\frac{1}{2}}\mathscr{D}_0 R_{-\frac{1}{2}} = (\lambda + i m_e r)\Delta^{\frac{1}{2}} R_{+\frac{1}{2}}, \tag{119}$$

$$\Delta^{\frac{1}{2}}\mathscr{D}_0^{\dagger}\Delta^{\frac{1}{2}} R_{+\frac{1}{2}} = (\lambda - i m_e r) R_{-\frac{1}{2}}, \tag{120}$$

$$\mathscr{L}_{\frac{1}{2}} S_{+\frac{1}{2}} = -(\lambda - a m_e \cos\theta) S_{-\frac{1}{2}}, \tag{121}$$

and

$$\mathscr{L}_{\frac{1}{2}}^{\dagger} S_{-\frac{1}{2}} = +(\lambda + a m_e \cos\theta) S_{+\frac{1}{2}}, \tag{122}$$

where it may be noted m_e is the mass of the particle expressed as the inverse of its Compton wavelength (i.e., $m_e c/\hbar$ is replaced by m_e).

We can eliminate $\Delta^{\frac{1}{2}}R_{+\frac{1}{2}}$ from equations (119) and (120) to obtain an equation for $R_{-\frac{1}{2}}$. Thus,

$$\left[\Delta \mathscr{D}_{\frac{1}{2}}^{\dagger}\mathscr{D}_0 - \frac{i m_e \Delta}{\lambda + i m_e r}\mathscr{D}_0 - (\lambda^2 + m_e^2 r^2)\right] R_{-\frac{1}{2}} = 0; \tag{123}$$

and $\Delta^{\frac{1}{2}}R_{+\frac{1}{2}}$ satisfies the complex-conjugate equation.

Similarly, we can eliminate $S_{+\frac{1}{2}}$ from equations (121) and (122) to obtain an equation for $S_{-\frac{1}{2}}$; thus,

$$\left[\mathscr{L}_{\frac{1}{2}}\mathscr{L}_{\frac{1}{2}}^{\dagger} + \frac{a m_e \sin\theta}{\lambda + a m_e \cos\theta}\mathscr{L}_{\frac{1}{2}}^{\dagger} + (\lambda^2 - a^2 m_e^2 \cos^2\theta)\right] S_{-\frac{1}{2}} = 0; \tag{124}$$

and $S_{+\frac{1}{2}}$ satisfies the 'adjoint' equation (obtained by replacing θ by $\pi - \theta$). Also, λ now appears as a characteristic-value parameter determined by the boundary conditions which require $S_{-\frac{1}{2}}$ (and, therefore, also $S_{+\frac{1}{2}}$) to be regular at $\theta = 0$ and $\theta = \pi$.

We shall return to these separated equations in §106 after considering the simpler case of the massless two-component neutrinos in §105.

105. Neutrino waves in Kerr geometry

The equations appropriate to the two-component neutrinos (satisfying Weyl's equations in Minkowski space) can be obtained by simply setting

$m_e = 0$ in equations (119)–(124). Also, by letting (cf. Ch. 8, equation (65) and Ch. 9, equation (72))

$$\Delta^{\frac{1}{2}} R_{+\frac{1}{2}} = P_{+\frac{1}{2}} \quad \text{and} \quad R_{-\frac{1}{2}} = P_{-\frac{1}{2}}, \tag{125}$$

we have the two pairs of equations

$$\Delta^{\frac{1}{2}} \mathscr{D}_0^\dagger P_{+\frac{1}{2}} = \lambda P_{-\frac{1}{2}}, \quad \Delta^{\frac{1}{2}} \mathscr{D}_0 P_{-\frac{1}{2}} = \lambda P_{+\frac{1}{2}}, \tag{126}$$

and

$$\mathscr{L}_{\frac{1}{2}} S_{+\frac{1}{2}} = -\lambda S_{-\frac{1}{2}}, \quad \mathscr{L}_{\frac{1}{2}}^\dagger S_{-\frac{1}{2}} = +\lambda S_{+\frac{1}{2}}. \tag{127}$$

The corresponding decoupled second-order equations are

$$(\Delta \mathscr{D}_{\frac{1}{2}} \mathscr{D}_0^\dagger - \lambda^2) P_{+\frac{1}{2}} = 0, \quad (\Delta \mathscr{D}_{\frac{1}{2}}^\dagger \mathscr{D}_0 - \lambda^2) P_{-\frac{1}{2}} = 0, \tag{128}$$

and

$$\mathscr{L}_{\frac{1}{2}}^\dagger \mathscr{L}_{\frac{1}{2}} S_{+\frac{1}{2}} = -\lambda^2 S_{+\frac{1}{2}}, \quad \mathscr{L}_{\frac{1}{2}} \mathscr{L}_{\frac{1}{2}}^\dagger S_{-\frac{1}{2}} = -\lambda^2 S_{-\frac{1}{2}}. \tag{129}$$

We observe that equations (128) are, indeed, special cases of Chapter 8, equations (96) and (97) appropriate for $|s| = \frac{1}{2}$. The general transformation theory of §§72 and 73 can accordingly be applied to transform equations (128) to the form of one-dimensional wave-equations. However, in the present instance, it is simpler to proceed directly from equations (128) and skirt the transformation theory.

As we have seen in §72, with the choice of r_* (defined in Chapter 8, equations (100) and (101)) as the independent variable, the operators $\mathscr{D}_0$ and $\mathscr{D}_0^\dagger$ take the simple forms

$$\mathscr{D}_0 = \frac{\varpi^2}{\Delta}\left(\frac{d}{dr_*} + i\sigma\right) \quad \text{and} \quad \mathscr{D}_0^\dagger = \frac{\varpi^2}{\Delta}\left(\frac{d}{dr_*} - i\sigma\right), \tag{130}$$

where it may be recalled that

$$\varpi^2 = r^2 + \alpha^2 = r^2 + a^2 + am/\sigma. \tag{131}$$

Equations (126) then become

$$\left(\frac{d}{dr_*} - i\sigma\right) P_{+\frac{1}{2}} = \lambda \frac{\Delta^{\frac{1}{2}}}{\varpi^2} P_{-\frac{1}{2}} \tag{132}$$

and

$$\left(\frac{d}{dr_*} + i\sigma\right) P_{-\frac{1}{2}} = \lambda \frac{\Delta^{\frac{1}{2}}}{\varpi^2} P_{+\frac{1}{2}}. \tag{133}$$

Letting

$$Z_\pm = P_{+\frac{1}{2}} \pm P_{-\frac{1}{2}}, \tag{134}$$

we can combine equations (132) and (133) to give

$$\left(\frac{d}{dr_*} - \lambda \frac{\Delta^{\frac{1}{2}}}{\varpi^2}\right) Z_+ = i\sigma Z_- \tag{135}$$

and

$$\left(\frac{d}{dr_*}+\lambda\frac{\Delta^{\frac{1}{2}}}{\varpi^2}\right)Z_- = i\sigma Z_+. \tag{136}$$

From these equations, we readily obtain the pair of one-dimensional wave-equations,

$$\left(\frac{d^2}{dr_*^2}+\sigma^2\right)Z_\pm = V_\pm Z_\pm, \tag{137}$$

where

$$V_\pm = \lambda^2\frac{\Delta}{\varpi^4}\pm\lambda\frac{d}{dr_*}\left(\frac{\Delta^{\frac{1}{2}}}{\varpi^2}\right). \tag{138}$$

(a) The problem of reflexion and transmission for $\sigma > \sigma_s$ $(= -am/2Mr_+)$

There is clearly no ambiguity in using the one-dimensional wave-equation (137) with either of the two potentials (138) so long as $\sigma > \sigma_s$: for the $r_*(r)$-relation is then single-valued and the potentials themselves are bounded and of short range. Besides, since the potentials $V_\pm$ given by equations (138) belong to the general class

$$V_\pm = \pm\beta\frac{df}{dr_*}+\beta^2 f^2+\kappa f, \tag{139}$$

considered in Chapter 4, §26 (with

$$\beta f = \lambda\frac{\Delta^{\frac{1}{2}}}{\varpi^2} \quad \text{and} \quad \kappa = 0), \tag{140}$$

it follows that the equations governing Z_+ and Z_- will yield the same reflexion and transmission coefficients. We shall see presently that this equality of the reflexion and transmission coefficients follows more directly from equations (135) and (136) according to which

$$Z_\pm \to \frac{1}{i\sigma}\frac{dZ_\mp}{dr_*} \quad \text{for} \quad r_*\to\pm\infty. \tag{141}$$

Returning to equations (137), we seek, as usual, solutions of the wave equation, with the potential V_+, for example, which have the asymptotic behaviours,

$$\left.\begin{aligned} Z_+ &\to e^{+i\sigma r_*}+A(\sigma)e^{-i\sigma r_*} \quad (r_*\to+\infty) \\ &\to \qquad\qquad B(\sigma)e^{+i\sigma r_*} \quad (r_*\to-\infty). \end{aligned}\right\} \tag{142}$$

With the aid of equation (135), we can derive from these solutions for Z_+, solutions Z_-, of the wave equation with the potential V_-, which have (by

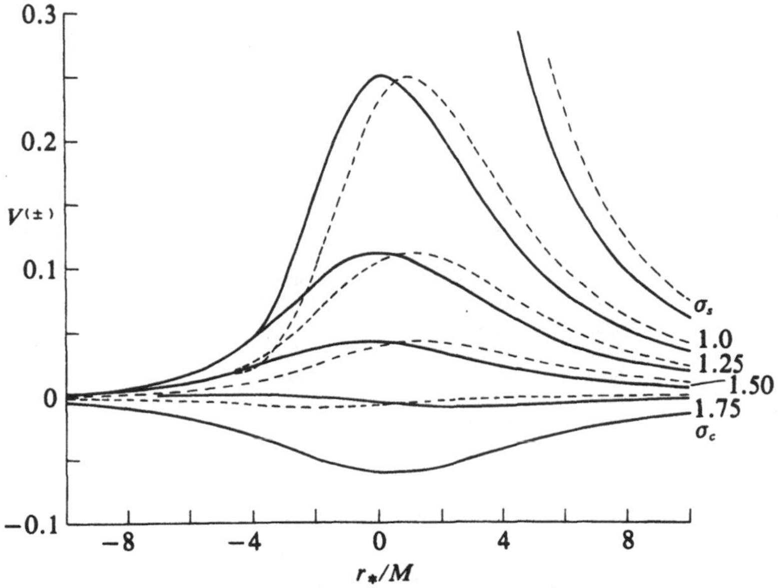

FIG. 46. The potential barriers, $V^{(+)}$ (the full-line curves) and $V^{(-)}$ (the dashed curves), surrounding a Kerr black-hole ($a = 0.95$) for the incidence of neutrino waves with $l = 2.5$, $m = -2.5$ for frequencies in the range of $\sigma^+_c \leqslant \sigma^+ < \sigma^+_s$. The curves are labelled by the values of σ^+ to which they belong.

equation (141)) the asymptotic behaviours

$$\left.\begin{aligned} Z_- &\to e^{+i\sigma r_*} - A(\sigma)e^{-i\sigma r_*} \quad (r_* \to +\infty) \\ &\to \qquad B(\sigma)e^{+i\sigma r_*} \quad (r_* \to -\infty). \end{aligned}\right\} \tag{143}$$

Hence the potentials $V_\pm$, will yield the same reflexion and transmission coefficients given by

$$\mathbb{R} = |A(\sigma)|^2 \quad \text{and} \quad \mathbb{T} = |B(\sigma)|^2, \tag{144}$$

satisfying the condition

$$\mathbb{R} + \mathbb{T} = 1. \tag{145}$$

We shall verify in §§106 and 107, in the more general context of Dirac's equation, that the reflexion and the transmission coefficients we have defined in terms of the solutions of the wave equations governing $Z_\pm$ are in accord with the physical definition of the flux of neutrinos across the event horizon.

In Fig. 46 we exhibit a family of the potentials $V_\pm$ for frequencies σ in the range, $\sigma_s < \sigma \leqslant \sigma_c$ $(= -m/a)$. In this range of frequencies, $\alpha^2 \leqslant 0$ but $0 \leqslant |\alpha| < r_+$; and $|\alpha| \to r_+ + 0$ as $\sigma \to \sigma_s + 0$. As a consequence of this last fact, the

TABLE XI

Reflexion coefficients for neutrinos incident on a Kerr black-hole with $a = 0.95$ ($l = 0.5$, $m = -0.5$)

σ	σ/σ_s	R	σ	σ/σ_s	R
0.181987	1.0055	0.97627	0.220987	1.2210	0.89187
0.182987	1.0111	0.97556	0.230987	1.2763	0.83457
0.183987	1.0166	0.97479	0.240987	1.3315	0.75356
0.184987	1.0221	0.97398	0.250000	1.3813	0.66005
0.185987	1.0276	0.97313	0.250987	1.3868	0.64885
0.186987	1.0332	0.97223	0.260987	1.4420	0.52777
0.187987	1.0387	0.97129	0.300000	1.6576	0.14125
0.188987	1.0442	0.97012	0.350000	1.9338	0.01622
0.189987	1.0497	0.96925	0.400000	2.2101	0.00197
0.200987	1.1105	0.95353	0.450000	2.4864	0.00027
0.210987	1.1658	0.92971			

potential barrier presented to the oncoming neutrino waves increases without bound as $\sigma \to \sigma_s + 0$. But the singularity of the potential, in this limit, at the horizon is weaker than it is for photons and gravitons. Thus, considering the integral over V as a measure of the barrier, we find that for both potentials given by equation (138)

$$\int_{-\infty}^{+\infty} V_{\pm}\, dr_* = -\frac{\lambda^2}{2|\alpha|}\lg\left[\frac{2Mr_+(1-\sigma_s/\sigma)}{(r_+ + |\alpha|)^2}\right]$$

$$(\text{for } \sigma_s < \sigma < \sigma_c \quad \text{and} \quad 0 \leqslant |\alpha| < r_+). \qquad (146)$$

This integral diverges logarithmically as $\sigma \to \sigma_s + 0$, in contrast to a divergence like $(1-\sigma_s/\sigma)^{-2}$ in the electromagnetic and the gravitational cases. It appears that this weaker divergence in the barrier for the neutrinos results in a *finite* transmission coefficient for them in the limit $\sigma \to \sigma_s + 0$. The numerical results given in Table XI strongly support this surmise, though it remains to be established rigorously.

(*b*) *The absence of super-radiance* ($0 < \sigma < \sigma_s$)

For $0 < \sigma < \sigma_s$, the $r_*(r)$-relation becomes double-valued; and, besides, the potentials, $V_{\pm}$, become singular at $r = |\alpha|$ ($> r_+$ when $\sigma < \sigma_s$). Thus, from the explicit expression for $V_{\pm}$, namely,

$$V_{\pm} = \lambda^2 \frac{\Delta}{\varpi^4} \pm \lambda \frac{\Delta^{\frac{1}{2}}}{\varpi^4}\left[(r-M) - \frac{2r\Delta}{\varpi^2}\right], \qquad (147)$$

we find that in the neighbourhood of $r = |\alpha|$, they have the behaviour

$$V_{\pm} \sim \mp\lambda \frac{\Delta_{|\alpha|}^{3/2}}{4|\alpha|^2}\frac{1}{(r-|\alpha|)^3}. \qquad (148)$$

This behaviour differs from the corresponding behaviours (Chapter 8, equation (193) and Chapter 9, equation (428)) of the potentials for $|s| = 1$ and 2 in an important respect—we now have a pole of order 3 instead of a pole of order 4.

Turning to the problem of reflexion and transmission, we must seek, as we have explained in Chapter 8, §§75(*c*) and 76(*c*), solutions of the wave equations for Z which satisfy the boundary conditions

$$Z \to e^{+i\sigma r_*} + A e^{-i\sigma r_*} \text{ along the branch } r \to \infty \quad \text{and} \quad r_* \to +\infty, \tag{149}$$

and

$$Z \to B e^{+i\sigma r_*} \text{ along the branch } r \to r_+ + 0 \quad \text{and} \quad r_* \to +\infty. \tag{150}$$

Also, by Chapter 8, equation (140), for the case $|s| = \frac{1}{2}$, which we are now considering, the Wronskian of Z and Z^* must remain the same on either side of the singularity at $r = |\alpha|$, in contrast to a change of sign demanded for integral spins. On this account, the conservation law for incident neutrino-waves will continue to be

$$\mathbb{R} + \mathbb{T} = 1 \quad \text{also for} \quad 0 < \sigma < \sigma_s. \tag{151}$$

In other words, *super-radiance will not be manifested*—a fact which emerges also from the numerical results of Table XI.

We shall now briefly indicate the behaviour of the solutions for $Z_\pm$ in the neighbourhood of $r = |\alpha|$.

Letting

$$x = r - |\alpha|, \tag{152}$$

we find that in the neighbourhood of $r = |\alpha|$, the equation governing $Z_\pm$ becomes

$$x\frac{d^2 Z_\pm}{dx^2} - \frac{dZ_\pm}{dx} \pm \frac{\lambda}{\Delta_{|\alpha|}^{1/2}} Z_\pm = 0. \tag{153}$$

And the solution of this equation is

$$Z_+ = y\mathscr{C}_2(\sqrt{y}) \quad \text{for } V_+ \text{ and} \quad Z_- = y\mathscr{C}_2(i\sqrt{y}) \quad \text{for } V_-, \tag{154}$$

where

$$y = 4\lambda x / \Delta_{|\alpha|}^{1/2}, \tag{155}$$

and $\mathscr{C}_2$ denotes the general solution of Bessel's equation of order 2. With this known behaviour of $Z_\pm$ at the singularity, there is no difficulty of principle in integrating the equation through $r = |\alpha|$ and completing the solution in the manner described in Chapter 8, §75(*c*) in the electromagnetic context.

The principal physical fact which emerges from the foregoing considerations is that particles of spin-$\frac{1}{2}$ do not show any super-radiance. (We shall confirm this fact also in the context of massive particles in §107.) On the analytical side, the origins of this circumstance are, *first*, that the potential barriers have a pole of order 3 (instead of a pole of order 4 as in the case of particles with integral spins) and *second* that the Wronskian, $[Z, Z^*]$, remains unchanged as the singularity in V is crossed (instead of changing sign as is demanded in the case of particles of integral spins).

106. The conserved current and the reduction of Dirac's equations to the form of one-dimensional wave-equations

We have shown in §104 that the solutions for the basic spinors, P^A and $\bar{Q}^{A'}$, in Kerr geometry, describing a spin-$\frac{1}{2}$ particle, are expressible in the forms

$$\left.\begin{aligned} P^0 &= \frac{1}{\bar{\rho}^*\sqrt{2}}R_{-\frac{1}{2}}(r)S_{-\frac{1}{2}}(\theta), \quad \bar{Q}^{0'} = -\frac{1}{\bar{\rho}\sqrt{2}}R_{-\frac{1}{2}}(r)S_{+\frac{1}{2}}(\theta), \\ P^1 &= R_{+\frac{1}{2}}(r)S_{+\frac{1}{2}}(\theta), \quad \text{and} \quad \bar{Q}^{1'} = R_{+\frac{1}{2}}(r)S_{-\frac{1}{2}}(\theta), \end{aligned}\right\} \tag{156}$$

where we have restored in the expressions for P^0 and $\bar{Q}^{0'}$ the factor $1/\sqrt{2}$ suppressed in passing from equations (116) and (117) to equations (119)–(122).

The equations governing the radial functions are

$$\Delta^{\frac{1}{2}}\left(\frac{\mathrm{d}}{\mathrm{d}r} - i\frac{r^2+\alpha^2}{\Delta}\sigma\right)P_{+\frac{1}{2}} = (\lambda - im_e r)P_{-\frac{1}{2}} \tag{157}$$

and

$$\Delta^{\frac{1}{2}}\left(\frac{\mathrm{d}}{\mathrm{d}r} + i\frac{r^2+\alpha^2}{\Delta}\sigma\right)P_{-\frac{1}{2}} = (\lambda + im_e r)P_{+\frac{1}{2}}, \tag{158}$$

where

$$P_{+\frac{1}{2}} = \Delta^{\frac{1}{2}}R_{+\frac{1}{2}} \quad \text{and} \quad P_{-\frac{1}{2}} = R_{-\frac{1}{2}}. \tag{159}$$

From equations (157) and (158), we readily find that

$$\frac{\mathrm{d}}{\mathrm{d}r}(|P_{+\frac{1}{2}}|^2 - |P_{-\frac{1}{2}}|^2) = 0. \tag{160}$$

An equivalent relation, involving the Wronskian, $[P_{-\frac{1}{2}}, P^*_{+\frac{1}{2}}]$, is

$$\frac{\Delta^{\frac{1}{2}}}{\lambda + im_e r}[P_{-\frac{1}{2}}, P^*_{+\frac{1}{2}}] = \text{constant}. \tag{161}$$

The quantity $(|P_{+\frac{1}{2}}|^2 - |P_{-\frac{1}{2}}|^2)$, whose constancy is guaranteed by equation (160), is, as we shall show presently, the conserved net current of particles given by

$$\frac{\partial N}{\partial t} = -\int_0^{2\pi}\int_0^{\pi} J^r(\sqrt{-g})\,\mathrm{d}\theta\,\mathrm{d}\varphi, \tag{162}$$

where

$$\sqrt{-g} = \rho^2\sin\theta = (r^2 + a^2\cos^2\theta)\sin\theta, \tag{163}$$

and

$$\frac{1}{\sqrt{2}} J^r = \sigma^r{}_{AB'} (P^A \bar{P}^{B'} + Q^A \bar{Q}^{B'}) \tag{164}$$

is the radial component of the Dirac current, J^i.

For the chosen basis vectors, $\boldsymbol{l}$, $\boldsymbol{n}$, $\boldsymbol{m}$, and $\bar{\boldsymbol{m}}$,

$$\sigma^r{}_{AB'} = \frac{1}{\sqrt{2}} \begin{vmatrix} l^r & m^r \\ \bar{m}^r & n^r \end{vmatrix} = \frac{1}{\sqrt{2}} \begin{vmatrix} 1 & 0 \\ 0 & -\Delta/2\rho^2 \end{vmatrix}. \tag{165}$$

Evaluating J^r with the aid of equations (156), (164), and (165), we find

$$J^r = \frac{1}{2\rho^2}\left(|R_{-\frac{1}{2}}|^2 - \Delta |R_{+\frac{1}{2}}|^2\right)\left(S^2_{-\frac{1}{2}} + S^2_{+\frac{1}{2}}\right). \tag{166}$$

Assuming that the angular functions, $S_{\pm\frac{1}{2}}(\theta)$ are normalized to unity, we find from equations (162), (163), and (166) that

$$\frac{\partial N}{\partial t} = 2\pi\left(|P_{+\frac{1}{2}}|^2 - |P_{-\frac{1}{2}}|^2\right) = \text{constant}. \tag{167}$$

(*a*) *The reduction of Dirac's equations to the form of one-dimensional wave-equations*

With our customary choice of r_* as the independent variable, equations (119) and (120) take the forms (cf. equations (132) and (133))

$$\left(\frac{\mathrm{d}}{\mathrm{d}r_*} - i\sigma\right) P_{+\frac{1}{2}} = \frac{\Delta^{\frac{1}{2}}}{\varpi^2}\left(\lambda - i m_e r\right) P_{-\frac{1}{2}} \tag{168}$$

and

$$\left(\frac{\mathrm{d}}{\mathrm{d}r_*} + i\sigma\right) P_{-\frac{1}{2}} = \frac{\Delta^{\frac{1}{2}}}{\varpi^2}\left(\lambda + i m_e r\right) P_{+\frac{1}{2}}. \tag{169}$$

Let

$$\vartheta = \tan^{-1}(m_e r/\lambda). \tag{170}$$

Then

$$\cos\vartheta = \frac{\lambda}{\sqrt{(\lambda^2 + m_e^2 r^2)}}, \quad \sin\vartheta = \frac{m_e r}{\sqrt{(\lambda^2 + m_e^2 r^2)}}, \tag{171}$$

and

$$(\lambda \pm i m_e r) = e^{\pm i\vartheta}\sqrt{(\lambda^2 + m_e^2 r^2)}; \tag{172}$$

and we can write equations (168) and (169) in the alternative forms

$$\left(\frac{\mathrm{d}}{\mathrm{d}r_*} - i\sigma\right) P_{+\frac{1}{2}} = \frac{\Delta^{\frac{1}{2}}}{\varpi^2}(\lambda^2 + m_e^2 r^2)^{1/2} P_{-\frac{1}{2}} \exp\left[-i\tan^{-1}\left(\frac{m_e r}{\lambda}\right)\right] \tag{173}$$

and

$$\left(\frac{\mathrm{d}}{\mathrm{d}r_*} + i\sigma\right) P_{-\frac{1}{2}} = \frac{\Delta^{\frac{1}{2}}}{\varpi^2}(\lambda^2 + m_e^2 r^2)^{1/2} P_{+\frac{1}{2}} \exp\left[+i\tan^{-1}\left(\frac{m_e r}{\lambda}\right)\right]. \tag{174}$$

We can eliminate the exponential factors on the right-hand sides of equations (173) and (174) by the substitutions

$$\left.\begin{aligned} P_{+\frac{1}{2}} &= \psi_{+\frac{1}{2}} \exp\left[-\tfrac{1}{2} i \tan^{-1}\left(\frac{m_e r}{\lambda}\right)\right] \\ \text{and} \qquad & \\ P_{-\frac{1}{2}} &= \psi_{-\frac{1}{2}} \exp\left[+\tfrac{1}{2} i \tan^{-1}\left(\frac{m_e r}{\lambda}\right)\right]. \end{aligned}\right\} \tag{175}$$

We are than left with

$$\frac{d\psi_{+\frac{1}{2}}}{dr_*} - i\sigma\left(1 + \frac{\Delta}{\varpi^2}\frac{\lambda m_e}{2\sigma}\frac{1}{\lambda^2 + m_e^2 r^2}\right)\psi_{+\frac{1}{2}} = \frac{\Delta^{\frac{1}{2}}}{\varpi^2}(\lambda^2 + m_e^2 r^2)^{1/2}\psi_{-\frac{1}{2}} \tag{176}$$

and

$$\frac{d\psi_{-\frac{1}{2}}}{dr_*} + i\sigma\left(1 + \frac{\Delta}{\varpi^2}\frac{\lambda m_e}{2\sigma}\frac{1}{\lambda^2 + m_e^2 r^2}\right)\psi_{-\frac{1}{2}} = \frac{\Delta^{\frac{1}{2}}}{\varpi^2}(\lambda^2 + m_e^2 r^2)^{1/2}\psi_{+\frac{1}{2}}. \tag{177}$$

We can obtain the same factor in front of the first term on the left-hand sides as there is in front of the second by changing the independent variable to

$$\hat{r}_* = r_* + \frac{1}{2\sigma}\tan^{-1}\frac{m_e r}{\lambda}, \tag{178}$$

in which case

$$d\hat{r}_* = \left(1 + \frac{\Delta}{\varpi^2}\frac{\lambda m_e}{2\sigma}\frac{1}{\lambda^2 + m_e^2 r^2}\right) dr_*. \tag{179}$$

Thus, by the change of variables (175) and (178), we obtain the equations

$$\left(\frac{d}{d\hat{r}_*} - i\sigma\right)\psi_{+\frac{1}{2}} = W\psi_{-\frac{1}{2}} \tag{180}$$

and

$$\left(\frac{d}{d\hat{r}_*} + i\sigma\right)\psi_{-\frac{1}{2}} = W\psi_{+\frac{1}{2}}, \tag{181}$$

where

$$W = \frac{\Delta^{\frac{1}{2}}(\lambda^2 + m_e^2 r^2)^{3/2}}{\varpi^2(\lambda^2 + m_e^2 r^2) + \lambda m_e \Delta/2\sigma}. \tag{182}$$

Now letting

$$Z_{\pm} = \psi_{+\frac{1}{2}} \pm \psi_{-\frac{1}{2}}, \tag{183}$$

we can combine equations (180) and (181) to give

$$\left(\frac{d}{d\hat{r}_*} - W\right)Z_+ = i\sigma Z_- \quad \text{and} \quad \left(\frac{d}{d\hat{r}_*} + W\right)Z_- = i\sigma Z_+. \tag{184}$$

From these equations, we readily obtain the pair of one-dimensional wave-equations,

$$\left(\frac{d^2}{d\hat{r}_*^2}+\sigma^2\right)Z_\pm = V_\pm Z_\pm, \tag{185}$$

where

$$\begin{aligned} V_\pm &= W^2 \pm \frac{dW}{d\hat{r}_*} \\ &= \frac{\Delta^{\frac{1}{2}}(\lambda^2+m_e^2r^2)^{3/2}}{[\varpi^2(\lambda^2+m_e^2r^2)+\lambda m_e\Delta/2\sigma]^2}\Big[\Delta^{\frac{1}{2}}(\lambda^2+m_e^2r^2)^{3/2} \pm \{(r-M)(\lambda^2+m_e^2r^2) \\ &\qquad\qquad + 3m_e^2 r\Delta\}\Big] \\ &\mp \frac{\Delta^{3/2}(\lambda^2+m_e^2r^2)^{5/2}}{[\varpi^2(\lambda^2+m_e^2r^2)+\lambda m_e\Delta/2\sigma]^3}\left[2r(\lambda^2+m_e^2r^2)+2m_e^2\,\varpi^2 r+\lambda m_e(r-M)/\sigma\right]. \end{aligned} \tag{186}$$

We observe that equations (184) and (185) arc of exactly the same forms as equations (137) and (138) appropriate to the massless neutrinos.

(*b*) *The separated forms of Dirac's equations in oblate-spheroidal coordinates in flat space*

It is clear that by setting

$$M = 0 \qquad \text{and} \qquad \Delta = r^2 + a^2 \tag{187}$$

in equations (157) and (158) and in the subsequent versions of these equations, we shall obtain the separated forms of Dirac's equation in oblate-spheroidal coordinates in flat space. For, when $M = 0$, Kerr's metric becomes

$$ds^2 = dt^2 - \left[\frac{r^2+a^2\cos^2\theta}{r^2+a^2}\,dr^2 + (r^2+a^2\cos^2\theta)\,d\theta^2 + (r^2+a^2)\sin^2\theta\,d\varphi^2\right], \tag{188}$$

and by letting

$$r = a\sinh\eta, \tag{189}$$

the metric reduces to the standard form for oblate-spheroidal coordinates, namely,

$$ds^2 = dt^2 - a^2[(\cosh^2\eta - \sin^2\theta)(d\eta^2 + d\theta^2) + \cosh^2\eta\sin^2\theta\,d\varphi^2]. \tag{190}$$

Since it does not seem to be generally known that Dirac's equation can be separated in oblate-spheroidal coordinates, we shall write down the explicit forms of the equations appropriate to this case.

With the same substitutions (156), the basic radial equations are

$$\left[\frac{\mathrm{d}}{\mathrm{d}r} - i\left(\sigma + \frac{am}{r^2+a^2}\right)\right]P_{+\frac{1}{2}} = \frac{\lambda - im_e r}{\sqrt{(r^2+a^2)}}P_{-\frac{1}{2}} \tag{191}$$

and

$$\left[\frac{\mathrm{d}}{\mathrm{d}r} + i\left(\sigma + \frac{am}{r^2+a^2}\right)\right]P_{-\frac{1}{2}} = \frac{\lambda + im_e r}{\sqrt{(r^2+a^2)}}P_{+\frac{1}{2}}, \tag{192}$$

where

$$P_{+\frac{1}{2}} = (r^2+a^2)^{1/2}R_{+\frac{1}{2}} \quad \text{and} \quad P_{-\frac{1}{2}} = R_{-\frac{1}{2}}. \tag{193}$$

By the further changes in the dependent and the independent variables,

$$Z_{\pm} = P_{+\frac{1}{2}}e^{i\vartheta_1/2} \pm P_{-\frac{1}{2}}e^{-i\vartheta_1/2} \tag{194}$$

and

$$\hat{r}_* = r + \frac{1}{\sigma}\left(\frac{1}{2}\vartheta_1 + m\vartheta_2\right) \tag{195}$$

where

$$\vartheta_1 = \tan^{-1}(m_e r/\lambda) \quad \text{and} \quad \vartheta_2 = \tan^{-1}(r/a), \tag{196}$$

the equations take the simple forms

$$\left(\frac{\mathrm{d}^2}{\mathrm{d}\hat{r}_*^2} + \sigma^2\right)Z_{\pm} = V_{\pm}Z_{\pm}, \tag{197}$$

where

$$V_{\pm} = W^2 \pm \frac{\mathrm{d}W}{\mathrm{d}\hat{r}_*}, \tag{198}$$

and

$$W = \frac{(r^2+a^2)^{1/2}(\lambda^2+m_e^2r^2)^{3/2}}{(r^2+a^2)(\lambda^2+m_e^2r^2)+[am(\lambda^2+m_e^2r^2)+\frac{1}{2}\lambda m_e(r^2+a^2)]\sigma^{-1}}. \tag{199}$$

And finally, it may be noted that the equations (121), (122), and (124) governing the angular functions $S_{\pm\frac{1}{2}}$ are unaffected by going to flat space.

107. The problem of reflexion and transmission

The most important respect in which the mass of the particle manifests itself in its interactions with the black holes is the behaviour of the potentials, $V_{\pm}$, for $r \to \infty$. In contrast to the potentials we have encountered hitherto for the massless particles, the potentials, $V_{\pm}$, given by equation (186), do not vanish with an r^{-2}-behaviour for $r \to \infty$; instead,

$$V_{\pm} \to m_e^2 - \frac{2Mm_e^2}{r} + O(r^{-2}) \quad (r \to \infty). \tag{200}$$

As a result, the asymptotic behaviour of the solutions, $Z_\pm$, of the wave equation (185), for $r \to \infty$, is given by

$$Z_\pm \to \exp\left\{\pm i\left[(\sigma^2 - m_e^2)^{1/2} r + \frac{M m_e^2}{(\sigma^2 - m_e^2)^{1/2}} \lg(r/2M)\right]\right\} \quad (r \to \infty). \tag{201}$$

This behaviour of the solutions implies that for real waves (incident or reflected) at infinity, σ *must exceed* m_e (remembering our convention that σ is always to be positive). This restriction on σ is consistent with the requirement that free particles at infinity must have energies exceeding their rest energies.

(*a*) *The constancy of the Wronskian*, $[Z_\pm, Z^*_\pm]$, *over the range of* r, $r_+ < r < \infty$

We observe that the potentials, $V_\pm$, defined in equation (186), become infinite on the horizon at r_+ when $\sigma = \sigma_s$, $\Delta = 0$, and

$$\varpi_+^2 = r_+^2 + a^2 + am/\sigma = 2Mr_+ + am/\sigma = 0. \tag{202}$$

We may have additional singularities in the potentials if λ should become negative (as it does for $m = -\frac{1}{2}$ and $\alpha\sigma \geqslant 1.8$; see Appendix, Table VII). While there is no difficulty of principle in integrating the equations governing $Z_\pm$ across such singularities (as we have explained in Chapter 8, §§75(*c*) and 76(*c*) and in §105 above), we yet need to know how the Wronskian, $[Z_\pm, Z^*_\pm]$, behaves as we cross the singularities. We shall now show that the Wronskian, $[Z_\pm, Z^*_\pm]$, in fact, retains the same constant value over the entire permissible range of r, namely, $r_+ < r < \infty$. Thus, in accordance with equations (167), (175), (183), and (184), we successively find

$$\begin{aligned}\frac{1}{2\pi}\frac{\partial N}{\partial t} &= |P_{+\frac{1}{2}}|^2 - |P_{-\frac{1}{2}}|^2 = |\psi_{+\frac{1}{2}}|^2 - |\psi_{-\frac{1}{2}}|^2 \\ &= \tfrac{1}{4}|Z_+ + Z_-|^2 - \tfrac{1}{4}|Z_+ - Z_-|^2 = Z_+^{(\text{real})} Z_-^{(\text{real})} + Z_+^{(\text{im})} Z_-^{(\text{im})} \\ &= \frac{1}{\sigma}\left[Z_+^{(\text{real})}\left(\frac{\mathrm{d}}{\mathrm{d}\hat{r}_*} - W\right) Z_+^{(\text{im})} - Z_+^{(\text{im})}\left(\frac{\mathrm{d}}{\mathrm{d}\hat{r}_*} - W\right) Z_+^{(\text{real})}\right] \\ &= \frac{1}{\sigma}\left[Z_+^{(\text{real})}\frac{\mathrm{d}}{\mathrm{d}\hat{r}_*} Z_+^{(\text{im})} - Z_+^{(\text{im})}\frac{\mathrm{d}}{\mathrm{d}\hat{r}_*} Z_+^{(\text{real})}\right] \\ &= -\frac{1}{2i\sigma}\left(Z_+\frac{\mathrm{d}Z_+^*}{\mathrm{d}\hat{r}_*} - Z_+^*\frac{\mathrm{d}}{\mathrm{d}\hat{r}_*} Z_+\right).\end{aligned} \tag{203}$$

Since $\partial N/\partial t$ retains the same constant value for $r_+ < r < \infty$, the constancy of the Wronskian, $[Z_+, Z_+^*]$, over the same range of r follows. The constancy of $[Z_-, Z_-^*]$ similarly follows.

Turning next to the problem of reflexion and transmission associated with equations (185), we must seek solutions for Z_+, for example, which have the asymptotic behaviours

$$Z_+ \to \exp\left\{+i\left[(\sigma^2-m_e^2)^{1/2}\hat{r}_* + \frac{Mm_e^2}{\sqrt{(\sigma^2-m_e^2)}}\int \mathrm{d}\hat{r}_*/r\right]\right\}$$
$$+R_+(\sigma)\exp\left\{-i\left[(\sigma^2-m_e^2)^{1/2}\hat{r}_* + \frac{Mm_e^2}{\sqrt{(\sigma^2-m_e^2)}}\int \mathrm{d}\hat{r}_*/r\right]\right\} \quad (r\to\infty) \tag{204}$$

and

$$Z_+ \to T_+(\sigma)\exp(i\sigma\hat{r}_*) \qquad (r\to r_+ + 0). \tag{205}$$

For solutions having these asymptotic behaviours,

$$\left(Z_+\frac{\mathrm{d}Z_+^*}{\mathrm{d}\hat{r}_*} - Z_+^*\frac{\mathrm{d}Z_+}{\mathrm{d}\hat{r}_*}\right)_{r=r_+} = -2i\sigma|T_+(\sigma)|^2 \tag{206}$$

and

$$\left(Z_+\frac{\mathrm{d}Z_+^*}{\mathrm{d}\hat{r}_*} - Z_+^*\frac{\mathrm{d}Z_+}{\mathrm{d}\hat{r}_*}\right)_{r\to\infty} = -2i\left[1-|R_+(\sigma)|^2\right]\sqrt{(\sigma^2-m_e^2)}. \tag{207}$$

Hence, by the constancy of the Wronskian that we have proved (in *all* cases),

$$1-|R_+(\sigma)|^2 = \frac{\sigma}{\sqrt{(\sigma^2-m_e^2)}}|T_+(\sigma)|^2. \tag{208}$$

The reflexion coefficient is, therefore, always less than unity and super-radiance cannot occur. Also, from a comparison of equations (203) and (206), we conclude that

$$\frac{1}{2\pi}\left(\frac{\partial N}{\partial t}\right)_{r_+} = |T_+(\sigma)|^2 > 0. \tag{209}$$

The net current of particles crossing the event horizon is, therefore, always positive.

(b) *The positivity of the energy flow across the event horizon*

The lack of super-radiance for spin-½ particles which we have demonstrated with the aid of the equations governing the problem can be traced directly to the energy-momentum spinor given by the quantum theory for these particles, namely,

$$T_{AA'BB'} = \tfrac{1}{2}i[P_A\bar{P}_{A';BB'} - \bar{P}_{A'}P_{A;BB'} + P_B\bar{P}_{B';AA'} - \bar{P}_{B'}P_{B;AA'}$$
$$- Q_A\bar{Q}_{A';BB'} + \bar{Q}_{A'}Q_{A;BB'} - Q_B\bar{Q}_{B';AA'} + \bar{Q}_{B'}Q_{B;AA'}]. \tag{210}$$

In terms of this energy-momentum spinor, we can directly evaluate (cf. Ch. 8, equations (247) and (253))

$$\left(\frac{\partial E}{\partial t}\right)_r = -\int_0^{2\pi}\int_0^{\pi} T^r{}_t(\sqrt{-g})\,d\theta\,d\varphi \tag{211}$$

and

$$\left(\frac{\partial L_z}{\partial t}\right)_r = -\int_0^{2\pi}\int_0^{\pi} T^r{}_\varphi(\sqrt{-g})\,d\theta\,d\varphi = \frac{m}{\sigma}\left(\frac{\partial E}{\partial t}\right)_r \tag{212}$$

in the coordinate system in which we have written the Kerr metric (Ch. 6, equations (134) and (135)). Since

$$T^r{}_t = -\frac{\Delta}{\rho^2}T_{rt}, \quad T^r{}_\varphi = -\frac{\Delta}{\rho^2}T_{r\varphi}, \tag{213}$$

and

$$\sqrt{-g} = \rho^2\sin\theta, \tag{214}$$

we may combine equations (211) and (212) to give

$$(r^2+a^2+am/\sigma)\left(\frac{\partial E}{\partial t}\right)_r = \varpi^2\left(\frac{\partial E}{\partial t}\right)_r$$

$$= 2\pi\int_0^{\pi}[\Delta(r^2+a^2)T_{rt}+a\Delta T_{r\varphi}]\sin\theta\,d\theta. \tag{215}$$

On the other hand, with the aid of the expressions for the basic vectors, ***l*** and ***n***, given in Chapter 6, equation (170), we readily verify that

$$\Delta(r^2+a^2)T_{rt}+a\Delta T_{r\varphi} = \tfrac{1}{4}\Delta^2 T_{ij}l^il^j - \rho^4 T_{ij}n^in^j. \tag{216}$$

Accordingly, we have

$$\varpi^2\left(\frac{\partial E}{\partial t}\right)_r = 2\pi\int_0^{\pi}(\tfrac{1}{4}\Delta^2 T_{ij}l^il^j - \rho^4 T_{ij}n^in^j)\sin\theta\,d\theta. \tag{217}$$

Now with the energy-momentum spinor given by equation (210), we find with the aid of equations (159) that as the horizon is approached

$$\begin{aligned}
T_{ij}l^il^j &= o^A\bar{o}^{A'}o^B\bar{o}^{B'}T_{AA'BB'}\\
&= io^A\bar{o}^{A'}o^B\bar{o}^{B'}[P_A\bar{P}_{A';BB'}-\bar{P}_{A'}P_{A;BB'}-Q_A\bar{Q}_{A';BB'}+\bar{Q}_{A'}Q_{A;BB'}]\\
&= i[P_0D\bar{P}_{0'}-\bar{P}_{0'}DP_0-Q_0D\bar{Q}_{0'}+\bar{Q}_{0'}DQ_0]\\
&= i[P^1D\bar{P}^{1'}-\bar{P}^{1'}DP^1-Q^1D\bar{Q}^{1'}+\bar{Q}^{1'}DQ^1]\\
&\to 2\sigma\frac{\varpi_+^2}{\Delta}|R_{+\frac{1}{2}}|^2(S^2_{+\frac{1}{2}}+S^2_{-\frac{1}{2}}).
\end{aligned} \tag{218}$$

Similarly, we find that

$$T_{ij}n^i n^j = \iota^A \bar{\iota}^{A'} \iota^B \bar{\iota}^{B'} T_{AA'BB'}$$

$$\rightarrow \tfrac{1}{2}\sigma \frac{\varpi_+^2}{\rho^4} |R_{-\frac{1}{2}}|^2 (S_{+\frac{1}{2}}^2 + S_{-\frac{1}{2}}^2). \tag{219}$$

Inserting the expressions (218) and (219) in equation (217) and remembering that we have normalized the functions $S_{+\frac{1}{2}}(\theta)$ and $S_{-\frac{1}{2}}(\theta)$ to unity, we obtain the formula

$$\left(\frac{\partial E}{\partial t}\right)_{r_+} = 2\pi\sigma[|P_{+\frac{1}{2}}|^2 - |P_{-\frac{1}{2}}|^2]. \tag{220}$$

The net flow of energy, across a surface of radius r per unit time and per unit solid-angle, is therefore given by

$$\left(\frac{\partial^2 E}{\partial t \partial \Omega}\right)_{r_+} = \tfrac{1}{2}\sigma[|P_{+\frac{1}{2}}|^2 - |P_{-\frac{1}{2}}|^2 = \frac{\sigma}{4\pi}\left(\frac{\partial N}{\partial t}\right)_{r_+}. \tag{221}$$

By equation (209), this flow, across the event horizon, is always positive.

(c) *The quantal origin of the lack of super-radiance*

The absence of super-radiance exhibited by spin-½ particles, in apparent violation of the area theorem of Hawking, is related to the fact that the energy-momentum spinor (210), provided by the quantum theory, violates the weak-energy condition,

$$T_{ij}k^i k^j > 0, \tag{222}$$

where k is any time-like vector, when $\sigma < \sigma_s$. Thus, with the choice,

$$k^i = \frac{1}{r^2}(\tfrac{1}{2}\Delta l^i + \rho^2 n^i) \quad \left(k^i k_i = \frac{\Delta\rho^2}{r^4} > 0 \quad \text{for} \quad r > r_+\right), \tag{223}$$

we have

$$T_{ij}k^i k^j = \frac{1}{r^4}(\tfrac{1}{4}\Delta^2 T_{ij}l^i l^j + \rho^4 T_{ij}n^i n^j + \rho^2 \Delta T_{ij}l^i n^j). \tag{224}$$

We have already evaluated $T_{ij}l^i l^j$ and $T_{ij}n^i n^j$ in equations (218) and (219). We similarly find that

$$T_{ij}l^i n^j = o^A \bar{o}^{A'} \iota^B \bar{\iota}^{B'} T_{AA'BB'}$$

$$\rightarrow \sigma \frac{\varpi_+^2}{\rho^2 \Delta}(|P_{+\frac{1}{2}}|^2 + |P_{-\frac{1}{2}}|^2)(S_{+\frac{1}{2}}^2 + S_{-\frac{1}{2}}^2). \tag{225}$$

Now combining equations (218), (219), (224), and (225), we obtain

$$T_{ij}k^i k^j \rightarrow 2\frac{\sigma\varpi_+^2}{r_+^4}(|P_{+\frac{1}{2}}|^2 + |P_{-\frac{1}{2}}|^2)(S_{+\frac{1}{2}}^2 + S_{-\frac{1}{2}}^2); \tag{226}$$

and this is negative for $\sigma < \sigma_s$ and violates the energy condition.

While the foregoing discussion clarifies the broad aspects of the phenomenon of reflexion and transmission of Dirac waves by the Kerr black-hole, there remain substantive questions which are unresolved. Thus, one may ask: does it happen that the potentials, $V_{\pm}$, have singularities, outside of the event horizon, for a frequency $\sigma > m_e$, derived from the term $\lambda m_e \Delta/2\sigma$ in the denominator of W when λ is negative? This does not happen in Schwarzschild geometry. If it should happen in Kerr geometry (perhaps for some $\sigma > \sigma_s$), then we should expect the reflexion coefficient to attain the maximum value unity at that frequency and decrease thereafter. In such an event, we may conclude that the reflexion of spin-$\frac{1}{2}$ particles by the Kerr black-hole exhibits the *Klein paradox*. Clearly, this is a matter of some consequence; but it cannot be resolved before we ascertain the dependence of the characteristic values, λ, on m_e, m, and σ; but this dependence is, as yet, unexplored. (It has since been shown that the paradox does not occur.)

BIBLIOGRAPHICAL NOTES

Dirac's equation in Kerr geometry was separated by:

1. S. Chandrasekhar, *Proc. Roy. Soc.* (*London*) A, 349, 571–5 (1976).

As is noted in this paper, by setting $M = 0$ in the derived equations, one obtains the separated form of Dirac's equation in oblate-spheroidal coordinates in flat space (see §106 (*b*)).

The method of paper (1) extends naturally and simply to any space-time of type D:

2. R. Güven, *Proc. Roy. Soc.* (*London*) A, 356, 465–70 (1977).

Weyl's equation governing the massless two-component neutrinos was separated earlier by:

3. W. G. Unruh, *Phys. Rev. Lett.*, 31, 1265–7 (1973).
4. S. A. Teukolsky, *Astrophys. J.*, 185, 635–47 (1973) (see Appendix B, p. 646).

§102. The spinor formalism for the study of the structure of space-time was initiated and developed by Penrose:

5. R. Penrose, *Ann. Phys.*, 10, 171–201 (1960).
6. ———, *An Analysis of the Structure of Space-Time*, Cambridge University Adams Prize Essay, Cambridge, England, 1966.
7. ——— in *Batelle Recontres* (*1967 Lectures in Mathematical Physics*), 121–35, edited by C. M. DeWitt and J. A. Wheeler, W. A. Benjamin Inc., New York, 1968.

The basic paper on the spinorial basis of the Newman–Penrose formalism is of course:

8. E. T. Newman and R. Penrose, *J. Math. Phys.*, 3, 566–79 (1962).

§§102 (*a*) and (*b*). The account of spinor analysis in the text is largely based on the author's notes of lectures given by Dirac in the spring of 1932 on spinor analysis and the relativistic theory of the electron. The style and content of the lectures do not seem to have faded in the intervening fifty years.

§§102 (*c*) and (*d*). The account in these sections owes much to discussions with J. Friedman. In particular, I have followed his manner of presentation of co-variant differentiation in §(*d*).

§103. Clearly no references are needed for Dirac's equation in flat space. However, the author found the treatment in "Landau and Lifshitz" most relevant for the purposes of translating Dirac's equation in the Newman–Penrose formalism:

9. V. B. BERESTETSKII, E. M. LIFSHITZ, and L. P. PITAEVSKII in *L. D. Landau and E. M. Lifshitz Courses in Theoretical Physics*, Vol. 4, Chap. 3, §20, 62–3, Pergamon Press, Oxford, 1971.

§104. This section is largely a repetition of the analysis of paper (1). For an alternative derivation of the same equations see:

10. B. CARTER and R. G. MCLENAGHAN in *Proc. 2d Marcel Grossman Meeting on General Relativity*, edited by R. Ruffini, North-Holland Publishing Co., Amsterdam, 1980.

§105. The analysis in this section is a simplified version of the treatment in:

11. S. CHANDRASEKHAR and S. DETWEILER, *Proc. Roy. Soc.* (*London*) A, 352, 325–38 (1977).

See also:

12. W. G. UNRUH, *Phys. Rev. D*, 10, 3194–205 (1974).

In the context of Schwarzschild's geometry the problem has been considered by:

13. D. R. BRILL and J. A. WHEELER, *Rev. Mod. Phys.*, 465–79 (1957).
14. J. B. HARTLE in *Magic without Magic: John Archibald Wheeler*, 259–75, edited by J. R. Klauder, W. H. Freeman and Co., San Francisco, 1972.

§106. This section is based on some unpublished work of the author. Equivalent results have been published by:

15. R. GÜVEN, *Phys. Rev. D*, 16, 1706–11 (1977).

See also Unruh (paper 12).

§107. The fact that super-radiance is absent in the reflexion of Dirac waves of the Kerr black-hole was explicitly demonstrated by Güven (paper 15) and by several other authors soon after Dirac's equation was separated:

16. M. MARTELLINI and A. TREVES, *Phys. Rev. D*, 15, 3060–4 (1977).
17. C. H. LEE, *Phys. Lett.*, 68B, 152–6 (1977).
18. B. R. IYER and A. KUMAR, *Phys. Rev. D*, 18, 4799–801 (1978).

For the Klein paradox mentioned in the concluding paragraph see:

19. O. KLEIN, *Z. f. Physik*, 157–65 (1929).
20. F. SAUTER, *Z. f. Physik*, 69, 742–64 (1931).

For aspects of the separated Dirac's equation not considered in the text see:

21. B. CARTER and R. G. MCLENAGHAN, *Phys. Rev. D*, 19, 1093–7 (1979).
22. B. R. IYER and A. KUMAR, *Pramāna*, 8, 500–11 (1977).
23. ———, ibid., 11, 171–85 (1978).
24. ———, ibid., 12, 103–20 (1979).

11

OTHER SOLUTIONS; OTHER METHODS

108. Introduction

The treatment of the spin-$\frac{1}{2}$ particles in the last chapter concludes our study of the Schwarzschild and the Kerr solutions—the digression into the Reissner–Nordström solution in Chapter 5 was largely dictated by the consideration that its study enables a deeper understanding of the Schwarzschild solution and provides a convenient bridge to the Kerr solution. On these accounts, one might have thought that the Kerr–Newman solution—which stands in the same relation to the Kerr solution as the Reissner–Nordström solution to the Schwarzschild solution—will enable a similar deeper understanding of the Kerr solution. While the derivation of the Kerr–Newman solution (in §110) along the lines of the Kerr solution (in §54) does provide some useful generalizations of the underlying concepts, it does not appear that the methods developed in Chapter 9 for the treatment of the gravitational perturbations of the Kerr black-hole can be extended in any natural way to the treatment of the coupled electromagnetic and gravitational perturbations of the Kerr–Newman black-hole. The origins of this apparently essential difference in the perturbed Kerr and Kerr–Newman space-times may lie deep in the indissoluble coupling of the spin-1 and spin-2 fields in the perturbed Kerr–Newman space-time—a coupling which it was possible to break only for very special reasons in the perturbed Reissner–Nordström space-time.

Besides the Kerr–Newman solution, we shall consider two other classes of solutions: axisymmetric black-hole solutions which are static but not asymptotically flat; and solutions which provide for an arbitrary number of isolated black-holes and which represent the relativistic analogue of the static equilibrium arrangement, possible in the Newtonian gravitational theory, of an arbitrary number of charged mass-points in which the mutual gravitational attractions are exactly balanced by the Coulomb repulsions.

The chapter (and the book) ends with a long section devoted to the still unsolved problem of the stability of the Kerr black-hole and to a preliminary consideration of an alternative method for treating perturbations of space-times.

109. The Einstein–Maxwell equations governing stationary axisymmetric space-times

As in Chapter 5, we start with a metric of the standard form

$$ds^2 = e^{2\nu}(dt)^2 - e^{2\psi}(d\varphi - \omega\, dt)^2 - e^{2\mu_2}(dx^2)^2 - e^{2\mu_3}(dx^3)^2, \tag{1}$$

where ν, ψ, ω, μ_2, and μ_3 are functions only of x^2 and x^3, with the freedom to impose a coordinate condition on μ_2 and μ_3.

Since we are now concerned with the Einstein–Maxwell equations, we shall consider, first, Maxwell's equations in a space-time with a metric of the chosen form. These equations can be readily written down by suitably specializing the equations of Chapter 2, §15. Thus, we find from Chapter 2, equations (95) (*a*)–(*h*), that in a stationary axisymmetric space-time

$$F_{01} = F_{23} = 0; \tag{2}$$

and that the remaining equations are

$$(e^{\psi+\mu_2}F_{12})_{,3} - (e^{\psi+\mu_3}F_{13})_{,2} = 0, \tag{3}$$

$$(e^{\psi+\mu_3}F_{02})_{,2} + (e^{\psi+\mu_2}F_{03})_{,3} = 0, \tag{4}$$

$$(e^{\nu+\mu_3}F_{12})_{,2} + (e^{\nu+\mu_2}F_{13})_{,3} = e^{\psi+\mu_3}F_{02}\omega_{,2} + e^{\psi+\mu_2}F_{03}\omega_{,3}, \tag{5}$$

$$(e^{\nu+\mu_2}F_{02})_{,3} - (e^{\nu+\mu_3}F_{03})_{,2} = e^{\psi+\mu_2}F_{12}\omega_{,3} - e^{\psi+\mu_3}F_{13}\omega_{,2}. \tag{6}$$

From equations (3) and (4), it is apparent that the two pairs of components, (F_{12}, F_{13}) and (F_{02}, F_{03}), of the Maxwell tensor, can be derived from two potentials, A and B, in the manner

$$\left.\begin{aligned} e^{\psi+\mu_2}F_{12} &= A_{,2}; \quad e^{\psi+\mu_2}F_{03} = -B_{,2}, \\ e^{\psi+\mu_3}F_{13} &= A_{,3}; \quad e^{\psi+\mu_3}F_{02} = +B_{,3}. \end{aligned}\right\} \tag{7}$$

In terms of the potentials, A and B, equations (5) and (6) take the forms

$$(e^{-\psi+\nu-\mu_2+\mu_3}A_{,2})_{,2} + (e^{-\psi+\nu+\mu_2-\mu_3}A_{,3})_{,3} = \omega_{,2}B_{,3} - \omega_{,3}B_{,2}, \tag{8}$$

$$(e^{-\psi+\nu-\mu_2+\mu_3}B_{,2})_{,2} + (e^{-\psi+\nu+\mu_2-\mu_3}B_{,3})_{,3} = \omega_{,3}A_{,2} - \omega_{,2}A_{,3}. \tag{9}$$

With the energy-momentum tensor of the Maxwell field given by

$$T_{ab} = \eta^{cd}F_{ac}F_{bd} - \tfrac{1}{4}\eta_{ab}F_{ef}F^{ef}, \tag{10}$$

the Einstein field-equation is given by

$$\begin{aligned} R_{ab} &= -2T_{ab} \\ &= -2\eta^{cd}F_{ac}F_{bd} - \eta_{ab}[(F_{02})^2 + (F_{03})^2 - (F_{12})^2 - (F_{13})^2]. \end{aligned} \tag{11}$$

Evaluating the various components of R_{ab} with the aid of equations (7), we find

$$R_{00} = R_{11} = e^{-2\psi-2\mu_2}[(A_{,2})^2+(B_{,2})^2]+e^{-2\psi-2\mu_3}[(A_{,3})^2+(B_{,3})^2], \quad (12)$$

$$R_{22} = -R_{33} = e^{-2\psi-2\mu_2}[(A_{,2})^2+(B_{,2})^2]-e^{-2\psi-2\mu_3}[(A_{,3})^2+(B_{,3})^2], \quad (13)$$

$$R_{10} = 2e^{-2\psi-\mu_2-\mu_3}(A_{,2}B_{,3}-A_{,3}B_{,2}), \quad (14)$$

$$R_{23} = 2e^{-2\psi-\mu_2-\mu_3}(A_{,2}A_{,3}+B_{,2}B_{,3}). \quad (15)$$

The required field equations now follow from equating the expressions on the right-hand sides of the foregoing equations with the corresponding expressions for the Ricci (or the Einstein) tensors in terms of the metric. The expressions for the latter can be directly read off from Chapter 6, equations (5)–(13). Thus, we find that Chapter 6, equations (12), (13), (7), (8), (9), and (10) are now replaced, respectively, by

$$\begin{aligned}(e^{\beta+\mu_3-\mu_2}\nu_{,2})_{,2}+(e^{\beta+\mu_2-\mu_3}\nu_{,3})_{,3} &= \tfrac{1}{2}e^{3\psi-\nu}[e^{\mu_3-\mu_2}(\omega_{,2})^2+e^{\mu_2-\mu_3}(\omega_{,3})^2]\\ &\quad +e^{-\psi+\nu}\{e^{\mu_3-\mu_2}[(A_{,2})^2+(B_{,2})^2]+e^{\mu_2-\mu_3}[(A_{,3})^2+(B_{,3})^2]\}, \quad (16)\end{aligned}$$

$$\begin{aligned}(e^{\beta+\mu_3-\mu_2}\psi_{,2})_{,2}+(e^{\beta+\mu_2-\mu_3}\psi_{,3})_{,3} &= -\tfrac{1}{2}e^{3\psi-\nu}[e^{\mu_3-\mu_2}(\omega_{,2})^2+e^{\mu_2-\mu_3}(\omega_{,3})^2]\\ &\quad -e^{-\psi+\nu}\{e^{\mu_3-\mu_2}[(A_{,2})^2+(B_{,2})^2]+e^{\mu_2-\mu_3}[(A_{,3})^2+(B_{,3})^2]\}, \quad (17)\end{aligned}$$

$$(e^{3\psi-\nu-\mu_2+\mu_3}\omega_{,2})_{,2}+(e^{3\psi-\nu+\mu_2-\mu_3}\omega_{,3})_{,3} = 4(A_{,2}B_{,3}-A_{,3}B_{,2}), \quad (18)$$

$$\begin{aligned}&(\psi+\nu)_{,2,3}-(\psi+\nu)_{,2}\,\mu_{2,3}-(\psi+\nu)_{,3}\,\mu_{3,2}+\psi_{,2}\,\psi_{,3}+\nu_{,2}\,\nu_{,3}\\ &\quad = \tfrac{1}{2}e^{2\psi-2\nu}\omega_{,2}\omega_{,3}-2e^{-2\psi}(A_{,2}A_{,3}+B_{,2}B_{,3}), \quad (19)\end{aligned}$$

$$\begin{aligned}&e^{-2\mu_3}[(\psi+\nu)_{,3,3}+(\psi+\nu)_{,3}(\nu-\mu_3)_{,3}+\psi_{,3}\psi_{,3}]\\ &+e^{-2\mu_2}[\nu_{,2}(\psi+\mu_3)_{,2}+\psi_{,2}\,\mu_{3,2}] = -\tfrac{1}{4}e^{2\psi-2\nu}[e^{-2\mu_2}(\omega_{,2})^2-e^{-2\mu_3}(\omega_{,3})^2]\\ &\quad +e^{-2\psi}\{e^{-2\mu_2}[(A_{,2})^2+(B_{,2})^2]-e^{-2\mu_3}[(A_{,3})^2+(B_{,3})^2]\}, \quad (20)\end{aligned}$$

and

$$\begin{aligned}&e^{-2\mu_2}[(\psi+\nu)_{,2,2}+(\psi+\nu)_{,2}(\nu-\mu_2)_{,2}+\psi_{,2}\psi_{,2}]\\ &+e^{-2\mu_3}[\nu_{,3}(\psi+\mu_2)_{,3}+\psi_{,3}\mu_{2,3}] = +\tfrac{1}{4}e^{2\psi-2\nu}[e^{-2\mu_2}(\omega_{,2})^2-e^{-2\mu_3}(\omega_{,3})^2]\\ &\quad -e^{-2\psi}\{e^{-2\mu_2}[(A_{,2})^2+(B_{,2})^2]+e^{-2\mu_3}[(A_{,3})^2+(B_{,3})^2]\}, \quad (21)\end{aligned}$$

where, as in Chapter 6,

$$\beta = \psi+\nu. \quad (22)$$

The sum and difference of equations (16) and (17) are

$$[e^{\mu_3-\mu_2}(e^{\beta})_{,2}]_{,2}+[e^{\mu_2-\mu_3}(e^{\beta})_{,3}]_{,3} = 0 \quad (23)$$

and

$$\begin{aligned}[e^{\beta+\mu_3-\mu_2}(\psi-\nu)_{,2}]_{,2}+[e^{\beta+\mu_2-\mu_3}(\psi-\nu)_{,3}]_{,3}\\ =-e^{3\psi-\nu}[e^{\mu_3-\mu_2}(\omega_{,2})^2+e^{\mu_2-\mu_3}(\omega_{,3})^2]\\ -2e^{-\psi+\nu}\{e^{\mu_3-\mu_2}[(A_{,2})^2+(B_{,2})^2]+e^{\mu_2-\mu_3}[(A_{,3})^2+(B_{,3})^2]\}. \quad (24)\end{aligned}$$

We observe that equation (23) is the same as Chapter 6, equation (14) obtained for the pure vacuum.

The addition of equations (20) and (21) yields the same equation (23), while subtraction gives

$$\begin{aligned}4e^{\mu_3-\mu_2}(\beta_{,2}\mu_{3,2}+\psi_{,2}\nu_{,2})-4e^{\mu_2-\mu_3}(\beta_{,3}\mu_{2,3}+\psi_{,3}\nu_{,3})\\ =2e^{-\beta}\{[e^{\mu_3-\mu_2}(e^{\beta})_{,2}]_{,2}-[e^{\mu_2-\mu_3}(e^{\beta})_{,3}]_{,3}\}\\ -e^{2\psi-2\nu}[e^{\mu_3-\mu_2}(\omega_{,2})^2-e^{\mu_2-\mu_3}(\omega_{,3})^2]\\ +4e^{-2\psi}\{e^{\mu_3-\mu_2}[(A_{,2})^2+(B_{,2})^2]-e^{\mu_2-\mu_3}[(A_{,3})^2+(B_{,3})^2]\}. \quad (25)\end{aligned}$$

(a) The choice of gauge and the reduction of the equations to standard forms

As in Chapter 6, §53, we can, by exercising the gauge freedom we have to impose a coordinate condition on μ_2 and μ_3 and by making further use of equation (23), arrange that the space-time we are considering has a smooth event horizon when a certain quadratic function in r, namely (cf. Ch. 6, equations (45) and (46)),

$$\Delta(r)=r^2-2Mr+M_0^2, \quad (26)$$

vanishes and

$$e^{2(\mu_3-\mu_2)}=\Delta(r) \qquad \text{and} \qquad e^{\beta}=\Delta^{1/2}\sin\theta. \quad (27)$$

In equation (26), M and M_0 are constants whose meanings will emerge later.

Returning to equations (8) and (9), we can combine them into a single equation, for the *complex potential*,

$$H=A+iB \quad (28)$$

in the form

$$(e^{-\psi+\nu-\mu_2+\mu_3}H_{,2})_{,2}+(e^{-\psi+\nu+\mu_2-\mu_3}H_{,3})_{,3}=i(\omega_{,3}H_{,2}-\omega_{,2}H_{,3}). \quad (29)$$

Next, writing equation (18) in the form

$$\begin{aligned}(e^{3\psi-\nu-\mu_2+\mu_3}\omega_{,2}-2AB_{,3}+2BA_{,3})_{,2}\\ +(e^{3\psi-\nu+\mu_2-\mu_3}\omega_{,3}+2AB_{,2}-2BA_{,2})_{,3}=0, \quad (30)\end{aligned}$$

we have, in terms of H,

$$(e^{3\psi-\nu-\mu_2+\mu_3}\omega_{,2}+2\,\mathrm{Im}\,HH^*_{,3})_{,2}+(e^{3\psi-\nu+\mu_2-\mu_3}\omega_{,3}-2\,\mathrm{Im}\,HH^*_{,2})_{,3}=0. \tag{31}$$

With the solutions for $e^{\mu_3-\mu_2}$ and e^{β} given in equations (27) and by using

$$\mu=\cos\theta, \tag{32}$$

instead of θ, as the variable indicated by the index '3', we can bring equations (29) and (31) to the forms

$$\left[\frac{\chi}{\sqrt{(\Delta\delta)}}\Delta H_{,2}\right]_{,2}+\left[\frac{\chi}{\sqrt{(\Delta\delta)}}\delta H_{,3}\right]_{,3}=i(\omega_{,2}H_{,3}-\omega_{,3}H_{,2}) \tag{33}$$

and

$$\left(\frac{\Delta}{\chi^2}\omega_{,2}-2\,\mathrm{Im}\,HH^*_{,3}\right)_{,2}+\left(\frac{\delta}{\chi^2}\omega_{,3}+2\,\mathrm{Im}\,HH^*_{,2}\right)_{,3}=0, \tag{34}$$

where we have written (again, as in Chapter 6)

$$\chi=e^{-\psi+\nu} \tag{35}$$

and

$$\delta=1-\mu^2=\sin^2\theta. \tag{36}$$

In similar fashion, equation (24), expressed in terms of χ and H, becomes

$$[\Delta(\lg\chi)_{,2}]_{,2}+[\delta(\lg\chi)_{,3}]_{,3}=\frac{1}{\chi^2}[\Delta(\omega_{,2})^2+\delta(\omega_{,3})^2]+\frac{2\chi}{\sqrt{(\Delta\delta)}}[\Delta|H_{,2}|^2+\delta|H_{,3}|^2]. \tag{37}$$

Equations (33), (34), and (37) are the basic equations of the theory: once these equations are solved, the solution can be completed by considering equations (19) and (25) which will determine $\mu_2+\mu_3$ by simple quadratures (see §110 below).

(*b*) *Further transformations of the equations*

We observe that equation (34) allows us to define a potential Φ by the equations (cf. Ch. 6, equation (87))

$$-\Phi_{,2}=\frac{\delta}{\chi^2}\omega_{,3}+2\,\mathrm{Im}\,HH^*_{,2},\quad +\Phi_{,3}=\frac{\Delta}{\chi^2}\omega_{,2}-2\,\mathrm{Im}\,HH^*_{,3}, \tag{38}$$

for equation (34) is no more than the integrability condition for these equations. Solving these equations for $\omega_{,2}$ and $\omega_{,3}$, we have

$$\omega_{,2} = \frac{\chi^2}{\Delta}(\Phi_{,3} + 2\,\mathrm{Im}\, H H^*_{,3}), \quad \omega_{,3} = -\frac{\chi^2}{\delta}(\Phi_{,2} + 2\,\mathrm{Im}\, H H^*_{,2}); \tag{39}$$

and the integrability condition for these equations is

$$\left[\frac{\chi^2}{\delta}(\Phi_{,2} + 2\,\mathrm{Im}\, H H^*_{,2})\right]_{,2} + \left[\frac{\chi^2}{\Delta}(\Phi_{,3} + 2\,\mathrm{Im}\, H H^*_{,3})\right]_{,3} = 0. \tag{40}$$

It is now convenient to define the function (cf. Ch. 6, equation (98)),

$$\Psi = \frac{\sqrt{(\Delta\delta)}}{\chi}. \tag{41}$$

In terms of Ψ, equation (40) takes the form

$$\left[\frac{\Delta}{\Psi^2}(\Phi_{,2} + 2\,\mathrm{Im}\, H H^*_{,2})\right]_{,2} + \left[\frac{\delta}{\Psi^2}(\Phi_{,3} + 2\,\mathrm{Im}\, H H^*_{,3})\right]_{,3} = 0, \tag{42}$$

or, expanding, we have

$$\begin{aligned}\Psi\{[\Delta(\Phi_{,2} + 2\,\mathrm{Im}\, H H^*_{,2})]_{,2} + [\delta(\Phi_{,3} + 2\,\mathrm{Im}\, H H^*_{,3})]_{,3}\} \\ = 2\Delta(\Phi_{,2} + 2\,\mathrm{Im}\, H H^*_{,2})\Psi_{,2} + 2\delta(\Phi_{,3} + 2\,\mathrm{Im}\, H H^*_{,3})\Psi_{,3}.\end{aligned} \tag{43}$$

Similarly, equation (37) written in terms of Ψ and Φ becomes

$$\begin{aligned}\Psi[(\Delta\Psi_{,2})_{,2} + (\delta\Psi_{,3})_{,3}] = \Delta(\Psi_{,2})^2 + \delta(\Psi_{,3})^2 - 2\Psi[\Delta|H_{,2}|^2 + \delta|H_{,3}|^2] \\ - \Delta(\Phi_{,2} + 2\,\mathrm{Im}\, H H^*_{,2})^2 - \delta(\Phi_{,3} + 2\,\mathrm{Im}\, H H^*_{,3})^2;\end{aligned} \tag{44}$$

while in place of equation (33), we have

$$\begin{aligned}\Psi[(\Delta H_{,2})_{,2} + (\delta H_{,3})_{,3}] = \Delta H_{,2}[\Psi_{,2} + i(\Phi_{,2} + 2\,\mathrm{Im}\, H H^*_{,2})] \\ + \delta H_{,3}[\Psi_{,3} + i(\Phi_{,3} + 2\,\mathrm{Im}\, H H^*_{,3})].\end{aligned} \tag{45}$$

We may note here for future reference the real and the imaginary parts of equation (33):

$$\left.\begin{aligned}\left(\frac{\Delta}{\Psi}A_{,2}\right)_{,2} + \left(\frac{\delta}{\Psi}A_{,3}\right)_{,3} &= \omega_{,3}B_{,2} - \omega_{,2}B_{,3}, \\ \left(\frac{\Delta}{\Psi}B_{,2}\right)_{,2} + \left(\frac{\delta}{\Psi}B_{,3}\right)_{,3} &= \omega_{,2}A_{,3} - \omega_{,3}A_{,2}.\end{aligned}\right\} \tag{46}$$

Finally, we shall state as lemmas two direct consequences of the foregoing equations.

LEMMA 1.

$$\Psi[(\Delta|H|^2_{,2})_{,2}+(\delta|H|^2_{,3})_{,3}] = 2\Psi[\Delta|H_{,2}|^2+\delta|H_{,3}|^2] + \Delta[|H|^2_{,2}\Psi_{,2}+2\,\mathrm{Im}\,HH^*_{,2}(\Phi_{,2}+2\,\mathrm{Im}\,HH^*_{,2})] + \delta[|H|^2_{,3}\Psi_{,3}+2\,\mathrm{Im}\,HH^*_{,3}(\Phi_{,3}+2\,\mathrm{Im}\,HH^*_{,3})]. \quad (47)$$

LEMMA 2.

$$2\Psi[(\Delta\,\mathrm{Im}\,HH^*_{,2})_{,2}+(\delta\,\mathrm{Im}\,HH^*_{,3})_{,3}] = 2\Delta\,\mathrm{Im}\,HH^*_{,2}\Psi_{,2}+2\delta\,\mathrm{Im}\,HH^*_{,3}\Psi_{,3} - [\Delta|H|^2_{,2}(\Phi_{,2}+2\,\mathrm{Im}\,HH^*_{,2})+\delta|H|^2_{,3}(\Phi_{,3}+2\,\mathrm{Im}\,HH^*_{,3})]. \quad (48)$$

(*c*) *The Ernst equations*

First, we observe that in view of the identity

$$(2i\,\mathrm{Im}\,HH^*_{,2})\Delta H_{,2} = (HH^*_{,2}+H^*H_{,2}-2H^*H_{,2})\Delta H_{,2} = \Delta|H|^2_{,2}-2H^*\Delta(H_{,2})^2, \quad (49)$$

equation (45) can be written in the form

$$\Psi[(\Delta H_{,2})_{,2}+(\delta H_{,3})_{,3}] = -2H^*[\Delta(H_{,2})^2+\delta(H_{,3})^2] + \Delta H_{,2}(\Psi+|H|^2+i\Phi)_{,2}+\delta H_{,3}(\Psi+|H|^2+i\Phi)_{,3}. \quad (50)$$

Next, eliminating the terms in $\mathrm{Im}\,HH^*_{,2}$ and $\mathrm{Im}\,HH^*_{,3}$ on the left-hand sides of equation (43) with the aid of lemma 2 (equation (48)), we are left with

$$\Psi[(\Delta\Phi_{,2})_{,2}+(\delta\Phi_{,3})_{,3}] = \Delta[\Phi_{,2}(2\Psi+|H|^2)_{,2}+2\,\mathrm{Im}\,HH^*_{,2}(\Psi+|H|^2)_{,2}] + \delta[\Phi_{,3}(2\Psi+|H|^2)_{,3}+2\,\mathrm{Im}\,HH^*_{,3}(\Psi+|H|^2)_{,3}]. \quad (51)$$

Similarly, combining equations (44) and (47), we obtain

$$\Psi\{[\Delta(\Psi+|H|^2)_{,2}]_{,2}+[\delta(\Psi+|H|^2)_{,3}]_{,3}\} = \Delta[(\Psi_{,2})^2+|H|^2_{,2}\Psi_{,2}-(\Phi_{,2})^2-2\,\mathrm{Im}\,HH^*_{,2}\Phi_{,2}] + \delta[(\Psi_{,3})^2+|H|^2_{,3}\Psi_{,3}-(\Phi_{,3})^2-2\,\mathrm{Im}\,HH^*_{,3}\Phi_{,3}]. \quad (52)$$

It is now clear that the left-hand sides of equations (51) and (52) can be combined into a single expression for the complex function,

$$Z = \Psi+|H|^2+i\Phi. \quad (53)$$

To show that the right-hand sides of the two equations can also be similarly combined as a single complex expression involving only Z and H, consider the combination

$$(\Psi_{,2})^2+|H|^2_{,2}\Psi_{,2}-(\Phi_{,2})^2+2i\Phi_{,2}\Psi_{,2}+i\Phi_{,2}|H|^2_{,2} + 2i\,\mathrm{Im}\,HH^*_{,2}(\Psi+|H|^2+i\Phi)_{,2} \quad (54)$$

which will occur on the right-hand side with the coefficient Δ. The term in $\operatorname{Im} HH^*_{,2}$ can be reduced in the manner

$$\begin{aligned}(2i\operatorname{Im} HH^*_{,2})Z_{,2} &= (HH^*_{,2}+H^*H_{,2}-2H^*H_{,2})Z_{,2}\\ &= (|H|^2_{,2}-2H^*H_{,2})Z_{,2}\\ &= |H|^2_{,2}(\Psi+|H|^2+i\Phi)_{,2}-2H^*H_{,2}Z_{,2};\end{aligned} \tag{55}$$

and the expression (54) becomes

$$\begin{aligned}&(\Psi_{,2})^2+2|H|^2_{,2}\Psi_{,2}+(|H|^2_{,2})^2-(\Phi_{,2})^2\\ &\qquad +2i(\Psi_{,2}+|H|^2_{,2})\Phi_{,2}-2H^*H_{,2}Z_{,2}=(Z_{,2})^2-2H^*H_{,2}Z_{,2}.\end{aligned} \tag{56}$$

The terms with the coefficient δ will similarly combine to give

$$(Z_{,3})^2-2H^*H_{,3}Z_{,3}. \tag{57}$$

Thus, equations (51) and (52) are equivalent to the single complex equation

$$\begin{aligned}\Psi[(\Delta Z_{,2})_{,2}+(\delta Z_{,3})_{,3}] &= \Delta(Z_{,2})^2+\delta(Z_{,3})^2\\ &\qquad -2H^*(\Delta Z_{,2}H_{,2}+\delta Z_{,3}H_{,3});\end{aligned} \tag{58}$$

and equation (50) with the terms in the second line written in terms of Z becomes

$$\begin{aligned}\Psi[(\Delta H_{,2})_{,2}+(\delta H_{,3})_{,3}] &= \Delta H_{,2}Z_{,2}+\delta H_{,3}Z_{,3}\\ &\qquad -2H^*[\Delta(H_{,2})^2+\delta(H_{,3})^2],\end{aligned} \tag{59}$$

where

$$\Psi = \operatorname{Re} Z-|H|^2. \tag{60}$$

Equations (58) and (59) are the *Ernst equations*: they combine in a convenient and a symmetric fashion the four equations governing χ, ω, A, and B into two complex equations; they replace the single equation, Chapter 6 (94), applicable to the pure vacuum.

(*d*) *The transformation properties of the Ernst equations*

Let the pair (Z, H) represent a solution of equations (58) and (59). Then the following pairs are also solutions:

(i) $(Z+i\alpha,\ H)$ where α is any real number;

(ii) $(\beta^2 Z,\ \beta e^{i\alpha}H)$ where α and β are two real numbers;

(iii) $(Z+2a^*H+aa^*,\ H+a)$ where a is any complex number; (61)

and

(iv) $(Z^{-1},\ HZ^{-1})$.

Of these transformations, (i) and (ii) follow 'trivially'; (iii) is readily verified; but the verification of (iv) requires some extensive reductions.

By applying successively one or the other of the transformations (i)–(iv), we can obtain other solution pairs. Thus

$$(Z,H)\Rightarrow\left(\frac{1}{Z},\frac{H}{Z}\right)\Rightarrow\left(\frac{1}{Z}+i\alpha,\frac{H}{Z}\right)\Rightarrow\left(\frac{Z}{1+i\alpha Z},\frac{H}{1+i\alpha Z}\right),\tag{62}$$

and

$$(Z,H)\Rightarrow\left(\frac{1}{Z},\frac{H}{Z}\right)\Rightarrow\left(\frac{1}{Z}+2a^*\frac{H}{Z}+aa^*,\frac{H}{Z}+a\right)$$
$$\Rightarrow\left(\frac{Z}{1+2a^*H+aa^*Z},\frac{H+aZ}{1+2a^*H+aa^*Z}\right).\tag{63}$$

These are, respectively, the *Ehlers* and the *Harrison* transformations. Gürses and Xanthopoulos have discovered a remarkable way of writing Ernst's equations which makes their transformation properties manifest. Thus, defining the Hermitian matrix,

$$\boldsymbol{P}=\left|\begin{matrix}\dfrac{1}{\Psi} & \sqrt{2}\dfrac{H}{\Psi} & i\left(\dfrac{Z}{\Psi}-1\right)\\[2ex] \sqrt{2}\dfrac{H^*}{\Psi} & 1+2\dfrac{|H|^2}{\Psi} & i\sqrt{2}\dfrac{H^*Z}{\Psi}\\[2ex] -i\left(\dfrac{Z^*}{\Psi}-1\right) & -i\sqrt{2}\dfrac{HZ^*}{\Psi} & \dfrac{|Z|^2}{\Psi}\end{matrix}\right|\tag{64}$$

and its inverse,

$$\boldsymbol{P}^{-1}=\left|\begin{matrix}\dfrac{|Z|^2}{\Psi} & -\sqrt{2}\dfrac{HZ^*}{\Psi} & +i\left(\dfrac{Z^*}{\Psi}-1\right)\\[2ex] -\sqrt{2}\dfrac{H^*Z}{\Psi} & 1+2\dfrac{|H|^2}{\Psi} & -i\sqrt{2}\dfrac{H^*}{\Psi},\\[2ex] -i\left(\dfrac{Z}{\Psi}-1\right) & +i\sqrt{2}\dfrac{H}{\Psi} & \dfrac{1}{\Psi}\end{matrix}\right|,\tag{65}$$

Gürses and Xanthopoulos have shown that Ernst's equations are included in the single matrix-equation

$$(\Delta\boldsymbol{P}^{-1}\boldsymbol{P}_{,2})_{,2}+(\delta\boldsymbol{P}^{-1}\boldsymbol{P}_{,3})_{,3}=0.\tag{66}$$

This matrix equation, written out explicitly in terms of its elements, represents nine equations, all of which are various combinations of Ernst's equations. It is an immediate consequence of equation (66) that if $\boldsymbol{P}$ represents a solution, then so does $\boldsymbol{APA}^{-1}$ where $\boldsymbol{A}$ is any constant invertible matrix. With different choices of $\boldsymbol{A}$, we can derive all the transformations applicable to Ernst's equations.

(e) The operation of conjugation

In Chapter 6, §52(*a*), it was shown that given a metric of the form (1), we can derive a *conjugate metric* of the same form by the transformation

$$t \to i\varphi \quad \text{and} \quad \varphi \to -it, \tag{67}$$

when χ and ω are replaced by

$$\tilde{\chi} = -\frac{\chi}{\chi^2 - \omega^2} \quad \text{and} \quad \tilde{\omega} = \frac{\omega}{\chi^2 - \omega^2}. \tag{68*}$$

To ascertain how the potential H transforms under conjugation, we first observe that by rewriting equation (33) in the form

$$\left(\chi\frac{\sqrt{\Delta}}{\sqrt{\delta}}H_{,2} - i\omega H_{,3}\right)_{,2} + \left(\chi\frac{\sqrt{\delta}}{\sqrt{\Delta}}H_{,3} + i\omega H_{,2}\right)_{,3} = 0, \tag{69}$$

we can define a potential $\tilde{H}$ by the equations

$$\left.\begin{aligned} -\tilde{H}_{,3} &= \chi\frac{\sqrt{\Delta}}{\sqrt{\delta}}H_{,2} - i\omega H_{,3}, \\ +\tilde{H}_{,2} &= i\omega H_{,2} + \chi\frac{\sqrt{\delta}}{\sqrt{\Delta}}H_{,3}. \end{aligned}\right\} \tag{70}$$

Solving these equations for $H_{,2}$ and $H_{,3}$, we obtain

$$\left.\begin{aligned} H_{,2} &= \tilde{\chi}\frac{\sqrt{\delta}}{\sqrt{\Delta}}\tilde{H}_{,3} + i\tilde{\omega}\tilde{H}_{,2}, \\ H_{,3} &= -\tilde{\chi}\frac{\sqrt{\Delta}}{\sqrt{\delta}}\tilde{H}_{,2} + i\tilde{\omega}\tilde{H}_{,3}, \end{aligned}\right\} \tag{71}$$

where $\tilde{\chi}$ and $\tilde{\omega}$ are defined as in equations (68); and the integrability condition of these equations is

$$\left(\tilde{\chi}\frac{\sqrt{\Delta}}{\sqrt{\delta}}\tilde{H}_{,2} - i\tilde{\omega}\tilde{H}_{,3}\right)_{,2} + \left(\tilde{\chi}\frac{\sqrt{\delta}}{\sqrt{\Delta}}\tilde{H}_{,3} + i\tilde{\omega}\tilde{H}_{,2}\right)_{,3} = 0. \tag{72}$$

This equation is exactly of the same form in the "tilded" variables as equation (69) is in the "untilded" variables. Accordingly, it would appear that conjugation results in replacing H by $\tilde{H}$. This inference can be confirmed by directly verifying that $\tilde{\chi}$, $\tilde{\omega}$, and $\tilde{H}$ do in fact satisfy equations of exactly the same forms as equations (30) and (37), though the actual verification is somewhat tedious.

From the fact that $\tilde{\chi}$, $\tilde{\omega}$, and $\tilde{H}$ satisfy the same equations as the "untilded"

* In the case of the pure vacuum considered in Chapter 6, $\tilde{\chi}$ can be chosen to be of either sign; but in the present context, the negative sign is the correct one, as it will emerge.

functions, it follows that with the further definitions,

$$\tilde{\Psi} = \frac{\sqrt{(\Delta\delta)}}{\tilde{\chi}} \tag{73}$$

and

$$\left.\begin{aligned} -\tilde{\Phi}_{,2} &= \frac{\delta}{\tilde{\chi}^2}\tilde{\omega}_{,3} + 2\,\mathrm{Im}\,\tilde{H}\tilde{H}^*{}_{,2}, \\ +\tilde{\Phi}_{,3} &= \frac{\Delta}{\tilde{\chi}^2}\tilde{\omega}_{,2} - 2\,\mathrm{Im}\,\tilde{H}\tilde{H}^*{}_{,3}, \end{aligned}\right\} \tag{74}$$

the functions

$$\tilde{Z} = \tilde{\Psi} + |\tilde{H}|^2 + i\tilde{\Phi} \qquad \text{and} \qquad \tilde{H} \tag{75}$$

will satisfy Ernst's equations of exactly the same forms as equations (58) and (59).

110. The Kerr–Newman solution: its derivation and its description in a Newman–Penrose formalism

As in the case of the Kerr solution (Chapter 6, §54), the Kerr–Newman solution follows, equally remarkably, from a consideration of the simplest solution of Ernst's equations for the conjugate metric.

Since we are now dealing with a pair of equations, the search for the "simplest solution" consists of two parts. *First*, is there a simple assumption which will reduce the pair of equations (58) and (59) to a single equation? And *second*, if such a reduced equation exists, what is its simplest solution?

With regard to the first question, the answer is that a linear relation of the form

$$H = Q(Z+1), \tag{76}$$

where Q is a complex constant, is consistent with the pair of equations (58) and (59). For, on this assumption,

$$\begin{aligned} \Psi &= \tfrac{1}{2}(Z+Z^*) - |Q(Z+1)|^2 \\ &= \tfrac{1}{2}[(1-2|Q|^2)(Z+Z^*) - 2|Q|^2(|Z|^2+1)]; \end{aligned} \tag{77}$$

and both equations (58) and (59) reduce to the same equation,

$$\begin{aligned} \tfrac{1}{2}[(1-2|Q|^2)(Z+Z^*) - 2|Q|^2(|Z|^2+1)][(\Delta Z_{,2})_{,2} + (\delta Z_{,3})_{,3}] \\ = [1-2|Q|^2(Z^*+1)][\Delta(Z_{,2})^2 + \delta(Z_{,3})^2]. \end{aligned} \tag{78}$$

The problem is thus effectively reduced to solving the Ernst equation for the vacuum. Accordingly, in seeking solutions of equation (78), we make the same transformation, Chapter 6, equation (95):

$$Z = \frac{1+E}{1-E}, \tag{79}$$

when

$$\Psi = \frac{1}{|1-E|^2}(1-4|Q|^2-|E|^2) \tag{80}$$

$$\Phi = \operatorname{Im} Z = -i\frac{E-E^*}{|1-E|^2}, \tag{81}$$

and the equation governing E is (cf. Chapter 6, equation (96)),

$$\begin{aligned}(1-4|Q|^2-|E|^2)[(\Delta E_{,2})_{,2}+(\delta E_{,3})_{,3}] \\ = -2E^*[\Delta(E_{,2})^2+\delta(E_{,3})^2].\end{aligned} \tag{82}$$

By changing the variable r to

$$\eta = (r-M)/(M^2-M_0^2)^{1/2}, \tag{83}$$

when

$$\Delta = (M^2-M_0^2)(\eta^2-1), \tag{84}$$

equation (82) takes the more symmetrical form

$$\begin{aligned}(1-4|Q|^2-|E|^2)\{[(\eta^2-1)E_{,\eta}]_{,\eta}+[(1-\mu^2)E_{,\mu}]_{,\mu}\} \\ = -2E^*[(\eta^2-1)(E_{,\eta})^2+(1-\mu^2)(E_{,\mu})^2].\end{aligned} \tag{85}$$

The derivation of the Kerr–Newman solution now proceeds as in §54. We consider equation (85) for the conjugate function, $\tilde{E}$; and we verify that it allows the simple solution

$$\tilde{E} = -p\eta - iq\mu, \tag{86}$$

where p and q are two real constants related in the manner

$$p^2+q^2 = 1-4|Q|^2. \tag{87}$$

For $\tilde{E}$ given by equation (86),

$$\frac{1+\tilde{E}}{1-\tilde{E}} = \frac{1}{(1+p\eta)^2+q^2\mu^2}[-p^2(\eta^2-1)+q^2(1-\mu^2)+4|Q|^2-2iq\mu]. \tag{88}$$

Therefore,

$$\operatorname{Re}\tilde{Z} = -\frac{p^2(\eta^2-1)-q^2(1-\mu^2)-4|Q|^2}{(1+p\eta)^2+q^2\mu^2}, \tag{89}$$

and

$$\operatorname{Im}\tilde{Z} = -\frac{2q\mu}{(1+p\eta)^2+q^2\mu^2}. \tag{90}$$

Also, by equation (80),

$$\tilde{\Psi} = -\frac{p^2(\eta^2-1)-q^2(1-\mu^2)}{(1+p\eta)^2+q^2\mu^2}; \tag{91}$$

and from the relation (60) it now follows that

$$|\tilde{H}|^2 = \operatorname{Re}\tilde{Z} - \tilde{\Psi} = \frac{4|Q|^2}{(1+p\eta)^2 + q^2\mu^2}. \tag{92}$$

Now, reverting to the variable r, we find that the foregoing solutions are much simplified by a particular choice of p and q compatible with the relation (87). Thus, the expression

$$-\operatorname{Re}\tilde{Z} = \left[\Delta - \frac{q^2}{p^2}(M^2 - M_0^2)\delta - 4|Q|^2\frac{(M^2-M_0^2)}{p^2}\right]$$
$$\div\left\{\left[(r-M) + \frac{(M^2-M_0^2)^{1/2}}{p}\right]^2 + \frac{q^2}{p^2}(M^2-M_0^2)\mu^2\right\} \tag{93}$$

is much simplified by the choice

$$p = (M^2 - M_0^2)^{1/2}/M \qquad \text{and} \qquad q = a/M, \tag{94}$$

where a is a constant; for, with this choice,

$$\left[(r-M) + \frac{(M^2-M_0^2)^{1/2}}{p}\right]^2 + \frac{q^2}{p^2}(M^2-M_0^2)\mu^2 = r^2 + a^2\mu^2 = \rho^2, \tag{95}$$

and

$$\Delta - \frac{q^2}{p^2}(M^2-M_0^2)\delta = \Delta - a^2\delta. \tag{96}$$

However, to be consistent with the relation (87), we must require that

$$(M^2 - M_0^2) + a^2 = M^2(1 - 4|Q|^2), \tag{97}$$

or

$$M_0^2 = a^2 + 4|Q|^2 M^2 = a^2 + Q_*^2 \quad \text{(say)}, \tag{98}$$

where

$$Q_*^2 = 4|Q|^2 M^2. \tag{99}$$

The expression for Δ now becomes

$$\Delta = r^2 - 2Mr + a^2 + Q_*^2. \tag{100}$$

Therefore, with the choices (94) and (99), the solutions (89)–(92) take the simple forms

$$\operatorname{Re}\tilde{Z} = -\frac{\Delta - a^2\delta - Q_*^2}{\rho^2}; \quad \operatorname{Im}\tilde{Z} = \tilde{\Phi} = -\frac{2aM\mu}{\rho^2}, \tag{101}$$

$$\tilde{\Psi} = \frac{\sqrt{(\Delta\delta)}}{\tilde{\chi}} = -\frac{\Delta - a^2\delta}{\rho^2}, \qquad \text{and} \qquad \tilde{H}^2 = \frac{Q_*^2}{\rho^2}. \tag{102}$$

Also, by equation (76),

$$\tilde{H} = Q(\tilde{Z}+1) = Q_*\frac{r - ia\mu}{\rho^2}; \tag{103}$$

and, therefore,

$$\tilde{A} = Q_* \frac{r}{\rho^2} \qquad \text{and} \qquad \tilde{B} = -aQ_* \frac{\mu}{\rho^2}. \tag{104}$$

From the solutions (104) we deduce:

$$\left.\begin{aligned} \tilde{A}_{,2} &= \frac{Q_*}{\rho^4}(-r^2 + a^2\mu^2), \quad & \tilde{A}_{,3} &= \frac{Q_*}{\rho^4}(-2a^2 r\mu), \\ \tilde{B}_{,3} &= a\frac{Q_*}{\rho^4}(-r^2 + a^2\mu^2); \quad & \tilde{B}_{,2} &= \frac{Q_*}{\rho^4}(+2ar\mu). \end{aligned}\right\} \tag{105}$$

With the aid of these equations, we find:

$$\left.\begin{aligned} 2\,\mathrm{Im}\,\tilde{H}\tilde{H}^*{}_{,2} &= 2(\tilde{B}\tilde{A}_{,2} - \tilde{A}\tilde{B}_{,2}) = -2\frac{Q_*^2}{\rho^4}a\mu, \\ 2\,\mathrm{Im}\,\tilde{H}\tilde{H}^*{}_{,3} &= 2(\tilde{B}\tilde{A}_{,3} - \tilde{A}\tilde{B}_{,3}) = +2\frac{Q_*^2}{\rho^4}ar. \end{aligned}\right\} \tag{106}$$

From equations (39), (101) and (106), it now follows that

$$\left.\begin{aligned} +\frac{\Delta}{\tilde{\chi}^2}\tilde{\omega}_{,2} &= \tilde{\Phi}_{,3} + 2\,\mathrm{Im}\,HH^*{}_{,3} = -\frac{2aM(r^2 - a^2\mu^2)}{\rho^4} + 2\frac{Q_*^2}{\rho^4}ar, \\ -\frac{\delta}{\tilde{\chi}^2}\tilde{\omega}_{,3} &= \tilde{\Phi}_{,2} + 2\,\mathrm{Im}\,HH^*{}_{,2} = \frac{4aMr\mu}{\rho^4} - 2\frac{Q_*^2}{\rho^4}a\mu. \end{aligned}\right\} \tag{107}$$

On the other hand, according to the solution for $\tilde{\Psi}$ given in equations (102),

$$\frac{\Delta}{\tilde{\chi}^2} = \frac{(\Delta - a^2\delta)^2}{\rho^4\delta} \qquad \text{and} \qquad \frac{\delta}{\tilde{\chi}^2} = \frac{(\Delta - a^2\delta)^2}{\rho^4\Delta}. \tag{108}$$

The equations which determine $\tilde{\omega}$ are, therefore,

$$\left.\begin{aligned} \tilde{\omega}_{,2} &= -\frac{2a\delta}{(\Delta - a^2\delta)^2}[M(r^2 - a^2\mu^2) - Q_*^2 r], \\ \tilde{\omega}_{,3} &= -\frac{2a\Delta\mu}{(\Delta - a^2\delta)^2}[2Mr - Q_*^2]. \end{aligned}\right\} \tag{109}$$

From these equations we readily find that

$$\tilde{\omega} = \frac{a\delta}{\Delta - a^2\delta}(2Mr - Q_*^2), \tag{110}$$

or, alternatively (cf. equations (68) and (100)),

$$\tilde{\omega} = \frac{\omega}{\chi^2 - \omega^2} = \frac{a\delta}{\Delta - a^2\delta}[(r^2 + a^2) - \Delta]; \tag{111}$$

and, similarly, from equations (68) and (102), we find

$$\tilde{\chi} = -\frac{\chi}{\chi^2-\omega^2} = -\frac{\rho^2}{\Delta - a^2\delta}\sqrt{(\Delta\delta)}. \tag{112}$$

With the aid of the solutions (111) and (112) for $\tilde{\omega}$ and $\tilde{\chi}$, we can solve for the metric functions, $e^{2\nu}$, $e^{2\psi}$, and ω exactly as we did in §54 (Ch. 6, equations (115)–(122)). We find:

$$e^{2\psi} = \frac{\delta\Sigma^2}{\rho^2}, \quad e^{2\nu} = \frac{\rho^2\Delta}{\Sigma^2}, \quad e^{2\beta} = e^{2(\psi+\nu)} = \Delta\delta, \tag{113}$$

$$\omega = \frac{a}{\Sigma^2}(r^2+a^2-\Delta) \qquad \text{and} \qquad \chi = e^{-\psi+\nu} = \frac{\rho^2\sqrt{\Delta}}{\Sigma^2\sqrt{\delta}}, \tag{114}$$

where

$$\Sigma^2 = (r^2+a^2)^2 - a^2\delta\Delta. \tag{115}$$

We observe that these solutions, apart from the definition of Δ, are exactly the same as the solutions (Ch. 6, equations (121)–(125)) for the Kerr metric (with the understanding that $2Mr$ in the solution for ω, in Chapter 6, equation (123), is to be replaced by $r^2+a^2-\Delta$).

Returning to equations (105), we can now obtain, by making use of equations (111) and (112), the solutions for the untilded quantities in accordance with equations (71); thus,

$$\left.\begin{aligned} A_{,r} &= +\frac{Q_*}{\rho^4}2a^2 r\sin^2\theta\cos\theta, \\ A_{,\theta} &= -\frac{Q_*}{\rho^4}(r^2+a^2)(-r^2+a^2\cos^2\theta)\sin\theta, \\ B_{,r} &= -\frac{Q_*}{\rho^4}a(-r^2+a^2\cos^2\theta)\sin^2\theta, \\ B_{,\theta} &= -\frac{Q_*}{\rho^4}2ar(r^2+a^2)\sin\theta\cos\theta, \end{aligned}\right\} \tag{116}$$

where, to avoid any ambiguity, we have explicitly indicated the variables to which the differentiations refer. For later use, we may note here that, according to the foregoing solutions,

$$A_{,r}A_{,\theta} + B_{,r}B_{,\theta} = 0, \tag{117}$$

and

$$\begin{aligned} (A_{,r})^2 + (B_{,r})^2 &= \frac{Q_*^2}{\rho^4}a^2\sin^4\theta, \\ (A_{,\theta})^2 + (B_{,\theta})^2 &= \frac{Q_*^2}{\rho^4}(r^2+a^2)^2\sin^2\theta. \end{aligned} \tag{118}$$

Finally, to determine $(\mu_2+\mu_3)$ and complete the solution, for the metric functions we turn to equations (19) and (25). By comparison with the corresponding equations, Chapter 6, (8) and (16), which obtain for the vacuum, we observe that, by virtue of the relation (117), equation (19) is the same as equation (8) while equation (25) has the additional term

$$4e^{-2\psi}\{e^{\mu_3-\mu_2}[(A_{,r})^2+(B_{,r})^2]-e^{\mu_2-\mu_3}[(A_{,\theta})^2+(B_{,\theta})^2]\}$$

$$=4e^{-2\psi}\left\{(\sqrt{\Delta})\frac{Q_*^2}{\rho^4}a^2\sin^4\theta-\frac{Q_*^2}{\rho^4\sqrt{\Delta}}(r^2+a^2)^2\sin^2\theta\right\}$$

$$=\frac{4\rho^2}{\Sigma^2\sin^2\theta}\cdot\frac{\sin^2\theta}{\sqrt{\Delta}}\frac{Q_*^2}{\rho^4}[a^2\Delta\sin^2\theta-(r^2+a^2)^2]=-\frac{4Q_*^2}{\rho^2\sqrt{\Delta}},\tag{119}$$

where, in the course of the simplification, we have made use of the relations (118). Therefore, in place of Chapter 6, equations (59) and (60), we now have

$$-\frac{\mu}{\delta}(\mu_2+\mu_3)_{,2}+\frac{r-M}{\Delta}(\mu_2+\mu_3)_{,3}=\frac{2}{(X+Y)^2}(X_{,2}Y_{,3}+Y_{,2}X_{,3}),\tag{120}$$

and

$$2(r-M)(\mu_2+\mu_3)_{,2}+2\mu(\mu_2+\mu_3)_{,3}=\frac{4}{(X+Y)^2}(\Delta X_{,2}Y_{,2}-\delta X_{,3}Y_{,3})$$
$$-3\frac{(r-M)^2-\Delta}{\Delta}+\frac{\mu^2+\delta}{\delta}-4\frac{Q_*^2}{\rho^2},\tag{121}$$

where

$$X=\chi+\omega\qquad\text{and}\qquad Y=\chi-\omega.\tag{122}$$

In view of the formal identity of the solutions for χ and ω (as written in equations (114)) with the corresponding solutions for the Kerr metric, the reduction of the right-hand side of equation (120) will lead to the same equation as in Chapter 6, (129). But the occurrence of the additional term $-4Q_*^2/\rho^2$ on the right-hand side of equation (121) requires that the reductions be carefully scrutinized. We find that

$$\frac{\Delta}{\chi^2}X_{,2}Y_{,2}=\frac{(r-M)^2}{\Delta}-\frac{4Mr}{\rho^2}+\frac{8a^2\mu^2rM}{\rho^4}+4\frac{Q_*^2r^2}{\rho^4}\tag{123}$$

and

$$\frac{\Delta}{\chi^2}X_{,2}Y_{,2}-\frac{\delta}{\chi^2}X_{,3}Y_{,3}=\frac{(r-M)^2}{\Delta}-\frac{4Mr}{\rho^2}+\frac{8a^2\mu^2rM}{\rho^4}+4\frac{Q_*^2r^2}{\rho^4}$$
$$-\frac{\mu^2}{\rho^4\delta}[(r^2+a^2+a^2\delta)^2-4a^2\Delta\delta]$$
$$=\frac{(r-M)^2}{\Delta}-\frac{4Mr}{\rho^2}-\frac{\mu^2}{\delta}+4\frac{Q_*^2}{\rho^2}.\tag{124}$$

By combining equations (121) and (124), we are again led to the same equation as in Chapter 6, (130). The solution for $(\mu_2+\mu_3)$ is, therefore, the same as for the Kerr metric, namely (cf. Ch. 6, equation (131)),

$$e^{\mu_2+\mu_3} = \rho^2/\sqrt{\Delta}. \tag{125}$$

The solution of the entire set of equations is now completed; and *the Kerr–Newman metric, except for the definition of* Δ, *is identical with the Kerr metric*. It follows that M, *as defined, does denote the inertial mass* of the black hole and *the parameter a is again to be interpreted as the angular momentum, per unit mass*, of the black hole.

(*a*) *The description of the Kerr–Newman space-time in a Newman–Penrose formalism*

From the formal identity of the Kerr and the Kerr–Newman metrics (apart from the definition of Δ) it follows that we can define a null-tetrad basis, for a description of the Kerr–Newman space-time in a Newman–Penrose formalism, exactly as in Chapter 6, §56 for the Kerr space-time. In particular, the required basis will continue to be given by Chapter 6, equations (170) and (173); and the spin coefficients with respect to the chosen basis will also be given by Chapter 6, equations (175). And we can conclude, from the vanishing of the spin coefficients, κ, σ, λ, and ν, that the Kerr–Newman space-time, like the Kerr space-time, is of Petrov type-D; and, further, that in the chosen basis,

$$\Psi_0 = \Psi_1 = \Psi_3 = \Psi_4 = 0. \tag{126}$$

Therefore, to complete the description, we need only to specify the Weyl scalar Ψ_2 and the Maxwell scalars ϕ_0, ϕ_1, and ϕ_2.

Considering first the Maxwell scalars, we can determine them by appropriately contracting F_{ij} with the basis vectors as given in Chapter 6, equation (177). Thus, remembering that F_{01} and F_{23} vanish, we have

$$\begin{aligned}2\phi_1 &= F_{ij}(l^i n^j + \bar{m}^i m^j)\\ &= F_{02}(l^0 n^2 - l^2 n^0) + F_{12}(l^1 n^2 - l^2 n^1)\\ &\quad + F_{03}(\bar{m}^0 m^3 - \bar{m}^3 m^0) + F_{13}(\bar{m}^1 m^3 - \bar{m}^3 m^1)\\ &= -\frac{1}{\rho^2}(r^2+a^2)e^{\nu+\mu_2}F_{02} - \frac{a\Delta}{\Sigma^2}e^{\psi+\mu_2}F_{12}\\ &\quad -\frac{i}{\rho^2}e^{\nu+\mu_3}(a\sin\theta)F_{03} - \frac{i}{\Sigma^2\sin\theta}e^{\psi+\mu_3}(r^2+a^2)F_{13}.\end{aligned} \tag{127}$$

Now substituting for the components of F_{ij} in accordance with equations (7) and (116), we find after some reductions that we are left with

$$\phi_1 = -i\frac{Q_*}{2(\bar{\rho}^*)^2}. \tag{128}$$

In similar fashion, we find that

$$\phi_0 = \phi_2 = 0, \tag{129}$$

so that ϕ_1 *is the only non-vanishing Maxwell scalar.*

To determine the Weyl scalar Ψ_2, we shall proceed somewhat differently by considering the Bianchi identities, Chapter 1, equations (321) (b) and (c) (with the "Ricci terms" given in Chapter 1, equations (339) (b) and (c)). Remembering that in the present context, the only non-vanishing Weyl and Maxwell scalars are Ψ_2 and ϕ_1 and that the spin coefficients κ, σ, λ, and ν vanish, we find that the Bianchi identities give

$$\left.\begin{aligned} -D\Psi_2 + 3\tilde{\rho}\Psi_2 + 4\tilde{\rho}\,\phi_1\phi_1^* &= 0 \\ \text{and} \qquad -\delta^*\Psi_2 - 3\pi\Psi_2 + 4\pi\phi_1\phi_1^* &= 0. \end{aligned}\right\} \tag{130}$$

Now by making use of the list of the spin coefficients in Chapter 6, equations (175), and of the definitions of the operators D and δ^* in Chapter 8, equations (3)–(5), we obtain the equations

$$\left.\begin{aligned} \frac{\partial \Psi_2}{\partial r} &= -\frac{3}{\bar{\rho}^*}\Psi_2 - \frac{Q_*^2}{4\rho^4\bar{\rho}^*}, \\ \text{and} \qquad \frac{\partial \Psi_2}{\partial \theta} &= -\frac{3ia\sin\theta}{\bar{\rho}^*}\Psi_2 + Q_*^2\frac{ia\sin\theta}{\rho^4\bar{\rho}^*}, \end{aligned}\right\} \tag{131}$$

after substituting for ϕ_1 its solution (128).

We readily verify that a particular integral of equations (131) is given by

$$\Psi_2 = \frac{Q_*^2}{\bar{\rho}(\bar{\rho}^*)^3}; \tag{132}$$

while the general solution of the homogeneous equations is

$$\Psi_2 = \frac{\text{constant}}{(\bar{\rho}^*)^3}. \tag{133}$$

By comparison with Chapter 6, equation (180), we conclude that the constant in the solution (133) is $-M$ since it must have the same value as for the Kerr solution (in the limit $Q_* = 0$). Thus, the required solution for Ψ_2 is

$$\Psi_2 = -\frac{M}{(\bar{\rho}^*)^3} + \frac{Q_*^2}{\bar{\rho}(\bar{\rho}^*)^3}. \tag{134}$$

111. The equations governing the coupled electromagnetic-gravitational perturbations of the Kerr–Newman space-time

As we have stated in the introductory section (§108), the methods that have proved to be so successful in treating the gravitational perturbations of the

Kerr space-time do not seem to be applicable (nor susceptible to easy generalizations) for treating the coupled electromagnetic-gravitational perturbations of the Kerr–Newman space-time. The principal obstacle is in finding separated equations. In this section, we shall briefly consider the origin of this apparent indissolubility of the coupling between the spin-1 and spin-2 fields in the perturbed space-time.

The basic equations for treating the perturbations of the Kerr–Newman space-time are again the "already linearized" equations provided by the four Bianchi identities (Chapter 1, equations (321) (a), (d), (e), and (h)), the two Ricci identities (Chapter 1, equations (310) (b) and (j)) and Maxwell's equations in the reduced form given in Chapter 5, equations (207) and (208). These latter equations suggest, as in the treatment of the perturbations of the Reissner–Nordström space-time, that we adopt a gauge in which ϕ_0 and ϕ_2 continue to vanish:

$$\phi_0 = \phi_2 = 0. \tag{135}$$

As we have explained in Chapter 5, §44(b), this choice of gauge is entirely permissible.* In this gauge, Maxwell's equations are replaced by the pair of equations,

$$(\triangle - 3\gamma - \gamma^* - 2\mu + \mu^*)\kappa - (\delta^* - 3\alpha + \beta^* - \tau^* - 2\pi)\sigma + 2\Psi_1 = 0, \tag{136}$$

$$(\delta + \pi^* + 2\tau - \alpha^* + 3\beta)\lambda - (D + 3\varepsilon + \varepsilon^* + 2\tilde{\rho} - \tilde{\rho}^*)\nu + 2\Psi_3 = 0, \tag{137}$$

for the spin coefficients. (Note that in the chosen gauge the Weyl scalars Ψ_1 and Ψ_3 cannot be assumed to vanish.)

Turning to the Bianchi identities (Chapter 1, equations (321) (a), (d), (e), and (h)), we must now replace the expressions on the right-hand sides of the first two of Chapter 9, equations (3) and (4) by (cf. Chapter 5, equations (220)),

$$\left.\begin{aligned}
+\kappa(3\Psi_2 - 4\phi_1\phi_1^*) &= -\frac{\kappa}{(\bar{\rho}^*)^3}\left[3\left(M - \frac{Q_*^2}{\bar{\rho}}\right) + Q_*^2\frac{\bar{\rho}^*}{\bar{\rho}^2}\right],\\
+\sigma(3\Psi_2 + 4\phi_1\phi_1^*) &= -\frac{\sigma}{(\bar{\rho}^*)^3}\left[3\left(M - \frac{Q_*^2}{\bar{\rho}}\right) - Q_*^2\frac{\bar{\rho}^*}{\bar{\rho}^2}\right],\\
-\lambda(3\Psi_2 + 4\phi_1\phi_1^*) &= +\frac{\lambda}{(\bar{\rho}^*)^3}\left[3\left(M - \frac{Q_*^2}{\bar{\rho}}\right) - Q_*^2\frac{\bar{\rho}^*}{\bar{\rho}^2}\right],\\
-\nu(3\Psi_2 - 4\phi_1\phi_1^*) &= +\frac{\nu}{(\bar{\rho}^*)^3}\left[3\left(M - \frac{Q_*^2}{\bar{\rho}}\right) + Q_*^2\frac{\bar{\rho}^*}{\bar{\rho}^2}\right].
\end{aligned}\right\} \tag{138}$$

Therefore, with the same definitions, Chapter 9 equations (6), we must now consider, in place of Chapter 9, equations (7), (8), and (9), the four equations

* It certainly appears the simplest: any other choice leads to a more complicated set of equations than equations (139)–(142).

(the last of which is the transformed version of equation (136)):

$$\left(\mathscr{L}_2-\frac{3ia\sin\theta}{\bar{\rho}^*}\right)\Phi_0-\left(\mathscr{D}_0+\frac{3}{\bar{\rho}^*}\right)\Phi_1 = -2k\left[3\left(M-\frac{Q_*^2}{\rho}\right)+Q_*^2\frac{\bar{\rho}^*}{\bar{\rho}^2}\right], \tag{139}$$

$$\Delta\left(\mathscr{D}_2^\dagger-\frac{3}{\bar{\rho}^*}\right)\Phi_0+\left(\mathscr{L}_{-1}^\dagger+\frac{3ia\sin\theta}{\bar{\rho}^*}\right)\Phi_1 = +2s\left[3\left(M-\frac{Q_*^2}{\bar{\rho}}\right)-Q_*^2\frac{\bar{\rho}^*}{\bar{\rho}^2}\right], \tag{140}$$

$$\left(\mathscr{D}_0+\frac{3}{\bar{\rho}^*}\right)s-\left(\mathscr{L}_{-1}^\dagger+\frac{3ia\sin\theta}{\bar{\rho}^*}\right)k = \frac{\bar{\rho}}{(\bar{\rho}^*)^2}\Phi_0, \tag{141}$$

$$\Delta\left(\mathscr{D}_2^\dagger-\frac{3}{\bar{\rho}^*}\right)k+\left(\mathscr{L}_2-\frac{3ia\sin\theta}{\bar{\rho}^*}\right)s = 2\frac{\bar{\rho}}{(\bar{\rho}^*)^2}\Phi_1. \tag{142}$$

We observe that equation (142) is the same equation as in Chapter 9, (29) which we supplied (ad hoc!) to restore the truncated symmetry of Chapter 9, equations (7)–(9).

We can now apply to equations (138)–(142) the same operations that led to Chapter 9, equations (17) and (31). Thus, with the aid of the identities

$$\Delta\left(\mathscr{D}_1+\frac{3}{\bar{\rho}^*}\right)\left(\mathscr{D}_2^\dagger-\frac{3}{\bar{\rho}^*}\right)+\left(\mathscr{L}_{-1}^\dagger+\frac{3ia\sin\theta}{\bar{\rho}^*}\right)\left(\mathscr{L}_2-\frac{3ia\sin\theta}{\bar{\rho}^*}\right)$$
$$= \Delta\mathscr{D}_1\mathscr{D}_2^\dagger+\mathscr{L}_{-1}^\dagger\mathscr{L}_2-6i\sigma\bar{\rho}+\frac{6\bar{\rho}}{(\bar{\rho}^*)^2}\left(M-\frac{Q_*^2}{\bar{\rho}}\right) \tag{143}$$

and

$$\Delta\left(\mathscr{D}_2^\dagger-\frac{3}{\bar{\rho}^*}\right)\left(\mathscr{D}_0+\frac{3}{\bar{\rho}^*}\right)+\left(\mathscr{L}_2-\frac{3ia\sin\theta}{\bar{\rho}^*}\right)\left(\mathscr{L}_{-1}^\dagger+\frac{3ia\sin\theta}{\bar{\rho}^*}\right)$$
$$= \Delta\mathscr{D}_2^\dagger\mathscr{D}_0+\mathscr{L}_2\mathscr{L}_{-1}^\dagger-6i\sigma\bar{\rho}+\frac{12\bar{\rho}}{(\bar{\rho}^*)^2}\left(M-\frac{Q_*^2}{\bar{\rho}}\right), \tag{144}$$

we find that, in place of the separated decoupled equations Chapter 9, (20), (21), (33), and (34), we are now left with the pair of coupled equations

$$(\Delta\mathscr{D}_1\mathscr{D}_2^\dagger+\mathscr{L}_{-1}^\dagger\mathscr{L}_2-6i\sigma\bar{\rho})\Phi_0 = -2Q_*^2\left(\mathscr{L}_{-1}^\dagger\frac{k\bar{\rho}^*}{\bar{\rho}^2}+\mathscr{D}_0\frac{s\bar{\rho}^*}{\bar{\rho}^2}\right) \tag{145}$$

and

$$(\Delta\mathscr{D}_2^\dagger\mathscr{D}_0+\mathscr{L}_2\mathscr{L}_{-1}^\dagger-6i\sigma\bar{\rho})\Phi_1 = +2Q_*^2\left(\Delta\mathscr{D}_2^\dagger\frac{k\bar{\rho}^*}{\bar{\rho}^2}-\mathscr{L}_2\frac{s\bar{\rho}^*}{\bar{\rho}^2}\right); \tag{146}$$

and all efforts to decouple (or separate) these equations were not successful. And numerous other alternative manipulations of the system of equations (139)–(142) were equally unsuccessful.

We may now ask why the system of equations (139)–(142) proves intractable in contrast to apparently similar systems of equations encountered in the treatment of the perturbations of the Reissner–Nordström and Kerr space-times. The experience with Chapter 5, equations (222)–(225), in the context of the Reissner–Nordström space-time, is perhaps apposite in this connection. In the first place, the angular functions, $S_{\pm 1}$ and $S_{\pm 2}$, are very simply related (cf. Chapter 5, equations (234)); and the characteristic values λ belonging to these functions are the same. Besides, the spherical symmetry of the background space assures the *separability* of the variables. But the *decoupling* of the system into a pair of second-order equations was not expected: it was enabled only by considering the peculiar combinations, Chapter 5 (239), of the radial functions belonging to the Weyl scalars and the spin coefficients, combinations in which the characteristic value parameter λ (common to both $S_{\pm 1}$ and $S_{\pm 2}$) occurs most non-linearly. In the general axisymmetric case, there is no *a priori* reason to expect that the angular functions belonging to the spin-1 and the spin-2 fields will be simply related: they are not, even in the Kerr background. Next, comparing equations (139) and (140) with the corresponding equations Chapter 9, (7) and (8), we observe that, in contrast to the simplicity of the terms on the right-hand sides of the latter equations, we now have the ugly combinations of $(M - Q_*^2/\bar{\rho})$ and $Q_*^2\bar{\rho}^*/\bar{\rho}^2$.

Altogether, then, one might be inclined to conclude that a decoupling of the system of equations (139)–(142) and a separation of the variables will be possible, if at all, only by contemplating equations of order 4 or higher.

112. Solutions representing static black-holes

We have seen in Chapter 6 (§55) that, compatible with the requirement of asymptotic flatness at infinity, the Schwarzschild solution is the unique static solution of Einstein's vacuum-equation with a smooth event horizon. This requirement of asymptotic flatness is equivalent to a restriction to black holes that are *isolated* in space. While it is natural that we restrict ourselves, in the first instance, to isolated black-holes, the question as to how they may be distorted by external distributions of mass is also of some interest. As far as the neighbourhood of such *distorted* black-holes is concerned, the assumption of the existence of an external distribution of mass is equivalent to relaxing the assumption of asymptotic flatness. In this section, we shall show how one can construct static axisymmetric solutions of Einstein's vacuum equations which may be considered as distorted Schwarzschild black-holes.

To construct static axisymmetric solutions of Einstein's equation, we start with a metric of the form

$$ds^2 = e^{2\nu}(dt)^2 - e^{2\psi}(d\varphi)^2 - e^{2\mu_2}(dx^2)^2 - e^{2\mu_3}(dx^3)^2, \tag{147}$$

i.e., of the same form that was considered in Chapter 6 (and in §109 of this chapter) except that the term in ω (representing the dragging of the inertial frame in stationary space-times) is now omitted. The consideration of the resulting field equations presents no formal difficulty.

With no loss of generality, we may, in the present context, assume that (cf. equations (26) and (27))

$$\left.\begin{aligned} e^{2(\mu_3-\mu_2)} &= \Delta(r) = r^2 - 2Mr \\ \text{and} \qquad e^{\beta} &= e^{\psi+\nu} = \Delta^{1/2}\sin\theta, \end{aligned}\right\} \tag{148}$$

consistently with the occurrence of an event horizon at $r = 2M$. The principal problem is then reduced to the consideration of the linear equation (cf. Chapter 6, equation (49)),

$$[\Delta(\psi-\nu)_{,2}]_{,2} + [\delta(\psi-\nu)_{,3}]_{,3} = 0, \tag{149}$$

where

$$\delta = 1-\mu^2 = \sin^2\theta, \tag{150}$$

and the index '3' refers to $\mu(=\cos\theta)$. Now letting

$$\eta = (r-M)/M \tag{151}$$

when

$$\Delta = M^2(\eta^2-1), \tag{152}$$

we can write equation (149) in the more symmetric form

$$[(\eta^2-1)(\lg\chi)_{,\eta}]_{,\eta} + [(1-\mu^2)(\lg\chi)_{,\mu}]_{,\mu} = 0, \tag{153}$$

where, as usual,

$$\chi = e^{-\psi+\nu}. \tag{154}$$

In our earlier considerations relative to the Kerr and the Kerr–Newman solutions, we found that the functions appropriate to the conjugate metric are, in some sense, more fundamental than the ones of the original metric. Accordingly, we shall consider (cf. Ch. 6, equations (97) and (98))

$$\Psi = \chi\sqrt{(\Delta\delta)} = \chi[(\eta^2-1)(1-\mu^2)]^{1/2}, \tag{155}$$

instead of χ. Since $\chi = (\eta^2-1)(1-\mu^2)$ is a solution of equation (153), Ψ satisfies the same equation:

$$[(\eta^2-1)(\lg\Psi)_{,\eta}]_{,\eta} + [(1-\mu^2)(\lg\Psi)_{,\mu}]_{,\mu} = 0. \tag{156}$$

It can now be verified that we recover the Schwarzschild metric with the solution

$$\Psi = \Psi_{\text{Sc}} = \frac{\eta-1}{\eta+1}. \tag{157}$$

The corresponding solution for χ is

$$\chi = \chi_{\text{Sc}} = \left[\frac{\eta-1}{(\eta+1)^3(1-\mu^2)}\right]^{1/2}. \tag{158}$$

Besides the "singular" solution (157), equation (156) allows separable solutions. Thus, writing

$$\lg \Psi = R(\eta)P(\mu), \tag{159}$$

we find that for $P(\mu)$ we must, in fact, choose* the Legendre function, $P_n(\mu)$. The radial function, $R(\eta)$, then, also satisfies Legendre's equation

$$\frac{d}{d\eta}\left[(\eta^2-1)\frac{dR}{d\eta}\right] - n(n+1)R = 0. \tag{160}$$

The solution for $R(\eta)$, that belongs with $P_n(\mu)$, is, therefore, a linear combination of the Legendre functions, $P_n(\eta)$ and $Q_n(\eta)$, of the first and the second kinds; thus,

$$R_n(\eta) = A_n P_n(\eta) + B_n Q_n(\eta), \tag{161}$$

where A_n and B_n are constants. The general solution for $\lg \Psi$ we obtain in this manner, is

$$\lg \Psi = \sum_{n=0}^{\infty} R_n(\eta) P_n(\mu) = S(\eta, \mu) \quad \text{(say)}. \tag{162}$$

Appropriate for a Schwarzschild black-hole distorted by an external distribution of mass, we shall assume for Ψ a solution of the form

$$\Psi = \Psi_{Sc}\, e^S = \frac{\eta-1}{\eta+1} e^S, \tag{163}$$

where S represents the general solution defined in equations (161) and (162). The corresponding solution for χ is

$$\chi = e^{-\psi+\nu} = \chi_{Sc}\, e^S = \left[\frac{\eta-1}{(\eta+1)^3(1-\mu^2)}\right]^{1/2} e^S. \tag{164}$$

From equation (164) and the further relation

$$e^\beta = e^{\psi+\nu} = [(\eta^2-1)(1-\mu^2)]^{1/2}, \tag{165}$$

we now find:

$$e^{2\nu} = \frac{\eta-1}{\eta+1} e^S \qquad \text{and} \qquad e^{2\psi} = (1-\mu^2)(\eta+1)^2 e^{-S}. \tag{166}$$

The resulting form of the metric is

$$ds^2 = \frac{\eta-1}{\eta+1} e^S (dt)^2 - (1-\mu^2)(\eta+1)^2 e^{-S} (d\varphi)^2$$
$$- \frac{e^{\mu_2+\mu_3}}{\sqrt{(\eta^2-1)}}[(d\eta)^2 + (\eta^2-1)(d\theta)^2]. \tag{167}$$

* Actually, the choice of $P_n(\mu)$, regular at the poles, is demanded by the requirement, we shall presently consider, (in § (*a*) below) of the local flatness of the metric for $\mu^2 = 1$ and $\eta \geqslant 1$. This same requirement, also, demands that the terms in the Legendre functions of the second kind, $Q_n(\eta)$, are not also included in the solution for $R(\eta)$ in equations (161) and (162).

To complete the solution, we must consider the equations which determine $(\mu_2+\mu_3)$. The equations are (cf. Ch. 6, equations (64) and (65))

$$-\frac{\mu}{1-\mu^2}(\mu_2+\mu_3)_{,\eta}+\frac{\eta}{\eta^2-1}(\mu_2+\mu_3)_{,\mu}=(\lg\chi)_{,\eta}(\lg\chi)_{,\mu} \tag{168}$$

and

$$2\eta(\mu_2+\mu_3)_{,\eta}+2\mu(\mu_2+\mu_3)_{,\mu}=(\eta^2-1)[(\lg\chi)_{,\eta}]^2-(1-\mu^2)[(\lg\chi)_{,\mu}]^2 -\frac{3}{\eta^2-1}+\frac{1}{1-\mu^2}, \tag{169}$$

since, for the problem on hand, $X=Y=\chi$. Solving equations (168) and (169) for $(\mu_2+\mu_3)_{,\eta}$ and $(\mu_2+\mu_3)_{,\mu}$, we obtain,

$$\frac{2(\eta^2-\mu^2)}{(\eta^2-1)(1-\mu^2)}(\mu_2+\mu_3)_{,\eta}=\frac{2(\eta^2-\mu^2)}{(\eta^2-1)(1-\mu^2)}\left[-\frac{2-\eta}{\eta^2-1}-S_{,\eta}\right] +\frac{4\eta}{\eta^2-1}S_{,\eta}-\frac{4\mu}{\eta^2-1}S_{,\mu}-2\mu S_{,\eta}S_{,\mu}+\eta(S_{,\eta})^2-\eta\frac{1-\mu^2}{\eta^2-1}(S_{,\mu})^2 \tag{170}$$

and

$$\frac{2(\eta^2-\mu^2)}{(\eta^2-1)(1-\mu^2)}(\mu_2+\mu_3)_{,\mu}=-\frac{2(\eta^2-\mu^2)}{(\eta^2-1)(1-\mu^2)}S_{,\mu} +\frac{4\mu}{1-\mu^2}S_{,\eta}+\frac{4\eta}{\eta^2-1}S_{,\mu}+2\eta S_{,\eta}S_{,\mu}+\mu\frac{\eta^2-1}{1-\mu^2}(S_{,\eta})^2-\mu(S_{,\mu})^2. \tag{171}$$

The integrability of these equations is, of course, guaranteed.

(a) The condition for the equilibrium of the black hole

It is clear that the equilibrium of the black hole requires that the space is locally flat along the θ-axis for all $\eta\geqslant 1$; for, otherwise, the black hole will be acted on by tidal forces which will propel it in one direction or the other. This condition for local flatness, stated differently, requires that the ratio of the circumference to the radius of an infinitesimal circle, drawn orthogonally to the axis, is 2π for all $\eta\geqslant 1$. For the metric written in the form (167), the requirement, then, is

$$\lim_{\theta\to 0 \text{ or } \pi}\left[\frac{e^{\mu_2+\mu_3}(\eta^2-1)^{1/2}(d\theta)^2}{(\eta+1)^2e^{-S}\sin^2\theta}\right]=1 \quad \text{for} \quad \eta\geqslant 1, \tag{172}$$

or

$$e^{\sigma}=\left[\frac{\eta-1}{(\eta+1)^3}\right]^{1/2}e^{\mu_2+\mu_3+S}=1 \quad \text{for} \quad \mu^2=1 \quad \text{and} \quad \eta\geqslant 1. \tag{173}$$

We now find that equations (170) and (171), expressed in terms of σ, simplify to give

$$\frac{2(\eta^2-\mu^2)}{(\eta^2-1)(1-\mu^2)}\sigma_{,\eta} = \frac{4\eta}{\eta^2-1}S_{,\eta} - \frac{4\mu}{\eta^2-1}S_{,\mu} - 2\mu S_{,\eta}S_{,\mu} + \eta(S_{,\eta})^2 - \eta\frac{1-\mu^2}{\eta^2-1}(S_{,\mu})^2 \tag{174}$$

and

$$\frac{2(\eta^2-\mu^2)}{(\eta^2-1)(1-\mu^2)}\sigma_{,\mu} = \frac{4\mu}{1-\mu^2}S_{,\eta} + \frac{4\eta}{\eta^2-1}S_{,\mu} + 2\eta S_{,\eta}S_{,\mu} - \mu(S_{,\mu})^2 + \mu\frac{\eta^2-1}{1-\mu^2}(S_{,\eta})^2. \tag{175}$$

A necessary (but not a sufficient) condition for equation (173) to be satisfied is that

$$\sigma_{,\eta} = 0 \quad \text{for} \quad 1-\mu^2 = 0. \tag{176}$$

Equation (174) will be consistent with this requirement, only provided $S(\eta, \mu)$ is regular for $\mu = \pm 1$. It is this requirement which restricts us to $P_n(\mu)$ when writing the solution (162) for $\lg \Psi$.

So long as $\sigma_{,\eta} = 0$ on the axis, we can certainly arrange that $\sigma = 0$ for $\theta = 0$ and $\eta \geqslant 1$. But to ensure that σ has the same value 0 for $\theta = \pi$, as well, we must, in addition, require

$$\int_0^\pi \sigma_{,\theta}(\eta = 1)\,d\theta = 0. \tag{177}$$

From equation (175) we conclude that the first requirement is that S be regular at $\eta = 1$. For this reason, we cannot include the Legendre functions of the second kind in the solution (161) for $R_n(\eta)$; and the solution for S reduces to

$$S(\eta, \mu) = \sum_{n=0}^{\infty} A_n P_n(\eta) P_n(\mu). \tag{178}$$

For this solution, regular at $\eta = 1$, equation (175) gives

$$\sigma_{,\mu} = 2S_{,\mu} \quad \text{and} \quad \eta = 1; \tag{179}$$

and the condition (177) yields

$$S(\eta = 1, \mu = +1) = S(\eta = 1, \mu = -1). \tag{180}$$

For S given by equation (178), local flatness on the axis requires that the odd coefficients, A_{2n+1}, are subject to the restriction

$$\sum_{n=0}^{\infty} A_{2n+1} P_{2n+1}(1) = 0; \tag{181}$$

but there are no restrictions on the even coefficients, A_{2n}.

Finally, we may note that the metric describing a static distorted black-hole is expressible in the form

$$ds^2 = \frac{\eta-1}{\eta+1}e^{+S}(dt)^2 - \frac{\eta+1}{\eta-1}e^{\sigma-S}(d\eta)^2$$
$$- (\eta+1)^2 e^{-S}[e^{\sigma}(d\theta)^2 + \sin^2\theta\,(d\varphi)^2], \quad (182)$$

where σ can be obtained from equations (174) and (175) by simple quadratures and the solution for S given by equation (178) is restricted by the condition (181) on the odd coefficients.

113. A solution of the Einstein–Maxwell equations representing an assemblage of black holes

In this section, we shall consider a solution of the Einstein–Maxwell equations which provides an analogue of the Newtonian arrangement of charged mass-points in which the mutual gravitational attractions are exactly balanced by the Coulomb repulsions. The solution was discovered by Majumdar and Papapetrou; but the correct interpretation of the solution as representing, in effect, an assemblage of extreme Reissner–Nordström black-holes (with $Q_* = \pm M$) is due to Hartle and Hawking.

The Majumdar–Papapetrou solution is obtained by considering static solutions of the Einstein–Maxwell equations for a space-time with the metric

$$ds^2 = e^{2\nu}(dt)^2 - e^{2\psi}[(dx^1)^2 + (dx^2)^2 + (dx^3)^2], \quad (183)$$

where ν and ψ are functions only of the spatial coordinates x^1, x^2, and x^3. The field-equations appropriate to this metric can be readily written down by suitably specializing the equations of Chapter 2. Thus, considering first Maxwell's equations, we find from Chapter 2, equations (95) (a)–(h):

$$(e^{2\psi}F_{12})_{,3} + (e^{2\psi}F_{23})_{,1} + (e^{2\psi}F_{31})_{,2} = 0, \quad (184)$$

$$\left.\begin{aligned}(e^{\psi+\nu}F_{01})_{,2} - (e^{\psi+\nu}F_{02})_{,1} &= 0,\\ (e^{\psi+\nu}F_{01})_{,3} - (e^{\psi+\nu}F_{03})_{,1} &= 0,\\ (e^{\psi+\nu}F_{12})_{,2} - (e^{\psi+\nu}F_{31})_{,3} &= 0;\end{aligned}\right\} \quad (185)$$

$$(e^{2\psi}F_{02})_{,2} + (e^{2\psi}F_{03})_{,3} + (e^{2\psi}F_{01})_{,1} = 0, \quad (186)$$

$$\left.\begin{aligned}(e^{\psi+\nu}F_{23})_{,2} - (e^{\psi+\nu}F_{31})_{,1} &= 0,\\ (e^{\psi+\nu}F_{23})_{,3} - (e^{\psi+\nu}F_{12})_{,1} &= 0,\\ (e^{\psi+\nu}F_{02})_{,3} - (e^{\psi+\nu}F_{03})_{,2} &= 0.\end{aligned}\right\} \quad (187)$$

From the two systems of equations (185) and (187) it follows that the components, (F_{01}, F_{02}, F_{03}) and (F_{12}, F_{23}, F_{31}), of the Maxwell tensor can be derived from two potentials A and B in the manner

$$\left.\begin{aligned} &e^{\psi+\nu}F_{0\alpha} = A_{,\alpha} \quad \text{and} \quad e^{\psi+\nu}F_{\alpha\beta} = B_{,\gamma} \\ &(\alpha,\ \beta, \text{ and } \gamma, \text{ a cyclical permutation of 1, 2, and 3}). \end{aligned}\right\} \tag{188}$$

(Here and in the sequel, Greek indices will be restricted to the spatial coordinates; and summation over α is to be understood as meaning summation over $\alpha = 1, 2$, and 3.)

Equations (184) and (186) now provide equations for A and B; thus,

$$\sum_\alpha (e^{\psi-\nu}A_{,\alpha})_{,\alpha} = 0 \quad \text{and} \quad \sum_\alpha (e^{\psi-\nu}B_{,\alpha})_{,\alpha} = 0. \tag{189}$$

With the energy-momentum tensor for the Maxwell field defined in equation (11), we find that in the present context,

$$R_{00} = e^{-2(\psi+\nu)}[|A_{,\alpha}|^2 + |B_{,\alpha}|^2], \tag{190}$$

$$R_{11} = e^{-2(\psi+\nu)}[|A_{,\alpha}|^2 + |B_{,\alpha}|^2 - 2(A_{,1})^2 - 2(B_{,1})^2], \tag{191}$$

$$R_{10} = +2e^{-2(\psi+\nu)}(A_{,2}B_{,3} - B_{,2}A_{,3}), \tag{192}$$

$$R_{12} = -2e^{-2(\psi+\nu)}(A_{,1}A_{,2} + B_{,1}B_{,2}), \tag{193}$$

where we have adopted the abbreviation,

$$|A_{,\alpha}|^2 = \sum_\alpha (A_{,\alpha})^2. \tag{194}$$

The expressions for the components R_{22}, etc., (not listed) can be obtained by cyclically permuting the indices 1, 2, and 3.

Turning to the Einstein field-equation, we find from Chapter 2, equations (75) (a)–(u) that the non-vanishing components of the Riemann tensor, for a metric of the form (183), are given by

$$R_{1212} = e^{-2\psi}(\psi_{,1,1} + \psi_{,2,2} + \psi_{,3}\psi_{,3}), \tag{195}$$

$$R_{1414} = e^{-2\psi}(\nu_{,1,1} + \nu_{,1}\nu_{,1} - \nu_{,1}\psi_{,1} + \nu_{,2}\psi_{,2} + \nu_{,3}\psi_{,3}), \tag{196}$$

$$R_{1442} = -e^{-2\psi}(\nu_{,1,2} + \nu_{,1}\nu_{,2} - \psi_{,1}\nu_{,2} - \psi_{,2}\nu_{,1}), \tag{197}$$

$$R_{1213} = e^{2\psi}(\psi_{,2,3} - \psi_{,2}\psi_{,3}), \tag{198}$$

and those which follow from these by a cyclical permutation of the indices 1, 2, and 3. The remaining components vanish:

$$\left.\begin{aligned} &R_{2114} = R_{3114} = R_{1224} = R_{3224} = R_{1334} = R_{2334} = 0, \\ \text{and} \quad &R_{1234} = R_{1423} = R_{1342} = 0. \end{aligned}\right\} \tag{199}$$

(It may be recalled that, according to the convention used in Chapter 2, the index 4 denotes *it*.)

(a) The reduction of the field equations

From the vanishing of the components of the Riemann tensor listed in (199), we conclude that

$$R_{14} = R_{1224} + R_{1334} = 0, \quad \text{i.e.,} \ R_{10} = 0. \tag{200}$$

Therefore, by equation (192),

$$A_{,2}B_{,3} - B_{,2}A_{,3} = 0. \tag{201}$$

From this equation and the similar ones which follow from the vanishing of R_{20} and R_{30}, we conclude that

$$\frac{A_{,1}}{B_{,1}} = \frac{A_{,2}}{B_{,2}} = \frac{A_{,3}}{B_{,3}}. \tag{202}$$

Accordingly, by a *duality transformation*, applied to the antisymmetric tensor F_{ij} and its dual $^*F_{ij}$ $(= \varepsilon_{ijkl}F^{kl})$, we can arrange for either A or B to vanish. Since A corresponds to the electrostatic potential, we shall find it convenient to set

$$B = 0. \tag{203}$$

Clearly, there is no loss of generality in making this assumption.

Since the energy-momentum tensor of a Maxwell field is traceless, the scalar curvature R must vanish, i.e.,

$$\tfrac{1}{2}R = R_{1212} + R_{2323} + R_{3131} + R_{1414} + R_{2424} + R_{3434} = 0. \tag{204}$$

Inserting for the components of the Riemann tensor, in accordance with equations (195) and (196), we find

$$\sum_\alpha \psi_{,\alpha,\alpha} + \sum_\alpha \psi_{,\alpha}\psi_{,\alpha} + \sum_\alpha (\psi + \nu)_{,\alpha,\alpha} + \sum_\alpha \nu_{,\alpha}(\psi + \nu)_{,\alpha} = 0. \tag{205}$$

The vanishing of the scalar curvature, R, implies, in particular, that R_{00} and G_{00} are equal. It is useful to have the corresponding components of the field equation written out explicitly. Since

$$+R_{00} = -R_{44} = R_{1414} + R_{2424} + R_{3434}, \tag{206}$$

and

$$-G_{00} = +G_{44} = R_{1212} + R_{2323} + R_{3131}, \tag{207}$$

we find from equations (190), (195), and (196) that

$$\sum_\alpha \nu_{,\alpha,\alpha} + \sum_\alpha \nu_{,\alpha}(\psi + \nu)_{,\alpha} = +e^{-2\nu}|A_{,\alpha}|^2 \tag{208}$$

and

$$2\sum_{\alpha}\psi_{,\alpha,\alpha}+\sum_{\alpha}\psi_{,\alpha}\psi_{,\alpha}=-e^{-2\nu}|A_{,\alpha}|^2, \tag{209}$$

remembering that we have set $B=0$.

By the addition of equations (208) and (209), we recover equation (205). An independent linear combination of these equations is

$$\sum_{\alpha}\psi_{,\alpha}\psi_{,\alpha}+2\sum_{\alpha}(\psi+\nu)_{,\alpha,\alpha}+2\sum_{\alpha}\nu_{,\alpha}(\psi+\nu)_{,\alpha}=e^{-2\nu}|A_{,\alpha}|^2. \tag{210}$$

Finally, considering the component

$$R_{12}=R_{1332}+R_{1442} \tag{211}$$

of the Ricci tensor, we obtain from equations (193), (197), and (198)

$$2\nu_{,1}\nu_{,2}+(\psi+\nu)_{,1,2}-(\psi+\nu)_{,1}(\psi+\nu)_{,2}=2e^{-2\nu}A_{,1}A_{,2}, \tag{212}$$

or, more generally,

$$2\nu_{,\alpha}\nu_{,\beta}+(\psi+\nu)_{,\alpha,\beta}-(\psi+\nu)_{,\alpha}(\psi+\nu)_{,\beta}=2e^{-2\nu}A_{,\alpha}A_{,\beta}. \tag{213}$$

The basic equations of the problem are provided by equations (189), (205), (210), and (213).

(b) The Majumdar–Papapetrou solution

It is plainly manifest from equations (205), (210), and (213) that the case

$$\psi+\nu=0 \tag{214}$$

is specially distinguished. For, when this is the case, the equations are drastically simplified and we are left with

$$\sum_{\alpha}\psi_{,\alpha,\alpha}+\sum_{\alpha}\psi_{,\alpha}\psi_{,\alpha}=0, \tag{215}$$

$$|\psi_{,\alpha}|^2=e^{2\psi}|A_{,\alpha}|^2 \quad\text{and}\quad \psi_{,\alpha}\psi_{,\beta}=e^{2\psi}A_{,\alpha}A_{,\beta}\quad(\alpha\neq\beta). \tag{216}$$

An alternative form of equation (215) is

$$\sum_{\alpha}(e^{\psi})_{,\alpha,\alpha}=0. \tag{217}$$

From equations (216) it follows that (cf. equation (188))

$$A_{,\alpha}=\mp e^{-\psi}\psi_{,\alpha}=\pm(e^{-\psi})_{,\alpha}\ (=F_{0\alpha}). \tag{218}$$

This solution for $A_{,\alpha}$ is consistent with equation (189) by virtue of equation (217) satisfied by e^{ψ}. The solution for the electrostatic potential, A, is now given by

$$A=\pm e^{-\psi}, \tag{219}$$

where the choice of sign is a matter of convention.

It is now convenient to use (x, y, z) in place of (x^1, x^2, x^3) and write U in place of e^{ψ}. The metric then takes the form,

$$ds^2 = \frac{1}{U^2}(dt)^2 - U^2(dx^2 + dy^2 + dz^2), \tag{220}$$

where U satisfies the Cartesian three-dimensional Laplace's equation,

$$\nabla^2 U = \left(\frac{\partial^2}{\partial x^2} + \frac{\partial^2}{\partial y^2} + \frac{\partial^2}{\partial z^2}\right) U = 0; \tag{221}$$

and the electrostatic potential is given by

$$A = U^{-1}. \tag{222}$$

This is the Majumdar–Papapetrou solution.

(c) *The solution representing an assemblage of black holes*

Any solution of Laplace's equation (221) can, of course, be used in conjunction with the metric (220). But the solution which is of particular concern to us as representing an assemblage of black holes, is the one which is derived from the Newtonian potential of a number of mass-points, $M_i (i = 1, \ldots, N)$, namely

$$U = 1 + \sum_{i=1}^{N} \frac{M_i}{r_i}, \tag{223}$$

where

$$r_i = [(x - x_i)^2 + (y - y_i)^2 + (z - z_i)^2]^{1/2}, \tag{224}$$

and the additive constant 1 in the solution for U is to ensure that the metric becomes Minkowskian in the absence of sources.

Two facts concerning the solution (223) require to be clarified.

First, from the form of the metric (220), when the different points (x_i, y_i, z_i) are widely separated, we may identify the constants M_i with the inertial masses enclosed by spheres of large radii surrounding the different points. We must, therefore, require that the M_i's are all positive:

$$M_i > 0 \quad (i = 1, \ldots, N). \tag{225}$$

Therefore, in the part of the manifold to which the coordinate chart (x, y, z) applies—as we shall see, it applies only to a part of the maximally extended manifold—$U(x, y, z)$ is positive and non-vanishing and the metric is everywhere regular except at the points r_i.

Second, the charge Q_i of the mass-point M_i, can be determined by evaluating the integral of the divergence of $F^{\alpha 0}\sqrt{-g}$ over a spherical volume, V_i, enclosing the point (x_i, y_i, z_i)*; thus,

$$\begin{aligned}4\pi Q_i &= \int_{V_i} dx^\alpha (F^{\alpha 0}\sqrt{-g})_{,\alpha} = \int_{\partial V_i} dS_\alpha F^{\alpha 0}\sqrt{-g}\\ &= \int_{\partial V_i} dS_\alpha U^2 F_{0\alpha} = \int_{\partial V_i} dS_\alpha U^2 (U^{-1})_{,\alpha}\\ &= -\int_{\partial V_i} \frac{\partial}{\partial x^\alpha}\left(\frac{M_i}{r_i}\right) dS^\alpha = 4\pi M_i. \end{aligned} \tag{226}$$

Therefore, in the units adopted, *the charge and the mass associated with the source at* (x_i, y_i, z_i) *are equal.* With the sign for A that has been chosen in equation (222), the charges $Q_i\,(= M_i)$ are all positive. But had we chosen the opposite sign for A in equation (222)—an option we had (cf. equation (219))—we should have found that all the charges are negative. In any event, the charges must all be of the same sign independently of choice or convention.

Returning to the solution (223), we first observe that when we have only one mass point and

$$U = 1 + M/r, \tag{227}$$

we recover the extreme Reissner–Nordström solution, in its standard form, by writing the metric in spherical polar-coordinates and replacing r by $r - M$.

The case of two mass-points, when

$$U = 1 + \frac{M_1}{r_1} + \frac{M_2}{r_2}, \tag{228}$$

is sufficiently illustrative of the general case that we shall consider it in some detail. In the coordinate chart (x, y, z) adopted, the solution for U and the metric (220) are not regular at $r = r_1$ and $r = r_2$. But the singularities at these points are only coordinate singularities and the geometry is entirely regular at r_1 and r_2. To make this fact manifest, we first transform to a system of polar coordinates centred at M_1 and with the θ_1-axis along the line joining the two mass points (see Fig. 47 *a*). Then,

$$U = 1 + \frac{M_1}{r_1} + \frac{M_2}{\sqrt{(r_1^2 + a^2 - 2ar_1\cos\theta_1)}}, \tag{229}$$

where a is the distance between M_1 and M_2; and the metric takes the form

$$ds^2 = \frac{(dt)^2}{U^2} - U^2[(dr_1)^2 + r_1^2(d\theta_1)^2 + (r_1^2\sin^2\theta_1)(d\varphi)^2]. \tag{230}$$

* This assertion follows from one of Maxwell's equations, namely,

$$(F^{\alpha 0}\sqrt{-g})_{,\alpha} = J^0\sqrt{-g}.$$

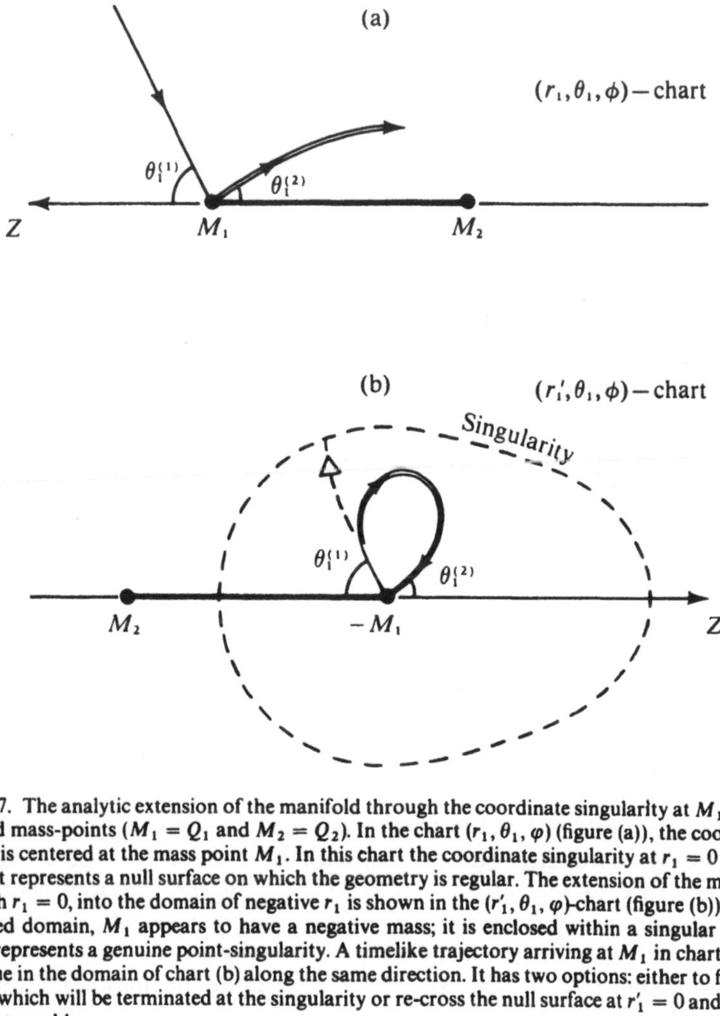

FIG. 47. The analytic extension of the manifold through the coordinate singularity at M_1 of two charged mass-points ($M_1 = Q_1$ and $M_2 = Q_2$). In the chart (r_1, θ_1, φ) (figure (a)), the coordinate system is centered at the mass point M_1. In this chart the coordinate singularity at $r_1 = 0$ is not a point: it represents a null surface on which the geometry is regular. The extension of the manifold through $r_1 = 0$, into the domain of negative r_1 is shown in the $(r_1', \theta_1, \varphi)$-chart (figure (b)). In this extended domain, M_1 appears to have a negative mass; it is enclosed within a singular surface which represents a genuine point-singularity. A timelike trajectory arriving at M_1 in chart (a) will continue in the domain of chart (b) along the same direction. It has two options: either to follow a course which will be terminated at the singularity or re-cross the null surface at $r_1' = 0$ and enter a different world.

When written in this form, it is evident that the surface area of a small sphere of radius r_1 surrounding the origin does not tend to zero as $r_1 \to 0$ but to the finite value $4\pi M_1^2$ (which is also the area of the event horizon of an extreme Reissner–Nordström black-hole of mass M_1). Therefore, $r_1 = 0$ *does not label a point; it represents a surface with the finite area* $4\pi M_1^2$.

To show that the geometry at $r_1 = 0$ is regular and that the surface it represents is in fact null, we make, following Hartle and Hawking, the further coordinate-transformation

$$t = u + F(r_1), \tag{231}$$

where

$$\frac{\mathrm{d}F}{\mathrm{d}r_1} = \left(1 + \frac{M_1}{r_1} + \frac{M_2}{a}\right)^2 = V^2(r_1) \quad \text{(say)}. \tag{232}$$

The metric then becomes

$$\mathrm{d}s^2 = \frac{(\mathrm{d}u)^2}{U^2} + 2\frac{V^2}{U^2}\mathrm{d}u\,\mathrm{d}r_1 - \frac{1}{U^2}(U^4 - V^4)(\mathrm{d}r_1)^2$$
$$- r_1^2 U^2[(\mathrm{d}\theta_1)^2 + (\sin^2\theta_1)(\mathrm{d}\varphi)^2]. \tag{233}$$

In the neighbourhood of $r_1 = 0$,

$$U(r_1) - V(r_1) = \frac{M_2\cos\theta_1}{a^2} r_1 + O(r_1^2), \tag{234}$$

and

$$\mathrm{d}s^2 \to \frac{r_1^2}{M_1^2}(\mathrm{d}u)^2 + 2\mathrm{d}u\,\mathrm{d}r_1 - \frac{4M_1 M_2\cos\theta_1}{a^2}(\mathrm{d}r_1)^2$$
$$- M_1^2[(\mathrm{d}\theta_1)^2 + (\sin^2\theta_1)(\mathrm{d}\varphi)^2] \quad (r_1 \to 0). \tag{235}$$

The regularity of the metric at $r_1 = 0$ is now manifest; and it is also manifest that $r_1 = 0$ is stationary *null-surface.*

Since the geometry is regular at $r_1 = 0$, the manifold can be extended to allow negative values of r_1: a time-like trajectory arriving at $r_1 = 0$, along some direction θ_1, will continue along the same direction into the domain of negative r_1. With the definitions (see Fig. 47 *b*),

$$r_1' = -r_1 \quad \text{and} \quad r_2' = (r_1'^2 + a^2 + 2ar_1'\cos\theta_1)^{1/2}, \tag{236}$$

the metric (in the domain of negative r_1) will continue to be of the same form (230) except that $U(r_1, \theta_1)$ will be replaced by

$$U'(r_1', \theta_1) = 1 - \frac{M_1}{r_1'} + \frac{M_2}{r_2'}. \tag{237}$$

It is clear that the function $U'(r_1', \theta_1)$ will vanish on some surface. A simple examination of the "equipotential" surfaces of $U'(r_1', \theta_1)$ shows that the surface on which U' vanishes completely encloses the origin at $r_1' = 0$. And on the surface $U' = 0$, the metric is singular. The singularity, now, is a genuine one; this can be confirmed by evaluating an invariant, such as $F_{ef}F^{ef}$, and demonstrating that it diverges on $U' = 0$. Thus (see equation (188)),

$$F_{ef}F^{ef} = -2|A_{,\alpha}|^2 = -2\left(\frac{\text{grad}\,U'}{U'^2}\right)^2 \tag{238}$$

clearly diverges when $U' = 0$. Moreover, *while in the chart* $(r'_1, \theta_1, \varphi)$, $U' = 0$ *appears as a surface; it is in reality a point.* For, by evaluating the area of a surface just inside $U' = 0$ and letting the surface tend to $U' = 0$ we find:

$$\iint [U'(r'_1, \theta_1)]^2 r'^2_1 \sin\theta_1 \, d\theta_1 \, d\varphi \to 0. \tag{239}$$

Therefore, the domain interior to $r'_1 = 0$ is very similar to the interior of the event horizon of an extreme Reissner–Nordström black-hole. In particular, a time-like trajectory, crossing the null surface at $r_1 = 0$, can follow one of two courses: *either* it can arrive at $U' = 0$ and be terminated, *or* it can turn around and emerge into a totally new world by recrossing $r_1 = 0$ (see Figs. 47 *a* and *b*).

All of the foregoing remarks, made with respect to M_1 and $r_1 = 0$, are, of course, equally applicable to M_2 and $r_2 = 0$. Indeed, quite generally, when U is given by equation (223), each of the points r_i represents a null surface with an area $4\pi M_i^2$; and the chart (x, y, z) can be extended into the interior of these null surfaces which enclose genuine singularities; and in all cases a maximal analytic extension of the manifold can be accomplished. We shall not pursue these matters further; but it is amply clear that we are justified in regarding the analytically extended space-time described by the solution (220) with U given by (223) as representing an assembly of N extreme Reissner–Nordström black-holes.

So far, our consideration of the metric (220) has been restricted to the simplest solution (223) of Laplace's equation. One might, at this stage, wish to consider similar superpositions of other more general solutions of Laplace's equation. But Hartle and Hawking have shown that in all these other cases, the space-time is characterized by naked singularities. On this account, these other solutions would not appear to hold the same interest. The same remark applies to a stationary generalization of the static Majumdar–Papapetrou solution discovered by Perjes and by Israel and Wilson.

114. The variational method and the stability of the black-hole solutions

Our considerations of the Schwarzschild and the Kerr solutions as representing the black holes of nature have, in very large measure, exploited their special algebraic character as grasped by the formalism of Newman and Penrose. But the very successes of the formalism, in providing separable solutions for the basic equations of mathematical physics and in revealing the analytic richness of the mathematical theory, have to some extent obscured the physical reality of the black holes behind a veil of geometrical appearance. For the physical reality of the black holes consists in a smooth event horizon, concealing a singularity, in an asymptotically flat space-time. And it is well to remember that the uniqueness of the Kerr solution does not derive from its special algebraic character; it derives rather from its physical reality: the algebraic speciality of the Kerr solution is not relevant to the reasonings of the Carter–Robinson theorems. Besides, the equations of the Newman–Penrose formalism have proved singularly inept in addressing, for example, the important physical problem of the stability of the Kerr space-time. In this last section, we shall describe an approach to these physical problems which departs, in some essential respects, from the ones that we have, in the main, adopted in the book.

The method of solution we shall describe is to treat the problem of perturbations by a direct linearization of Einstein's equation about the stationary solution and expressing the solution of the equations governing the perturbations in terms of a variational principle. This method, common enough in other branches of mathematical physics, must incorporate features unique to the general theory of relativity. They concern questions of gauge and the role of the initial-value equations and of the Bianchi identities. We shall provide an introduction to these matters. The considerations we shall present will, however, be restricted to perturbations of axisymmetric vacuum-solutions of Einstein's equation which preserve their axisymmetry (though the considerations themselves are of more general applicability).

We consider then a stationary axisymmetric solution of Einstein's equation with a metric of the standard form, namely,

$$ds^2 = e^{2\nu}(dt)^2 - e^{2\psi}(d\varphi - \omega\, dt)^2 - e^{2\mu_2}(dx^2)^2 - e^{2\mu_3}(dx^3)^2, \qquad (240)$$

where the gauge freedom in the choice of the functions μ_2 and μ_3 has already been exercised. As we have shown in Chapter 2, §12, a non-stationary metric, which will suffice to describe axisymmetric perturbations of a space-time with the metric (240), is given by (Ch. 2, equation (38))

$$ds^2 = e^{2\nu}(dt)^2 - e^{2\psi}(d\varphi - q_2\, dx^2 - q_3\, dx^3 - \omega\, dt)^2 - e^{2\mu_2}(dx^2)^2 - e^{2\mu_3}(dx^3)^2, \qquad (241)$$

where we no longer have the freedom to impose a coordinate condition on μ_2 and μ_3. In other words, a perturbation of the metric (240), which preserves its axisymmetry, will require us to consider that ν, ψ, ω, μ_2, *and* μ_3 are all affected to the first order so that they become

$$\nu + \delta\nu,\ \psi + \delta\psi,\ \omega + \delta\omega,\ \mu_2 + \delta\mu_2, \quad \text{and} \quad \mu_3 + \delta\mu_3. \qquad (242)$$

In addition, q_2 and q_3 must be allowed to be non-vanishing (to the first order) and subject to the restrictions shown in Chapter 2, equations (42). It was, however, shown by Friedman and Schutz that, for considerations of stability, it is important that we do *not* effect any change in $(\mu_2 + \mu_3)$ i.e., we must ensure that

$$\delta(\mu_2 + \mu_3) = 0. \qquad (243)$$

Clearly, one cannot accomplish this without considering a more general form for the metric than the one given by (241). To allow ourselves the requisite freedom to impose a restriction such as (243), we shall generalize (241) to the form

$$ds^2 = e^{2\nu}(dt)^2 - e^{2\psi}(d\varphi - \omega\,dt - q_2\,dx^2 - q_3\,dx^3)^2 - e^{2\mu_2}(dx^2 + \chi\,dx^3)^2 - e^{2\mu_3}(dx^3)^2, \quad (244)$$

where χ is a further function dependent on t, x^2, and x^3. This metric is *more general* than the one considered in Chapter 2, §13, in that we have an additional function χ, but is also *less general* in that we are not now allowing the metric functions to depend on φ.

In evaluating the components of the Riemann tensor for a metric of the form (244) by Cartan's calculus, we shall adopt (as in Ch. 2, §13) the complex notation (Ch. 2, equation (44))

$$dt = -i\,dx^4, \quad \partial_t = i\partial_4, \quad \nu = \mu_4, \quad \text{and} \quad \omega = iq_4 \quad (245)$$

and use, in place of Chapter 2, (46), the basis one-forms,

$$\left.\begin{aligned} \omega^1 &= e^{\psi}\Big(d\varphi - \sum_A q_A\,dx^A\Big); & d\varphi - \sum_A q_A\,dx^A &= \omega^1 e^{-\psi}, \\ \omega^2 &= e^{\mu_2}(dx^2 + \chi\,dx^3); & dx^2 &= e^{-\mu_2}\omega^2 - \chi e^{-\mu_3}\omega^3, \\ \omega^3 &= e^{\mu_3}\,dx^3; & dx^3 &= e^{-\mu_3}\omega^3, \\ \omega^4 &= e^{\mu_4}\,dx^4; & dx^4 &= e^{-\mu_4}\omega^4, \end{aligned}\right\} \quad (246)$$

where (as in Ch. 2, §13) the capital Roman indices are restricted to 2, 3, and 4.

We find that much of the analysis of Chapter 2, §13 can be taken over in the present context, if we define a '*colon*' and a '*colon-prime*' derivative, with respect to the index 3, of a function $f(x^2, x^3, x^4)$ by

$$\left.\begin{aligned} f_{:3} &= f_{,3} - \chi f_{,2} \\ \text{and} \qquad f_{:3'} &= f_{,3} - (\chi f)_{,2} = f_{:3} - \chi_{,2} f. \end{aligned}\right\} \quad (247)$$

(It should be noted that: 3-derivative satisfies the Leibnitz rule; but: 3′-derivative does not.) Thus, in place of Chapter 2, equation (60), we have

$$\begin{aligned} \omega^1{}_2 &= e^{-\mu_2}\psi_{,2}\omega^1 + \tfrac{1}{2}e^{\psi-\mu_2-\mu_3}Q_{23}\omega^3 + \tfrac{1}{2}e^{\psi-\mu_2-\mu_4}Q_{24}\omega^4, \\ \omega^1{}_3 &= e^{-\mu_3}\psi_{:3}\omega^1 + \tfrac{1}{2}e^{\psi-\mu_3-\mu_4}\bar{Q}_{34}\omega^4 + \tfrac{1}{2}e^{\psi-\mu_2-\mu_3}Q_{32}\omega^2, \\ \omega^1{}_4 &= e^{-\mu_4}\psi_{,4}\omega^1 + \tfrac{1}{2}e^{\psi-\mu_2-\mu_4}Q_{42}\omega^2 + \tfrac{1}{2}e^{\psi-\mu_3-\mu_4}\bar{Q}_{43}\omega^3, \\ \omega^2{}_3 &= -\tfrac{1}{2}e^{\psi-\mu_2-\mu_3}Q_{23}\omega^1 + e^{-\mu_2-\mu_3}(e^{\mu_2})_{:3'}\omega^2 - e^{-\mu_2}\mu_{3,2}\omega^3 \\ &\quad - \tfrac{1}{2}e^{\mu_2-\mu_3-\mu_4}\chi_{,4}\omega^4, \end{aligned}$$

$$\omega^3{}_4 = -\tfrac{1}{2}e^{\psi-\mu_3-\mu_4}\bar{Q}_{34}\omega^1 + e^{-\mu_4}\mu_{3,4}\omega^3 - e^{-\mu_3}\mu_{4:3}\omega^4 + \tfrac{1}{2}e^{\mu_2-\mu_3-\mu_4}\chi_{,4}\omega^2,$$

$$\omega^4{}_2 = -\tfrac{1}{2}e^{\psi-\mu_2-\mu_4}Q_{42}\omega^1 + e^{-\mu_2}\mu_{4,2}\omega^4 - e^{-\mu_4}\mu_{2,4}\omega^2 - \tfrac{1}{2}e^{\mu_2-\mu_3-\mu_4}\chi_{,4}\omega^3, \tag{248}$$

where (cf. Ch. 2, equation (58)),

$$Q_{AB} = q_{A,B} - q_{B,A}, \tag{249}$$

and

$$\bar{Q}_{34} = Q_{34} - \chi Q_{24} = -\bar{Q}_{43}; \tag{250}$$

and in place of Chapter 2, equations (62) and (63), we have

$$\begin{aligned}
\mathrm{d}(F\omega^1) = {} & e^{-\psi-\mu_2}(Fe^{\psi})_{,2}\omega^2\wedge\omega^1 + e^{-\psi-\mu_3}(Fe^{\psi})_{:3}\omega^3\wedge\omega^1 \\
& + e^{-\psi-\mu_4}(Fe^{\psi})_{,4}\omega^4\wedge\omega^1 + Fe^{\psi}[e^{-\mu_2-\mu_3}Q_{23}\omega^2\wedge\omega^3 \\
& + e^{-\mu_3-\mu_4}\bar{Q}_{34}\omega^3\wedge\omega^4 + e^{-\mu_2-\mu_4}Q_{42}\omega^4\wedge\omega^2], \\
\mathrm{d}(F\omega^2) = {} & e^{-\mu_2-\mu_3}(Fe^{\mu_2})_{:3'}\omega^3\wedge\omega^2 + e^{-\mu_2-\mu_4}(Fe^{\mu_2})_{,4}\omega^4\wedge\omega^2 \\
& + e^{\mu_2-\mu_3-\mu_4}F\chi_{,4}\omega^4\wedge\omega^3, \\
\mathrm{d}(F\omega^3) = {} & e^{-\mu_2-\mu_3}(Fe^{\mu_3})_{,2}\omega^2\wedge\omega^3 + e^{-\mu_3-\mu_4}(Fe^{\mu_3})_{,4}\omega^4\wedge\omega^3, \\
\mathrm{d}(F\omega^4) = {} & e^{-\mu_2-\mu_4}(Fe^{\mu_4})_{,2}\omega^2\wedge\omega^4 + e^{-\mu_3-\mu_4}(Fe^{\mu_4})_{:3}\omega^3\wedge\omega^4.
\end{aligned} \tag{251}$$

The components of the Riemann tensor can now be evaluated with the aid of Cartan's second equation of structure. We find:

$$\begin{aligned}
-R_{1212} = {} & -e^{-\psi-\mu_2}(e^{\psi-\mu_2}\psi_{,2})_{,2} - e^{-2\mu_3-\mu_2}(e^{\mu_2})_{:3'}\psi_{:3} \\
& - e^{-2\mu_4}\mu_{2,4}\psi_{,4} + \tfrac{1}{4}e^{2\psi-2\mu_2}(e^{-2\mu_3}Q_{23}^2 + e^{-2\mu_4}Q_{24}^2), && \text{(a)} \\
-R_{1313} = {} & -e^{-\psi-\mu_3}(e^{\psi-\mu_3}\psi_{:3})_{:3} - e^{-2\mu_2}\psi_{,2}\mu_{3,2} - e^{-2\mu_4}\psi_{,4}\mu_{3,4} \\
& + \tfrac{1}{4}e^{2\psi-2\mu_3}(e^{-2\mu_2}Q_{32}^2 + e^{-2\mu_4}\bar{Q}_{34}^2), && \text{(b)} \\
-R_{1414} = {} & -e^{-\psi-\mu_4}(e^{\psi-\mu_4}\psi_{,4})_{,4} - e^{-2\mu_2}\psi_{,2}\mu_{4,2} - e^{-2\mu_3}\psi_{:3}\mu_{4:3} \\
& + \tfrac{1}{4}e^{2\psi-2\mu_4}(e^{-2\mu_2}Q_{42}^2 + e^{-2\mu_3}\bar{Q}_{43}^2), && \text{(c)} \\
-R_{2323} = {} & -e^{-\mu_2-\mu_3}\{[e^{-\mu_3}(e^{\mu_2})_{:3'}]_{:3'} + (e^{\mu_3-\mu_2}\mu_{3,2})_{,2}\} \\
& - e^{-2\mu_4}\mu_{2,4}\mu_{3,4} - \tfrac{3}{4}e^{2\psi-2\mu_2-2\mu_3}Q_{23}^2 + \tfrac{1}{4}(e^{\mu_2-\mu_3-\mu_4}\chi_{,4})^2, && \text{(d)} \\
-R_{2424} = {} & -e^{-\mu_2-\mu_4}[(e^{\mu_2-\mu_4}\mu_{2,4})_{,4} + (e^{\mu_4-\mu_2}\mu_{4,2})_{,2}] \\
& - e^{-\mu_2-2\mu_3}(e^{\mu_2})_{:3'}\mu_{4:3} - \tfrac{3}{4}e^{2\psi-2\mu_2-2\mu_4}Q_{24}^2 + \tfrac{1}{4}(e^{\mu_2-\mu_3-\mu_4}\chi_{,4})^2, && \text{(e)} \\
-R_{3434} = {} & -e^{-\mu_3-\mu_4}[(e^{\mu_3-\mu_4}\mu_{3,4})_{,4} + (e^{\mu_4-\mu_3}\mu_{4:3})_{:3}] \\
& - e^{-2\mu_2}\mu_{3,2}\mu_{4,2} - \tfrac{3}{4}e^{2\psi-2\mu_3-2\mu_4}\bar{Q}_{34}^2 - \tfrac{3}{4}(e^{\mu_2-\mu_3-\mu_4}\chi_{,4})^2, && \text{(f)}
\end{aligned}$$

$$-R_{1213} = -e^{-\psi-\mu_3}(e^{\psi-\mu_2}\psi_{,2})_{:3} + e^{-\mu_2-\mu_3}\psi_{:3}\mu_{3,2}$$
$$+\tfrac{1}{4}e^{2\psi-\mu_2-\mu_3-2\mu_4}Q_{42}\bar{Q}_{43} - \tfrac{1}{2}e^{\mu_2-\mu_3-2\mu_4}\psi_{,4}\chi_{,4}, \quad \text{(g)}$$

$$-R_{1214} = -e^{-\psi-\mu_2}(e^{\psi-\mu_4}\psi_{,4})_{,2} + e^{-\mu_2-\mu_4}\psi_{,2}\mu_{2,4}$$
$$+\tfrac{1}{4}e^{2\psi-\mu_2-2\mu_3-\mu_4}Q_{32}\bar{Q}_{34} + \tfrac{1}{2}e^{\mu_2-2\mu_3-\mu_4}\psi_{:3}\chi_{,4}, \quad \text{(h)}$$

$$-R_{1314} = -e^{-\psi-\mu_3}(e^{\psi-\mu_4}\psi_{,4})_{:3} + e^{-\mu_3-\mu_4}\psi_{:3}\mu_{3,4}$$
$$+\tfrac{1}{4}e^{2\psi-2\mu_2-\mu_3-\mu_4}Q_{23}Q_{24} + \tfrac{1}{2}e^{-\mu_3-\mu_4}\psi_{,2}\chi_{,4}, \quad \text{(i)}$$

$$-R_{1223} = \tfrac{1}{2}e^{\psi-2\mu_2-\mu_3}[Q_{23,2} + Q_{23}(3\psi-\mu_2-\mu_3)_{,2}]$$
$$+\tfrac{1}{2}e^{\psi-\mu_3-2\mu_4}\bar{Q}_{43}\mu_{2,4} - \tfrac{1}{4}e^{\psi-\mu_3-2\mu_4}Q_{42}\chi_{,4}, \quad \text{(j)}$$

$$-R_{1224} = \tfrac{1}{2}e^{\psi-2\mu_2-\mu_4}[Q_{24,2} + Q_{24}(3\psi-\mu_2-\mu_4)_{,2}]$$
$$+\tfrac{1}{2}e^{\psi-\mu_2-2\mu_3-\mu_4}\bar{Q}_{34}(e^{\mu_2})_{:3'} + \tfrac{1}{4}e^{\psi-2\mu_3-\mu_4}Q_{32}\chi_{,4}, \quad \text{(k)}$$

$$-R_{1334} = \tfrac{1}{2}e^{\psi-2\mu_3-\mu_4}[\bar{Q}_{34:3} + \bar{Q}_{34}(3\psi-\mu_3-\mu_4)_{:3}]$$
$$+\tfrac{1}{2}e^{\psi-2\mu_2-\mu_4}Q_{24}\mu_{3,2} + \tfrac{1}{4}e^{\psi-2\mu_3-\mu_4}Q_{23}\chi_{,4}, \quad \text{(l)}$$

$$-R_{1332} = \tfrac{1}{2}e^{\psi-2\mu_3-\mu_2}[Q_{32:3} + Q_{32}(3\psi-\mu_2-\mu_3)_{:3}]$$
$$+\tfrac{1}{2}e^{\psi-\mu_2-2\mu_4}Q_{42}\mu_{3,4} - \tfrac{1}{4}e^{\psi+\mu_2-2\mu_3-2\mu_4}\bar{Q}_{43}\chi_{,4}, \quad \text{(m)}$$

$$-R_{1442} = \tfrac{1}{2}e^{\psi-\mu_2-2\mu_4}[Q_{42,4} + Q_{42}(3\psi-\mu_2-\mu_4)_{,4}]$$
$$+\tfrac{1}{2}e^{\psi-\mu_2-2\mu_3}Q_{32}\mu_{4:3} - \tfrac{1}{4}e^{\psi+\mu_2-2\mu_3-2\mu_4}\bar{Q}_{43}\chi_{,4}, \quad \text{(n)}$$

$$-R_{1443} = \tfrac{1}{2}e^{\psi-\mu_3-2\mu_4}[\bar{Q}_{43,4} + \bar{Q}_{43}(3\psi-\mu_3-\mu_4)_{,4}]$$
$$+\tfrac{1}{2}e^{\psi-2\mu_2-\mu_3}Q_{23}\mu_{4,2} + \tfrac{1}{4}e^{\psi-\mu_3-2\mu_4}Q_{42}\chi_{,4}, \quad \text{(o)}$$

$$-R_{2334} = e^{-\mu_2-\mu_4}[\mu_{3,2,4} + \mu_{3,2}(\mu_3-\mu_2)_{,4} - \mu_{3,4}\mu_{4,2}]$$
$$-\tfrac{3}{4}e^{2\psi-\mu_2-2\mu_3-\mu_4}Q_{23}\bar{Q}_{34} - \tfrac{1}{2}e^{-\mu_2-\mu_3}(e^{2\mu_2-\mu_3-\mu_4}\chi_{,4})_{:3'}$$
$$-\tfrac{1}{2}e^{-2\mu_3-\mu_4}(e^{\mu_2})_{:3'}\chi_{,4}, \quad \text{(p)}$$

$$-R_{3224} = e^{-\mu_2-\mu_3}(e^{\mu_2-\mu_4}\mu_{2,4})_{:3'} - e^{-\mu_2-\mu_3-\mu_4}(e^{\mu_2})_{:3'}\mu_{3,4}$$
$$-\tfrac{3}{4}e^{2\psi-2\mu_2-\mu_3-\mu_4}Q_{32}Q_{24} - \tfrac{1}{2}e^{-\mu_2-\mu_3}(e^{\mu_2-\mu_4}\chi_{,4})_{,2}$$
$$-\tfrac{1}{2}e^{-\mu_3-\mu_4}\mu_{3,2}\chi_{,4}, \quad \text{(q)}$$

$$-R_{3442} = e^{-\mu_3-\mu_4}(e^{-\mu_2+\mu_4}\mu_{4,2})_{:3} - e^{-\mu_2-\mu_3}\mu_{4:3}\mu_{3,2}$$
$$-\tfrac{3}{4}e^{2\psi-\mu_2-\mu_3-2\mu_4}\bar{Q}_{34}Q_{42} + \tfrac{1}{2}e^{-\mu_2-\mu_4}(e^{2\mu_2-\mu_3-\mu_4}\chi_{,4})_{,4}$$
$$+\tfrac{1}{2}e^{\mu_2-\mu_3-2\mu_4}\mu_{2,4}\chi_{,4}, \quad \text{(r)}$$

$$-R_{1234} = \tfrac{1}{2}e^{\psi-\mu_2-\mu_3-\mu_4}\{Q_{24}[\psi_{:3} - e^{-\mu_2}(e^{\mu_2})_{:3'}] - Q_{23}(\psi-\mu_2)_{,4}\}$$
$$+\tfrac{1}{2}e^{-\psi-\mu_2}(e^{2\psi-\mu_3-\mu_4}\bar{Q}_{34})_{,2}, \quad \text{(s)}$$

$$-R_{1423} = \tfrac{1}{2}e^{\psi-\mu_2-\mu_3-\mu_4}[\bar{Q}_{43}(\psi-\mu_4)_{,2} - Q_{42}(\psi-\mu_4)_{:3}] + \tfrac{1}{2}e^{-\psi-\mu_4}(e^{2\psi-\mu_2-\mu_3}Q_{23})_{,4} \quad \text{(t)}$$

$$-R_{1324} = \tfrac{1}{2}e^{\psi-\mu_2-\mu_3-\mu_4}[\bar{Q}_{34}(\psi-\mu_3)_{,2} - Q_{32}(\psi-\mu_3)_{,4}] + \tfrac{1}{2}e^{-\psi-\mu_3}(e^{2\psi-\mu_2-\mu_4}Q_{24})_{:3}. \quad \text{(u)}$$

(252)

It should perhaps be noted that for particular components of the Riemann tensor, one obtains expressions which are not manifestly the same. Thus, one finds

$$R^1{}_{314} = -e^{-\psi-\mu_4}(e^{\psi-\mu_3}\psi_{:3})_{,4} + e^{-\mu_3-\mu_4}\psi_{,4}\mu_{4:3} + \tfrac{1}{4}e^{2\psi-2\mu_2-\mu_3-\mu_4}Q_{24}Q_{23} - \tfrac{1}{2}e^{-\mu_3-\mu_4}\psi_{,2}\chi_{,4} \quad (253)$$

and

$$R^1{}_{413} = -e^{-\psi-\mu_3}(e^{\psi-\mu_4}\psi_{,4})_{:3} + e^{-\mu_3-\mu_4}\psi_{:3}\mu_{3,4} - \tfrac{1}{4}e^{2\psi-2\mu_2-\mu_3-\mu_4}Q_{23}Q_{42} + \tfrac{1}{2}e^{-\mu_3-\mu_4}\psi_{,2}\chi_{,4} \quad (254)$$

by using, respectively, the equations

$$\Omega^1{}_3 = d\omega^1{}_3 + \omega^1{}_2 \wedge \omega^2{}_3 + \omega^1{}_4 \wedge \omega^4{}_3 \quad (255)$$

and

$$\Omega^1{}_4 = d\omega^1{}_4 + \omega^1{}_3 \wedge \omega^3{}_4 + \omega^1{}_2 \wedge \omega^2{}_4. \quad (256)$$

Nevertheless, by identities involving the colon derivatives, one can show that the two expressions (253) and (254) are, in fact, the same (though the establishment of similar equalities is not always as simple).

Now by combining appropriately the components of the Riemann tensor that we have listed, we can obtain the components of the Ricci and the Einstein tensors; the resulting field equations are

$$[\psi_{,2,4} + \psi_{,2}(\psi-\mu_2)_{,4} - \psi_{,4}\mu_{4,2} + \mu_{3,2,4} + \mu_{3,2}(\mu_3-\mu_2)_{,4} - \mu_{3,4}\mu_{4,2}] - \tfrac{1}{2}e^{2\psi-2\mu_3}Q_{23}\bar{Q}_{34} = \tfrac{1}{2}e^{2\mu_2-2\mu_3}[\psi_{:3} + e^{-\mu_2}(e^{\mu_2})_{:3'}]\chi_{,4} + \tfrac{1}{2}e^{\mu_4-\mu_3}(e^{2\mu_2-\mu_3-\mu_4}\chi_{,4})_{:3'}, \quad (R_{42}) \quad (257)$$

$$[\psi_{,4:3} + \psi_{:3}(\psi-\mu_3)_{,4} - \psi_{,4}\mu_{4:3} + e^{\mu_4-\mu_2}(e^{\mu_2-\mu_4}\mu_{2,4})_{:3'} - e^{-\mu_2}(e^{\mu_2})_{:3'}\mu_{3,4}] - \tfrac{1}{2}e^{2\psi-2\mu_2}Q_{32}Q_{24} = \tfrac{1}{2}(\psi+\mu_3)_{,2}\chi_{,4} + \tfrac{1}{2}e^{-\mu_2+\mu_4}(e^{\mu_2-\mu_4}\chi_{,4})_{,2}, \quad (R_{43}) \quad (258)$$

$$(e^{3\psi-\mu_2-\mu_3+\mu_4}Q_{32})_{:3} + (e^{3\psi-\mu_2+\mu_3-\mu_4}Q_{42})_{,4} = -e^{3\psi+\mu_2-\mu_3-\mu_4}\bar{Q}_{34}\chi_{,4}, \quad (R_{12}) \quad (259)$$

$$(e^{3\psi-\mu_2-\mu_3+\mu_4}Q_{23})_{,2} + (e^{3\psi+\mu_2-\mu_3-\mu_4}\bar{Q}_{43})_{,4} = 0, \quad (R_{13}) \quad (260)$$

$$(e^{3\psi-\mu_2+\mu_3-\mu_4}Q_{24})_{,2}+(e^{3\psi+\mu_2-\mu_3-\mu_4}\bar{Q}_{34})_{:3}$$
$$=e^{3\psi+\mu_2-\mu_3-\mu_4}\bar{Q}_{34}\chi_{,2}, \quad (R_{41}) \quad (261)$$

$$e^{\mu_3-\mu_2}[\psi_{,2,2}+\psi_{,2}\psi_{,2}+(\psi+\mu_3)_{,2}(\mu_3-\mu_2)_{,2}+\mu_{3,2,2}]$$
$$+e^{\mu_2-\mu_3}\{\psi_{:3:3}+\psi_{:3}\psi_{:3}+\psi_{:3}[e^{-\mu_2}(e^{\mu_2})_{:3'}-\mu_{3:3}]$$
$$+e^{\mu_3-\mu_2}[e^{-\mu_3}(e^{\mu_2})_{:3'}]_{:3'}\}+\tfrac{1}{4}e^{2\psi-\mu_2-\mu_3}Q_{23}^2$$
$$-\tfrac{1}{4}e^{2\psi-2\nu}(e^{\mu_3-\mu_2}Q_{24}^2+e^{\mu_2-\mu_3}\bar{Q}_{34}^2)-\tfrac{1}{4}e^{\mu_2+\mu_3}(e^{\mu_2-\mu_3-\mu_4}\chi_{,4})^2$$
$$+e^{-2\nu+\mu_2+\mu_3}[\psi_{,4}(\mu_2+\mu_3)_{,4}+\mu_{2,4}\mu_{3,4}]=0, \quad (G_{44}) \quad (262)$$

$$e^{\mu_3-\mu_2}[\nu_{,2,2}+\nu_{,2}\nu_{,2}+(\nu+\mu_3)_{,2}(\mu_3-\mu_2)_{,2}+\mu_{3,2,2}]$$
$$+e^{\mu_2-\mu_3}\{\nu_{:3:3}+\nu_{:3}\nu_{:3}+\nu_{:3}[e^{-\mu_2}(e^{\mu_2})_{:3'}-\mu_{3:3}]$$
$$+e^{\mu_3-\mu_2}[e^{-\mu_3}(e^{\mu_2})_{:3'}]_{:3'}\}+\tfrac{3}{4}e^{2\psi-\mu_2-\mu_3}Q_{23}^2$$
$$+\tfrac{3}{4}e^{2\psi-2\nu}(e^{\mu_3-\mu_2}Q_{24}^2+e^{\mu_2-\mu_3}\bar{Q}_{34}^2)+\tfrac{1}{4}e^{\mu_2+\mu_3}(e^{\mu_2-\mu_3-\mu_4}\chi_{,4})^2$$
$$+e^{-2\nu+\mu_2+\mu_3}[\mu_{3,4,4}+\mu_{2,4,4}+\mu_{2,4}(\mu_2-\mu_4)_{,4}+\mu_{3,4}(\mu_3-\mu_4)_{,4}$$
$$+\mu_{3,4}\mu_{2,4}]$$
$$=0, \quad (G_{11}) \quad (263)$$

$$e^{-2\mu_4}[\psi_{,4,4}+\psi_{,4}(\psi-\mu_4+\mu_3)_{,4}+\mu_{3,4,4}+\mu_{3,4}(\mu_3-\mu_4)_{,4}]$$
$$+e^{-2\mu_3}[\psi_{:3:3}+\psi_{:3}(\psi-\mu_3+\mu_4)_{:3}+\mu_{4:3:3}+\mu_{4:3}(\mu_4-\mu_3)_{:3}]$$
$$+e^{-2\mu_2}[\psi_{,2}\mu_{4,2}+\mu_{3,2}(\psi+\mu_4)_{,2}]+\tfrac{1}{4}e^{2\psi-2\mu_3-2\mu_4}\bar{Q}_{34}^2$$
$$-\tfrac{1}{4}e^{2\psi-2\mu_2}(e^{-2\mu_3}Q_{32}^2+e^{-2\mu_4}Q_{42}^2)+\tfrac{3}{4}(e^{\mu_2-\mu_3-\mu_4}\chi_{,4})^2=0, \quad (G_{22}) \quad (264)$$

$$e^{-2\mu_4}[\psi_{,4,4}+\psi_{,4}(\psi-\mu_4+\mu_2)_{,4}+\mu_{2,4,4}+\mu_{2,4}(\mu_2-\mu_4)_{,4}]$$
$$+e^{-2\mu_2}[\psi_{,2,2}+\psi_{,2}(\psi-\mu_2+\mu_4)_{,2}+\mu_{4,2,2}+\mu_{4,2}(\mu_4-\mu_2)_{,2}]$$
$$+e^{-2\mu_3}[\psi_{:3}\mu_{4:3}+e^{-\mu_2}(e^{\mu_2})_{:3'}(\psi+\mu_4)_{:3}]+\tfrac{1}{4}e^{2\psi-2\mu_2-2\mu_4}Q_{24}^2$$
$$-\tfrac{1}{4}e^{2\psi-2\mu_3}(e^{-2\mu_2}Q_{23}^2+e^{-2\mu_4}\bar{Q}_{43}^2)-\tfrac{1}{4}(e^{\mu_2-\mu_3-\mu_4}\chi_{,4})^2=0 \quad (G_{33}) \quad (265)$$

$$(\psi+\mu_4)_{,2:3}-(\psi+\mu_4)_{,2}\mu_{2:3}-(\psi+\mu_4)_{:3}\mu_{3,2}+\psi_{:2}\psi_{:3}+\mu_{4,2}\mu_{4:3}$$
$$+\tfrac{1}{2}e^{2\psi-2\mu_4}\bar{Q}_{43}Q_{42}+\tfrac{1}{2}e^{2\mu_2-2\mu_4}(\psi+\mu_2)_{,4}\chi_{,4}$$
$$+\tfrac{1}{2}e^{\mu_3-\mu_4}(e^{2\mu_2-\mu_3-\mu_4}\chi_{,4})_{,4}=0. \quad (R_{23}) \quad (266)$$

$$e^{\mu_3-\mu_2}[\psi_{,2,2}+\psi_{,2}(\psi+\mu_4+\mu_3-\mu_2)_{,2}]$$
$$+e^{\mu_2-\mu_3}\{\psi_{:3:3}+\psi_{:3}[(\psi+\mu_4-\mu_3)_{:3}+e^{-\mu_2}(e^{\mu_2})_{:3'}]\}$$
$$+e^{\mu_2+\mu_3-2\mu_4}[\psi_{,4,4}+\psi_{,4}(\psi-\mu_4+\mu_2+\mu_3)]$$
$$-\tfrac{1}{2}e^{2\psi-2\mu_4}(e^{\mu_3-\mu_2}Q_{24}^2+e^{\mu_2-\mu_3}\bar{Q}_{34}^2)-\tfrac{1}{2}e^{2\psi-\mu_2-\mu_3}Q_{23}^2=0. \quad (R_{11}) \quad (267)$$

(a) The linearization of the field equations about a stationary solution; the initial-value equations

We now suppose that, as the result of a perturbation, preserving axisymmetry, a stationary axisymmetric space-time, with the metric (240), is no longer stationary and takes the more general form (244). The various functions satisfying equations (257)–(267) will then be

$$\begin{gathered} \nu+\delta\nu,\ \psi+\delta\psi,\ \omega+\delta\omega,\ \mu_2+\delta\mu_2,\ \mu_3+\delta\mu_3 \\ q_2,q_3,\quad\text{and}\quad \chi, \end{gathered} \tag{268}$$

where $\delta\nu$, $\delta\psi$, $\delta\omega$, $\delta\mu_2$, $\delta\mu_3$, q_2, q_3, and χ are all quantities of the first order of smallness and dependent on time, while ν, ψ, ω, μ_2, and μ_3 are functions, independent of time, which satisfy the field equations appropriate to a stationary axisymmetric space-time, namely those listed in Chapter 6, equations (5)–(16).

Since q_2 and q_3 are non-vanishing only on account of the perturbation, we shall find it convenient to write

$$q_{2,0}\quad\text{and}\quad q_{3,0}\quad\text{in place of}\quad q_2\quad\text{and}\quad q_3. \tag{269}$$

Also, since χ is now assumed to be a quantity of the first order of smallness, we need not distinguish the colon derivatives of combinations of functions in which χ occurs as a factor from the normal comma derivatives. We need not also distinguish between $\bar{Q}_{43}$ and Q_{43} when it occurs multiplied by a quantity of the first order of smallness such as Q_{23}.

The equations governing the perturbations can now be obtained by linearizing equations (257)–(267) about the stationary solution. The resulting equations are of two kinds: *initial-value equations* and *dynamical equations.* Initial-value equations are those which are linear and homogeneous in the time derivative; and dynamical equations are those which are of the second order in the time derivative and inhomogeneous. The initial-value equations can accordingly be integrated (with respect to the time), as they stand, and provide simple time-independent linear relations among the various perturbations: as such, the relations which follow from these equations may be interpreted as determining the initial conditions of the perturbation. The dynamical equations, then, determine their time-evolution.

In view of the basic role they play in the entire theory, we shall consider first the initial-value equations; these are provided by the (4, 2)-, (4, 3)-, (1, 2)-, and the (1, 3)-components of the field equations.

Considering the (4, 2)-component given by equation (257), and remembering that in accordance with our present convention (269),

$$Q_{23}=q_{2,3,4}-q_{3,2,4}, \tag{270}$$

we can integrate the equation with respect to x^4 to obtain

$$\delta\psi_{,2}+\psi_{,2}(\delta\psi-\delta\mu_2)-\delta\psi\nu_{,2}+\delta\mu_{3,2}+\mu_{3,2}\delta(\mu_3-\mu_2)-\nu_{,2}\delta\mu_3 + \tfrac{1}{2}e^{2\psi-2\mu_3}(q_{2,3}-q_{3,2})\omega_{,3}=\tfrac{1}{2}e^{2\mu_2-2\mu_3}(\psi+\mu_2)_{,3}\chi + \tfrac{1}{2}e^{\nu-\mu_3}(e^{2\mu_2-\mu_3-\nu}\chi)_{,3}. \quad (271)$$

Now defining

$$Q=e^{3\psi+\nu-\mu_2-\mu_3}(q_{2,3}-q_{3,2}), \quad (272)$$

we can rewrite equation (271) in the form

$$e^{\mu_3-\mu_2+\beta}[\delta\psi_{,2}+(\psi-\nu)_{,2}\delta\psi+\delta\mu_{3,2}-(\nu-\mu_3)_{,2}\delta\mu_3-(\psi+\mu_3)_{,2}\delta\mu_2] + \tfrac{1}{2}Q\omega_{,3}-\tfrac{1}{2}e^{2\nu-2\mu_2}(e^{\psi-\nu+3\mu_2-\mu_3}\chi)_{,3}=0, \quad (273)$$

where $\beta=\psi+\nu$ (cf. Ch. 6, equation (11)). Similarly, the (4, 3)-component of the equation gives

$$e^{\mu_2-\mu_3+\beta}[\delta\psi_{,3}+(\psi-\nu)_{,3}\delta\psi+\delta\mu_{2,3}-(\nu-\mu_2)_{,3}\delta\mu_2-(\psi+\mu_2)_{,3}\delta\mu_3] - \tfrac{1}{2}Q\omega_{,2}-\tfrac{1}{2}e^{2\nu-2\mu_3}(e^{\psi-\nu+\mu_2+\mu_3}\chi)_{,2}=0. \quad (274)$$

Alternative forms of equations (273) and (274) which we shall find useful are

$$e^{\mu_3-\mu_2}[(e^{\beta}\delta\psi)_{,2}-(e^{\beta})_{,2}\delta\mu_3+e^{\beta}\delta\mu_{3,2}-2e^{\beta}\nu_{,2}\delta\psi + e^{\beta}(\psi+\mu_3)_{,2}\delta(\mu_3-\mu_2)]+\tfrac{1}{2}Q\omega_{,3}-\tfrac{1}{2}e^{2\nu-2\mu_2}(e^{\psi-\nu+3\mu_2-\mu_3}\chi)_{,3}=0 \quad (275)$$

and

$$e^{\mu_2-\mu_3}[(e^{\beta}\delta\psi)_{,3}-(e^{\beta})_{,3}\delta\mu_2+e^{\beta}\delta\mu_{2,3}-2e^{\beta}\nu_{,3}\delta\psi + e^{\beta}(\psi+\mu_2)_{,3}\delta(\mu_2-\mu_3)]-\tfrac{1}{2}Q\omega_{,2}-\tfrac{1}{2}e^{2\nu-2\mu_3}(e^{\psi-\nu+\mu_2+\mu_3}\chi)_{,2}=0. \quad (276)$$

Considering next the (1, 2)-component of the equation given by equation (259), and ignoring terms which are manifestly of the second order, we have

$$(e^{3\psi-\mu_2-\mu_3+\nu}Q_{23})_{,3}+(e^{3\psi-\mu_2+\mu_3-\nu}Q_{02})_{,0}=+e^{3\psi+\mu_2-\mu_3-\nu}Q_{03}\chi_{,0}, \quad (277)$$

where we have reverted to the variable x^0 in accordance with the definitions (245). In view of equations (270) and (272), equation (277) integrates to give

$$Q_{,3}+e^{3\psi-\mu_2+\mu_3-\nu}[(\delta\omega_{,2}-q_{2,0,0})+\omega_{,2}(3\delta\psi-\delta\mu_2+\delta\mu_3-\delta\nu)] = e^{3\psi+\mu_2-\mu_3-\nu}\omega_{,3}\chi \quad (278)$$

or, alternatively,

$$\delta\omega_{,2}-q_{2,0,0}=-\omega_{,2}(3\delta\psi-\delta\nu-\delta\mu_2+\delta\mu_3)-e^{-3\psi+\nu+\mu_2-\mu_3}Q_{,3} + e^{2\mu_2-2\mu_3}\omega_{,3}\chi. \quad (279)$$

Similarly, the (1, 3)-component of the field equation gives

$$\delta\omega_{,3} - q_{3,0,0} = -\omega_{,3}(3\delta\psi - \delta\nu + \delta\mu_2 - \delta\mu_3) + e^{-3\psi+\nu-\mu_2+\mu_3}Q_{,2} + \omega_{,2}\chi. \quad (280)$$

While equations (279) and (280) relate to the initial conditions on ω, they also provide a dynamical equation for Q. Thus the integrability condition for ω yields for Q the inhomogeneous wave-equation:

$$(e^{-3\psi+\nu+\mu_2-\mu_3}Q_{,3})_{,3} + (e^{-3\psi+\nu-\mu_2+\mu_3}Q_{,2})_{,2}$$
$$= e^{-3\psi+\nu+\mu_2+\mu_3}Q_{,0,0} + (e^{2\mu_2-2\mu_3}\omega_{,3}\chi)_{,3} - (\omega_{,2}\chi)_{,2}$$
$$-[\omega_{,2}(3\delta\psi - \delta\nu - \delta\mu_2 + \delta\mu_3)]_{,3} + [\omega_{,3}(3\delta\psi - \delta\nu + \delta\mu_2 - \delta\mu_3)]_{,2}. \quad (281)$$

This equation, in fact, governs the emission of gravitational radiation.

We shall state as a lemma two relations which follow from equations (279) and (280) and which are important for the linearization of the remaining field equations.

LEMMA.

$$\delta[e^{2\psi-2\nu}(e^{\mu_3-\mu_2}Q_{02}^2 + e^{\mu_2-\mu_3}Q_{03}^2)]$$
$$= -e^{2\psi-2\nu}[4X\delta\psi + Y\delta(\mu_3-\mu_2)] + 2S + 4e^{2\psi-2\nu-\mu_3+\mu_2}\omega_{,2}\omega_{,3}\chi \quad (282)$$

and

$$\delta[e^{2\psi-2\nu}(e^{\mu_3-\mu_2}Q_{02}^2 - e^{\mu_2-\mu_3}Q_{03}^2)]$$
$$= -e^{2\psi-2\nu}[4Y\delta\psi + X\delta(\mu_3-\mu_2)] + 2D, \quad (283)$$

where

$$X = e^{\mu_3-\mu_2}(\omega_{,2})^2 + e^{\mu_2-\mu_3}(\omega_{,3})^2, \quad (284)$$

$$Y = e^{\mu_2-\mu_3}(\omega_{,2})^2 - e^{\mu_2-\mu_3}(\omega_{,3})^2, \quad (285)$$

$$S = e^{-\beta}(Q_{,2}\omega_{,3} - Q_{,3}\omega_{,2}), \text{ and } D = -e^{-\beta}(Q_{,2}\omega_{,3} + Q_{,3}\omega_{,2}). \quad (286)$$

Proof: Clearly,

$$\delta(e^{2\psi-2\nu+\mu_3-\mu_2}Q_{02}^2)$$
$$= e^{2\psi-2\nu+\mu_3-\mu_2}[2\omega_{,2}(\delta\omega_{,2} - q_{2,00}) + (\omega_{,2})^2\delta(2\psi - 2\nu + \mu_3 - \mu_2)]; \quad (287)$$

and making use of equation (279), we have

$$\delta(e^{2\psi-2\nu+\mu_3-\mu_2}Q_{02}^2) = -[4\delta\psi + \delta(\mu_3-\mu_2)]e^{2\psi-2\nu+\mu_3-\mu_2}(\omega_{,2})^2$$
$$-2e^{-\psi-\nu}Q_{,3}\omega_{,2} + 2e^{2\psi-2\nu-\mu_3+\mu_2}\omega_{,2}\omega_{,3}\chi. \quad (288)$$

Similarly,

$$\delta(e^{2\psi-2\nu+\mu_2-\mu_3}Q_{03}^2) = -[4\delta\psi-\delta(\mu_3-\mu_2)]e^{2\psi-2\nu+\mu_2-\mu_3}(\omega_{,3})^2 + 2e^{-\psi-\nu}Q_{,2}\omega_{,3}+2e^{2\psi-2\nu-\mu_3+\mu_2}\omega_{,2}\omega_{,3}\chi. \quad (289)$$

And by addition and subtraction of the two equations, we obtain the relations stated in the lemma.

Finally, we may note the further relation which follows from equations (279) and (280):

$$\omega_{,3}(\delta\omega_{,2}-q_{2,0,0})+\omega_{,2}(\delta\omega_{,3}-q_{3,0,0}) = -2\omega_{,2}\omega_{,3}(3\delta\psi-\delta\nu) + e^{-3\psi+\nu}(e^{\mu_3-\mu_2}Q_{,2}\omega_{,2}-e^{\mu_2-\mu_3}Q_{,3}\omega_{,3})+e^{\mu_2-\mu_3}X\chi. \quad (290)$$

(b) The Bianchi identities

It will be observed that if quantities manifestly of the second order are neglected, the expressions for R_{41} and G_{44} given in equations (261) and (262) do not involve any time derivatives. They give the impression of being initial-value equations; but their status, in reality, is different. For reverting to the variable x^0, the (0, 1)-component of the field equation is

$$(e^{3\psi-\mu_2+\mu_3-\nu}Q_{02})_{,2}+(e^{3\psi+\mu_2-\mu_3-\nu}\bar{Q}_{03})_{:3} = e^{3\psi+\mu_2-\mu_3-\nu}\bar{Q}_{03}\chi_{,2}. \quad (291)$$

To the first order in χ,

$$\begin{aligned}(e^{3\psi+\mu_2-\mu_3-\nu}\bar{Q}_{03})_{:3} &= [e^{3\psi+\mu_2-\mu_3-\nu}(Q_{03}-\chi Q_{02})]_{:3}\\ &= (e^{3\psi+\mu_2-\mu_3-\nu}Q_{03})_{:3}-(e^{3\psi+\mu_2-\mu_3-\nu}\chi\omega_{,2})_{,3}+O(\chi^2)\\ &= (e^{3\psi+\mu_2-\mu_3-\nu}Q_{03})_{,3}-\chi(e^{3\psi+\mu_2-\mu_3-\nu}\omega_{,3})_{,2}\\ &\quad -(e^{3\psi+\mu_2-\mu_3-\nu}\chi\omega_{,2})_{,3}+O(\chi^2).\end{aligned} \quad (292)$$

Therefore, the linearized version of equation (291) is

$$\delta[(e^{3\psi-\mu_2+\mu_3-\nu}Q_{02})_{,2}+(e^{3\psi+\mu_2-\mu_3-\nu}Q_{03})_{,3}] = \chi_{,2}\omega_{,3}e^{3\psi+\mu_2-\mu_3-\nu}+\chi(e^{3\psi+\mu_2-\mu_3-\nu}\omega_{,3})_{,2} + (e^{3\psi+\mu_2-\mu_3-\nu}\chi\omega_{,2})_{,3}. \quad (293)$$

On evaluating the right-hand side of this equation we find

$$e^{3\psi+\mu_2-\mu_3-\nu}\{2\chi\omega_{,2,3}+\chi_{,2}\omega_{,3}+\chi_{,3}\omega_{,2} + \chi[\omega_{,2}(3\psi-\nu+\mu_2-\mu_3)_{,3}+\omega_{,3}(3\psi+\mu_2-\mu_3-\nu)_{,2}]\}; \quad (294)$$

while the left-hand side gives

$$\{e^{3\psi-\mu_2+\mu_3-\nu}[(\delta\omega_{,2}-q_{2,0,0})+\omega_{,2}(3\delta\psi-\delta\mu_2+\delta\mu_3-\delta\nu)]\}_{,2} + \{e^{3\psi+\mu_2-\mu_3-\nu}[(\delta\omega_{,3}-q_{3,0,0})+\omega_{,3}(3\delta\psi+\delta\mu_2-\delta\mu_3-\delta\nu)]\}_{,3}. \quad (295)$$

Now substituting for the quantities in square brackets in (295) from equations (279) and (280), we have

$$(-Q_{,3}+e^{3\psi-\nu+\mu_2-\mu_3}\omega_{,3}\chi)_{,2}+(Q_{,2}+e^{3\psi-\nu+\mu_2-\mu_3}\omega_{,2}\chi)_{,3}; \qquad (296)$$

and this expression clearly reduces to (294). Therefore, the (0, 1)-component of the equation provides no new information: it is an expression of a Bianchi identity of the linearized equations.

Considering next the (0, 0)-component of the field equation given by equation (262), we have

$$\begin{aligned}
&e^{\mu_3-\mu_2}[\psi_{,2,2}+\psi_{,2}\psi_{,2}+\mu_{3,2,2}+(\psi+\mu_3)_{,2}(\mu_3-\mu_2)_{,2}]\\
&\quad+e^{\mu_2-\mu_3}\,\{\psi_{:3:3}+\psi_{:3}\psi_{:3}+\psi_{:3}[e^{-\mu_2}(e^{\mu_2})_{:3'}-\mu_{3:3}]\\
&\quad+e^{\mu_3-\mu_2}[e^{-\mu_3}(e^{\mu_2})_{:3'}]_{:3'}\} = -\tfrac{1}{4}e^{2\psi-2\nu}(e^{\mu_3-\mu_2}Q_{20}^2+e^{\mu_2-\mu_3}Q_{30}^2)\\
&\qquad\qquad+\tfrac{1}{2}e^{2\psi-2\nu+\mu_2-\mu_3}\omega_{,2}\omega_{,3}\chi+O(\chi^2). \qquad (297)
\end{aligned}$$

With the aid of equation (282), we find that the linearization of the right-hand side gives

$$\tfrac{1}{4}e^{2\psi-2\nu}[4X\delta\psi+Y\delta(\mu_3-\mu_2)]-\tfrac{1}{2}S-\tfrac{1}{2}e^{2\psi-2\nu+\mu_2-\mu_3}\omega_{,2}\omega_{,3}\chi. \quad (298)$$

In linearizing the left-hand side, we must first replace the colon derivatives by the equivalent partial derivatives. By this procedure, we find

$$\begin{aligned}
&e^{\mu_3-\mu_2}\{[\delta\psi_{,2,2}+2\psi_{,2}\delta\psi_{,2}+\delta\mu_{3,2,2}+(\delta\psi+\delta\mu_3)_{,2}(\mu_3-\mu_2)_{,2}\\
&\quad+(\psi+\mu_3)_{,2}\,\delta(\mu_3-\mu_2)_{,2}]+\delta(\mu_3-\mu_2)[\psi_{,2,2}+\psi_{,2}\psi_{,2}+\mu_{3,2,2}\\
&\quad+(\psi+\mu_3)_{,2}(\mu_3-\mu_2)_{,2}]\}\\
&\quad+e^{\mu_2-\mu_3}\{2\leftrightarrow 3\}-\tfrac{1}{4}e^{2\psi-2\nu}[4X\delta\psi+Y\delta(\mu_3-\mu_2)] = -\tfrac{1}{2}S\\
&\quad-\tfrac{1}{2}\omega_{,2}\omega_{,3}e^{2\psi-2\nu+\mu_2-\mu_3}\chi+e^{\mu_2-\mu_3}\{\chi_{,2,3}+\chi_{,2}(\psi+2\mu_2-\mu_3)_{,3}\\
&\quad+\chi_{,3}(\psi+\mu_2)_{,2}+\chi[2(\psi+\mu_2)_{,2,3}+2\psi_{,2}\psi_{,3}+(\psi+\mu_2)_{,2}(\mu_2-\mu_3)_{,3}\\
&\qquad\qquad+(\psi+\mu_2)_{,3}(\mu_2-\mu_3)_{,2}]\}. \qquad (299)
\end{aligned}$$

Equation (299) combined with the initial-value equations (273) and (274) provides an important new relation. Thus, from the latter equations, we have

$$\begin{aligned}
&\tfrac{1}{2}[-Q\omega_{,3}+e^{2\nu-2\mu_2}(e^{\psi-\nu+3\mu_2-\mu_3}\chi)_{,3}]_{,2}\\
&\quad+\tfrac{1}{2}[+Q\omega_{,2}+e^{2\nu-2\mu_3}(e^{\psi-\nu+\mu_2+\mu_3}\chi)_{,2}]_{,3}\\
&\quad=\{e^{\mu_3-\mu_2+\beta}[(\delta\psi+\delta\mu_3)_{,2}+(\psi-\nu)_{,2}\delta\psi+\mu_{3,2}\,\delta(\mu_3-\mu_2)\\
&\qquad\qquad-\nu_{,2}\delta\mu_3-\psi_{,2}\delta\mu_2]\}_{,2}+\{e^{\mu_2-\mu_3+\beta}[2\leftrightarrow 3]\}. \qquad (300)
\end{aligned}$$

The left-hand side of this equation reduces to give

$$\begin{aligned}
&-\tfrac{1}{2}Se^{\beta}+\tfrac{1}{2}[e^{2\nu-2\mu_2}(e^{\psi-\nu+3\mu_2-\mu_3}\chi)_{,3}]_{,2}\\
&\quad+\tfrac{1}{2}[e^{2\nu-2\mu_3}(e^{\psi-\nu+\mu_2+\mu_3}\chi)_{,2}]_{,3}. \qquad (301)
\end{aligned}$$

While, on evaluating the right-hand side of equation (300), we find after some considerable reductions that, apart from a common factor e^{β}, we are left with precisely the terms on the left-hand side of equation (299). We thus obtain the equality

$$e^{\beta+\mu_2-\mu_3}\Big[\!\Big[-\tfrac{1}{2}e^{2\psi-2\nu}\omega_{,2}\omega_{,3}\chi+\{\chi_{,2,3}+\chi_{,2}(\psi+2\mu_2-\mu_3)_{,3}$$
$$+\chi_{,3}(\psi+\mu_2)_{,2}+\chi[2(\psi+\mu_2)_{,2,3}+2\psi_{,2}\psi_{,3}+(\psi+\mu_2)_{,2}(\mu_2-\mu_3)_{,3}$$
$$+(\psi+\mu_2)_{,3}(\mu_2-\mu_3)_{,2}]\}\Big]\!\Big]$$
$$=\tfrac{1}{2}[e^{2\nu-2\mu_2}(e^{\psi-\nu+3\mu_2-\mu_3}\chi)_{,3}]_{,2}+\tfrac{1}{2}[e^{2\nu-2\mu_3}(e^{\psi-\nu+\mu_2+\mu_3}\chi)_{,2}]_{,3}. \qquad (302)$$

This is an important equation governing χ: it is again an expression of a Bianchi identity which in this context provides non-trivial information.

(c) *The linearized versions of the remaining field equations*

Finally, we list below the result of linearizing the remaining field equations.

$$e^{\mu_3-\mu_2}\{[\delta\nu_{,2,2}+2\nu_{,2}\delta\nu_{,2}+\delta\mu_{3,2,2}+(\delta\nu+\delta\mu_3)_{,2}(\mu_3-\mu_2)_{,2}$$
$$+(\nu+\mu_3)_{,2}\delta(\mu_3-\mu_2)_{,2}]$$
$$+\delta(\mu_3-\mu_2)[\nu_{,2,2}+\nu_{,2}\nu_{,2}+\mu_{3,2}+(\nu+\mu_3)_{,2}(\mu_3-\mu_2)_{,2}]\}$$
$$+e^{\mu_2-\mu_3}\{2\leftrightarrow 3\}=e^{-2\nu+\mu_2+\mu_3}\delta(\mu_3+\mu_2)_{,0,0}+\tfrac{3}{2}S$$
$$-\tfrac{3}{4}[4X\delta\psi+2Y\delta(\mu_3-\mu_2)]$$
$$+e^{\mu_2-\mu_3}\Big[\!\Big[\tfrac{3}{2}\omega_{,2}\omega_{,3}e^{2\psi-2\nu}\chi+\{\chi_{,2,3}+\chi_{,2}(\nu+2\mu_2-\mu_3)_{,3}+\chi_{,3}(\nu+\mu_2)_{,2}$$
$$+\chi[2(\nu+\mu_2)_{,2,3}+2\nu_{,2}\nu_{,3}+(\nu+\mu_2)_{,2}(\mu_2-\mu_3)_{,3}$$
$$+(\nu+\mu_2)_{,3}(\mu_2-\mu_3)_{,2}]\}\Big]\!\Big], \qquad (\delta G_{11}=0), \quad (303)$$

$$e^{\mu_3-\mu_2}[\delta\beta_{,2,2}+2\beta_{,2}\delta\beta_{,2}+(\mu_3-\mu_2)_{,2}\delta\beta_{,2}]+e^{\mu_2-\mu_3}[2\leftrightarrow 3]$$
$$=e^{-2\nu+\mu_2+\mu_3}[2\delta\psi_{,0,0}+\delta(\mu_3+\mu_2)_{,0,0}]-e^{\mu_3-\mu_2}\beta_{,2}\delta(\mu_3-\mu_2)_{,2}$$
$$+e^{\mu_2-\mu_3}\beta_{,3}\delta(\mu_3-\mu_2)_{,3}-(2e^{-\beta}W+\tfrac{1}{2}e^{2\psi-2\nu}Y)\delta(\mu_3-\mu_2)$$
$$+e^{\mu_2-\mu_3}\{\beta_{,2}\chi_{,3}+\beta_{,3}\chi_{,2}$$
$$+\chi[2\beta_{,2,3}+2\beta_{,2}\beta_{,3}+\beta_{,2}(\mu_2-\mu_3)_{,3}+\beta_{,3}(\mu_2-\mu_3)_{,2}]\},$$
$$[\delta(G_{33}+G_{22})=0], \quad (304)$$

$$e^{\mu_3-\mu_2}[\delta\beta_{,2,2}+2\psi_{,2}\delta\psi_{,2}+2\nu_{,2}\delta\nu_{,2}-\delta\beta_{,2}(\mu_3+\mu_2)_{,2}-\beta_{,2}\delta(\mu_3+\mu_2)_{,2}]$$
$$-e^{\mu_2-\mu_3}[2\leftrightarrow 3]=2e^{-\beta}U\,\delta(\mu_3-\mu_2)-e^{-2\nu+\mu_2+\mu_3}\,\delta(\mu_3-\mu_2)_{,0,0}$$
$$-\tfrac{1}{2}e^{2\psi-2\nu}[4Y\delta\psi+X\,\delta(\mu_3-\mu_2)]+D-e^{\mu_2-\mu_3}\Big[\!\Big[-\omega_{,2}\omega_{,3}e^{2\psi-2\nu}\chi$$
$$+\beta_{,2}\chi_{,3}-\beta_{,3}\chi_{,2}+\chi[2\beta_{,2,3}+2\psi_{,2}\psi_{,3}+2\nu_{,2}\nu_{,3}-\beta_{,2}(\mu_2+\mu_3)_{,3}$$
$$-\beta_{,3}(\mu_2+\mu_3)_{,2}]\Big]\!\Big]\quad[\delta(G_{33}-G_{22})=0],\qquad(305)$$

and

$$\delta\beta_{,2,3}-\delta\beta_{,2}\mu_{2,3}-\delta\beta_{,3}\mu_{3,2}-\beta_{,2}\delta\mu_{2,3}-\beta_{,3}\delta\mu_{3,2}+\psi_{,2}\delta\psi_{,3}+\psi_{,3}\delta\psi_{,2}$$
$$+\nu_{,2}\delta\nu_{,3}+\nu_{,3}\delta\nu_{,2}=-2e^{2\psi-2\nu}\omega_{,2}\omega_{,3}\delta\psi+\tfrac{1}{2}e^{-2\nu+2\mu_2}\chi_{,0,0}$$
$$+\tfrac{1}{2}e^{-\psi-\nu}(e^{\mu_3-\mu_2}Q_{,2}\omega_{,2}-e^{\mu_2-\mu_3}Q_{,3}\omega_{,3})+\tfrac{1}{2}e^{2\psi-2\nu+2\mu_2-2\mu_3}(\omega_{,3})^2\chi$$
$$+\chi[\beta_{,2,2}-\beta_{,2}(\mu_2+\mu_3)_{,2}+\psi_{,2}\psi_{,2}+\nu_{,2}\nu_{,2}],\qquad(\delta R_{23}=0)\qquad(306)$$

where in equations (304) and (305) we have used the abbreviations

$$\left.\begin{aligned}U&=e^{\beta}[e^{\mu_3-\mu_2}(\beta_{,2}\mu_{3,2}+\psi_{,2}\nu_{,2})+e^{\mu_2-\mu_3}(\beta_{,3}\mu_{2,3}+\psi_{,3}\nu_{,3})],\\W&=e^{\beta}[e^{\mu_3-\mu_2}(\beta_{,2}\mu_{3,2}+\psi_{,2}\nu_{,2})-e^{\mu_2-\mu_3}(\beta_{,3}\mu_{2,3}+\psi_{,3}\nu_{,3})].\end{aligned}\right\}\qquad(307)$$

It may be noted that in deriving equations (304) and (305) use has been made of the following two equations which obtain in the stationary case (cf. Ch. 6, equations (14) and (16)):

$$\left.\begin{aligned}&[e^{\mu_3-\mu_2}(e^{\beta})_{,2}]_{,2}+[e^{\mu_2-\mu_3}(e^{\beta})_{,3}]_{,3}=0\\\text{and}\quad&[e^{\mu_3-\mu_2}(e^{\beta})_{,2}]_{,2}-[e^{\mu_2-\mu_3}(e^{\beta})_{,3}]_{,3}=2W+\tfrac{1}{2}e^{3\psi-\nu}Y.\end{aligned}\right\}\qquad(308)$$

This completes the formal developments.

(*d*) *Equations governing quasi-stationary deformations; Carter's theorem*

To illustrate the purposes to which the linearized equations we have derived can be put, we shall first consider the limit of infinitely slow deformation when the system (subject to the deformation) is effectively in a state of equilibrium at all times. Deformations effected in this manner are said to be *quasi-stationary* or *adiabatic*. Quasi-stationary deformations are useful for determining whether (or not) along a sequence of stationary systems (labelled by a parameter) a *point of bifurcation* occurs. Along the Kerr sequence (labelled by a), Carter's theorem assures that no such bifurcation can occur. We shall now show how a proof of Carter's theorem can be constructed by considering quasi-stationary deformations of stationary axisymmetric space-times external to the event horizon of a black hole.

A quasi-stationary deformation of an axisymmetric space-time with the metric (240) will effect infinitesimal changes in ν, ψ, ω, μ_2, and μ_3 while preserving the coordinate condition we may have imposed on μ_2 and μ_3. On this latter account, we shall find it convenient to start with a metric initially of the form

$$ds^2 = e^{2\nu}(dt)^2 - e^{2\psi}(d\varphi - \omega dt)^2 - e^{2\mu}[(dx^2)^2 + (dx^3)^2], \qquad (309)$$

i.e., with $\mu_2 = \mu_3$. A quasi-stationary deformation of this space-time will effect, infinitely slowly, changes in ν, ψ, ω and μ by amounts, say, $\delta\nu$, $\delta\psi$, $\delta\omega$ and $\delta\mu$. The equations governing these changes can be obtained directly from the equations we have derived in §§(*b*) and (*c*), by setting the time-derivatives in all the dynamical equations equal to zero and in the *integral forms* of the initial-value equations. Besides, for quasi-stationary deformations of a space-time with a metric of the adopted form (309),

$$\chi = 0 \qquad \text{and} \qquad \delta(\mu_3 - \mu_2) = 0. \qquad (310)$$

We observe first that the consideration of equation (304) provides an immediate simplification. For quasi-stationary deformations, this equation gives

$$(\delta\beta_{,2,2} + 2\beta_{,2}\delta\beta_{,2}) + (\delta\beta_{,3,3} + 2\beta_{,3}\delta\beta_{,3}) = 0; \qquad (311)$$

or, in the notation of two-dimensional Cartesian tensors (which we shall adopt),

$$(e^{\beta}\delta\beta)_{,\alpha,\alpha} = 0, \quad (\alpha = 2, 3) \qquad (312)$$

since the equation governing equilibrium guarantees that

$$(e^{\beta})_{,\alpha,\alpha} = 0. \qquad (313)$$

From equation (312) it follows that

$$\delta\beta \equiv 0. \qquad (314)$$

For by multiplying equation (312) by $e^{\beta}\delta\beta$ and integrating over all of the 3-space (x^1, x^2, x^3), external to the event horizon, we find, after an integration by parts,

$$\iiint (e^{\beta}\delta\beta)_{,\alpha}(e^{\beta}\delta\beta)_{,\alpha}\, d\mathbf{x} = 0, \qquad (315)$$

the integrated part vanishing on account of the boundary conditions (cf. Ch. 6, §§53 and 55)

$$\left.\begin{aligned} &e^{\beta} = 0 \quad \text{on the horizon} \\ \text{and} \quad &e^{\beta} = O(r) \quad \text{and} \quad \delta\beta = O(r^{-1}) \quad \text{for} \quad r \to \infty \end{aligned}\right\} \qquad (316)$$

at the boundaries. The integrand in equation (315) being positive-definite, the identical vanishing of $\delta\beta$ follows. We conclude then that for quasi-stationary deformations, besides the vanishing of χ and $\delta(\mu_3-\mu_2)$,

$$\delta\psi = -\delta\nu. \tag{317}$$

In view of the requirements (310) and (317), the initial-value equations (275), (276), (279), and (280) now take the forms

$$\left.\begin{aligned}(e^\beta\delta\psi)_{,2}-(e^\beta)_{,2}\delta\mu+e^\beta\delta\mu_{,2}-2e^\beta\nu_{,2}\delta\psi &= -\tfrac{1}{2}Q\omega_{,3},\\ (e^\beta\delta\psi)_{,3}-(e^\beta)_{,3}\delta\mu+e^\beta\delta\mu_{,3}-2e^\beta\nu_{,3}\delta\psi &= +\tfrac{1}{2}Q\omega_{,2},\end{aligned}\right\} \tag{318}$$

and

$$\left.\begin{aligned}\delta\omega_{,2} &= -4\omega_{,2}\delta\psi-e^{-3\psi+\nu}Q_{,3},\\ \delta\omega_{,3} &= -4\omega_{,3}\delta\psi+e^{-3\psi+\nu}Q_{,2},\end{aligned}\right\} \tag{319}$$

where

$$Q = e^{3\psi+\nu-2\mu}(q_{2,3}-q_{3,2}); \tag{320}$$

while equations (281) and (282) give

$$(e^{-3\psi+\nu}Q_{,\alpha})_{,\alpha} = 4(\omega_{,3}\delta\psi_{,2}-\omega_{,2}\delta\psi_{,3}). \tag{321}$$

and

$$\delta(e^{2\psi-2\nu}X) = -4e^{2\psi-2\nu}X\,\delta\psi+2S, \tag{322}$$

where

$$X = \omega_{,\alpha}\omega_{,\alpha} \quad \text{and} \quad Se^\beta = Q_{,2}\omega_{,3}-Q_{,3}\omega_{,2}. \tag{323}$$

We now consider the linearization of the R_{11}-equation (cf. Ch. 6, equation (6)),

$$(e^\beta\psi_{,\alpha})_{,\alpha}+\tfrac{1}{2}e^{3\psi-\nu}X = 0. \tag{324}$$

Making use of equations (314) and (322), we obtain

$$(e^\beta\delta\psi_{,\alpha})_{,\alpha} = 2e^{3\psi-\nu}X\,\delta\psi-Se^\beta. \tag{325}$$

(This equation also follows directly from linearizing equation (267).)

Now multiplying equation (325) by $\delta\psi$ and integrating over the whole of the 3-space external to the horizon, we find, after an integration by parts,

$$\iiint[e^\beta\delta\psi_{,\alpha}\delta\psi_{,\alpha}+2e^{3\psi-\nu}X(\delta\psi)^2-Se^\beta\delta\psi]\,\mathrm{d}x = 0 \tag{326}$$

with the integrated part vanishing by virtue of the conditions (316) together with the requirement (which follows from an examination of equation (325))

$$\delta\psi = O(r^{-1}) \quad \text{for} \quad r\to\infty. \tag{327}$$

A more symmetric form of equation (326) can be obtained by observing that

$$\begin{aligned}\iiint S e^{\beta}\delta\psi\,\mathrm{dx} &= +\iiint (Q_{,2}\omega_{,3}-Q_{,3}\omega_{,2})\delta\psi\,\mathrm{dx}\\ &= -\iiint Q(\omega_{,3}\delta\psi_{,2}-\omega_{,2}\delta\psi_{,3})\,\mathrm{dx}\\ &= -\tfrac{1}{4}\iiint Q(e^{-3\psi+\nu}Q_{,\alpha})_{,\alpha}\,\mathrm{dx}\\ &= +\tfrac{1}{4}\iiint e^{-3\psi+\nu}Q_{,\alpha}Q_{,\alpha}\,\mathrm{dx},\end{aligned} \tag{328}$$

where an integration by parts has been effected in passing from line 1 to 2, equation (321) has been used in passing from line 2 to 3, and a further integration by parts has been effected in passing from line 3 to 4. The integrated parts, in both instances, vanish by virtue of the conditions (316), (327), the known r^{-3}-behaviour of ω at infinity, and the further requirement that $Qe^{-2\psi}$ be smooth on the horizon (which requirement follows from equations (316) and (318)). Making use of the final result of the reduction (328), we can rewrite equation (326) in the form

$$\iiint [e^{\beta}\delta\psi_{,\alpha}\delta\psi_{,\alpha}+2e^{3\psi-\nu}X(\delta\psi)^2-\tfrac{1}{4}e^{-3\psi+\nu}Q_{,\alpha}Q_{,\alpha}]\,\mathrm{dx}=0. \tag{329}$$

To prove the impossibility of a quasi-stationary deformation and Carter's theorem, we must transform the integrand in equation (329) to a positive-definite form. To achieve this end, we need the following two lemmas.

LEMMA 1.

$$\begin{aligned}Q_{,2}\omega_{,3}-Q_{,3}\omega_{,2} &= e^{2\psi}[(e^{-2\psi}Q)_{,2}\omega_{,3}-(e^{-2\psi}Q)_{,3}\omega_{,2}]\\ &\quad +2Q(\psi_{,2}\omega_{,3}-\psi_{,3}\omega_{,2});\end{aligned} \tag{330}$$

and

LEMMA 2.

$$\begin{aligned}\iiint [\tfrac{1}{4}e^{\beta}(Qe^{-2\psi})_{,\alpha}(Qe^{-2\psi})_{,\alpha}+e^{-3\psi+\nu}Q^2\psi_{,\alpha}\psi_{,\alpha}\\ -\tfrac{1}{4}e^{-3\psi+\nu}Q_{,\alpha}Q_{,\alpha}+\tfrac{1}{4}e^{-\beta}Q^2X]\,\mathrm{dx}=0.\end{aligned} \tag{331}$$

Lemma 1 is directly verified. To prove Lemma 2, we multiply equation (324) by $\frac{1}{2}Q^2e^{-4\psi}$ and integrate over the 3-space external to the event horizon. After an integration by parts of the first term in the integrand, we obtain

$$\iiint [-\tfrac{1}{2}(Q^2)_{,\alpha}e^{-3\psi+\nu}\psi_{,\alpha}+2Q^2e^{-3\psi+\nu}\psi_{,\alpha}\psi_{,\alpha}+\tfrac{1}{4}Q^2Xe^{-\beta}]\,\mathrm{dx}=0, \tag{332}$$

the integrated parts vanishing by virtue of the boundary conditions obtaining at infinity and on the horizon. On rearranging the terms in the integrand of equation (332), we obtain equation (331).

To prove Carter's theorem, we first verify by making use of relations included in (328) that

$$\iiint e^{\beta}(\delta\psi_{,\alpha}\delta\psi_{,\alpha} - S\delta\psi)\,\mathrm{d}\mathbf{x}$$

$$= \iiint [e^{\beta}\delta\psi_{,\alpha}\delta\psi_{,\alpha} + Q(\omega_{,3}\delta\psi_{,2} - \omega_{,2}\delta\psi_{,3})]\,\mathrm{d}\mathbf{x}$$

$$= \iiint \{e^{\beta}[(\delta\psi_{,2} + \tfrac{1}{2}e^{-\beta}Q\omega_{,3})^2 + (\delta\psi_{,3} - \tfrac{1}{2}e^{-\beta}Q\omega_{,2})^2]$$

$$- \tfrac{1}{4}e^{-\beta}XQ^2\}\,\mathrm{d}\mathbf{x}. \quad (333)$$

Next, eliminating the form $-\frac{1}{4}e^{-\beta}XQ^2$ in equation (333) with the aid of Lemma 2, we can write equation (329) in the form

$$\iiint \{e^{\beta}[(\delta\psi_{,2} + \tfrac{1}{2}e^{-\beta}Q\omega_{,3})^2 + (\delta\psi_{,3} - \tfrac{1}{2}e^{-\beta}Q\omega_{,2})^2]$$

$$+ \tfrac{1}{4}e^{\beta}(Qe^{-2\psi})_{,\alpha}(Qe^{-2\psi})_{,\alpha} + e^{-3\psi+\nu}Q^2\psi_{,\alpha}\psi_{,\alpha}$$

$$- \tfrac{1}{4}e^{-3\psi+\nu}Q_{,\alpha}Q_{,\alpha} + 2e^{3\psi-\nu}\omega_{,\alpha}\omega_{,\alpha}(\delta\psi)^2\}\,\mathrm{d}\mathbf{x} = 0. \quad (334)$$

On the other hand, by equations (328) and (330),

$$\frac{1}{4}\iiint e^{-3\psi+\nu}Q_{,\alpha}Q_{,\alpha}\,\mathrm{d}\mathbf{x} = \iiint e^{\beta}S\delta\psi\,\mathrm{d}\mathbf{x}$$

$$= \iiint (Q_{,2}\omega_{,3} - Q_{,3}\omega_{,2})\delta\psi\,\mathrm{d}\mathbf{x}$$

$$= \iiint \delta\psi\{e^{2\psi}[(e^{-2\psi}Q)_{,2}\omega_{,3} - (e^{-2\psi}Q)_{,3}\omega_{,2}]$$

$$+ 2Q(\psi_{,2}\omega_{,3} - \psi_{,3}\omega_{,2})\}\,\mathrm{d}\mathbf{x}. \quad (335)$$

Inserting this last relation in equation (334) and rearranging the terms, we obtain

$$\iiint \Big[\!\Big[e^{\beta}\{(\tfrac{1}{2}e^{-\beta}Q\omega_{,3} + \delta\psi_{,2})^2 + (\tfrac{1}{2}e^{-\beta}Q\omega_{,2} - \delta\psi_{,3})^2$$

$$+ [(\tfrac{1}{2}e^{-2\psi}Q)_{,2} - e^{\psi-\nu}\omega_{,3}\delta\psi]^2 + [(\tfrac{1}{2}e^{-2\psi}Q)_{,3} + e^{\psi-\nu}\omega_{,2}\delta\psi]^2\}$$

$$+ e^{-3\psi+\nu}[(Q\psi_{,2} - e^{3\psi-\nu}\omega_{,3}\delta\psi)^2 + (Q\psi_{,3} + e^{3\psi-\nu}\omega_{,2}\delta\psi)^2]\Big]\!\Big]\,\mathrm{d}\mathbf{x} = 0.$$

(336)

We have thus reduced the integrand of equation (329) to a sum of squares. And the only way in which the integral can vanish is for each of the summands to vanish identically; and it can be readily verified that this is possible if and only if

$$\delta\psi \equiv Q \equiv 0. \tag{337}$$

A non-trivial quasi-stationary deformation is therefore impossible; and Carter's theorem follows.

(*e*) *A variational formulation of the perturbation problem*

We begin by restating the problem we are considering. We start with a stationary solution of Einstein's vacuum-equation satisfying certain well-defined boundary conditions such as are required of the solutions external to the event horizons of black holes. We subject the solution to a *variation* which effects infinitesimal first-order changes, $\delta\nu$, $\delta\psi$, $\delta\omega$, Q, $\delta\mu_2$, $\delta\mu_3$, and χ, in the various metric functions. These variations satisfy equations of two kinds: the initial-value equations and the dynamical equations. We also require that the variations satisfy certain boundary conditions at the horizon and at infinity. If we now suppose that the variations have a time-dependence, $e^{i\sigma t}$ (where σ is allowed to be complex), then the dynamical equations satisfied by the variations, together with the boundary conditions imposed on them, define a characteristic-value problem for σ.

In a variational formulation of the characteristic-value problem, we associate with the *proper variations* (which satisfy both the initial-value and the dynamical equations), *trial variations* $\bar{\delta}\nu$, $\bar{\delta}\psi$, $\bar{\delta}\omega$, $\bar{Q}$, $\bar{\delta}\mu_2$, $\bar{\delta}\mu_3$, and $\bar{\chi}$ which we require to be consistent with the initial-value equations and satisfy the same boundary conditions as the proper variations, but which are arbitrary otherwise. In other words, the barred (trial) variations differ from the unbarred (proper) variations only in that the latter are required to satisfy the dynamical equations while the former are not. In particular, the barred quantities will have the same time-depenence $e^{i\sigma t}$ as the unbarred quantities.

The first step in a variational formulation of the underlying characteristic-value problem is to obtain a formula for σ^2 by multiplying an initial-value equation satisfied by a barred quantity by one or other of the unbarred quantities (or a dynamical equation satisfied by an unbarred quantity, by a barred quantity) and integrating over the entire 3-space external to the event horizon; and to arrange, if possible, by a suitable number of integrations by parts (as in §(*d*) in the proof of Carter's theorem), making use of the initial-value and the dynamical equations satisfied by the unbarred quantities and the initial-value equations satisfied by the barred quantities, that, in the integrals which occur in the formula for σ^2, the barred and the unbarred quantities appear entirely symmetrically. If such a reduction can be achieved, then we say that the differential operators defining the characteristic-value problem are *self-adjoint*; and a variational formulation then becomes possible. But first we shall indicate how such a formula for σ^2 can be obtained for the problem on hand.

Considering equations (275) and (276) for the barred quantities—we are justified in considering these equations for the barred quantities since they are initial-value equations—and multiplying them by $\delta\psi_{,2}$ and $\delta\psi_{,3}$, respectively, and adding, we integrate over the 3-space included between two 2-surfaces S_1 and S_2. We shall eventually let S_1 tend to the event horizon and S_2 to infinity; but for the present, we shall leave them unspecified. After several integrations by parts and making use (only) of the (0, 0)-component of the linearized field-equation, namely (299), for the proper variations *and* the relation (302) which follows from it, we obtain.

$$
\begin{aligned}
0 = \iiint \Big[\!\Big[& -2e^{\beta}(e^{\mu_3-\mu_2}\delta\psi_{,2}\bar{\delta}\psi_{,2} + e^{\mu_2-\mu_3}\delta\psi_{,3}\bar{\delta}\psi_{,3}) \\
& + e^{\beta}\{(e^{\mu_3-\mu_2}\delta\mu_{3,2})_{,2}\bar{\delta}\psi + (e^{\mu_3-\mu_2}\bar{\delta}\mu_{3,2})\delta\psi \\
& - e^{\mu_3-\mu_2}(\psi+\mu_3)_{,2}[\delta\psi_{,2}\bar{\delta}(\mu_3-\mu_2) + \bar{\delta}\psi_{,2}\delta(\mu_3-\mu_2)]\} \\
& + e^{\beta}\{2 \leftrightarrow 3\} \\
& - e^{3\psi-\nu}X\,\delta\psi\bar{\delta}\psi - (W + \tfrac{1}{4}e^{3\psi-\nu}Y)[\delta\psi\,\bar{\delta}(\mu_3-\mu_2) + \bar{\delta}\psi\,\delta(\mu_3-\mu_2)] \\
& + \tfrac{1}{2}e^{\beta}(S\bar{\delta}\psi + \bar{S}\delta\psi) \\
& - \tfrac{1}{2}\{[e^{-2\mu_2+2\nu}(e^{\psi-\nu+3\mu_2-\mu_3}\chi)_{,3}]_{,2} \\
& + [e^{-2\mu_3+2\nu}(e^{\psi-\nu+\mu_2+\mu_3}\chi)_{,2}]_{,3}\}\,\bar{\delta}\psi \\
& - \tfrac{1}{2}\{[e^{-2\mu_2+2\nu}(e^{\psi-\nu+3\mu_2-\mu_3}\bar{\chi})_{,3}]_{,2} \\
& + [e^{-2\mu_3+2\nu}(e^{\psi-\nu+\mu_2+\mu_3}\bar{\chi})_{,2}]_{,3}\}\,\delta\psi \Big]\!\Big]\, d\varphi\, dx^2\, dx^3 \\
+ \iint \Big[\!\Big[& e^{\mu_3-\mu_2+\beta}[\delta\psi_{,2}\bar{\delta}\psi + (\psi+\mu_3)_{,2}\delta(\mu_3-\mu_2)\bar{\delta}\psi + (\psi+\nu)_{,2}\delta\psi\,\bar{\delta}\mu_3 \\
& - \delta\psi\,\bar{\delta}\mu_{3,2}] - \tfrac{1}{2}\bar{Q}\omega_{,3}\delta\psi \\
& + \tfrac{1}{2}e^{-2\mu_2+2\nu}(e^{\psi-\nu+3\mu_2-\mu_3}\bar{\chi})_{,3}\,\delta\psi \Big]\!\Big]_{[x^2]} d\varphi\, dx^3 \\
+ \iint \Big[\!\Big[& e^{\mu_2-\mu_3+\beta}[\delta\psi_{,3}\bar{\delta}\psi + (\psi+\mu_2)_{,3}\delta(\mu_2-\mu_3)\bar{\delta}\psi + (\psi+\nu)_{,3}\delta\psi\,\bar{\delta}\mu_2 \\
& - \delta\psi\,\bar{\delta}\mu_{2,3}] + \tfrac{1}{2}\bar{Q}\omega_{,2}\delta\psi \\
& + \tfrac{1}{2}e^{-2\mu_3+2\nu}(e^{\psi-\nu+\mu_2+\mu_3}\bar{\chi})_{,2}\,\delta\psi \Big]\!\Big]_{[x^3]} d\varphi\, dx^2. \qquad (338)
\end{aligned}
$$

In equation (338) the symbol

$$\Big[\!\Big[\cdots\Big]\!\Big]_{[x^{\alpha}]}$$

in the integrands of the surface integrals (that result from the various integrations by parts) has the following meaning. For a fixed $x^{\beta}\,(\beta \neq \alpha)$ let the appropriate limits of x^{α} be $x^{\alpha}(1)$ and $x^{\alpha}(2)$ and $x^{\alpha}(2) > x^{\alpha}(1)$; the symbol then stands for the difference in the values of the quantity enclosed in double brackets at $x^{\alpha}(2)$ and $x^{\alpha}(1)$.

Equation (338) has been derived from the initial-value equations only: this fact is evident from the absence of σ^2 in this equation. We also observe that the volume integral in equation (338) is symmetric in the barred and the unbarred quantities.

Next we multiply by $\delta\nu_{,2}$ and $\delta\nu_{,3}$ the same equations (275) and (276); and we also multiply by $\bar{\chi}$ the (2, 3)-component of the linearized field-equation (cf. equation (306)),

$$\begin{aligned}
&\tfrac{1}{2}e^{\psi-\nu+3\mu_2-\mu_3}\chi_{,0,0}+\tfrac{1}{2}e^{3\psi-\nu+3\mu_2-3\mu_3}(\omega_{,3})^2\chi\\
&\quad-2e^{3\psi-\nu+\mu_2-\mu_3}\omega_{,2}\omega_{,3}\delta\psi+\tfrac{1}{2}e^{\mu_2-\mu_3}(e^{\mu_3-\mu_2}Q_{,2}\omega_{,2}-e^{\mu_2-\mu_3}Q_{,3}\omega_{,3})\\
&\quad+e^{\mu_2-\mu_3+\beta}\chi[\beta_{,2,2}-\beta_{,2}(\mu_3+\mu_2)_{,2}+\psi_{,2}\psi_{,2}+\nu_{,2}\nu_{,2}]\\
&\quad-e^{\mu_2-\mu_3+\beta}[\delta\nu_{,2,3}+\delta\nu_{,2}(\nu-\mu_2)_{,3}+\delta\nu_{,3}(\nu-\mu_3)_{,2}-\beta_{,2}\delta\mu_{2,3}\\
&\qquad\qquad+\delta\psi_{,2,3}+\delta\psi_{,2}(\psi-\mu_2)_{,3}+\delta\psi_{,3}(\psi-\mu_3)_{,2}-\beta_{,3}\delta\mu_{3,2}]=0. \qquad (339)
\end{aligned}$$

(Since equation (339), unlike equations (275) and (276), is a dynamical equation, we are considering it for a proper variation χ and multiplying it by a trial variation $\bar{\chi}$.) We now integrate the sum of the three equations, obtained by the respective multiplications, over the 3-space included between S_1 and S_2. After numerous integrations by parts and extensive rearrangement of the terms, we find that we are left with

$$\begin{aligned}
0=\iiint\Big[\!\Big[&Xe^{3\psi-\nu}[3\delta\psi\bar{\delta}\psi+\tfrac{1}{4}\delta(\mu_3-\mu_2)\bar{\delta}(\mu_3-\mu_2)]\\
&+(\tfrac{3}{4}e^{3\psi-\nu}Y-W)[\delta\psi\bar{\delta}(\mu_3-\mu_2)+\bar{\delta}\psi\,\delta(\mu_3-\mu_2)]\\
&\qquad\qquad-U\delta(\mu_3-\mu_2)\bar{\delta}(\mu_3-\mu_2)\\
&-e^{\mu_3-\mu_2+\beta}\{[\beta_{,2}\delta\mu_{3,2}+(\nu+\mu_3)_{,2}\delta\psi_{,2}]\bar{\delta}(\mu_3-\mu_2)\\
&\qquad\qquad+[\beta_{,2}\bar{\delta}\mu_{3,2}+(\nu+\mu_3)_{,2}\bar{\delta}\psi_{,2}]\delta(\mu_3-\mu_2)\}\\
&-e^{\mu_2-\mu_3+\beta}\{2\leftrightarrow 3\}\\
&+e^{\beta}[(e^{\mu_3-\mu_2}\delta\mu_{3,2})_{,2}\bar{\delta}\psi+(e^{\mu_3-\mu_2}\bar{\delta}\mu_{3,2})\delta\psi]+e^{\beta}[2\leftrightarrow 3]
\end{aligned}$$

$$+2\sigma^2 e^{-2\nu}\sqrt{-g}\cdot[\tfrac{1}{4}\delta(\mu_3+\mu_2)\bar{\delta}(\mu_3+\mu_2)+\tfrac{1}{2}\delta\psi\,\bar{\delta}(\mu_3+\mu_2)$$
$$+\tfrac{1}{2}\bar{\delta}\psi\,\delta(\mu_3+\mu_2)$$
$$-\tfrac{1}{4}\delta(\mu_3-\mu_2)\bar{\delta}(\mu_3-\mu_2)]-\tfrac{1}{2}\sigma^2 e^{-3\psi-\nu+\mu_2+\mu_3}Q\bar{Q}$$
$$-\tfrac{1}{2}e^{\beta}[3S\bar{\delta}\psi+3\bar{S}\delta\psi+D\bar{\delta}(\mu_3-\mu_2)+\bar{D}\delta(\mu_3-\mu_2)]$$
$$+\tfrac{1}{2}e^{-3\psi+\nu}(e^{\mu_3-\mu_2}Q_{,2}\bar{Q}_{,2}+e^{\mu_2-\mu_3}Q_{,3}\bar{Q}_{,3})$$
$$-\tfrac{1}{2}\sigma^2 e^{\psi-\nu+3\mu_2-\mu_3}\chi\bar{\chi}-2e^{3\psi-\nu+\mu_2-\mu_3}\omega_{,2}\omega_{,3}(\bar{\chi}\delta\psi+\chi\bar{\delta}\psi)$$
$$+\tfrac{1}{2}e^{\mu_2-\mu_3}(e^{\mu_3-\mu_2}Q_{,2}\omega_{,2}-e^{\mu_2-\mu_3}Q_{,3}\omega_{,3})\bar{\chi}$$
$$+\tfrac{1}{2}e^{\mu_2-\mu_3}(e^{\mu_3-\mu_2}\bar{Q}_{,2}\omega_{,2}-e^{\mu_2-\mu_3}\bar{Q}_{,3}\omega_{,3})\chi$$
$$-e^{\mu_2-\mu_3+\beta}\{[\delta\psi_{,2,3}+\delta\psi_{,2}(\psi-\mu_2)_{,3}+\delta\psi_{,3}(\psi-\mu_3)_{,2}$$
$$-\beta_{,2}\delta\mu_{2,3}-\beta_{,3}\delta\mu_{3,2}]\bar{\chi}$$
$$+[\bar{\delta}\psi_{,2,3}+\bar{\delta}\psi_{,2}(\psi-\mu_2)_{,3}+\bar{\delta}\psi_{,3}(\psi-\mu_3)_{,2}$$
$$-\beta_{,2}\bar{\delta}\mu_{2,3}-\beta_{,3}\bar{\delta}\mu_{3,2}]\chi$$
$$-[\beta_{,2,2}-\beta_{,2}(\mu_3+\mu_2)_{,2}+\psi_{,2}\psi_{,2}+\nu_{,2}\nu_{,2}]\chi\bar{\chi}$$
$$-\tfrac{1}{2}e^{2\psi-2\nu+2\mu_2-2\mu_3}(\omega_{,3})^2\chi\bar{\chi}\}\Big]\!\Big]\,d\varphi\,dx^2\,dx^3$$
$$+\iint\Big[\!\Big[e^{\mu_3-\mu_2+\beta}[\delta\nu_{,2}\bar{\delta}\psi+(\nu+\mu_3)_{,2}\delta(\mu_3-\mu_2)\bar{\delta}\psi-\delta\psi\,\bar{\delta}\mu_{3,2}$$
$$+(\psi+\nu)_{,2}\,\delta\psi\,\bar{\delta}\mu_3+\delta(\psi+\nu)_{,2}\bar{\delta}\mu_3+\beta_{,2}\delta(\mu_3-\mu_2)\bar{\delta}\mu_3]$$
$$+\tfrac{1}{2}\bar{Q}\omega_{,3}[3\delta\psi-\delta(\mu_3-\mu_2)]-\tfrac{1}{2}e^{-3\psi+\nu+\mu_3-\mu_2}Q\bar{Q}_{,2}$$
$$-\tfrac{1}{2}\bar{Q}\omega_{,2}\chi-\tfrac{1}{2}e^{-2\mu_2+2\nu}(e^{\psi-\nu+3\mu_2-\mu_3}\bar{\chi})_{,3}\delta\nu$$
$$+e^{\beta+\mu_2-\mu_3}[(\psi+2\mu_2-\mu_3)_{,3}\bar{\chi}\delta\nu-(\nu+2\mu_2-\mu_3)_{,3}\chi\bar{\delta}\psi$$
$$-\beta_{,3}\bar{\delta}\mu_3\chi+\bar{\chi}_{,3}\delta\nu-\chi_{,3}\bar{\delta}\psi]\Big]\!\Big]_{[x^2]}\,d\varphi\,dx^3$$
$$+\iint\Big[\!\Big[e^{\mu_2-\mu_3+\beta}[\delta\nu_{,3}\bar{\delta}\psi+(\nu+\mu_2)_{,3}\delta(\mu_2-\mu_3)\bar{\delta}\psi-\delta\psi\,\bar{\delta}\mu_{2,3}$$
$$+(\psi+\nu)_{,3}\delta\psi\,\bar{\delta}\mu_2+\delta(\psi+\nu)_{,3}\bar{\delta}\mu_2+\beta_{,3}\delta(\mu_2-\mu_3)\bar{\delta}\mu_2]$$
$$-\tfrac{1}{2}\bar{Q}\omega_{,2}[3\delta\psi+\delta(\mu_3-\mu_2)]-\tfrac{1}{2}e^{-3\psi+\nu+\mu_2-\mu_3}Q\bar{Q}_{,3}$$
$$+\tfrac{1}{2}e^{2\mu_2-2\mu_3}\bar{Q}\omega_{,3}\chi-\tfrac{1}{2}e^{-2\mu_3+2\nu}(e^{\psi-\nu+\mu_2+\mu_3}\bar{\chi})_{,2}\delta\nu$$
$$+e^{\beta+\mu_2-\mu_3}[(\psi+\mu_2)_{,2}\bar{\chi}\delta\nu-(\nu+\mu_2)_{,2}\chi\bar{\delta}\psi-\beta_{,2}\bar{\delta}\mu_2\chi]$$
$$-(e^{\beta+\mu_2-\mu_3}\delta\nu)_{,2}\bar{\chi}-(e^{\beta+\mu_2-\mu_3}\bar{\delta}\psi)_{,2}\chi\Big]\!\Big]_{[x^3]}\,d\varphi\,dx^2. \qquad (340)$$

In the course of the reductions leading to equation (340), use has been made of the further dynamical equations (303), (304), and (305), again, only for the unbarred quantities.

We observe that the volume integral in equation (340) is symmetric in the barred and the unbarred quantities. By subtracting equation (338) from equation (340), we obtain the somewhat simpler equation

$$\begin{aligned}
\sigma^2 \iiint & e^{-2\nu}[2\delta\tau\bar{\delta}\tau + 2\delta\psi\bar{\delta}\psi - 2\delta(\psi+\mu)\bar{\delta}(\psi+\mu) \\
&\quad + \tfrac{1}{2}e^{-4\psi}Q\bar{Q} + \tfrac{1}{2}e^{2\mu_2-2\mu_3}\chi\bar{\chi}]\sqrt{-g}\,d\varphi\,dx^2\,dx^3 \\
= \iiint \Big[\!\Big[& Xe^{3\psi-\nu}(4\delta\psi\bar{\delta}\psi + \delta\tau\bar{\delta}\tau) + 2Ye^{3\psi-\nu}(\delta\psi\bar{\delta}\tau + \bar{\delta}\psi\delta\tau) \\
& - 4U\,\delta\tau\bar{\delta}\tau + 2e^{\beta}(e^{\mu_3-\mu_2}\delta\psi_{,2}\bar{\delta}\psi_{,2} + e^{\mu_2-\mu_3}\delta\psi_{,3}\bar{\delta}\psi_{,3}) \\
& - 2e^{\mu_3-\mu_2+\beta}\{[\beta_{,2}\delta\mu_{3,2} - (\psi-\nu)_{,2}\delta\psi_{,2}]\bar{\delta}\tau \\
& + [\beta_{,2}\bar{\delta}\mu_{3,2} - (\psi-\nu)_{,2}\bar{\delta}\psi_{,2}]\delta\tau\} \\
& + 2e^{\mu_2-\mu_3+\beta}\{[\beta_{,3}\delta\mu_{2,3} - (\psi-\nu)_{,3}\delta\psi_{,3}]\bar{\delta}\tau \\
& + [\beta_{,3}\bar{\delta}\mu_{2,3} - (\psi-\nu)_{,3}\bar{\delta}\psi_{,3}]\delta\tau\} \\
& - e^{\beta}(2S\bar{\delta}\psi + 2\bar{S}\delta\psi + D\bar{\delta}\tau + \bar{D}\delta\tau) \\
& + \tfrac{1}{2}e^{-3\psi+\nu}(e^{\mu_3-\mu_2}Q_{,2}\bar{Q}_{,2} + e^{\mu_2-\mu_3}Q_{,3}\bar{Q}_{,3}) \\
& + \tfrac{1}{2}\{[e^{-2\mu_2+2\nu}(e^{\psi-\nu+3\mu_2-\mu_3}\chi)_{,3}]_{,2} \\
& \quad + [e^{-2\mu_3+2\nu}(e^{\psi-\nu+\mu_2+\mu_3}\chi)_{,2}]_{,3}\}\bar{\delta}\psi \\
& + \tfrac{1}{2}\{[e^{-2\mu_2+2\nu}(e^{\psi-\nu+3\mu_2-\mu_3}\bar{\chi})_{,3}]_{,2} \\
& \quad + [e^{-2\mu_3+2\nu}(e^{\psi-\nu+\mu_2+\mu_3}\bar{\chi})_{,2}]_{,3}\}\delta\psi \\
& + \tfrac{1}{2}e^{\mu_2-\mu_3}(e^{\mu_3-\mu_2}Q_{,2}\omega_{,2} - e^{\mu_2-\mu_3}Q_{,3}\omega_{,3})\bar{\chi} \\
& + \tfrac{1}{2}e^{\mu_2-\mu_3}(e^{\mu_3-\mu_2}\bar{Q}_{,2}\omega_{,2} - e^{\mu_2-\mu_3}\bar{Q}_{,3}\omega_{,3})\chi \\
& - 2e^{3\psi-\nu+\mu_2-\mu_3}\omega_{,2}\omega_{,3}(\bar{\chi}\delta\psi + \chi\bar{\delta}\psi) \\
& - e^{\mu_2-\mu_3+\beta}\{[\delta\psi_{,2,3} + \delta\psi_{,2}(\psi-\mu_2)_{,3} + \delta\psi_{,3}(\psi-\mu_3)_{,2} \\
& \qquad - \beta_{,2}\delta\mu_{2,3} - \beta_{,3}\delta\mu_{3,2}]\bar{\chi} \\
& \qquad + [\bar{\delta}\psi_{,2,3} + \bar{\delta}\psi_{,2}(\psi-\mu_2)_{,3} + \bar{\delta}\psi_{,3}(\psi-\mu_3)_{,2} \\
& \qquad - \beta_{,2}\bar{\delta}\mu_{2,3} - \beta_{,3}\bar{\delta}\mu_{3,2}]\chi \\
& \qquad - [\beta_{,2,2} - \beta_{,2}(\mu_3+\mu_2)_{,2} + \psi_{,2}\psi_{,2} + \nu_{,2}\nu_{,2}]\chi\bar{\chi} \\
& \qquad - \tfrac{1}{2}e^{2\psi-2\nu+2\mu_2-2\mu_3}(\omega_{,3})^2\chi\bar{\chi}\}\Big]\!\Big]\,d\varphi\,dx^2\,dx^3 \\
& + \text{surface integrals,}
\end{aligned} \tag{341}$$

where for the sake of brevity, we have not explicitly written out the surface integrals. Also, in equation (341) we have introduced the abbreviations

$$\delta\mu = \tfrac{1}{2}\delta(\mu_3 + \mu_2) \qquad \text{and} \qquad \delta\tau = \tfrac{1}{2}\delta(\mu_3 - \mu_2). \tag{342}$$

(i) *A variational principle*

In deriving equation (341), we supposed that the trial variations represented by the barred quantities satisfied only the initial-value equations. In view of the symmetry of the integrand of the volume integral in the barred and the unbarred quantities, we now formally identify them to obtain

$$\begin{aligned}
\sigma^2 \iiint & e^{-2\nu}\{2(\delta\tau)^2 + 2(\delta\psi)^2 - 2[\delta(\psi+\mu)]^2 \\
& + \tfrac{1}{2}e^{-4\psi}Q^2 + \tfrac{1}{2}e^{2\mu_2 - 2\mu_3}\chi^2\}\sqrt{-g}\,d\varphi\,dx^2\,dx^3 \\
= \iiint \Big[\!\Big[& Xe^{3\psi-\nu}[4(\delta\psi)^2 + (\delta\tau)^2] + 4Ye^{3\psi-\nu}\delta\psi\,\delta\tau - 4U(\delta\tau)^2 \\
& + 2e^{\beta}[e^{\mu_3-\mu_2}(\delta\psi_{,2})^2 + e^{\mu_2-\mu_3}(\delta\psi_{,3})^2] \\
& - 4e^{\beta}\{e^{\mu_3-\mu_2}[\beta_{,2}\delta\mu_{3,2} - (\psi-\nu)_{,2}\delta\psi_{,2}] \\
& \qquad - e^{\mu_2-\mu_3}[\beta_{,3}\delta\mu_{2,3} - (\psi-\nu)_{,3}\delta\psi_{,3}]\}\,\delta\tau \\
& - 2e^{\beta}(2S\delta\psi + D\delta\tau) + \tfrac{1}{2}e^{-3\psi+\nu}[e^{\mu_3-\mu_2}(Q_{,2})^2 + e^{\mu_2-\mu_3}(Q_{,3})^2] \\
& + \{[e^{-2\mu_2+2\nu}(e^{\psi-\nu+3\mu_2-\mu_3}\chi)_{,3}]_{,2} + [e^{-2\mu_3+2\nu}(e^{\psi-\nu+\mu_2+\mu_3}\chi)_{,2}]_{,3}\}\delta\psi \\
& + e^{\mu_2-\mu_3}(e^{\mu_3-\mu_2}Q_{,2}\omega_{,2} - e^{\mu_2-\mu_3}Q_{,3}\omega_{,3})\chi - 4e^{3\psi-\nu+\mu_2-\mu_3}\omega_{,2}\omega_{,2}\chi\delta\psi \\
& + e^{\mu_2-\mu_3+\beta}\{[\beta_{,2,2} - \beta_{,2}(\mu_3+\mu_2)_{,2} + \psi_{,2}\psi_{,2} + \nu_{,2}\nu_{,2}]\chi^2 \\
& + \tfrac{1}{2}e^{2\psi-2\nu+2\mu_2-2\mu_3}(\omega_{,3})^2\chi^2\} - 2e^{\beta+\mu_2-\mu_3}[\delta\psi_{,2,3} \\
& + \delta\psi_{,2}(\psi-\mu_2)_{,3} + \delta\psi_{,3}(\psi-\mu_3)_{,2} - \beta_{,2}\delta\mu_{2,3} - \beta_{,3}\delta\mu_{3,2}]\chi\Big]\!\Big]\,d\varphi\,dx^2\,dx^3 \\
& + \text{surface integrals.}
\end{aligned} \tag{343}$$

We now consider equation (343) as a *formula* for σ^2 in which the variations are assumed to be consistent only with the initial-value equations and the proper boundary conditions. Suppose now that we evaluate σ^2, successively, with two sets of trial variations, $\delta\psi$, $\delta\nu$, Q, χ, etc., and $\delta\psi + \tfrac{1}{2}\delta^2\psi$, $\delta\nu + \tfrac{1}{2}\delta^2\nu$, $Q + \tfrac{1}{2}\delta Q$, $\chi + \tfrac{1}{2}\delta\chi$, etc., where $\tfrac{1}{2}\delta^2\psi$, $\tfrac{1}{2}\delta^2\nu$, $\tfrac{1}{2}\delta Q$, $\tfrac{1}{2}\delta\chi$, etc., are infinitesimal increments. In other words, we consider the effect on σ^2, given by equation (343), of arbitrary infinitesimal increments in the trial variations that are consistent only with the initial-value equations and the proper boundary conditions. We can readily write down an explicit expression for $\delta\sigma^2$ by subjecting the integrand in equation (343) to the desired variations. We start with the expression for $\delta\sigma^2$ so obtained and trace backwards the reductions that led to equation (341) starting from equations (275), (276), and (339), but with one essential difference: we are not now entitled to use any of the dynamical equations. We find

$$\begin{aligned}\delta\sigma^2 \iiint & e^{-2\nu}\{2(\delta\tau)^2 + 2(\delta\psi)^2 - 2[\delta(\psi+\mu)]^2 \\ & + \tfrac{1}{2}e^{-4\psi}Q^2 + \tfrac{1}{2}e^{2\mu_2-2\mu_3}\chi^2\}\sqrt{-g}\,\mathrm{d}\varphi\,\mathrm{d}x^2\,\mathrm{d}x^3 \\ = -\iiint \Big[\!\Big[& \delta\{[G^{(3)(3)} + G^{(2)(2)}]\sqrt{-g}\}\delta^2\mu + \delta\{[G^{(3)(3)} - G^{(2)(2)}]\sqrt{-g}\}\delta^2\tau \\ & + \delta\{R^{(2)(3)}e^{\mu_2-\mu_3}\sqrt{-g}\}\delta\chi + \delta\{e^{\mu_3+\mu_2}G^{(1)(1)}\}e^{\beta}\delta^2\psi \\ & + \tfrac{1}{2}\{\sigma^2 e^{-3\psi-\nu+\mu_2+\mu_3}Q + (e^{-3\psi+\nu-\mu_2+\mu_3}Q_{,2})_{,2} \\ & + (e^{-3\psi+\nu+\mu_2-\mu_3}Q_{,3})_{,3} + (\omega_{,2}\chi)_{,2} - (e^{2\mu_2-2\mu_3}\omega_{,3}\chi)_{,3} \\ & + [\omega_{,2}(3\delta\psi - \delta\nu + \delta\mu_3 - \delta\mu_2)]_{,3} \\ & - [\omega_{,3}(3\delta\psi - \delta\nu + \delta\mu_2 - \delta\mu_3)]_{,2}\}\,\delta Q \Big]\!\Big]\,\mathrm{d}\varphi\,\mathrm{d}x^2\,\mathrm{d}x^3. \qquad (344)\end{aligned}$$

From this equation it follows that *if we subject a trial variation to arbitrary infinitesimal increments, consistent only with the initial-value equations and the proper boundary conditions, and require that $\delta\sigma^2$ vanish for all such increments, then the dynamical equations of the problem must be satisfied.* In other words, the trial variation must represent a true solution of the problem. This is the *variational principle* governing the solution of the linearized equations.

(ii) *The stability of the Kerr solution to axisymmetric perturbations*

We shall now show how the stability of the Kerr solution to axisymmetric perturbations can be deduced from the variational expression (343) for σ^2. The choice of gauge in which

$$\delta\mu = \tfrac{1}{2}\delta(\mu_3 + \mu_2) = 0 \qquad (345)$$

is of crucial importance in this connection as was first shown by Friedman and Schutz. For, had we worked with a metric of the form (241), the coefficient of σ^2 on the left-hand side of the equation would have been

$$\begin{aligned}\iiint & e^{-2\nu}\{2(\delta\tau)^2 + 2(\delta\psi)^2 - 2[\delta(\psi+\mu)]^2 \\ & + \tfrac{1}{2}e^{-4\psi}Q^2\}\sqrt{-g}\,\mathrm{d}\varphi\,\mathrm{d}x^2\,\mathrm{d}x^3; \qquad (346)\end{aligned}$$

and this expression is not positive-definite. On the other hand, working with the more general metric (244) we can now let $\delta\mu = 0$ by allowing $\chi \neq 0$. With the choice of gauge,

$$\delta\mu = 0 \qquad \text{and} \qquad \chi \neq 0, \qquad (347)$$

the coefficient of σ^2 in the variational expression becomes

$$\iiint e^{-2\nu}[2(\delta\tau)^2+\tfrac{1}{2}e^{-4\psi}Q^2+\tfrac{1}{2}e^{2\mu_2-2\mu_3}\chi^2]\sqrt{-g}\,d\varphi\,dx^2\,dx^3; \quad (348)$$

and this expression *is* positive-definite. In our further considerations, we shall suppose that the gauge (347) has been chosen.

The stability of the Kerr solution to axisymmetric perturbations follows from three basic facts:

(1) a zero-frequency mode is not allowed along the Kerr sequence;
(2) instability, if it sets in, can only be by a purely imaginary mode: $\sigma = -ik$, $k > 0$; and
(3) the imaginary frequencies of the unstable modes (if they exist) must depend continuously on the Kerr parameter a.

That a zero-frequency mode is not allowed, is, of course, assured by Carter's theorem. And as we have seen in § (*d*) the present approach to the perturbation problem provides an independent proof.

To deduce the reality of σ^2 from the variational expression (even with the choice of gauge (347)), we must, first, ensure that the surface integrals, when we let S_1 tend to the horizon and S_2 tend to infinity (cf. remarks in the paragraph preceding equation (338)), do not make any contributions to the expression on the right-hand side of the equation (343). This, however, follows from the general considerations of Chapter 9 which show that in the r_*- variable (which represents a single-valued transformation when $m = 0$ and $r > r_+$) the solutions have the character of outgoing and ingoing waves. This fact, together with the divergence of the coefficient (348) resulting from the occurrence of $e^{-2\nu}$ in the integrand) guarantees that, for unstable modes with a complex σ, the surface integrals can make no contributions. (It should be noted that the ratio of the integral expressions on the two sides of equation (343) is constant for all r.) Therefore, the positive definiteness of the coefficient (348) ensures that if instability sets in, it can only be by a finite, purely imaginary mode (since σ^2 is real).

Since $\sigma = 0$ is forbidden, the spectrum of σ must lie entirely on the positive or entirely on the negative side of the imaginary axis. But we know that for the Schwarzschild solution, with $a = 0$, there are no unstable exponentially growing modes. Consequently, if σ can be shown to depend continuously on the parameter a, then instability for *any* value of a can be excluded.

It would appear that there are theorems which can be used to establish the continuity of $\sigma(a)$; for, the differential operators we are presently considering are self-adjoint and 'densely defined' on the space of square-integrable functions. And, moreover, these operators are analytic in the parameter a. These facts are sufficient to ensure the required continuity of $\sigma(a)$.

We conclude, then, that *the Kerr solution is stable to axisymmetric perturbations.*

The question occurs whether the present preliminary considerations of this section can be usefully extended to the solution of other problems in the mathematical theory of black holes. "This might be the subject of a new story; but our present story has ended."

BIBLIOGRAPHICAL NOTES

§§ 109, 110. What has now come to be called the 'Kerr–Newman' solution was discovered by:

1. E. T. NEWMAN, E. COUCH, K. CHINNAPARED *et al.*, *J. Math. Phys.*, 6, 918–9 (1965).

The solution was obtained by these authors by applying a complex coordinate-transformation to the Reissner–Nordström solution in the manner Newman and Janis (in reference 2 below) had earlier obtained the Kerr solution by applying a similar complex coordinate-transformation to the Schwarzschild solution.

2. E. T. NEWMAN and A. I. JANIS, *J. Math. Phys.*, 6, 915–7 (1965).

The *raison d'etre* for what appeared at first as a 'curious procedure' was later supplied by:

3. E. T. NEWMAN, *J. Math. Phys.*, 14, 774–6 (1972).

The derivation of the Kerr–Newman solution in the text is patterned after the derivation of the Kerr solution in Chapter 6 (§§52–54) and in:

4. S. CHANDRASEKHAR, *Proc. Roy. Soc.* (*London*) A, 358, 405–20 (1978).

The derivations in § 109, in the main, are devoted to obtaining the Ernst equations directly from the Einstein–Maxwell equations written out explicitly for a stationary axisymmetric space-time. The original derivation by Ernst with a different motivation, is contained in:

5. F. J. ERNST, *Phys. Rev.*, 168, 1415–7 (1968).

For the transformation properties of the Ernst equations in §109(*d*) see:

6. B. C. XANTHOPOULOS, *Phys. Letters*, 98B, 377–80 (1981).
7. M. GÜRSES and B. C. XANTHOPOULOS, *J. Phys. A* (in press).

Also:

8. B. K. HARRISON, *J. Math. Phys.*, 9, 1744–52 (1968).

The effect of conjugation on the potential *H*, considered in §109(*e*), was obtained in collaboration with B. C. Xanthopoulos to whom I am grateful for intensive discussions pertaining to the presentation in these sections.

It is, of course, obvious from the formal identity of the Kerr and the Kerr–Newman metrics (apart from the definition of the 'horizon function' Δ) that the null tetrad-basis adopted in Chapter 6, §56, for the description of the Kerr space-time in a Newman–Penrose formalism, and the various expressions for the spin-coefficients obtained in Chapter 6, equations (175) can be used, without any alterations, for the Kerr–Newman space-time. See:

9. S. K. BOSE, *J. Math. Phys.*, 16, 772–5 (1975).

§111. The perturbations of the Kerr–Newman space-time have been considered widely in the literature, for example, in:

10. C. H. LEE, *J. Math. Phys.*, 17, 1226–35 (1976).
11. D. M. CHITRE, *Phys. Rev. D*, 13, 2713–9 (1976).

The treatment in the text follows:

12. S. CHANDRASEKHAR, *Proc. Roy. Soc.* (*London*) A, 358, 421–39 (Appendix B, pp. 437–8) (1978).

But to the extent that the basic equations have not been decoupled or separated, to that extent it may be stated that there has been no progress in this subject.

However, problems which do *not* require the space-time to be perturbed can be treated exactly as they were in the Kerr space-time. This applies in particular to the discussion of the geodesics (in which some additional cases have to be considered because of the different definitions of Δ) and to the separation of Dirac's equation (for which, literally, no changes in the analysis of Chapter 10 are needed). See:

13. M. Calvani, F. de Felice, and L. Nobili, *J. Phys. A*, 13, 3213–9 (1980).
14. F. de Felice, L. Nobili, and M. Calvani, ibid., 3635–41 (1980).
15. D. N. Page, *Phys. Rev. D*, 14, 1509–10 (1976).

§112. The reduction of Einstein's equation, for static axisymmetric vacuum space-times, to Laplace's equation in Euclidian 3-space, was first accomplished by:

16. H. Weyl, *Ann. Phys.*, 54, 117–45 (1917).

For an account, see:

17. J. L. Synge, *Relativity: The General Theory*, 309–17, North-Holland Publishing Co., Amsterdam, 1964.

The resulting solutions have generally been called *Weyl's solutions* in the literature. It is, of course, clear that the reduction of the equations to Laplace's equation can be accomplished in a variety of ways by different choices of gauge and coordinates. In Weyl's original reduction (commonly used), Schwarzschild's solution appears with its event horizon as a coordinate line-singularity along the axis. For the purposes of treating distorted Schwarzschild black-holes, one might wish for a reduction in which Schwarzschild's solution appears as Schwarzschild's solution. A gauge and a coordinate system appropriate for this consummation are adopted in the text.

Distorted static black-holes, derived from Weyl's general solution, have been considered in the literature by several authors, including:

18. W. Israel and K. A. Khan, *Nuovo Cim.*, 33, 331–44 (1964).

and

19. P. C. Peters, *J. Math. Phys.*, 20, 1481–5 (1979).

But the most complete discussion is due to:

20. R. Geroch and J. B. Hartle, *J. Math. Phys.*, 23, 680–92 (1982).

This last paper clarifies many aspects concerning these black holes including those related to Hawking radiation. The author is indebted to James Hartle for discussions pertaining to many of these issues, and most particularly for his emphasizing the importance of the condition of local flatness along the axis.

§113. The solution considered in this section was discovered independently by:

21. S. D. Majumdar, *Phys. Rev.*, 72, 390–8 (1947).

and

22. A. Papapetrou, *Proc. Roy. Irish Acad.*, 51, 191–205 (1947).

The correct interpretation of these solutions is due to:

23. J. B. Hartle and S. W. Hawking, *Commun. Math. Phys.*, 26, 87–101 (1972).

The discussion in §(*c*) is, in the main, taken from reference 23.

The stationary generalization of the Majumdar–Papapetrou solution was discovered, independently, by:

24. W. Israel and G. A. Wilson, *J. Math. Phys.*, 13, 865–7 (1972).

and

25. Z. Perjés, *Phys. Rev. Letters*, 27, 1668–70 (1971).

These solutions are discussed by Hartle and Hawking (reference 23).

§114. The basic elements of the theory outlined in this section were developed by Chandrasekhar and Friedman in a series of papers in the larger context of the stability of uniformly rotating stars in the framework of general relativity:

26. S. Chandrasekhar and J. L. Friedman, *Astrophys. J.*, 175, 379–405 (1972).

27. Ibid., 176, 745–68 (1972).

28. Ibid., 177, 745–56 (1972).

29. Ibid., 181, 481–95 (1973).

One of these papers (Paper 28) was devoted to vacuum solutions and Carter's theorem. But all of them were based on a non-stationary metric of the form (241).

The crucial importance of including a term in g_{23} in the metric for inferring the reality of σ^2 for the unstable modes belonging to axisymmetric perturbations was first recognized by Friedman and Schutz. Indeed, it provided the basis for their proof of the stability of the Kerr solution for axisymmetric perturbations:

30. J. L. Friedman and B. F. Schutz, Jr., *Phys. Rev. Letters*, 32, 243–5 (1973).

Since Friedman and Schutz have not published any details of their analysis the theory is developed *ab initio* in the text. In the main, the analysis follows along the lines of Papers 26 and 27; but it is carried out for the more general metric (244).

The proof of Carter's theorem in §(*d*) is the same as that given in Paper 28; and the basic arguments of §(*e*) for the stability of the Kerr solution (§(*e*), 2) for axisymmetric perturbations are those of Friedman and Schutz (Paper 30). However, Friedman and Schutz inferred the continuity of σ on the Kerr parameter a by appealing to scattering theory as applied to the Teukolsky equations, and proving a theorem less general than (but sufficient for their purposes) the ones proved by Hartle and Wilkins:

31. J. B. Hartle and D. C. Wilkins, *Commun. Math. Phys.*, 38, 47–63 (1974).

As we have indicated in the text, one can avoid appealing to theorems in scattering theory by going directly to the theorems which establish that the characteristic values of a self-adjoint differential operator, depending analytically on a parameter and defined densely on the space of square-integrable functions, are continuous in the parameter; see:

32. F. Riesz and B. Sz.-Nagy, *Functional Analysis*, §136, 373–9, Blackie & Son Limited, London, 1956.

I am greatly indebted to John L. Friedman for advice and helpful discussions on all aspects of the matter dealt with in this section.

APPENDIX

TABLES OF TEUKOLSKY AND ASSOCIATED FUNCTIONS

In Chapters 7, 9, and 10, we have seen how the reaction of the Kerr black-hole to incident electromagnetic, gravitational, and massless neutrino-waves can be fully described in terms of Teukolsky's radial and angular functions belonging to the different spins. Apart from the papers by Breuer, Ryan, Jr., and Waller and by Fackerell and Crossman, there have been no serious attempts to study these functions as befits their place among the special functions of mathematical physics; and none to the study of the functions which emerge from the separation of Dirac's equation with the additional feature of the characteristic-value parameter depending explicitly on the mass of the particle.

Eventually, one hopes to understand the origin of the remarkable identities and properties that were found during the course of solving the linearized Newman–Penrose equations, in their entirety, in Chapter 9. But as yet, there are no clues.

Since most of the identities and relations derived in Chapter 9 were the results of very lengthy reductions, it was felt useful to check them (as they were) by direct numerical evaluations to ensure against inadvertent algebraic or other errors. For these and other purposes, Dr. Steven Detweiler kindly undertook the task of tabulating the radial and the angular functions belonging to $s = \pm 2$ and $s = \pm 1$ for two typical cases, $a = 0.95$ and $\sigma = 0.5$ and 0.25, and for $a = 0$. These tables are included in this Appendix, since they may be similarly useful to others who may become interested in this area of analysis.

The radial and the angular functions belonging to spin-2, included in Tables I and II, are in the normalizations used in the text: the radial functions, belonging to $s = +2$ and $s = -2$, so as to be consistent with complex-conjugate Starobinsky relations (Chapter 9, equations (41) and (42)) and the angular functions normalized to unity. The functions $\mathscr{R}$ and $\mathscr{S}$ are defined in Chapter 9, equations (178), (179), (183), and (184); and the various bracket expressions in §81(b). The various constants (Chapter 9, equation (161)) associated with the tabulated solutions are:

$$\begin{array}{lll} a = & 0.95; & l = -m = 2 \\ \sigma = & 0.5 & 0.25 \\ \lambda\!\!\!^{-} = & 0.88693 & 2.4308 \end{array}$$

TABLE I

The radial functions for $s = \pm 2$ ($a = 0.95$, $\sigma = 0.5$; $l = -m = 2$)

r/M	$[P]^+$	$[DP]^+$	$[DDP]^+$	$[P]^-$	$[DP]^-$	$[DDP]^-$	$\mathscr{R}$
2.00	2.00000	−1.07940	5.38789	0.000000	4.75963	25.1627	0.865586
2.10	1.87877	−0.926641	5.46889	0.727147	7.29748	26.9098	1.02641
2.20	1.70012	−0.990398	5.40319	1.71762	10.0176	28.8345	1.25339
2.30	1.41550	−1.30720	5.15534	2.97871	12.9180	30.8465	1.54821
2.40	0.973712	−1.91092	4.70475	4.51446	15.9838	32.8803	1.91043
2.50	0.321416	−2.83272	4.04018	6.32375	19.1910	34.8845	2.33813
2.60	−0.596200	−4.10067	3.15689	8.39906	22.5081	36.8177	2.82829
2.70	−1.83470	−5.73938	2.05492	10.7257	25.8982	38.6449	3.37695
2.80	−3.44959	−7.76962	0.738136	13.2813	29.3198	40.3359	3.97938
2.90	−5.49545	−10.2079	−0.786504	16.0357	32.7278	41.8645	4.63018
3.00	−8.02499	−13.0665	−2.50942	18.9503	36.0743	43.2076	5.32336
3.10	−11.0882	−16.3526	−4.41878	21.9788	39.3091	44.3451	6.05242
3.20	−14.7313	−20.0688	−6.50082	25.0668	42.3805	45.2592	6.81043
3.30	−18.9959	−24.2125	−8.74010	28.1522	45.2362	45.9345	7.59012
3.40	−23.9181	−28.7763	−11.1197	31.1658	47.8229	46.3580	8.38391
3.50	−29.5276	−33.7473	−13.6215	34.0314	50.0880	46.5186	9.18399
3.60	−35.8469	−39.1078	−16.2264	36.6664	51.9795	46.4074	9.98243
3.70	−42.8904	−44.8347	−18.9142	38.9830	53.4463	46.0175	10.7712
3.80	−50.6637	−50.9003	−21.6643	40.8882	54.4393	45.3439	11.5421
3.90	−59.1632	−57.2717	−24.4554	42.2851	54.9116	44.3837	12.2873
4.00	−68.3750	−63.9116	−27.2658	43.0735	54.8187	43.1359	12.9987
4.10	−78.2751	−70.7783	−30.0737	43.1511	54.1192	41.6011	13.6686
4.20	−88.8282	−77.8255	−32.8571	42.4146	52.7752	39.7819	14.2894
4.30	−99.9879	−85.0035	−35.5943	40.7602	50.7526	37.6828	14.8537
4.40	−111.697	−92.2585	−38.2636	38.0854	48.0216	35.3097	15.3545
4.50	−123.884	−99.5336	−40.8436	34.2896	44.5567	32.6704	15.7852
4.60	−136.470	−106.769	−43.3135	29.2756	40.3373	29.7742	16.1394
4.70	−149.362	−113.902	−45.6531	22.9506	35.3479	26.6320	16.4114
4.80	−162.455	−120.869	−47.8427	15.2274	29.5783	23.2561	16.5958
4.90	−175.634	−127.602	−49.8635	6.02563	23.0237	19.6602	16.6878
5.00	−188.773	−134.035	−51.6977	−4.72704	15.6853	15.8593	16.6831
5.10	−201.737	−140.099	−53.3283	−17.0940	7.56960	11.8797	16.5780
5.20	−214.379	−145.725	−54.7395	−31.1282	−1.31055	7.70867	16.3695
5.30	−226.545	−150.844	−55.9167	−46.8715	−10.9365	3.39469	16.0551
5.40	−238.072	−155.388	−56.8463	−64.3534	−21.2837	−1.05287	15.6329
5.50	−248.791	−159.289	−57.5163	−83.5902	−32.3215	−5.61379	15.1018
5.60	−258.525	−162.482	−57.9157	−104.584	−44.0132	−10.2671	14.4613
5.70	−267.094	−164.904	−58.0353	−127.321	−56.3166	−14.9912	13.7117
5.80	−274.312	−166.491	−57.8670	−151.774	−69.1834	−19.7639	12.8537
5.90	−279.992	−167.188	−57.4045	−177.897	−82.5601	−24.5625	11.8891
6.00	−283.945	−166.937	−56.6427	−205.629	−96.3875	−29.3643	10.8199
6.10	−285.982	−165.687	−55.5784	−234.890	−110.601	−34.1459	9.64910
6.20	−285.917	−163.392	−54.2096	−265.583	−125.133	−38.8841	8.38033
6.30	−283.564	−160.008	−52.5362	−297.593	−139.907	−43.5556	7.01777
6.40	−278.745	−155.498	−50.5593	−330.788	−154.848	−48.1373	5.56628
6.50	−271.285	−149.830	−48.2820	−365.016	−169.872	−52.6061	4.03132
6.60	−261.018	−142.976	−45.7085	−400.109	−184.895	−56.9393	2.41895
6.70	−247.788	−134.917	−42.8450	−435.880	−199.827	−61.1148	0.735764
6.80	−231.449	−125.637	−39.6987	−472.126	−214.577	−65.1107	−1.01110
6.90	−211.867	−115.128	−36.2788	−508.626	−229.052	−68.9059	−2.81401

The quantities $[P]^-$, $[DP]^-$ and $[DDP]^-$ are all purely imaginary and the entries in the respective columns should be understood to have the factor i.

TABLE I, Continued
($a = 0.95$, $\sigma = 0.25$; $l = -m = 2$)

r/M	$[P]^+$	$[DP]^+$	$[DDP]^+$	$[P]^-$	$[DP]^-$	$[DDP]^-$	$\mathcal{R}$
2.00	2.00000	15.9355	45.1217	0.000000	−2.95809	23.9078	−5.59002
2.10	3.80931	20.6317	51.7891	−0.402230	−1.48548	28.1807	−6.18269
2.20	6.11835	26.1438	59.0316	−0.658754	0.583079	32.9755	−6.77236
2.30	9.02331	32.4815	66.7997	−0.676982	3.32893	38.3038	−7.33757
2.40	12.6134	39.6531	75.0438	−0.356858	6.83718	44.1730	−7.85760
2.50	16.9716	47.6631	83.7152	0.411260	11.1946	50.5885	−8.31243
2.60	22.1750	56.5116	92.7676	1.74745	16.4885	57.5543	−8.68270
2.70	28.2943	66.1930	102.156	3.78248	22.8062	65.0734	−8.94952
2.80	35.3937	76.6962	111.838	6.65774	30.2342	73.1480	−9.09449
2.90	43.5302	88.0043	121.770	10.5250	38.8584	81.7790	−9.09959
3.00	52.7529	100.094	131.913	15.5461	48.7633	90.9666	−8.94717
3.10	63.1025	112.938	142.226	21.8926	60.0320	100.710	−8.61998
3.20	74.6100	126.500	152.670	29.7451	72.7458	111.007	−8.10110
3.30	87.2969	140.740	163.206	39.2928	86.9840	121.855	−7.37402
3.40	101.173	155.613	173.797	50.7332	102.824	133.250	−6.42263
3.50	116.239	171.067	184.406	64.2710	120.340	145.186	−5.23120
3.60	132.480	187.044	194.996	80.1179	139.604	157.659	−3.78450
3.70	149.872	203.483	205.531	98.4917	160.685	170.660	−2.06771
3.80	168.374	220.317	215.976	119.615	183.648	184.181	−0.0665556
3.90	187.934	237.471	226.297	143.717	208.557	198.214	2.23276
4.00	208.485	254.871	236.458	171.027	235.469	212.748	4.84347
4.10	229.943	272.433	246.428	201.782	264.440	227.772	7.77822
4.20	252.210	290.071	256.172	236.217	295.521	243.274	11.0491
4.30	275.174	307.695	265.660	274.571	328.757	259.240	14.6675
4.40	298.702	325.209	274.859	317.081	364.193	275.657	18.6443
4.50	322.650	342.517	283.738	363.987	401.863	292.510	22.9895
4.60	346.852	359.515	292.269	415.523	441.803	309.783	27.7126
4.70	371.129	376.097	300.422	471.925	484.040	327.458	32.8223
4.80	395.282	392.155	308.168	533.421	528.597	345.519	38.3264
4.90	419.095	407.577	315.479	600.239	575.491	363.947	44.2323
5.00	442.336	422.249	322.330	672.597	624.735	382.722	50.5462
5.10	464.753	436.053	328.693	750.710	676.336	401.825	57.2738
5.20	486.079	448.870	334.545	834.784	730.294	421.235	64.4199
5.30	506.028	460.578	339.860	925.017	786.606	440.930	71.9883
5.40	524.296	471.055	344.616	1021.59	845.261	460.888	79.9822
5.50	540.563	480.175	348.789	1124.70	906.243	481.085	88.4037
5.60	554.493	487.814	352.360	1234.48	969.529	501.500	97.2543
5.70	565.731	493.843	355.307	1351.11	1035.09	522.107	106.534
5.80	573.907	498.135	357.610	1474.72	1102.90	542.881	116.243
5.90	578.635	500.563	359.252	1605.43	1172.90	563.797	126.379
6.00	579.513	500.997	360.215	1743.35	1245.06	584.830	136.941
6.10	576.125	499.311	360.483	1888.57	1319.33	605.954	147.925
6.20	568.038	495.375	360.039	2041.15	1395.63	627.140	159.326
6.30	554.809	489.063	358.871	2201.17	1473.91	648.363	171.141
6.40	535.979	480.248	356.964	2368.64	1554.10	669.595	183.361
6.50	511.077	468.806	354.307	2543.58	1636.11	690.808	195.982
6.60	479.622	454.612	350.887	2725.99	1719.87	711.973	208.994
6.70	441.119	437.545	346.697	2915.82	1805.28	733.062	222.389
6.80	395.066	417.486	341.726	3113.02	1892.24	754.047	236.157
6.90	340.951	394.316	335.967	3317.52	1980.66	774.898	250.288

The quantities $[P]^-$, $[DP]^-$ and $[DDP]^-$ are all purely imaginary and the entries in the respective columns should be understood to have the factor i.

Table II
The normalized angular functions for $s = \pm 2$
$(a = 0.95, \sigma = 0.5; l = -m = 2)$

$\cos\theta$	S_{+2}	S_{-2}	$[S]^+$	$[\mathscr{L}S]^+$	$[\mathscr{L}\mathscr{L}S]^+$	$[S]^-$	$[\mathscr{L}S]^-$	$[\mathscr{L}\mathscr{L}S]^-$	$\mathscr{S}$
0.96	1.66329	0.00037	1.66366	−0.87863	0.55703	1.66292	−0.89925	0.12516	0.13581
0.92	1.57437	0.00149	1.57586	−1.17445	1.08767	1.57287	−1.23281	0.23460	0.18038
0.88	1.48894	0.00340	1.49234	−1.35688	1.59237	1.48553	−1.46413	0.32906	0.20711
0.84	1.40690	0.00612	1.41303	−1.47481	2.07158	1.40077	−1.63997	0.40925	0.22376
0.80	1.32816	0.00970	1.33785	−1.54843	2.52573	1.31846	−1.77926	0.47591	0.23358
0.76	1.25261	0.01414	1.26675	−1.58874	2.95522	1.23847	−1.89212	0.52974	0.23833
0.72	1.18017	0.01949	1.19966	−1.60267	3.36043	1.16068	−1.98481	0.57143	0.23915
0.68	1.11074	0.02578	1.13652	−1.59501	3.74172	1.08496	−2.06159	0.60167	0.23680
0.64	1.04423	0.03304	1.07728	−1.56926	4.09942	1.01119	−2.12551	0.62116	0.23185
0.60	0.98056	0.04132	1.02188	−1.52815	4.43387	0.93924	−2.17890	0.63056	0.22475
0.56	0.91965	0.05064	0.97028	−1.47384	4.74535	0.86901	−2.22354	0.63053	0.21583
0.52	0.86140	0.06104	0.92244	−1.40809	5.03415	0.80036	−2.26086	0.62173	0.20538
0.48	0.80574	0.07256	0.87830	−1.33241	5.30053	0.73319	−2.29204	0.60483	0.19360
0.44	0.75260	0.08524	0.83784	−1.24806	5.54472	0.66737	−2.31801	0.58046	0.18072
0.40	0.70190	0.09912	0.80101	−1.15615	5.76695	0.60278	−2.33959	0.54927	0.16687
0.36	0.65356	0.11424	0.76779	−1.05767	5.96741	0.53932	−2.35743	0.51189	0.15222
0.32	0.60751	0.13064	0.73815	−0.95350	6.14629	0.47687	−2.37209	0.46895	0.13687
0.28	0.56368	0.14837	0.71205	−0.84443	6.30375	0.41531	−2.38405	0.42108	0.12093
0.24	0.52200	0.16747	0.68948	−0.73120	6.43992	0.35453	−2.39369	0.36891	0.10450
0.20	0.58242	0.18799	0.67041	−0.61449	6.55494	0.29442	−2.40134	0.31306	0.08767
0.16	0.44485	0.20998	0.65483	−0.49493	6.64890	0.23487	−2.40725	0.25414	0.07051
0.12	0.40925	0.23348	0.64273	−0.37313	6.72189	0.17577	−2.41164	0.19277	0.05310
0.08	0.37554	0.25855	0.63410	−0.24968	6.77398	0.11699	−2.41466	0.12956	0.03550
0.04	0.34368	0.28524	0.62892	−0.12513	6.80522	0.05844	−2.41643	0.06513	0.01778

Table II, Continued
$(a = 0.95, \sigma = 0.25; l = -m = 2)$

$\cos\theta$	S_{+2}	S_{-2}	$[S]^+$	$[\mathscr{L}S]^+$	$[\mathscr{L}\mathscr{L}S]^+$	$[S]^-$	$[\mathscr{L}S]^-$	$[\mathscr{L}\mathscr{L}S]^-$	$\mathscr{S}$
0.96	1.59016	0.00049	1.59064	−0.86289	0.63892	1.58967	−0.89008	0.06868	0.13895
0.92	1.51587	0.00196	1.51782	−1.15641	1.25074	1.51391	−1.23295	0.12890	0.18550
0.88	1.44380	0.00443	1.44823	−1.33936	1.83560	1.43937	−1.47926	0.18102	0.21407
0.84	1.37393	0.00792	1.38185	−1.45925	2.39362	1.36601	−1.67351	0.22539	0.23241
0.80	1.30621	0.01246	1.31867	−1.53558	2.92490	1.29375	−1.83339	0.26239	0.24375
0.76	1.24061	0.01805	1.25866	−1.57897	3.42958	1.22256	−1.96822	0.29238	0.24983
0.72	1.17709	0.02473	1.20182	−1.59610	3.90774	1.15237	−2.08370	0.31570	0.25177
0.68	1.11563	0.03250	1.14813	−1.59156	4.35949	1.08313	−2.18360	0.33272	0.25032
0.64	1.05618	0.04140	1.09758	−1.56876	4.78492	1.01478	−2.27064	0.34380	0.24606
0.60	0.99872	0.05144	1.05016	−1.53031	5.18413	0.94729	−2.34681	0.34929	0.23941
0.56	0.94321	0.06263	1.00585	−1.47832	5.55719	0.88058	−2.41365	0.34954	0.23072
0.52	0.88962	0.07502	0.96464	−1.41451	5.90419	0.81461	−2.47239	0.34491	0.22015
0.48	0.83792	0.08860	0.92653	−1.34036	6.22519	0.74932	−2.52400	0.33575	0.20827
0.44	0.78808	0.10341	0.89149	−1.25712	6.52026	0.68467	−2.56927	0.32241	0.19496
0.40	0.74006	0.11947	0.85954	−1.16592	6.78947	0.62059	−2.60887	0.30525	0.18049
0.36	0.69384	0.13681	0.83065	−1.06775	7.03287	0.55704	−2.64333	0.28460	0.16503
0.32	0.64939	0.15543	0.80482	−0.96350	7.25052	0.49396	−2.67313	0.26083	0.14870
0.28	0.60667	0.17538	0.78205	−0.85400	7.44245	0.43129	−2.69863	0.23428	0.13163
0.24	0.56566	0.19666	0.76232	−0.74002	7.60871	0.36900	−2.72015	0.20530	0.11394
0.20	0.52633	0.21931	0.74564	−0.62227	7.74933	0.30702	−2.73795	0.17424	0.09572
0.16	0.48865	0.24335	0.73199	−0.50144	7.86435	0.24530	−2.75224	0.14145	0.07708
0.12	0.45259	0.26880	0.72138	−0.37818	7.95378	0.18379	−2.76319	0.10726	0.05781
0.08	0.41813	0.29568	0.71381	−0.25310	8.01765	0.12244	−2.77092	0.07204	0.03887
0.04	0.38523	0.32404	0.70927	−0.12683	8.05596	0.06119	−2.77553	0.03612	0.01948

TABLE III

The radial functions for $a = 0$, $s = \pm 2$

($\sigma = 0.5$; $l = -m = 2$)

r/M	$[P]^+$	$[DP]^+$	$[DPP]^+$	$[P]^-$	$[DP]^-$	$[DDP]^-$	$\mathfrak{R}$
2.10	1.66046	3.05695	−1.64363	1.25666	−1.31413	−4.21285	−0.836984
2.20	0.691547	3.23681	1.12446	2.14365	0.848822	−1.11410	−0.366649
2.30	−0.0598288	2.72551	2.51969	2.44870	2.22828	1.09792	0.0329922
2.40	−0.672384	2.09362	3.23429	2.60223	3.24104	2.69108	0.357504
2.50	−1.22137	1.43940	3.56968	2.71122	4.05816	3.89029	0.629729
2.60	−1.75141	0.779221	3.67008	2.80879	4.75533	4.82288	0.864907
2.70	−2.28981	0.111032	3.61248	2.90446	5.36838	5.56265	1.07231
2.80	−2.85426	−0.570955	3.44184	2.99841	5.91439	6.15443	1.25756
2.90	−3.45673	−1.27231	3.18648	3.08650	6.40083	6.62671	1.42407
3.00	−4.10555	−1.99734	2.86547	3.16232	6.82987	6.99828	1.57389
3.10	−4.80645	−2.74888	2.49250	3.21803	7.20064	7.28197	1.70817
3.20	−5.56319	−3.52837	2.07795	3.24485	7.51049	7.48683	1.82749
3.30	−6.37787	−4.33594	1.63012	3.23337	7.75577	7.61947	1.93205
3.40	−7.25114	−5.17060	1.15600	3.17363	7.93227	7.68486	2.02181
3.50	−8.18227	−6.03031	0.661616	3.05535	8.03553	7.68697	2.09654
3.60	−9.16922	−6.91210	0.152402	2.86801	8.06112	7.62902	2.15593
3.70	−10.2087	−7.81218	−0.366673	2.60095	8.00472	7.51381	2.19960
3.80	−11.2960	−8.72601	−0.890978	2.24353	7.86229	7.34388	2.22716
3.90	−12.4253	−9.64835	−1.41615	1.78522	7.63019	7.12160	2.23823
4.00	−13.5895	−10.5734	−1.93804	1.21572	7.30517	6.84927	2.23244
4.10	−14.7801	−11.4948	−2.45271	0.525111	6.88455	6.52920	2.20947
4.20	−15.9876	−12.4057	−2.95639	−0.296069	6.36618	6.16373	2.16905
4.30	−17.2009	−13.2990	−3.44548	−1.25666	5.74853	5.75525	2.11101
4.40	−18.4082	−14.1670	−3.91656	−2.36476	5.03072	5.30626	2.03520
4.50	−19.5962	−15.0019	−4.36636	−3.62761	4.21255	4.81933	1.94161
4.60	−20.7507	−15.7958	−4.79182	−5.05148	3.29452	4.29717	1.83027
4.70	−21.8564	−16.5404	−5.19003	−6.64151	2.27787	3.74255	1.70135
4.80	−22.8971	−17.2276	−5.55829	−8.40164	1.16455	3.15838	1.55508
4.90	−23.8558	−17.8490	−5.89407	−10.3345	−0.0427275	2.54766	1.39181
5.00	−24.7147	−18.3965	−6.19508	−12.4411	−1.34051	1.91347	1.21198
5.10	−25.4554	−18.8619	−6.45920	−14.7213	−2.72459	1.25898	1.01616
5.20	−26.0588	−19.2375	−6.68454	−17.1729	−4.19003	0.587458	0.804980
5.30	−26.5058	−19.5154	−6.86944	−19.7923	−5.73117	−0.0977922	0.579209
5.40	−26.7766	−19.6885	−7.01245	−22.5740	−7.34165	−0.793399	0.339700
5.50	−26.8516	−19.7496	−7.11234	−25.5107	−9.01439	−1.49595	0.0874053
5.60	−26.7112	−19.6923	−7.16814	−28.5930	−10.7417	−2.20202	−0.176624
5.70	−26.3358	−19.5104	−7.17908	−31.8100	−12.5151	−2.90813	−0.451246
5.80	−25.7065	−19.1983	−7.14466	−35.1485	−14.3259	−3.61086	−0.735232
5.90	−24.8047	−18.7512	−7.06459	−38.5935	−16.1643	−4.30674	−1.02727
6.00	−23.6124	−18.1645	−6.93883	−42.1279	−18.0205	−4.99236	−1.32596
6.10	−22.1129	−17.4348	−6.76755	−45.7328	−19.8839	−5.66436	−1.62985
6.20	−20.2901	−16.5589	−6.55119	−49.3874	−21.7436	−6.3194	−1.93741
6.30	−18.1295	−15.5346	−6.29038	−53.0689	−23.5884	−6.95425	−2.24704
6.40	−15.6177	−14.3606	−5.98600	−56.7528	−25.4066	−7.56572	−2.55711
6.50	−12.7430	−13.0363	−5.63914	−60.4128	−27.1864	−8.15074	−2.86593
6.60	−9.49563	−11.5617	−5.25110	−64.0211	−28.9157	−8.70633	−3.17179
6.70	−5.86740	−9.93821	−4.82339	−67.5483	−30.5824	−9.22965	−3.47291
6.80	−1.85241	−8.16761	−4.35771	−70.9635	−32.1743	−9.71797	−3.76755
6.90	2.55311	−6.25284	−3.85597	−74.2347	−33.6790	−10.1687	−4.05390
7.00	7.35054	−4.19766	−3.32025	−77.3286	−35.0845	−10.5795	−4.33020

The quantities $[P]^-$, $[DP]^-$ and $[DDP]^-$ are all purely imaginary and the entries in the respective columns should be understood to have the factor i.

TABLE III
The radial functions for $a = 0$, $s = \pm 2$
($\sigma = 0.5$; $l = -m = 2$)

r/M	$[P]^+$	$[DP]^+$	$[DPP]^+$	$[P]^-$	$[DP]^-$	$[DDP]^-$	$\mathscr{R}$
2.10	1.66046	3.05695	−1.64363	1.25666	−1.31413	−4.21285	−0.836984
2.20	0.691547	3.23681	1.12446	2.14365	0.848822	−1.11410	−0.366649
2.30	−0.0598288	2.72551	2.51969	2.44870	2.22828	1.09792	0.0329922
2.40	−0.672384	2.09362	3.23429	2.60223	3.24104	2.69108	0.357504
2.50	−1.22137	1.43940	3.56968	2.71122	4.05816	3.89029	0.629729
2.60	−1.75141	0.779221	3.67008	2.80879	4.75533	4.82288	0.864907
2.70	−2.28981	0.111032	3.61248	2.90446	5.36838	5.56265	1.07231
2.80	−2.85426	−0.570955	3.44184	2.99841	5.91439	6.15443	1.25756
2.90	−3.45673	−1.27231	3.18648	3.08650	6.40083	6.62671	1.42407
3.00	−4.10555	−1.99734	2.86547	3.16232	6.82987	6.99828	1.57389
3.10	−4.80645	−2.74888	2.49250	3.21803	7.20064	7.28197	1.70817
3.20	−5.56319	−3.52837	2.07795	3.24485	7.51049	7.48683	1.82749
3.30	−6.37787	−4.33594	1.63012	3.23337	7.75577	7.61947	1.93205
3.40	−7.25114	−5.17060	1.15600	3.17363	7.93227	7.68486	2.02181
3.50	−8.18227	−6.03031	0.661616	3.05535	8.03553	7.68697	2.09654
3.60	−9.16922	−6.91210	0.152402	2.86801	8.06112	7.62902	2.15593
3.70	−10.2087	−7.81218	−0.366673	2.60095	8.00472	7.51381	2.19960
3.80	−11.2960	−8.72601	−0.890978	2.24353	7.86229	7.34388	2.22716
3.90	−12.4253	−9.64835	−1.41615	1.78522	7.63019	7.12160	2.23823
4.00	−13.5895	−10.5734	−1.93804	1.21572	7.30517	6.84927	2.23244
4.10	−14.7801	−11.4948	−2.45271	0.525111	6.88455	6.52920	2.20947
4.20	−15.9876	−12.4057	−2.95639	−0.296069	6.36618	6.16373	2.16905
4.30	−17.2009	−13.2990	−3.44548	−1.25666	5.74853	5.75525	2.11101
4.40	−18.4082	−14.1670	−3.91656	−2.36476	5.03072	5.30626	2.03520
4.50	−19.5962	−15.0019	−4.36636	−3.62761	4.21255	4.81933	1.94161
4.60	−20.7507	−15.7958	−4.79182	−5.05148	3.29452	4.29717	1.83027
4.70	−21.8564	−16.5404	−5.19003	−6.64151	2.27787	3.74255	1.70135
4.80	−22.8971	−17.2276	−5.55829	−8.40164	1.16455	3.15838	1.55508
4.90	−23.8558	−17.8490	−5.89407	−10.3345	−0.0427275	2.54766	1.39181
5.00	−24.7147	−18.3965	−6.19508	−12.4411	−1.34051	1.91347	1.21198
5.10	−25.4554	−18.8619	−6.45920	−14.7213	−2.72459	1.25898	1.01616
5.20	−26.0588	−19.2375	−6.68454	−17.1729	−4.19003	0.587458	0.804980
5.30	−26.5058	−19.5154	−6.86944	−19.7923	−5.73117	−0.0977922	0.579209
5.40	−26.7766	−19.6885	−7.01245	−22.5740	−7.34165	−0.793399	0.339700
5.50	−26.8516	−19.7496	−7.11234	−25.5107	−9.01439	−1.49595	0.0874053
5.60	−26.7112	−19.6923	−7.16814	−28.5930	−10.7417	−2.20202	−0.176624
5.70	−26.3358	−19.5104	−7.17908	−31.8100	−12.5151	−2.90813	−0.451246
5.80	−25.7065	−19.1983	−7.14466	−35.1485	−14.3259	−3.61086	−0.735232
5.90	−24.8047	−18.7512	−7.06459	−38.5935	−16.1643	−4.30674	−1.02727
6.00	−23.6124	−18.1645	−6.93883	−42.1279	−18.0205	−4.99236	−1.32596
6.10	−22.1129	−17.4348	−6.76755	−45.7328	−19.8839	−5.66436	−1.62985
6.20	−20.2901	−16.5589	−6.55119	−49.3874	−21.7436	−6.3194	−1.93741
6.30	−18.1295	−15.5346	−6.29038	−53.0689	−23.5884	−6.95425	−2.24704
6.40	−15.6177	−14.3606	−5.98600	−56.7528	−25.4066	−7.56572	−2.55711
6.50	−12.7430	−13.0363	−5.63914	−60.4128	−27.1864	−8.15074	−2.86593
6.60	−9.49563	−11.5617	−5.25110	−64.0211	−28.9157	−8.70633	−3.17179
6.70	−5.86740	−9.93821	−4.82339	−67.5483	−30.5824	−9.22965	−3.47291
6.80	−1.85241	−8.16761	−4.35771	−70.9635	−32.1743	−9.71797	−3.76755
6.90	2.55311	−6.25284	−3.85597	−74.2347	−33.6790	−10.1687	−4.05390
7.00	7.35054	−4.19766	−3.32025	−77.3286	−35.0845	−10.5795	−4.33020

The quantities $[P]^-$, $[DP]^-$ and $[DDP]^-$ are all purely imaginary and the entries in the respective columns should be understood to have the factor i.

$$\alpha^2 = -2.8975 \qquad -6.6975$$
$$\sigma(\lambda\alpha^2 - 6a^2) = -3.99244 \qquad -5.42382$$
$$\Gamma_1 = 11.2530 \qquad 15.7935$$
$$\mathscr{C}_1 = 12.4482 \qquad 17.3831$$
$$\mathscr{C}_2(= -\Gamma_2) = -6 \qquad -3$$

In Tables III and IV the functions for $a = 0$, for the Schwarzschild limit, are tabulated.

Tables V and VI similarly provide the radial and the angular functions (in their standard normalizations) for spin-1. The constants associated with the tabulated solutions are:

$$a = 0.95; \qquad l = -m = +2$$
$$\sigma = 0.5 \qquad 0.25$$
$$\lambda = 3.9317 \qquad 4.9294$$
$$\alpha^2 = -2.8975 \qquad -6.6925$$
$$\mathscr{C} = D = 4.2844 \qquad 5.0964$$

TABLE IV

The angular functions for $a = 0$, $s = \pm 2$
($l = -m = 2$)

$\cos\theta$	S_{+2}	S_{-2}	$[S]^+$	$[\mathscr{L}S]^+$	$[\mathscr{L}\mathscr{L}S]^+$	$[S]^-$	$[\mathscr{L}S]^-$	$\mathscr{S}$
0.96	1.51853	0.00063	1.51916	−0.85002	0.74377	1.51789	−0.88544	0.14166
0.92	1.45718	0.00253	1.45971	−1.14021	1.45718	1.45465	−1.23935	0.19003
0.88	1.39709	0.00569	1.40279	−1.32176	2.14023	1.39140	−1.50200	0.22029
0.84	1.33828	0.01012	1.34840	−1.44128	2.79292	1.32816	−1.71581	0.24021
0.80	1.28072	0.01581	1.29653	−1.51789	3.41526	1.26491	−1.89737	0.25298
0.76	1.22443	0.02277	1.24720	−1.56198	4.00724	1.20167	−2.05524	0.26033
0.72	1.16941	0.03099	1.20040	−1.58007	4.56886	1.13842	−2.19454	0.26334
0.68	1.11565	0.04048	1.15613	−1.57666	5.10012	1.07517	−2.31862	0.26277
0.64	1.06316	0.05123	1.11439	−1.55508	5.60102	1.01193	−2.42981	0.25918
0.60	1.01193	0.06325	1.07517	−1.51789	6.07157	0.94868	−2.52982	0.25298
0.56	0.96196	0.07653	1.03849	−1.46716	6.51176	0.88544	−2.61992	0.24453
0.52	0.91327	0.09107	1.00434	−1.40458	6.92159	0.82219	−2.70111	0.23410
0.48	0.86583	0.10688	0.97272	−1.33160	7.30106	0.75895	−2.77417	0.22193
0.44	0.81966	0.12396	0.94362	−1.24948	7.65018	0.69570	−2.83972	0.20825
0.40	0.77476	0.14230	0.91706	−1.15931	7.96893	0.63246	−2.89827	0.19322
0.36	0.73112	0.16191	0.89303	−1.06209	8.25733	0.56921	−2.95025	0.17701
0.32	0.68874	0.18278	0.87152	−0.95872	8.51538	0.50596	−2.99600	0.15979
0.28	0.64763	0.20492	0.85255	−0.85002	8.74306	0.44272	−3.03579	0.14167
0.24	0.60779	0.22832	0.83611	−0.73676	8.94039	0.37947	−3.06985	0.12279
0.20	0.56921	0.25298	0.82219	−0.61968	9.10735	0.31623	−3.09839	0.10328
0.16	0.53189	0.27891	0.81081	−0.49945	9.24396	0.25298	−3.12154	0.08324
0.12	0.49584	0.30611	0.80195	−0.37673	9.35022	0.18974	−3.13943	0.06279
0.08	0.46106	0.33457	0.79563	−0.25217	9.42611	0.12649	−3.15214	0.04203
0.04	0.42754	0.36429	0.79183	−0.12639	9.47165	0.06325	−3.15975	0.02106

TABLE V

The radial functions for $s = \pm 1$

$(a = 0.95, \sigma = 0.5; l = -m = 2)$

r/M	Real parts $P_{\pm 1}$	Real parts $dP_{\pm 1}/dr$	Imaginary parts $P_{\pm 1}$	Imaginary parts $dP_{\pm 1}/dr$
2.10	1.2003	2.1833	−0.10873	−1.2468
2.20	1.4350	2.5017	−0.24999	−1.5818
2.30	1.6992	2.7757	−0.42574	−1.9361
2.40	1.9887	3.0079	−0.63781	−2.3080
2.50	2.2994	3.1991	−0.88783	−2.6946
2.60	2.6272	3.3492	−1.1771	−3.0926
2.70	2.9679	3.4575	−1.5066	−3.4985
2.80	3.3173	3.5234	−1.8770	−3.9087
2.90	3.6711	3.5463	−2.2884	−4.3194
3.00	4.0251	3.5258	−2.7407	−4.7270
3.10	4.3748	3.4617	−3.2336	−5.1279
3.20	4.7160	3.3540	−3.7660	−5.5185
3.30	5.0442	3.2031	−4.3368	−5.8953
3.40	5.3551	3.0095	−4.9445	−6.2549
3.50	5.6447	2.7741	−5.5871	−6.5940
3.60	5.9086	2.4980	−6.2625	−6.9095
3.70	6.1430	2.1827	−6.9681	−7.1984
3.80	6.3439	1.8298	−7.7012	−7.4579
3.90	6.5077	1.4412	−8.4586	−7.6853
4.00	6.6310	1.0191	−9.2371	−7.8781
4.10	6.7105	0.56571	−10.033	−8.0341
4.20	6.7432	0.83637	−10.843	−8.1510
4.30	6.7264	−0.042439	−11.662	−8.2272
4.40	6.6576	−0.095550	−12.487	−8.2608
4.50	6.5346	−1.5066	−13.313	−8.2505
4.60	6.3557	−2.0746	−14.135	−8.1951
4.70	6.1192	−2.6562	−14.950	−8.0937
4.80	5.8241	−3.2480	−15.752	−7.9455
4.90	5.4694	−3.8466	−16.538	−7.7500
5.00	5.0547	−4.4484	−17.301	−7.5071
5.10	4.5797	−5.0499	−18.037	−7.2168
5.20	4.0448	−5.6475	−18.743	−6.8793
5.30	3.4505	−6.2376	−19.412	−6.4952
5.40	2.7977	−6.8167	−20.040	−6.0653
5.50	2.0876	−7.3812	−20.623	−5.5906
5.60	1.3220	−7.9277	−21.157	−5.0722
5.70	0.50282	−8.4527	−21.636	−4.5118
5.80	−0.36768	−8.9528	−22.058	−3.9110
5.90	−1.2868	−9.4250	−22.417	−3.2716
6.00	−2.2516	−9.8659	−22.711	−2.5960
6.10	−3.2589	−10.273	−22.935	−1.8863
6.20	−4.3049	−10.642	−23.087	−1.1451
6.30	−5.3860	−10.972	−23.163	−0.37505
6.40	−6.4980	−11.260	−23.161	0.42089
6.50	−7.6365	−11.503	−23.078	1.2397
6.60	−8.7970	−11.699	−22.913	2.0784
6.70	−9.9747	−11.847	−22.662	2.9334
6.80	−11.165	−11.944	−22.326	3.8015
6.90	−12.362	−11.989	−21.902	4.6791
7.00	−13.561	−11.980	−21.390	5.5627

TABLE V, continued

The radial functions for $s = \pm 1$

$(a = 0.95, \sigma = 0.25; l = -m = 2)$

r/M	Real parts $P_{\pm 1}$	Real parts $dP_{\pm 1}/dr$	Imaginary parts $P_{\pm 1}$	Imaginary parts $dP_{\pm 1}/dr$
2.10	0.94906	−0.28975	0.070474	0.56174
2.20	0.93987	0.096678	0.11579	0.35698
2.30	0.96682	0.43572	0.14371	0.20822
2.40	1.0259	0.74175	0.15841	0.089098
2.50	1.1144	1.0253	0.16201	−0.015377
2.60	1.2305	1.2936	0.15552	−0.11405
2.70	1.3728	1.5518	0.13922	−0.21232
2.80	1.5406	1.8035	0.11295	−0.31361
2.90	1.7334	2.0510	0.076319	−0.42012
3.00	1.9507	2.2959	0.028707	−0.53332
3.10	2.1925	2.5392	−0.030602	−0.65420
3.20	2.4585	2.7814	−0.10241	−0.78344
3.30	2.7488	3.0229	−0.18758	−0.92149
3.40	3.0631	3.2638	−0.28701	−1.0687
3.50	3.4015	3.5039	−0.40163	−1.2252
3.60	3.7639	3.7431	−0.53238	−1.3913
3.70	4.1501	3.9812	−0.68020	−1.5668
3.80	4.5600	4.2177	−0.84606	−1.7520
3.90	4.9935	4.4522	−1.0309	−1.9467
4.00	5.4504	4.6844	−1.2357	−2.1509
4.10	5.9303	4.9137	−1.4614	−2.3645
4.20	6.4330	5.1397	−1.7089	−2.5874
4.30	6.9582	5.3619	−1.9792	−2.8195
4.40	7.5053	5.5797	−2.2731	−3.0606
4.50	8.0739	5.7927	−2.5916	−3.3105
4.60	8.6636	6.0004	−2.9355	−3.5691
4.70	9.2738	6.2021	−3.3057	−3.8361
4.80	9.9038	6.3975	−3.7030	−4.1113
4.90	10.553	6.5859	−4.1282	−4.3944
5.00	11.221	6.7670	−4.5821	−4.6851
5.10	11.906	6.9401	−5.0655	−4.9832
5.20	12.609	7.1049	−5.5790	−5.2884
5.30	13.327	7.2607	−6.1234	−5.6003
5.40	14.060	7.4073	−6.6993	−5.9186
5.50	14.808	7.5441	−7.3073	−6.2430
5.60	15.569	7.6706	−7.9481	−6.5731
5.70	16.342	7.7864	−8.6221	−6.9084
5.80	17.126	7.8912	−9.3299	−7.2487
5.90	17.920	7.9845	−10.072	−7.05935
6.00	18.722	8.0659	−10.849	−7.9424
6.10	19.532	8.1351	−11.661	−8.2949
6.20	20.349	8.1916	−12.508	−8.6507
6.30	21.170	8.2353	−13.391	−9.0093
6.40	21.995	8.2656	−14.310	−9.3701
6.50	22.823	8.2824	−15.265	−9.7329
6.60	23.651	8.2853	−16.256	−10.097
6.70	24.480	8.2741	−17.284	−10.462
6.80	25.306	8.2485	−18.349	−10.827
6.90	26.129	8.2082	−19.450	−11.192
7.00	26.947	8.1532	−20.587	−11.557

TABLE VI

The angular functions for $s = \pm 1$

	($a = 0.95$, $\sigma = 0.5$)					($a = 0.95$, $\sigma = 0.25$)			
$\cos\theta$	S_{+1}	$dS_{+1}/d\theta$	S_{-1}	$dS_{-1}/d\theta$	$\cos\theta$	S_{+1}	$dS_{+1}/d\theta$	S_{-1}	$dS_{-1}/d\theta$
0.96	0.484864	1.566731	0.007161	0.074963	0.96	0.457337	1.491370	0.007979	0.083382
0.92	0.659681	1.363899	0.020161	0.147218	0.92	0.624867	1.317715	0.022405	0.163014
0.88	0.776809	1.171990	0.036860	0.216699	0.88	0.738909	1.151512	0.040852	0.238847
0.84	0.861868	0.990669	0.056467	0.283339	0.84	0.823242	0.992645	0.062411	0.310830
0.80	0.925258	0.819615	0.078507	0.347065	0.80	0.887456	0.841005	0.086531	0.378909
0.76	0.972562	0.658514	0.102647	0.407804	0.76	0.936665	0.696481	0.112820	0.443033
0.72	1.007246	0.507061	0.128628	0.465477	0.72	0.974032	0.558966	0.140974	0.503147
0.68	1.031673	0.364960	0.156242	0.520003	0.68	1.001698	0.428355	0.170746	0.559195
0.64	1.047558	0.231925	0.185310	0.571297	0.64	1.021213	0.304541	0.201921	0.611121
0.60	1.056202	0.107677	0.215674	0.619269	0.60	1.033749	0.187424	0.234314	0.658866
0.56	1.058631	−0.008053	0.247191	0.663826	0.56	1.040229	0.076902	0.267754	0.702373
0.52	1.055674	−0.115529	0.279731	0.704873	0.52	1.041399	−0.027125	0.302086	0.741582
0.48	1.048019	−0.215005	0.313168	0.742307	0.48	1.037877	−0.124754	0.337162	0.776430
0.44	1.036245	−0.306726	0.347384	0.776025	0.44	1.030183	−0.216082	0.372844	0.806858
0.40	1.020847	−0.390932	0.382264	0.805917	0.40	1.018765	−0.301203	0.408999	0.832799
0.36	1.002255	−0.467855	0.417693	0.831871	0.36	1.004013	−0.380210	0.445494	0.854192
0.32	0.980847	−0.537719	0.453558	0.853768	0.32	0.986270	−0.453195	0.482202	0.870968
0.28	0.956955	−0.600743	0.489747	0.871487	0.28	0.965842	−0.520248	0.518996	0.883063
0.24	0.930879	−0.657136	0.526143	0.884900	0.24	0.943003	−0.581457	0.555748	0.890406
0.20	0.902887	−0.707103	0.562631	0.893877	0.20	0.918004	−0.636910	0.592331	0.892930
0.16	0.873222	−0.750843	0.599087	0.898280	0.16	0.891073	−0.686692	0.628612	0.890563
0.12	0.842106	−0.788545	0.635389	0.897970	0.12	0.862419	−0.730888	0.664460	0.883233
0.08	0.809744	−0.820397	0.671403	0.892799	0.08	0.832239	−0.769579	0.699739	0.870867
0.04	0.776324	−0.846577	0.706995	0.882617	0.04	0.800714	−0.802849	0.734306	0.853390
0.00	0.742020	−0.867258	0.742020	0.867266	0.00	0.768015	−0.830776	0.768015	0.830727

TABLE VII

The characteristic values λ for $s = 2$

$a\sigma$	$l = 2$ $m = 2$	$l = 2$ $m = 1$	$l = 2$ $m = 0$	$l = 2$ $m = -1$	$l = 2$ $m = -2$
0	4.0000	4.0000	4.0000	+4.0000	+4.0000
0.2	5.3446	4.6839	4.0186	+3.3503	+2.6767
0.4	6.7137	5.4035	4.0740	+2.7330	+1.3717
0.6	8.1108	6.1600	4.1645	+2.1463	+0.08242
0.8	9.5395	6.9542	4.2873	+1.5889	−1.1935
1.0	11.004	7.7856	4.4387	+1.0601	−2.4580
1.2	12.508	8.6528	4.6136	+0.56000	−3.7124
1.4	14.056	9.5528	4.8056	+0.08978	−4.9576
1.6	15.654	10.481	5.0071	−0.34779	−6.1937
1.8	17.305	11.432	5.2089	−0.74819	−7.4199
2.0	19.014	12.397	5.4000	−1.1047	−8.6344
2.2	20.787	13.364	5.5682	−1.4081	−9.8344
2.4	22.630	14.321	5.6990	−1.6460	−11.016
2.6	24.546	15.252	5.7764	−1.8028	−12.173
2.8	26.542	16.136	5.7821	−1.8588	−13.298
3.0	28.623	16.949	5.6960	−1.7899	−14.383
3.2	30.794	17.665	5.4954	−1.5673	−15.417
3.4	33.059	18.252	5.1558	−1.1563	−16.387
3.6	35.425	18.673	4.6500	−0.5166	−17.278
3.8	37.896	18.885	3.9484	+0.3992	−18.072
4.0	40.476	18.842	3.0187	+1.6454	−18.750

TABLE VII, continued

The characteristic values λ for $s = 1$ and $s = \frac{1}{2}$

$a\sigma$	$l = 1.0$ $m = 1.0$	$l = 1$ $m = 0$	$l = 1.0$ $m = -1.0$	$l = 2$ $m = -2$	$l = 0.5$ $m = 0.5$	$l = 0.5$ $m = -0.5$
0	2.0000	2.0000	+2.0000	+6.0000	1.0000	1.0000
0.2	2.6189	2.0227	+1.4173	+5.0934	1.2908	+0.75658
0.4	3.2775	2.0905	+0.86643	+4.2389	1.6311	+0.55754
0.6	3.9787	2.2003	+0.3428	+3.4341	2.0228	+0.39903
0.8	4.7243	2.3477	−0.15866	+2.6762	2.4671	+0.27622
1.0	5.5150	2.5266	−0.64375	+1.9618	2.9658	+0.18321
1.2	6.3497	2.7294	−1.1187	+1.2872	3.5210	+0.11286
1.4	7.2255	2.9470	−1.5904	+0.6480	4.1362	+0.056690
1.6	8.1368	3.1685	−2.0664	+0.0392	4.8170	+0.0047234
1.8	9.0754	3.3816	−2.5547	−0.5451	5.5713	−0.054695
2.0	10.029	3.5724	−3.0641	−1.1114	6.4113	−0.13508
2.2	10.984	3.7253	−3.6040	−1.6671	7.3535	−0.25199
2.4	11.919	3.8232	−4.1844	−2.2206	8.4204	−0.42325
2.6	12.809	3.8478	−4.8163	−2.7814	9.6416	−0.66921
2.8	13.624	3.7791	−5.5110	−3.3602	11.0553	−1.01293
3.0	14.327	3.5960	−6.2811	−3.9688	12.709	−1.4805
3.2	14.874	3.2762	−7.1397	−4.6205	14.663	−2.1012
3.4	15.216	2.7961	−8.1007	−5.3300	16.990	−2.9080
3.6	15.291	2.1316	−9.1791	−6.1138	19.776	−3.9377
3.8	15.032	1.2574	−10.390	−6.9900	23.127	−5.2311
4.0	14.360	0.1478	−11.752	−7.9785	27.165	−6.8339

Finally, in Table VII, we list the characteristic values λ belonging to the angular functions for the different spins and l and m values: those for $s = 2$ and $s = 1$ were evaluated with data provided by Press and Teukolsky, and those for $s = \frac{1}{2}$, with similar data provided by Page.

BIBLIOGRAPHICAL NOTES

Papers dealing with the spin-weighted spherical and spheroidal harmonics are:

1. J. N. GOLDBERG, A. J. MACFARLANE, E. T. NEWMAN, *et al.*, *J. Math. Phys.*, 8, 2155–61 (1967).
2. R. A. BREUER, M. P. RYAN, JR., and S. WALLER, *Proc. Roy. Soc.* (*London*) A, 71–86 (1977).
3. E. D. FACKERELL and R. G. CROSSMAN, *J. Math. Phys.*, 18, 1849–54 (1977).

Press and Teukolsky have tabulated the coefficients of series expansions for λ in powers of $a\sigma$ both for $s = 2$ (reference 4) and 1 (reference 5):

4. W. H. PRESS and S. A. TEUKOLSKY, *Astrophys. J.*, 185, Appendix C, 668 (1973).
5. S. A. TEUKOLSKY and W. H. PRESS, ibid., 193, Table 2, 454 (1974).

The coefficients of similar power-series expansions for λ for $s = \frac{1}{2}$ were provided to the author by D. N. Page.

EPILOGUE

There is no excellent beauty that hath not some strangeness in the proportion.
Francis Bacon

Beauty is the proper conformity of the parts to one another and to the whole.
Werner Heisenberg

The author had occasion to ask Henry Moore how one should view sculptures: from afar or from near by. Moore's response was that the greatest sculptures can be viewed—indeed, should be viewed—from all distances, since new aspects of beauty will be revealed at every scale. Moore cited the sculptures of Michelangelo as examples: from the excellence of their entire proportion to the graceful delicacy of the fingernails. The mathematical perfectness of the black holes of Nature is, similarly, revealed at every level by some strangeness in the proportion in conformity of the parts to one another and to the whole.

INDEX

Milton Keynes UK
Ingram Content Group UK Ltd.
UKHW020402220124
436454UK00002B/6

9 780198 503705